D0521493

Accessing AutoCAD® Architecture 2012

Accessing AutoCAD® Architecture 2012

autodesk Press

WILLIAM G. WYATT

DELMAR
CENGAGE Learning™

Australia • Brazil • Japan • Korea • Mexico • Singapore • Spain • United Kingdom • United States

Accessing AutoCAD® Architecture 2012
William G. Wyatt

Vice President, Editorial: Dave Garza

Director of Learning Solutions: Sandy Clark

Acquisitions Editor: Stacy Masucci

Managing Editor: Larry Main

Senior Product Manager: John Fisher

Editorial Assistant: Andrea Timpano

Vice President, Marketing: Jennifer Baker

Marketing Director: Deborah Yarnell

Marketing Manager: Katie Hall

Associate Marketing Manager: Jillian Borden

Senior Production Director: Wendy Troeger

Senior Content Project Manager: Glenn Castle

Senior Art Director: David Arsenault

Technology Project Manager: Joe Pliss

Cover image: ist2_11955229-architectural-abstract
© Kasper Christiansen/iStockphoto

Library of Congress Control Number: 2011924519

ISBN-13: 978-1-111-64831-2

ISBN-10: 1-111-64831-X

Delmar
5 Maxwell Drive
Clifton Park, NY 12065-2919
USA

Cengage Learning is a leading provider of customized learning solutions with office locations around the globe, including Singapore, the United Kingdom, Australia, Mexico, Brazil, and Japan. Locate your local office at: **international. cengage.com/region**

Cengage Learning products are represented in Canada by Nelson Education, Ltd.

To learn more about Delmar, visit **www.cengage.com/delmar**

Purchase any of our products at your local college store or at our preferred online store **www.cengagebrain.com**

Notice to the Reader

Printed in the United States of America
1 2 3 4 5 6 7 15 14 13 12 11

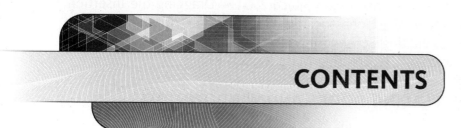

CONTENTS

INTRODUCTION

PREFACE

Accessing AutoCAD Architecture 2012, 10th edition, is a comprehensive presentation of the tools included in AutoCAD Architecture 2012. The format of the text includes the introduction of tools followed by an explanation of the options of the command and how it is used in the development of drawings. A tool access table is provided for each command presented. The text includes screen captures of the dialog boxes associated with each tool. Tutorials are included throughout each chapter to show you step by step how to use the tools of the chapter. In addition, the tutorials provide you a written guide to refer to when doing the independent projects located at the end of each chapter. Projects and review questions are also provided for each chapter to reinforce your knowledge of the software. The AutoCAD Architecture 2012 software is state-of-the-art software for architectural design. Recently, programs of study in architectural technology and architecture have integrated computer-aided design within the curriculum. With this in mind, *Accessing AutoCAD Architecture 2012,* 10th edition, has been written in order to provide detailed and systematic explanations of the application of the tools to create architectural working drawings. The book includes techniques and tutorials that apply the software to the creation of drawings for residential and commercial buildings.

The first twelve chapters of the book provide the necessary instruction regarding the use of AutoCAD Architecture 2012 for creating architectural working drawings. The beginning chapters will step you through the placement and editing of doors, walls, windows, and roofs. The tutorials track the development of working drawings from the beginning of a residence. Specific tutorials are included to demonstrate how the software is used to create floor plans, foundation plans, elevations, and sections. Each tutorial focuses on how to access a command and how to use it to create architectural working drawings. This format allows the reader to immediately begin drawing floor plans.

Chapter 4 explains how to develop space plans and mass models to represent the shape of a building. Chapter 13 explains how the tools of AutoCAD Architecture 2011 can be used to create drawings for commercial buildings. This chapter includes the structural catalog, layout curves, ceiling grids, camera views, and animations.

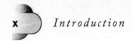

FEATURES OF THIS EDITION

Included in *Accessing AutoCAD Architecture 2012* are tutorials, which provide step-by-step instruction in the use of AutoCAD Architecture 2012 to develop architectural working drawings for residential and commercial buildings. Short tutorials are included throughout the chapter and project-based tutorials are placed at the end of the chapters. Command access tables and the online Student Companion contents together help the reader identify the appropriate commands and techniques to develop the working drawings. The features of this edition are summarized below:

- Command access tables are provided for each new command introduced. They describe in table format how to access the commands from the menu bar, command line, tool palette, and shortcut menu.
- Instructional video files, etutorials, and eprojects are located at the Student Companion site of http://www.cengagebrain.com.
- Imperial and Metric tutorials are included to introduce and reinforce the options of the commands. They include both residential and commercial building applications.
- Development of complex roofs, roof slabs, roof intersections, and dormers for residential and commercial buildings are included.

STYLE CONVENTIONS

Throughout the book you are requested to select commands and respond to the prompts in the workspace. Metric values for the tutorials are included in brackets adjacent to the Imperial values (e.g., 24'-0" [7300]). The text style conventions are used systematically to enhance the understanding and recall of the commands. The style conventions of the text are as follows:

Element	Example
Commands	WallAdd or WALLADD
Menu	Format > Style Manager
Dialog box elements	Select the Edit button
Workspace prompts	Line start point:
Keyboard input	Press ENTER to end the command.
User Input	Type OFFICE--CAD in the field.
File and directory names	*C:\Documents and Settings\All users\Application Data\Autodesk*

HOW TO USE THIS BOOK

The design of each chapter of the text is to introduce the commands of AutoCAD Architecture 2012 and allow you, through the tutorials, to gain the skills and understanding of the software needed to sufficiently create your own architectural working drawings. Each chapter includes an introduction and objectives. Read the objectives of each chapter carefully to determine the commands and types of drawings created in the chapter. The purpose of each tutorial is not to finish quickly but to gain hands-on experience in using the commands to create drawings. You may find it most helpful to repeat the tutorial after completing the chapter to gain recognition of the commands on the tool palette and of the content of the Properties palette. The book includes screen captures of the tool palettes and Properties palettes to allow you to study the commands when you might not have access to the software. However, having access to the software as you read the book greatly enhances the learning process. A summary and a set of review questions are provided at the end of each chapter, as are additional project exercises.

Command Access Tables

The commands included in the text are summarized and presented in command access tables. Included in each table is a list of methods used to access the command from the menu bar, command prompt, tool palette, and shortcut menu. Commands to create objects are usually selected from the ribbon and tool palettes, while commands to edit objects are selected from shortcut menus or contextual tabs of the ribbon for the selected object.

The command access table for the **Wall** tool (**WallAdd command**) is shown below as an example.

Command prompt	**WALLADD**
Tool palette	Select the Wall tool from the Design palette as shown in Figure 002.

Organizing Tutorial Directories

Drawing files for all tutorials are located at the Student Companion site from CengageBrain.

Accessing a Student Companion site from CengageBrain:

1. GO TO: http://www.cengagebrain.com
2. TYPE author, title, or ISBN in the **Search** window
3. LOCATE the desired product and click on the title
4. When you arrive at the Product Page, CLICK on the **Access Now** button.
5. USE the "**Click Here**" link to be brought to the Companion site
6. Click on the Student Resources link in the left navigation pane to access the resources; click on Dataset Files. Download and extract the zip file.

The drawing files are sorted by chapter and located in the *Accessing Tutor* folder at the site. When you perform a tutorial, you will be directed to create a drawing or project in your Accessing Student folder by chapter; therefore, create a student folder (e.g., *C:\Accessing Student*) and folders for each of the 13 chapters. Copy the *Accessing Tutor* and all its subdirectories from the CengageBrain site to the root directory of your computer, placing the *Accessing Tutor* and its contents at the C:\ location.

Create your student folder now.	**STOP**

INSTRUCTOR SITE

An Instructor Companion Website containing supplementary material is available. This site contains an Instructor Guide, testbank, instructional video files, image gallery of text figures, and chapter presentations done in PowerPoint. Contact Delmar Cengage Learning or your local sales representative to obtain an instructor account.

Accessing an Instructor Companion Website from SSO Front Door

1. GO TO: http://login.cengage.com and login using the Instructor email address and password.
2. ENTER author, title, or ISBN in **the Add a title to your bookshelf** search box; click on **Search** button.
3. CLICK **Add to My Bookshelf** to add Instructor Resources.
4. At the Product page click on the **Instructor Companion site** link.

New Users

If you're new to Cengage.com and do not have a password, contact your sales representative.

WE WANT TO HEAR FROM YOU

We welcome your comments and suggestions regarding the contents of this text. Your input will result in the improvement of future publications. Please forward your comments and questions to:

The CADD Team
C/O Autodesk Press
Executive Woods
5 Maxwell Drive
Clifton Park, NY 12065-8007

ABOUT THE AUTHOR

William G. Wyatt, Sr., is an instructor at John Tyler Community College in Chester, Virginia. He has taught architectural drafting and related technical courses in the Architectural Engineering Technology program since 1972. He earned his doctor of education degree from Virginia Tech and his master of science and bachelor of science degrees in industrial technology from Eastern Kentucky University. He earned his associate's degree in applied science in architectural technology from John Tyler Community College. He is a certified Architectural and Building Construction Technician and Autodesk Certified Instructor for the Autodesk Training Center at Tidewater Community College.

DEDICATION

The 2012 edition is dedicated to H. Barry Edwards, long-time friend and colleague, who died in April 2011. Barry was a dynamic professor and Department Head of Mechanical Engineering Technology at John Tyler Community College from 1972 to 1990. It was under Barry's leadership the drafting courses converted from manual procedures to CAD. Barry and his dear wife, Janet, went home to Shelby, N.C. after retiring. It was over those 21 years in Shelby where their family grew. Both their children Chris and Sherry married and grandchildren followed. Barry was devoted to his family but he never neglected to reach out to his community and friends. We spoke often on the phone. Barry always displayed interest and concern for my professional endeavors and remembered each member of my family.

I am so fortunate to be surrounded by people that care about me both personally and professionally.

The first people that come to mind are my parents, Leslie (died 1989) and Catherine (died 1981) Wyatt, who supported me throughout their lives with their encouragement and unconditional love. They will always be loved and remembered. They would have appreciated the contribution of this book towards learning.

Our children's lives continue active, productive, and give us reason to celebrate.

Our daughter, Sarah, is a Graduate Research Assistant in the Mathematics Department at Virginia Tech. She never ceases to amaze us with her strong work ethic, sheer determination, and love for learning. We are so proud of Sarah.

Two of our children married this year. Our oldest daughter, Leslie, married Sean Jansen. Sean graduated from University of Virginia Law School in May, 2011. Leslie taught her second year of English Literature at Culpepper High School in Virginia. Our son, Will, married Sabrina Schumaker. Sabrina graduated from Virginia Commonwealth University, School of Nursing in May, 2011. Will continues his full time employment as an ALS Paramedic for Dinwiddie Fire and EMS. Both Sabrina and Sean are heartily welcomed into our family.

My mother-in-law, Helen Hedahl, is completing her fourth year as a Virginian which was a big switch from being a Montanan. Helen continues to meet the health challenges that the aging process brings with grace and patience. She is a constant reminder how valuable each person is in the family mosaic.

But most of all, I want to remember my wife, Bevin Hedahl Wyatt, who remains the cornerstone of all my projects and brings joy into the ordinary. Through the daily writing process, she provides the momentum and encouragement needed to sustain the writing and editing processes. Her quiet love and support have made all the difference.

ACKNOWLEDGMENTS

The author would like to thank and acknowledge the team of reviewers who reviewed the chapters and provided guidance and direction to the work. A special thanks to the following reviewers, who reviewed the chapters in detail:

Paul Adams, Denver Technical College, Denver, Colorado

Deanna Blickham, Moberly Area Community College, Moberly, Missouri

David Braun, Spokane Community College, Spokane, Washington

Paul N. Champigny, New England Institute of Technology, Warwick, Rhode Island

James Freygang, Ivy Tech State College, South Bend, Indiana

Lynn A. Gurnett, York County Technical College, Wells, Maine

Donald W. Hain, Orleans/Niagara BOCES, Sanborn, New York

Christopher LeBlanc, Porter & Chester Institute, Chicopee, Massachusetts

Jeff Levy, Pulaski County High School, Dublin, Virginia

Joseph M. Liston, University of Arkansas Westark, Fort Smith, Arkansas

Jeff Porter, Porter & Chester Institute, Watertown, Connecticut

Charles T. Walling, Silicon Valley College, Walnut Creek, California

Special thanks go to Chris Lucas for his careful technical editing of the manuscript. Chris Lucas is employed as a CADD Designer I for Tectonic Engineering & Surveying Consultants, P.C. of Richmond Virginia. His technical expertise has allowed this text to include many practical details and tips that will benefit the reader.

I would like to thank Ronald A. Williams of Ronald A. Williams, LTD, Autodesk Education Representative of Virginia for his support and encouragement of this project.

The author would like to acknowledge and thank the following staff from Delmar Cengage Learning:

Sandy Clark, Editorial Director; John Fisher, Senior Product Manager; Glenn Castle, Senior Content Project Manager; and Stacy Masucci, Acquisitions Editor. The author would like to acknowledge and thank Dewanshu Ranjan and Aravinda Kulasekar Doss, Project Managers, and the staff at PreMediaGlobal.

Introduction to AutoCAD Architecture

INTRODUCTION

AutoCAD Architecture 2012 is intended to assist the designer in developing architectural working drawings, as well as preliminary schematics and computer models of a building. AutoCAD Architecture 2012 evolved from Autodesk Architectural Desktop, which was first released as a vertical product of AutoCAD Release 14 in 1998. AutoCAD Architecture 2012 utilizes the new features of AutoCAD 2012.

AutoCAD Architecture 2012 creates components of a building as 3D objects. Objects consist of walls, doors, windows, stairs, and roofs. Because 3D objects are used to develop drawings, the drawings can be put together to create a three-dimensional model of a building. In addition, AutoCAD Architecture 2012 includes tools for creating schedules that allow you to extract a comprehensive list of doors, windows, furniture, rooms, and wall types from the drawing and models.

OBJECTIVES

After completing this chapter, you will be able to:

- Describe how to start AutoCAD Architecture 2012 and identify the components of the workspace.

- Identify the purpose and resources of the tabs of the Ribbon.

- Identify the purpose of AutoCAD Architecture templates and Drawing Setup.

- Identify the display options of Tool palettes and the Properties palette.

- Describe how to use display configurations and the Object Viewer.

- Create revision and demolition plans.

- Describe how to use the Project Browser and Project Navigator to create a project.

ADVANTAGES OF OBJECT TECHNOLOGY

Using AutoCAD Architecture 2012 requires a basic knowledge of AutoCAD. You can create a sketch using lines and arcs and convert this geometry to walls, slabs, roofs, and mass elements. Doors and windows as well as the walls are inserted in a drawing through ObjectARX technology. These objects consist of several entities that have associative interaction with other objects, when inserted. When doors and windows are placed in the wall, they behave as magnets within the wall, and the wall is correctly edited for the window or door. Using ObjectARX technology reduces the need to edit the drawing. AutoCAD Architecture 2012 objects are three dimensional and consist of several components. For instance, a door object consists of a door panel, frame, stop, and swing components.

NOTE   The content of this book is based upon the Typical installation, which installs the Imperial and Metric content. The recommended screen resolution is 1280 × 1024 with a 32-bit color video adapter. Selection of lower resolutions may hide screen options.

STARTING AUTOCAD ARCHITECTURE 2012

Start AutoCAD Architecture 2012 by choosing the **Start** button of Windows Vista/ Windows XP and selecting **All Programs > Autodesk > AutoCAD Architecture 2012 > AutoCAD Architecture 2012** or by double-clicking the **AutoCAD Architecture 2012** icon on the desktop. Once AutoCAD Architecture 2012 is launched, it opens to the Autodesk Exchange window if the computer is connected to the internet, as shown in Figure 1.1. The **Autodesk Exchange window** provides announcements and access to Help and resources from Autodesk. Click the Help button as shown at p1 in Figure 1.1 to access the Help file.

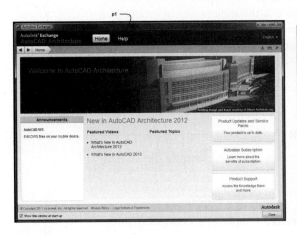

FIGURE 1.1 *Autodesk Exchange*

When you launch AutoCAD Architecture, the Info Center located in the upper-right corner of the workspace (see Figure 1.2) provides quick access to Help and the Exchange for the Autodesk community. You can type a topic in the window and help will be displayed in the Autodesk Exchange window.

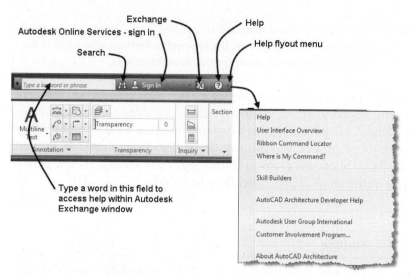

FIGURE 1.2 *Info Center*

The program launches to Drawing 1 based upon the *Aec Model (Imperial Stb).dwt [Aec Model (Metric Stb).dwt]* template. To save a non-project drawing with a specific name, choose **Save** from the Quick Access toolbar shown in Figure 1.3 (located in the upper-left corner of the screen), or use Ctrl + S and type a new drawing name in the File name field.

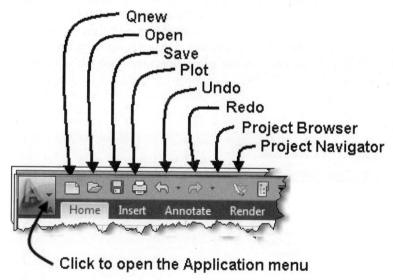

FIGURE 1.3 *Quick Access toolbar*

To create a new drawing, choose **QNew** from the Quick Access toolbar shown in Figure 1.3, and the program creates a new drawing based upon the *Aec Model (Imperial Stb).dwt [Aec Model (Metric Stb).dwt]* template. If you select **New > Drawing** from the Application menu, the **Select Template** dialog box will open, allowing you to specify a template.

TIP

The default template used with the QNew command is specified in the Template Settings of the Files tab in the Options dialog box. Therefore, the Imperial or Metric profile listed in the Profiles tab of Options includes the specification for the default template.

The default template of the Imperial profile is *Aec Model (Imperial Stb).dwt*, whereas the default template of the Metric profile is *Aec Model (Metric Stb).dwt*. New drawings should utilize one of the templates listed in Table 1.1. Drawings developed as part of a project are printed or plotted from drawings created, based upon the AEC Sheet templates. All templates are located in the *(Vista- C:\ProgramData\ Autodesk\ ACA 2012\enu\Template)* folder.

TABLE 1.1 *AutoCAD Architecture 2012 templates*

Template	Purpose
Aec Model (Imperial Ctb)	To create drawings using object styles based on imperial units and color-dependent plotting style tables for plotting.
Aec Model (Imperial Stb)	To create drawings using object styles based on imperial units and named plot style tables for plotting.
Aec Model (Metric Ctb)	To create drawings using object styles based on metric units and color-dependent plotting style tables for plotting.
Aec Model (Metric Stb)	To create drawings using object styles based on metric units and named plot style tables for plotting.

TIP

The **Select template** dialog box includes a Places panel at the left. The **Content** option of the Places panel will navigate to Vista- C:\ProgramData\Autodesk\ACA 2012\enu. The settings in Windows Explorer may suppress the display of selected hidden files and directories. To view hidden files and folders, launch Windows Explorer as shown in Figure 1.4 and select **Organize** > **Folder and Search Options** from the menu. Select the **View** tab and select **Show hidden files and folders**.

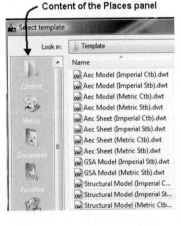

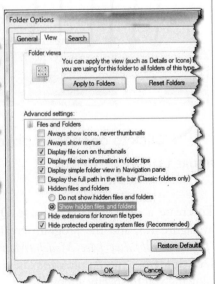

FIGURE 1.4 *Display of hidden files and folders*

Properties of Templates

Templates specifically designed for AutoCAD Architecture 2012 include preset values for annotation, dimensioning, grid, layer standards, limits, snap, and units. The settings for the imperial and metric templates are shown in Table 1.2.

TABLE 1.2 *Settings of the AutoCAD Architecture 2012 templates*

Setting	Imperial	Metric
Annotation Plot Size	3/32	3.5 mm
Grid	10'	3000 mm
Limits	0,0 to 288',192'	0,0 to 53600,41400
Scale Factor	96.000	100.000
Ltscale	0.5	0.5
MsLtscale	1	1
PsLtscale	1	1
Snap	1/16	1 mm
Units	Architectural	Decimal
Global Cut Plane Height	3'-6"	1400 mm

AUTOCAD ARCHITECTURE 2012 SCREEN LAYOUT

The AutoCAD Architecture 2012 screen is shown in Figure 1.5. The screen consists of the Application menu and Quick Access Toolbar at top left. The InfoCenter is located at top right of the graphics screen. The Ribbon is located immediately below the Quick Access toolbar. The Ribbon consists of eight tabs. When you select a tab, a group of panels is displayed, which includes tools related to the tab group. The View Cube and Navigation bar shown at right consist of view, zoom, and pan tools. The Tool Palettes and Properties palette are shown at left in the Drawing window of Figure 1.5. The Tool Palettes have been resized in Figure 1.5 to expose the In Canvas Controls. The In Canvas Controls menu includes flyout menus for Viewport, View, and Visual Styles controls. When a project is open, the Project Navigator may be displayed as shown at right in the Drawing window. The project name and project drawing file names are shown in the Drawing Window status bar immediately below the Drawing window. The current annotation scale, display configuration, and cut plane for a drawing are located in the Drawing Window status bar.

When you select a tool from the Tool Palette, the program displays information regarding the settings of the current tool in the **Properties** palette. If you select an existing object, its properties are displayed and can be edited as shown in the Design tab of the Properties palette on the left in Figure 1.5.

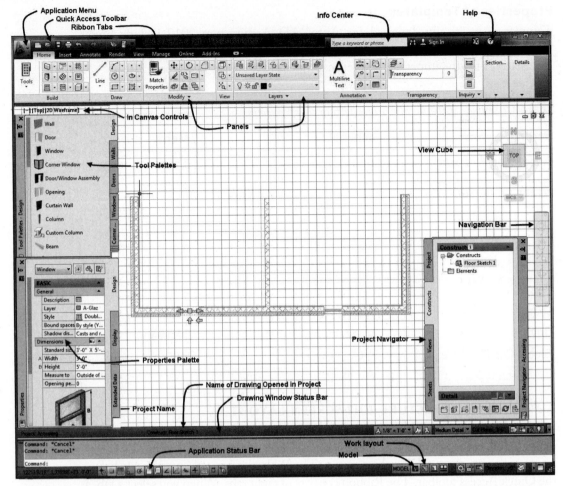

FIGURE 1.5 *AutoCAD Architecture 2012 screen layout*

Using Workspaces

The content of the screen display is defined by the Architecture workspace. When you open a session of AutoCAD Architecture 2012, the Architecture workspace is current. You can use the Workspace Switching toggle located in the Application status bar to select or create new workspaces as shown in Figure 1.6. The workspace defines the content of the ribbon and screen settings. Therefore, if you modify the ribbon, you can create an additional workspace if you choose **Save Current As** from the drop-down menu of Workspace Switching as shown in Figure 1.6. Selecting the Customize option of the drop-down menu opens the Customize User Interface dialog box shown in Figure 1.6, which lists the content and allows customization of the workspace.

FIGURE 1.6 *Accessing Workspace Switching from the Application status bar*

If you choose the **Workspace Settings** option from the flyout menu, the **Workspace Settings** dialog box opens as shown in Figure 1.6. In the Workspace Settings dialog box, you can choose **Automatically save workspace changes** to retain any changes you have made to the workspace when you switch to a different workspace. If you choose the radio button **Do not save changes to workspace** as shown in Figure 1.6, the screen display will be preserved as defined when the workspace was created or last saved.

Preserving the content of a workspace may assist in learning the content of the workspace. After learning the original content of the workspaces that were shipped with the software, toggle on **Automatically save workspace changes** to develop a custom workspace.

Steps to Creating Additional Workspaces:

1. Arrange palettes and customize the ribbon.
2. Choose **Save Current As** from the Workspace Switching flyout menu.
3. Type the name of the workspace in the **Save Workspace** dialog box.
4. Choose **Save** to save the workspace and dismiss the dialog box.

Accessing the Drawing Window Status Bar

The Drawing Window status bar is located above the command window, as shown in Figure 1.7, for Model and Work layouts. The status bar includes options to set the annotation scale or viewport scale and display configuration. If the current drawing is part of a project, the drawing title and the project name are displayed at the left. The Floor Sketch 1 drawing is a construct drawing of the *Accessing* project as shown in Figure 1.7.

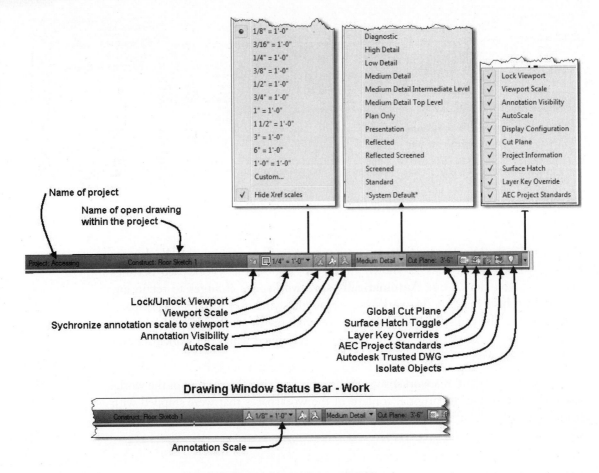

Drawing Window Status Bar - Work

Drawing Window Status Bar - Model

FIGURE 1.7 *Drawing Window status bar*

Accessing the Application Status Bar

The **Application** status bar located below the command line includes additional toggles to control the display of the workspace. The **Dynamic Input** toggle shown in Figure 1.8 should be turned on when you use AutoCAD Architecture 2012 since it controls the display of dynamic dimensions that become active when you are drawing walls, doors, and windows. The **Model** and **Layout** toggles provide access to Model and Paper Space. You may right-click over the toggles to display Model and Layout tabs. The **Elevation** field displays the current elevation of drawings associated with the levels of a project. The **Tray Settings** flyout located in the lower-right corner allows you to turn on or off the display of the Drawing Window status bar and the content of the Application status bar.

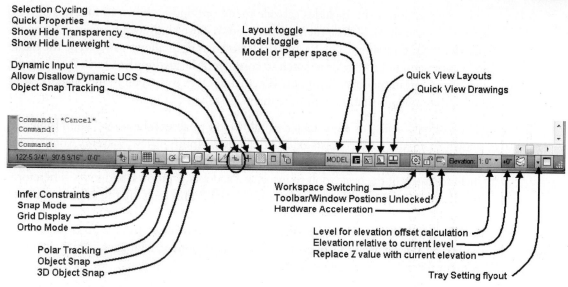

FIGURE 1.8 *Application status bar*

Using the Application Menu and Ribbon

The Application menu shown in Figure 1.9 includes a menu of file-related operations. This menu includes the following menu items including New, Open, Save, Save As, Export, Publish, Print, Drawing Utilities, and Close. You can choose the Recent Documents or Open Documents button shown in Figure 1.9 to open or switch to a drawing. Project names are also included in the list; therefore, you can choose a project name from the Recent Documents list to open a project.

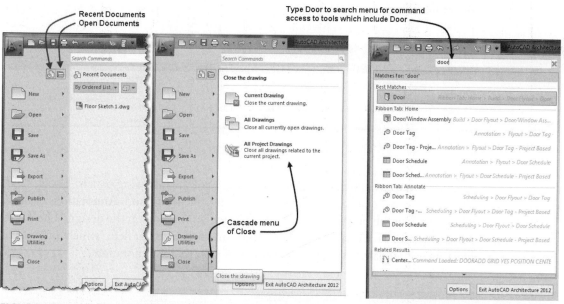

FIGURE 1.9 *Application Menu*

The Application menu includes cascade menu options as shown in Figure 1.9. The cascade menu for the Close menu allows you to close all drawings when you choose **Close > All Drawings**. The Search field shown at right allows you to type the name of a command to browse for a match to commands located in the Quick Access toolbar, Application menu, and Ribbon. The name Door was typed in the Search box to display commands that match in the menu as shown in Figure 1.9.

The ribbon consists of eight tabs that allow you to access the tools. The Home tab shown in Figure 1.10 consists of Build, Draw, Modify, View, Layers, Annotation, Transparency, Inquiry, Section & Elevation, and Details panels. The Build panel includes many of the basic tools to insert walls, doors, windows, and roofs. Therefore, the Build panel provides tools to start a design. The tools of the Build panel are included on the Design palette. When you choose the Tool button shown at left, the Design palette is displayed. In addition, you can choose the Detail Components button at the far right to open the Detail Component Manager, which includes content for the development of details.

FIGURE 1.10 *Home tab of the Ribbon*

The Insert tab consists of Reference, Block, Attributes, Import, Content and Seek panels as shown in Figure 1.11. The Content panel provides the tools for placing content from the Content Browser, Building Components, and Design Center, which include symbols and design components. The Seek panel includes a window that allows you to type search names for drawing content. The Content Browser will be discussed later in this chapter.

FIGURE 1.11 *Insert tab of the Ribbon*

The Annotate tab shown in Figure 1.12 consists of the following panels: Tools, Text, Dimensions, Scheduling, Callouts, Keynoting, Markup, and Annotation Scaling. Therefore, this tab provides the tools for placing annotation on the design. The Tools panel includes an Annotation Tools button located at far left, which opens the Document tool palette group that consists of palettes for annotation.

FIGURE 1.12 *Annotate tab of the Ribbon*

The Render tab shown in Figure 1.13 consists of the following panels: Tools, Render, Materials, Sun & Location, Lights, Camera, and Animations. You can choose the Render Tools button shown at far left to open the Visualization tool palette group.

FIGURE 1.13 *Render tab of the Ribbon*

The View tab (see Figure 1.14) consists of the following panels: Navigate, Appearance, Coordinates, Visual Styles, Viewports, and Windows.

FIGURE 1.14 *View tab of Ribbon*

The Manage tab consists of the following panels: Action Recorder, CAD Standards, Project Standards, Style & Display, Applications, and Customization. The Style & Display panel shown in Figure 1.15 provides access to the Style Manager and Display Manager. The Style Manager includes tools to create and edit styles. Each object inserted such as walls, doors, and windows are defined by a named style. The Style Manager allows you to edit the styles of the objects. The Display Manager provides global control of how objects are displayed in the drawing. The settings for the Display Configuration are selected from the Drawing Window status bar and are defined within the Display Manager.

FIGURE 1.15 *Manage tab of the Ribbon*

The Online and Add-Ins tab are created to the ribbon when the AutoCAD WS plug in is installed. The Online tab provides tools for downloading and uploading files to the AutoCAD WS. The Add-Ins tab includes the Content panel, which allows you to open the Content Explorer. The Content Explorer provides indexing of content such as blocks, layers, linetypes, and annotation styles that are located in network folders, your computer, and Autodesk Seek.

Online tab

Add Ins tab

FIGURE 1.16 *Online and Add Ins tabs*

When you select an object, a contextual tab is displayed, which consists of panels with tools necessary to edit the selected object. The Wall tab shown in Figure 1.17 is displayed when a wall is selected. This tab consists of the edit tools for a selected wall. The edit tools are also included in the shortcut menu displayed when you right-click after selecting the wall. In addition, when you select an object such as a wall, the properties are displayed in the Properties palette.

FIGURE 1.17 *Wall tab for edit of a wall*

You can click the title of a panel and drag it from the ribbon to create a sticky panel. The sticky panel will display continuously although you have changed tabs of the ribbon. The Build panel has been dragged into the workspace from the Home tab as shown in Figure 1.18 at left. You can move the cursor over a sticky panel to display its menu options as shown in Figure 1.18 at left. The menu options allow you to display the ribbon in either a horizontal or a vertical orientation. Choose Return to Panel from the menu to return the panel to the ribbon.

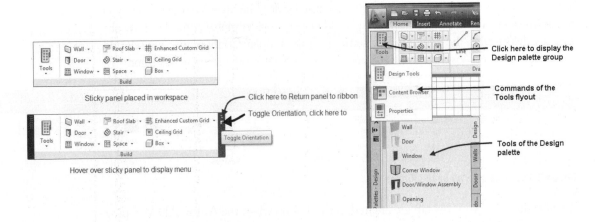

FIGURE 1.18 *Creating a sticky panel and accessing Design tool palettes*

Tools of the Tool Palette

The tools of the tool palette include the basic tools for creating AutoCAD Architecture 2012 objects. The tools include commands for adding doors, windows, walls, and the roof. The Tools flyout provides access to the Design Tools palette, Content Browser, and Properties. Tool palettes can be accessed as shown in Table 1.3.

TABLE 1.3 *Accessing palette groups*

Palette	Ribbon
Design	Home tab > Build panel
Document	Annotate tab > Tools panel
Visualization	Render > Tools panel

Select commands by clicking the icon of the tool palette as shown at right in Figure 1.18. When the command is executed, the Properties palette opens; review the settings in the Design tab and respond to the prompts in the command line or the Dynamic Prompts in the workspace. The tool palettes can be customized to include additional tools needed by the designer. You can drag styles of objects created in the **Style Manager** to a tool palette. You can click and drag content such as plumbing fixtures and furniture from the DesignCenter and **Content Browser** onto tool palettes. These palettes are grouped into four palette sets: Design, Document, Detailing, and Visualization. The content of the palette sets consists of a sample of tools available from the Content Browser. You can right-click over the title bar of the tool palettes and choose the palette set as shown in Figure 1.19. The Design, Document, and Visualization palette groups can be accessed from the Ribbon as shown in Figure 1.19.

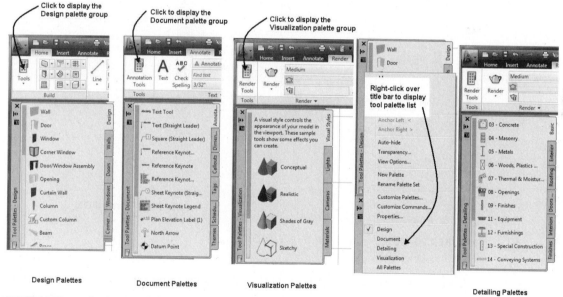

FIGURE 1.19 *Tool palette groups*

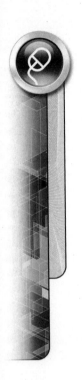

TUTORIAL 1-0: ACCESSING TOOL PALETTES FROM THE RIBBON

1. Launch AutoCAD Architecture 2012 by selecting the shortcut from the desktop. Close the Welcome Screen.

2. Choose the Annotate tab. Click twice the **Annotation Tools** button of the Tools panel at left on the tab to display the Document palette set as shown in Figure 1.19. The Document palette includes tool palettes for dimensioning, annotating, tagging, placing callouts, and placing schedules.

3. Choose the Render tab. Choose the **Render Tools** button of the Tools panel located at left on the tab to display the Visualization palette set as shown in Figure 1.19. The Visualization palette includes Visual Styles, Lights, Camera, and Materials palettes.

4. Choose the Home tab. Choose the **Tools** button located at left of the Build panel to display the Design palette set. The Design palette includes tools for placing walls, doors, windows, door/window assemblies, openings, curtain walls, structural columns, structural beams, structural braces, slabs, roof slabs, roofs, stairs, railings, structural column grids, ceiling grids, and spaces.

5. To display the Detailing palette set, right-click the title bar of the Design palette set and choose Detailing from the context menu. The Detailing palette includes detail tools from each of the Master Format divisions.

6. Right-click over the title bar of tool palette, and choose Auto-hide as shown in Figure 1.20. When Auto-hide is selected, a check will precede the Auto-hide option. Move the cursor from the tool palette to turn off palette display.

Auto-hide Turned ON

FIGURE 1.20 *Accessing Auto-hide for a palette*

7. Move the cursor to the title bar of the tool palette to return the display of palettes.

8. In this step, you will toggle off the Auto-hide option. Right-click over the title bar of the tool palette, and choose Auto-hide.

9. Right-click over the title bar of the tool palettes, and choose View Options from the shortcut menu. The View Options dialog box allows you to change the display of the palettes. Choose the Icon with text radio button. Verify that the Apply to option is set to Current Palette as shown in Figure 1.21; click OK to change the display of the palette.

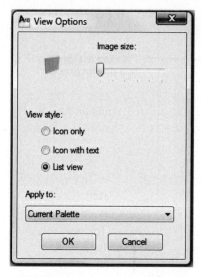

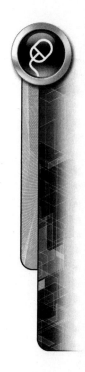

FIGURE 1.21 *Accessing View Options for palettes*

10. Right-click over the title bar of the tool palettes, and choose **View Options**. Choose the List view radio button; choose OK to return the palette display to List view.

11. To change the size of the tool palettes, move the cursor to the edge of the palette. When a two-way arrow is displayed, left-click and drag the palette to change its size as shown in Figure 1.22.

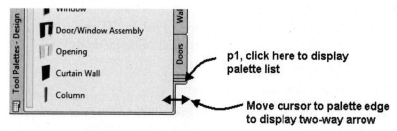

FIGURE 1.22 *Resizing a palette set*

12. You can resize the palettes to hide the tabs of the palette. To display the hidden tabs, click the palette stack at **p1** as shown in Figure 1.22 and choose the name of a palette from the flyout list.

13. To view the palette settings, right-click the tool palette title bar and choose **Customize Palettes** from the shortcut menu. The palettes of the palette groups are shown in the right window in Figure 1.23. You may choose a palette shown in the left window and drag the palette to the right window to assign that palette to a group.

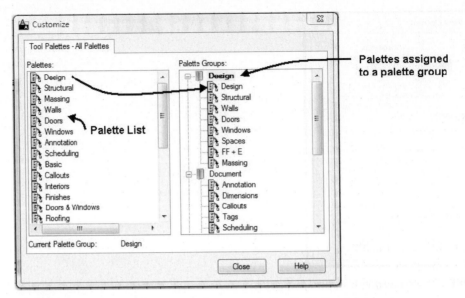

FIGURE 1.23 *Assigning palettes to a palette group in the Customize dialog box*

14. Save and close the drawing.

Customizing Tool Palettes

Sample styles of objects are included in the Design tool palette. However, you can create additional tools for a palette by using one of the following methods:

- Drag an instance of an object onto a palette.
- Drag an object style from Style Manager onto a palette.
- Drag the i-drop of an object style in the Content Browser onto a palette.
- Drag detail content from the Detail Component Manager onto a palette.
- Drag blocks from the DesignCenter onto a palette.

When you place tools on a tool palette, the definition of the tool is specified in its source. Therefore, if you drag an instance of an object onto a palette, the source drawing must remain available to your computer. You can identify the drawing that is the source for the style of each tool if you right-click over the tool on the tool palette and choose Properties to open the **Tool Properties** dialog as shown in Figure 1.24. The source for the style of the tool is shown in Figure 1.24. If you drag an instance of an object to a tool palette, for example, the tool functions from the style definition specified by the tool. If the style source drawing is moved, the tool will not function. If the style source points to the Content Browser, then the tool is not dependent upon a specific drawing and will continue to function in other drawings. The tool catalog generator command introduced in Chapter 10 allows you to create a catalog of all tools used in a drawing or project that can be accessed through the Content Browser.

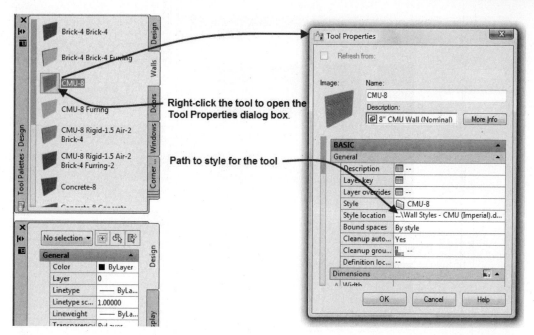

FIGURE 1.24 *Properties of a tool*

The **Content Browser** consists of a library of tool catalogs shipped with the software. The catalogs are shown in Figure 1.25. The metric and imperial Design catalogs consist of object styles for doors, windows, and walls. The Content and Plug-ins Catalog provides links to Online resources such as Kohler and Marvin Windows. The metric and imperial Document catalogs include tools for annotations, callouts, and schedule tables. The sample palettes for metric and imperial tools are also included in the Content Browser. A tool catalog stores pointers to the physical tool catalogs instead of storing the tool definitions in a file.

The Content Browser consists of the following catalogs:

- Stock Tool Catalog
- Sample Palette Catalog - Imperial and Sample Palette Catalog - Metric
- Design Tool Catalog Imperial and Design Tool Catalog Metric
- Documentation Tool Catalog Imperial and Documentation Tool Catalog Metric
- Visualization Catalog
- Content & Plug-ins Catalog
- My Tool Catalog

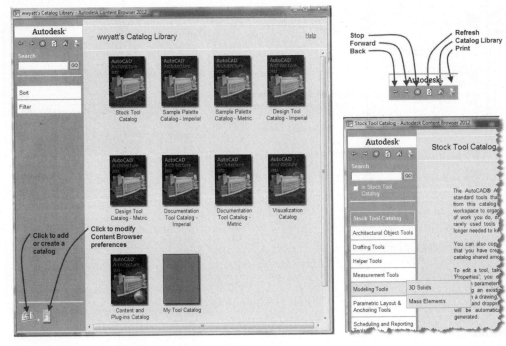

FIGURE 1.25 *Content Browser and Stock Tool Catalog*

Access the **Content Browser** as shown in Table 1.4.

TABLE 1.4 *Content Browser command access*

Command prompt	**CONTENTBROWSER**
Ribbon	Home tab > Tools panel, as shown in Figure 1.26

When you access the **Content Browser**, it opens as shown in Figure 1.26.

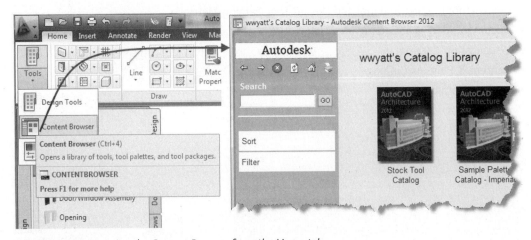

FIGURE 1.26 *Accessing the Content Browser from the Home tab*

The Content Browser consists of two panes. The left pane has navigation controls at the top and bottom. When a catalog is opened, the categories will display in the left pane. Tool catalogs may include the following:

Tools—Used to place or edit objects. A tool can be used to place a wall in the drawing.

Tool palette—Multiple tools displayed as a group because they are used for similar tasks. The tool palette is considered a single object.

Tool package—A collection of tools that you can select for your tool palette.

The right pane displays the contents of tool catalog categories. If you move the pointer over the categories listed in the left pane, the subcategories of the category will expand as shown in Figure 1.27. You can select the subcategory by clicking its title while its subtitle is expanded. The Mass Elements subcategory is shown opened in Figure 1.27. The right pane displays the tools of the Mass Elements subcategory. You can return to the previous state of the Content Browser by selecting the **Back** arrow shown at the top of Figure 1.27.

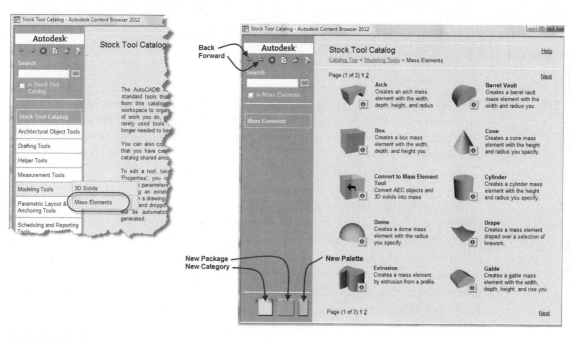

FIGURE 1.27 *Mass Elements of the Stock Tool Catalog*

You can open a catalog in a new window if you right-click over a catalog image of the **Content Browser** and choose Open in **New Window**. In addition, press CTRL + N to open an additional window of the Content Browser. The Content Browser will minimize when you click on the workspace. However, you can right-click the title bar of the Content Browser and choose **Always on Top** to retain the display of the Content Browser when you click on the workspace.

The **My Tool Catalog** shown in Figure 1.28 is provided to store custom tool palettes. It starts out as an empty catalog located in the Content Browser. You can open the My Tool Catalog and other tool catalogs in separate Content Browser windows, then use the i-drop to drop tools into the My Tool Catalog. You can use the i-drop feature to insert tools from palettes of your drawing into the My Tool Catalog. To drag a tool from a tool palette to the My Tool Catalog, open the My Tool Catalog, right-click over the title bar, and choose Always on Top. Drag the i-drop from a tool palette to the My Tool Catalog. The buttons at the bottom in the left pane are used to create new palettes, packages, or categories. Type CTRL + N to open an additional Content Browser window, then you can navigate to other catalogs in the new window and drag content into your catalog. The i-drop shown in the right pane allows you to drag a tool or a tool palette into the current workspace. In Figure 1.28, the **Basic Legend** tool was inserted with i-drop from the AutoCAD Architecture 2012 Stock Tool Catalog shown on the left to the My Tool Catalog shown on the right.

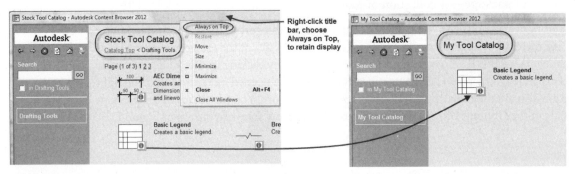

FIGURE 1.28 *Adding tools to the My Tool Catalog*

TIP	The Sample Palette Catalog-Imperial and Sample Palette Catalog-Metric of the Content Browser consist of palette sets included when the software is installed. Therefore, you can click and drag the i-drop for a palette from the Content Browser to restore tool content of the original sample palettes.

Using the Properties Palette

When you choose a tool from the tool palette, the **Properties** palette opens to display the current settings for the new object. The current settings for a tool are inherited from the last insertion of that tool. Open the Properties palette as shown in Table 1.5. The Properties palette is used to set the properties of new objects and to modify existing objects. The palette includes three tabs: **Design**, **Display**, and **Extended Data**. The settings of a wall object as displayed during placement are shown in Figure 1.29.

TABLE 1.5 *Accessing the Properties palette*

Command prompt	**PROPERTIES or** CTRL + 1
Ribbon	Home tab > Tools panel, choose Properties from the drop down

The most basic properties are listed in the **Basic** section at the top. If you are drawing a wall, the Basic section lists the general information such as style, dimensions of the wall, and location, while the **Advanced** section includes cleanups, style overrides, and worksheets. Depending on the object created, the Properties palette may include graphical illustrations of features of the object. Selecting the **Illustration** toggle as shown in Figure 1.29 will turn off/on the display of the illustration feature. Each category of the Properties palette can be collapsed by selecting the **Close** toggle or opened by selecting the **Open** toggle. The entire Basic or Advanced category can be opened or closed by selecting the **Open** or **Close** category toggle. The Display tab allows you to edit the display settings for the object. The display settings of objects are presented in this chapter. The **Extended Data** tab of the Properties palette provides space for recording documentation files that support the object and may include property data for schedules.

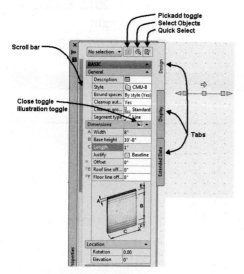

FIGURE 1.29 *Properties palette*

Access Instructional Video 1.1—Introduction to the Workspace located **in the Instructional Video category of the Student Companion site of CengageBrain** http://www.cengagebrain. com described in the Preface.

TUTORIAL 1-1: ACCESSING THE RIBBON AND THE CONTENT BROWSER

1. Launch AutoCAD Architecture 2012 by selecting the shortcut from the desktop. If AutoCAD Architecture is open, choose User Interface Overview from the Help drop-down menu shown in Figure 1.30.

FIGURE 1.30 *Accessing the User Interface Overview*

2. Select the Maximize button in the AutoCAD Architecture User Interface Overview window. Move the cursor over each of the workspace components shown in Figure 1.31. Read the description displayed at left for the component.

3. Choose Close on the title bar to dismiss the AutoCAD Architecture User Interface Overview window.

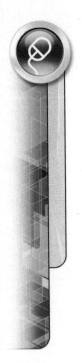

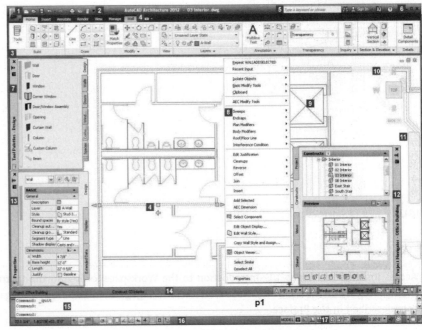

FIGURE 1.31 *User Interface Overview screen*

4. Choose **QNew**, shown in Figure 1.32, from the Quick Access toolbar.

5. Verify that the **Model** is selected from the Application status bar as shown at **p1** in Figure 1.31.

6. In the following steps, you will display the Design, Document, and Visualization palettes using the Ribbon.

7. Choose the Annotate tab of the Ribbon. The Annotate tab consists of panels for adding text and dimensions to a drawing.

8. Click twice the Annotation Tools button of the Tools panel to display the Document tool palette set as shown at left in Figure 1.32.

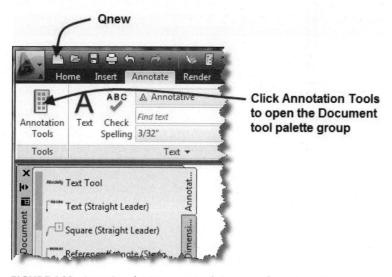

FIGURE 1.32 *Accessing the Document palette group from the Ribbon*

9. Choose the Render tab of the Ribbon. The Render tab consists of panels that support rendering operations such as adding lights and materials.

10. Click twice the Render Tools button of the Tools panel to display the Visualization tool palette set as shown at right in Figure 1.33.

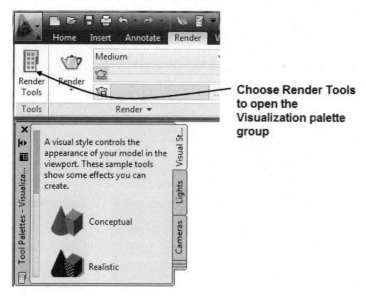

FIGURE 1.33 *Accessing the Visualization palette group from the Ribbon*

11. Choose the Home tab of the Ribbon. The Home tab consists of tools for placing walls, doors, and windows.

12. Click the Tools button of the Build panel to display the Design tool palette group as shown in Figure 1.34.

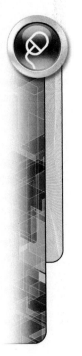

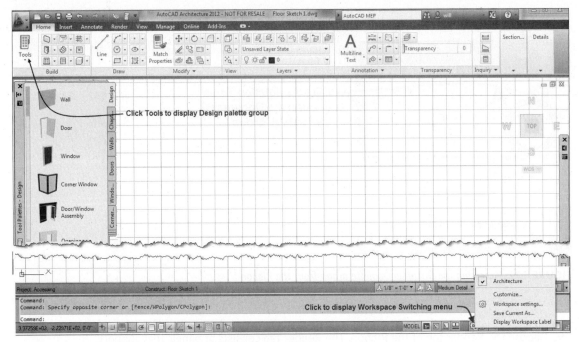

FIGURE 1.34 *Accessing the Design palette group from the Ribbon*

13. Choose the flyout of the Workspace Switching toggle shown in Figure 1.34 located at right in the Application status bar. Select the Architecture workspace. The screen will flash to display the content of the Architecture workspace. Verify that the Home tab is current and the Design palette group is displayed at left.

14. To create a new palette in the Design palette set, right-click over the title bar of the tool palette and choose **New Palette**. Overtype **Chapter 1** as the name of the new palette.

15. Choose the flyout of the Tools button in the Build panel of the Home tab; choose **Content Browser** tool as shown in Figure 1.35.

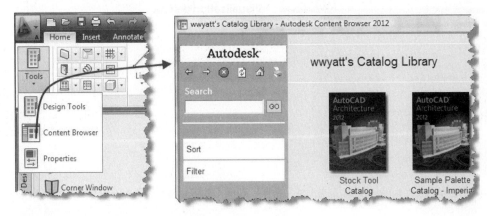

FIGURE 1.35 *Accessing the Content Browser from the Ribbon*

16. Double-click the Design Tool Catalog - Imperial [Design Tool Catalog - Metric]. Click the **Walls > Stud**. Choose **Next** in the lower-right corner to navigate to page 3 [2]. Select and drag the i-drop of the **Stud-3.5 Sheathing-0.5 Siding [Stud-064 Siding-018]** tool onto the new palette as shown in Figure 1.36. To minimize the Content Browser, left-click in the workspace.

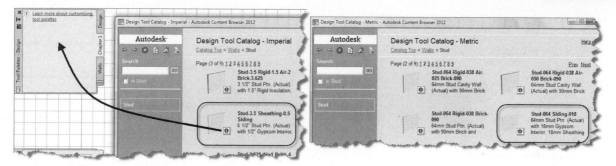

FIGURE 1.36 *Placing tools on a tool palette*

17. Toggle ON Snap Display, Grid Display, Polar Tracking, Object Snap, Object Snap Tracking, Dynamic Input, and Show/Hide Transparency toggles of the Application status bar as shown in Figure 1.37.

18. Verify that the Dynamic Input is toggled ON in the Application status bar as shown in Figure 1.37. Right-click over the Dynamic Input icon toggle and choose Settings to open the Drafting Settings dialog box. Check Enable Dimension Input where possible and Show command prompting and command input near crosshairs.

19. Choose the Snap and Grid tab as shown in Figure 1.37. In this step, you will set your cursor movement to intervals of 1'-4" [400] and major grid lines will be displayed every **4'-0" [1200]**. Edit Snap x spacing = **16 [400]**, Snap Y spacing = **16 [400]**, Grid X spacing = **16 [400]**, Grid Y spacing = **16 [400]**, and Major line every = **3**.

20. Click OK to dismiss the dialog box.

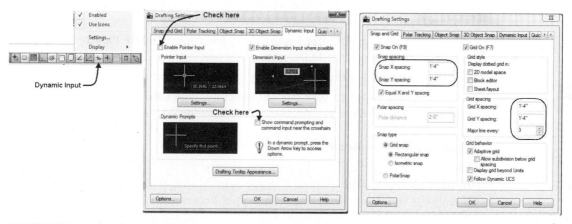

FIGURE 1.37 *Dynamic Input settings*

21. Choose **Stud-3.5 Sheathing-0.5 Siding [Stud-064 Siding-018]** tool from the new palette. Move the pointer to the workspace. Respond to the workspace prompts as follows:

Start point or: *(Select a point near a grid line p1 as shown in Figure 1.38. Move the cursor right to display the dynamic dimension 20'-0" [6000].)*

End point or: *(Roll the mouse wheel forward to magnify and display grid lines 16" [400] apart, and move the cursor right to display 21'-4" [6400] in the dynamic dimension as shown at right in Figure 1.38. Left-click to end the wall segment. Press Esc to end the command.)*

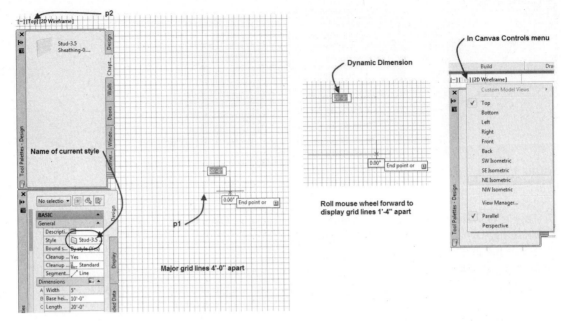

FIGURE 1.38 *Placing a wall in the workspace*

22. Move the cursor to the edge of the tool palettes to display the resize arrows, left click, and drag the top edge of the tool palettes down to expose the In Canvas Controls as shown in Figure 1.38.

23. Choose Top from the menu of the In Canvas Controls shown at p2 in Figure 1.38 to display the View menu and choose SE Isometric.

24. Choose 2D Wireframe from the menu of the In Canvas Controls to display the visual styles flyout list; choose Realistic to view the wall in the Realistic visual style.

25. Right-click over the **Chapter 1** tool palette and choose **Delete Palette**. Click **OK** to dismiss the Confirm Palette Deletion dialog box.

26. Choose **Save As > AutoCAD Drawing** from the Application menu.

27. Edit the **Save in** list to your student directory, type **Lab 1-1** in the **File name** edit field, and choose **Save** to dismiss the **Save Drawing As** dialog box.

Accessing Design Resources

The resources for creating a drawing are located in the **Style Manager**, **Content Browser**, **DesignCenter**, **Detail Component Manager**, and **Structural Catalog**. Throughout your work with AutoCAD Architecture 2012, you will open each of these resources to insert objects or annotations. Wall, door, and window objects are inserted into the drawing using the tools of the Home tab and Design tool palette set. Each object is defined in a style definition. You can access the styles of objects in the **Style Manager**, which allows you to create, edit, import, and export styles. Therefore, to create complex walls that include brick veneer and concrete masonry units, you can create a wall style that includes brick and concrete masonry unit wall components. The wall style definition controls the shape and appearance of the wall. Refer to Table 1.6 to access the Style Manager. Wall styles can be dragged from the Style Manager, shown at the left in Figure 1.39, to a tool palette. Tools located on the tool palette can be dragged to catalogs of the Content Browser.

TABLE 1.6 *Accessing the Style Manager*

Command prompt	**AECSTYLEMANAGER**
Ribbon	Manage > Style & Display panel choose Style Manager

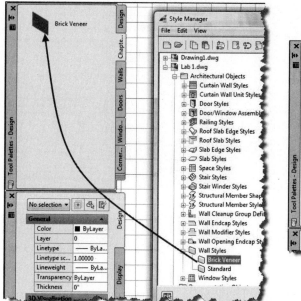
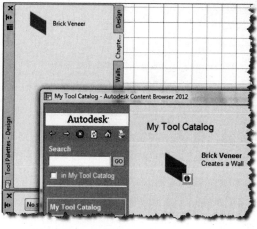

FIGURE 1.39 *Copying styles from the Style Manager to tool palettes and from tool palettes to My Tool Catalog of the Content Browser*

Resources of the Style Manager

Styles define doors, walls, windows, and other objects included in AutoCAD Architecture 2012. The style definitions are accessed in the **Style Manager**. The Style Manager, shown in Figure 1.39, organizes object styles into Architectural Objects, Documentation Objects, and Multi-Purpose Objects folders. The **Architectural Objects** folder includes such styles as doors, windows, stairs, and railings. The **Documentation Objects** folder includes such styles as elevation styles, section styles, and schedule tables. The **Multi-Purpose Objects** folder includes layer key styles, profiles, and material definitions. Styles are not included in the template; therefore, the Style Manager is used to copy a style from other drawings to the current drawing.

Inserting Symbols

Symbols are inserted to represent furniture, appliances, and plumbing fixtures. Symbols are multi-view blocks that are inserted in the drawing from the FF + E palette of the Design palette group. The FF + E palette includes Fixtures, Furnishings, and Equipment. These components are also located in the Design Tool Catalog-Imperial and Design Tool Catalog-Metric catalogs of the Content Browser. This content can also be accessed from the DesignCenter. Refer to Table 1.7 to access the DesignCenter. The AEC Content view of the DesignCenter is shown in Figure 1.40. Additional content is available on the Internet. Type in the Seek window to open Autodesk Seek Internet window as shown in Figure 1.40.

TABLE 1.7 *Accessing the DesignCenter*

Command prompt	**ADCENTER or** CTRL + 2
Ribbon	Insert tab > Content panel choose DesignCenter as shown in Figure 1.40

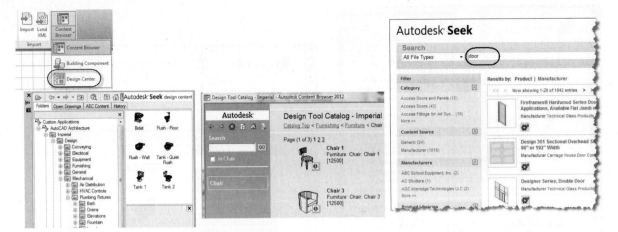

FIGURE 1.40 *DesignCenter and Design Tool Catalog of the Content Browser*

Inserting Documentation

Dimensions, schedules, and symbols can be placed in the drawing from tool palettes of the Documentation palette group. The Documentation folder of the DesignCenter or the Documentation Tool Catalog-Metric and the Documentation Tool Catalog-Imperial catalogs of the Content Browser includes additional documentation tools such as break marks, callouts, dimensions, keynotes, schedules, and tags.

Creating Detail Components

Components for details such as brick, concrete masonry units, and 2 × 4s can be accessed from the Detailing palette group. The **Detailing** tool palettes allow access to the **Detail Component Manager** (see Figure 1.41) and its content. Refer to Table 1.8 to access the Detail Component Manager. (The Detail Component Manager will be presented in Chapter 12.)

TABLE 1.8 *Accessing the Detail Component Manager*

Command prompt	**AECDTLCOMPMANAGER**
Ribbon	Home tab > Details panel choose Detail Components

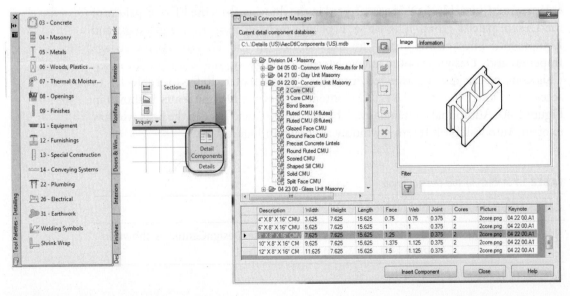

FIGURE 1.41 *Detail Component Manager*

Inserting Structural Components

Structural components such as precast concrete, steel, and wood components can be inserted in the drawing from the **Structural Member Catalog** (see Figure 1.42). Refer to Table 1.9 to access the Structural Member Catalog. (The Structural Member Catalog is presented in detail in Chapter 13.)

TABLE 1.9 *Accessing the Structural Member Catalog*

Command prompt	**AECSMEMBERCATALOG**
Ribbon	Manage tab > Style & Display panel choose Structural Member Catalog

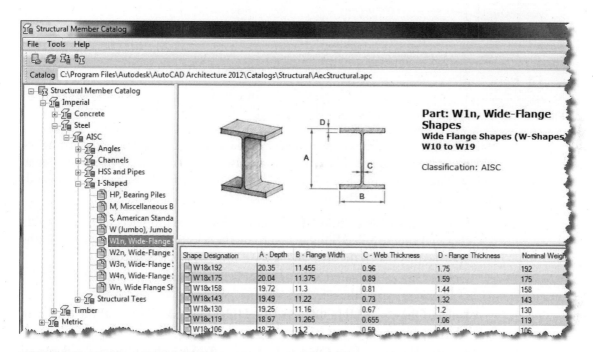

FIGURE 1.42 *Structural Member Catalog*

Using Contextual Tabs for Editing

Contextual tabs of the ribbon are displayed when you select an object. The content of the contextual tab includes the commands for editing the object. Therefore, after placing a wall, you can select the wall and choose commands for the edit of the wall. Access the Wall tab of the ribbon as shown in Figure 1.43.

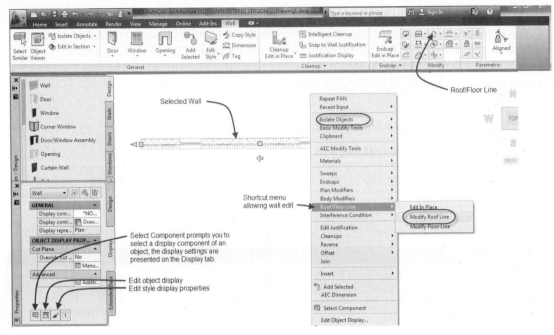

FIGURE 1.43 *Editing objects from the ribbon and shortcut menu*

The General panel of the contextual tab includes options for viewing and isolating the objects in the workspace. The edit tools are also located on the shortcut menu of an object. The shortcut menu and contextual tabs also include the following commands to select and control the display of objects:

Select Similar (SelectSimilar)—This tool allows you to select an object; right-click and choose **SelectSimilar** to turn on the selection of all other objects in the drawing that match the style and layer of the initial object selected.

Isolate Objects > Isolate Objects (AecIsolateObjects)—Turns off the display of all objects except the selection.

Isolate Objects > Hide Objects (AecHideObjects)—Turns off the display of selected objects.

Isolate Objects > End Objects Isolation (AecUnisolateObjects)—Turns off control of Isolate Objects and Hide Objects.

Isolate Objects > Edit in Section (AecEditInSection)—Provides a series of prompts to select objects and generate a section view of the selected objects that can be edited while viewed in section.

Isolate Objects > Edit in Elevation (AecEditInElevation)—Provides a series of prompts to select objects and generate an elevation view of the selected objects for editing.

Isolate Objects > Edit in Plan (AecEditInPlan)—Provides a series of prompts to select one or more objects and perform an edit in place from the plan orientation.

Add Selected—Select a wall and choose Add Selected from the General panel to repeat the WallAdd tool and place a wall of the same style and properties of the selected wall.

TIP

Use the SelectSimilar command and the Properties palette to quickly select all windows in a drawing, and edit one or more properties in the Properties palette.

CONFIGURING THE DRAWING SETUP

The **Drawing Setup** command allows you to set the units, scale, layering, and display. Access the Drawing Setup command as shown in Table 1.10.

TABLE 1.10 *Drawing Setup command access*

Command prompt	**AECDWGSETUP**
Application menu	Utilities > Drawing Setup

Selecting the **Drawing Setup** command opens the **Drawing Setup** dialog box shown in Figure 1.44. This dialog box consists of the **Units, Scale, Layering,** and **Display** tabs. The settings selected from the Annotation Scale flyout of the Drawing Window status bar are reflected in the Scale tab of the Drawing Setup dialog box. The layer standard used by AutoCAD Architecture 2012 is set in the **Layering** tab. Symbols inserted in the drawing from a tool palette will be placed on layers according to the layer standard specified in the Drawing Setup dialog box. The **Units** tab shown in Figure 1.44 allows you to set the drawing units.

Setting Units

The **Units** tab allows you to set the units of the AutoCAD Architecture 2012 symbols. The Units tab of the Drawing Setup dialog box includes a **Drawing Units** drop-down list and **Length, Area, Angle,** and **Volume** sections, as shown in Figure 1.44.

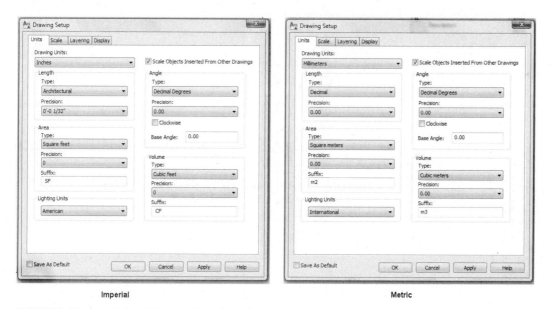

Imperial   Metric

FIGURE 1.44 *Imperial and Metric Units tab of the Drawing Setup dialog box*

Setting Scale

The **Scale** of the drawing is specified by selecting a scale from the Annotation Scale of the Drawing window. The Scale tab of the Drawing Setup dialog box lists the current setting of the Annotation Scale. The changes in scale executed in the Scale tab of the Drawing Setup dialog box do not change the scale of the drawing. Also, a display configuration can be linked to a scale in the Scale tab. When you select an annotation scale from the Drawing Window status bar and the association is specified, the display configuration is changed. If you change the display configuration for a scale, the Drawing Setup—Update Display Configuration dialog box will open as shown in Figure 1.45. Choose the *Update to match my changes* option to change the setting in the current drawing. No link is established if you choose None from the Display Configuration list.

The Scale tab lists the current scale factor of the scale specified by Annotation Scale. Symbols and annotation inserted in the drawing will be scaled according to the scale factor specified. Annotative tools, presented in Chapter 11, include scale representations for one or more annotative scales selected for a drawing. Annotative tools include AEC Dimensions, tags, and text, which are located in the Document palette set. The scale listed in Figure 1.45 indicates the drawing is set for 1/8″ = 1′-0″ [1:100]; the resulting scale factor is shown inactive below the scale list. The Scale Value is set to 96 [100] for the 1/8″ = 1′-0″ [1:100] scale. For example, if a receptacle symbol is placed in a drawing with the scale set to 1/4″ = 1′-0″ [1:50], it is drawn with a 6 [150] diameter circle. This symbol, if placed in a drawing with the 1/8″ = 1′-0″ [1:100] annotation scale, would be drawn with a 12 [300] diameter circle because the scale factor has changed from 48 to 96 [50 to 100]. The scale of a drawing sets the scale factor, which is used as a multiplier for selected symbols and annotation. The scale factor is the ratio between the size of the AutoCAD entity and its size when printed on paper. The Edit Scale List button allows you to add custom scales.

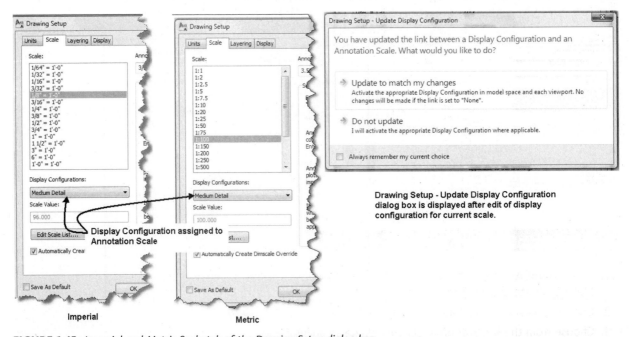

FIGURE 1.45 *Imperial and Metric Scale tab of the Drawing Setup dialog box*

Defining the Layer Standard

The **Layering** tab shown in Figure 1.46 allows you to select a predefined layer standard for the drawing. The imperial or metric layer standard is set when AutoCAD Architecture is launched from the desktop shortcut AutoCAD Architecture 2012 (US Imperial) or AutoCAD Architecture 2012 (US Metric). The Layering tab includes a **Layer Standards/Key File to Auto Import** edit field and a **Default Layer Standard Layer Key Style** list. Using the *AecLayerStd* file and the AIA (256 color)* current drawing will place architectural features such as walls, doors, and windows on the layer as defined in the AIA standard. This standard can be verified in this tab. The tutorials included in this book utilize the AIA (256 Color) [BS1192 Cisfb (256 Color)] layer key style of the *AecLayerStd.dwg* drawing.

> New layers created that do not comply with the standard will plot monochrome if the Normal plot style is not used.

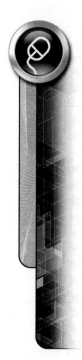

NOTE

Controlling Display in Drawing Setup

The **Display** tab lists the AutoCAD Architecture 2012 objects in the left window (see Figure 1.46). The display representations available for each object are in the right window. The display representations are shown in the figure for a door object. The default display configuration can be set as shown in Figure 1.46. Display configuration is presented in more detail later in this chapter.

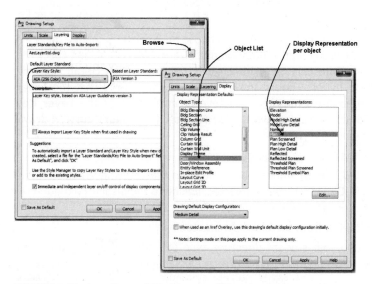

FIGURE 1.46 *Layering and Display tabs of Drawing Setup*

TUTORIAL 1-2: SETTING ANNOTATION SCALE

1. Select AutoCAD Architecture 2012 (US Imperial) [AutoCAD Architecture 2012 (US Metric)] shortcut from the desktop.

2. Choose **QNew** from the Quick Access toolbar.

3. Choose from the Application menu **Utilities > Drawing Setup**.

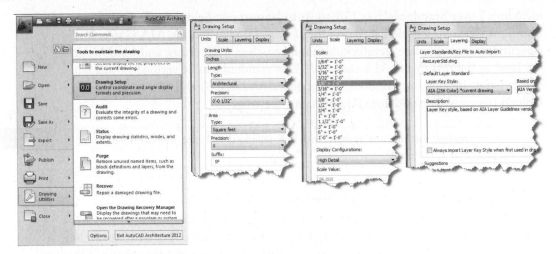

FIGURE 1.47 *Applying Drawing Setup*

4. To verify the units of the drawing, choose the **Units** tab and verify that the units are set to inches [millimeters].

5. To set the scale and display configuration link, choose the **Scale** tab, change the scale to **1/4" = 1'-0" [1:50]**, verify that Annotation plot size = **3/32" [3.5]**, and check the **Automatically Create Dimscale Override** checkbox. Choose **High Detail** from the Display Configuration flyout. Choose the **Layering** tab and verify that the **Layer Standards/Key File to Auto-Import** is *AecLayerStd.dwg* and that the **Layer Key Style** is **AIA (256 Color)*current drawing [BS1192 Cisfb (256 color)*current drawing]**, as shown in Figure 1.47.

6. Click **OK** to dismiss the Drawing Setup dialog box.

7. Choose the ¼" = 1'-0" **[1:50]** Annotation Scale from the Drawing Window status bar. Since the ¼" = 1'-0" **[1:50]** scale is selected, linking to the display configuration changes the Display Configuration to High Detail.

8. Choose **Save** or **CTRL + S** from the Quick Access toolbar.

9. Edit the **Save in** list to your student directory, type **Lab 1-2** in the **File name** edit field, and click **Save** to dismiss the Save Drawing As dialog box.

10. Close the drawing.

DEFINING DISPLAY CONTROL FOR OBJECTS

Display control in AutoCAD Architecture 2012 consists of three levels: display representations, display representation sets, and display configurations. These levels can be viewed and edited in the Display Manager dialog box and the Display tab of the Properties palette. The **AecDisplayManager** command opens the Display Manager dialog box. The three methods of defining display are:

- Drawing Default—displays properties defined in the Display Manager
- Object Override—displays properties defined uniquely per object
- Style—displays properties defined in the style definition of an object

The Display Manager lists the Drawing default settings for a drawing. Access the AecDisplayManager command as shown in Table 1.11.

TABLE 1.11 *Display Manager command access*

Command prompt	AECDISPLAYMANAGER
Ribbon	Manage tab > Style & Display panel choose Display Manager

If this command is accessed while in paper space, you will be prompted to select a viewport and the **Display Manager** will open, presenting information relative to the selected viewport. If model space is current for a viewport when you select this command, the Display Manager will open, presenting information relative to the active viewport.

The **Display Manager** dialog box shown in Figure 1.48 has combined all three levels of the display system into one dialog box. The Display Manager consists of three categories: **Configurations**, **Sets**, and **Representations by Object**. Each of these categories can be opened to view the available options in the right pane. Selecting the plus (+) sign of the Configurations category within the tree in the left pane will open the category and display the available configurations. In Figure 1.48, the Medium Detail configuration is shown bold because it is the configuration applied to the current viewport.

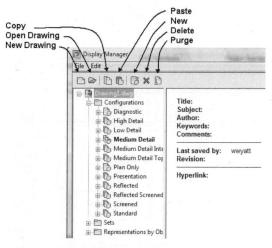

FIGURE 1.48 *Display Configurations in the Display Manager*

Selecting Display Representations

Display representations control how objects are displayed. One or more display representations can be applied to define the display of the object. The display representations define how much detail to display. If you are creating a drawing for $\frac{1}{4}'' = 1'\text{-}0''$ [1:50] objects, they should be displayed with greater detail than $\frac{1}{16}'' = 1'\text{-}0''$ [1:200]. Objects consist of several display components; the number of display components varies according to the detail desired for the display representation. Each component of an object can be controlled for visibility, layer, color, and linetype in the display representation. You can view the properties of a display representation in the Display Manager as shown in Figure 1.49.

In Figure 1.49, the plus (+) symbol of the **Representations by Object** category has been selected to display a list of all objects. If you scroll down the list of objects and select an object in the left pane, the right pane of the Display Manager will display a list of the display representations for that object (see Figure 1.49). The list of **Sets** in which the display representations can be used is listed across the top of the right pane. The display representation for an object varies according to the desired representation of that object in the drawing. The Display Representations column shown in Figure 1.49 lists the display representations for a door. Each of these representations will turn on or off the display of specific components of the object. For example, a door object consists of the following display components when the **Plan** display representation is used: Door Panel, Frame, Stop, Swing, Direction, Door Panel Above Cut Plane, Frame Above Cut Plane, Stop Above Cut Plane, Swing Above Cut Plane, Door Panel Below Cut Plane, Frame Below Cut Plane, Stop Below Cut Plane, and Swing Below Cut Plane. A door when shown for a floor plan could use the **Plan** set, which uses the **Plan** and **Threshold Plan** representations as shown in Figure 1.49. Certain display components of the door are visible in the Plan representation, while others common to other display representations are not included. The Threshold Plan representation allows the threshold display component to be turned on. The door shown in Figure 1.50 is displayed according to the three different display configurations. The door shown at the left in the figure is defined by the **Reflected** display configuration that assigns the Reflected display representation to objects. The Reflected display is used for developing reflected ceiling plans. The door shown at right in Figure 1.50 is a simplified representation of the door using Low Detail, which is typical for 1/16" = 1'-0" scale.

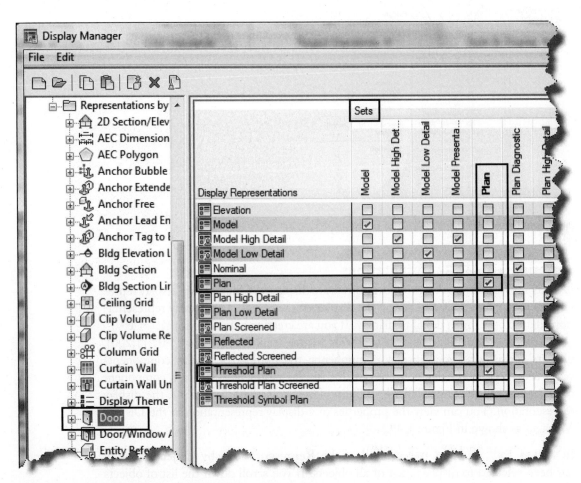

FIGURE 1.49 *Door display representations in the Display Manager*

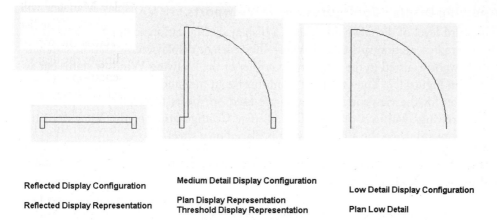

Reflected Display Configuration

Reflected Display Representation

Medium Detail Display Configuration

Plan Display Representation
Threshold Display Representation

Low Detail Display Configuration

Plan Low Detail

FIGURE 1.50 *Display configurations applied to the display of a door*

Contents of Display Representation Sets

A display representation set groups display representations to obtain an appropriate object representation for a drawing. The display representation sets are shown across the top of Figure 1.49. Notice that the Plan and Threshold display representations are checked for the Plan set. You can view the available sets by expanding the **Sets** folder of the **Display Manager** as shown in Figure 1.51. In this figure, the left pane lists the fifteen sets. The Plan set is selected in the left pane and the **Display Representation Control** tab is selected in the right pane. The Objects column in the right pane lists all objects and the top row lists all display representations available. If a box is checked, that object and its display representation are applied for the Plan set.

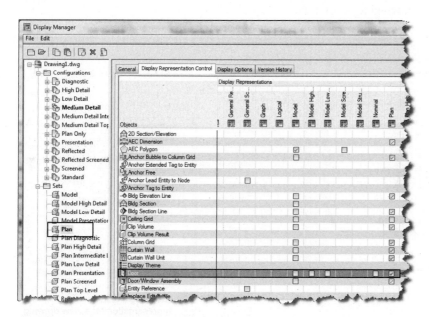

FIGURE 1.51 *Sets folder of the Display Manager*

A display representation set therefore controls the visibility of the components according to the needs of the drawing.

Applying Display Configurations to Viewports

The third level of display control uses a display configuration to apply one or more display representation sets to a viewport. The current display configuration assigned to a viewport is listed in the lower-right corner of the Drawing Window status bar as shown in Figure 1.7. You can select from this flyout menu to change the display configuration for the viewport. Viewports are used on sheets to display the content of View drawings within a project. The Display Configuration of the View drawing comes forward when the View drawing is placed on a sheet.

To view the content of a display configuration, choose from the ribbon the **Manage** tab > **Style & Display panel** > **Display Manager**. Expand the **Configurations** folder of the left pane as shown in Figure 1.52. The right pane of this tab describes the display configuration.

The purpose of each tab in the right pane is as follows:

General—Allows you to edit the name and description of the display configuration.

Configuration—Allows you to assign display representation sets to define the display representation set for a given view direction. In Figure 1.52, the Configuration tab is shown for the **Medium Detail** display configuration and the view directions with their respective display representation sets are shown in the right pane. Therefore, when this display configuration is used and the objects are viewed from the top, the Plan display representation set will be applied. The available display representation sets are shown at the right in Figure 1.52. The **Section_Elev** display representation set will be applied when the design is viewed from the left, right, front, or back. The **Model** display representation set will be used when a pictorial view is applied.

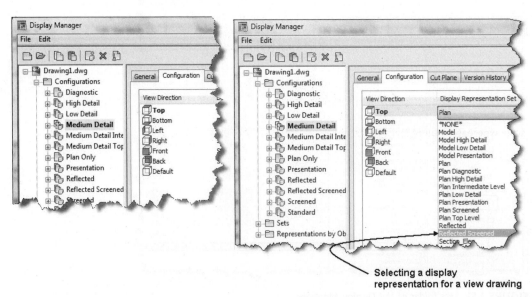

FIGURE 1.52 *Configuration of the Medium Detail display configuration*

Cut Plane—This tab, shown in Figure 1.53, allows you to change the cut plane height and the range of display. The default cut plane height is 3'-6" [1400]. You can modify the cut plane height here or edit the Global Cut Plane value in the Drawing Window status bar as shown in Figure 1.7.

You can change the display representation set assigned to a view direction by selecting the **View Direction** of the **Configuration** tab, clicking the current name in the **Display Representation Set** column, and selecting from the list, as shown in Figure 1.53.

The display configuration shown in Figure 1.53 is **View Directional Dependent**. A View Directional Dependent display configuration will control the display of an object dependent upon the view. You can check the **Override View Direction** checkbox of the **Configuration** tab and assign a **Fixed View Direction** for the viewport by selecting from the **Fixed View Direction** list as shown at the right.

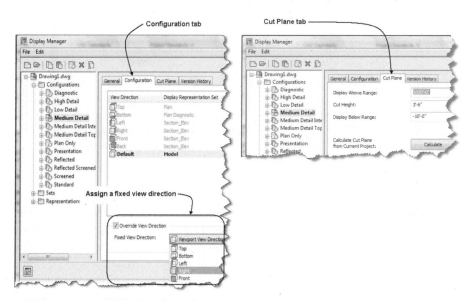

FIGURE 1.53 *Overriding the view direction*

Although display configurations can be edited, the display configurations available in the AEC templates are usually adequate for creating architectural working drawings.

> Display configurations can cause objects to be visible in one viewport but not necessarily in all viewports. In later tutorials, display configurations will be set to display objects and dimensions based upon the intended scale. High Detail is assigned for the 1/4″ = 1′-0″ [1:50] scale whereas Medium Detail is assigned for the 1/8″ = 1′-0″ [1:100] scale.

NOTE

VIEWING OBJECTS WITH THE OBJECT VIEWER

One or more objects can be selected and viewed in the **Object Viewer** window. The Object Viewer allows you to view the objects using display configurations from various view directions. The Object Viewer shown in Figure 1.54 includes a listing of the commands on the toolbars of the Object Viewer. Access the Object Viewer as shown in Table 1.12.

TABLE 1.12 *Object Viewer command access*

Command Prompt	**OBJECTVIEWER**
Ribbon	Select an AEC object, and choose Object Viewer from the General panel of the objects contextual tab.
Shortcut menu	Select an AEC object, right-click, and choose Object Viewer.

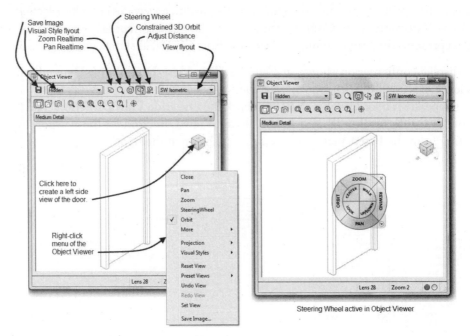

FIGURE 1.54 *Object Viewer display of a door*

When the Object Viewer opens, the object is displayed in the window. The toolbar and shortcut menu allow you to adjust the view within the window. If you move your pointer over the view and right-click, a shortcut menu allows you to choose other viewing options.

WARNING

> To terminate a real-time zoom or pan, click outside of the Viewer. Pressing the ESC key to end real-time zoom while in the Viewer will close the Viewer.

After creating an isometric view, you can create a perspective view if you select **Projection > Perspective** from the shortcut menu in the graphics window of the viewer. The image can also be saved as a *.png*, *.jpg*, *.bmp*, or *.tif* file if you choose the **Save Image** command as shown in Figure 1.54. The **Set View** option of the shortcut menu sets the view in the drawing area to be equal to that displayed in the Object Viewer.

EDITING OBJECTS WITH OBJECT DISPLAY

The display of each object can also be controlled as an exception to that defined in the style or drawing default. This level of individual control is accessed by **ObjectDisplay**. Access **ObjectDisplay** as shown in Table 1.13.

TABLE 1.13 *ObjectDisplay command access*

Command prompt	OBJECTDISPLAY
Ribbon	Select an object and choose Object Viewer from the General panel the contextual tab.
Shortcut menu	Select an object, right-click, and choose Edit Object Display.

When you select an object, right-click, and select **Edit Object Display**, the **Object Display** dialog box opens as shown in Figure 1.55. The Object Display dialog box consists of **Materials**, **Display Properties**, and **General Properties** tabs. The **Materials** tab shown in Figure 1.55 is for a door. You can add or edit the materials assigned for the components. If you check the **Object Override** box for a component, such as the Frame shown in Figure 1.55, you can modify the **Material Definition** assigned to that component. Material Definition provides realistic appearance to simulate materials, whereas the Display Properties tab is used just to view the current setting of the display representation. If you check the **Object Override** checkbox or click on the **Properties** button, the **Display Properties** dialog box opens, which includes tabs for editing the display. You can choose the **Layer/Color/Linetype** tab and turn on or off the display of components for the object. If you turn off the light bulb for the **Stop**, the stop component will not be displayed for the selected display representation.

You can use this technique to change the linetype and color of components within an object rather than modifying its object style. The display settings specified in object styles will be presented in Chapter 3.

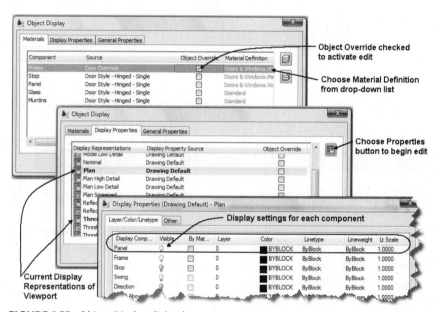

FIGURE 1.55 *Object Display dialog box*

USING THE DISPLAY TAB OF THE PROPERTIES PALETTE

The Display tab of the Properties palette lists the settings controlling the display of a selected object. You can modify the display as an object override, style change, or drawing default within the Display tab. If you select a door, right-click, and choose Properties, the display settings for the door are shown in the Display tab. The options of the **Display controlled by** flyout include control as follows: This object, object style, and drawing default settings. If you choose the Display controlled by **This object**, the **Add Object Override** dialog box opens as shown in Figure 1.56. If you choose OK, you can create an object override when you modify the display components of the object listed in the drop-down list as shown at the right. After creating an object override, you can remove any overrides when you reselect the **This object** option. Display settings for the door style or drawing default can also be specified in the Display tab.

TIP	Right-click the mouse with no objects selected and choose Select Component. The cursor is transformed to a pick box. When you select an object, the Display tab of the Properties palette will open, allowing you to edit the display of the selected components.

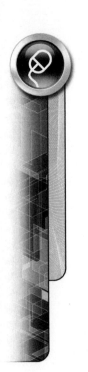

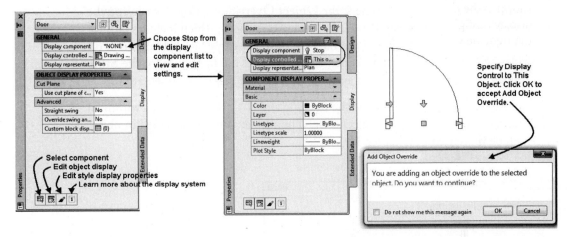

FIGURE 1.56 *Display tab of the Properties palette*

TUTORIAL 1-3: VIEWING THE DISPLAY REPRESENTATION OF OBJECTS

1. Right-click over **Start** in the Windows taskbar and choose Explore to open the Windows Explorer.

2. Verify that the Accessing Tutor content of the CengageBrain has been downloaded to your Accessing Tutor folder on your computer as described in Organizing Tutorial Directories in the Preface.

3. Choose **Open** from the **Quick Access** toolbar. Select Ex 1-3 from your *Accessing Tutor\ Imperial\Ch1* or *Accessing Tutor\Metric\Ch 1* directory.

4. Choose **Save As > AutoCAD Drawing** from the Application menu and save the drawing as **Lab 1-3** in your student directory.

5. Verify that the **Model** toggle is selected from the **Application** status bar. The annotation scale is currently set to 1/8″ = 1′-0″ [1:100].

6. To link the display configuration for the annotation scale, choose from the Application menu **Utilities > Drawing Setup**. Choose the Scale tab. Choose ¼″ = 1′-0″ [1:50] scale. Choose High Detail from the Display Configuration flyout. Choose OK to dismiss the Drawing Setup dialog box.

7. Choose ¼″ = 1′-0″ [1:50] from the Annotation Scale flyout. As a result of step 6, the display configuration changes to **High Detail** as shown in the **Drawing Window** status bar. When the display configuration changes to High Detail, the density of the hatch pattern representing the space object increases.

8. Select the **Work** layout from the **Application** status bar as shown in Figure 1.57, and click on the right viewport. Choose ¼″ = 1′-0″ [1:50] from the Viewport Scale flyout of the Drawing Window status bar.

9. Click on the left viewport. Choose ¼″ = 1′-0″ [1:50] from the Viewport Scale flyout.

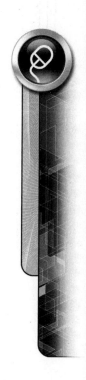

10. In this step, you will use Object Isolation to view only the windows of the drawing. To select all windows of the same style, select a window in the left viewport, and choose **Select Similar** from the Window tab > General panel of the Ribbon. Windows of similar style in the drawing are now selected. To isolate the windows, choose from the **Window** tab > **General** panel > **Isolate Objects**.

11. There are eight windows shown in the left viewport; however, only four windows are shown in the plan view of the right viewport. If you adjust the cutting plane height of display configuration, the missing windows of the right viewport will display. Click on the right viewport and choose **Cut Plane** in the Drawing Window status bar to open the **Global Cut Plane** dialog box shown in Figure 1.57. Type **5'** **[1500]** in the Cut Height field. Click OK to dismiss the Global Cut Plane dialog box.

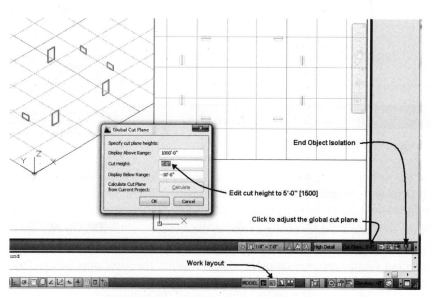

FIGURE 1.57 *Global Cut Plane modified for display configuration*

12. To end the isolation state, choose **End Object Isolation** from the Drawing Window status bar as shown in Figure 1.58.

13. To view the building with the Reflected display configuration, verify that the right viewport is current. Choose the **Reflected** display configuration from the **Display Configuration** flyout menu of the Drawing Window status bar. The viewport should display the ceiling grid as shown in Figure 1.58.

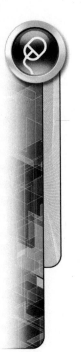

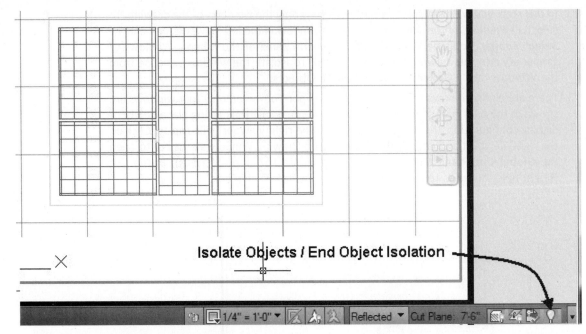

Isolate Objects / End Object Isolation

1/4" = 1'-0"   Reflected   Cut Plane: 7-6"

FIGURE 1.58 *Reflected display configuration of Work layout tab*

14. To view the settings of the **Reflected** display configuration, choose **Manage tab > Style & Display > Display Manager** from the ribbon. Choose the plus (**+**) sign of the **Configurations** folder. Note that the Reflected display configuration is shown bold.

15. Choose the **Reflected** display configuration in the left pane of the Display Manager.

16. Choose the **Configuration** tab of the right pane to display the **View Directions** list. The Reflected display representation is assigned when the view direction is **Top**. Choose the **Cut Plane** tab; the Cut Height is set to **7'-6" [2300]**. The cut height defines the cutting plane position, which allows view of the ceiling grid.

17. Choose the plus (**+**) of the **Sets** folder in the left pane and choose **Reflected** in the left pane. Choose the **Display Representation Control** tab in the right pane as shown in Figure 1.59.

18. Scroll down the list of objects to **Ceiling Grid** and scroll to the right to the **Reflected** display representation as shown in Figure 1.59.

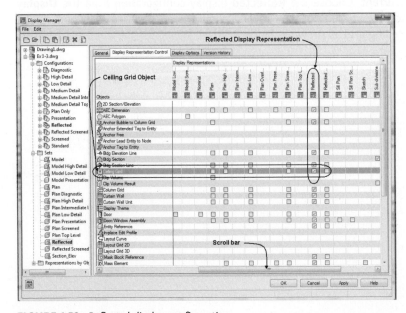

FIGURE 1.59 *Reflected display configuration*

19. The **Ceiling Grid** object is toggled ON for the Reflected display representation as shown in Figure 1.59. Select OK to close the Display Manager.

20. You can modify the display representation of doors in the Display Manager; however, in this step, you will use the **Display** tab of the Properties palette to modify the door display in the Reflected display configuration. Click on the left viewport. Zoom as necessary to select a door. The grips of the selected door are displayed in each viewport.

21. Choose the Display tab of the Properties palette. Verify that the display representation listed in the General section of the Properties palette for the door in the left viewport is Model High Detail.

22. Click on the right viewport. Verify that the display representation of the door is Reflected in the Display tab of the Properties palette. Verify that Display controlled by is set to **Drawing default setting** in the Properties palette. Choose the drop-down list for the Display Component and choose the **Direction** component. Click on the light bulb icon to display the component as shown in Figure 1.60. Since this setting deviates from the drawing default, the **Modify Display Component at Drawing Default Level** dialog box opens. Choose **OK** to accept the change.

23. Choose the drop-down list for the Display Component and choose the **Panel Below Cut Plane** component. Click on the light bulb icon to display the component. The panel of the door is displayed in the closed position in the right viewport.

24. Choose the drop-down list for the Display Component and choose the **Swing Below Cut Plane** component. Click on the light bulb to display the component. Move the cursor from the Properties palette, and press ESC to end edit.

25. Doors are not shown in the open position as defined in settings for the Reflected display representation. From the Advanced section of the Display tab shown in Figure 1.60, choose **No** for the Override swing angle. Move the cursor from the palette to view the doors in the open position.

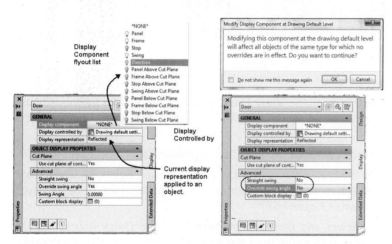

FIGURE 1.60 *Setting door display for the Reflected display representation*

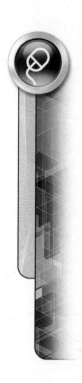

26. Select **Model** in the Application status bar. Choose Reflected from the Display Configuration flyout of the Drawing Window status bar.

27. Choose the High Detail display configuration to display the spaces.

28. Choose the Reflected display configuration.

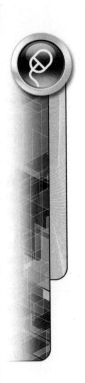

29. To turn off the display of spaces in the drawing, select from the ribbon **Home** tab > **Layer** panel and then click to display the layer drop-down list. Click on the light bulb of the A-Area-Spce [Aec-Space] layer to turn this layer off.

30. To view the building in the Object Viewer, select the entire building using a window selection set from **p1** to **p2** as shown in Figure 1.61 and choose **Object Viewer** from the ribbon **Multiple Objects** > **General** panel.

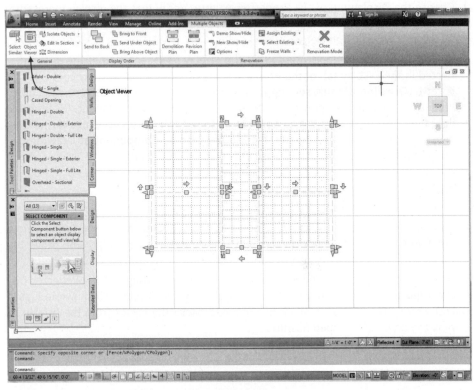

FIGURE 1.61 *Building selected for view*

31. Choose **SW Isometric** from the **View** flyout menu of the Object Viewer. Choose **Realistic** from the Visual Styles drop-down list. Select **Maximize** on the Object Viewer title bar.

32. Verify that **Reflected** is selected from the **Configuration** list of the Object Viewer.

33. Move the cursor to the graphics screen of the Object Viewer, right-click, and choose **Projection** > **Perspective** from the shortcut menu. The building should be displayed in the viewer as shown in Figure 1.62.

34. Close the Object Viewer and then choose **Save** from the Quick Access toolbar. Choose **Close** > **Current Drawing** from the Application menu to close the drawing.

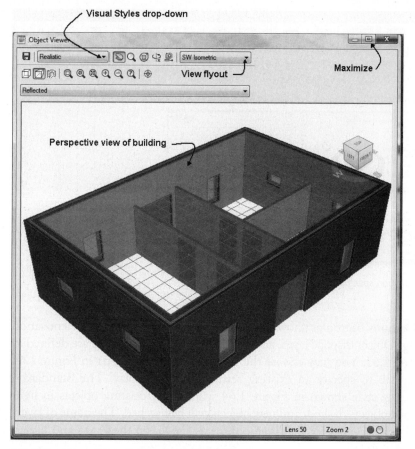

Visual Styles drop-down

Maximize

View flyout

Perspective view of building

FIGURE 1.62 *Building displayed in Object Viewer*

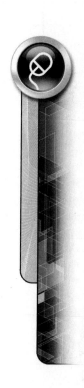

CREATING RENOVATION DRAWINGS

A new feature of AutoCAD Architecture 2012 is the ability to quickly create renovation, demolition, and revision drawings. The Renovation Mode toggle creates a renovation display configuration, which creates a display configuration based upon the current display configuration. The new renovation display configuration when activated will change the display of existing objects and apply display representation for demolished and new objects. The style name of existing and demolished objects is appended to create unique names. Access the Renovation mode as shown in Table 1.14.

TABLE 1.14 *Renovation command access*

Command prompt	AECRenovation
Ribbon	Select Renovation Mode from the Style & Display panel of the Manage tab.

When you toggle on the Renovation Mode, the First Activation of Renovation Mode dialog is displayed as shown in Figure 1.63. This dialog box allows you to modify the name of the new display configuration and set the options for the display. During the renovation session, objects added are classified as new. Existing objects of the drawing can be deleted or changed, and are classified as demolished. When the edits of the drawing are complete, choose **Close Renovation Mode** of the Renovation panel.

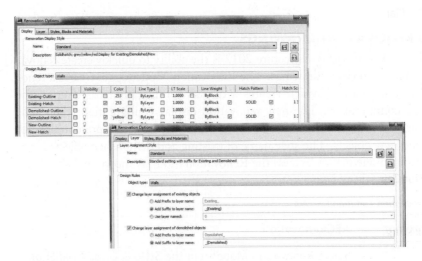

FIGURE 1.63 *Display Configuration specified for Renovation Mode*

Choose the **Options** button as shown at **p1** in Figure 1.63 to open the Renovation Options dialog. Object display, layer, and style naming for renovations are defined as shown in Figure 1.63. You may choose the Name flyout shown at p1 in Figure 1.64 of the Display tab to specify an existing setting for the display. The Standard - renovation display style shown in Figure 1.64 will display existing objects in light grey, demolished objects dashed, and new objects will be hatched. The name of layers and styles will be appended for modified objects as shown on the Layer and Styles, Blocks and Materials tabs.

FIGURE 1.64 *Renovation Options*

The edit tools to assign and select components to the demolition, new, and existing categories are located on the Renovation panel as shown in Figure 1.65 and Table 1.15.

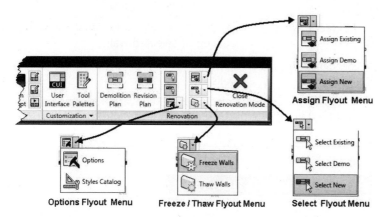

FIGURE 1.65 *Edit tools of the Renovation panel*

TABLE 1.15 *Commands of the Renovation Panel*

Command Prompt	Ribbon	Function
AecCreateDemoPlan	Demolition Plan	Shows the existing components and shows as dashed the objects to be demolished.
AecCreateRevPlan	Revision Plan	Shows the existing components to remain and shows new objects to the plan.
AecDemOff	Demolished Show/Hide	Toggles on or off objects included in the demolished category.
AecNewOff	New Show/Hide	Toggles on or off objects included in the new category of the plan.
AecRenovationOptions	Options	Opens the Renovation Options dialog box allowing the edit of renovation display settings.
AecRenovationCatalog	Styles Catalog	Opens the Renovation Styles Catalog Settings dialog box for the import of renovation styles.
AecAddToExisting	Assign Existing	Allows user to select an object for assignment to the existing category.
AecAddToDemolition	Assign Demo	Allows user to select an object for assignment to the demolish category.
AecAddToNew	Assign New	Allows user to select an object for assignment to the new category.
AecSelectExisting	Select Existing	Selects objects of the plan defined to the existing category.
AecSelectDemolition	Select Demo	Selects objects of the plan defined to the demolished category.

Continued

TABLE 1.15 *Continued*

Command Prompt	Ribbon	Function
AecSelectNew	Select New	Selects objects of the plan defined in the new category of renovation.
AecFreezeWalls	Freeze Walls	Suspends the renovation mode allowing cleanup of selected walls.
AecResetFrozenWalls	Thaw Walls	Resumes renovation mode for selected walls.
AecRenovation	Close Renovation Mode	Selects Close Renovation Mode from the Renovation panel to end the renovation session.

TUTORIAL 1-4: CREATING A RENOVATION PLAN

1. Verify that the Accessing Tutor content of the CengageBrain has been downloaded to your Accessing Tutor folder on your computer as described in Organizing Tutorial Directories in the Preface.

2. Choose **Open** from the **Quick Access** toolbar. Select Ex 1-4 from your *Accessing Tutor \Imperial\Ch1* or *Accessing Tutor\Metric\Ch 1* directory.

3. Choose Save As > AutoCAD Drawing from the Application menu and save the drawing as Lab 1-4 in your student folder.

4. Choose the **Manage** tab. Choose **Renovation Mode** of the Style & Display panel.

5. Verify that **Renovation_Medium Detail** is the name of the new Display Configuration in the First Activation of Renovation Mode dialog box.

6. Choose the Options button of the First Activation of Renovation Mode dialog box. Verify that **Standard** is the name of the Renovation Display Style of the Renovation Options dialog. Choose **OK** to dismiss all dialog boxes.

7. Select the door at **p1** as shown in Figure 1.66. The door includes grips to edit the hinge, swing, size, and location of the door. Choose the square grip at p1 to change the door location, and drag the cursor toward **p2**.

8. Press Escape to view door in a new position.

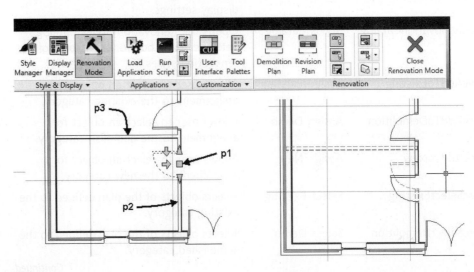

FIGURE 1.66 *Creating content for demolition and new categories*

9. Select the wall at **p3** as shown in Figure 1.66, right-click, and choose Basic Modify Tools > Delete. The display of the wall is represented by a dashed line.

10. Choose the Demolition Plan of the Renovation panel shown at **p1** in Figure 1.67 to open the **Activation of a Demolition Plan**. Verify that the **Renovation_Medium Detail** is specified for the Name of Renovation Display Configuration to activate. Choose **OK** to dismiss the dialog box and view the demolition plan as shown at left in Figure 1.67.

11. Choose **Revision Plan** of the Renovation panel. Select Medium Detail of the flyout of the Name of Display Configuration to activate flyout. Choose OK to dismiss the dialog box.

12. Choose **Close Renovation Mode** of the **Renovation** panel as shown at right in Figure 1.67.

13. Save and close the drawing.

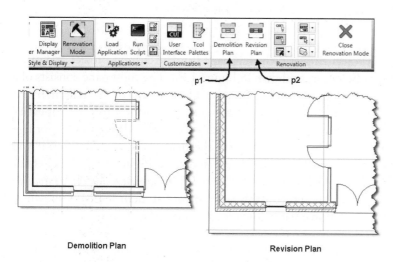

FIGURE 1.67 *Demolition and Revision Plans created with the Renovation Mode*

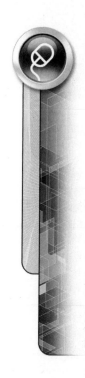

CREATING PROJECTS

AutoCAD Architecture 2012 includes a **Project Browser** and **Project Navigator** to create and define projects. Drawings for single-story buildings can be developed without the use of projects. However, the drawings for multistory buildings should be developed using a project since project tools provide organization and collaboration of the levels and horizontal divisions within the building. Project tools allow you to insert one or more levels to create a digital model of the building, which can be used for the development of elevations and section drawings. The project tools also include techniques for placing View drawings onto sheets for plotting or printing. These model management tools automate the design, data, and model presentation. The **Project Navigator** allows you to manage the use of external reference files and extract data across several drawings. Drawings are referenced from within the Project Navigator using the tools of the Project Navigator. When using projects, do not attach project drawings independent of the Project Navigator using the XREF command. Open, create, and reference all drawings from within the Project Navigator. Drawings are developed and classified as **Elements**, **Constructs**, **Views**, and **Sheets** in the **Project Navigator**. Drawings can be developed specifically for a project, or existing drawings can be converted to components of a project.

When AutoCAD Architecture is installed, sample projects are provided for reference. These projects are located in the *Vista– C:\Users\Login name\Documents\Autodesk\My Projects* folder. The sample projects are for commercial buildings; however, they provide excellent examples of how the final documents are created.

Using the Project Browser to Create the Project

The first step in developing a model is to create a project name in the Project Browser. Select **Project Browser** from the Quick Access toolbar (see Table 1.16 for command access) to open the Project Browser.

TABLE 1.16 *Project Browser access*

Command prompt	PROJECTBROWSER
Quick Access toolbar	Choose Project Browser

When you open the Project Browser, the left pane lists the current project and other projects in the **Project Selector**, as shown in Figure 1.63. The right pane consists of a project bulletin board connected to an HTM page. The contents of the bulletin board are specified in the HTM file, and its location is specified in the **Project Properties** dialog box shown in Figure 1.68. You can open this file in Word and add text or figures to the bulletin board. The current path to the project can be identified if you select the drop-down list of the Project Selector shown in Figure 1.68. The default path to the My Projects folder is *Vista– C:\Users\Login name\Documents\Autodesk\My Projects*. Therefore, you should edit this path to your student folder as shown in Figure 1.68 from the drop-down list before creating a project.

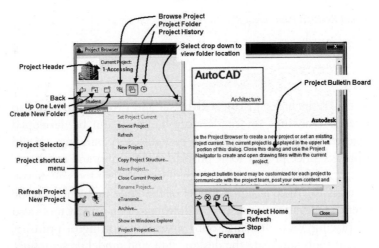

FIGURE 1.68 *Navigating the Project Browser*

To create a new project, select the **New Project** button located in the lower-left corner of the Project Browser shown in Figure 1.68 and open the **Add Project** dialog box. The Add Project dialog box is shown in Figure 1.69. To create a new project, you must enter a name; project number is optional. The project number and name appear in the Project Header of the Project Browser. The **1-Accessing** project number and name are shown in Figure 1.68. You can check the box **Create from template project** shown in Figure 1.69 and choose Browse to specify the file path to a template. Any previous project may be used as a template to create a new project. The tutorials of this text will require that you download the projects located in the **Accessing Tutor** of the Student Companion site of CengageBrain and specify as a template for your projects. If you clear the checkbox for **Create from template project**, the project created will not include folders necessary for project standards. When you create a project, it becomes the current project. Only one project may be current.

You can check the **Create from template project** checkbox shown in Figure 1.69 to create a new project that includes a copy of the levels and drawing files of any existing project.

TIP

The content of the **Add Project** dialog box can be revised for an existing project by selecting the name of the project in the left pane of the **Project Browser**, right-clicking, and choosing **Project Properties**. The **Project Properties** command opens the **Project Properties** dialog box, as shown in Figure 1.69. The paths to the default templates for the Sheet, View, Construct, and Element drawings are listed in the **Project Properties** dialog box shown in Figure 1.69. The folder location of default templates used for creating new drawings within the project is specified in the Project Properties dialog box.

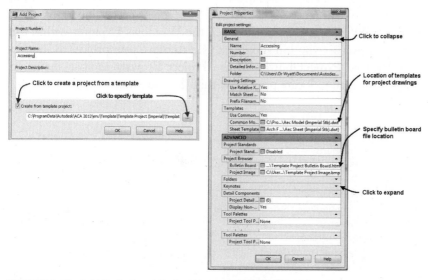

FIGURE 1.69 *Add Project and Project Properties dialog boxes*

Projects may be started from the project templates shipped with the software, or from previous projects. Project templates shipped with the software for metric and imperial projects are located in the Template folder of *Vista- C:\ProgramData\ Autodesk\ACA 2012\enu*. The content of projects created from templates is described in Table 1.17 and Figure 1.70. You can create a project from the **New 2012** *Project.apj* template located in the *Vista- C:\Users\Login name\Documents\Autodesk\My Projects* folder; however, this project is similar to a project created without a template. The templates shipped with the software are as follows:

TABLE 1.17 *Project Templates*

Template	Purpose
Commercial Template Project (Imperial).apj or Commercial Template Project (Metric).apj	Designed for the development of commercial buildings and includes sample empty drawings, preset values for levels, and folders for project standards.
Template Project (Imperial).apj or Template Project (Metric).apj	Designed for the development of residential or commercial buildings and includes folders for the project standards.
New 2012 Project.apj	Designed for residential or commercial buildings with no provision for project standards.

To set existing projects as current, double-click the project name in the Project Selector of the Project Browser. Additional tools for archiving, copying, closing, and browsing a project are included in the shortcut menu of the selected project in the Project Selector of the Project Browser, as shown in Figure 1.68. The Archiving option allows you to create a zip or executable file that includes all support files for a project. A new feature of this release is the ability to select a project name, right-click, and choose Show in Windows Explorer to open the Windows Explorer within the Project Selector. The **Archive** option should be used to move your project to another computer. The **Browse Project** button of the Project Browser toolbar shown in Figure 1.68 can be selected to locate projects within the computer.

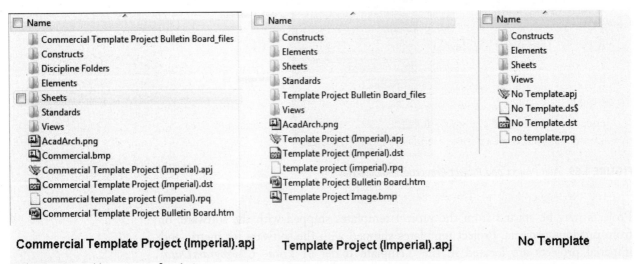

Commercial Template Project (Imperial).apj   **Template Project (Imperial).apj**   **No Template**

FIGURE 1.70 *Folder content of projects*

When a project is created, a directory is created within the folder listed in the Project Selector. The project folder includes a project file (*.*apj* extension), sheet set file (*.*dst* extension), and the following folders: **Constructs**, **Elements**, **Sheets**, **Standards**, **Views**, and **Template Project Bulletin Board_files**. The project file (*.*apj*) is created, which contains such data as project name, number, levels, templates, and project detail text information. Drawing files created in the project are saved to one of the subdirectories of the project directory. The folders are empty when the project is created; however, as the project develops, the drawing files of the project are automatically placed in the Constructs, Elements, Sheets, and Views folders. An XML file is created with each drawing file created or used in the project. The XML file retains such supporting data as level, divisions, viewport scale, and display configuration. Therefore, the XML file is essential to the function of the drawing file in the project. Do **not** delete or move files within the project with the Windows Explorer.

You can double-click a project file *.apj in Windows Explorer to open the project.	NOTE

Creating Project Content with the Project Navigator

The Project Browser is used to create the project name and data supporting the project. However, the **Project Navigator** is used to create the drawings of the project. You can return to the Project Browser from the **Project Navigator** by selecting the **Project Browser** button in the lower-left corner of the Project tab in the **Project Navigator** as shown in Figure 1.71. Access the **Project Navigator** by selecting **Project Navigator Palette** from the Quick Access toolbar; refer to Table 1.18 for command access.

TABLE 1.18 *Project Navigator Palette access*

Command prompt	**PROJECTNAVIGATORTOGGLE or** CTRL + 5
Quick Access toolbar	Choose Project Navigator

The **Project Navigator** includes four tabs: **Project**, **Constructs**, **Views**, and **Sheets**, as shown in Figure 1.71, and the tabs are described as follows.

Project—This tab defines the divisions and levels of the project. The floor elevations are specified in the level definition as shown in Figure 1.71.

Constructs—Drawings classified as elements and constructs are the building blocks of floor plans and are created from this tab.

Views—Views are created in this tab by attaching Construct drawings as reference files. The View drawing identifies level, divisions, and Construct drawings. Dimensions and annotation are placed in the View drawing.

Sheets—Drawings classified as Sheets are created on this tab. The Sheet drawing includes the title block and border for plotting. View drawings are attached to the sheet to create the document for plotting.

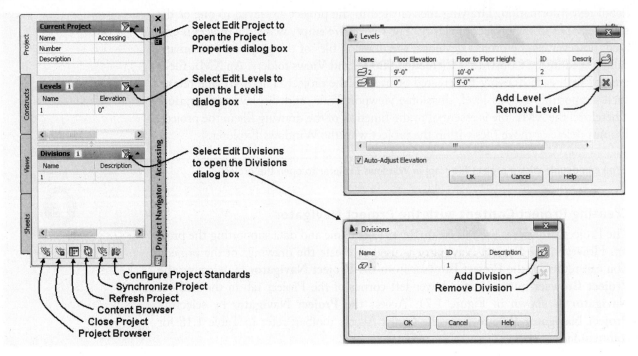

FIGURE 1.71 *Creating levels and divisions in the Project Navigator*

NOTE Access Instructional Video 1.2—Creating a Project located in the Instructional Video category of the Student Companion site of CengageBrain http://www.cengagebrain.com described in the Preface.

Using Project Components to Develop Construction Documents

Project content is created to ultimately create the working drawings. The development of a floor plan is described in Figure 1.72. As shown at the left, the first step is to define the levels and divisions of a project. After specifying levels, you create a floor plan as a Construct drawing, which includes the walls, doors, and windows. The Construct drawing is a unique drawing developed as part of the project and associated with a level and division. This construct is attached to a View drawing. The view completes the plan when dimensions, notes, and schedules are added. A View drawing is created for each floor plan. The floor plan View drawing is then referenced into a Sheet drawing to place the drawing in a viewport within the sheet. The display configuration and cut plane height are specified for the viewport of the Sheet drawing. The Sheet drawing includes a title block and border as defined in the sheet template. The Sheets tab of the Project Navigator provides the functions of the Sheet Set Manager of AutoCAD 2012.

NOTE According to the anticipated scale, set the Annotation Scale and Display Configuration in the Drawing Window status bar prior to adding dimensions and annotation in a View drawing.

Elevations and sections are created by attaching each floor plan or other Construct drawings to a model View drawing as shown in Figure 1.72. The model drawing includes all levels and divisions of the project. A model View drawing should be created early in the project to view vertical alignment of features per floor level. Elevation and section drawings are developed from the model as View drawings using the elevation and section tools of the Callout tool palette. The elevation and section View drawings are placed on a sheet created in the Sheets tab of the Project Navigator.

Development of Floor Plans and Elevations Using a Project

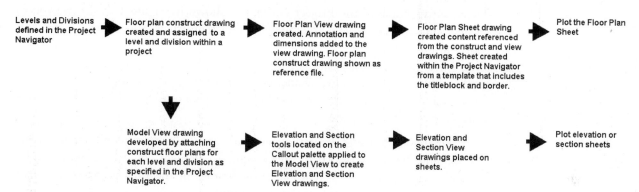

FIGURE 1.72 *Development of plans and elevations using a project*

Defining Divisions and Levels of the Project

The Project tab lists the project data, levels, and divisions of the project. The level definition defines the floor to floor height and floor elevations, while divisions specify wings or horizontal divisions of the building.

The **Edit Project**, **Edit Levels**, and **Edit Divisions** buttons located at the right in Figure 1.71 allow you to specify the content of the project. Selecting the **Edit Project** button of the **Project** tab opens the **Project Properties** dialog box, which allows you to edit the project data, such as project templates and titleblock data.

Levels of the project are defined by selecting the **Edit Levels** button to open the **Levels** dialog box. The **Levels** dialog box shown in Figure 1.71 includes level name, floor elevation, floor to floor height, and description. Each name of the level must be unique. Click on the columns of the **Levels** dialog box to edit the floor elevation and floor to floor height. Drawing files created for a level are inserted as reference files based on the floor elevation value.

Additional levels can be created by selecting the **Add Level** button, or you can select the name of a level, right-click, and choose **Add Level Above** or **Add Level Below**. The shortcut menu of a level name also includes **Copy Level** and **Copy Level and Contents** options for creating a new level from an existing level. The Copy Level and Contents option can be used to copy the first floor plan of a house to the second level; the second level can then be edited to retain the bearing walls common to each level. The **Auto-Adjust Elevation** checkbox located in the lower-left corner, if checked, will assign the floor elevation of the new level equal to the sum of the previous floor level and floor elevation. When you add a level between levels, the floor elevation of the levels above the new level is automatically adjusted. The **Floor Elevation** and **Floor to Floor Height** values may be edited at any point during the development of drawings for the project.

The **Divisions** section includes the **Edit Divisions** button, shown in Figure 1.71, which opens the **Divisions** dialog box to create divisions. Select the **Add Division** button to open the **Divisions** dialog box and define division names and descriptions. To edit the **Divisions** dialog box, click on the **Name**, **ID**, and **Description** columns.

Creating Constructs

The Constructs tab of the Project Navigator, shown in Figure 1.73, allows you to create Construct and Element drawings. The Constructs tab includes a toolbar, shown at the bottom of the tab in Figure 1.73, which consists of commands for creating categories, elements, and constructs. The concept of the project is to create small unique drawings, such as elements and constructs, which can be inserted as references into View drawings to create the desired plan or model. Elements and constructs are defined as follows:

> **Elements**—Drawings created as elements are the simplest components of the building that are repeated in the design. An element could be a drawing of a bathroom layout that is repeated several times in the final drawing. Element drawings are attached to Construct drawings.
>
> **Constructs**—Drawings of unique content in the building are constructs. A construct can be a floor plan. Element drawings can be referenced into Construct drawings to complete the plan.

Prior to creating a new drawing, you can create a category within the Construct tab to organize the drawings. Categories can be created to classify the drawings in such categories as architectural, plans, details, or site work. Element or Construct categories are created by selecting the **Element** or **Construct** folder and then selecting the **Add Category** command from the toolbar of the **Constructs** tab, as shown in Figure 1.73. To create a new drawing in a category, select the category name and then select the **Add Element** or the **Add Construct** command from the toolbar on the **Constructs** tab.

When you select the **Add Element** command, the **Add Element** dialog box opens, as shown in Figure 1.74. To obtain a template for the Element drawing, select in the **Drawing Template** edit field.

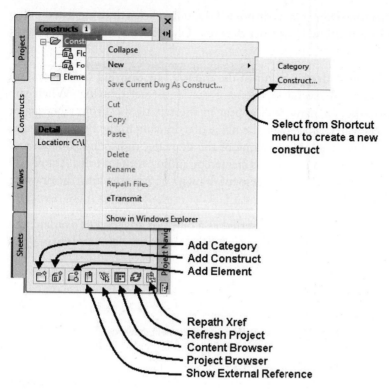

FIGURE 1.73 *Constructs tab*

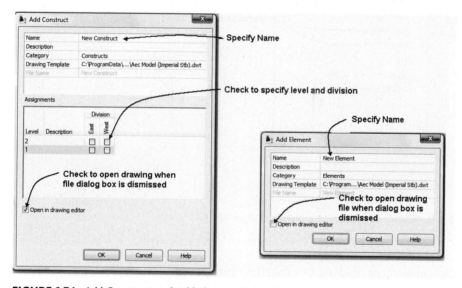

FIGURE 1.74 *Add Construct and Add Element dialog boxes*

When you select the **Add Construct** command, the **Add Construct** dialog box opens as shown at the left in Figure 1.74. The **Add Construct** dialog box lists the levels and divisions defined for the project. If you select the checkbox to assign a division to a level, the Construct drawing for the division will be inserted as an external reference file for the level specified in the View drawing. The drawing is attached and recorded in the project data files. You can choose the *Open in drawing editor* checkbox to open the file when the dialog box is dismissed. Double-click a drawing listed in the Construct tab to open the new Construct drawing. Click Save on the Quick Access toolbar to save the file.

Elements are the simplest drawing unit of the model. The drawing classified as an element can be inserted in several locations of the Construct drawing. When you add an element drawing, its name will appear in the tree of the Project Navigator (see Figure 1.75). You can double-click the name of a drawing in the Project Navigator to open the drawing or select the name of the drawing, right-click, and choose **Open**. Content of the element drawing is created by adding content from AutoCAD Architecture 2012. Save and Close the element drawing before attaching the drawing to a Construct drawing.

The content of other files can also be inserted as a block or attached or overlaid as a reference file to the construct file. The procedure to attach an element drawing to a construct file is shown below:

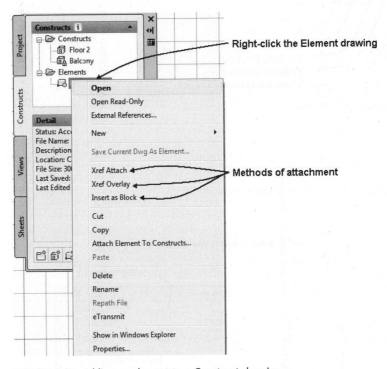

FIGURE 1.75 *Adding an element to a Construct drawing*

To attach an element drawing to a construct file, complete the following steps:

1. Open the **Project Navigator**, select the Construct drawing, right-click, and choose **Open**.

2. Select the name of the element drawing in the **Project Navigator** tree, right-click, and choose one of the following: **Xref Attach**, **Xref Overlay**, **Insert as Block**, or **Attach Element to Constructs**, as shown in Figure 1.75.

Drawings that were created as elements can be converted to constructs by selecting the name of the element in the **Project Navigator** and dragging it to the Construct category of the tree of the **Constructs** tab. When you drag an element to the

Construct category, the **Add Construct** dialog box opens, allowing you to specify the levels and divisions of the construct. Constructs can also be converted to element drawings by dragging them into the Element category. When Construct drawings are converted to elements, the level and division data are lost.

Existing drawings can be converted to element or Construct drawings in the Project Navigator. To convert a drawing, open the drawing and the **Project Navigator**. Select the category of the construct or element, right-click, and choose **Save Current Dwg As Element** or **Save Current Dwg As Construct**. The **Add Element** or **Add Construct** dialog box opens, allowing you to specify the name and other data. When existing drawings are converted to element or Construct drawings, an accompanying XML file is also created. The existing drawing is not linked when converted to an element or construct. Therefore, an existing floor plan can be converted to a Construct drawing to take advantage of the Project Navigator management tools.

Creating Views

Views are created to specify which elements and constructs are used to define the drawing outcome of the model. View drawings can be created as **General Views, Section/ Elevation Views,** or **Detail Views,** as shown in the shortcut menu in Figure 1.76. You can create a general view that includes only constructs of the first floor or a composite view of all floors. A view is created by selecting the **Views** tab. A category can be created by selecting the **Add Category** button from the toolbar at the bottom of the **Views** tab. You can also create a view by selecting the **Add View** command from the toolbar at the bottom of the tab. When you select **Add View**, the **Add View** dialog box opens, which prompts you to specify the type as **General, Section/ Elevation,** or **Detail**. After you specify the **General** type, the **Add General View** dialog box opens, as shown in Figure 1.77. The name and description are defined on the **General** page.

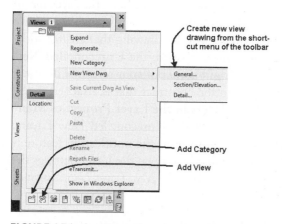

FIGURE 1.76 *Creating a View drawing*

The **Context** page, shown in Figure 1.77, lists the levels and divisions of the project. The check placed in the box for Division East of Level 2 indicates that the content specified for Level **2** and division **East** will be included in the view. If you are creating a model view of the composite components, all levels should be checked for all divisions.

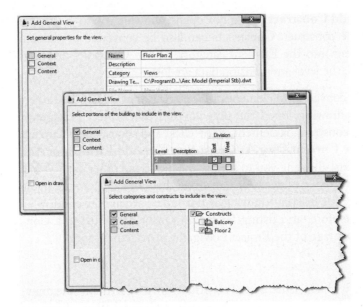

FIGURE 1.77 *Add General View dialog box*

The constructs included in the view are specified in the **Content** page. The **Content** page shown in Figure 1.77 includes two Construct drawings. The check to the left of the construct indicates that the construct will be displayed. The **Balcony** drawing checkbox is clear; therefore, it will not be displayed.

Reconciling Layers in Reference Files

AutoCAD Architecture includes the new layer notification feature. Since projects utilize reference files, you may notice the Unreconciled New Layers message box in the lower-right corner of your screen. The layer notification tracks the layer list of each file attached as a reference file and will notify the user if additional layers have been added. Therefore, if you add content to a construct drawing that creates additional layers when the associated View drawing is opened, the **Unreconciled New Layers** message box will display in the lower-right corner of your screen. This message box opens since layers have been added to the construct that is attached to the View drawing. Click **View unreconciled new layers in the Layer Properties Manager** hyperlink to open the Layer Manager. Select the Unreconciled New Layers, right-click, and choose Reconcile Layer. You can turn off the notification feature if you choose the **Settings** button of the Layer Manager to open the Layer Settings dialog box as shown in Figure 1.78. Clear the checkbox for *New Layer Notification* of the Layer Settings dialog box.

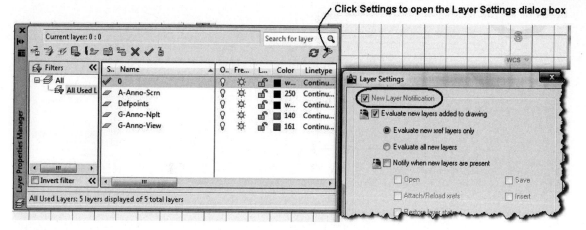

FIGURE 1.78 *Editing New Layer Notification in the Layer Properties Manager*

TIP

A toggle exists within the Project Properties dialog box that allows you to choose **Match Sheet View Layers** to View. If Match Sheet View Layers to View is set to Yes, the settings of a View drawing will be retained when placed on a sheet. Click on the Edit Project button of the Project tab within the Project Navigator to open the Project Properties dialog.

Creating Model Space Views within a View Drawing

A model space view can be created within an existing View drawing. The saved view specifies the scale. A model space view should be created for detail and plan View drawings prior to placing the View drawing on a sheet. Title marks located on the Callouts palette are used to place the title for each View drawing as part of the document management.

To create a model space view within a View drawing, complete the following steps:

1. Open a View drawing in the **Project Navigator**.

2. Select the View drawing name in the **View** tab of the **Project Navigator**, right-click, and choose **New Model Space View** from the shortcut menu to open the **Add Model Space View** dialog box shown in Figure 1.79.

3. Edit the name and description edit fields. Choose the **Define View Window** button to return to the workspace and select from point **p1** to **p2** to specify the content of the view. Select **OK** to dismiss the **Add Model Space View** dialog box.

4. Verify that the model space view is listed below the View drawing in the **Project Navigator**, as shown at the right in Figure 1.79.

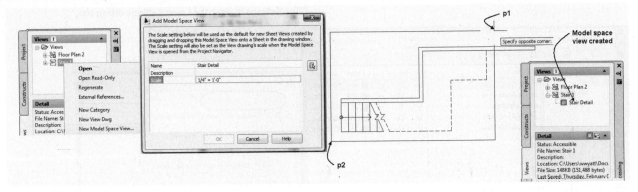

FIGURE 1.79 *Creating model space views in a View drawing*

Creating Sheets

Sheet drawings are created in the Sheets tab. Sheets can be created using the default template provided or you can specify a custom template. The template can be specified for all sheets or the subsets. The following steps specify a sheet template and create a new sheet:

1. Select the **Sheet Set** title at p1 in Figure 1.80 on the **Sheets** tab in the **Project Navigator**, right-click, and choose **Properties** to open the **Sheet Set Properties** dialog box.

2. To edit the sheet creation template path, select the **Browse** button shown at **p2** in Figure 1.80, which opens the **Select Layout as Sheet Template** dialog box.

3. If necessary, click on the **Browse** button of the **Select Layout as Sheet Template** dialog box as shown at **p3** in Figure 1.80 to select an Aec Sheet template drawing that includes layouts. Select a layout that will be used as the default for new sheets. Select **OK** to dismiss all dialog boxes. If this step changed the template, the Sheet Set—Confirm Changes dialog box opens. Select **Apply changes to all nested subsets** to apply a template setting for all nested subsets.

4. To create a new sheet, select a subset, right-click, and choose **New > Sheet** from the shortcut menu to open the **New Sheet** dialog box as shown in Figure 1.81.

5. Type the sheet number and name in the **New Sheet** dialog box and select **OK** to dismiss the New Sheet dialog box.

TIP

Templates can be specified in the **Properties** section of each subset, shown at p1 in Figure 1.81 of the sheet set.

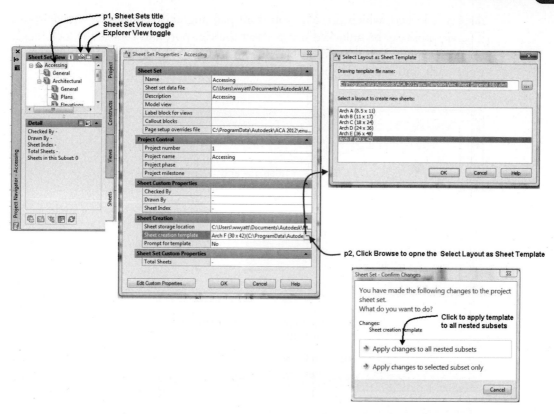

FIGURE 1.80 *Selecting a sheet template*

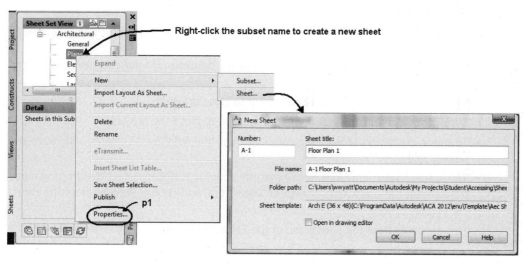

FIGURE 1.81 *Creating a new sheet*

If you check the *Open in drawing editor* checkbox, the sheet will open when you close the New Sheet dialog box. After a sheet has been created, double-click the sheet name to open the drawing from the Sheets tab. To place content on the sheet, select the **Views** tab of the **Project Navigator**, select a view, and drag the view attached to the pointer into the workspace of the sheet as shown in Figure 1.82. A phantom version of the view will be attached to the pointer as you drag the view onto the sheet. You can right-click to edit the scale of the view prior to specifying the location of the view on the sheet. The viewport graphics boundary on the sheet will not plot and its display is locked, which prevents you from panning and zooming in the viewport. The viewport can be unlocked if you select the viewport and choose the lock toggle on the Drawing Window status bar, shown in Figure 1.82, or select the viewport, right-click, and choose **Display Locked > No**.

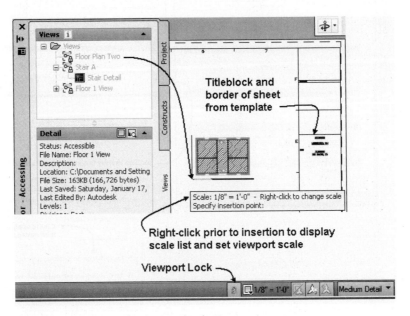

FIGURE 1.82 *Placing the view on the sheet*

Project Rules

- Create new drawings from the Constructs, Views, and Sheets tabs of the Project Navigator.
- Click **Save** on the Quick Access toolbar to save drawings. Do not choose Save As from the Application menu to save a drawing.
- Choose **Close** from the Application menu to close drawings.
- Click **Close Current Project** on the Project tab to close the project.
- To move a drawing to another computer, use the Windows Explorer and move the project folder.
- To rename project drawings, verify that the drawing is closed, select the drawing name from within the Constructs, Views, or Sheets tab, and choose Rename from the shortcut menu.
- Plot drawings from the Sheets tab of the Project Navigator.

Keeping the Project Navigator up to Date

During the process of defining the components of a project in a networked environment, the project tree of the Project Navigator may not be displayed up to date. Select the **Refresh Project** command located on the bottom toolbar of each tab to recalculate the settings of the project and display the current drawing with the latest information. The **Repath Xref** command, also located on the toolbar of the Constructs and Views tabs, opens the External References palette. The External References palette allows you to view the path to the files referenced and reload the files.

When reference files change, a balloon notification appears in the lower-right corner of the Drawing Window status bar. Select the hyperlink of the notification window to reload the file. However, do not use the Xref Manager to attach or overlay files.

TIP

When you complete your drawing session in a project, choose Close Current Project button of the Project tab to close all drawings associated with the project. When you choose the Close Current Project button, you can choose to close all project drawings and close the project without closing the project drawings. In addition, you can close the project from the **Project Browser**, choose the project title, right-click, and choose **Close Current Project** from the shortcut menu. You can move the project folder in Windows Explorer or select the project name in the Project Browser and choose **Move**. The Archive option creates a zip or executable file for moving the project.

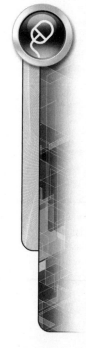

When you open a project that has been moved, you will be prompted to repath a project. A Project Browser – Project Location Changed warning box, shown in Figure 1.83, will prompt you with options to repath the project. Select **Repath the project now** to repath the project to the drive and directory of the new location or a new computer.

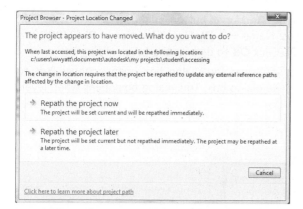

FIGURE 1.83 *Repathing a project*

TUTORIAL 1-5: CREATING A PROJECT

1. Choose **Project Browser** from the Quick Access toolbar.

2. Choose the drop-down list of the **Project Selector** and set the path to your *Student\ Ch 1* folder. Refer to Organizing Tutorial Directories located in the Preface.

3. Select the **New Project** button of the Project Browser shown in Figure 1.84 to open the **Add Project** dialog box.

4. Type **1** in the **Number** edit field and **Ex 1-5** in the **Project Name** edit field. Verify that **Create from template project** is checked. Click on the **Browse** button to open the **Select Project** dialog box. Click on the **Content** button of the Places panel as shown in Figure 1.84. Navigate to *Template\Template Project (Imperial)\Template Project (Imperial).apj* or *Template\Template Project (Metric)\Template Project (Metric).apj*. Select **Open** to choose the template file and dismiss the **Select Project** dialog box.

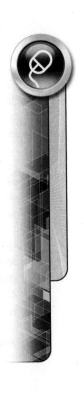

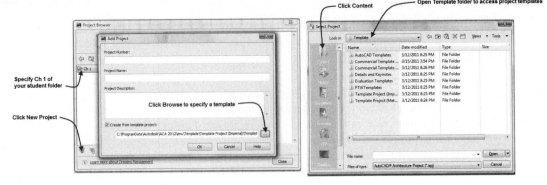

FIGURE 1.84 *Creating a project in the Project Browser*

5. Click OK to dismiss the Add Project dialog box and return to Project Browser. Choose Close to close the Project Browser and open the **Project Navigator** in the workspace.

6. Select the **Project** tab of the Project Navigator and choose **Edit Levels** as shown in Figure 1.85. Click on the edit fields to edit level 1 as follows: Name = **1 [G]**, Floor Elevation = **0**, Floor to Floor Height = **9' [2700]**. Verify that the **Auto-Adjust Elevation** checkbox is checked.

7. To create an additional level, click the **Add Level** button, as shown in Figure 1.85. Edit the properties of level 2 as follows: Name = **2 [1]**, Floor Elevation = **9' [2700]**, and Floor to Floor Height = **9' [2700]**. Click **OK** to dismiss the Levels dialog box.

8. After the AutoCAD Architecture 2012 warning dialog box opens, choose **Yes** to regenerate the views and dismiss the dialog box. This dialog box opens when levels are changed or added.

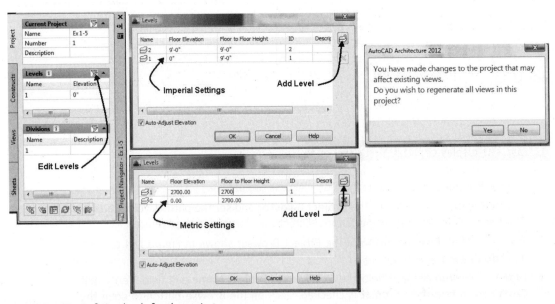

FIGURE 1.85 *Defining levels for the project*

9. The floor plan of the house will be started in a Construct drawing. This project will consist of a basement; therefore, the main floor will be developed first as level 2. To create a Construct drawing, choose the Constructs tab in the Project Navigator. Choose the Constructs category, right-click, and choose New > Construct to open the Add Construct dialog box. Edit the Add Construct dialog box as follows: Name = Floor 2 and check level 2 [1] for Division 1 as shown in Figure 1.86. Select Open in drawing editor checkbox. The content of level 1 [G] will be developed in future tutorials as either a foundation plan or a basement plan. Click OK to dismiss the Add Construct dialog box.

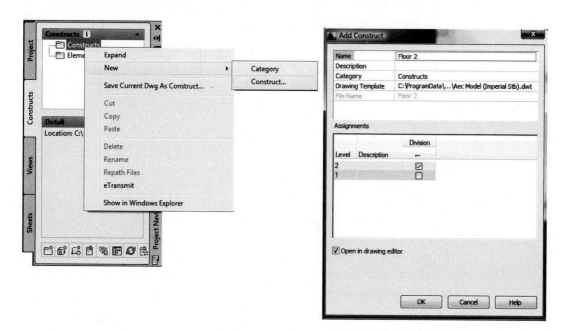

FIGURE 1.86 *Creating a Construct drawing*

10. Choose Utilities > Drawing Setup from the Application menu to open the Drawing Setup dialog box. In this step, you will link the scale and display configuration. Choose the Scale tab. Select ¼" = 1'-0" [1:50] scale. Choose **High Detail** from the display configuration list. Choose OK to dismiss the dialog box. Choose ¼" = 1'-0"[1:50] scale from the Annotation Scale flyout of the Drawing window status bar. The display configuration changes to High Detail.

11. Choose the Work layout from the Application status bar and verify that the right viewport is current. Choose ¼" = 1'-0" [1:50] from the Viewport Scale flyout of the Drawing Window status bar for the right viewport. Click on the left viewport and set the Viewport Scale to ¼" = 1'-0" [1:50].

12. Click Model in the Application status bar.

13. In future tutorials, dimensions will be added in a View drawing for the floor plan. To create a View drawing, choose the **Views** tab. Choose the **Views** category, right-click, and choose **New View Dwg** > **General** to open the **Add General View** dialog box.

14. Type **Floor 2 View** in the **Name** edit field of the General page of the **Add General View** dialog box, as shown in Figure 1.87. Choose *Open in the drawing editor* checkbox. Click the **Next** button to open the **Context** page; check level **2 [1]** for **Division 1**. Click **Next** to open the **Content** page. Verify that the **Floor 2** construct is checked for this construct. Choose **Finish** to dismiss the **Add General View** dialog box and open the drawing. Verify that **View: Floor 2 View** is the drawing title listed in the Drawing Window status bar.

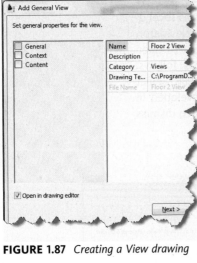
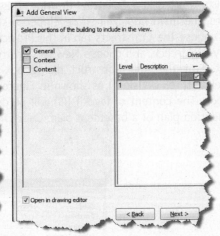
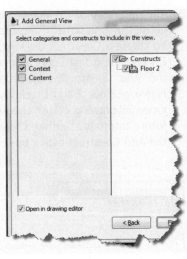

FIGURE 1.87 *Creating a View drawing*

15. Choose from the Application menu Utilities > Drawing Setup. Choose the Scale tab. Choose ¼" = 1'-0" [1:50] from the scale list and High Detail from the display configurations flyout list. Click OK to dismiss the dialog box.

16. Choose ¼" = 1'-0" [1:50] from the Annotation Scale flyout of the Drawing Window status bar; the display configuration is changed to High Detail.

17. Choose the Work layout and verify that the right viewport is current. Choose ¼" = 1'-0" [1:50] from the Viewport Scale flyout of the Drawing Window status bar for the right viewport. Click on the left viewport and set Viewport Scale to ¼" = 1'-0" [1:50].

18. Click Model in the Application status bar.

19. To create a sheet, choose the **Sheets** tab of the Project Navigator. Expand Architectural category. Choose the **Plans** category, right-click, and choose **New > Sheet** to open the **New Sheet** dialog box. Type **A-2** in the **Number** field and **Floor Plan 2** in the **Sheet title** edit field, as shown in Figure 1.88. Verify that Arch F (30 × 42) **[ISO A0 841 × 1189]** is the sheet template. Select the *Open in drawing editor* checkbox. Select **OK** to dismiss the New Sheet dialog box. The title block and border should be displayed at the right.

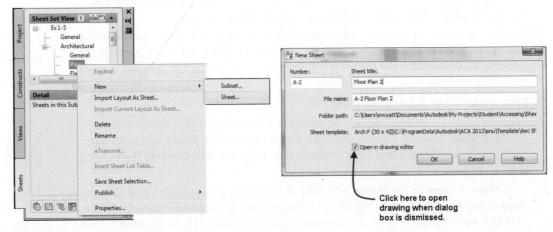

FIGURE 1.88 *Creating a Sheet*

20. Choose the **Project** tab of the Project Navigator. Select the **Close Current Project** button, shown in Figure 1.89, to open the **Project Browser – Close Project Files** dialog. Choose Close all project files to close all drawings and close the project. Click Yes in the AutoCAD message boxes to Save change to drawings.

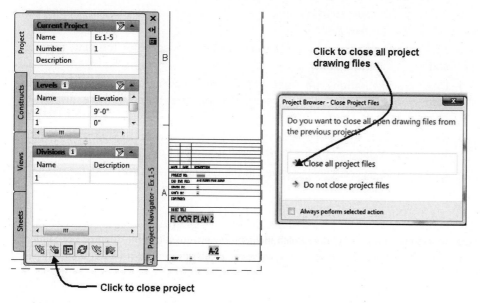

FIGURE 1.89 *Closing a project*

SUMMARY

- AutoCAD Architecture 2012 objects are created by selecting tools from the tool palette.
- The Home tab provides access to tools to place typical building components such as doors, windows, and roof.
- When you select an object, an edit tab for the selected object is displayed in the ribbon, which includes typical tools for the edit of the object.
- The shortcut menu of a selected AutoCAD Architecture 2012 object includes the commands for editing the object.
- Doors, windows, and stairs are inserted in the drawing as objects.
- Workspaces can be created to retain the display of the ribbon and palettes.
- The resource files for AutoCAD Architecture 2012 are located in the *Vista-C:\ProgramData\Autodesk\ACA 2012\enu* directory.
- The Content Browser is used to create and customize tool palettes.
- The name of the project associated with a drawing is listed on the Drawing Window status bar.
- Projects are created using the Project Browser, whereas the Project Navigator is used to create drawings assigned to the project.
- Units, scale, layer, and display settings are listed in the Drawing Setup dialog box.

- Templates for creating a floor plan using imperial units are Aec Model (Imperial Stb) and Aec Model (Imperial Ctb). Templates for creating a floor plan using metric units are Aec Model (Metric Stb) and Aec Model (Metric Ctb).

- A display representation specifies which components of an object are displayed.

- A display representation set identifies one or more display representations that are used to display objects.

- Toggle on the Renovation mode to create object display for the new, existing, and demolition categories.

- Display configurations are selected from the Drawing Window status bar to identify one or more display representation sets to govern the display of objects.

- The Display tab of the Properties of an object allows you to view and override the display as defined by the drawing default, object override, or object style.

- One or more objects can be viewed in the Object Viewer.

NOTE

Refer to Review Questions folder, which includes review questions for Chapter 1 of Student Companion Site of **CengageBrain** http://www.cengagebrain.com described in the Preface.

Creating Floor Plans

INTRODUCTION

This chapter introduces the commands to draw and modify walls using the Standard wall style. The focus of this chapter is the placement of the wall and the tools available for precise placement of walls for the creation of a floor plan. Options of the **WallAdd** command and how to use properties for editing will be explained. The tools to edit walls with precision, such as **Justification**, **WallOffsetCopy**, **WallOffsetMove**, **OffsetSetFrom**, Intelligent Cleanup, and constraint tools, will be presented.

OBJECTIVES

After completing this chapter, you will be able to

- Draw straight or curved walls using the WallAdd command and Properties palette.

- Use the Close and Ortho close options to close a polygon shape.

- Identify the function of the wall grips.

- Set the cleanup radius and display of graph linesUse the WallOffsetCopy, WallOffsetMove, WallOffsetSet, and Intelligent Cleanup commands for adding and modifying walls.

- Apply geometric and dimensional 2D constraints to walls.

- Merge walls with Join and Wall Merge commands.

- Reverse the direction of a wall using the Reverse command.

- Use Add Selected to execute the WallAdd command.

SETTING UP FOR DRAWING WALLS

When a new file is created to draw a floor plan, the imperial template Aec Model (Imperial Stb) or Aec Model (Imperial Ctb) is used. Metric templates are Aec Model (Metric Stb) and Aec Model (Metric Ctb). The file can be created as a new Construct of a project or created independently of a project. The Model or

Work layout can be used. Prior to placing walls the following settings should be reviewed: Dynamic Input, AutoSnap, Object Snap, and Justification. The Annotation Scale and Display Configuration should be set in the Drawing Window status bar as follows:

- 1/16" = 1'-0" [1:200] Annotation Scale use Low Detail Display Configuration
- 1/8" = 1'-0" [1:100] Annotation Scale use Medium Detail Display Configuration
- 1/4" = 1'-0" [1:50] Annotation Scale use High Detail Display Configuration

CONNECTING WALLS USING OBJECT SNAP AND AUTOSNAP

When placing walls the object snap modes can be applied to wall components or the wall justification line. The Object Snap tab of the Drafting Settings dialog box includes the object snap modes and the AutoCAD Architecture toggle: Allow general object snap settings to act upon wall justification line. When the Allow general object snap settings to act upon wall justification line is toggled ON, the modes of object snap will snap to the wall justification line rather than to components or the linework of hatching. If Allow general object snap settings to act upon wall justification line is toggled ON when placing walls, wall endpoints automatically connect to adjoining wall justification lines as shown at Wall "A" in Figure 2.1.

The **Node** object snap mode can be used to snap to the endpoint and midpoint of the wall along its justification. The points selected using the Node object snap mode are located at the bottom and top of the wall along the justification line. The **Node** object snap mode should be selected when you draw walls or manipulate mass elements. The Node object snap allows you to snap to the justification line of walls.

The Autosnap toggle located on the AEC Object Settings tab in the Options dialog box will assist you in placing walls. The default setting of Autosnap is controlled by toggling ON Autosnap New Wall Justification Lines in the AEC Object Settings tab of the Options dialog box (see Figure 2.1). The Autosnap feature detects other wall baselines within a specified radius of the current wall and will snap the two walls together. The Autosnap radius is set to 6" [75 mm in metric drawings] by default. The Autosnap feature can be toggled ON or OFF for grip editing by checking Autosnap Grip Edited Wall Justification Lines. Turn OFF the Autosnap toggle to edit walls with lengths less than 6" [75 mm].

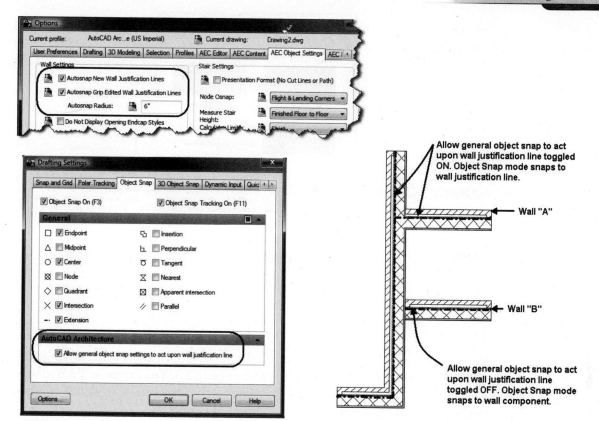

FIGURE 2.1 *AEC Object Settings and Object Snap tabs*

USING DYNAMIC INPUT

When Enabled, the Dynamic Input toggle located on the Application status bar shown in Figure 2.2 allows you to view dynamic dimensions and respond to the command prompts in the workspace. Dynamic Input should be enabled and will be applied throughout this text. Dynamic Input consists of pointer input, dimensional input, and dynamic prompts. Right-click over the Dynamic Input toggle, and choose **Settings** to display the Dynamic Input tab of the Drafting Settings dialog box. Choose the Settings button of the Enable Pointer Input section to view the Pointer Input Settings as shown in Figure 2.2. If Enable Pointer Input is checked and you choose a command that requires input, the pointer tooltip displays the typed values. When you type @, the pointer input format changes to relative Polar Tracking. If you type the # key or a number, the absolute format is applied as shown in Figure 2.2. The lock icon in the pointer prompt indicates that a value has been manually entered. If you press the TAB key, you can toggle to edit the value.

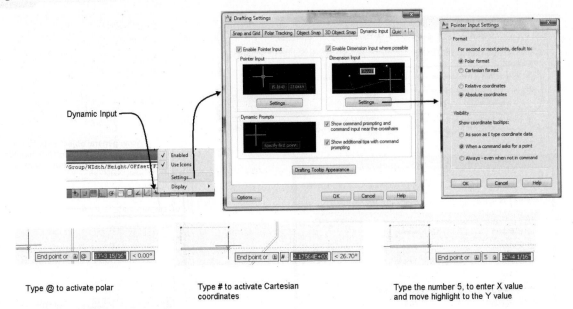

Dynamic Input

/Group/WIdth/Height/OFfset/F

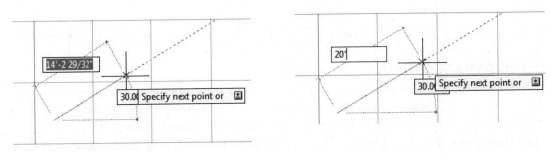

Type @ to activate polar

Type # to activate Cartesian coordinates

Type the number 5, to enter X value and move highlight to the Y value

FIGURE 2.2 *Displays of Pointer Input*

When you type values in the keyboard, the values override the value listed in the highlighted dynamic dimension, as shown in Figure 2.3. You can press TAB to toggle the edit of dynamic dimensions. Pressing TAB allows you to toggle the edit between linear and angular dimensions as shown in Figure 2.3.

Dynamic Dimension Displayed for Line

Value Typed in Keyboard to Edit Dynamic Dimension

FIGURE 2.3 *Display and edit of dynamic dimensions*

NOTE

Throughout this text, toggle ON the Enable Pointer Input, Enable Dimension Input, and Show command prompting and command input near the crosshairs options so the tutorials and resulting workspace prompts will be listed for the command description.

DRAWING WALLS

Drawing walls is the beginning point for the creation of floor plans for residential or commercial structures. Walls created in AutoCAD Architecture 2012 are objects that have three-dimensional properties of width, height, and length. Therefore, the wall can be viewed in plan to create a floor plan or viewed as an isometric to create a pictorial drawing of the building (see Figure 2.4). Walls placed in a drawing have attributes defined by their style. The simplest style of wall is the Standard wall style that consists of only two wall lines. The style of the wall is listed in the Basic section within the Design tab of the Properties palette shown in Figure 2.4.

When developing a floor plan using the Standard wall style, the representation of walls should be kept simple and the width of a wall component should be set equal to the wall's structural component. If you are drawing 2 × 4 [50 × 100] wood frame walls, the wall width for the wall should be set to 3 1/2″ [89], the actual width of a 2 × 4 [38 × 89]. Setting the width to the actual size of the structural component allows you to snap to the lines representing the wall for specific location dimensions. Wall components could be added to represent the sheathing or gypsum wall board in a wall style. Walls created using the Standard wall style can be converted to complex walls such as brick veneer frame. Complex wall styles, discussed in Chapter 3, consist of multiple wall components.

Because walls are inserted as three-dimensional objects using object technology, the placement of doors and windows in a wall will cause the wall to be edited correctly to display the other objects in the wall. Wall anchors exist that lock other objects such as doors to the anchor. Therefore, you can select a door, select its grips, and slide it down the wall without moving the door out of the wall. When a wall is placed, it is automatically placed on the **A-Wall** layer if the AIA (256 color) Layer Key Style [A210G if the BS1192 Cisfb Layer Key Style] is used. The default color for the A-Wall and [A210G] layers is color 113 (a hue of green).

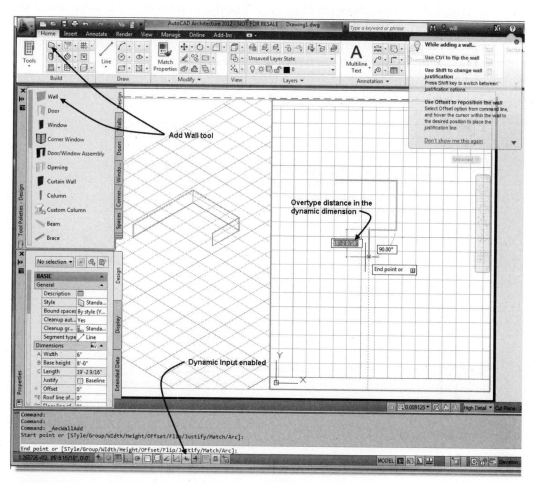

FIGURE 2.4 *Plan and pictorial view of walls in Work layout*

Creating Walls with the WallAdd Command

The **Wall** tool (**WallAdd** command) is accessed from the Build panel of the Home tab or the Design tool palette as shown in Figure 2.4. Access the **WallAdd** command as shown in Table 2.1.

TABLE 2.1 *WallAdd command access*

Command prompt	WALLADD
Ribbon	Select Wall from the Build panel of the Home tab
Tool palette	Select the Wall tool from the Design palette as shown in Figure 2.4.

When you select the Wall tool from the Design tool palette, the Properties palette opens as shown in Figure 2.4 and you are prompted for a start point to begin the wall. Prior to specifying the start point you should review the settings in the Design tab of the Properties palette. The options of the WallAdd command can be selected in the Design tab of the Properties palette or by choosing the down arrow in the workspace prompt. The command line options are shown below:

```
Start point or [STyle/Group/WIdth/Height/OFfset/Flip/
Justify/Match/Arc]:
```

The settings of a wall are retained from the last instance of the command. The start point and end point of the wall can be specified just as you would specify the length of an AutoCAD line as shown in Figure 2.3. The dynamic inputs for drawing a wall are shown in Figure 2.5. Once the wall is begun, the distance from the start point to the current cursor location is dynamically displayed in the Length edit field of the Design tab in the Properties palette as shown in Figure 2.5.

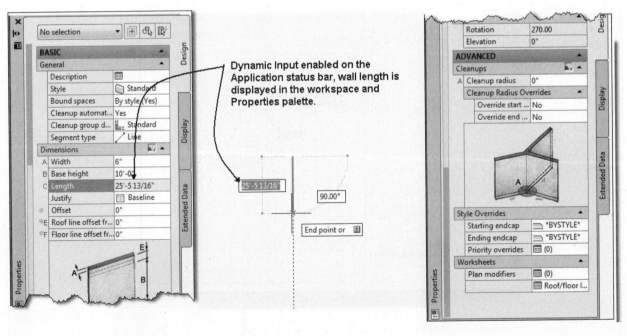

FIGURE 2.5 *Wall Properties*

Using Add Selected

After you have placed a wall in a drawing you can add walls using the WallAdd or WallAddSelected command. The **WallAddSelected** command allows you to select a wall, right-click, choose Add Selected, and then begin drawing a wall. The properties of the new wall are inherited from the selected wall. Access WallAddSelected as shown in Table 2.2 and Figure 2.6.

TABLE 2.2 *Add Selected command access*

Command prompt	**WALLADDSELECTED**
Ribbon	Select a wall and choose Add Selected from the Wall tab > General panel
Shortcut menu	Select a wall, right-click, and choose Add Selected.

> The Add Selected command can be applied to insert other AEC objects. Therefore, if you select a door, right-click and choose **Add Selected**. The DoorAdd command is active.

TIP

Using Match Properties

The **Match Properties** button shown in the Modify panel of the Home tab of Figure 2.6 allows you to match the style and object display properties of objects. When you choose the Match Properties button, you are prompted to select a wall to obtain the properties from the source object. The style of the source object is assigned to the destination objects. The Settings option of the command opens the Property Settings dialog box, which allows you to control the match for the following properties: color, layer, ltype, ltscale, lineweight, thickness, plotstyle, and additional special properties. After selecting the source object you can select one or more destination objects to define the source object properties to the destination objects as shown in the following workspace prompt.

> Select source object: *(Select a wall.)*
>
> Select destination object(s) or [Settings]: *(Select the wall for edit).*

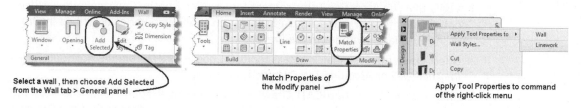

Select a wall , then choose Add Selected from the Wall tab > General panel

Match Properties of the Modify panel

Apply Tool Properties to command of the right-click menu

FIGURE 2.6 *Add Selected and Match Properties commands of the ribbon*

> To create additional walls with properties identical to the walls of the drawing, select a wall and drag it to a tool palette. Tool palette content created from a drawing will allow you to create additional objects with identical properties.

TIP

Changing Walls with the Apply Tool Properties to Command

The **Apply Tool Properties to** command is accessed from the shortcut menu of the **Wall** tool on the Design tool palette as shown at right in Figure 2.6. This command allows you to change one or more walls to the settings of the tool. The settings of the tool are as specified in the Design tab of the Properties palette. The **Apply Tool Properties to** can convert linework to walls. Access the **Apply Tool Properties to** command as shown in Table 2.3.

TABLE 2.3 *Apply Tool Properties to command access*

Shortcut menu	Select the Wall tool of the Design palette, right-click, and choose Apply Tool Properties to > Wall or Apply Tool Properties to > Linework.

When you select the **Wall** tool, right-click, and choose **Apply Tool Properties to**, a submenu appears, and you can select either Wall or Linework option. You can select one or more walls and then edit the Design tab of the Properties palette; the selected walls will match the properties specified in the Design tab of the Properties palette. If you select the linework option and then select lines, arcs, or circles, the linework will be converted to walls.

TIP   The SelectSimilar command discussed in Chapter 1 can assist in selecting walls with the same layer and style property. When the walls have a mixture of properties, Varies will be displayed in the property field of the Properties palette.

Contents of the Wall Properties Palette

The Design tab of the **Properties** palette consists of **Basic** and **Advanced** sections to describe a wall being placed. The Properties palette is changed by clicking in the edit field to the right of the property and then typing a new value or selecting from the options of the drop-down list. The content of the Design tab of the Properties palette is presented in Tutorial 2.0.

TUTORIAL 2-0: EXPLORING WALL PROPERTIES

1. Verify that the Accessing Tutor content for Chapter 2 of the CengageBrain has been downloaded to your Accessing Tutor folder on your computer as described in Organizing Tutorial Directories in the Preface.

2. Select **Open** from the Quick Access toolbar. Choose Ex 2-0 from the *Accessing Tutor\ Imperial\Ch 2* or *Accessing Tutor\Metric\Ch 2* folders.

3. Choose **Save As > AutoCAD Drawing** from the Application menu, and save the drawing as **Lab 2-0** in your student directory. Click Save to dismiss the Save Drawing As dialog box.

4. Select wall **A** as shown in Figure 2.7; choose **Add Selected** from the General panel of the Wall contextual tab. The Properties palette consists of the settings to create the wall based upon the settings of the selected wall as shown at left in Figure 2.7.

5. The General section lists the style as **CMU-8 [CMU 190]**. The Bound spaces field is set to **By style (Yes)**; therefore a space object could be created from the boundary of the four walls shown.

6. Press Escape, and select wall B as shown in Figure 2.7. The Properties palette for this wall shown at right includes additional fields since it describes an existing wall. The Bound spaces field, shown at **p1**, for this wall is set to **No**.

7. Retain the selection of wall B, and choose the **Cleanup automatically field** (refer to **p2** in Figure 2.7). The Yes and No options control the automatic cleanup when walls intersect. The Cleanup group definition lists the **Standard** cleanup group style. Cleanup group definitions can be used to selectively clean up walls.

8. The Segment type is Arc for wall B as shown at **p3**. The wall was created with the Segment type set to **Arc**. Press Escape to end selection.

9. Select wall **A**; the Segment type is **Line** for this wall as shown at left in the Properties palette.

10. Retain the selection of wall **A**; the Width field of the wall shown at **p4** is inactive since its width is preset in the style definition. Press Escape.

11. Select wall B. Notice the Width field of this wall is 6 [240]; you can click in this field to change the Width value.

12. Retain the selection of wall **B**. The Base height of this wall is **8'-0"** [2500] as shown at **p5**. Press Escape to clear selection.

13. Select wall **A**; this wall has a Base height of **10'-0"** [**3000**].

14. Verify that wall **A** is selected, and choose **Add Selected** from the General panel of the Wall contextual tab. Click in the workspace, as you move the cursor; the Length field will dynamically change as you place the wall. If you choose an existing wall, you can edit this field to change wall length. Press Escape to end the command.

15. Select wall **A**; the Justify field setting for the wall is currently set to Baseline as shown at **p6** in Figure 2.7. However you can click at p6 to view the following additional justifications from the flyout: Center, Right, and Left.

16. Retain the selection of wall A, and choose **Add Selected** from the General panel Wall tab. The Offset field is currently set to 0; click at **p7** and type **12"** [**300**]. Click anywhere in the drawing to begin the wall; notice the wall is dynamically displayed 12" [300] from the cursor as you move the cursor. Press Escape to end the command and wall selection.

17. Select wall **A**, and choose **Add Selected** from the General panel Wall tab. The E Roof line offset from baseline and F Floor line offset from baseline fields allow you to enter a distance to extend the wall above the base height or below the baseline of the wall as shown in the graphic at **p8** in Figure 2.7.

18. Retain the selection of wall **A**; the Rotation and Elevation fields display the current settings for placing the wall.

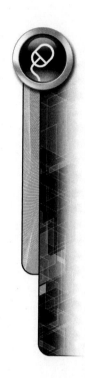

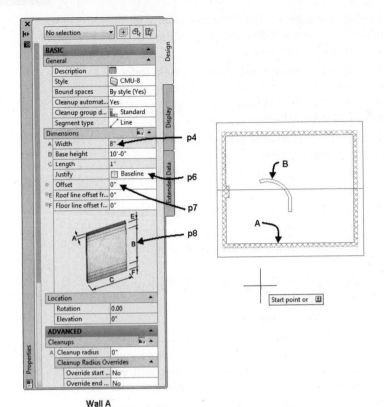

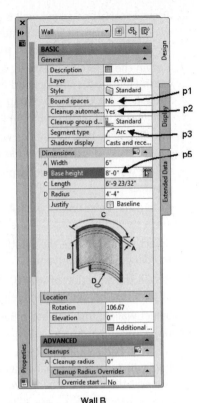

Wall A   Wall B

FIGURE 2.7 *Basic section of the Properties palette of a wall*

19. The Advanced section of the Properties palette consists of settings regarding cleanup radii and endcap settings. Select the wall **A** as shown in Figure 2.8 and toggle ON **Justification Display** of the Cleanup panel in the Wall contextual tab. This toggle will allow you to view the justification lines and the cleanup radii of all walls.

20. In the Advanced section click the Cleanup radius field at **p1** shown in Figure 2.8, type **12″ [300]**. The Cleanup radius discussed in this chapter can be used to merge components of walls. The Cleanup radius is graphically displayed by circles at each end of the wall as shown in Figure 2.8. The default cleanup radius for the remaining walls is 0; therefore the cleanup radius is not displayed. The Override start cleanup radius and Override end cleanup radius allow you to edit the cleanup radius for each end.

21. The Style Overrides category of the Advanced section allows you to specify the endcap and wall priorities for each end of the wall. Endcaps and wall component priorities will be discussed in Chapter 3.

22. The Worksheets section includes Plan modifiers and Roof/floor line worksheets. Click the Plan modifier worksheet at **p2** of Figure 2.8 to view the settings for plan modifiers. A plan modifier is located near **B** as shown in Figure 2.8. Plan modifiers are vertical projections such as pilasters that are placed along the wall. Choose Cancel to dismiss the dialog box.

23. Retain the selection of wall **A** shown in Figure 2.8. To view the gable of the wall, choose **SW Isometric** from the View cube. Choose the Roof/floor line worksheet at **p3** in Figure 2.8. The Edit Roof Line radio button is toggled ON in the Roof and Floor Line worksheet; therefore the gable of the wall is defined in the upper pane.

24. Save and close the drawing.

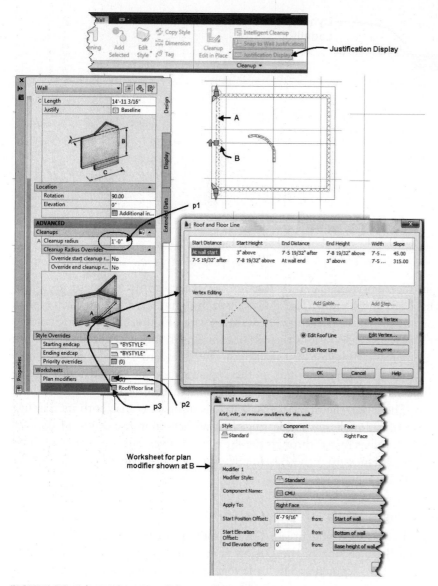

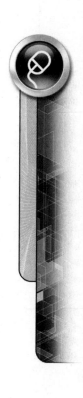

FIGURE 2.8 *Advanced section of the Properties palette for a wall*

Setting Justification for Precision Drawing

When you draw a wall, a justification line is displayed in which the wall components follow, as shown in Figure 2.9. This justification line becomes the handle for drawing the wall; it begins and ends at the coordinates specified for the wall. The grips of the wall are located on the justification line. The justification line is similar to an axis line in which the wall is displayed. When you select an existing wall, the justification line is displayed as well as the grips of the wall, as shown in Figure 2.9. The justification line is shown as a dashed line. A directional arrow is shown indicating the direction the wall was drawn. The walls in Figure 2.9 were drawn from left to right. The justification of a wall sets the location of the justification line relative to the wall width. A wall drawn with right justification has its handle on the right when you orient yourself at the start point and face toward the end of the wall.

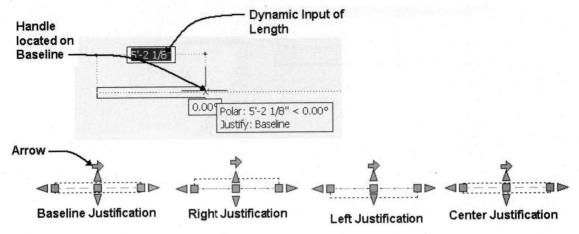

FIGURE 2.9 *Wall justifications*

Types of justifications available in the Design tab of the Properties palette are **Baseline**, **Right**, **Left**, and **Center**. The walls shown in Figure 2.9 illustrate each type of justification and its handle position. The walls drawn in Figure 2.9 used the Standard wall style. This wall style positions the justification line in the center of the wall width for both Center and Baseline justifications. The Standard style consists of only two lines, which are offset one-half of the wall width value on each side of the Baseline. The value entered for the wall width is balanced one-half on each side of the Baseline. Therefore, the Standard wall style is unique because both its Baseline and Center justifications result in locating the wall handle in the center of the wall width. Complex walls created through the use of wall styles can consist of wall lines offset a specific distance from the Baseline. Complex walls created using wall styles will be presented in Chapter 3.

Creating Straight Walls

Walls can be drawn straight or curved. The **Segment** property located in the General category of the Basic section of the Design tab in the Properties palette is used to toggle to Line or Arc segments. The Line toggle is the default wall type that is retained if arc walls were previously drawn. Therefore, segment type will be **Line** unless you toggle **Arc** in the Segment edit field of the Basic section of the Design tab of the Properties palette. To draw a straight wall, select the Wall tool from the Build panel of the Home tab, edit the Properties palette, and then specify the beginning and ending points of the wall. If you toggle ON **Ortho Mode** or **Polar Tracking**, the direction of the wall will be controlled. A wall 8'-0" [2440] long is created in Figure 2.10 when **Dynamic Input** is toggled ON in the Application status bar. The typed value of 8' [2440] will edit the dynamic dimension. When the dynamic prompt option of Dynamic Input is ON, values typed will edit the dynamic dimension in the workspace; however, you can click on the command line to respond in the command line. The workspace prompts are shown below:

```
(Select the Wall tool from the Build panel of the Home tab or
Design palette.)
Start point or: (Click a location for p1 in the graphics
screen.)
(Move the cursor to the right.)
End point or: 8'[2440] ENTER (The p2 point is specified as
shown in Figure 2.10.)
```

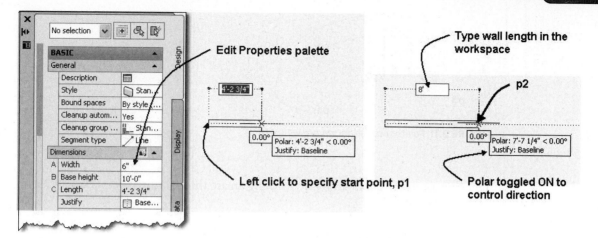

FIGURE 2.10 *Creating a straight wall*

The endpoint can be established by any of the following four coordinate entry methods: absolute coordinate, relative coordinate, Polar Tracking coordinate, or direct distance entry. In the earlier example, direct distance entry was used to draw the wall 8' [2440] long.

Drawing walls in AutoCAD Architecture 2012 is very similar to drawing a line in AutoCAD. Notice in the command sequence that the Undo option is added to command options once the endpoint of the wall is established. If you are drawing a series of wall segments, you can undo each segment by continuing to select the Undo option in the **WallAdd** command prompt. To exit the command, press ENTER; the null response will cause the WallAdd command to be terminated.

Designing Curved Walls

Curved walls are drawn by selecting **Arc** in the **Segment** edit field of the Basic section in the Design tab of the Properties palette. Three points establish the curve of the wall: the beginning point, a point along the curve, and the endpoint of the curved wall. The following workspace prompts and entries illustrate the method to create the curved wall shown in Figure 2.11.

> (Select the Wall tool from the Build panel of the Home tab and select Arc from the Segment type in the Design tab of the Properties palette.)
>
> Start point or: (Click a location in the graphics screen to locate **p1**, move the pointer left as shown in Figure 2.11.)
>
> Mid point or: @-5',2' [@-1500,600] (Specifies **p2** as shown in Figure 2.11.)
>
> End point or: @-5',-2' [@-1500,-600] (Specifies **p3** as shown in Figure 2.11.)
>
> Mid point or: ENTER (Ends the WallAdd command.)

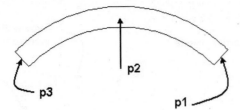

FIGURE 2.11 *Points specified for an arc wall*

Drawing a curved wall is similar to drawing an arc through three points with the Arc command of AutoCAD.

Using the Close Option

The **WallAdd** command includes options of **Close** and **Ortho close**. You can access these options from the command line or from the shortcut menu displayed while you are drawing the wall. The Close option is used to draw a final wall segment from the endpoint of the last wall segment back to the start point of the first wall segment of the wall segment series as shown in Figure 2.11. This option is similar to the Close option when AutoCAD lines are drawn. The Close option of the AutoCAD **Line** command allows you to draw several line segments and enter a C for Close to close the last line segment to the beginning of the first line segment. The Close option for the wall becomes active after two wall segments have been drawn. Select the Close option from the workspace prompts or choose **Close** from the shortcut menu displayed after two segments are drawn. The following command sequence was used to close the last wall segment:

    Start point or: (Select p1 as shown in Figure 2.12.)
    End point or: (Select p2 as shown in Figure 2.12.)
    End point or: (Select p3 as shown in Figure 2.12.)
    End point or: (Right-click and choose Close from the shortcut menu.)
    (Final wall segment is drawn to p1.)

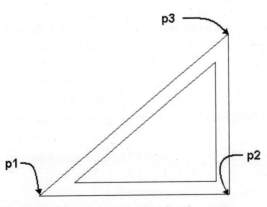

FIGURE 2.12 *Walls drawn using the Close option*

TUTORIAL 2-1: DRAWING WALLS AND USING CLOSE

1. Select New from the Quick Access toolbar.

2. Choose **Save As > AutoCAD Drawing** from the Application menu, and save the drawing as **Lab 2-1** in your student directory. Click Save to dismiss the Save Drawing As dialog box.

3. Right-click over the Object Snap toggle, and choose Settings to open the Drafting Settings dialog box. Choose the Endpoint object snap mode and Extension. Verify that Allow general object snap settings to act upon wall justification line is checked as shown in Figure 2.13.

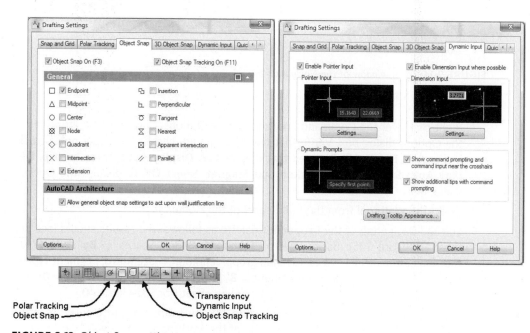
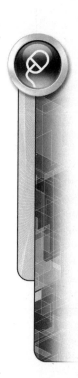

FIGURE 2.13 *Object Snap settings*

4. Verify that only Polar Tracking, Object Snap, Object Snap Tracking, Dynamic Input, and Transparency are toggled ON in the status bar as shown in Figure 2.13. Right-click over the Dynamic Input toggle, and choose Enable.

5. Right-click over the Dynamic Input toggle, and choose Settings. Verify that the Enable Pointer Input, Enable Dimensional Input where possible, and Show command prompting and command input near the crosshairs checkboxes are checked as shown in Figure 2.13. Select OK to close Drafting Settings.

6. Select **Zoom Window** from the Zoom flyout of the **Navigation** bar as shown in Figure 2.14. Respond as follows to the following workspace prompts:

Specify first corner: **100',0'** [800, 6000] ENTER

Specify opposite corner: **200',75'** [25000, 25000] ENTER

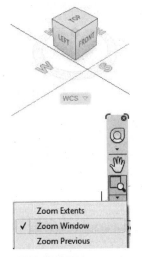

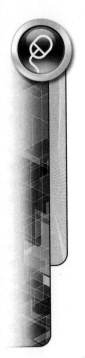

FIGURE 2.14 *Zoom window of the Navigation bar*

7. Right-click over the Command line, and choose Options. Choose the Drafting tab of the Options dialog box. Verify that the following are checked: **Display polar tracking vector**, **Display full-screen tracking vector**, **Display Auto Track tooltip**, **Display Auto Snap tooltip**, **Marker**, and **Magnet**. Select **OK** to dismiss the Options dialog box.

8. Select the **Wall** tool from the Build panel of the Home tab.

9. Edit the Design tab of the **Properties** palette as follows (the results should look like those in Figure 2.15): Basic: Style = **Standard**, Bound Spaces = **By style (Yes)**, Cleanup automatically = **Yes**, Cleanup group definition = **Standard**, Segment type = **Line**, A-Width = **3 1/2 [89]**, Base height = **8' [2440]**, Justify = **Left**, Offset = **0**, Roof line offset from base height = **0**, and Floor line offset from baseline = **0**.

10. Begin drawing the walls as shown in Figure 2.15 using direct distance entry. Type the distances in the dynamic dimension field of the workspace as follows:

> Start point or: 150',50' [13750,20000] ENTER *(Point* p1 *established as shown in Figure 2.15.)*
>
> (Move the pointer right; verify Polar Tracking angle in Auto Track tooltip = 0.)
>
> End point or: **24' [7200]** ENTER (p2)
>
> (Move the pointer down; verify Polar Tracking angle in Auto Track tooltip = 270.)
>
> End point or: **24' [7200]** ENTER (p3)
>
> (Move the pointer left; verify Polar Tracking angle in Auto Track tooltip = 180.)
>
> End point or: **12' [3600]** ENTER (p4)
>
> (Move the pointer down; verify Polar Tracking angle in Auto Track tooltip = 270.)
>
> End point or: **2'[600]** ENTER (p5)
>
> (Move the pointer left; verify Polar Tracking angle in Auto Track tooltip = 180.)
>
> End point or: **12'[3600]** ENTER (p6)

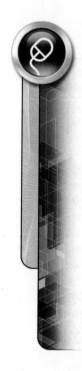

> (Move the pointer up; verify Polar Tracking angle in Auto
> Track tooltip = 90.)
>
> End point or: 2' [600] ENTER (p7)
>
> (Move the pointer left; verify Polar Tracking angle in Auto
> Track tooltip = 180.)
>
> End point or: 12' [3600] ENTER (p8)
>
> (Move the pointer up; verify Polar Tracking angle in Auto
> Track tooltip = 90.)
>
> End point or: 12' [3600] ENTER (p8)
>
> (Right-click and choose Close from shortcut menu.)
>
> (Wall segments will be drawn as shown in Figure 2.15.)

11. Click **Work layout** in the **Application** status bar, and click on the left viewport. Select **Zoom Extents** from the Zoom flyout of the **Navigation** bar.

12. Click on the right viewport, and select **Zoom Extents** from the Zoom flyout of the **Navigation** bar.

13. Verify that the right viewport is current, and select **1/4" = 1'-0" [1:50]** from the **Viewport Scale** flyout menu of the Drawing Window status bar. Select **High Detail** from the display configuration flyout of the Drawing Window status bar.

14. Choose Save from the Quick Access toolbar. Choose **Close > Current Drawing** from the Application menu.

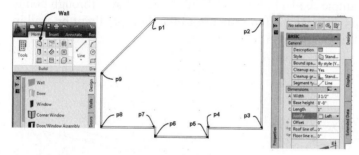

FIGURE 2.15 *Close option used to draw the last wall segment*

Using the Ortho Close Option

The **Ortho close** option draws the final two wall segments of a series of walls to connect at an angle of 90 degrees to the beginning of the first wall segment. Execute this option by selecting **Ortho close** from the shortcut menu or typing OR in the command line. If Ortho Mode is toggled ON in the Application status bar, this option allows you to draw a rectangle after establishing the length and width distances by the first two wall segments. The following workspace prompts were used to create a rectangle using the Ortho close option. Note in the following workspace prompts that when you choose Ortho close you are prompted to select a "Point on wall in direction of close." This point sets the direction of completing the wall segment.

> Start point or: (Select **p1** as shown in Figure 2.16.)
>
> End point or: (Select **p2** as shown in Figure 2.16.)
>
> End point or: (Select **p3** as shown in Figure 2.16.)
>
> End point or: (Right-click and choose Ortho close from the
> shortcut menu.)
>
> Point on wall in direction of close: *(Select a location near*
> **p4** *as shown in Figure 2.16.)*

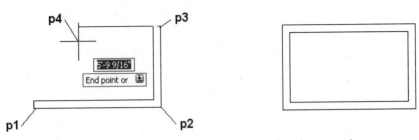

FIGURE 2.16 *Identifying a point for Ortho close to complete the rectangle*

The **Ortho close** option is not active until two wall segments are drawn. In the above workspace prompts Ortho Mode was toggled ON prior to beginning the WallAdd command; therefore, when you choose the "Point on wall in direction of close": the point will be perpendicular to the last wall segment, and close the polygon to form a rectangle.

If the first two wall segments were not perpendicular to each other when Ortho close was used, the final wall segment will still be close to the first wall segment at an angle of 90 degrees. In the example shown in Figure 2.17 walls were drawn through points **p1**, **p2**, and **p3**. Point **p4** was specified in response to the prompt "Point on wall in direction of close": to close the polygon.

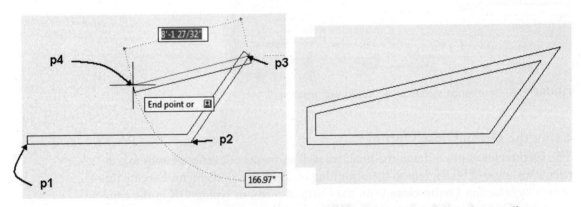

Point p4 selected for Ortho close option

Walls drawn using Ortho close option

FIGURE 2.17 *Ortho close used to complete the walls of a polygon*

Using the Offset Option to Move the Wall Handle

The **Offset** edit field of the Properties palette allows you to enter a distance to offset the handle for the wall relative to the graph line of the wall. If you enter an offset of 2 [50] using a right wall justification, the handle for the wall will be located 2 [50] beyond the right wall face, as shown on the left in Figure 2.18. The wall shown is 8 [200] wide and is being drawn from left to right.

If a negative value is entered for the offset distance, the handle for the same wall will move up 2 [50] from the right face, as shown on the right in Figure 2.18. The ability to shift the handle relative to the justification of a wall is useful in creating a foundation plan. In residential construction drawings, the masonry walls are offset 3/4 [19] from a wood stud partition line, as shown in Figure 2.19. This feature would allow you to trace an existing AutoCAD floor plan to create a foundation plan in AutoCAD Architecture 2012. Based on this arrangement of the wall elements, the masonry foundation wall would be created with a 7 5/8 [194] wall width and a baseline offset of 3/4 [19]. These settings would allow you to trace the first floor plan geometry attached as a reference file to create a foundation plan using the Endpoint Object Snap mode.

After specifying the start point of a wall, type O or choose the Offset option from the shortcut menu to display assistant lines. If you left-click when the assistant line is displayed, you specify an offset value equal to the distance from the wall justification to the assistant line in the workspace. Assistant lines are displayed for the center and face of each wall component and the overall center of the wall. As shown in Figure 2.18 a dashed assistant line is displayed when the cursor is positioned over the center of a wall component to specify an offset of 3″, whereas the center linetype is displayed for assistant lines of the center of the wall.

In commercial construction the offset distance could be set according to the distance a wall is offset from the column centerline. This would allow you to trace the column line to draw a wall offset the desired distance between the column line and the wall.

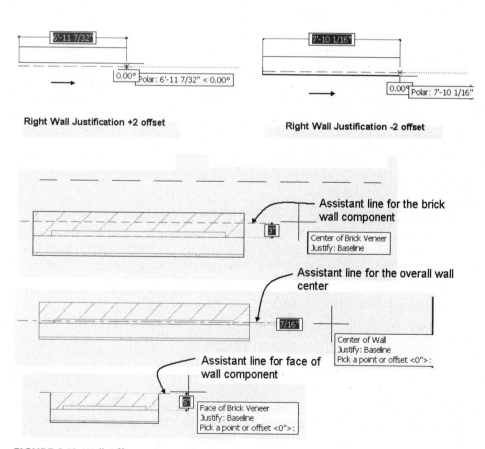

FIGURE 2.18 *Wall Offset options and assistant lines*

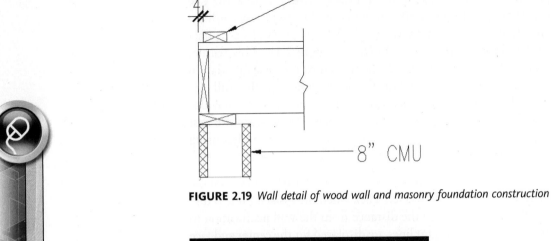

FIGURE 2.19 *Wall detail of wood wall and masonry foundation construction*

TUTORIAL 2-2: USING WALL OFFSET

1. Verify that the Accessing Tutor Imperial or Accessing Tutor Metric content for Chapter 2 of the CengageBrain has been downloaded to your Accessing Tutor folder on your computer as described in Organizing Tutorial Directories in the Preface.

2. Select **Open** from the Quick Access toolbar. Choose Ex 2-2 from the *Accessing Tutor\ Imperial\Ch* 2 or *Accessing Tutor\Metric\Ch* 2 folders.

3. Choose **SaveAs > AutoCAD Drawing** from the Application menu, and save the drawing as **Lab 2-2** in your student directory. Click Save to dismiss the Save Drawing As dialog box.

4. Verify that **Model** is toggled ON in the **Application** status bar.

5. Right-click over the **Object Snap** toggle, and choose **Settings**. Clear all modes except **Intersection**. Verify that the **Allow general object snap settings to act upon wall justification line** checkbox is checked. Choose **OK** to dismiss the **Drafting Settings** dialog box. Verify that **Objects Snaps** are Enabled in the Application status bar.

6. Select CMU-8 Rigid-1.5 Air-2 Brick-4 Furring-2 [CMU-190 Rigid-038 Air-050 Brick-090 Furring] from the Walls tool palette. Edit the Properties palette as follows: set Justification = Baseline, Height = 10'. Respond to the workspace prompts as follows:

   ```
   Start point or: (Choose a point at p1 to start the wall as
   shown in Figure 2.20.)
   End point or: o (Type o to choose the Offset option.)
   Pick a point or offset <0">: (Move the cursor right. Move the
   cursor up to point p2 to display the assistant line and click
   to specify the face of the Air Gap wall component as shown in
   Figure 2.20.)
   End point or: (Select a point at p3 as shown in Figure 2.20.)
   End point or: (Select a point at p4 as shown in Figure 2.20.)
   End point or: (Select a point at p5 as shown in Figure 2.20.)
   End point or: (Right click and choose Close from the shortcut
   menu.)
   ```

7. In this step you will add walls using Apply tool properties to > Linework. Click the Brick-4 Brick-4 [Brick-090 Brick-090] tool from the Walls tool palette, right-click, and choose Apply Tool Properties to > Linework. Respond to the workspace as follows:

Select lines, arcs, circles, or polylines to convert into
walls: (Select the polyline at **p6** as shown in Figure 2.20.)
1 found

Select lines, arcs, circles, or polylines to convert into
walls: ENTER (Press ENTER to end selection.)

Erase layout geometry? [Yes/No] <No>: Y (Choose Yes to erase
the layout geometry.)

4 new wall(s) created.

(Press ESC to clear selection.)

8. Select Wall A as shown in Figure 2.20; choose Add Selected from the General panel in the Wall tab. Respond to the workspace prompts as follows:

Start point or: (Choose a point at **p7** to start the wall.)

End point or: **o** (Type o to choose the Offset options.)

Pick a point or offset <O>: (Move the cursor until the Center
of Wall tool tip is displayed; left click to specify the
offset.

End point or: Move the cursor to choose the intersection at **p8**.)

(Press ESC to end the command.)

9. Choose **Save** from the Quick Access toolbar. Choose **Close** > **Current Drawing** from the Application menu.

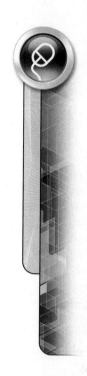

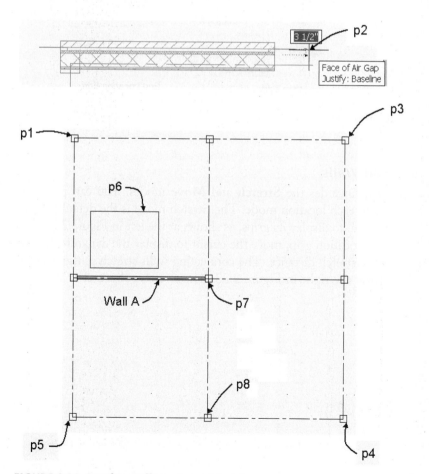

FIGURE 2.20 *Specifying Offset using assistant lines*

Grips of the Wall

You can use AutoCAD grips to lengthen, move, or increase the height or width of a wall. The functions of the grips of a wall are described in Figure 2.21. The grips are located at the endpoints and midpoint of the wall justification line. When you select a grip, it becomes hot; movement of the pointer will stretch the grip to new location. Hot grips allow you to directly manipulate the size and location of a wall. Movement of a hot grip displays one or more dynamic dimensions. When a dynamic dimension is displayed with blue highlight, you can type a value in the keyboard and the wall will stretch to the dimension specified. When a grip is hot, you can press TAB to cycle through and highlight other dynamic dimensions.

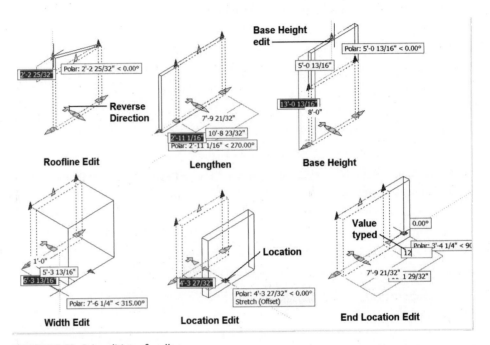

FIGURE 2.21 *Grip editing of walls*

Using Grips to Edit Walls

The wall location grip includes the **Stretch** and **Move** modes. You can press the CTRL key to toggle to each location mode. The Stretch mode is the default; therefore, if you select the wall to display its grips, as shown at the left in Figure 2.22, you can choose the middle location grip, move the cursor to display the dynamic dimension, and then type the stretch distance. The connecting walls stretch as the selected wall moves as shown at the right in Figure 2.22. The stretch is limited to the connected walls.

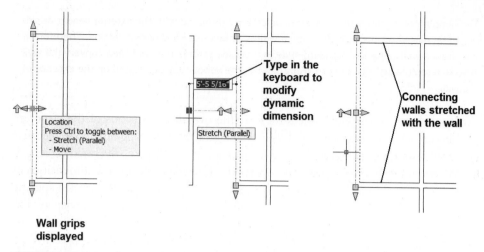

Wall grips
displayed

FIGURE 2.22 *Using the Stretch mode of a location grip to edit walls*

The Move mode can be applied if you select the location grip, press the CTRL key to toggle to Move, and then move the cursor to display the dynamic dimension. When you type a value in the keyboard, you edit the dynamic dimension and move the wall independent of the adjoining walls.

Steps to Copying a Wall Using Grips

1. Select the wall to display the grips.
2. Select the Location grip at the midpoint to make it hot.
3. Right-click and choose **Copy** from the shortcut menu.
4. Move the pointer in the direction for the new wall.
5. Type the distance for the Stretch (Offset) in the dynamic dimension.

Setting Justification

This section explains the wall justifications when using the Standard wall style. The justification of complex walls should be set to baseline, as discussed in Chapter 3. If you are working from a sketch that identifies the overall dimensions to the outer wall surface, the left or right wall justification should be used. The justification of the wall sets the location of the handle for placing the wall. This handle is located at the bottom of the wall or at the zero Z coordinate. Therefore, as the wall is drawn, the bottom of the wall is placed at $Z = 0$ for all wall vertices. If a wall is drawn with center justification, the handle for the wall is located at the bottom of the wall and is centered between the sides of the wall.

When drawing a floor plan, you often start with a basic shape such as a rectangle. The overall dimensions of the basic shape are set by the outer wall surfaces. To draw walls using the Standard wall style dimensioned to the outer surfaces, set the justification to left and draw the walls in a **clockwise** direction.

Access Instructional Video **2.1 Floor Plan Setup** from the **Instructional Video** category of the **Student Companion** site of **CengageBrain** http://www.cengagebrain.com described in the Preface.

NOTE

> **TIP** The sample wall styles of the Walls tool palette are designed with the exterior veneer or siding to be placed on the outside when a wall is drawn in a clockwise direction. Therefore if you draw a wall using the Standard style in a clockwise direction and then convert the wall later to a sample wall style in the Walls palette, the veneer will be located on the exterior of the wall.

If you are drawing a wood frame structure, the overall dimensions of the structure should usually be a multiple of 16" [400] or 24" [600] for framing efficiency. Because studs and joists are usually placed 16" [400] o.c. or 24" [600] o.c., the building length should be a multiple of 16" [400] or 24" [600]. Therefore, the overall dimensions of the building should not be a random distance but rather a distance that is a specific multiple of 16" [400] or 24" [600]. Setting the Snap Mode and Grid Display distance to a modular distance will aid in controlling wall lengths. With Snap Mode, Grid Display, and the appropriate wall justification, the outer wall surfaces will be snapped to dimensions that are multiples of 16 inches [400 mm] as the cursor is moved. In Figure 2.23 the Grid Display and Snap Mode were set to 16" [400]. The left justification of the walls is shown by the location of the grips. The exterior walls were drawn clockwise.

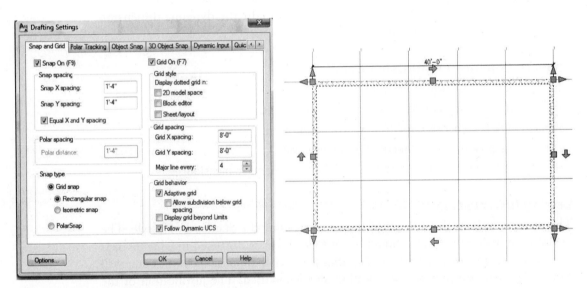

FIGURE 2.23 *Wall justification displayed with grips*

Using Edit Justification

You can edit the justification of a wall in the Basic section in the Design tab of the Properties palette. However, the **Edit Justification** command can be used to specifically change the justification. Access this command as shown in Table 2.4. When this command is executed, trigger grips are displayed on the wall, which indicate the justification options, as shown in Figure 2.24.

The current justification is shaded. If you move your pointer over a justification, a tip will display the name of the justification and dynamically display how the wall will change if this justification is selected. The justification can be changed by clicking the justification trigger grip. You may press the CTRL key to toggle options to maintain wall position or maintain justification line position. After you click the justification trigger grip, the markers are cleared and only the wall is displayed.

TABLE 2.4 *WallEditJustification command access*

Command prompt	WALLEDITJUSTIFICATION
Ribbon	Select the wall to display the Wall tab. Choose Edit Justification from the slide out panel of the Cleanup panel. Refer to Figure 2.24.
Shortcut menu	Select the wall, right-click, and choose Edit Justification.

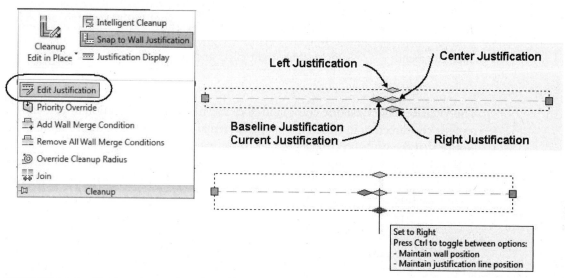

FIGURE 2.24 *Justification trigger grips of a wall*

After you specify the start point of the wall you can toggle through wall justification options when you press the Shift key.	**TIP**

CHANGING WALL DIRECTION WITH WALLREVERSE

The **Reverse (WallReverse)** and WallReverseBaseline (WallReverseBaseline) commands allow you to change the direction in which a wall is drawn. The commands are most useful in flipping the thickness of a wall to the other side of the justification line. Access the WallReverse command as shown in Table 2.5 and Figure 2.25.

TABLE 2.5 *WallReverse command access*

Command prompt	WALLREVERSE
Ribbon	Select a wall, and choose WallReverse from the Reverse flyout of the Wall Reverse flyout shown in Figure 2.25.
Shortcut menu	Select a Wall, right-click, and choose Reverse > In Place.

When choosing **WallReverse** from the Ribbon, you can change the direction of the wall in place without changing its position in the drawing. The WallReverse command can also be chosen from the shortcut menu by choosing the Reverse > **In Place**. The **In Place** option will reverse the direction and retain wall location as shown at the left in Figure 2.25. The **Baseline** option reverses the justification line;

therefore, when the wall is drawn with left or right justification, the thickness of the wall will flip to the other side. Access the Wall Reverse Baseline command as shown in Table 2.6.

TABLE 2.6 *WallReverseBaseline command access*

Command prompt	WALLREVERSEBASELINE
Ribbon	Select a wall, choose WallReverseJustificationLine from the Reverse flyout of Modify panel flyout in the Wall tab, and refer to Figure 2.25.
Shortcut menu	Select a Wall, right-click, and choose Reverse > Baseline.

The **WallReverse Baseline** command does not move the justification line; it simply reverses the direction in which the wall is drawn as shown at the right in Figure 2.25. Notice that the wall-directional indicator is reversed from the top walls. The wall directional indicator is a trigger grip, which, when selected, will apply the **Wall Reverse** command. If you hold down CTRL when you select the trigger grip, the **Wall Reverse Baseline** command is applied to the wall.

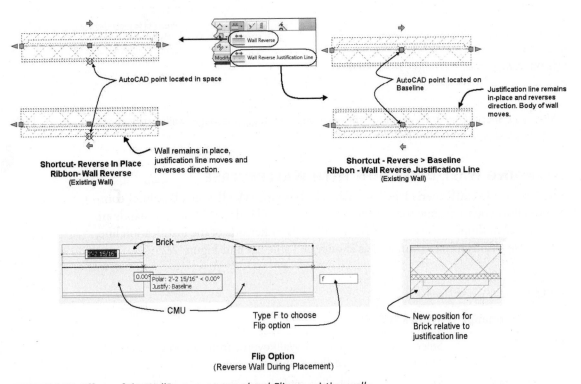

FIGURE 2.25 *Effects of the WallReverse command and Flip on existing walls*

The Flip option can be selected when placing a wall to reverse the direction of the wall. The Flip option can also be selected by pressing the CTRL key to toggle segment orientation or reverse the direction. Select the Flip option after specifying the start point of a wall to flip the body of the wall about the justification line as shown in Figure 2.25.

USING WALL CLEANUP

The **Cleanup** properties of a wall allow you to control whether walls clean up when they intersect. If you are drawing walls representing the same type of construction, the walls should clean up. However, if you are drawing a floor plan with masonry and wood frame walls, the walls should not clean up or merge together. Control over the merge and cleanup of walls is governed in the cleanup settings of the wall, and additional controls can be defined in the wall style definition. Because Autosnap is toggled ON and *Allow general object snap settings to act upon wall justification line* is checked, the cleanup radius of walls can be set to zero. The method of setting the cleanup radius is discussed in the "Setting the Cleanup Radius" section.

Cleanup automatically can be set to Yes or No in the Basic section in the Design tab of the Properties palette, as shown in Figure 2.26. When cleanup is turned on by selecting Yes, walls with the same cleanup group definition and with connecting justification lines will clean up, as shown in Figure 2.26. Therefore walls should clean up if they share the same Standard wall style and Standard Cleanup Group Definition with connecting justification lines. The name of the Cleanup Group Definition of a wall is listed in the Basic section in the Design tab of Properties palette as shown in Figure 2.26 (Cleanup Group Definitions are discussed later in this chapter). The cleanup setting of the wall is changed by editing the cleanup setting in the Basic section in the Design tab of the Properties palette of the selected wall.

If Cleanup automatically is set to No, walls will not clean up as they are drawn. The default for cleanup is Yes; therefore, each time the WallAdd command is selected, the cleanup is set to Yes unless you change it.

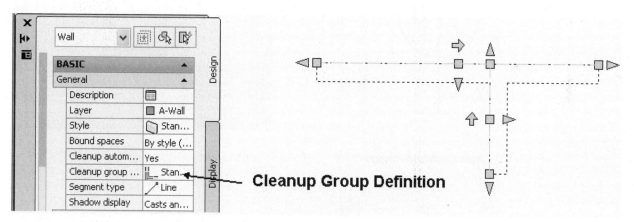

FIGURE 2.26 *Cleanup settings in the Design tab of the Properties palette of a wall*

Setting the Cleanup Radius

The cleanup radius of the wall governs which walls are selected for cleanup. Walls will clean up with other walls if the cleanup radius crosses the justification line, as shown in Figure 2.27. This radius is centered by default along the Justification wall line. The **Cleanup Radius** is set in the Advanced section in the Design tab of the Properties palette as shown in Figure 2.27. The default cleanup radius for the Standard Wall is 0. If you need to edit the cleanup radius in response to a Solution Tip, set the cleanup to one-half the wall width. The Justification Display toggle (WallJustificationDisplayToggle) located on the Cleanup panel of the Wall contextual tab allows you to view the wall components, justification line, and cleanup radii as shown in Figure 2.27 and Table 2.7. When Wall Justification has been toggled on cleanup radius, grips are displayed allowing you to grip edit the cleanup radius.

TABLE 2.7 *WallJustificationDisplayToggle command access*

Command prompt	**WALLJUSTIFICATIONDISPLAYTOGGLE**
Ribbon	Select a wall to display the Wall tab. Choose Justification Display from the Cleanup panel, and refer to Figure 2.27.
Shortcut menu	Select a Wall, right-click, and choose Cleanups > Toggle Wall Justification Display.

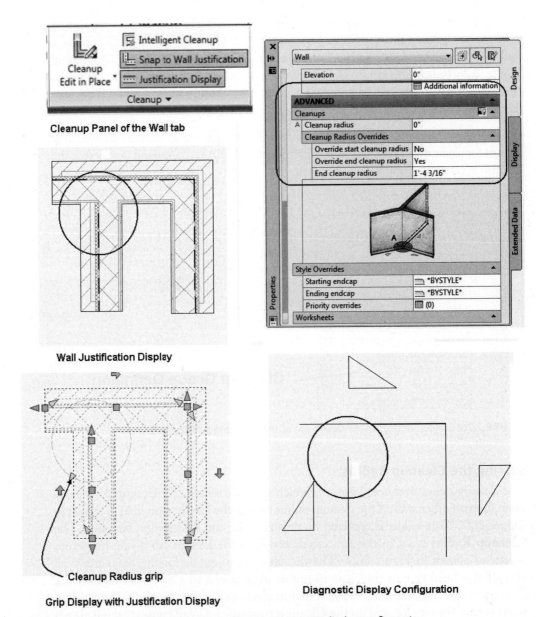

Cleanup Panel of the Wall tab

Wall Justification Display

Cleanup Radius grip

Grip Display with Justification Display

Diagnostic Display Configuration

FIGURE 2.27 *Cleanup radii displayed for walls using Diagnostic display configuration*

If you select the **Diagnostic** display configuration on the Drawing Window status bar, you can view the justification lines and cleanup radii of the walls. The Diagnostic display configuration applies the Graph display representation, which includes a wall-directional arrow or triangle that points to the endpoint of the wall. The grips shown in Figure 2.27 include a cleanup radius grip that allows you to drag the grip and increase the cleanup radius. The width of the wall is not shown in the Diagnostic display configuration. Compare the wall cleanup radii shown at the right in Figure 2.27 with the actual wall cleanup shown at the left.

The cleanup radius can be set as an exception or overridden for the start or end of the wall segment. This allows you to edit the value of the cleanup radius for a specific end of the wall. The override is set in the Advanced section in the Design tab of the Properties palette. Additional information regarding cleanup of wall components is presented in Chapter 3.

Using Priority to Control Component Merger

Complex walls created using Wall styles can have more than one wall component. The wall components merge with other components based upon the priority value assigned to the component. A wall can be created that consists of a concrete masonry unit component and a brick component. Each of the components can be assigned a priority value. Wall components with the same numerical priority value will merge together. Components with the lowest value priority override the display of components with higher priority values. In Figure 2.28 the brick veneer component has a priority of 810, and the concrete masonry unit component has a priority of 300. Therefore, the concrete masonry unit components merge together. Additional information regarding setting priorities of wall components is presented in Chapter 3.

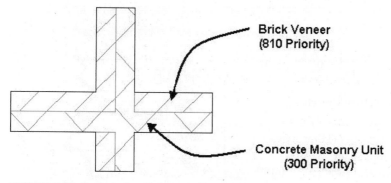

Brick Veneer
(810 Priority)

Concrete Masonry Unit
(300 Priority)

FIGURE 2.28 *Control of wall cleanup using priority*

Creating Cleanup Group Definitions

Cleanup Group Definitions can be defined to control wall cleanup. If walls have different cleanup group definitions, they will not clean up. The walls can be of the same style, but if they have different cleanup group definitions, the cleanup will not occur. Cleanup groups are created to control the merging of walls during automatic cleanup. Regardless of style name, walls can merge with other walls, and the intersections are cleaned up according to wall component priority number. The priority number of wall components controls how the walls clean up. Walls with components of equal priority numbers merge together if they share the same cleanup group definition.

Walls with different cleanup group definitions will not blend together regardless of wall component priority number. When cleanup group definitions are assigned to walls in the Design tab of the Properties palette, the walls that share the same cleanup group definition will merge based upon the priority number of the wall component. Walls that do not share the same cleanup group will not merge, regardless of the wall component priority number.

Cleanup Groups are created in the **Style Manager**. Wall **Cleanup Group Definitions** are located in the Architectural Objects folder of the left pane in the Style Manager. Shown below are the steps to creating a cleanup group definition.

Steps to Create and Apply a Cleanup Group

1. Select **Manage** tab, and choose **Style Manager** from the Style & Display panel in the ribbon.
2. Select the **(+)** to expand the Architectural Objects folder in the left pane.
3. Select **Wall Cleanup Group Definitions** in the left pane.
4. Select the **New Style** command on the **Style Manager** toolbar, and type **Mystyle** as the name of the new style, as shown in Figure 2.29.
5. Select **OK** to dismiss the Style Manager.
6. Select a wall in the current drawing, right-click, and choose **Properties** from the shortcut menu.
7. To assign cleanup groups to a wall, click on the edit field of the **Cleanup Group Definitions** in the Design tab of the **Properties** palette. Select **Mystyle** from the drop-down list, as shown in Figure 2.30. Walls do not merge as shown at the right in Figure 2.30 because they have different cleanup group definitions.

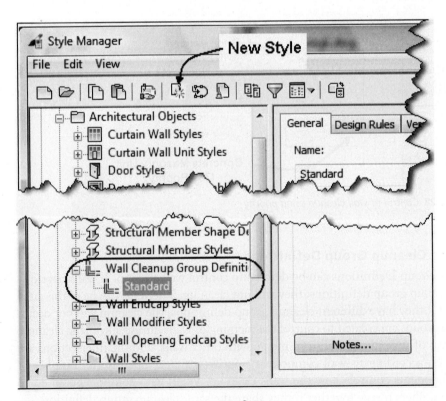

FIGURE 2.29 *Creating Wall Cleanup Group Definitions*

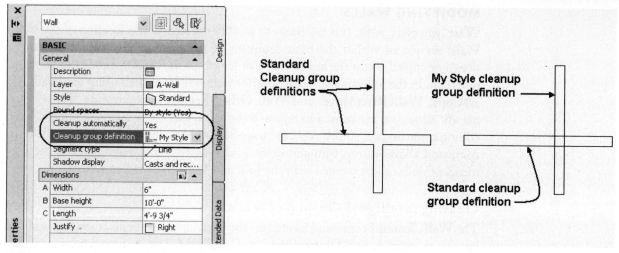

FIGURE 2.30 *New Cleanup Group Definition in the Design tab of the Properties palette*

Solution Tips

A solution tip, as shown in Figure 2.31, will be displayed if a wall fails to clean up. A defect marker was displayed in previous versions, whereas the new Solution Tip provides an explanation and possible solutions.

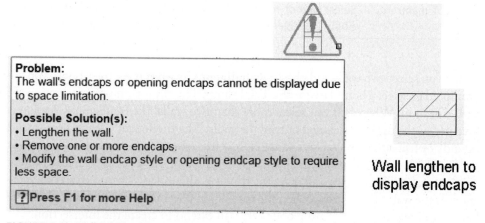

Wall lengthen to display endcaps

FIGURE 2.31 *Wall Solution Tip*

Solution tips are displayed when the Justification lines do not intersect. When you toggle ON the Justification Display, the justification line is shown as well as other wall components. Table 2.8 lists the reasons and solutions for removing solution tips.

TABLE 2.8 *Reasons Wall Solution Tips are displayed*

Reason	Solution
Multiple walls are intersecting at angles.	Select the walls, and add a cleanup radius to each wall less than the wall width.
The wall is coincident with another wall.	Select the Solution tip and choose Erase from the Modify panel of the Home tab.
Cleanup radius may be too large.	Select the wall and reduce the cleanup radius to the wall width in the Design tab of the Properties palette.
Walls drawn are too short to display wall endcap.	Increase the wall length or modify the wall endcap.

MODIFYING WALLS

When you place walls, it is necessary to modify the length and location of the walls. Walls are placed within the basic footprint of the building. Interior walls can be drawn or copied across the entire width or length of the building and later trimmed to length. In the following sections, interior walls will be created using the **WallOffsetCopy**, **WallOffsetMove**, and **Wall Offset Set From** commands. These commands allow you to copy and move walls with precision. Although most basic editing commands of AutoCAD can be applied to the walls, the **Intelligent Cleanup** command allows you to trim and extend walls based upon wall length. Short segments of walls can be merged with the **Join** and **Wall Merge** commands.

Creating Intelligent Cleanup of Walls

The **WallCleanupI** command located on the Cleanups panel walls is selected and can be used to create L and T intersections and multiple L and T intersections using the Intelligent Cleanup option. Access the WallCleanupI command as shown in Table 2.9.

TABLE 2.9 *Intelligent Cleanup command access*

Command prompt	WALLCLEANUPI
Ribbon	Select one or more walls to display the Wall tab. Choose Intelligent Cleanup from the Cleanup panel (refer to Figure 2.32).
Shortcut menu	Select one or more walls, right-click, and choose Cleanups > Apply Intelligent Cleanup.

When you select two walls and choose **Intelligent Cleanup** from the ribbon, you are prompted to select a boundary for the T-Cleanup. The command provides option for an L-Cleanup. Whereas if you select more than two walls, the Intelligent Cleanup is executed as the default to extend or trim the walls. The Intelligent Cleanup command extends or trims the two walls that are closest to each other. Therefore, if you select the four walls from **p4** and **p5** as shown in Figure 2.32, the walls are extended to the adjacent walls. The shorter wall segment crossing a boundary is trimmed in the operation. If the walls do not cross, when the Intelligent Cleanup command is applied, the longer wall segment crossing the boundary will extend to create the intersection.

```
(Select a wall for editing at p1 as shown in Figure 2.32 and
choose Intelligent Cleanup from the Cleanup panel of ribbon.)

Select the boundary wall for T-Cleanup or [Create L-Cleanup]
<Create L-Cleanup>: (Select the wall at p2.)

A T intersection is created as shown in Figure 2.32.
```

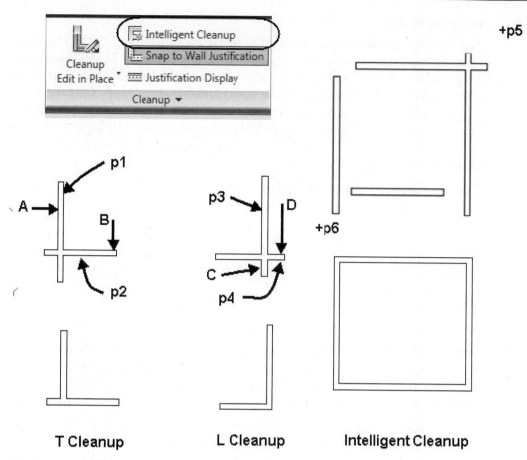

FIGURE 2.32 *Editing walls with Wall Cleanup L and Wall Cleanup T*

The Create L-Cleanup option of Intelligent Cleanup deletes the shorter wall segment to create an L cleanup. In Figure 2.32 shorter wall segments C and D are removed to create the L cleanup. The following command line sequence created an L intersection of walls.

```
(Select walls at p3 and p4 for editing, choose Intelligent
Cleanup from the Cleanup panel of the Wall tab in the ribbon.)

Select the boundary wall for T-Cleanup or [Create L-Cleanup]
<Create L-Cleanup>: (Press Enter to choose the Create
L-Cleanup option.)

(The shortest wall trimmed to create the L-Cleanup as shown
in Figure 2.32.)
```

The following command sequence performs an Intelligent Cleanup.

```
Create a Crossing selection from p5 to p6 to select the walls
as shown in Figure 2.32.

Choose Intelligent Cleanup from the Cleanup panel of the Wall
tab in the ribbon.

Select boundary wall for T-Cleanup or [Intelligent Cleanup]
<Intelligent Cleanup>: (Press Enter to choose the
Intelligent Cleanup option. Walls are trimmed and extended
as shown in Figure 2.32.)
```

Merging Walls

The **WallMergeAdd** command is used to merge two intersecting walls. The Wall-MergeRemove command is used to remove the merger of walls. The WallMergeAdd command is useful for creating short walls or projections of a wall, or for cleaning up walls that intersect perpendicular to the current wall. If several wall components are merged together, each wall component can also be removed from the merge using this command. Access the **WallMergeAdd** command as shown in Table 2.10.

TABLE 2.10 *Wall Merge command access*

Command prompt	WALLMERGEADD
Ribbon	Select a wall to display the Wall tab. Choose Add Wall Merge Condition from the slide out of the Cleanup panel (refer to Figure 2.33).
Shortcut menu	Select a wall, right-click, and choose Cleanups > Add Wall Merge Condition.

When you select a wall and then choose **Add Wall Merge Condition** from the Cleanup panel, you are prompted in the workspace to select the walls to merge with the selected wall. Walls selected in response to this prompt will merge together. The workspace prompts of the WallMergeAdd command are shown below. Multiple components can be merged while remaining in the WallMergeAdd command.

(Select wall as shown at p1 in Figure 2.33 choose from the
Cleanup panel Add Wall Merge Condition.)

Select walls to merge with: (Select the wall at p2 as shown in
Figure 2.33.) 1 found

Select walls to merge with: (Press ENTER to end the command.)

(Walls are merged as shown at right in Figure 2.33.)

If the results of the merger are not satisfactory, you can remove the merger using the **WallMergeRemove** command. Access the **WallMergeRemove** command as shown in Table 2.11.

TABLE 2.11 *WallMergeRemove command access*

Command prompt	WALLMERGEREMOVE
Ribbon	Select a wall to display the Wall tab. Choose the Remove All Wall Merge Conditions from the Cleanup panel (refer to Figure 2.33).
Shortcut menu	Select the wall, right-click, and choose Cleanups > Remove All Wall Merge Conditions.

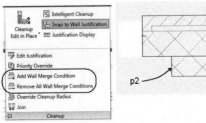
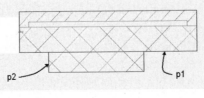
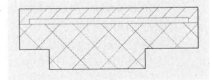

FIGURE 2.33 *Walls selected for Wall Merge*

When you select a wall that has been merged, and choose **Remove All Wall Merge** Conditions from the Cleanup panel, the merger will be removed and the command line will report the results as shown in the following command line sequence.

```
Command:WallMergeRemove
(1) wall merge conditions removed.
```

> *The commands WallMergeAdd and WallMergeRemove can also be accessed by entering WALLMERGE in the command line. The options of this command when entered in the command line are shown below.*

```
Command: WallMerge
Wall Merge [Add/Remove/remoVe all]: (Type Add to Wall Merge
Add or R to Wall Merge
Remove. Note that if you type V, wall merge will be applied to
the selected walls.)
```

Access Instructional **Video 2.2 Placing Walls with Precision** from the **Instructional Video** category of the **Student Companion** site of **CengageBrain** http://www.cengagebrain.com described in the Preface.

Joining Walls

When constructing a floor plan, you can draw short segments of walls that connect to other short walls, as shown in Figure 2.34. The two segments when unselected appear as a single wall because of the cleanup function. However, each segment is treated uniquely when you place doors or windows in the wall. If you insert a door centered in the wall, the door will be centered in a wall segment. Therefore, the two wall segments should be joined together or joined to make one wall. The **Join** command will convert the two walls to one. Access the **WallJoin** command as shown in Table 2.12.

TABLE 2.12 *WallJoin command access*

Command prompt	**WALLJOIN**
Ribbon	Select the walls to display the Wall contextual tab. Choose the Join command from the slide out of the Cleanup panel.
Shortcut menu	Select a wall segment, right-click, and choose Join.

```
(Select the wall segments to display the Wall tab, choose
Join from the Cleanup panel.)
Command:WallJoin
Select first wall:Select second wall: (Select wall at p1 as
shown in Figure 2.34.)
(Walls segments joined.)
```

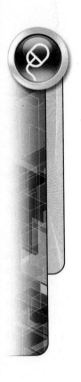

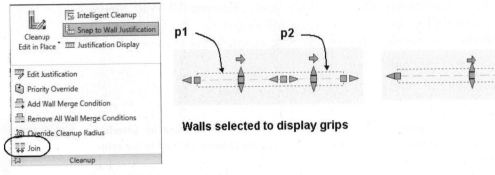

Walls selected to display grips

FIGURE 2.34 *Wall selected for Join command*

TUTORIAL 2-3: EDITING WALLS USING EDIT JUSTIFICATION, REVERSE, JOIN, AND MERGE WALL

1. Verify that the Accessing Tutor content for Chapter 2 of the CengageBrain has been downloaded to your Accessing Tutor folder on your computer as described in Organizing Tutorial Directories in the Preface.

2. Select **Open** from the Quick Access toolbar.

3. Select Accessing Tutor\Imperial\Ch 2\Ex 2-3 or *Accessing Tutor\Metric\Ch 2\Ex 2-3* from the **Select File** dialog box, and select **Open** to dismiss the **Select File** dialog box.

4. Choose **SaveAs > AutoCAD Drawing** from the Application menu, and save the drawing as **Lab 2-3** in your student directory.

5. Verify that Polar Tracking, Object Snap, Object Snap Tracking, Allow Disallow Dynamic UCS, Dynamic Input, and Quick Properties are toggled ON in the status bar. Toggle OFF Infer Constraints, Snap Mode, Grid Display, Ortho Mode, 3D Object Snap, Show Hide Lineweight, Show Hide Transparency, and Selection Cycling on the Application status bar.

6. Right-click over the Dynamic Input toggle of the Application status bar, and choose Settings. Verify that the **Enable Pointer Input, Enable Dimensional Input where possible**, and **Show command prompting and command input near the crosshairs** checkboxes are checked.

7. Move the pointer over the **Object Snap** (Object Snap) button on the status bar, right-click, and choose **Settings**. Edit the **Drafting Settings, Object Snap** tab as follows: Clear all Osnap modes except **Endpoint** and **Node**. Verify that the **Allow general object snap settings to act upon wall justification line** checkbox is checked. Select OK to dismiss the dialog box.

8. Select **Wall A**, as shown in Figure 2.35, and choose **Edit Justification** from the Cleanup slide out panel. Move the pointer over the gray justification marker to identify the justification of the wall, as shown in Figure 2.35.

9. Move the pointer over the **Left** justification marker to display the tip and then click on the **Left** marker to change the justification as shown at the right in Figure 2.35.

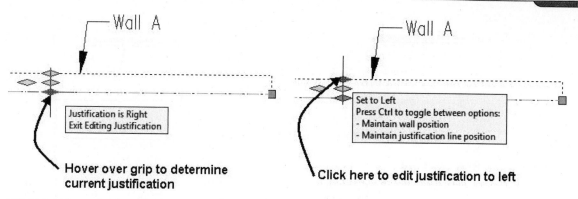

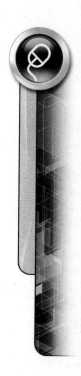

FIGURE 2.35 *Justification markers displayed for the Edit Justification command*

10. Retain selection of Wall A, and verify that the Wall tab is current. From the Reverse flyout of Modify panel choose **Wall Reverse Justification Line** (WallReverseBaseline) command.

11. Select **Wall A** and verify that the wall has left justification and that the directional trigger grip is reversed to point left. Press ESC to clear grip display of the wall.

12. In this step you will perform an L-Cleanup using Intelligent Cleanup. Select **Wall B** at **p1**, as shown in Figure 2.36, and choose **Intelligent Cleanup** from the Cleanup panel of the Wall tab. Respond to the workspace prompts as shown below:

    Select boundary wall for T-Cleanup or [Create L-Cleanup] <Create L-Cleanup>: (Press Enter to choose L-Cleanup.)

    Select the second wall: (Select the wall at **p2** as shown in Figure 2.36.)

    (Wall trimmed as shown at the right in Figure 2.36. Press ESC to clear selection.)

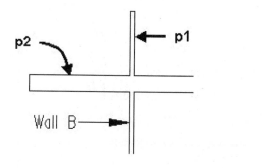

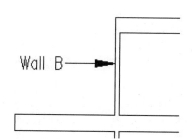

L-Cleanup Created with Intelligent Cleanup

FIGURE 2.36 *Applying Intelligent Cleanup to walls*

13. Select all walls except Wall D and E shown in Figure 2.37, and choose **Intelligent Cleanup** from the Cleanup panel of the Wall tab. Respond to the workspace prompts as shown below:

    Select boundary wall for T-Cleanup or [Intelligent Cleanup] <Intelligent Cleanup>: (Press Enter to choose Intelligent Cleanup.)

    (Walls trimmed and extended as shown at the right in Figure 2.37. Press ESC to clear selection.)

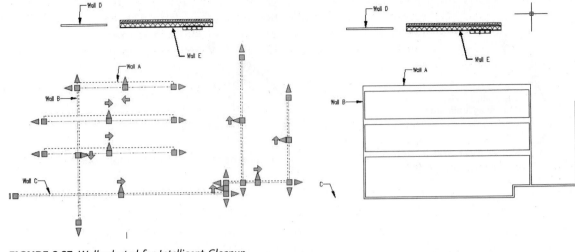

FIGURE 2.37 *Wall selected for Intelligent Cleanup*

14. Select **Wall D** at p1, as shown in Figure 2.38, and choose **Join** from Cleanup slide out panel of the Wall tab. Respond to the workspace prompts as shown below:

    Select second wall: (Select adjoining wall segment at p2 in Figure 2.38.)

    (Select the wall and verify that only one trigger grip is displayed. Press ESC to clear selection.)

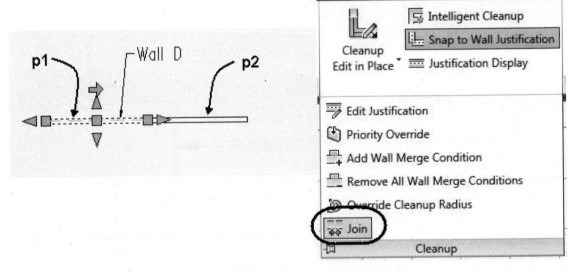

FIGURE 2.38 *Wall segment selected for Join operation*

15. Select **Wall E** at p1, as shown in Figure 2.39, and choose **Add Wall Merge Condition** from the Cleanup slide out panel. Respond to the workspace prompts as shown below:

    Select walls to merge with: 1 found (Select wall at p2 as shown in Figure 2.39.)

    Select walls to merge with: ENTER (Press ENTER to end selection.)

    Merged with 1 wall(s).

16. In this step you will apply Match Properties to change the style of a wall. The Match Properties command does not modify the justification. Retain the selection of the wall at **p1**, as shown in Figure 2.39. Choose the Home tab. Choose **Match Properties** located in the Modify panel as shown in Figure 2.39. Respond to the workspace prompts as shown below:

> Select destination object(s) or [Settings]: *(Select wall A in the drawing.)*
>
> Select destination object(s) or [Settings]: *ENTER (Press ENTER to end selection.)*

17. Save and close your drawing.

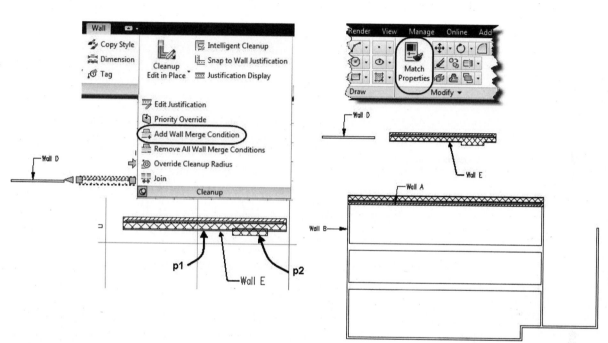

FIGURE 2.39 *Wall selected for Merge Wall*

Using WallOffsetCopy to Draw Interior Walls

The **OFFSET** command of AutoCAD is frequently used to copy an existing entity a specified distance. The AutoCAD OFFSET command copies a wall; however, it does not compensate for the wall width as shown at A in Figure 2.40. The **Wall Offset** command performs a similar function; however, the wall width is considered in placing the wall. The shortcut menu of a wall has the following three wall offset options: **Offset > Copy (WallOffsetCopy), Offset > Move (WallOffsetMove),** and **Offset > Set From (WallOffsetSet).** The **WallOffsetCopy** command copies the selected wall a specified distance from the wall. The specified distance can include the wall width, as shown at B in Figure 2.40, or the specified distance can place walls a specified distance from center to center, as shown at C in Figure 2.40. The wall shown at B in Figure 2.40 is offset by selecting the wall at **p1**, right-clicking, and choosing **Offset > Copy.** A red assistant line is displayed near the face of the GWB wall component. When the assistant line is displayed, click the mouse to specify the reference surface. The assistant line at B in Figure 2.40 sets the reference surface to the face of the GWB. Move the pointer to a point near **p2** to specify the direction, and type the offset distance in the dynamic dimension field of the workspace. The offset wall is positioned to set the overall dimension measured to the outer surface of the wall at **p1**. The wall surface selected becomes the reference point for the dimension.

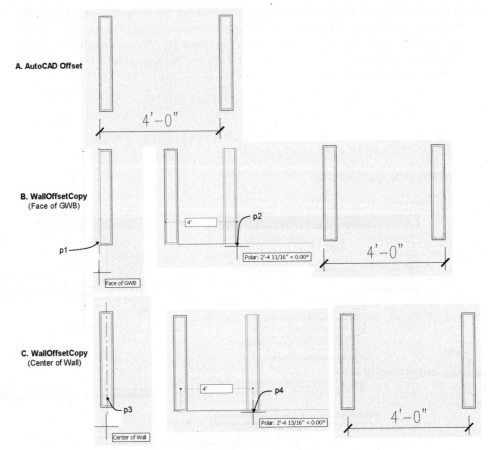

FIGURE 2.40 *Effects of wall component selected by the WallOffsetCopy command*

The wall offset shown at C in Figure 2.40 is created by selecting the wall at **p3**, right-clicking, and choosing **Offset > Copy**. Move your pointer near the center of the wall to display a red centerline, and click to specify the center as the reference. Move the pointer to near **p4** to set the direction, and type the offset distance in the dynamic dimension edit field. The offset distance becomes the dimension between the centers of the two walls. Selecting a wall by the surface nearest to the direction of the offset will create another wall relative to the surface selected and exclusive of the wall width. The WallOffsetCopy command allows you to specify stud surface within complex wall styles as the reference surface. This command continues after you copy a wall, allowing you to add additional walls with specified interior dimensions. Access the **WallOffsetCopy** command as shown in Table 2.13.

TABLE 2.13 *WallOffsetCopy command access*

Command prompt	**WALLOFFSETCOPY**
Ribbon	Select a wall to display the Wall tab, and choose Copy from the Offset flyout of the Modify palette shown in Figure 2.41.
Shortcut menu	Select a wall, right-click, and choose Offset > Copy.

When you select a wall, choose Copy from the Offset flyout of the Modify panel. You are prompted in the workspace as shown below.

(Select a wall as shown at **p1** in Figure 2.41.)

Select the component to offset from: (Move the pointer over the wall component to display the assistant line at p2. The tip displays the name of the wall component under the pointer unless Object Snap is toggled ON; Move the pointer down to set the direction in the workspace as shown in Figure 2.42.)

Select a point to offset to: 3' [914] ENTER (Type 3' in the dynamic dimension.)

Select a point to offset to ENTER (Press ENTER to end the command.)

> **NOTE**
>
> The tooltip described below specifying the component is displayed if you Enable Dynamic Input in the Application status bar, toggle ON the Enable Dimension Input in the Dynamic Input tab of Drafting Settings, and toggle ON Display *AutoSnap tooltip in the Drafting tab of the Options* dialog box. If the tip is not displayed, toggle OFF Object Snap to avoid the display of the Osnap mode.

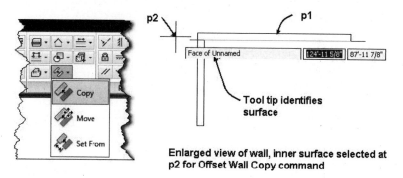

FIGURE 2.41 *Selection of wall face*

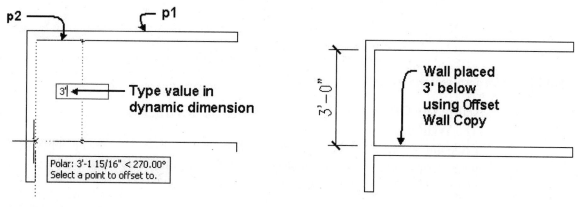

FIGURE 2.42 *Type offset distance in workspace*

Using WallOffsetMove

The **WallOffsetMove** command will allow you to move an existing wall offset from a specified point on the selected wall. Access the **WallOffsetMove** command as shown in Table 2.14.

TABLE 2.14 *WallOffsetMove command access*

Command prompt	WALLOFFSETMOVE
Ribbon	Select a wall to display the Wall contextual tab. Choose Move from the Offset flyout of the Modify panel (refer to Figure 2.41).
Shortcut	Select a wall, right-click, and choose Offset > Move.

When you select a wall and then select **Move** from the Offset flyout of the Modify panel, you can move walls as shown in the following workspace prompts.

(Select a wall as shown at **p1** in Figure 2.43; select Move from the Offset flyout of the Modify panel.)

Select the component to offset from:

(Move the pointer over the wall component to display the assistant line at **p2** and click to specify the reference. The tip displays the name of the wall component under the pointer.)

Select a point to offset to: **5' [1524]** ENTER

(Move the pointer down to set the direction. Distance is dynamically displayed; then type the offset distance in the workspace as shown in Figure 2.43.)

(Wall is moved 5' [1524 mm] as specified.)

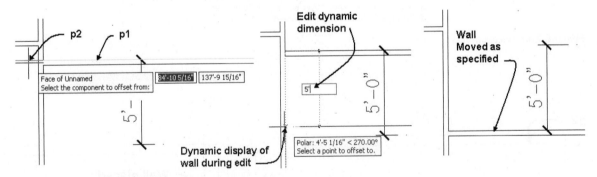

FIGURE 2.43 *Wall component selected for WallOffsetMove*

Using the Offset Set from Command

The **WallOffsetSet** command will allow you to move an existing wall from a specified point on the selected wall and relative to existing walls in the drawing. This command allows you to measure distances from walls and enter a new distance. Access the **Offset Set From** command as shown in Table 2.15.

TABLE 2.15 *WallOffsetSet command access*

Command prompt	WALLOFFSETSET
Ribbon	Select a wall to display the Wall contextual tab. Choose **Set From** from the Offset flyout of the Modify panel (refer to Figure 2.41).
Shortcut	Select a wall, right-click, and choose Offset > Set From.

When you select a wall, and then select **Set From** from the Offset flyout of the Modify panel, you can move walls as shown in the following workspace prompts and in Figure 2.44.

> (Toggle ON running Endpoint Object Snap.)
>
> (Select a wall as shown at **p1** in Figure 2.44; select **Set From** from the Offset flyout of the Modify panel.)
>
> Select the component to offset from: (Move the pointer over the wall component to display the assistant line at **p2** and click as shown in Figure 2.44.)

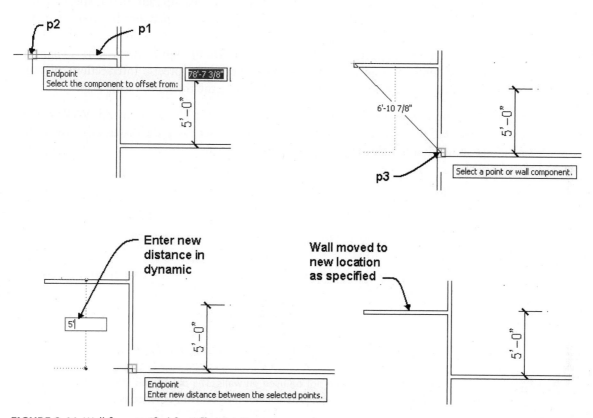

FIGURE 2.44 *Wall face specified for Offset Set From command*

> Select a point or wall component: (Move the pointer down to display the assistant line at **p3**. Click to measure the distance between the two walls. The distance is displayed as a dynamic dimension as shown at the right in Figure 2.44)
>
> Enter new distance between the selected points. **5'[1524]** ENTER *(Value typed in the workspace.)*
>
> (Selected wall surface is moved to a point 5' [1524] from the wall at **p3** as shown at the right in Figure 2.44.)

The AutoCAD **Offset** command can be used to offset walls. However, it does not compensate for wall widths or allow you to specify a reference point for the offset distance. Therefore, when you use the AutoCAD **Offset** command to create a clear distance between walls, the offset distance must equal the thickness of a wall plus the clear distance between the walls.

To create a tub enclosure using 3 1/2" [89] walls that have a 5' [1524] inside clear dimension between partitions, make the offset distance 5'–3 1/2" [1613] when using the AutoCAD Offset command.

Editing Walls with AutoCAD Editing Commands

Most other AutoCAD commands can be used to edit the walls of a drawing. Walls can be edited with the following AutoCAD commands: **Align**, **Array**, **Break**, **Copy**, **Chamfer**, **Fillet**, **Erase**, **Extend**, **Mirror**, **Move**, **Offset**, **Rotate**, **Scale**, **Stretch**, and **Trim**. When the **Extend** and **Trim** commands are used, the justification line of the wall serves as the boundary or cutting edge. Wall cleanup radii may display a wall as connected to an adjacent wall, yet the justification line may not be connected. You can apply AutoCAD editing commands to wall justification lines as displayed with the Diagnostic display configuration as shown in Figure 2.45. The Diagnostic display configuration allows you to easily verify the precise intersections of walls. The **Align** and **Scale** commands do not change the thickness of the wall.

Walls may be arrayed using the Array option of the Copy command. Walls may be arrayed to fit between two points or arrayed by specifying the distance between walls as shown in the following workspace prompts:

```
Select Copy from the Modify panel in the Home tab.
Select objects: (Select wall at p1 as shown in Figure 2.45.)
Select objects: (Press ENTER to end selection.)
Specify base point or: (Select the endpoint of the wall at p2
as shown in Figure 2.45.)
Specify second point or [Array]: A (Press the down arrow key,
choose Array.)
Enter number of items to array: 3 enter
Specify second point or [Fit]: (Select a point at p3 to
specify the array distance as shown in Figure 2.45.)
```

The **Explode** command should not be used on walls. The object definition is lost when you explode objects; therefore, the objects will not interact with other objects.

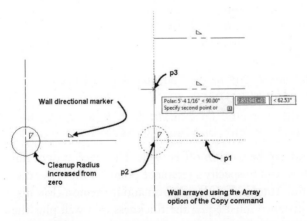

FIGURE 2.45 *Array option of Copy command applied to walls displayed with the Diagnostic display configuration*

TUTORIAL 2-4: EDITING WALLS USING WALLOFFSETCOPY AND WALLOFFSETSET

1. Verify that the Accessing Tutor content for Chapter 2 of the CengageBrain has been downloaded to your Accessing Tutor folder on your computer as described in Organizing Tutorial Directories in the Preface.

2. Select **Open** from the **Quick Access** toolbar.

3. Select *Accessing Tutor\Imperial\Ch 2\ Ex 2-4* or *Accessing Tutor\Metric\Ch 2 \Ex 2-4* from the **Select File** dialog box and select **OK** to dismiss the **Select File** dialog box.

4. Choose **Save As > AutoCAD Drawing** from the Application menu, and save the drawing as **Lab 2-4** in your student directory.

5. If tool palettes are not displayed, choose the Home tab. Choose **Tools** from the Tools flyout.

6. Verify in the Application status bar that Polar Tracking, Object Snap, Object Snap Tracking, Allow Disallow Dynamic UCS, Dynamic Input, and Quick Properties are toggled ON in the status bar. Verify that Constraints, Snap Mode, Grid Display, Ortho Mode, 3D Object Snap, Show Hide Lineweight, Show Hide Transparency, and Selection Cycling are toggled OFF.

7. Right-click over the Dynamic Input toggle, and choose Settings. Verify that the Enable Pointer Input, Enable Dimensional Input where possible, and Show command prompting and command input near the crosshairs checkboxes are checked.

8. Select the Object Snap tab of the **Drafting Settings** dialog; edit as follows: clear all Osnap modes except **Endpoint**. Verify that the **Allow general object snap settings to act upon wall justification line** checkbox is checked. Select **OK** to dismiss the Drafting Settings dialog box.

9. Select **Wall A** shown in Figure 2.46 to display the Wall tab. Choose **Set From** from the **Offset** flyout of the **Modify** panel. Respond to the following workspace prompts to move it **24' [7315]** from Wall B.

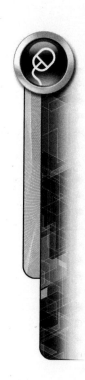

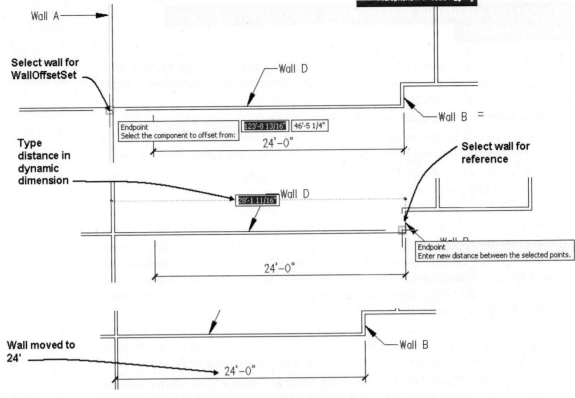

FIGURE 2.46 *Offset component selected for Offset Set From command*

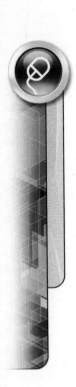

Select the component to offset from: (Move the cursor to the left face of Wall **A** and click to select as shown in Figure 2.46.)

Select a point or wall component: (Move the cursor to the right wall face of wall **B** to display the **Endpoint** object snap; click to select the right face as shown in Figure 2.46.)

Enter new distance between the selected points: **24'** [7315] ENTER

Type distance in the dynamic dimension on screen and press ENTER. The wall moves 24' [7315] from Wall B as shown in Figure 2.46. (Press ESC to clear selection.)

10. Right-click over the graphics screen, and choose **Basic Modify Tools > Trim** from the shortcut menu. Respond to the following workspace prompts as shown below:

(**Trim** command selected from graphics screen shortcut.)

Select objects or <select all>: 1 found (*Select wall at* **p1** *shown in Figure 2.47.*)

Select objects: ENTER (Press ENTER to end selection of boundaries.)

Select object to trim or shift-select to extend or: (*Select wall at* **p2** *shown in Figure 2.47.*)

Select object to trim or shift-select to extend or: ENTER (*Press* ENTER *to end selection.*)

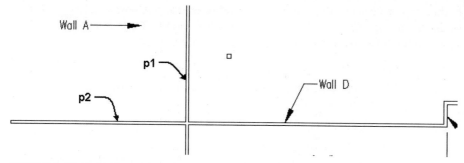

FIGURE 2.47 *Wall selected for the Extend command*

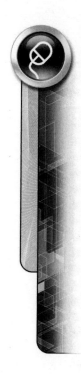

11. This step will create three interior room spaces with 5' [1524] clear dimension between partitions. Select **Wall D**, and choose **Copy** from the Offset flyout of the Modify panel. Respond to the following workspace prompts to create the rooms:

```
Command: WallOffsetCopy
```

Select the component to offset from: (Move the pointer to
display the assistant line at wall surface X shown in Figure
2.48, and click to specify the surface; move the pointer above
Wall D to a point near **p1.**)

Select a point to offset to: **5' [1524]** ENTER *(Move the pointer
to p1.)*

Select a point to offset to: **5' [1524]** ENTER *(Retain pointer
at p1 location.)*

Select a point to offset to: **5' [1524]** ENTER *(Retain pointer
at p1 location.)*

Select a point to offset to: *(Press* ENTER *to end the command.
Press ESC to clear selection.)*

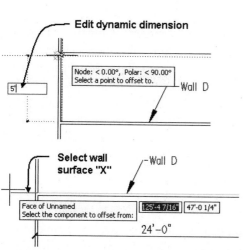

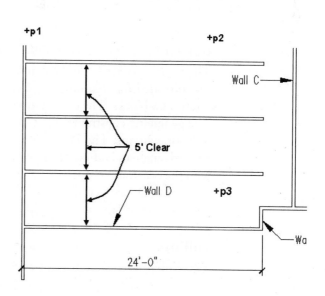

FIGURE 2.48 *Wall drawn using the WallOffsetCopy*

12. In this step you will use the Extend command to extend the walls created in the previous step to Wall C. Right-click and choose **Basic Modify Tools** > **Extend**; then respond to the following workspace prompts to create the rooms:

    Select boundary edges…

    Select objects or <select all>: 1 found *(Select Wall C as shown in Figure 2.48.)*

    Select objects: ENTER *(Press ENTER to end selection.)*

    Select object to extend or shift-select to trim or: (Click the down arrow in the workspace prompt; select **Fence** option).

    Specify first fence point: (Select a point near **p2** in Figure 2.48.)

    Specify next fence point or: (Select a point near **p3** in Figure 2.48.)

    Specify next fence point or: ENTER (Press ENTER to end selection. Walls extended as shown in Figure 2.49.)

13. Choose Save from the Quick Access toolbar. From the Application menu choose Close > Current Drawing to close your drawing.

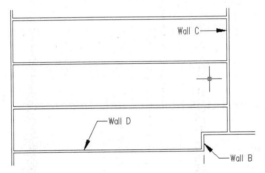

FIGURE 2.49 *Walls extended to Wall C*

USING PARAMETRIC CONSTRAINTS

The location of walls can be constrained using 2D constraints. Constraint tools are also provided for column grids, beams, columns, curtain walls, mass elements, and multi-view blocks. The constraints consist of geometric and dimensional parameters. The geometric constraints allow you to control the relative location of walls to other walls or columns. The Dimensional constraints position the location of walls using aligned, angular, and radial dimensions. A constraint marker will be placed on walls with the constraint property as shown at right in Figure 2.50. The right-click menu of the constraint marker allows you to hide the marker from the drawing. Constraint markers do not print with the drawing.

Collinear

The Collinear constraint can be used to align two walls. The edges or center of the wall can be used as the reference. Walls may also be aligned with the profile edges, center edges, and boundary edges of column objects. The selection of the reference surface is done using the Tab key.

Access the **Collinear** command as shown in Table 2.16.

TABLE 2.16 *Collinear command access*

Command prompt	**GCCOLLINEAR**
Ribbon	Select a wall to display the Wall contextual tab. Choose **Collinear** from the Parametric panel (refer to Figure 2.50).

Select wall at **p1** as *shown in Figure 2.50. Choose Collinear from the Parametric panel of the Wall contextual tab.*

Select first object or [Multiple]:(Choose the left edge of the wall at **p2**.)

Select second object: *(Choose the wall to make collinear to the first, choose the left edge of the wall at p3 in Figure 2.50.)*

The wall at **p3** is shifted to align with the wall at **p2**.

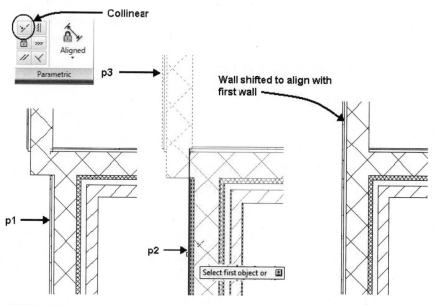

FIGURE 2.50 *Accessing the Collinear constraint*

Fix

The Fix constraint can be used to restrain a point or an object relative to the World Coordinate System. When the Fix command is applied to a wall or column grid, the objects are locked in position relative to the World Coordinate System.

Access the **Collinear** command as shown in Table 2.17.

TABLE 2.17 *Fix command access*

Command prompt	**GCFIX**
Ribbon	Select a wall to display the Wall contextual tab. Choose **Fix** from the Parametric panel (refer to Figure 2.51).

(Select wall at **p1** as *shown in Figure 2.51. Choose Fix from the Parametric panel of the Wall contextual tab.*)

Select point or [Object] <Object>:(Choose the wall at **p2** as shown in Figure 2.51, Node object snap tip displayed on wall.)

The wall is locked to the world coordinate system and will not allow location edit with grips or modify commands.

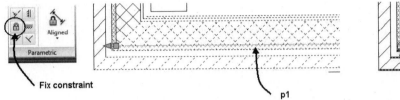

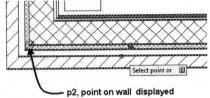

Fix constraint

p1

p2, point on wall displayed

FIGURE 2.51 *Creating a Fix constraint*

Parallel

The Parallel constraint can be used to restrain two points or wall to retain the same angle relative to the World Coordinate System. When the Parallel command is applied to a wall, the angle of the first wall controls the angle for the second wall.

Access the **Parallel** command as shown in Table 2.18.

TABLE 2.18 *Parallel command access*

Command prompt	**GCPARALLEL**
Ribbon	Select a wall to display the Wall contextual tab. Choose **Parallel** from the Parametric panel (refer to Figure 2.52).

(Select wall at p1 as shown in Figure 2.52. Choose Parallel from the Parametric panel of the Wall contextual tab.)

Select first object:*(Select wall at p1.)*

Select second object: *(Select wall at p2.)*

The wall located at **p2** is modified to the angle of the wall at **p1**.

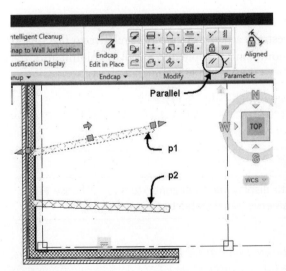

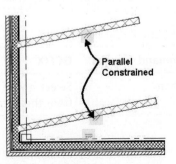

FIGURE 2.52 *Accessing the Parallel constraint*

Vertical

The Vertical constraint can be used to constrain two points or a wall to vertical relative to the current UCS. When the Vertical constraint is applied to points, the second point is made vertical relative to the first.

Access the **Vertical** constraint shown in Table 2.19.

TABLE 2.19 *Vertical command access*

Command prompt	**GCVERTICAL**
Ribbon	Select a wall to display the Wall contextual tab. Choose **Vertical** from the Parametric panel (refer to Figure 2.53).

(Select wall at p1 as shown in Figure 2.53. Choose Vertical from the Parametric panel of the Wall contextual tab.)

Select an object or [2Points] <2Points>: *(Select wall at p1.)*

The wall is positioned to vertical.

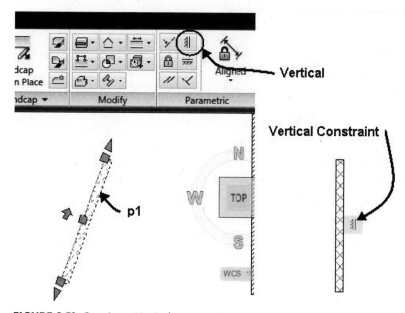

FIGURE 2.53 *Creating a Vertical constraint*

Horizontal

The Horizontal constraint can be used to constrain two points or a wall to horizontal relative to the current UCS. When the Horizontal constraint is applied to points, the second point is made vertical relative to the first.

Access the **Horizontal** constraint shown in Table 2.20.

TABLE 2.20 *Horizontal command access*

Command prompt	GCHORIZONTAL
Ribbon	Select a wall to display the Wall contextual tab. Choose **Horizontal** from the Parametric panel (refer to Figure 2.54).

*(Select wall at **p1** as shown in Figure 2.20. Choose Horizontal from the Parametric panel of the Wall contextual tab.)*

Select an object or [2Points] <2Points>: *(Select wall at **p1**.)*

The wall is positioned to horizontal.

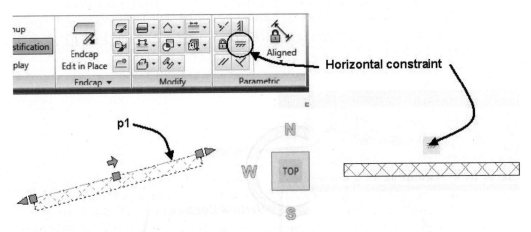

FIGURE 2.54 *Creating a Horizontal constraint*

Perpendicular

The Perpendicular constraint can be used to constrain two lines or walls to maintain a 90 angle to each other. When the Perpendicular constraint is applied to walls the second wall is made perpendicular to the first wall.

Access the **Perpendicular** constraint as shown in Table 2.21.

TABLE 2.21 *Perpendicular command access*

Command prompt	GCPERPENDICULAR
Ribbon	Select a wall to display the Wall contextual tab. Choose **Perpendicular** from the Parametric panel (refer to Figure 2.55).

*(Select wall at **p1** as shown in Figure 2.55. Choose Perpendicular from the Parametric panel of the Wall contextual tab.)*

Select first object: *(Select wall at **p1**.)*

Select second object: *(Select wall at **p2**.)*

The wall at **p2** is positioned to perpendicular to the wall at **p1**.

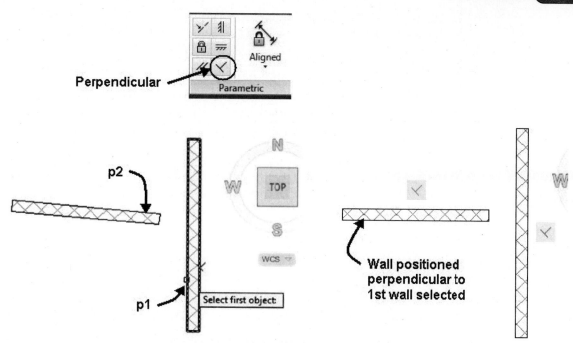

FIGURE 2.55 *Creating a Perpendicular constraint*

Dimensional Constraints

The Aligned, Angular, and Radial dimensional constraints are accessed from the fly-out of the Parametric panel. The Aligned dimensional constraint positions the second object parallel to the first the distance specified. The distance may be specified as a number or a relationship. The wall shown in Figure 2.56 is position 1/2 the dimension value of d1. The distance may be specified to the edges or centerline of the wall. In addition if positioned to a column object the dimension may be specified to the profile edges, center edges, or boundary edges.

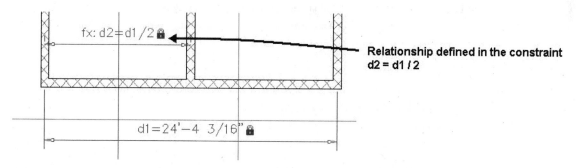

FIGURE 2.56 *Relationship defined in the constraint*

Access the **Aligned** dimensional constraint as shown in Table 2.22.

TABLE 2.22 *DcAligned command access*

Command prompt	DCALIGNED
Ribbon	Select a wall to display the Wall contextual tab. Choose **Aligned** from the Parametric panel (refer to Figure 2.57).

(Select wall at p1 as shown in Figure 2.57. Choose Aligned from the Parametric panel of the Wall contextual tab.)

Select first line: *(Select wall at p1.)*

Select second line to make parallel: *(Select the wall at p2.)*

Specify dimension line location: *(Select a point near p3.)*

Specify text = 2'2 1/16" *(Overtype dimension value to specify the distance.)*

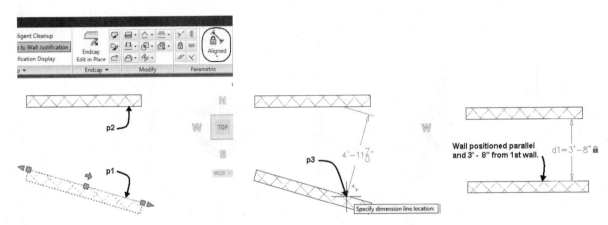

FIGURE 2.57 *Creating an Aligned constraint*

The Angular constraint allows you to modify the angle between two walls. The second wall selected is modified to form the angle from the first wall selected. Access the **Angular** dimensional constraint as shown in Table 2.23.

TABLE 2.23 *DcAngular command access*

Command prompt	DCANGULAR
Ribbon	Select a wall to display the Wall contextual tab. Choose **Angular** from the dimensional constraint flyout of the Parametric panel (refer to Figure 2.58).

(Select wall at p1 as shown in Figure 2.58. Choose Aligned from the Parametric panel of the Wall contextual tab.)

Select first line or arc or [3Point]<3Point>: *(Select wall at p1.)*

Select second line: *(Select the wall at p2.)*

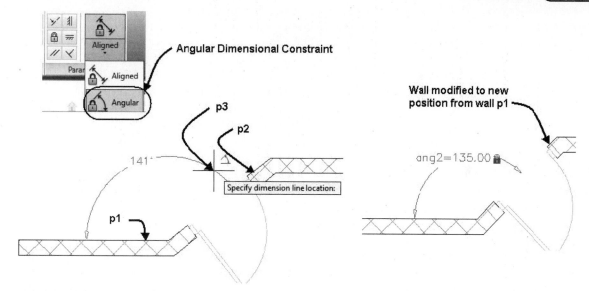

FIGURE 2.58 *Creating an Angular dimensional constraint*

The Radial constraint allows you to modify the radius of the wall. Access the **Radial** dimensional constraint as shown in Table 2.24.

TABLE 2.24 *DcRadial command access*

Command prompt	DCRADIAL
Ribbon	Select a wall to display the Wall contextual tab. Choose **Radial** from the dimensional constraint flyout of the Parametric panel (refer to Figure 2.59).

*(Select wall at **p1** as shown in Figure 2.59. Choose Radial from the Parametric panel of the Wall contextual tab.)*

Select arc or circle: *(Select the wall at **p1** as shown in Figure 2.59.)*

Dimension text = 13'9 5/8"

Specify dimension line location: *(Select a location at **p2**.)*

(Overtype 12' to specify the new dimension.)

Select second line: *(Select the wall at **p2**.)*

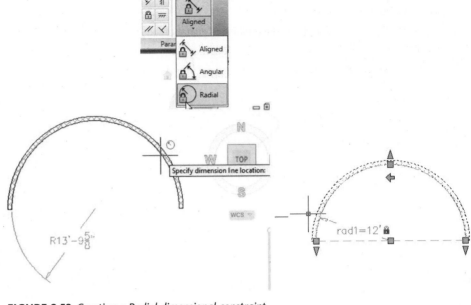

FIGURE 2.59 *Creating a Radial dimensional constraint*

TUTORIAL 2-5: APPLYING WALL CONSTRAINTS

1. Verify that the Accessing Tutor content for Chapter 2 of the CengageBrain has been downloaded to your Accessing Tutor folder on your computer as described in Organizing Tutorial Directories in the Preface.

2. Select **Open** from the **Quick Access** toolbar.

3. Select *Accessing Tutor\Imperial\Ch 2\ Ex 2-5* or *Accessing Tutor\Metric\Ch 2 \Ex 2-5* from the **Select File** dialog box, and select **OK** to dismiss the **Select File** dialog box.

4. Choose **Save As** > **AutoCAD Drawing** from the Application menu, and save the drawing as **Lab 2-5** in your student directory.

5. Verify that the **Model** is current in the **Application** status bar.

6. If tool palettes are not displayed, choose the Home tab. Choose **Tools** from the Tools flyout.

7. Verify that only the following toggles are ON on the Application status bar: Polar Tracking, Object Snap, Object Snap Tracking, Allow Disallow Dynamic UCS, Dynamic Input, and Quick Properties.

8. Right-click over the Dynamic Input toggle, and choose Enable. Right-click over the Dynamic Input toggle, and choose Settings. Verify that the Enable Pointer Input, Enable Dimensional Input where possible, and Show command prompting and command input near the crosshairs checkboxes are checked.

9. In this step you will use the Collinear constraint to move the interior wall to align its GWB component to the GWB component of the exterior wall. Select the exterior wall at p1 as shown in Figure 2.60 to display the Wall contextual tab.

Choose Collinear from the Parametric panel. Respond to the
workspace prompts as follows.

Select first object or [Multiple]: *(Select the wall at p1 as
shown in Figure 2.60.)*

Select second object: *(Select the wall at p2 as shown in
Figure 2.60.)*

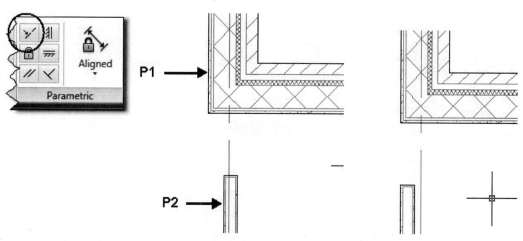

FIGURE 2.60 *Applying the Collinear constraint*

10. The interior wall shown at p1 shown in Figure 2.61 is currently rotated 5 degrees. In
    this step you will use the Horizontal constraint to modify the interior wall to the hor-
    izontal position. Select the interior wall at p1 as shown in Figure 2.61 to display the
    Wall contextual tab.

    Choose **Horizontal** from the Parametric panel. Respond to the
    workspace prompts as follows.

    Select an object or [2Points]<2Points>: *(Select the wall at
    p1 as shown in Figure 2.61.)*

    Wall is positioned to horizontal as shown in Figure 2.61.

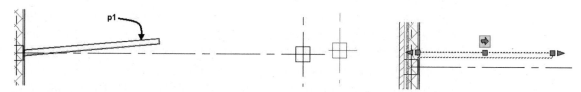

FIGURE 2.61 *Horizontal constraint applied to wall*

11. This step will begin the process of using Aligned constraints to position the interior
    walls shown at left to create room dimension with 4 equal spaces. Select the exterior
    wall at **p1** to display the wall contextual tab. Choose Aligned command from the
    Parametric panel. Respond to the workspace prompts as follows.

    Select first line: *(Select the edge of the wall at p1 as shown
    in Figure 2.62.)*

Select second line to make parallel: *(Select the edge of the wall at p2.)*

Specify dimension line location: *(Select a point near p3 to place the dimension.)*

Dimension text = 38' - 10" [11858] *(Press Enter to accept default value.)*

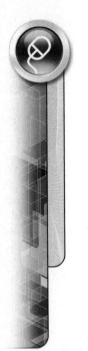
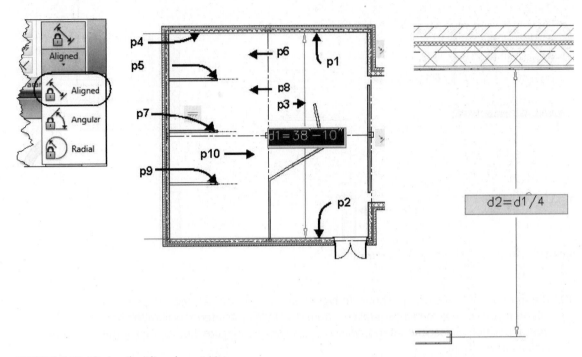

FIGURE 2.62 *Placing the Aligned constraints*

12. In this step you will use the Aligned constraint tool to begin the process of using Aligned constraints to position the center of the 3 interior walls shown at left into 4 equal spaces. Select the exterior wall at p1 as shown in Figure 2.62 to display the Wall contextual tab. Choose Aligned command from the Parametric panel. Respond to the workspace prompts as follows:

    Select first line: *(Select the edge of the wall at p4 as shown in Figure 2.62.)*

    Select second line to make parallel: *(Press Tab to toggle to the center of the wall at p5 as shown in Figure 2.62.)*

    Specify dimension line location: *(Select a point near p6 to place the dimension.)*

    Dimension text = 8' - 7 9/16" [2770] *(Overtype the dimension as follows:* d2=d1/4 ENTER*).*

13. Select the exterior wall at p1 as shown in Figure 2.62 to display the wall contextual tab. Choose Aligned command from the Parametric panel. Respond to the workspace prompts as follows:

    Select first line: *(Press Tab to toggle to center of wall at p5 as shown in Figure 2.62.)*

    Select second line to make parallel: *(Select center of the wall at p7 as shown in Figure 2.62.)*

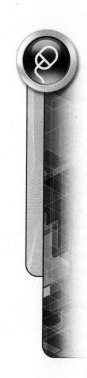

> Specify dimension line location: (*Select a point near p8 to place the dimension.*)
>
> Dimension text = 7' - 7 13/16" [2843] (*Overtype the dimension as follows:* d3=d1/4 ENTER).

14. Select the exterior wall at p1 as shown in Figure 2.62 to display the wall contextual tab. Choose Aligned command from the Parametric panel. Respond to the workspace prompts as follows:

> Select first line: (*Press Tab to toggle to center of wall at p7 as shown in Figure 2.62.*)
>
> Select second line to make parallel: (*Select center of the wall at p9 as shown in Figure 2.62.*)
>
> Specify dimension line location: (*Select a point near p10 as shown in Figure 2.62 to place the dimension.*)
>
> Dimension text = 9' - 0" [3554] (*Overtype the dimension as follows:* d4=d1/4 ENTER).

15. In this step you will use the Perpendicular constraint tool to modify the position of the wall at p2 as shown in Figure 2.63. Select the interior wall at p1 to display the wall contextual tab. Choose the Perpendicular command from the Parametric panel. Respond to the workspace prompts as follows:

> Select first object: (*Select the wall at p1 as shown in Figure 2.63.*)
>
> Select second object: (*Select the wall at p2 as shown in Figure 2.63. The wall at p2 is now perpendicular to p1. Press Escape to end the command.*)

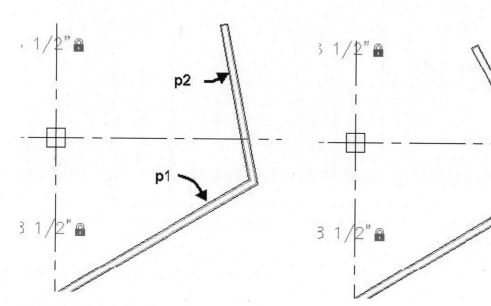

FIGURE 2.63 *Placing the Perpendicular constraints*

16. Save and close the drawing.

ACCESSING AUTOCAD WS

AutoCAD WS allows you to view your drawing in an AutoCAD WS browser window as shown in Figure 2.76. AutoCAD WS is accessed when you choose the Upload tool of the Upload panel in the Online tab as shown in Figure 2.64. To use AutoCAD WS you must create an account and a login. If you choose the Upload tool, you will be prompted to login or create an account. Since AutoCAD Architecture drawings are based upon AutoCAD, you can upload the file to Auto-CAD WS and perform viewing and edits using AutoCAD commands in mobile applications.

A summary of the commands accessed from the Online tab shown in Figure 2.64 follow:

Upload Panel

Upload—save the current drawing, then choose Upload to upload the current drawing to AutoCAD WS.

Manage Uploads—choose Manage Uploads to resume or stop uploads in Auto-CAD WS.

Upload Multiple Files to open the Choose Files to Upload. This tool allows you to upload AutoCAD drawings, Adobe PDF, or Microsoft Office documents.

Content Panel

Open Online—Opens the current drawing in the AutoCAD WS for online edits, sharing, and collaboration.

Online Drawings—Opens the Drawings pane of the AutoCAD WS for selection of uploaded drawings.

Timeline—Displays the revision history of uploaded drawings.

Share Panel

Share Drawing—Opens AutoCAD WS which allows you to invite your contacts per email addresses to share in the edit or download of the drawing. A link to the drawing is sent within the email to participants.

Get Link—Generates a link to the drawing. The link may be copied to allow others to view the drawing.

Messages—Displays a list of the activity in your account

AutoCAD WS provides you with editing tools for AutoCAD; therefore you can measure, erase, and view properties of drawings created in AutoCAD Architecture. You cannot modify the AEC Objects such as walls, doors, windows, roofs from within AutoCAD WS. If you add AutoCAD entities to the drawing while online, AutoCAD WS does not preset layer and scale for the entity based upon the layer standard of AutoCAD Architecture.

You may begin a drawing from within AutoCAD Architecture and upload the file into AutoCAD WS. If you add edits to the file and download it, you may open it and convert it to a construct of a project. When you upload a view drawing, any referenced construct files will also upload; however, you will be able edit only the content of the view drawing using AutoCAD commands.

PROJECT TUTORIAL 2-6: STARTING THE FLOOR PLAN AND USING AUTOCAD WS

1. Verify that the Accessing Tutor content for Chapter 2 of the CengageBrain has been downloaded to your Accessing Tutor folder on your computer as described in Organizing Tutorial Directories in the Preface.

2. Open AutoCAD Architecture 2012, and verify that a drawing is open. Select **Project Browser** from the Quick Access toolbar. Use the Project Selector drop-down list to navigate to your *Student\Ch2* directory, as shown in Figure 2.64. Select **New Project** to open the **Add Project** dialog box. Type **2** in the **Project Number** field, and type **Ex 2-6** in the **Project Name** field as shown in Figure 2.64. Check **Create** from template project, choose **Browse**, and edit the path to *Accessing Tutor\Imperial\Ch 2\ Ex 2-6 \Ex 2-6.apj* [Accessing Tutor\Metric\Ch 2\Ex 2-6\Ex 2-6.apj] in the **Look in** edit field. Choose **Open** to close the **Select Project** dialog box. Choose **OK** to dismiss the **Add Project** dialog box. Select **Close** to dismiss the Project Browser.

3. Select the **Constructs** tab of the **Project Navigator**. Double-click on **Floor 2** to open the Floor 2 drawing. The Floor 2 drawing was created in Tutorial 1.5 and is empty and ready for you to start the floor plan.

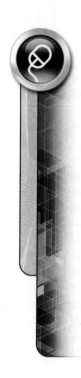

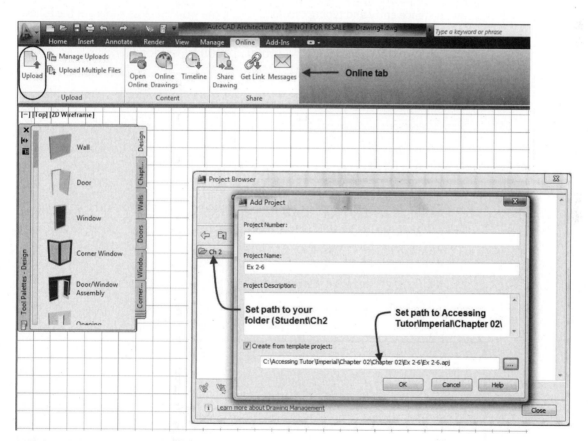

FIGURE 2.64 *Creating the Ex 2-6 project*

4. Verify that only Polar Tracking, Object Snap, Object Snap Tracking, Allow Disallow Dynamic UCS, Dynamic Input, and Quick Properties are toggled ON in the Application status bar.

5. Right-click over the **Dynamic Input** toggle, and choose **Enable**. Right-click over the **Dynamic Input** toggle, and choose **Settings**. Verify that the **Enable Pointer Input**, **Enable Dimensional Input where possible**, and **Show command prompting and command input near the crosshairs** checkboxes are checked. Choose **OK** to dismiss all dialog boxes.

6. Move the pointer over the Object Snap button on the status bar, right-click, and choose **Settings**. Edit the **Drafting Settings**, **Object Snap** tab as follows: clear all Osnap modes except **Endpoint** and **Node**. Verify that the **Allow general object snap settings to act upon wall justification line** checkbox is checked. Select **OK** to dismiss the **Drafting Settings** dialog box.

7. Choose the **Wall** tool from the Build panel of the Home tab.

8. Edit the Design tab of the **Properties** palette as follows: Style = **Standard**, Cleanup automatically = **Yes**, Cleanup group definition = **Standard**, Segment = **Line**, A-Width = **3.5 [89]**, B-Base height = **8' [2440]**, Justify = **Left**, Offset = **0**, Roof line offset from base height = **0**, and Floor line offset from baseline = **0**.

9. Move the pointer from the Properties palette, and begin drawing the walls using direct distance entry and typing the distances in the workspace as shown in the following workspace prompts and in Figure 2.65.

> Start point or: 75',40'[24000,12000] ENTER *(Point p1 is established as shown in Figure 2.65.)*
>
> *(Move the pointer up to Auto Track tooltip Polar Tracking angle = 90°.)*
>
> End point or: 28'[8534] ENTER (p2)
>
> *(Move the pointer right to Auto Track tooltip Polar Tracking angle = 0°.)*
>
> End point or: 68'-3" [20803] ENTER (p3)
>
> *(Right-click and choose* Ortho close *from the shortcut menu.)*
>
> Point on wall in direction of close: *(Move the pointer down to Auto Track tooltip Polar Tracking angle = 270° and select a point near* p4 *as shown in Figure 2.65.)*
>
> *(Walls are drawn to create the rectangle as shown in Figure 2.66.)*

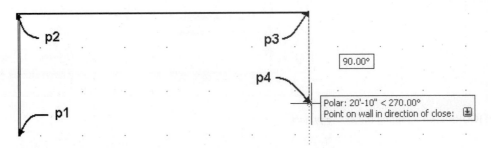

FIGURE 2.65 *Location of points for wall segments*

10. Select **Zoom Window** from the **Zoom** flyout of the Navigation bar. Click a location near **p1** and **p2** as shown in Figure 2.66.

11. Select the vertical wall at **p3** shown in Figure 2.66; choose **Copy** from the Offset flyout Modify panel. Respond as shown to the following workspace prompts to create interior walls using the **WallOffsetCopy** command.

> Select the component to offset from: (Move the pointer to display the assistant line at the endpoint of the wall as shown in Figure 2.67 and click to specify the wall component face.)
>
> (Move the pointer to the right with a Polar Tracking angle of 0°.)
>
> Select a point to offset to: 31'[9449] ENTER (Type the distance dynamically displayed in the Auto Track tooltip, as shown in Figure 2.68.)
>
> (Move the pointer to the right of the last wall segment with a Polar Tracking angle of 0°.)
>
> Select a point to offset to: 12'[3658] ENTER (Type the distance in the command line dynamically displayed in the Auto Track tooltip.)
>
> (Move the pointer to the right of the last wall segment with a Polar Tracking angle of 0°.)
>
> Select a point to offset to: 8' [2438] ENTER (Type the distance in the command line dynamically displayed in the Auto Track tooltip.)
>
> (Move the pointer to the right of the last wall segment with a Polar Tracking angle of 0.)
>
> Select a point to offset to: 3' [914] ENTER (Type the distance in the command line dynamically displayed in the Auto Track tooltip.)
>
> Select a point to offset to: ENTER *(Press ENTER to end the command. Press ESC to end selection of wall.)*

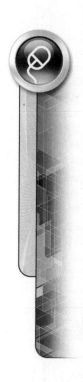

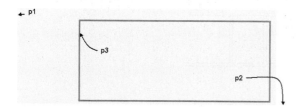

FIGURE 2.66 *Rectangular polygon drawn with Ortho close option*

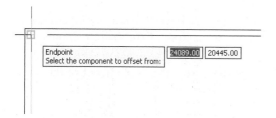

FIGURE 2.67 *Wall component specified for WallOffsetCopy*

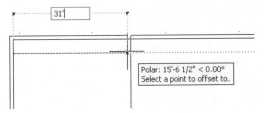

FIGURE 2.68 *Distance specified for WallOffsetCopy*

12. Select **Zoom Extents** from the **Zoom** flyout of the Navigation bar. Select **Wall A** as shown in Figure 2.69.

13. To edit the length of the wall using grips, select the Start Grip (bottom square grip) of Wall A to make it hot, move the pointer up, and type **12'6 [3810]** ENTER, dynamically displayed in the tooltip, to stretch the wall up as shown in Figure 2.69.

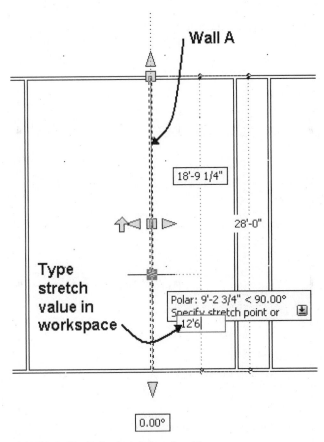

FIGURE 2.69 *Wall selected for grip edit*

14. Press ESC to end grip edit and selection of wall.

15. To view the building in isometric and plan views, click Work layout from the Application status bar and click on the left viewport. Select **Zoom Extents** from Zoom flyout of the Navigation bar.

16. Click in the right viewport. Select **Zoom Extents** from the Zoom flyout of the Navigation bar.

17. Choose the Model toggle of the Application status bar. To verify that **Autosnap** is toggled ON in the **Options** dialog box, right-click over the command line, and choose **Options**.

18. Select the AEC Object Settings tab, and verify that the following are checked ON as shown in Figure 2.70: Autosnap New Wall Justification Lines, Autosnap Grip Edited Wall Justification Lines, and Autosnap Radius = 6" [75].

19. Select **OK** to dismiss the Options dialog box.

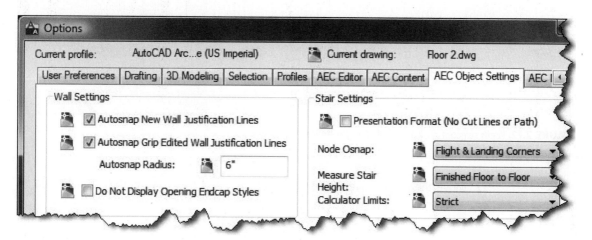

FIGURE 2.70 *AEC Object Settings tab of the Options dialog box*

20. To view the justification line of the walls, click the **Display Configuration** menu of the **Drawing Window** status bar, and select **Diagnostic** as shown in Figure 2.71.

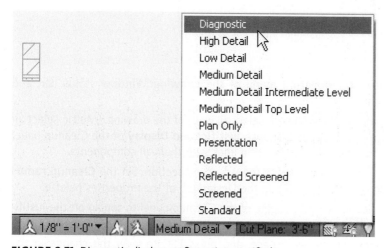

FIGURE 2.71 *Diagnostic display configuration specified*

21. To edit the cleanup radius, type Ctrl A to select all walls in the drawing. [Metric-select walls using a crossing selection.] Edit the **Cleanup radius** to **12 [300]** in the **Advanced** section in the Design tab of the **Properties** palette as shown in Figure 2.72.

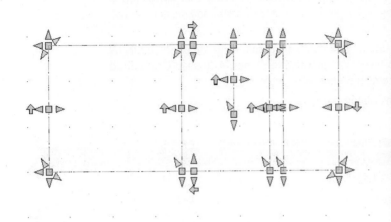

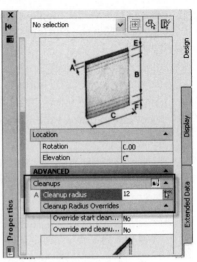

FIGURE 2.72 *Walls selected for change of cleanup radius*

22. Move the pointer over the workspace, and press ESC to view the cleanup radii, as shown in Figure 2.73.

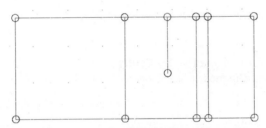

FIGURE 2.73 *Cleanup radius displayed and graph line of walls*

23. Click the **Display Configuration** menu of the **Drawing Window** status bar, and select **High Detail**.

24. Type Ctrl +A in the command line to select all walls of the drawing. [Metric-select all walls using a crossing selection.] Choose **Justification Display** of the Cleanup panel to view the cleanup radii and justification lines with the wall components.

25. Scroll down the Properties palette to the Advanced section. Set the **Cleanup radius** equal to **0** in the Advanced section in the Design tab of the Properties palette.

26. Choose the Justification Display toggle of the Cleanup panel to toggle off the Justification Display.

27. Move the pointer over the workspace, and press ESC to view the change in the cleanup radii.

28. Use **WallOffsetCopy** to add wall **A** 12'-7" [3835] from the top wall and wall **B** 12'-4" [3759] from the bottom wall, as shown in Figure 2.74.

29. Choose **Zoom Extents** from the Navigation bar. Choose Save from the Quick Access Toolbar.

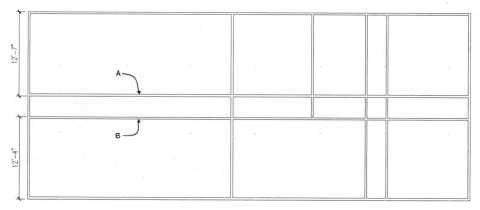

FIGURE 2.74 *Dimensions for additional walls*

30. Verify Floor 2 is the current drawing. In this step you will upload the Floor 2 drawing to AutoCAD WS. Choose the Online tab. Choose the **Upload** command of the Upload panel to launch AutoCAD WS as shown in Figure 2.75. If you have an account, type your Autodesk user name and password. Choose the Log In button; open your session of AutoCAD WS. If you do not have an account, choose the Create An Account button to create the account.

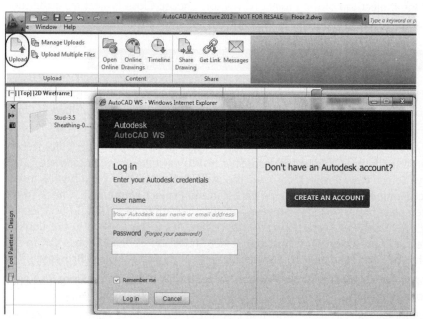

FIGURE 2.75 *Uploading a drawing to AutoCAD WS*

31. Verify the Auto-upload changes is toggled ON when Floor 2 is uploaded to your account as shown in Figure 2.75. When Auto-upload is toggle ON, as you work on your drawing the changes will be automatically uploaded to AutoCAD WS. Choose Close to dismiss the AutoCAD WS dialog box.

32. Although you have closed the project, you can view the Floor 2 drawing online from AutoCAD WS accessed from AutoCAD Architecture 2012 or a mobile device.

33. To view the Floor 2 drawing online, choose the Online tab. Choose **Online Drawings** from the Content panel in the Online tab to open AutoCAD WS to the Drawings pane. Double click on **Floor 2** drawing to open it in AutoCAD WS as shown in Figure 2.76. Choose the Home tab; you may use the measure tools and add AutoCAD entities to the file while online as shown at right in Figure 2.76.

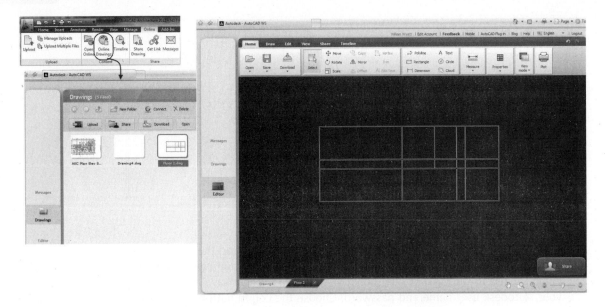

FIGURE 2.76 *Floor 2 shown in AutoCAD WS*

34. Select the Project tab of the Project Navigator. Choose Close Project from the tool-bar located on the Project tab. Close and Save all project drawings.

PROJECT TUTORIAL 2-7: ADDING WALLS TO A FLOOR PLAN

1. Verify that the Accessing Tutor content for Chapter 2 of the CengageBrain has been downloaded to your Accessing Tutor folder on your computer as described in Organizing Tutorial Directories in the Preface.

2. Open AutoCAD Architecture 2012. Choose **Project Browser** from the Quick Access toolbar. Use the Project Selector drop-down list to navigate to your *Student\Ch 2* directory. Select **New Project** to open the **Add Project** dialog box. Type **3** in the **Project Number** edit field and type **Ex 2-7** in the **Project Name** field. Check create from template project, choose **Browse**, and edit the path to *\Accessing Tutor\Imperial\ Ch 2\Ex 2-7\Ex 2-7.apj [Accessing Tutor\Metric\Ch 2\Ex 2-7\Ex 2-7.apj]* in the **Look in** edit field. Choose **Open** to dismiss the Select Project dialog box. Choose **OK** to dismiss the **Add Project** dialog box. Select **Close** to dismiss the **Project Browser**.

3. Select the **Constructs** tab of the **Project Navigator**. Double-click on **Floor 2** to open the Floor 2 drawing.

4. Add additional walls as dimensioned in Figure 2.77 (Imperial) or Figure 2.78 (Metric).

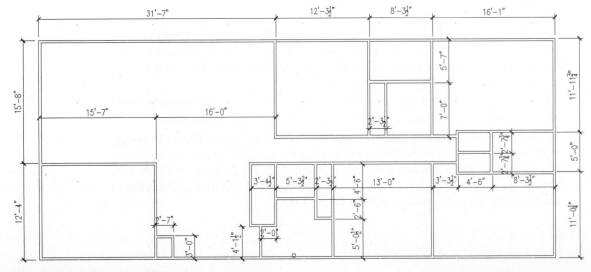

FIGURE 2.77 *Wall placement dimensions (Imperial)*

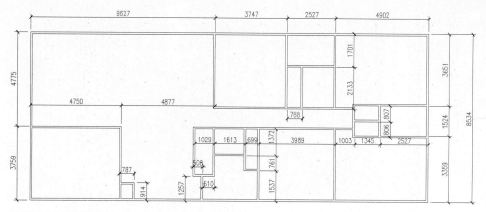

FIGURE 2.78 *Wall placement dimensions (Metric)*

SUMMARY

- Walls are created with the **WallAdd** command as objects with properties of width, height, and justification.

- Draw curved walls by toggling ON **Arc** for Segment type in the Properties palette and identifying a start point, a point on the curve, and the endpoint of the curve.

- If the Close option is used, the last wall segment is drawn to connect to the beginning of the first wall segment.

- The Ortho close option of the **WallAdd** command can be used to create walls in the shape of a rectangle.

- The Baseline Offset option shifts the justification handle off the justification line a specified distance.

- The handle for placing a wall is located at the $Z = 0$ location along the justification line.

- Object snap nodes are located at the beginning, middle, and end of the justification line along the top and bottom of the wall.

- Grips of a wall are located at the beginning, middle, and end of the bottom of the justification line.

- The Cleanup Radius value determines how close to an existing wall a new wall can begin and be automatically connected to the wall.

- AutoCAD edit commands such as **Chamfer**, **Fillet**, **Extend**, **Offset**, **Stretch**, and **Trim** can be used to create and modify walls.

- The grips of a wall allow easy editing of the wall to copy, stretch, and move walls.

Refer to review questions for Chapter 2 in the Review Questions folder of the Student Companion Site of **CengageBrain** http://www.cengagebrain.com described in the Preface.

NOTE

CHAPTER 3

Advanced Wall Features

INTRODUCTION

Autodesk AutoCAD Architecture 2012 provides additional tools to enhance the creation of walls that consist of multiple wall components. The use of wall styles, wall cleanup groups, wall endcaps, wall surface modifiers, and body modifiers increases the speed of design and the detail of wall representation.

OBJECTIVES

After completing this chapter, you will be able to

- Create and modify a wall style.

- Change the style of a wall.

- Control the display properties of wall components to include materials.

- Create wall endcap styles, wall modifier styles, and body modifiers to modify the properties of walls.

- Use Edit In Place to edit wall endcaps and wall components, wall modifiers, and body modifiers.

- Use the Project Navigator to develop additional floor plans and a foundation plan.

ACCESSING WALL STYLES

Although the Standard wall style may be adequate in many design applications, it consists of only two wall lines. Other wall styles can be used to create wall representations for complex walls that consist of several wall components. Wall styles allow the designer to create a wall style for each wall type. Wall styles created in the drawing are listed in the **Style** drop-down list in the Design tab of the Properties palette shown in Figure 3.2. Other defined wall styles can be selected from the Walls tool palette as shown in Figure 3.1. Additional styles can be i-dropped into the drawing or onto tool palettes from the Walls category of the Design Tool Catalog-Imperial or Design

Tool Catalog-Metric as shown in Figure 3.1. The Walls category consists of Brick, Casework, CMU, Concrete, and Stud subcategories. The Casework category includes a wall style for countertops for cabinets.

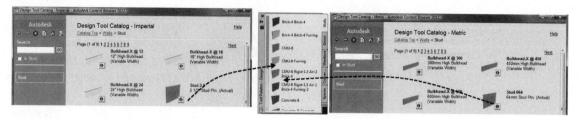

FIGURE 3.1 *Walls tool palette and Design Tool Catalog*

When you click a style from the Walls palette, the **WallAdd** command begins with the style preset in the Properties palette. After you place a wall style, the style is added to the styles listed in the Style Manager and in the Style drop-down list in the Design tab of the Properties palette as shown at the left in Figure 3.2. Therefore, the Properties palette retains a running list of the styles used in the drawing. If you want to change the style of a wall, select the wall and select a style from the Style list in the Properties palette. You can use the Style Manager to bring additional styles into the drawing, which are not listed on the Walls tool palette. When you import or create a new style, it is listed in the Style list in the Design tab of the Properties palette.

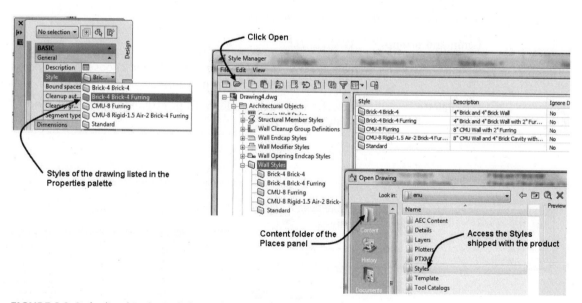

FIGURE 3.2 *Styles listed in the Style list in the Design tab of the Properties palette*

You can choose the *Content* folder of the *Places* panel in the Open Drawing dialog box as shown at the right in Figure 3.2 to access wall styles within the Style Manager from the enu folder of *C:\ProgramData\Autodesk\ACA 2012\enu\Styles*.

TIP

CREATING AND EDITING WALL STYLES

The **Style Manager** is used to create, edit, import, and export wall styles in drawings. Unused styles of the drawing can be purged or deleted from the drawing within the Style Manager. Access the **Style Manager** as shown in Table 3.1.

TABLE 3.1 *Style Manager access*

Command prompt	**AECSTYLEMANAGER**
Ribbon	Select Manage tab < Style & Display panel, choose Style Manager.

When you choose Style Manager from the Manage tab **> Style & Display** panel, the **Style Manager** dialog box opens as shown in Figure 3.3. Styles of the drawing are listed in one of the following three categories: Architectural Objects, Documentation Objects, and Multi-Purpose Objects folders. Wall styles are included in the Wall Styles folder of the Architectural Objects folder.

You can open the Style Manager directly to the Walls folder if you type **WallStyle** in the command line or choose **Wall Styles** from the Edit Style flyout (refer to Table 3.2). You can access the information of the Style Manager within the Wall Style Properties dialog box if you select **Edit Style** button of the General panel of the Wall contextual tab as shown in Figure 3.3 and Table 3.3.

TABLE 3.2 *Accessing Wall Styles in the Style Manager*

Command prompt	**WALLSTYLE**
Ribbon	Select a wall to display the Wall tab, and choose Wall Styles from the Edit Style flyout menu of General panel.
Tool Palette	Select a wall tool on the Walls palette, right-click, and choose Wall Styles.

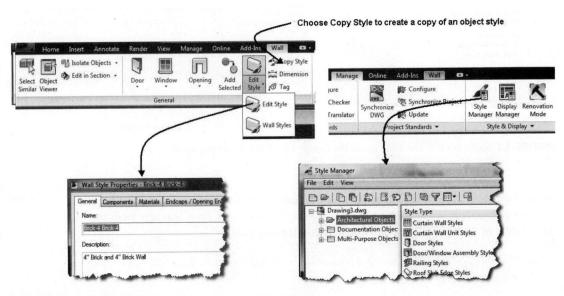

FIGURE 3.3 *Accessing the Wall Style Properties and Style Manager*

When you expand the style categories of the left pane in the Style Manager, the properties of the selected item are displayed in the right pane. When you select the plus (**+**) sign, the style type expands, and the styles used in the drawing are shown in the left pane. When you select a style category in the left pane, the right pane lists the styles of the drawing that apply to the category as shown in Figure 3.4.

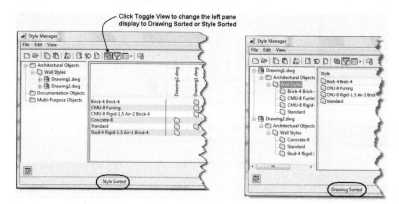

FIGURE 3.4 *Style Manager dialog boxes with styles sorted by drawing and by type*

When you select a wall style in the left pane, the right pane may consist of either the description of the style or the edit style properties tabs. You can choose the **Inline Edit Toggle**, shown in the toolbar of Figure 3.5, to switch between the editing view and the list/view as shown in Figure 3.6. The content of the description pane consists of the Viewer and List tabs shown in Figures 3.5 and 3.6. The **Viewer** tab includes a viewing toolbar and display representations list as shown in Figure 3.5. The **Viewing** toolbar and the shortcut menu inside the **Viewer** tab include options to change the view direction and shading of the wall style. The **List** tab will display the properties of the wall style in text format as shown in Figure 3.6.

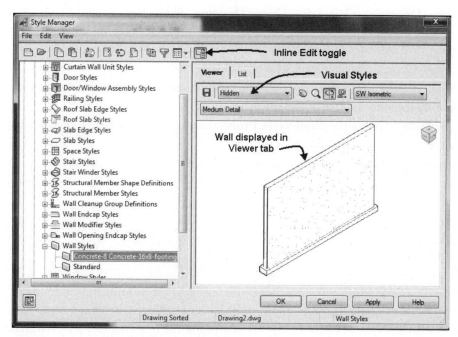

FIGURE 3.5 *Viewer tab of the Style Manager*

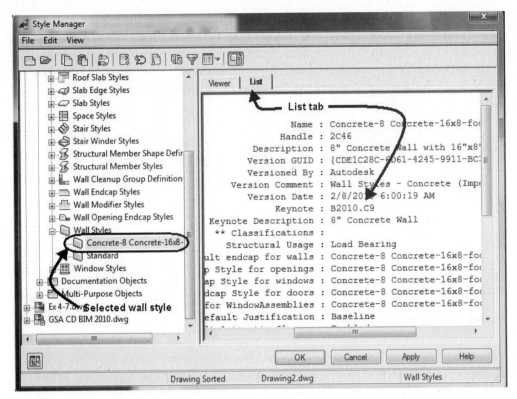

FIGURE 3.6 *List tab of the Style Manager*

The commands of the toolbar in the Style Manager shown in Figure 3.7 include the following:

New Drawing—Creates a new drawing file in which you can place new styles rather than create the styles in the current or existing drawings.

Open Drawing—Opens the **Open Drawing** dialog box, which allows you to select drawings that are not currently opened. Open Drawing allows you to open other drawings and access their styles in the Style Manager.

Copy—Copies a selected wall style or wall style folder to the clipboard.

Paste—Pastes a wall style or wall style folder from the clipboard into another drawing listed in the tree of the left pane.

Edit Style—Opens the **Wall Style Properties** dialog box, which allows you to define the detail of the selected wall style.

New Style—Creates a field to type a new name for the wall style.

Set From—Specifies geometry for style definitions such as profiles and wall endcaps.

Purge Styles—Opens the **Select Wall Style** dialog box, which lists the wall styles not used in the drawing. When a wall style is purged from the drawing, its definition is deleted from the drawing.

Toggle View—Toggles between sorting styles by drawing or sorting styles by style type. The Style Manager on the left in Figure 3.7 is set to sorting styles by styles type, while the Style Manager on the right is set to sorting styles by drawing.

Filter Style Type—Limits the display of styles in the tree view to a style selected by the filter as shown in Figure 3.7.

Views—Controls the format for displaying the contents in the right pane. Options for Icon Format include Small Icons, Large Icons, List, and Details.

Inline Edit Toggle—Toggles between the Viewer and List to wall style properties.

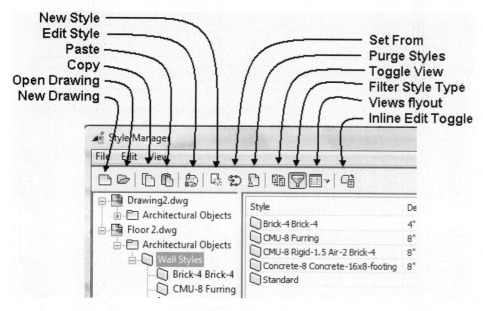

FIGURE 3.7 *Filter applied in the Style Manager*

Creating a New Wall Style Name

The first step to create a new wall style is to establish a new name. Names for wall styles should be kept short and descriptive. Spaces are permitted in the wall style name; however, the inch symbol (") is not accepted. When you select the Wall Styles category and click the **New Style** button shown in Figure 3.8 on the Style Manager toolbar, the default name **"New Style"** is assigned and is highlighted in the left pane. You can overtype to enter a desired name. Specific properties of a wall style are defined in the Wall Style Properties pane as shown in Figure 3.9. Once a name has been created, it is listed in the style list of the left pane or tree view.

You can select *New Style* from the shortcut menu of the Wall Styles category within the Style Manager to create a new style.	**TIP**

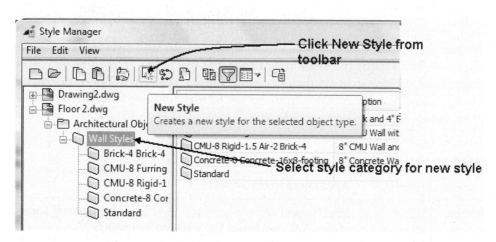

FIGURE 3.8 *New Style command of the Style Manager*

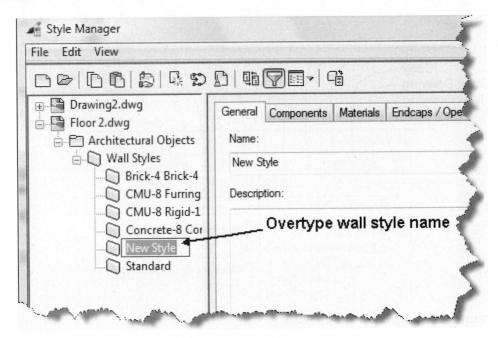

FIGURE 3.9 *New wall style name overtyped in the left pane*

Steps to Create a New Wall Style

1. Choose Style Manager from the Style & Display panel of the Manage tab from the ribbon.
2. Select the plus (+) sign of the **Architectural Objects** folder in tree view of the left pane.
3. Select the **Wall Styles** folder.
4. Click the **New Style** button in the **Style Manager** toolbar as shown in Figure 3.9.
5. Overtype a new name in the left pane as shown in Figure 3.9.

Creating a New Style from an Existing Style

The **CopyAndAssignStyle** command allows you to select an object and create a new style from the style definition of the selected object. The new style definition will be assigned to the selected object. The **CopyAndAssignStyle** command can be accessed from contextual tab of an object as shown in Table 3.3 and Figure 3.3. This command, when applied to a wall, creates a copy of the wall style assigned to the selected wall and opens the **Wall Style Properties** dialog box. The name and other properties of the new style can then be changed in the **Wall Style Properties** dialog box. The CopyAndAssignStyle command can be used to create additional styles for other objects such as windows and doors.

TABLE 3.3 *AecCopyAndAssignStyle command access*

Command prompt	**AECCOPYANDASSIGNSTYLE**
Ribbon	Select a wall and choose Copy Style from the General panel of the Wall contextual tab (refer to Figure 3.3).
Shortcut menu	Select a wall, right-click, and choose Copy Wall Style and Assign Style.

Defining the Properties of a Wall Style

Editing a wall style can occur prior to or during the use of the wall style. The current definition for the wall style is applied to all instances of the wall in the drawing. Therefore, if a wall style is defined initially to consist of three wall lines with a total width of 12" [300] and the style is later edited to a width of 10" [250], all previous walls placed in the drawing using this style will automatically be updated to 10" [250] wide walls. Using wall styles allows you to globally edit all the walls of an entire building by changing the original wall style definition.

To edit a wall style, select the wall and choose Edit Style from the General panel of the Wall contextual tab; the **Wall Style Properties** dialog box opens as shown in Figure 3.10. The Wall Style Properties dialog box is identical in content to the right pane of the Style Manager. The right pane of the Style Manager or the Wall Style Properties dialog box consists of the following tabs: **General**, **Components**, **Materials**, **Endcaps/Opening Endcaps**, **Classifications**, **Display Properties**, and **Version History**. Note that the Version History tab is not displayed for new styles. The Version History tab is created and utilized to track style versions during project synchronization, which will be discussed in Chapter 10. A description of the options in each tab of the Wall Style Properties dialog box is explained in the following tutorial.

TUTORIAL 3-0: EXPLORING WALL STYLE PROPERTIES

1. Verify that the Accessing Tutor content for Chapter 3 of the CengageBrain has been downloaded to your Accessing Tutor folder on your computer as described in Organizing Tutorial Directories in the Preface.

2. Open *Accessing Tutor\Imperial\Ch 3\Ex 3-0.dwg* or *Accessing Tutor\Metric\Ch 3\Ex 3-0.dwg*. Choose **Save As > AutoCAD Drawing** from the **Application** menu and save the drawing as **Lab 3-0** to your student directory.

3. Select wall number 1 to display the Wall contextual tab. Choose Edit Style from the Edit Style flyout of the General panel of the Wall contextual tab to review the settings of the Concrete-8 Concrete-16x8 footing [Concrete-200 Concrete-400 x 200-footing].

4. Choose the General panel. The Keynote field shown in Figure 3.10 lists the metric or imperial keynote. Choose the Select Keynote button to display the Select Keynote dialog box. The keynote database file includes a list of keynotes. Choose OK to dismiss the Select Keynote dialog box.

5. You may choose the Property Sets button of the General tab shown in Figure 3.10 to open the **Edit Property Set Data** dialog box. This dialog box allows you to edit or add the property sets associated with a wall style. Property data are used to develop schedules from the drawing and will be presented in Chapter 11. Click OK to dismiss the Edit Property Set Data dialog.

6. The checkbox **Objects of this style may act as a boundary for automatic spaces** located in the lower-left corner of the General tab, which is checked; therefore, walls created with this style will function as a boundary for spaces created by the **Space-Generate** command.

7. Choose the Viewer button located in the lower-left corner as shown in Figure 3.10 to preview the wall style. The Viewer is shown in Figure 3.11. Choose SW Isometric from the View flyout at p1 to view the wall in pictorial. Choose the X in the upper-right corner to close the Viewer.

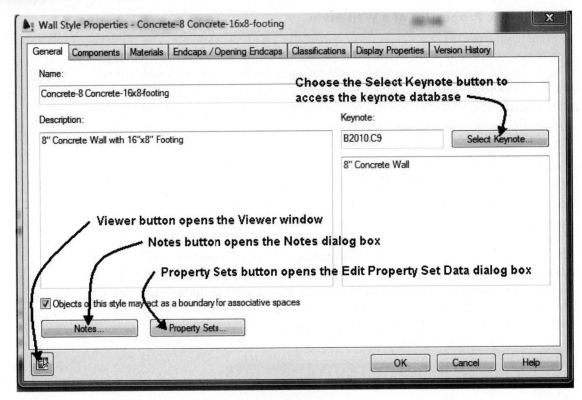

FIGURE 3.10 *General tab of the Wall Style Properties dialog box*

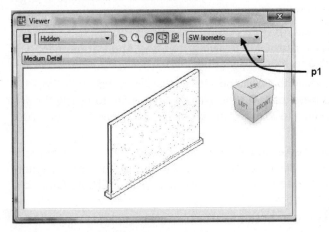

FIGURE 3.11 *Wall displayed in the Viewer*

8. Choose the Components tab to review the settings of the components, which define the wall style. The left pane provides a dynamic view of the wall style as shown in Figure 3.12.

9. Move the cursor over the left pane and right-click to display a shortcut menu. Choose **Zoom** from the shortcut menu to activate real-time zoom within the window. Left-click and continue to hold down the left button and drag the mouse forward to magnify the image. To end real-time zoom, move the cursor to the right pane and left-click; do not press Escape.

10. In this step, you will move a component up within the list of components. Choose Index 2 column to select the Concrete (Footing) component, and choose the Move Component Up in List button as shown at right in Figure 3.12.

11. Choose Index 1 component to reselect Concrete (Footing), then choose Move Component Down button as shown in Figure 3.12.

12. Note the wall style is defined with the exterior of the wall components located at left or in the positive direction as shown in the lower-left corner of Figure 3.12.

13. Choose the Open Wall Style Browser button located at right in Figure 3.12 to open the Wall Style Component Browser as shown in Figure 3.13.

14. You may choose the **Open drawing** button shown in Figure 3.13 of the Wall Style Browser to open files for the import of wall components into the current style. The Wall Style Components Browser of Figure 3.13 lists the wall styles of the current drawing. Therefore, to import a component, choose the + sign of the left pane to expand the CMU-8 Rigid-1.5 Air-2 Brick-4 Furring [CMU-190 Rigid-038 Air-050 Brick-090 Furring] wall style. Select the Air Gap component, right-click, and choose **Copy**. Move the cursor to right pane of the Components tab of the Wall Style Properties palette, right-click, and choose Paste to add this component into the current wall style.

15. To remove the Air Gap component, select the component and then choose Remove Component button shown at right in the Components tab in Figure 3.12.

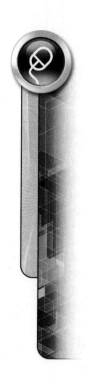

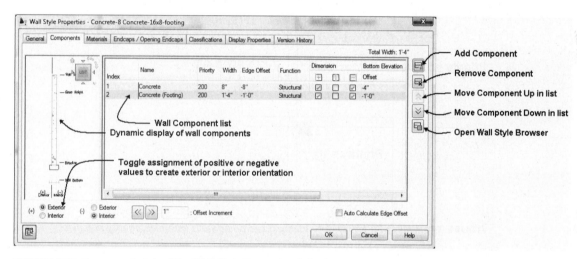

FIGURE 3.12 *Components tab of the Wall Style Properties dialog box*

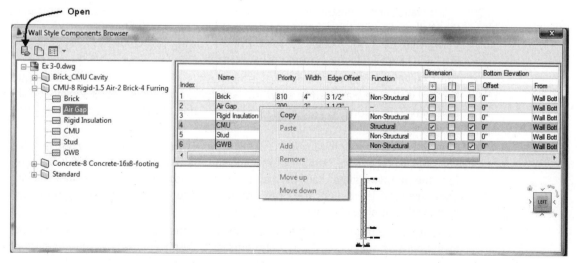

FIGURE 3.13 *Components from wall styles listed in the Wall Style Components Browser*

16. Select Index 1 to choose the Concrete component, and choose the Add Component button shown at right in Figure 3.12 to copy the Concrete component. There are now three Index numbers that identify each component whereas the remaining columns define the size and location of the component.

17. Click the Concrete name of the Name column for Index 2 and overtype Brick to change the name of the new component.

18. The Priority field allows you to assign a priority number to control how the wall component will clean up with other walls. Wall components with the same priority number will merge when walls intersect. Components with low priority numbers will override the display of wall components with higher priority numbers as shown in Figure 3.14.

The walls shown in Figure 3.14 were drawn with priorities of 5, 10, and 20. The horizontal wall and the left wall have low component priority; therefore, they override the display of the wall component with a priority of 20. Assigning priority numbers to wall components would be useful for controlling the display of components of fire-rated walls. Fire-rated walls are displayed without penetrations from other non-rated walls. Therefore, in the example shown in Figure 3.14, the right vertical wall could correctly represent a non-rated wall that intersects the horizontal fire-rated wall.

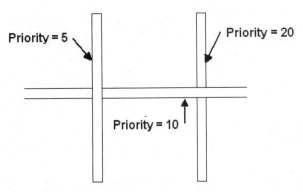

FIGURE 3.14 *Walls with low priority numbers overriding walls with high priority numbers*

19. Press ESCAPE to dismiss the dialog box without saving any changes to the wall style. Press ESCAPE to clear the grips from wall 1.

20. Select wall 2, and choose Edit Style from the Edit Style flyout of the General panel in the Wall contextual tab. Choose the General tab of the Wall Style Properties dialog. Verify that the CMU-8 Rigid-1.5 Air-2 Brick-4 Furring [CMU-190 Rigid-038 Air-050 Brick-090 Furring] is the current wall style name.

21. The CMU-8 Rigid-1.5 Air-2 Brick-4 Furring [CMU-190 Rigid-038 Air-050 Brick-090 Furring] wall style was used to draw the horizontal and vertical walls shown at right from wall 2. Position the Wall Style Properties dialog in the workspace allowing you to view the wall intersection during the next step.

22. Choose the Components tab. The CMU is assigned the lowest priority number, 300. The Brick has a priority number of 810 and the Air Gap has a priority of 700; therefore, the priority number controls display of wall components. Because the CMU priority number is the lowest, the CMU display overrides that of the Brick and Air Gap components as shown at right in the drawing.

23. Choose the Help button located in the lower-right corner of the Components tab. Help opens to Specifying the Components of a Wall Style. Review the contents of

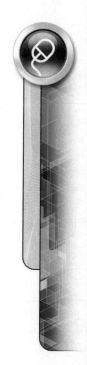

this topic. Scroll to the end of the document, then choose Default Cleanup Priorities of Wall Components.

The priority index values used in AutoCAD Architecture 2012 are shown in Figure 3.15. Notice the priority assignment in the styles systematically assigns low priority to components placed early in construction and a higher priority value is assigned to finish components. Therefore, a concrete footing component has a priority of 200, whereas the gypsum wallboard has a priority of 1200.

24. Close the Help window to return to the Wall Style Properties dialog box for the CMU-8 Rigid-1.5 Air-2 Brick-4 Furring [CMU-190 Rigid-038 Air-050 Brick-090 Furring].

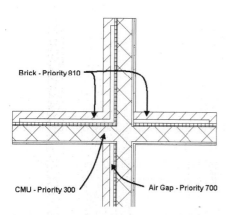

Component	Index
Air Gap	700
Air Gap (Brick/Brick)	805
Air Gap (CMU/CMU)	305
Air Gap (Stud/Stud)	505
Brick	800
Brick Veneer	810
Bulkhead	1800
Casework - Backsplash	2030
Casework - Base	2010
Casework - Counter	2020
Casework - Upper	2000
CMU	300
CMU Veneer	350
Concrete	200

Component	Index
Concrete (Footing)	200
Glass	1200
GWB	1200
GWB (X)	1200, 1210, 1220, 1230
Insulation (CMU/Brick, Stud/Brick)	600
Metal Pane	1000
Precast Panel	400
Rigid Insulation (Brick)	404
Siding	900
Stucco	1100
Stud	500
Toilet Partiton	3000

FIGURE 3.15 *Priority Index Numbers of AutoCAD Architecture 2012 wall components*

The width of a wall component is the distance between the faces of the component. A wall component is placed within the wall relative to the baseline of the wall style. The Edge Offset value defines the position of a component from the baseline. The baseline is typically aligned within a wall style to place a component such as a wood stud or masonry unit placed at the beginning of the construction of a wall. The position of the baseline may also be defined to permit the vertical alignment of wall components for floors above or below.

The design of all Autodesk wall styles places the exterior veneer component toward the positive; therefore, if wall segments are placed in a clockwise direction, the exterior veneer is placed on the positive side as shown in Figure 3.16. The width of components is assigned positive values. A positive edge offset value is assigned to place the component on the exterior side of the wall relative to the baseline. A negative edge offset value is assigned to place components on the interior side of the baseline.

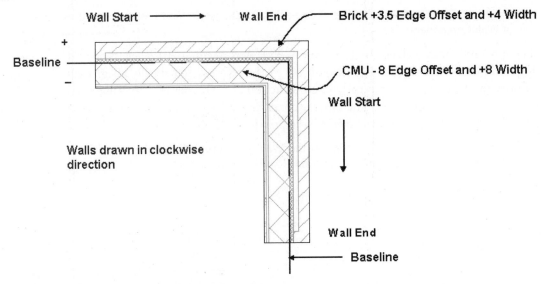

FIGURE 3.16 *Components positioned with edge offset*

25. Return to the CMU-8 Rigid-1.5 Air-2 Brick-4 Furring [CMU-190 Rigid-038 Air-050 Brick-090 Furring] Wall Style Properties dialog.

26. The Brick component is positioned with a positive 3 1/2" [88] edge offset and its width is positive 4" [90]. The CMU component has a negative 8" [–190] edge offset and a positive 8" [190] width.

27. Click in the Width column for the Brick component to display the drop-down list as shown in Figure 3.17. The width value of 4 creates a component 4 units wide in the positive or exterior direction as defined from the Edge Offset value.

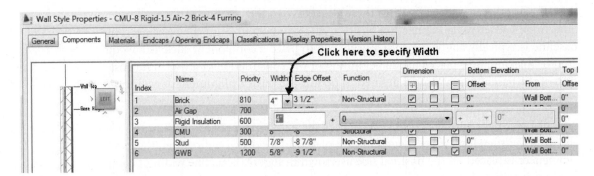

FIGURE 3.17 *Setting wall width*

Note, if you create a new wall style from the Standard wall style, the Base Width is specified by default as the variable value. Failure to select the drop-down hides the variable value setting. The drop-down list defines a formula, which may consist of a fixed value, variable value, operator, and operand.

A fixed value is set by typing a value in the fixed value field and setting the variable value to 0 as shown at the left in Figure 3.18. Selecting the **Base Width** or other values from the variable drop-down list applies the variable value in a formula.

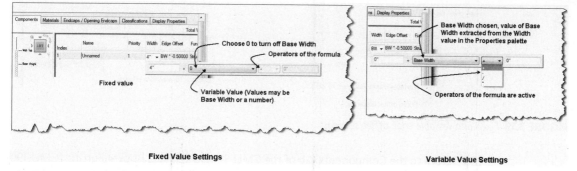

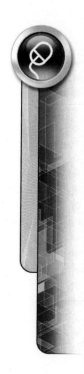

FIGURE 3.18 *Setting component width*

The Base Width value is the width extracted from the **Width** field in the Design tab of the Properties palette of the wall. The operator and operand are then applied to the base value. The operator options can be add, subtract, divide, or multiply. Positive or negative distances can be entered in the Width field as shown in Figure 3.18.

28. Return to the Components tab of the CMU-8 Rigid-1.5 Air-2 Brick-4 Furring [CMU-190 Rigid-038 Air-050 Brick-090 Furring] wall style. Click in the Edge Offset column for the Brick component to display the drop-down fields. The Edge Offset value of 3 1/2" [88] specifies the start point for the development of the Brick component width in the exterior or positive direction from the Baseline.

The Edge Offset column is used to define the location of a wall component relative to the **Baseline**; therefore, to position a veneer component in the CMU-8 Rigid-1.5 Air-2 Brick-4 Furring [CMU-190 Rigid-038 Air-050 Brick-090 Furring] wall style, a positive 3 1/2" [88] offset is specified. The CMU component has an Edge Offset of −8" [−190], which positions this component to the interior direction from baseline, and it is developed using an 8" [190] width.

To set the Edge Offset distance, click in the **Edge Offset** column and select options from the drop-down list. The drop-down list consists of the same formula of fixed and variable values as specified in the Width option described above. Base Width is specified by default as the variable value for Edge Offset in Standard wall style. Failure to select the drop-down hides the setting of the variable value; therefore, review the drop-down when setting the Edge Offset. The edge offset can be set to a fixed or variable value as shown in Figure 3.19.

Typing a value in the fixed offset field and setting the variable offset to 0 will specify a fixed offset as shown at the left in Figure 3.19. A variable offset is specified by selecting the **Base Width** option as shown in the drop-down list. The Base Width option is used in the Standard wall style. The **Base Width** value is extracted from the **A-Width** of the wall in the Properties palette. The Base Width is applied to the operator and operand as shown in Figure 3.19. The Operator options include add, subtract, multiply, and divide. The operand is a number applied to the operator and the base width. The Standard wall applies this method and specifies the edge offset as BW × −1/2. This statement will multiply −1/2 times the base width to specify the offset. Therefore, if you were drawing with the Standard wall style and set the Width to 6" in the Properties palette, the edge offset distance would be 3" ($6 \times -1/2 = -3$). The variable offset method can also specify a fixed value, which is added to arithmetic operations of the Base Width.

FIGURE 3.19 *Fixed and variable edge offset settings*

29. Return to the Components tab of the CMU-8 Rigid-1.5 Air-2 Brick-4 Furring [CMU-190 Rigid-038 Air-050 Brick-090 Furring] wall style. Click in the Function column for the Brick component. The drop-down includes the following options: Unspecified, Structural, and Non-structural. The options allow you to define in the wall style which component surfaces will be dimensioned as referenced in an AEC Dimension.

30. The positive checkbox is checked; therefore, the AEC Dimension style will include an extension line to the positive surface of the brick.

The Function and Dimension columns, shown in Figure 3.20, allow you to define in the wall style which component surfaces will be dimensioned as referenced in an AEC Dimension. The settings specified in an AEC Dimension are specified in the Content tab of the Display Properties tab as shown at the right in Figure 3.20. If the **Structural by Style** is specified in an AEC dimension style as shown in Figure 3.20, the AEC Dimension tool will only dimension components specified with this structural function. The positive, center, and negative surfaces of the structural component will be dimensioned if checked. If the **By Style** option is specified in the AEC dimension style, the components are dimensioned as specified in the checkboxes: positive, center, or negative, regardless of the structural function. When AEC Dimensions are applied to a wall, the wall style definition and the AEC Dimension style jointly determine how objects are dimensioned. The **Exterior** and **Interior** toggles control how the wall component is treated for the dimension. AEC Dimensions are presented in Chapter 11. The positive and negative surfaces of a component within a wall are shown in Figure 3.20. The positive surface is the top surface when a horizontal wall is drawn from left to right or in a clockwise direction.

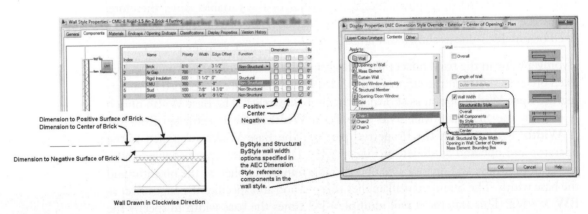

FIGURE 3.20 *Structural Function and Dimension Settings*

31. Press ESCAPE twice to close the dialog box and clear the selection of the wall.

32. Select wall 3, and choose the Edit Style button of the General panel of the Wall contextual tab for the **Brick_CMU Cavity** wall. If necessary, resize the dialog box to display the Bottom Elevation and Top Elevation columns at right.

33. You can set the bottom elevation of each component relative to the Wall Top, Base Height, Baseline, and Wall Bottom as shown in the left pane of Figure 3.21. The bottom of the wall component will move up if the distance is assigned a positive value from the Wall Bottom and Baseline, or negative value from the Wall Top and Base Height. The Bottom Elevation for all components except the Footing and 10" [250] CMU is set to 5'-0" [1500] above the Baseline to create cavity wall components in the upper half of the wall as shown in Figure 3.21.

34. The 10" [250] CMU component has a Bottom Elevation of 10 [250] from the Wall Bottom and a Top Elevation of 5'-0" [1500] from the Baseline. The top elevation restricts the height of this component. Controlling the elevation of wall components allows the wall style to simulate actual construction.

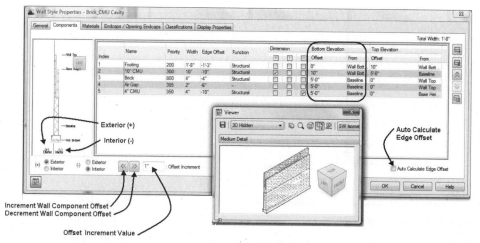

FIGURE 3.21 *Effects of setting Bottom Elevation 5'-0" [1500] above Wall Bottom for cavity components*

35. Choose the Footing component to display the footing component shown in green in the left pane. The bottom elevation of the footing is specified as 0" from Wall Bottom, and its top elevation is defined as 10" [250] from the Wall Bottom as shown in Figure 3.21. The location of the Wall Bottom is specified by the FloorLine command presented in this chapter.

36. Choose the 10" CMU [250 CMU] component. The 10" CMU [250] component starts 10" [250] above the Wall Bottom as defined in the Offset value of the Bottom Elevation section.

Components such as the footing can be projected below the baseline to the Wall Bottom location by the **FloorLine** command presented in this chapter. Components may be projected from the Base Height to the Wall Top by the RoofLine command as shown in Figure 3.22. The RoofLine command may be used to cover the wood framing between floors. If the brick veneer component is specified as 0" from Wall Top and the remaining components are restricted to 0" from Base Height, the brick will project to the wall top based upon the RoofLine Offset.

CAUTION

Setting the bottom elevation with a positive value relative to the wall top or base height will cause the component to disappear. The component is defined but not visible.

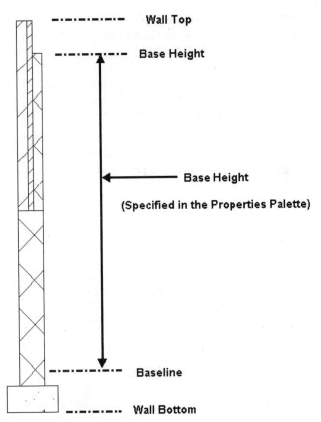

Wall Top

Base Height

Base Height

(Specified in the Properties Palette)

Baseline

Wall Bottom

Brick CMU Cavity Wall Style

FIGURE 3.22 *Editing the Top Elevation and Bottom Elevation*

CAUTION Setting the top elevation with a negative value relative to the wall bottom or baseline causes the component to disappear. The component is defined but not visible.

37. Return to Components tab of the Brick_CMU Cavity wall style, and choose the 10″ CMU [250] component. The Offset Increment value is shown as 1″ [1]; therefore, when you choose the **Increment Wall Component Offset** arrow as shown in Figure 3.21, the edge offset will step over 1″ [1] from the baseline in the positive direction changing the Edge Offset from –10″ [250] to –9″ [–249].

38. Choose the Decrement Wall Component Offset to step the component in the negative direction 1″ [1] changing the Edge Offset from –9″ [–249] to –10″ [–250].

39. The AutoCalculate Edge Offset checkbox shown in Figure 3.21 located in the lower-right corner, if selected, will adjust the offset equal to the previous width to avoid the overlap of components. Therefore, check the Auto Calculate Edge Offset check box, choose the Brick component, and then choose the Add Component button at right. A new Brick component is created with an Edge Offset of 0″, an increase of 4 [90] units.

40. Choose the Materials tab. This tab allows you to assign material definitions to a component. Click the Material Definition column as shown in Figure 3.23 to choose a material definition. A material definition displays the material in section and elevation.

The material definition is a style that you access in the Multi-Purpose Object category of the Style Manager. The **Standard** material definition is the only material definition in a drawing unless material definitions have been imported or other objects with material definitions have been inserted into the drawing. You can choose **Content** in the **Places** panel of the **Style Manager** to import material definitions from the *Material Definitions (Imperial).dwg* or *Material Definitions (Metric).dwg* of Imperial and Metric folders in *Vista–C:\ProgramData\Autodesk\ACA 2012\enu\Styles.* directory.

The **Material** tool of the **Design** tool palette allows you to assign materials from the respective Material Definition file to wall components. The Material tool is presented later in this chapter. Material definitions can also be created from materials placed with the i-drop tool from the Visualization Catalog in the Content Browser.

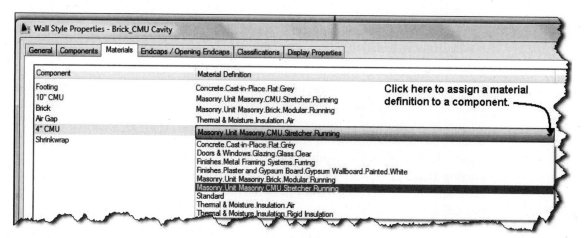

FIGURE 3.23 *Materials tab of the Wall Style Properties dialog box*

41. Press ESCAPE twice to dismiss the Wall Style Properties dialog box and clear the selection of the wall.

42. Select wall 2. Choose Edit Style from the Edit Style flyout to open the Wall Style Properties-CMU-8 Rigid-1.5 Air-2 Brick-4 Furring [CMU-190 Rigid-038 Air-050 Brick-090 Furring] dialog box. Choose the Endcaps/Opening Endcaps tab. This tab allows you to define an endcap style as shown in Figure 3.24.

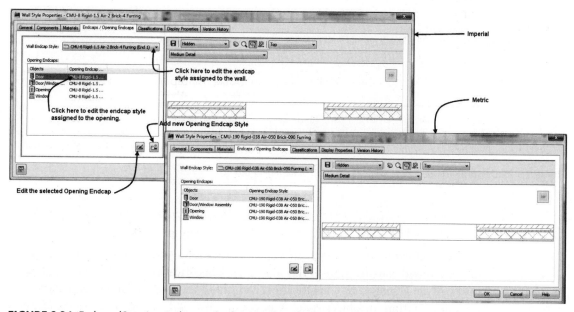

FIGURE 3.24 *Endcaps/Opening Endcaps tab of the Wall Style Properties dialog box*

In Figure 3.24, the CMU-8 Rigid-1.5 Air-2 Brick-4 Furring (End 1) [CMU-190 Rigid-038 Air-050 Brick-090 Furring (End 1)] endcap style is assigned to cap the ends of the wall and the openings. Creating endcap styles from polylines will be discussed later in this chapter. To change the endcap style assigned, click in the endcap list and select from the list of available endcap styles in the drawing. The left pane of this tab includes the following buttons:

> **Edit the Selected Opening Endcap Style**—This button opens the **Opening Endcap Style** dialog box, which allows you to edit the existing opening endcap style.
>
> **Add a New Opening Endcap Style**—Clicking this button opens the **Opening Endcap Style** dialog box to create a new opening endcap style.

The Opening Endcap Style dialog box has three tabs: **General**, **Design Rules**, and **Version History**. The General tab allows you to change the name and description and add notes. The Design Rules tab allows you to assign endcaps to the start, end, sill, and head locations of the opening as shown in Figure 3.25. The Version History tab lists the revision history of the style. The wall opening endcap style specified in Figure 3.25 applies the jamb endcap named CMU-8 Rigid-1.5 Air-2 Brick-4 Furring (End 1) [CMU-190 Rigid-038 Air-050 Brick-090 Furring (End 1)] to each jamb of the opening. You can specify different endcap styles for each of the four edges of the opening in the opening endcap style.

You can create and edit an opening endcap style from the Style Manager by selecting a style located in the *Architectural Objects\Wall Opening Endcaps* folder.

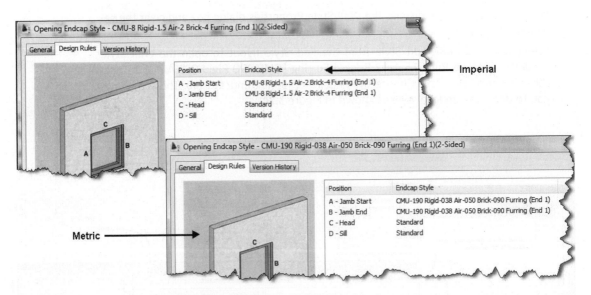

FIGURE 3.25 *Design Rules tab of the Opening Endcap Style dialog box*

43. Choose the Classifications tab. This tab allows you to assign a classification definition to a wall style.

The Classifications tab shown in Figure 3.26 allows you to assign a classification definition and classification to the wall style. **Classification Definitions** are created in the Style Manager. A Classification Definition includes classifications that create categories within an object style. You could create a Building Status classification definition that would consist of New and Existing classifications. Therefore, the New and Existing classification can be assigned to a wall style name. A display representation set could then be edited to show or hide all objects with the New classification. Classifications also allow the development of schedules restricted to one classification of an object style. A wall schedule could be developed that includes one classification within a wall style. Classifications allow you to sort objects with the same object style according to their classification. The Classifications tab shown in Figure 3.26 includes **Load Bearing** and **Non-Load Bearing** classification definitions, which can be selected from the drop-down list.

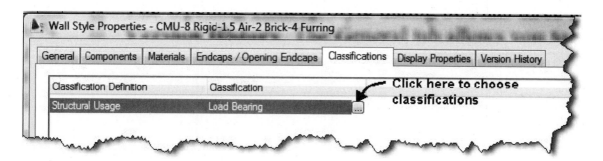
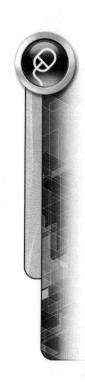

FIGURE 3.26 *Classifications tab of the Wall Style Properties dialog box*

If no classifications are defined in the Style Manager, this tab will not contain any information.

44. Choose the Display Properties tab as shown in Figure 3.27. The display method for the wall components can be specified for each of the display representations. The display representation that is current in the drawing is shown as bold.

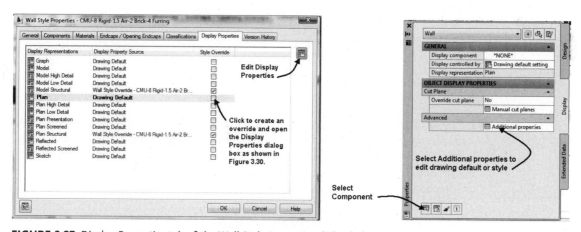

FIGURE 3.27 *Display Properties tab of the Wall Style Properties dialog box*

45. Clicking the **Edit Display Properties** button shown in Figure 3.27 shown at the right allows you to access the Display Properties (Drawing Default) Plan dialog box shown in Figure 3.28. This dialog box includes four tabs that describe the default settings of a wall.

46. Click OK to dismiss the Display Properties (Drawing Default) Plan dialog box. Retain the display of the Display Properties tab of the Wall Style Properties dialog.

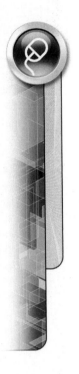

47. A display representation can be controlled by a **Style Override** or the **Drawing Default**. In this step, you will create a style override and view the settings of display defined by the wall style. Check the **Style Override** checkbox, and the **Display Properties** (Wall Style Override-CMU-8 Rigid-1.5 Air-2 Brick-4 Furring [CMU-190 Rigid-038 Air-050 Brick-090 Furring]) dialog box opens as shown at the right in Figure 3.28. A **Layer/Color/Linetype** tab of a Style Override defines visibility, color, and linetype for each of the wall components defined by the wall style. Notice that the Display Components are listed by boundary number with the specific component name.

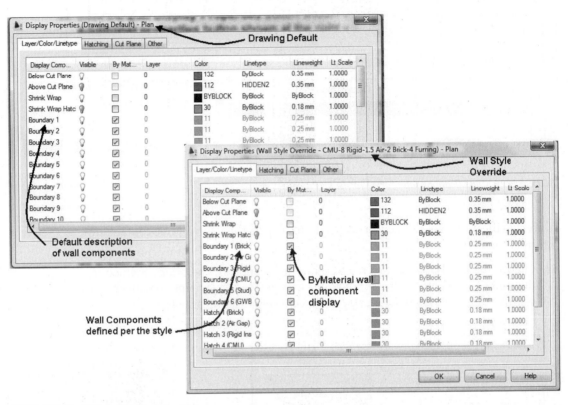

FIGURE 3.28 *Editing Display Properties*

48. The Boundary display components consist of the entities that represent the width of the wall components. In Figure 3.28, Boundary 1 (Brick) and Boundary 2 (Air Gap) are the wall components created as Index 1 and Index 2 on the Components tab. You can turn the display of each component off by turning off the visibility of the component.

49. The Hatch component applies a hatch pattern for each wall component. The Hatch 1 (Brick) and Hatch 2 (Air Gap) display components will apply the specified hatch pattern to each of the wall components. The hatch pattern is determined by material assigned to a component. If the Material checkbox is clear, the hatch pattern is specified in the **Hatching** tab. To display hatching of a component, turn ON the light bulb in the **Visible** column.

50. The Below Cut Plane display component shown in Figure 3.28 consists of the representation of the wall located at an elevation below the cutting plane. The default cutting plane elevation is 3'-6" [1060] for the Medium Detail display configuration defined in the Display Manager. You can alter the global height of the cutting plane by editing the Display Manager. The cut plane height can be defined in the wall style independent of the Display Manager if you override the cutting plane

height in the **Cut Plane** tab. Turning on visibility turns on entities to represent objects below the cutting plane.

51. The Above Cut Plane components consist of the component representation at an elevation above the cutting plane. In the dialog box shown in Figure 3.28, the visibility of the Above Cut Plane display component is turned off.

52. The Shrink Wrap display component is displayed when the **Wall Interference** command is applied to a wall that intersects with other objects. Shrink-wrap entities add emphasis to the intersection of objects.

53. Choose the Hatching tab to specify the hatch pattern, scale, orientation, and offsets for the hatch. To change the hatch properties, click in the respective column and edit the hatch pattern, hatch scale factor, angle, orientation, X offset, and Y offset. If the **By Material** checkbox is clear in the Layer/Color/Linetype tab for a display component, the default **User Single** pattern is applied as shown in Figure 3.29. You can select other hatch patterns by selecting the **User Single** option of the **Pattern** column to display the **Hatch Pattern** dialog box shown in Figure 3.29.

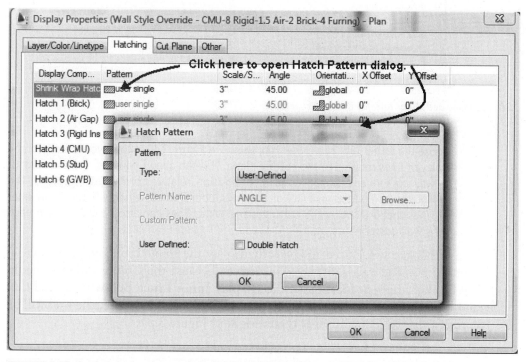

FIGURE 3.29 *Setting the hatch pattern in the Hatching tab*

54. The Cut Plane tab shown in Figure 3.30 allows you to set the cut plane height for the wall style. Cut plane elevations can be defined for the display configuration in the Display Manager or within the wall style. Since the check box is clear, the wall style will be displayed with the cut plane defined by the display configuration settings.

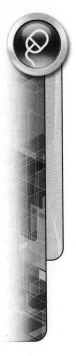

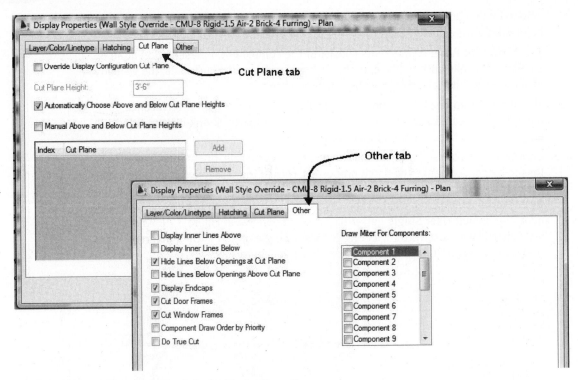

FIGURE 3.30 *Cut Plane and Other tabs for Display Properties*

55. Choose the Other tab. The Other tab, shown in Figure 3.30, allows you to control the display of lines relative to the cutting plane. This tab allows you additional control of the wall components based upon their position relative to the cutting plane. The checkboxes turn on the visibility of openings, door frames, window frames, and inner wall lines that are cut by the cutting plane. Miter lines can also be checked for each of the wall components.

The Other tab was edited to correctly display the foundation vent as shown in Figure 3.31. The wall is displayed incorrectly because the footing and wall should not break at the vent location. If the **Display Inner Lines Below** is toggled ON and **Hide Lines Below Openings at Cut Plane** is toggled OFF, the wall will be correctly displayed as shown at the far right in Figure 3.31.

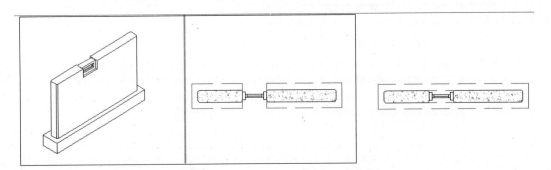

FIGURE 3.31 *Control of lines beneath openings to create an accurate foundation vent display*

56. Choose ESCAPE twice to dismiss all dialog boxes.

57. The content of the Other tab shown in Figure 3.30 can be displayed from Display tab of the Properties palette. Verify that wall 2 is selected, and choose the Display tab of the Properties palette. Choose the Additional properties button shown in Figure 3.32 to display the Wall Additional Properties dialog. Choose OK to dismiss the dialog box.

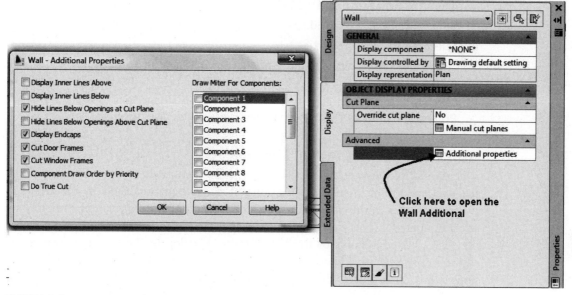

FIGURE 3.32 *Accessing Display properties from the Display tab of the Properties palette*

58. Choose Edit Style of the General panel of the ribbon. Choose the Version History tab. The Version History tab shown in Figure 3.33 lists the date-modified history of a style. This information is used during synchronization of a project to determine if the style matches a project standard. The synchronization of a project will be presented in Chapter 10.

59. Save and Close the drawing.

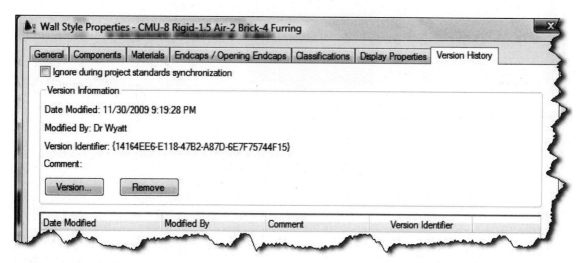

FIGURE 3.33 *Version History tab*

NOTE **Access Instructional Video 3.1—Creating a Wall Style** from the Instructional Video category of the **Student Companion site of CengageBrain** http://www.cengagebrain.com described in the Preface.

Steps to Editing the Drawing Default Display Representation for a Wall Style

1. Select a wall and choose **Edit Style** from the Wall tab of the ribbon.

2. Select the **Display Properties** tab.

3. Select the current display representation set and choose **Edit Display Properties** button at right.

4. Select the **Layer/Color/Linetype**, **Hatching**, or **Cut Plane** tabs to edit the **Display Properties** (Drawing Default) dialog box. Click OK to dismiss all dialog boxes.

Assigning Materials to Wall Components

Material definitions are assigned to each wall component in the Materials tab of the wall style. The material definitions of a drawing are listed in the Material Definition styles in the Style Manager. The material definitions can be imported by selecting material tools from the Materials palette of the Visualization palette set or the AEC Material Tools of the Visualization Catalog in the Content Browser as shown in Figure 3.34. The sample material tools can be used to apply material definitions to any AEC object with display components.

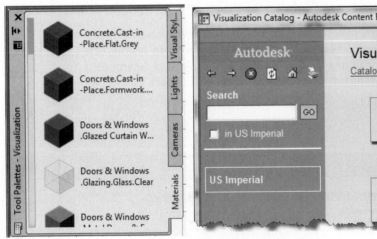
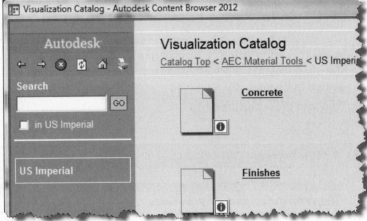

FIGURE 3.34 *Sample material tools for material definitions*

The **Material** tool (**MaterialApply**), located on the Design tool palette shown in Figure 3.34, can be selected to assign a material definition from the master list of material definitions. The master list of material definitions is located in the *Material Definition (Imperial).dwg* or *Material Definition (Metric).dwg* of the Imperial and Metric folders in *Vista–C:\ProgramData\Autodesk\ACA 2012\enu\Styles* directory.

Prior to selecting the tool, right-click the Material tool and choose Properties to open the Tool Properties dialog box. Material definitions can be specified in the Tool Properties dialog box as shown at right in Figure 3.35. When you choose the Material tool, you are prompted to select a component. Moving the cursor over a wall component will display the name of the wall component as defined in the wall style.

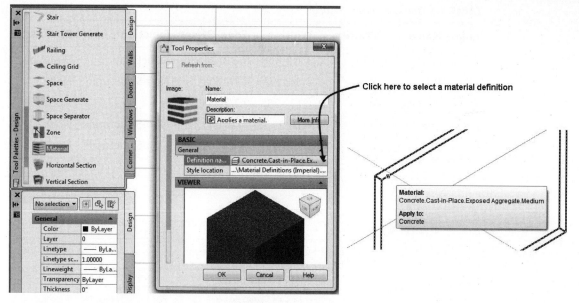

FIGURE 3.35 *Using the Material tool to assign a material definition*

After assigning materials to a wall, the **MaterialList** command allows you to determine the volume of a material used in a wall. If you assign materials to wall components, the command will list the material definition and the total volume of the material contained in the wall as shown in the following command sequence:

```
Command: MaterialList
Select Objects: Specify opposite corner: 2 found (Select two
concrete walls.)
Select Objects: ENTER
Material Name Volume
- - - - - - - - - -
Concrete.Cast-in-Place.Flat.Grey 109 CF
```

TUTORIAL 3-1: CREATING A WALL STYLE

1. Open AutoCAD Architecture 2012 and choose **QNew** from the Quick Access toolbar. Choose **SaveAs > AutoCAD Drawing** from the Application menu and save the drawing as Lab 3-1 in your student directory.

2. Select **Style Manager** from the Style & Display panel of the Manage tab.

3. To create a new wall style, expand the Architectural Objects folder in the left pane.

4. Select the **Wall Styles** category in the left pane, as shown in Figure 3.36, right-click, and choose **New** from the shortcut menu. Overtype **Frame Brick** as the name of the new wall style.

5. Select the name **Frame Brick** in the left pane of the **Style Manager**.

6. Select the **General** tab of the right pane and type **Wood frame brick veneer** in the **Description** field.

7. In the following steps, you will create components as described in Table 3.4. To create the Wood component, select the **Components** tab and select Index 1. Select **Unnamed** in the **Name** field and type **Wood**.

TABLE 3.4 *Wall Component Setting*

Index	Name	Priority	Width	Edge Offset	Function	Dimension
1	Wood	500	3 1/2 [89]	−3 1/2 [−89]	Structural	Positive
2	Brick	810	3 5/8 [92]	1 1/2 [38]	Non-Structural	None
3	Air Gap	700	−1 [−25]	1 1/2 [38]	—	None
4	Sheathing	900	−1/2 [−13]	1/2 [13]	Non-Structural	None
5	Gypsum	1200	1/2 [13]	−4 [−102]	Non-Structural	None

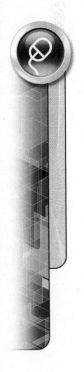

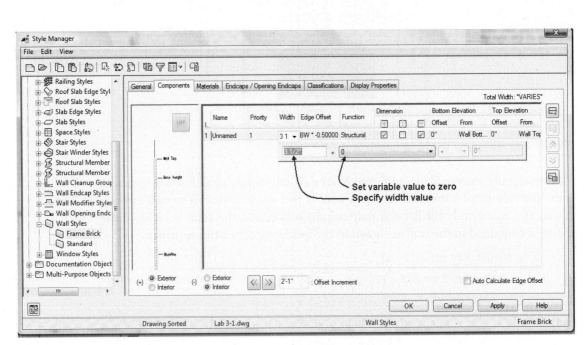

FIGURE 3.36 *Creating a new style in the Style Manager*

8. Click in the **Priority** column and type **500**.

9. Click in the **Width** column and select the arrow to expand the **Width** field. Set the fixed value to **3 1/2 [89]** and variable value to **0** as shown in Figure 3.36.

10. Click in the **Edge Offset** column and select the arrow to expand the **Edge Offset** field. Set the fixed value to **−3 1/2 [−89]** and variable value to **0**.

11. To set the function of the component, click the **Function** drop-down and choose **Structural**. Check the positive surface of the component in the **Dimension** field. Clear the **negative surface** checkbox.

12. Click the **Add Component** button in the right margin of the dialog box.

13. To create the Brick component, select **2** in the **Index** column and then click in the **Name** field and type **Brick**.

14. Click in the **Priority** column and type **810**.

15. Click in the **Width** column and select the arrow to expand the **Width** field. Set the fixed value to **3 5/8 [92]** and variable value to **0**.

16. Select in the **Edge Offset** column and select the arrow to expand the **Edge Offset** field. Set the fixed value to **1 1/2 [38]** and variable value to **0**.

17. To set the function of the component, click the **Function** drop-down list and choose **Non-Structural**. Clear all **Dimension** checkboxes.

18. Click the **Add Component** button in the right margin of the dialog box.

19. To create the Air Gap component, select **3** in the **Index** column and then click in the **Name** field and type **Air Gap**.

20. Click in the **Priority** column and type **700**.

21. Click in the **Width** column and select the arrow to expand the **Width** field. Set the fixed value to **−1 [−25]** and variable value to **0**.

22. Click in the **Edge Offset** column and select the arrow to expand the **Edge Offset** field. Set the fixed value to **1 1/2 [38]** and variable value to **0**.

23. To set the function of the component, click the **Function** drop-down list and choose **Undefined (- -)**. Clear all Dimension checkboxes.

24. To create the Sheathing component, click the **Add Component** button in the right margin. Select **4** in the **Index** column and then click in the **Name** field and type **Sheathing**.

25. Click in the **Priority** column and type **900**.

26. Click in the **Width** column and select the arrow to expand the **Width** field. Set the fixed value to **−1/2 [−13]** and variable value to **0**.

27. Click in the **Edge Offset** column and select the arrow to expand the **Edge Offset** field. Set the fixed value to **1/2 [13]** and variable value to **0**.

28. To set the function of the component, click the Function drop-down and choose **Non-Structural**. Clear all **Dimension** checkboxes.

29. Click the **Add Component** button in the right margin of the dialog box.

30. To create the Gypsum component, select **5** in the **Index** column and then click in the **Name** field and type **Gypsum**.

31. Click in the **Priority** column and type **1200**.

32. Click in the **Width** column and select the arrow to expand the **Width** field. Set the fixed value to **1/2 [13]** and variable value to **0**.

33. Click in the **Edge Offset** column and select the arrow to expand the **Edge Offset** field. Set the fixed value to **−4 [−102]** and variable value to **0**.

34. To set the function of the component, click the **Function** drop-down list and choose **Non-Structural**. Clear all **Dimension** checkboxes.

35. Verify that the **Name**, **Priority**, **Edge Offset**, and **Width** values are as shown in Figure 3.37.

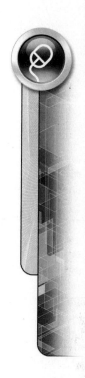

FIGURE 3.37 *Wall component settings*

36. Click **OK** to dismiss the **Style Manager**.

37. To insert an instance of the wall, choose the **Wall** tool of the **Design** palette and verify that **Frame Brick** is the current style in the **Properties** palette. Edit **Justify** in the **Properties** palette to **Baseline**.

38. Respond to the workspace prompts as follows:

> Start point or: *(Click to start the wall, move the cursor to the right, and set the polar tooltip to* **0**.*)*
>
> Endpoint or: **10' [3000]** *(Type* **10' [3000]** *in dynamic dimension to specify the length of the wall.)*
>
> Endpoint or: ENTER *(Press* ENTER *to end the command.)*

39. Choose **NE Isometric** from the **View Cube**.

40. Right-click the **Material** tool from the **Design** tool palette and choose properties, as shown in Figure 3.38. Choose the **Masonry.Unit Masonry.Brick.Modular.Running.Brown** material definition as shown in Figure 3.38. Choose **OK** to dismiss the Tool Properties dialog box.

41. Select the Material tool from the design tool palette. Select the **Brick** wall component, right-click, and choose **this Object**.

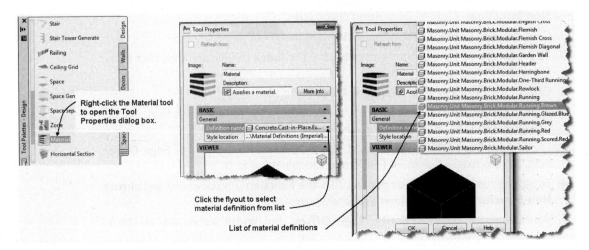

FIGURE 3.38 *Assigning materials to a wall component*

42. The material is assigned to the Brick component as shown in Figure 3.39. Save and close the drawing.

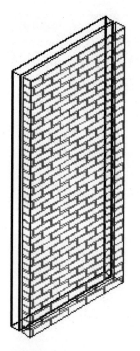

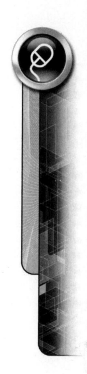

FIGURE 3.39 *Materials applied to the Brick component*

EXPORTING AND IMPORTING WALL STYLES

Wall styles can be applied using the i-drop tool from the Design Tool Catalog–Imperial or Design Tool Catalog–Metric. The wall styles that have been applied with the i-drop tool or selected from the Walls tool palette are displayed within the Wall Style folder of the Style Manager. The wall styles of the Design Tool Catalog–Imperial and Design Tool Catalog–Metric are located in the Metric and Imperial folders of *Vista-C:\ProgramData\Autodesk\ACA 2012\enu\Styles* directory.

The commands of the Style Manager toolbar shown in Figure 3.7 are useful in importing and exporting styles. The **New Drawing, Open Drawing, Copy**, and **Paste** commands are used to import and export wall styles. The New Drawing button opens the **New Drawing** dialog box to create a new file for placing wall styles. The Open Drawing button opens the **Open Drawing** dialog box, allowing you to open files other than the existing file. The **Copy** and **Paste** commands of the toolbar allow you to copy a style from one drawing and paste it in another drawing in the left pane. If you have created a style in a drawing that is opened, the drawing will be listed in the left pane of the Style Manager. To copy or export a file to another drawing, copy the style to the clipboard and paste it into other drawings within the left pane. In addition to using the Copy and Paste buttons, you can click on a style, drag it to another drawing name, and release it in the left pane over the file name. You can drag styles from the Style Manager to tool palettes to import the style to other drawings.

If wall styles of a drawing are selected for export to a file that contains the same named wall styles, the **Import/Export-Duplicate Names Found** dialog box will open as shown in Figure 3.40. This dialog box displays the duplicate names. There are three options: **Leave Existing, Overwrite Existing**, and **Rename To Unique**.

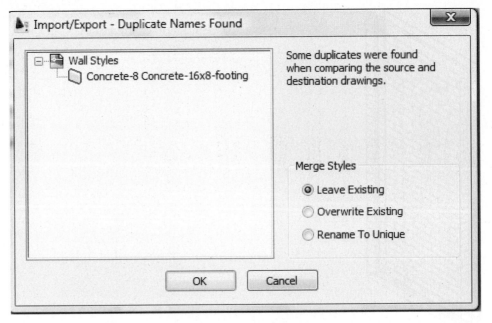

FIGURE 3.40 *Import/Export-Duplicate Names Found dialog box*

The **Leave Existing** option does not replace the wall style of the destination drawing with that of the source file. The **Overwrite Existing** option revises the wall style definition of the destination file with that of the wall style of the source file. The **Rename To Unique** option appends a number such as 2, 3, or 4 to the wall style name from the source file to distinguish it from the wall style of the destination drawing.

Steps to Import a Wall Style in the Style Manager

1. Select **Style Manager** from the Style & Display panel of the Manage tab in the ribbon.
2. Click the **Open Drawing** button on the **Style Manager** toolbar.
3. Select the source drawing that includes the desired wall style from the **Open Drawing** dialog box.
4. Select the source drawing in the left pane and expand the **Architectural Objects\ Wall Styles** folder.
5. Select the name of the desired wall style of the source file in the left pane, right-click, and choose **Copy**.
6. Select the name of the target file or the current drawing in the left pane, right-click, and choose **Paste**.

The wall style is imported into the target or current drawing. The procedure to import wall styles can be applied to exporting styles to other drawings by copying the wall style to the clipboard in the Style Manager of the current drawing and pasting the style into the target drawing.

PURGING WALL STYLES

Wall styles can also be purged from a drawing to decrease the list of wall styles in the Style Manager. Only wall styles that have not been used in the drawing can be deleted. Wall styles can be purged by opening the **Style Manager**, expanding the **Wall Styles** folder of the current drawing in the left pane, selecting the Wall Styles category, right-clicking, and choosing **Purge** from the shortcut menu. The **Select Wall Styles** dialog box opens as shown in Figure 3.41.

This dialog box lists all the wall styles of the drawing that have not been used. You can deselect a wall style for purging by clearing the checkbox for that wall style in the **Select Wall Styles** dialog box. Click the **OK** button, and all unused wall styles will be deleted from the drawing.

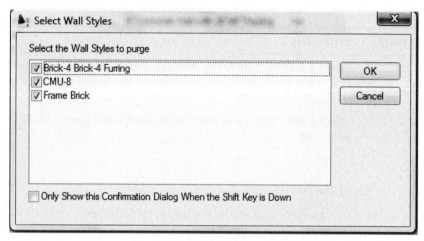

FIGURE 3.41 *Wall styles listed in the Select Wall Styles dialog box*

IDENTIFYING WALL STYLES USING ROLLOVER TOOLTIPS

The object name, object style name, layer, and style description are displayed in a rollover tooltip that is displayed when you hover the cursor over an architectural object. The rollover tooltip feature is turned on as a default setting in Options. The content displayed in the rollover tooltip can be modified in the Customize User Interface dialog box. Access the **Rollover Tooltip** command as shown in Table 3.5 and in Figure 3.42.

TABLE 3.5 *Rollover Tooltip command access*

Command prompt	OPTIONS
Shortcut Menu	Right-click over the command line and choose Options. Check Show rollover ToolTips checkbox of the Display tab of the Options dialog box.

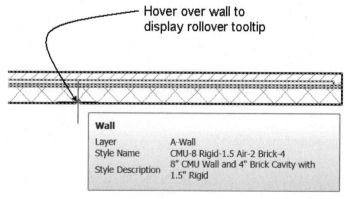

FIGURE 3.42 *Rollover Tooltip*

EXTENDING A WALL WITH ROOFLINE AND FLOORLINE

The **RoofLine** and **FloorLine** commands allow you to extend the wall beyond the base height and baseline. The commands can be useful in creating walls that extend to other objects such as gables or stairs. The **Offset** option of the command allows you to extend a wall component uniformly a specified distance along its length. Access the **RoofLine** command as shown in Table 3.6.

TABLE 3.6 *RoofLine command access*

Command prompt	ROOFLINE
Ribbon	Select a wall to display the Wall tab and choose Modify Roof Line from the Roof/Floor Line flyout of the Modify panel as shown in Figure 3.43.
Shortcut menu	Select a wall, right-click, and choose Roof/Floor Line > Modify Roof Line.

The options of the **RoofLine** command include the following:

Offset—Allows you to specify the distance the wall is projected up.

Project to polyline—Allows you to select the walls and project the walls to the intersection of a selected polyline. The polyline is used as the boundary instead of the roof object.

Generate polyline—Creates a polyline from the wall; the polyline is created at the top of the wall.

Auto project—Projects the wall to the roof, roof slab, or stair.

Reset—Removes previous modifications to the roof line or floor line.

The **Offset** option extends the wall the specified offset distance toward the roof. If the brick component Top Elevation is equal to 0 from Wall Top and the Top Elevation of other components are restrained to 0 from Base Height, the brick can be extended up beyond the base height as shown in Figure 3.43. The **RoofLine** command and restraints specified in the wall style allow you to offset the brick veneer component above the framing components to cover the floor system when the height needed for a floor system is determined. To modify a roof line, select a wall, right-click, choose **Roof/Floor Line > Modify Roof Line**, choose the **Offset** option in the workspace, and then respond to the workspace prompt as follows:

```
Enter offset <0>: 12 [300] ENTER
(Press ENTER to end the command.)
```

Extending the Wall with FloorLine

The **FloorLine** command can be used to extend the wall up or down about the baseline of the wall. In Figure 3.43, the wall was extended down below the baseline to create the footing. A negative offset of the floor line will extend the wall down below the baseline. The floor line extends the wall down to create the wall bottom. Access the **FloorLine** command as shown in Table 3.7.

TABLE 3.7 *FloorLine command access*

Command prompt	FLOORLINE
Ribbon	Select a wall to display the Wall tab and choose Modify Floor Line from the Roof/Floor Line flyout of the Modify panel as shown in Figure 3.43
Shortcut menu	Select a wall, right-click, and choose Roof/Floor Line > Modify Floor Line.

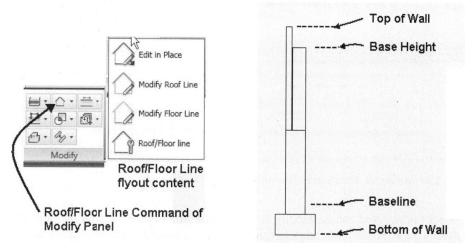

FIGURE 3.43 *Wall components extended with RoofLine and FloorLine*

The base height of the wall specified in the B-Base height field in the Design tab of the Properties palette is the distance between the baseline and the base height features of a wall. The FloorLine command allows you to extend wall components below the baseline to represent the footing or wall bottom, as shown in the following sequence.

To modify a floor line, select a wall to display the Wall tab, choose **Modify Floor-Line** from the **Roof/Floor Line** flyout of the **Modify** panel, choose the **Offset** option in the workspace, and then respond to the workspace prompt as follows:

```
Enter offset <1'-0"> [300]: -16 [-400] ENTER (Specifies the
offset distance down.)
(Press ENTER to end the command.)
```

TUTORIAL 3-2: CREATING A WALL STYLE FOR ROOFLINE EDIT

1. Verify that the Accessing Tutor content for Chapter 3 of the CengageBrain has been downloaded to your Accessing Tutor folder on your computer as described in Organizing Tutorial Directories in the Preface.

2. Open *Accessing Tutor\Imperial\Ch 3\Ex 3-2.dwg* or *Accessing Tutor\Metric\Ch 3\Ex 3-2.dwg*. Choose **Save As** > **AutoCAD Drawing** from the **Application** menu and save the drawing as **Lab 3-2** to your student directory.

3. Select the Walls tool palette. Move the cursor over the **CMU-8 Rigid-1.5 Air-2 Brick-4 Furring** [CMU-190 Rigid-038 Air-050 Brick-090 Furring] tool, right-click, and

choose **Apply tool properties to > Wall**. Select the wall of the drawing. Press ENTER to end the command.

4. Select **West** from the **View Cube**.

5. Choose the wall and choose **Edit Style** from General panel in the **Wall** tab.

6. Choose the **Components** tab of the Wall Style Properties dialog box.

7. In this step, you will modify the top elevation of the components to restrain all components except the Brick veneer component. Edit the **Top Elevation** to **Base Height** for the following components as shown in Figure 3.44: **Air Gap**, **Rigid Insulation**, **CMU, Stud, and GWB**. The **Base Height** setting will restrict the components to the height of the wall when the **RoofLine** command is applied to the wall. Click **OK** to dismiss the **Wall Style Properties** dialog box.

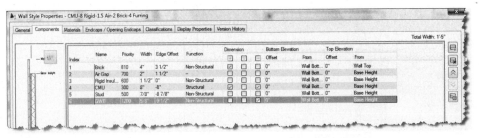

FIGURE 3.44 *Setting top elevations to Base Height for wall components*

8. Select the wall and choose Modify Roof Line from the Roof/Floor Line flyout of the Modify panel in the Walls tab. Choose the Offset option from the workspace option list. Respond to the workspace prompts as follows:

```
Enter offset: 12 [300] ENTER
```

(Press Esc to end the command. RoofLine edit applied to wall as shown in Figure 3.45.)

9. Save and close the drawing.

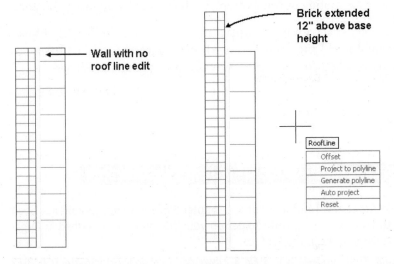

FIGURE 3.45 *Brick component extended with RoofLine command*

CREATING WALL ENDCAPS

When openings are created in a wall or when a wall terminates, the treatment of the ends is controlled through **Endcap** styles as shown in Figure 3.46. **Wall Endcap** styles enclose the end of each component of the wall and **Wall Opening Endcap** styles close the edges of walls where doors, windows, or openings are placed. The **Wall Opening Endcap** styles can be assigned to the sill, start jamb, end jamb, and header of the opening. The Wall Endcap and Wall Opening Endcaps are defined with the wall style. The Standard wall style shown at A in Figure 3.46 uses a Standard wall endcap style and a Standard opening endcap style. The Standard endcap styles are simple straight lines connecting the two faces of the wall. The endcap style for the Concrete-8 Concrete-16×8-footing wall style is shown at B in Figure 3.46. Endcap geometry is applied for each of the wall components. Therefore, six endcaps have been applied to the CMU-8 Rigid-1.5 Air-2 Brick-4 Furring wall style as shown at C in Figure 3.46.

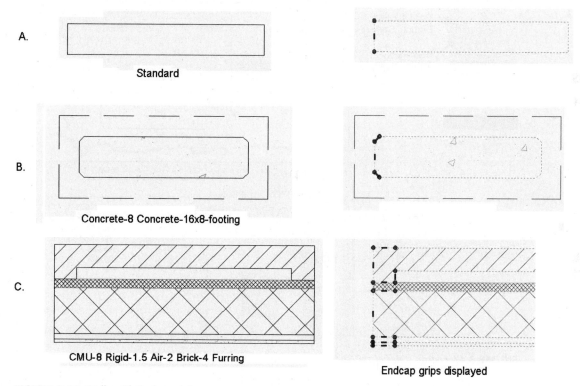

A. Standard

B. Concrete-8 Concrete-16x8-footing

C. CMU-8 Rigid-1.5 Air-2 Brick-4 Furring

Endcap grips displayed

FIGURE 3.46 *Walls with Endcap styles*

You can view the properties on a wall endcap style in the Style Manager as shown in Figure 3.47. The tools to edit and creating an endcap are located on the Endcap panel in the Wall tab for a selected wall. Access the Edit Wall Endcap Style command as shown from the ribbon in Table 3.8.

TABLE 3.8 *WallEndcapStyle command access*

Command prompt	**WALLENDCAPSTYLE**
Ribbon	Select a wall to display the Wall tab and choose Edit Wall Endcap Style from the Endcap panel as shown in Figure 3.47.

The endcap treatment around the opening of doors and windows can be edited when you choose the Edit Opening Endcap Style command as shown in Table 3.9. Wall endcap styles and Wall Opening Endcap Styles are listed in the Style Manager for walls of the current drawing as shown in Figure 3.47.

TABLE 3.9 *OpeningEndcapStyleEdit command access*

Command prompt	**OpeningEndcapStyleEdit**
Ribbon	Select a wall to display the Wall tab and choose Edit Opening Endcap Style from the Endcap panel. Next, choose a door, window, opening, or door/window assembly within the wall.

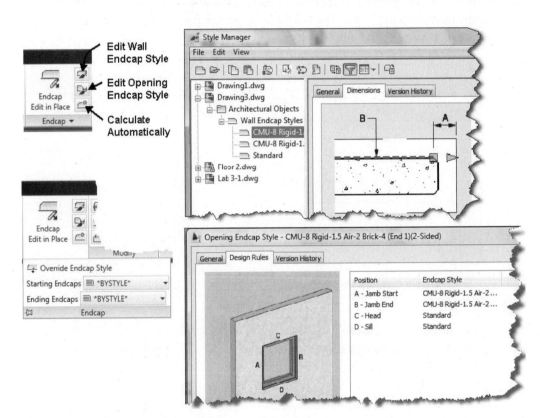

FIGURE 3.47 *Endcap styles of the Style Manager dialog box*

Creating Wall Endcap Styles

Wall endcap styles can be created and edited in the Style Manager or in the drawing editor using Calculate Automatically. Creating an endcap style in the Style Manager is awkward since the polyline is drawn independent of the wall. Previous versions of AutoCAD Architecture allowed the polyline to be created as a sketch that was scaled and rotated to create the endcap. The Calculate Automatically tool simplifies the procedure since it allows you to draw the polyline in the workspace in the shape and size necessary to cap each wall component. The polyline should contact the wall component and include the specific shape and size necessary to cap the wall component.

Access **Calculate Automatically** as shown in Table 3.10.

TABLE 3.10 *Automatically calculate endcap from polyline command access*

Command prompt	WALLAUTOENDCAP
Ribbon	Select a wall to display the Wall tab and choose the Calculate Automatically command from the Endcap panel.
Shortcut menu	Select a wall and then select Endcaps > Calculate Automatically.

When you select a wall, the Wall tab is displayed. Choose **Calculate Automatically** from the Endcap panel of the Wall tab. You are prompted as shown in the following workspace sequence:

> Select polylines: *(Select a polyline at p1 as shown in* Figure 3.48.*)* 1 found
>
> Select polylines: ENTER *(Press* ENTER *to end selection of polylines and open the Wall Endcap Style dialog.)*
>
> The highlighted polyline can be applied to more than one component.
>
> Press ENTER to apply to highlighted component or Tab to go to next possible component: *(Press* ENTER *to apply the endcap to the component shown at p2.)*
>
> *(Edit the Wall Endcap Style dialog box as shown in Figure 3.48.)*

Before you select **Calculate Automatically**, you must draw in the workspace a polyline that you want to use as an endcap. The first prompt of the command is to select the polyline for the new endcap application. Viewing the wall in isometric allows you to view the highlight of the polyline and the wall components. You can press Tab to toggle through the wall components. The Wall Endcap Style dialog box opens and provides you with the option to name and erase the layout geometry. The Wall Endcap Style dialog box also includes toggles to apply the endcap style as the default for wall style or as an override to the specified end of the wall.

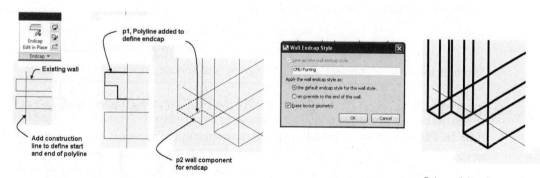

FIGURE 3.48 *Creating an endcap style using Calculate Automatically*

The selected polyline is assigned to a wall component in the wall style definition. The start and end of the polyline should align with a similar x coordinate as shown in Figure 3.49. If the polyline has different coordinates, the cap will, in essence, bend the end of the wall when the endcap style is attached.

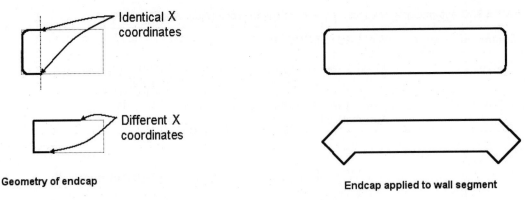

Geometry of endcap

Endcap applied to wall segment

FIGURE 3.49 *Application of endcaps to walls*

If a wall consists of two wall components, draw two polylines—one for each wall component. The index number of the wall component is linked to the polyline selected in the process of defining wall endcap styles.

When wall endcaps are edited for a wall, the results are displayed in the Advanced section in the Design tab of the Properties palette. You can select from the endcap drop-down list in the Advanced section in the Design tab of the Properties palette to create an override for the endcap. The endcaps listed are the endcap styles used by the walls inserted in the drawing or created in the Style Manager.

Using Edit In Place and Override Endcap Style

The endcap of a wall can be edited in the workspace using the commands of the **Endcap** panel of the Wall tab. The **Edit In Place** and **Override Endcap Style** allow you to customize the application of the endcap styles. The **Edit In Place** option provides access to commands for editing the endcap and using grips to edit the endcap. Access **Edit In Place** as shown in Table 3.11.

TABLE 3.11 *WallEndcapEdit command access*

Command prompt	**WALLENDCAPEDIT**
Ribbon	Select a wall to display the Wall tab and choose Endcap Edit in Place from the Endcap panel shown at left in Figure 3.50.
Shortcut menu	Select a wall, right-click, and choose Endcaps > Edit In Place.

When you select a wall, the Wall contextual tab is displayed. Next, choose Endcap **Edit In Place** from the Endcap panel; you are prompted to select near the endcap to specify the endcap for editing. After you select a point near the endcap, the grips of the endcap are displayed and the Edit In Place: Wall Endcap tab is displayed as shown in Figure 3.50. The Profile panel of this tab consists of tools to modify the profile used in the endcap style as follows:

> **Zoom To**—Executes the **Zoom** command to enlarge the view of the endcap.
>
> **Add Vertex**—Executes the **InplaceEditAddVertex** command. You are prompted to select a point on the endcap to create an additional vertex.
>
> **Remove Vertex**—Executes the **InplaceEditRemoveVertex** command. You are prompted to select a point to remove the vertex.

Hide Edge—Executes the **InplaceEditHideEdge** command. You are prompted to select an edge to hide in the plan view of the workspace. The selected profile segments are turned off.

Show Edge—Executes the **InplaceEditShowEdge** command. You are prompted to select an edge of profile segments that were turned off with the **InplaceEditHideEdge** command to display.

Replace Endcap—Executes the **InplaceEditReplaceEndcap** command. You are prompted to select an existing profile in the drawing to be used as the new endcap.

Remove Endcap—Executes the **InplaceEditRemoveEndcap** command, removing the endcap of the wall.

The Edits panel consists of the following commands:

Save As—Executes the **InplaceEditSaveAs** command to save the editing of the endcap using a new name in the Wall Endcap Style dialog box shown in Figure 3.50.

Cancel—Executes the **InplaceEditDiscard** command to exit the editing of the profile without saving any changes.

Finish—Executes the **InplaceEditSaveAll** command to save the edited profile as a new style or updates the existing style in the **Wall Endcap In-Place Edit – Save Changes** dialog box.

You can select the grips and stretch the grips of the profile to change the shape of the endcap. When the grips are displayed, right-click and the shortcut menu that is displayed includes the following additional options for editing the vertices:

Add Vertex—Executes the **InplaceEditAddVertex** command. You are prompted to select a point to add in the workspace. The existing profile will be extended to include the vertex.

Remove Vertex—Executes the **InplaceEditRemoveVertex** command. You are prompted to select a point to remove in the workspace.

The AEC Modify Tools are included in the Modify panel. When the grips of a profile are displayed, the shortcut menu and the Modify panel include the following AEC Modify Tools:

Fillet—Executes the **AecFillet** command during the Inplace Edit, which allows you to add a radius to the lines defining the endcap.

Chamfer—Executes the **AecChamfer** command during the Inplace Edit.

Merge—Executes the **AecLineworkMerge** command, allowing you to merge a closed polyline into the endcap geometry.

Subtract—Executes the **AecLineworkSubtract** command, which allows you to subtract a closed polyline shape from the existing endcap geometry during the inplace edit.

Trim—Executes the **AecLineworkTrim** command. When you execute the command, you are prompted to specify two points to define a trim line to remove a section of the endcap.

Extend—Executes the **AecLineworkExtend** command, which allows you to extend the endcap geometry to a boundary edge defined by linework that is added to the drawing beyond the endcap.

When you select options from the shortcut menu, the grips are cleared, and you are prompted to edit the polyline. Prior to saving or discarding the edit, you can select the endcap to display the grips and edit the grips or make additional selections from the shortcut menu.

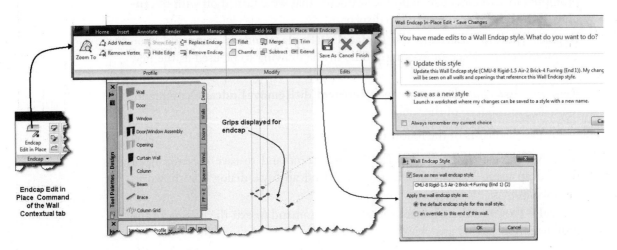

FIGURE 3.50 *Editing endcap style with Edit In Place*

Using Override Endcap Style

The **Override Endcap Style** command allows you to select from a list of endcaps to use at the end of a wall. The default endcap specified in the style is defined in the **Endcaps/Opening Endcaps** tab of the **Wall Style Properties** dialog box. The **Override Endcap Style** command allows you to change the endcap used at one end of the wall. Access the **Override Endcap Style** command as shown in Table 3.12.

TABLE 3.12 *Endcap Style command access*

Command prompt	**WALLAPPLYENDCAP**
Ribbon	Select a wall to display the Wall tab, expand the slide out panel of Endcap, and choose Override Endcap Style.
Shortcut menu	Select a wall, right-click, and choose Endcaps > Override Endcap Style.

When you select a wall, choose **Override Endcap Style** from the slide out panel of Endcap; you are prompted to select a point near the endcap for edit. Selecting the point specifies the wall endcap for edit and opens the **Select an Endcap Style** dialog box, as shown in Figure 3.51. Select a style from the dialog box and the endcap is changed.

FIGURE 3.51 *Select an Endcap Style dialog box*

Modifying Wall Component in Place

The WallCleanUpEdit command allows you to edit the profile of wall components when they intersect other walls. The command provides convenient wall component edits that are not global. You can merge components if the components share the same priority. Access the WallCleanupEdit command as shown in Table 3.13.

TABLE 3.13 *WallCleanupEdit command access*

Command prompt	**WALLCLEANUPEDIT**
Ribbon	Select a wall to display the Wall tab and choose Cleanup Edit in Place from the Cleanup panel shown in Figure 3.52.

When you choose the **Cleanup Edit in Place** tool, the Edit in Place: Wall Cleanup tab opens and the grips of a wall component are displayed. You can edit the selected wall component or press ESC to clear selection and then select a component for edit. The selected wall component can be edited using the commands of the Modify or Profile panels shown in Figure 3.52.

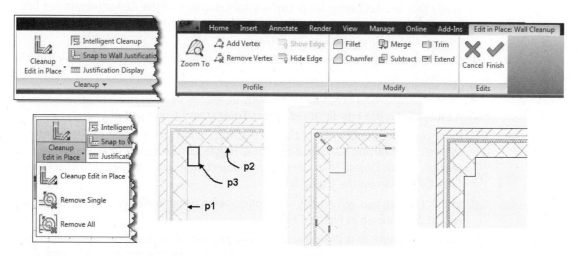

FIGURE 3.52 *Accessing the Edit in Place: Wall Cleanup command*

The **WallCleanupEdit** command allows you to merge a component to a closed polyline. The following workspace prompts were applied to modify the wall components shown in the following workspace prompts and at right in Figure 3.52.

(Select walls at p1 and p2, choose Cleanup Edit in Place from the Wall tab.)

(Select the CMU components at p1 and p2 to display the grips.)

(Choose Merge from the Modify panel of Edit in Place: Wall Cleanup tab.)

Select Object(s) to merge or None to pick rectangle. (Select the polyline at p3.)

Select Object(s) to merge or None to pick rectangle. (Press ENTER to end selection.)

Erase selected linework? [Yes/No] <No>: Y (Enter Y to choose the Yes option.)

(Choose Finish from the Edits panel of Edit in Place: Wall Cleanup tab.)

Inserting an Endcap as a Polyline for Edit

If an existing endcap style needs to be modified, the **WallEndcap** command can be used to insert the polyline of the endcap style in the drawing for editing. This command is selected as shown in Table 3.14.

TABLE 3.14 *WallEndcap command access*

Command prompt	WALLENDCAP

When this command is typed in the command line, the command has two options: **Styles** and **Pline**. When you select the Pline option, the Endcap Styles dialog opens, allowing you to select and insert the endcap in the drawing. The Pline option is useful since the polyline is inserted in the drawing for the edit and future applications of endcap styles.

TUTORIAL 3-3: CREATING AND EDITING ENDCAPS WALL COMPONENTS

1. Verify that the Accessing Tutor content for Chapter 3 of the CengageBrain has been downloaded to your Accessing Tutor folder on your computer as described in Organizing Tutorial Directories in the Preface.

2. Open \Accessing Tutor\Imperial\Ch3\Ex3-3.dwg or Accessing Tutor\Metric\Ch3\Ex 3-3.dwg.

3. Save the drawing as **Lab 3-3** in your student directory.

4. If tool palettes are not displayed, select **Tools** from the Home tab.

5. Right-click over the **OSNAP** option on the Application status bar and choose Settings to open the Drafting Settings dialog box. Toggle ON Endpoint, Intersection, and Extension. Check Object Snap On (F3) and Object Snap Tracking On (F11) checkboxes. Click OK to dismiss the dialog box.

6. In this tutorial, you will modify wall components using **WallCleanupEdit**. Select the wall at p1 as shown in Figure 3.53.

7. Choose Cleanup Edit in Place from the **Cleanup** panel of the Wall Contextual tab to open the Edit in Place: Wall Cleanup tab. Respond to the workspace prompts as follows:

   Pick a point at wall joint to edit wall cleanup results: (Select the wall at p1 in Figure 3.53.)

8. Choose **Trim** from the Modify panel of the Edit in Place: Wall Cleanup tab. Respond to the workspace prompts as follows:

   Select the first point of the trim line or ENTER to pick on screen. (Select the wall at p2 shown in Figure 3.53.)

   Select the second point of the trim line. (Select the wall at p3 shown in Figure 3.53.)

   Select the side to trim. (Select a point near p4 shown in Figure 3.53.)

   Choose **Finish** from the Edits panel to complete the edit.

9. In this step, you will begin the process of merging a component to a closed polyline. Select the wall at **p5** as shown in Figure 3.53. Choose Cleanup Edit in Place from the Cleanup panel. To change the component for edit press ESCAPE to clear selection and select the brick component at **p5** as shown in Figure 3.53. Verify the Edit in Place grips are now displayed for the brick component.

10. Choose Merge from the Modify panel. Respond to the workspace prompts as follows:

    Select object(s) to merge or None to pick rectangle. *(Select the closed polyline at p6 as shown in Figure 3.53.)*

    Select object(s) to merge or None to pick rectangle: ENTER. *(Press ENTER to end selection.)*

    Erase selected linework? Y. *(Choose Yes. Choose **Finish** of the Edits panel in the Edit in Place: Wall Cleanup contextual tab.)*

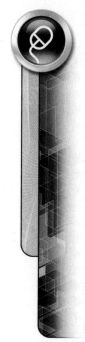

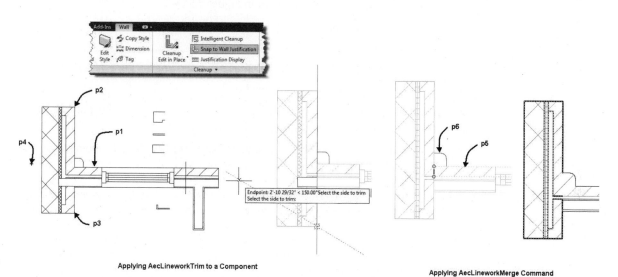

Applying AecLineworkTrim to a Component

Applying AecLineworkMerge Command

FIGURE 3.53 *Edit in Place of wall components*

11. In the next steps, you will create an endcap. The drawing consists of four polylines that will be applied to the sample wall.

12. Verify that no entity or object is selected; right-click over the workspace and choose Basic Modify Tools > Move from the shortcut menu. Respond to the workspace prompts as follows:

    Command: _move

    Select objects: 1 found (Select the polyline at p1 as shown in Figure 3.54.)

    Select objects: (Press ENTER to end selection.)

    Specify base point or [Displacement] <Displacement>: (Select the polyline at p1 using the endpoint object snap as shown in Figure 3.54.)

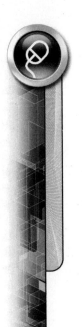

Specify second point or <use first point as displacement>: (Select a point using the intersection object snap at p2 as shown in Figure 3.54.)

(Press the space bar to repeat the last command.)

Command: _move

Select objects: 1 found (Select the polyline at p3 as shown in Figure 3.54.)

Select objects: (Press ENTER to end selection.)

Specify base point or [Displacement] <Displacement>: (Select the polyline at p3 using the endpoint object snap as shown in Figure 3.54.)

Specify second point or <use first point as displacement>: (Select a point using the intersection object snap at p4 as shown in Figure 3.54.)

(Press the space bar to repeat the last command.)

Command: _move

Select objects: 1 found (Select the polyline at p5 as shown in Figure 3.54.)

Select objects: (Press ENTER to end selection.)

Specify base point or [Displacement] <Displacement>: (Select the polyline at p5 using the endpoint object snap as shown in Figure 3.54.)

Specify second point or <use first point as displacement>: (Select a point using the intersection object snap at p6 as shown in Figure 3.54.)

(Press the space bar to repeat the last command.)

Command: _move

Select objects: 1 found (Select the polyline at p7 as shown in Figure 3.54.)

Select objects: (Press ENTER to end selection.)

Specify base point or [Displacement] <Displacement>: (Select the polyline at p7 using the endpoint object snap as shown in Figure 3.54.)

Specify second point or <use first point as displacement>: (Select a point using the intersection object snap at p8 as shown in Figure 3.54.)

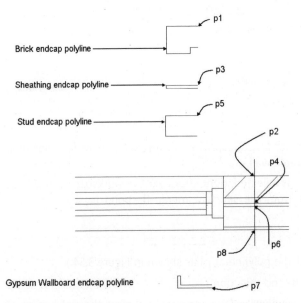

FIGURE 3.54 *Polylines for wall component endcaps new*

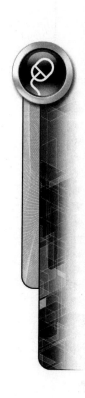

13. Select the wall at p1 as shown in Figure 3.55 and then select Calculate Automatically from the Endcap panel. Respond to the workspace prompts as follows:

Select polylines: (Select a point near p2 making a crossing window to p3 as shown in Figure 3.55.)

Specify opposite corner. ENTER (Press ENTER to end selection.)

(Type FrameBrick in the Save as new opening endcap style and the Save as new wall endcap style field. Select the radio buttons and checkboxes as shown in Figure 3.55.)

Select OK to dismiss the Opening Endcap Style dialog box.

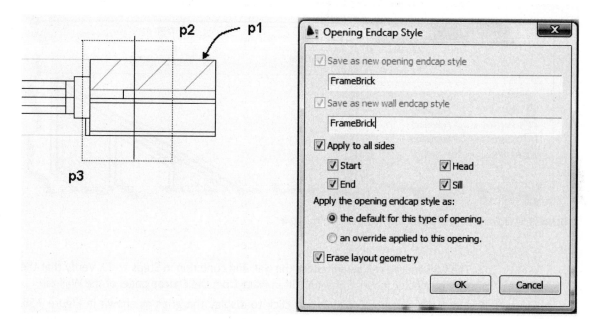

FIGURE 3.55 *Polylines selected for FrameBrick endcap style*

14. The opening endcap style is applied as shown in Figure 3.56.

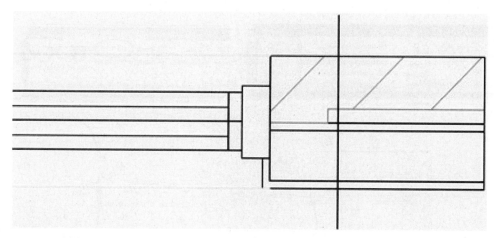

FIGURE 3.56 *Opening endcap style applied to wall*

15. To apply the new endcap style to the end of the wall and other opening applications, select the wall as shown in Figure 3.56 and choose Edit Style from the Wall tab.

16. Choose the Endcaps/Opening Endcaps tab. Select the down arrow of the Wall Endcap Style field and choose FrameBrick as shown in Figure 3.57. Choose FrameBrick from the Opening Endcap Style drop-down for the Door, Door/WindowAssembly, and Opening fields as shown in Figure 3.57. Select OK to dismiss the dialog box.

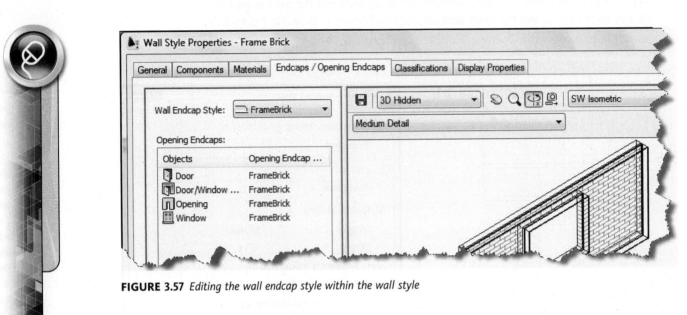

FIGURE 3.57 *Editing the wall endcap style within the wall style*

17. The wall endcap will be edited for the wall end condition in Steps 17–19. Verify that the wall is selected; choose Endcap Edit in Place from the Endcap panel of the Wall tab.

18. Move the cursor to point A and click to display the grips as shown in Figure 3.58. Choose Trim from the Edit In Place: Wall Endcap tab. Respond to the command prompts as follows:

Select the first point of the trim line or ENTER to pick on screen: (Select the end of the wall at p2.)

Select the second point of the trim line: (Select the end of the wall at p1.)

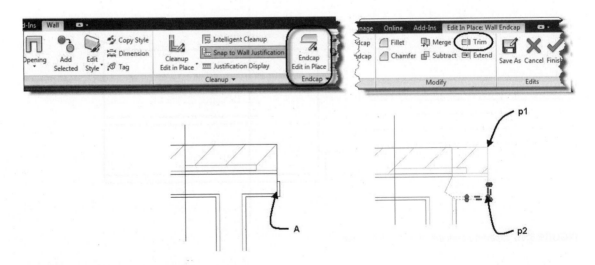

FIGURE 3.58 *In-Place Edit of wall endcap style new*

19. Choose the **Save As** from the Edits panel of the Edit In Place Wall Endcap tab. Respond to the workspace prompts as follows:

Select an in-place edit profile: (Choose the endcap at p2 as shown in Figure 3.58.)

The Wall Endcap Style dialog opens; check Save as new wall endcap style. Overtype FrameBrickEnd as the name of the new style. Verify that the default endcap style for this wall style radio button is checked as shown in Figure 3.59.

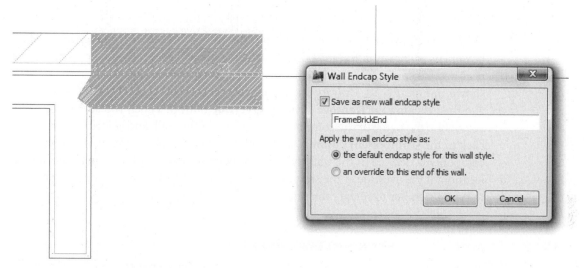

FIGURE 3.59 *Endcap style applied to wall*

20. Choose OK to dismiss the dialog box.
21. Choose View tab. Choose SW Isometric from the View Cube.
22. Choose Realistic from the Visual Styles flyout of the Visual Styles panel.
23. The opening endcap does not display in the model view. To display the endcap in model view, select the wall near the window and choose Edit Style from the General panel of the Wall tab.
24. Choose the Display Properties tab. Verify that the Model display representation is current. Check the Style Override checkbox. Choose the Other tab of the Display Properties (Wall Style Override – Frame Brick) – Model dialog box. Choose the *Display Opening Endcaps* checkbox. Choose OK to dismiss all dialog boxes.
25. Press ESC to end selection of wall. Save and Close the drawing.

OVERRIDING THE PRIORITY OF COMPONENTS IN A WALL STYLE

The priority of the wall components can be edited as an exception to the priorities defined in the wall style. **Wall Style Overrides** are listed in the Advanced section in the Design tab of the Properties palette shown in Figure 3.60. You can edit the wall priorities from the Properties palette or select WallPriorityOverrides from the ribbon as shown in Table 3.15.

TABLE 3.15 *WallEndcap command access*

Command prompt	WALLPRIORITYOVERRIDES
Ribbon	Select a wall and expand the slide out for the Cleanup panel to choose Priority Override.
Properties palette	Click in the Priority Overrides button of the Style Overrides section in the Advanced section on the Design tab of the Properties palette.

When you choose Priority Override from the ribbon, the **Priority Overrides** dialog box opens as shown in Figure 3.60. This dialog box will be empty if no overrides have been set. To create an override, click the **Add Priority Override** button at the right as shown in Figure 3.60. Priority overrides can be removed by selecting the component and then clicking the **Remove Priority Override** button. When you add an override, you define the name of the component, location on the wall, and priority value. If you click in the **Component column,** a drop-down list will display the names of the wall components defined in the wall style. The **Override** column allows you to specify the location of the override at the start or end of the wall. The **Priority** column allows you to type a number to specify the priority for the wall component.

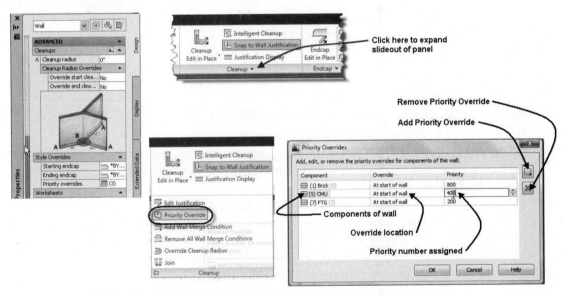

FIGURE 3.60 *Endcap Override listed in the Design tab and Priority Override Dialog box*

CREATING WALL MODIFIERS AND STYLES

Wall modifiers can be created to add projections to a wall at the beginning, end, or at a specified distance from the beginning, end, or midpoint of a wall. Wall modifiers can consist of rectangular projections or the shape defined by a polyline saved as a wall modifier style. You can create pilasters or other decorative projections of the wall using wall modifier styles. Wall modifiers are placed on existing walls by clicking the **Plan modifier** button of the Advanced section in the Design tab of the Properties palette for a selected wall. Wall modifiers can also be added in the workspace from the shortcut menu of a selected wall or the ribbon. The Plan Modifier flyout of the

Modify panel for a selected wall includes the following wall modifier options: **Add, Edit Profile in Place, Convert Polyline to Wall Modifier, Remove, Plan Modifier,** and **Edit Wall Modifier Style** as shown in Figure 3.61. Access the **WallModifierAdd** command as shown in Table 3.16.

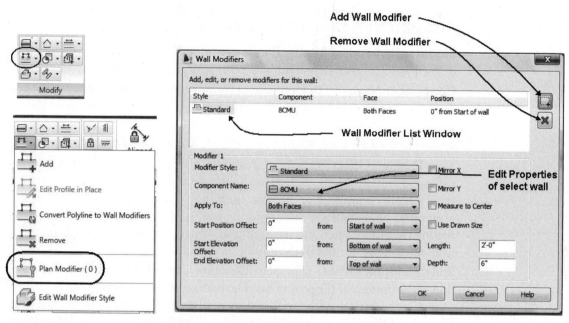

FIGURE 3.61 *Wall Modifiers dialog box*

TABLE 3.16 *WallModifierAdd command access*

Command prompt	**WALLPLANMODIFIERS**
Ribbon	Select a wall to display the Wall tab, and expand Plan Modifiers flyout from the Modify panel. Choose Plan Modifier button as shown in Figure 3.61.
Properties palette	Click the Plan Modifiers worksheet of the Advanced section in the Design tab of the Properties palette.

You can insert a wall modifier by clicking the **Plan Modifier** button on the ribbon, to open the **Wall Modifiers** dialog box as shown in Figure 3.61. This dialog box allows you to specify the style, wall component, and location of the wall modifier.

The top portion of the **Wall Modifiers** dialog box lists the wall modifiers that are currently attached to the selected wall. The bottom section of the dialog box is designed to edit the modifier that is selected in the top portion. The options of the **Wall Modifiers** dialog box include:

Add Wall Modifier—The Add Wall Modifier button activates the dialog box, allowing you to assign a wall modifier style to the selected wall. The Standard wall modifier style will be used as the default unless other wall modifier styles are selected and defined in the drawing.

Remove Wall Modifier—If you select a wall modifier and then click the Remove Wall Modifier button, the wall modifier will be removed.

Wall Modifier List window—The Wall Modifier List Window lists the wall modifiers that are defined for the selected wall.

The lower portion of the dialog box is used to edit properties and conditions of attachment to the wall. The purpose of each of the fields of the lower portion of the **Wall Modifiers** dialog box is described as follows:

Modifier Style—The Modifier Style drop-down list allows you to select from defined wall modifier styles that have been created in the drawing.

Component Name—The Component Name drop-down list allows you to specify the wall component to apply the wall modifier.

Apply To—The Apply To drop-down list allows the wall modifier style to be applied to the Left Face, Right Face, or Both Faces of the wall.

Start Position Offset—The Start Position Offset field allows you to set the position horizontally along the wall to locate the wall modifier. The distance entered can be a positive or negative number. The **from** list allows the start position to be established relative to the wall start, wall end, or wall midpoint. Setting the distance to 0 and measuring the distance relative to the wall midpoint will locate the projection in the middle of the wall.

Start Elevation Offset—The Start Elevation Offset field establishes the elevation of the bottom of the projection. The bottom of the projection can be defined to start at the bottom of the wall or at some distance from the bottom. The **Start Elevation Offset** distance and the **from** list allow the distance to be defined from the wall top, wall base height, wall baseline, or wall bottom.

End Elevation Offset—The End Elevation Offset section establishes the elevation of the top of the projection. The **End Elevation Offset** distance and the **from** list define the distance from the wall top, wall base height, wall baseline, or wall bottom for the top of the projection.

Note that if the start elevation offset and end elevation offset are set to the same distance and reference surface, the modifier will become invisible because the projection is starting and ending at the same elevation. Setting a distance in the start elevation offset from the wall bottom will lift the projection up from the bottom. Editing the start position offset, start elevation offset, and end elevation offset would allow the creation of the wall modifiers as shown in Figure 3.62. A projection can be positioned relative to the beginning, midpoint, and end of the wall. In Figure 3.63, the start position offset is set to 0 relative to the end of wall. This projection has a start elevation of 1'-0" from the Bottom of wall and an end elevation offset of a negative 1'-0" from Top of Wall.

Use Drawn Size—The Use Drawn Size checkbox turns OFF the size established in the **Length** and **Depth** fields. If this box is checked, the wall modifier will be created according to the actual size of the polyline geometry used to create the wall modifier style.

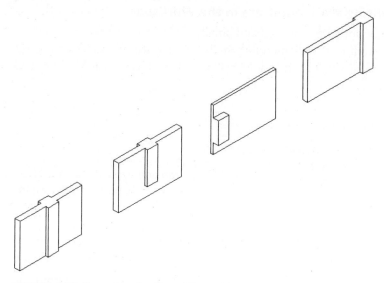

FIGURE 3.62 *Examples of wall modifier styles*

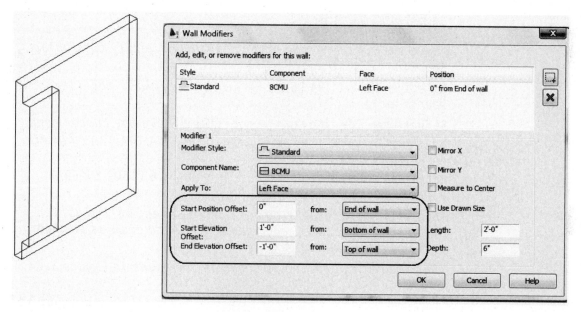

FIGURE 3.63 *Start and End Elevation set for wall modifier style*

Length—The Length field allows you to establish the length of the wall modifier. The distance entered in the Length field can be the same as the wall length.

Depth—The Depth field sets the distance the wall modifier will project from the wall.

Mirror X—The Mirror X checkbox is used to mirror the wall modifier in the X direction.

Mirror Y—The Mirror Y checkbox is used to mirror the wall modifier in the Y direction.

Measure to Center—The Measure to Center checkbox will position the wall modifier relative to its center along the wall.

Inserting and Editing Wall Modifiers in the Workspace

Wall modifiers can be added or removed and polylines can be converted to modifiers in the workspace. Adding a wall modifier in the workspace allows you to select points in the workspace to begin and end the wall modifier. Access WallModifierAdd as shown in Table 3.17.

TABLE 3.17 *WallModifierAdd command access*

Command prompt	**WALLMODIFIERADD**
Ribbon	Select a wall to display the Wall tab and expand Plan Modifier flyout from the Modify panel. Choose Add button as shown in Figure 3.61.
Shortcut menu	Select a wall, right-click, and choose Plan Modifiers > Add.

When you select a wall and choose **Add** from the Plan Modifier flyout within the Modify panel, you are prompted as follows in the workspace:

Select start point: *(Select a point near* p1 *as shown at the left in Figure 3.64.)*

(Move the cursor to the right near p2 *and verify that polar tracking tooltip angle* = 0.)

Select endpoint: 24 [600] ENTER *(Width of wall modifier specified.)*

Select the side to draw the modifier: *(Move the cursor down and select a point near* p3, *polar tracking tooltip* = 270 *as shown in Figure 3.64.)*

Enter wall modifier depth <8 5/8 [220]>: 6 [150] ENTER *(Specify distance for projection.)*

(Add Wall Modifier dialog box opens; select wall modifier style and elevations, and then click OK to dismiss the Add Wall Modifier dialog box.)

(Wall modifier created as shown at the right in Figure 3.64.)

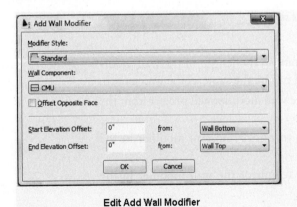

Edit Add Wall Modifier

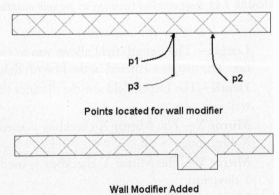

Points located for wall modifier

Wall Modifier Added

FIGURE 3.64 *Creating a wall modifier*

Removing a Wall Modifier

A wall modifier can be removed from a wall by accessing the **WallModifierRemove** command as shown in Table 3.18.

TABLE 3.18 *WallModifierRemove command access*

Command prompt	WALLMODIFIERREMOVE
Ribbon	Select the wall to display the Wall tab and choose Remove from the Plan Modifier flyout of the Modify panel as shown in Figure 3.61.
Shortcut menu	Select the wall, right-click, and choose Plan Modifiers > Remove.

When you access **WallModifierRemove** from ribbon, you are prompted to select the wall modifier as follows:

```
Command: WallModifierRemove

Select a modifier: (Select a wall modifier at p1 as shown in
Figure 3.65.)

Convert removed modifier to a polyline? [Yes/No] <No>: N
ENTER (Select N to not retain the polyline geometry in the
workspace.)

(Wall modifier is removed as shown in Figure 3.65.)
```

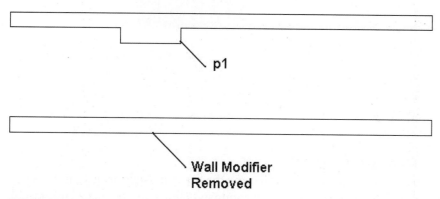

FIGURE 3.65 *Wall modifier removed from wall*

Edit In Place

A wall modifier can be edited using the **Edit In Place** option of the shortcut menu. **Edit In Place** allows you to change the geometry of the wall modifier in the workspace without accessing the **Style Manager**. Access **Edit In Place** as shown in Table 3.19.

TABLE 3.19 *Edit In Place of a Wall Modifier command access*

Command prompt	WALLMODIFIEREDIT
Ribbon	Select the wall to display the Wall tab and choose Edit Profile in Place from the Plan Modifier flyout of the Modify panel as shown in Figure 3.61.
Shortcut menu	Select the wall, right-click, and choose Plan Modifiers > Edit In Place.

If the wall modifier was created from a polyline drawn to actual size, grips are immediately displayed on the boundary of the wall modifier. Wall modifiers not drawn to actual size must first be converted. An AutoCAD Architecture message box, as shown in Figure 3.66, will open if the wall modifier was not drawn to actual size. A Yes response to this dialog box will convert the wall modifier and display grips along the boundary of the plan modifier as shown in Figure 3.66.

The wall modifier can then be edited by selecting grips or choosing from the shortcut menu or the Profile panel of the ribbon. Refer to the Edit In Place options presented earlier in the Using Edit In Place and Override Endcap Style section in this chapter.

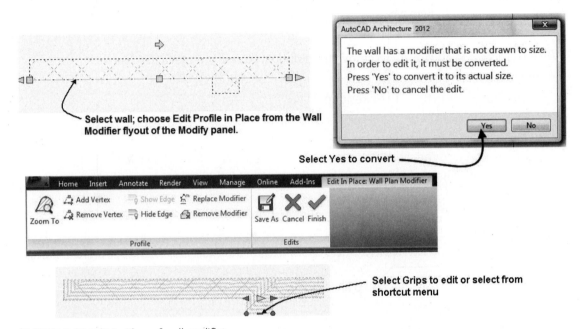

FIGURE 3.66 *Edit In Place of wall modifier*

Creating Wall Modifier Styles

Wall modifiers can be added to the wall using the Standard style or custom styles that you can create. The **WallModifierStyle** command allows you to create custom wall modifier shapes. Access the **WallModifierStyle** command as shown in Table 3.20.

TABLE 3.20 *WallModifierStyle command access*

Command prompt	**WALLMODIFIERSTYLE**
Ribbon	Select a wall to display the Wall tab and choose Edit Wall Modifier Style from the Plan Modifier flyout of the Modify panel as shown in Figure 3.61.

Creating Wall Modifier Styles with the Style Manager

Prior to selecting the **WallModifierStyle** command, create an open polyline to define the geometry for the wall modifier style. To open the Style Manager, select a wall and choose Edit Wall Modifier Style from the Plan Modifier flyout of the Modify panel. When the Style Manager opens, the **Wall Modifier Styles** category is selected in the left pane; right-click and choose **New**. Overtype the name of the new style in the left pane. The content of the wall modifier consists of an open polyline. Examples of polylines used as wall modifiers are shown in Figure 3.67. To define the polyline for a wall modifier, select the wall modifier name in the **Style Manager**, right-click, and choose **Set From**. The Style Manager temporarily closes, allowing you to select the polyline.

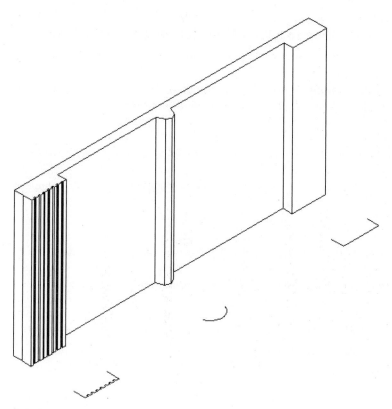

FIGURE 3.67 *Examples of Wall Modifier Styles*

The polyline used to define the wall modifier can be drawn the actual size of the desired projection or drawn as a sketch and applied with specified length and depth dimensions in the **Wall Modifiers** dialog box. If a wall modifier style is created from geometry drawn to actual size, it can be applied to a wall using the actual dimensions of the original geometry by checking the **Use Drawn Size** checkbox in the **Wall Modifiers** dialog box as shown in Figure 3.61.

If a polyline has been drawn in the workspace, you can convert this geometry to a wall modifier by selecting the wall and choosing **Convert Polyline to Wall Modifier** from the Plan Modifier flyout of the Modify panel from the ribbon. This process allows you to create a wall modifier for a wall without accessing the **Style Manager**. Access the **Convert Polyline to Wall Modifier** command as shown in Table 3.21.

TABLE 3.21 *WallModifierConvert command access*

Command prompt	WALLMODIFIERCONVERT
Ribbon	Select a wall to display the Wall tab and choose Convert Polyline to Wall Modifiers from the Plan Modifier flyout of the Modify panel as shown in Figure 3.61.
Shortcut menu	Select a wall, right-click, and choose Plan Modifiers > Convert Polyline to Wall Modifier.

Steps to Convert a Polyline to a Wall Modifier for a Selected Wall

1. Open a drawing that consists of an open polyline and a wall.
2. Select the wall at **p1** as shown in Figure 3.68 to display the Wall tab and choose **Convert Polyline to Wall Modifier** from the Plan Modifier flyout of the Modify panel. Respond to the following command line prompts as shown below:

   Command: WallModifierConvert

   Select a polyline: *(Select the arc at* p2 *as shown in Figure 3.68.)*

   Erase layout geometry? [Yes/No] <No>: Y ENTER *(Type Y, press* ENTER *to erase the polyline.)*

FIGURE 3.68 *Creating a Wall Modifier style*

3. Type a name in the **New Wall Modifier Style Name** dialog box as shown in Figure 3.68. Click **OK** to dismiss the dialog box and open the **Add Wall Modifier** dialog box.
4. Edit the **Add Wall Modifier** dialog box, specify style and start and end elevations, and click **OK**.

A wall modifier is placed on the wall as shown in Figure 3.68. Note that the wall modifier is placed on the wall near the location where the polyline is drawn. You can specify the location of the wall modifier in the **Wall Modifiers** dialog box by selecting **Plan Modifiers** in the Advanced section in the Design tab of the Properties palette.

CREATING WALL SWEEPS USING PROFILES

The **WallSweep** command is used to generate a wall that has the shape defined from an AEC Profile. The AEC Profile is created from a closed polyline. The closed polyline should be identical to the outline of a typical vertical section of the wall. The shape and size of the profile is then swept along the length of an existing wall. The profile is substituted as a wall component of the wall. Prior to sweeping the wall, the AECPROFILEDEFINE command must be used to create the AEC Profile. Access the **Profile Definition** command as shown in Table 3.22.

TABLE 3.22 *Accessing the Profile Definition command*

Command prompt	AECPROFILEDEFINE
Ribbon	Select Manage tab, and from the Style & Display panel, choose Profile Definitions.
Shortcut menu	Select a closed polyline, right-click, and choose Convert to Profile Definition.

Profiles can be created from the shortcut menu of a closed polyline or in the Style Manager. An AEC Profile is created in the following procedure.

Select a closed polyline at p1 shown in Figure 3.69, right-click, and choose Convert to > Profile Definition.

Insertion Point or [Add ring/Centroid]: *(Select the lower-right corner at* p2 *with the endpoint object snap.)*

Profile Definition [New/Existing] <New>: N *(Choose the New option, the New Profile Definition dialog box opens, and type the name of the new profile. Click OK to dismiss the dialog box.)*

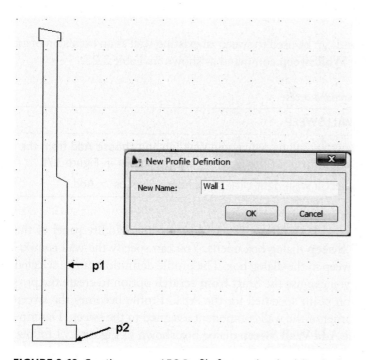

FIGURE 3.69 *Creating a new AEC Profile from a closed polyline in the workspace*

Profiles of the drawing can be created and edited in the Style Manager. Profiles are listed in the Multi-Purpose Objects folder as shown in Figure 3.70. A new style is created by selecting the **Profiles** category and then clicking the **New Style** button on the **Style Manager** toolbar. The new style is named by overtyping the name in the left pane. Select the profile name, right-click, and choose **Set From**; then, select a closed polyline and define the insertion point in the workspace to define the geometry of the AEC Profile.

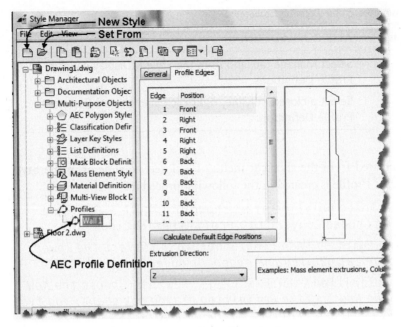

FIGURE 3.70 *Creating a new AEC Profile in the Style Manager*

Creating the Sweep

The **WallSweep** command can be used to sweep an existing wall as an extrusion of an AEC Profile. Access the **WallSweep** command as shown in Table 3.23.

TABLE 3.23 *Sweep Profile command access*

Command prompt	**WALLSWEEP**
Ribbon	Select a wall to display the Wall tab and choose Add from the Sweep flyout of the Modify panel as shown in Figure 3.71.
Shortcut menu	Select a wall, right-click, and choose Sweeps > Add.

When you select a wall and choose **Sweeps > Add** from the Modify panel of the Wall tab, the **Add Wall Sweep** dialog box opens. You can specify the wall component and profile for the sweep in the dialog box. The profile definition can be selected from existing profiles or you can use the **Start from scratch** option to create the profile in place. The insertion point specified for the AEC Profile becomes the sweep point for the lower left corner of the wall component assigned to the sweep. Descriptions of the options of the **Add Wall Sweep** dialog box shown in Figure 3.71 follow.

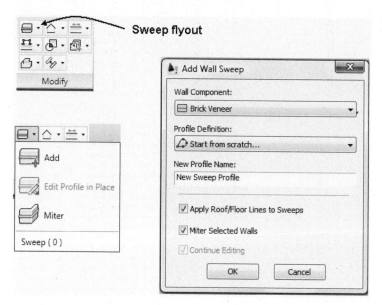

FIGURE 3.71 *Creating an AEC Profile for a wall sweep*

Wall Component—The Wall Component list displays the wall components of the selected wall. The sweep will be attached to the specified wall component.

Profile Definition—The Profile Definition allows you to select from **Start from scratch** or the defined profiles of the drawing.

New Profile Name—The New Profile Name is assigned to the wall when a sweep has been performed.

Apply Roof/Floor Lines to Sweeps—The Apply Roof/Floor Lines to Sweeps checkbox allows the sweep to include the roof and the floor line of the wall. Therefore, if the wall has been extended up for the roof line or down for the floor line, the sweep will also be extended accordingly.

Miter Selected Walls—When connecting walls are selected for the sweep, the Miter Selected Walls checkbox, when checked, miters the intersection of the connecting walls.

Continue Editing—Selecting the Continue Editing checkbox allows you to edit in place and grip edit the profile prior to sweeping the wall.

Steps to Create a Sweep

1. Open a drawing that consists of an AEC Profile and a wall.
2. Select the wall and choose **Add** from the Sweep flyout of the Modify panel in the Wall tab.
3. Edit the **Add Wall Sweep** dialog box, specify the wall component and profile definition, and check **Apply Roof/Floor Lines to Sweeps** and **Miter Selected Walls**.
4. Click **OK** to dismiss the **Add Wall Sweep** dialog box.

The wall is swept as shown in Figure 3.72.

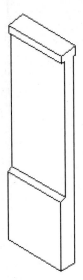

FIGURE 3.72 *Wall created with wall sweep*

Modifying Swept Walls

The intersection of two walls can be swept without miter. The **Miter** options will apply a miter to connecting walls. The **WallSweepMiterAngles** command extends the planes of each wall to a common intersection. The walls shown on the left in Figure 3.73 have not been mitered, whereas the walls on the right have been mitered. Access the **WallSweepMiterAngles** command as shown in Table 3.24.

TABLE 3.24 *WallSweepMiterAngles command access*

Command prompt	WALLSWEEPMITERANGLES
Ribbon	Select a wall to display the Wall tab and choose Miter from the Sweep flyout of the Modify panel.
Shortcut menu	Select a wall, right-click, and choose Sweeps > Miter.

When you select walls to apply the miter, choose **Miter** from the Sweep flyout of the Modify panel; the walls are mitered as shown on the right in Figure 3.73. If you type the command in the command line, you are prompted to select the walls for the miter.

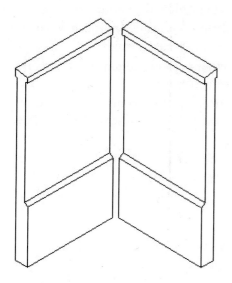 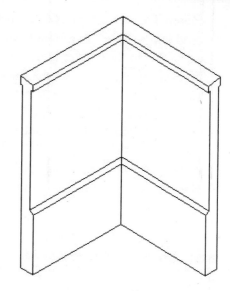

Walls before Miter   **Walls Mitered**

FIGURE 3.73 *Application of Sweep Profile Miter Angles command to walls*

Using Edit In Place with Wall Sweep

The profile of a swept wall can be edited in place using the **WallSweepProfileEdit** command shown in Table 3.25. If you select a wall that has a sweep operation, you can choose Edit Profile in Place from the Sweep flyout of the Modify panel. The grips of the profile will be displayed and can be edited as shown in Figure 3.74. The Edit In Place: Wall Sweep contextual tab includes options to edit the profile and save your changes. The following AEC Modify tools are included on the Modify panel: Merge, Subtract, Trim, and Crop. You can select a **Vertex** or an **Edge grip** and stretch to a new location to redefine the shape of the profile.

TABLE 3.25 *WallSweepProfileEdit command access*

Command prompt	**WALLSWEEPPROFILEEIDT**
Ribbon	Select a wall to display the Wall tab and choose Edit Profile in Place from the Sweep flyout of the Modify panel.
Shortcut menu	Select a wall, right-click, and choose Sweeps > Edit Profile in Place.

The commands of the Edit In Place: Wall Sweep include the Aec Modify tools of Merge, Subtract, Trim, and Crop. The profile may be modified with the following commands:

Add Vertex—Executes the **InplaceEditAddVertex** command available from the shortcut menu of a selected profile. You are prompted to select a point to add in the workspace. The existing polyline will be extended to include the vertex.

Remove Vertex—Executes the **InplaceEditRemoveVertex** command available from the shortcut menu of a selected profile. You are prompted to select a point to remove in the workspace.

Zoom To—Executes the **Zoom** command to magnify the profile.

Add Ring—Executes the **InplaceEditAddRing** command, which allows you to add a ring to the profile. Additional rings can be created as voids.

Remove Ring—Executes the **InplaceEditRemoveRing** command. The workspace prompts allow you to select a ring entity to remove.

Replace Ring—Executes the **InplaceEditReplaceRing** command. The command line prompts for this command allow you to select a closed polyline, spline, ellipse, or circle.

Finish—Saves the changes made to the profile.

Save As—Saves the changes in the profile using a new name.

Cancel—Ends the in-place editing without saving the changes to the profile.

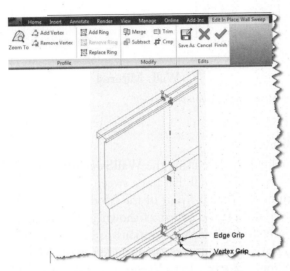

FIGURE 3.74 *Edit In Place of a wall sweep profile*

TUTORIAL 3-4: CREATING A MOLDING SWEEP

1. Verify that the Accessing Tutor content for Chapter 3 of the CengageBrain has been downloaded to your Accessing Tutor folder on your computer as described in Organizing Tutorial Directories in the Preface.

2. Choose Open from the Quick Access toolbar, select *Accessing Tutor\Imperial\Ch 3\Ex3-4.dwg* or *Accessing Tutor\Metric\Ch 3\Ex 3-4.dwg* from the **Select File** dialog box, and click **Open** to dismiss the **Select File** dialog box. Save the drawing to your student folder as *Lab 3-4.dwg*

3. In this step, you will create a wall component 4 1/8″ [80] wide to represent the molding. Select the wall and choose **Edit Style** from the General panel of the Wall tab. Choose the **Components** tab. Select Index 3 and click the **Add Component** button at the right. Overtype **Molding** as the name of the new component as shown in Figure 3.75, **Edit Priority = 1200, Width = 4 1/8″ [80], Edge Offset = 5/8″ [18]**. Click **OK** to dismiss the **Wall Style Properties** dialog box.

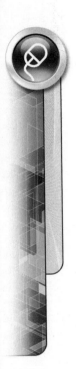

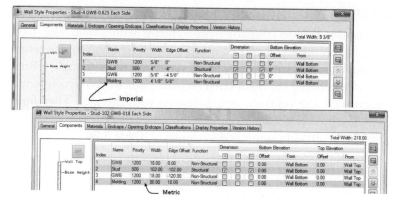

FIGURE 3.75 *Imperial and Metric molding component settings for molding*

4. The drawing consists of a polyline shape drawn to represent the crown molding. Click in the right viewport, select the polyline, right-click, and choose Convert to > Profile Definition. Respond to the workspace prompts as follows:

    Insertion point: *(Click in the left viewport, choose the endpoint of the wall component at p1 as shown in Figure 3.76.)*

    The **New Profile Definition** dialog box opens.

    Type **Crown** to specify the name of the profile. Click **OK** to dismiss the New Profile Definition dialog box.

5. Select the wall and choose **Add** from the Sweep flyout of the Modify panel in the Wall tab. The **Add Wall Sweep** dialog box opens. Select from the drop-down fields to edit the **Add Wall Sweep** dialog box as follows: **Wall Component = Molding** and **Profile Definition = Crown**. Click **OK** to dismiss the dialog box.

6. Save and close the drawing.

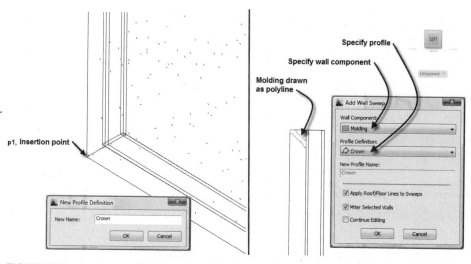

FIGURE 3.76 *Creating a wall sweep*

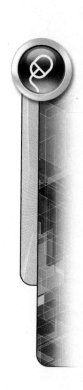

ADDING MASS ELEMENTS USING BODY MODIFIERS

The **WallBody** command is used to combine mass elements with a wall. The mass elements, covered in Chapter 4, can be used to model custom building components. Mass elements such as cylinders, boxes, arches, barrel vault, and right-triangle prisms can be combined to create building components. The mass element can be combined using Boolean operations of add, subtract, or intersection with the wall. Therefore, this command can be used to create projections in the wall or to cut out a portion from the wall. Access the **WallBody** command as shown in Table 3.26.

TABLE 3.26 *WallBody command access*

Command prompt	**AECWALLBODY**
Ribbon	Select a wall to display the Wall tab and choose Add from the Body Modifier flyout shown in Figure 3.77.
Shortcut menu	Select a wall, right-click over the drawing area, and choose Body Modifiers > Add.

When you select a wall, right-click, and choose **Body Modifiers > Add**, you are prompted to select the mass element. When you select the mass element, the **Add Body Modifier** dialog opens as shown in Figure 3.77. The cylinder mass element is combined with the wall since the operation is set to **Additive - Cut Openings** in the **Add Body Modifier** dialog box.

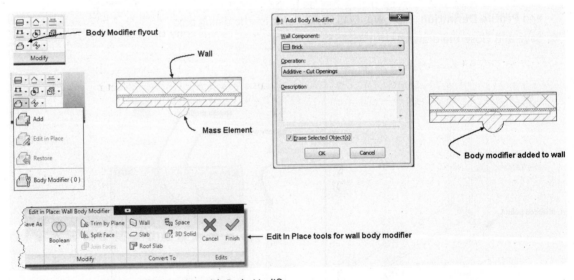

FIGURE 3.77 *Creating a wall projection with Body Modifier*

The options of the Add Body Modifier dialog box include:

Wall Component—The Wall Component list allows you to select a wall component to which to attach the mass element.

Operation—The Operation options allow you to attach the mass element using one of the following Boolean operations: **Additive - Cut Openings**, **Additive**, **Subtractive**, and **Replace**.

Erase Selected Object—The Erase Selected Object checkbox, if checked, will erase the mass element after it has been added to the wall.

After a mass element is added to the wall, select the wall and choose Edit in Place from the Body Modifier flyout to access **Restore** and **Edit in Place** options for editing the mass element. When you select the Restore option, you are prompted in the workspace: Remove Body Modifier? Respond Yes to this prompt to remove the body modifier. Choosing **Edit in Place** allows you to edit the size of the mass element. While **Edit in Place** is active, the shortcut menu and the Edit in Place: Wall Body Modifier tab include the following additional editing options: **Boolean**, **Trim by Plane**, **Split Face**, and **Join Faces**. Upon completion of the editing of the wall, choose Finish to save changes to the body modifier.

> **Boolean**—Allows you to change the Boolean operation applied to Union, Subtract, and Intersect.
>
> **Trim by Plane**—Executes the **MassElementTrim** command, allowing you to trim the mass element by specifying a plane through three points.
>
> **Split Face**—Executes the **MassElementFaceDivide** command to divide faces of the mass element.
>
> **Join Face**—Executes the **MassElementFaceJoin** command to combine faces of the mass element.

CREATING ADDITIONAL FLOORS

The main level or first floor is usually the starting point for the development of the working drawings. After you develop the majority of the first-floor plan, additional floors can be developed based on its shape. The additional floors can be created by creating a construct for the second floor in the **Project Navigator**. The shortcut menu of the **Constructs** tab shown in Figure 3.78 allows you to copy the Construct drawing of the floor plan assigned to level 2 to level 3. Therefore, the second-floor plan construct consists of exterior walls with the same XY coordinates, ensuring the vertical alignment of the floors. The second-floor plan can be developed by deleting non-load-bearing walls and creating the rooms of the second floor.

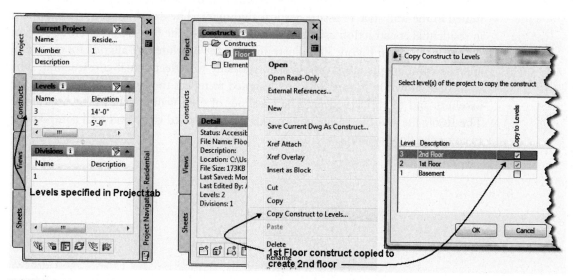

FIGURE 3.78 *Using the Project Navigator to copy a construct to another level*

Floor plans can be developed without starting the first floor as part of a project. If the first floor has been developed, you can create a project and save a current drawing as a construct of the project. When you are ready to create a model to check vertical alignment of features, you can convert each plan view drawing to a Construct drawing as part of the project. To convert a file to a construct, open the file and create or open a project. Select the **Constructs** tab of the **Project Navigator**, right-click, and choose **Save Current Dwg As Construct**. The **Add Construct** dialog box opens, allowing you to specify the construct name and level, as shown in Figure 3.79. Each drawing for a floor is created as a construct and assigned to a level defined in the **Project** tab.

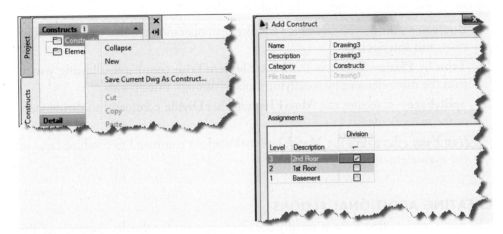

FIGURE 3.79 *Using the Project Navigator to convert a drawing to a construct*

After you have created constructs for each level, you can create a model drawing in the **Views** tab. The Construct drawings for each level can be attached as reference files to the model drawing in the Project Navigator. The model drawing allows you to check the vertical alignment of bearing walls and exterior walls.

If the exterior walls of one floor differ in construction from those of another floor, the wall styles of each floor can be created to permit the vertical alignment of the walls. If exterior walls of each floor are baseline justified, the wall components can be positioned in the wall style to assure vertical alignment. The exterior walls of a basement in residential construction usually differ from those of the first floor. The wall section shown at the left in Figure 3.80 illustrates how the baselines are aligned for each level and the components are positioned horizontally to reflect actual construction. The wall styles shown in Figure 3.81 are designed to reflect the vertical alignment of the baselines and the brick veneer. The baselines of each wall are shown by bold lines. The RoofLine and FloorLine commands can be applied to project wall components to cover the floor system.

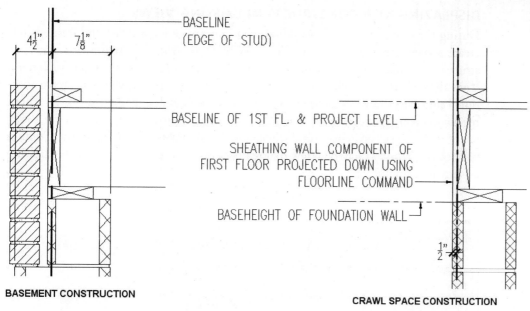

FIGURE 3.80 *Wall section for basement and crawl space construction*

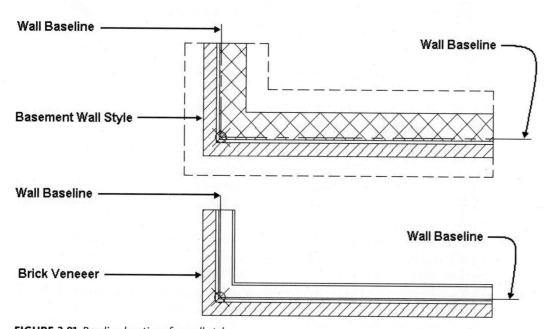

FIGURE 3.81 *Baseline locations for wall styles*

Therefore, if the first-floor plan is copied to create the basement level, you can modify the wall style of the basement exterior walls to ensure vertical wall alignment of the baselines.

> You can check the vertical alignment of levels by creating a model view in the *Project Navigator* with constructs of floor plan 1 and the basement floor plan attached. A model view of a residence that includes a basement is presented in Chapter 12. The View drawing assists in creating a model for checking vertical alignment of exterior walls and bearing walls.

NOTE

DISPLAYING NEW CONSTRUCTS IN EXISTING VIEWS

During the development of a building, you may create a construct and immediately create a view drawing for the construct. When you create a view drawing, the construct is assigned to the View drawing on the Content page of the **Modify General View** dialog box as shown in Figure 3.82. Later in the project, you may create additional construct drawings and review construct drawings assigned per level. To edit or review the construct drawings assigned, choose the view drawing name in the Views tab of the Project Navigator, right-click, and choose Properties. Choose the Content page of the Modify General View dialog box and edit the constructs for the drawing as shown in Figure 3.82.

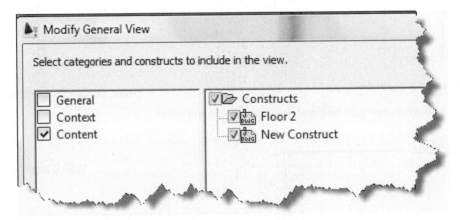

FIGURE 3.82 *Specifying construct drawings for an existing view*

CREATING A FOUNDATION PLAN

A foundation plan for a house that does not have a basement can be developed by copying the first-floor construct to a level below the first-floor plan. The vertical alignment of the structural components is established through controlling the baseline locations of the walls of each floor. The wall height is reduced to the anticipated height of the foundation wall. The levels defined in the project for a foundation plan are adjusted based upon the foundation wall height. The foundation walls should be baseline justified with the components shifted horizontally as shown in Figure 3.81. Piers can be inserted as walls or columns using layout curves to space the piers along a centerline. Layout curves, presented in Chapter 13, allow you to equally space the piers along a centerline. You will create a wall style for a foundation plan in Tutorial 3-8, which includes a concrete footing wall component in addition to the concrete masonry unit.

NOTE

To represent double joists and girders, change interior walls to a DOUBLE_JOIST wall style. The components of a DOUBLE_JOIST wall style can be assigned 0 width, 0 offset, and the linetype set to center. Change the wall style of interior walls, which require double joists, to this DOUBLE_JOIST joist wall style.

PLOTTING SHEETS

This chapter has provided you with basic commands for creating floor plans. The document workflow occurs in the Project Navigator and begins with a Construct drawing being attached to a View drawing. Dimensions and annotations are added in View drawings for each floor plan. The View drawing with its attached Construct drawing is attached to a Sheet drawing. The Sheet drawing consists of a viewport placed in paper space in which you view the content of the View and Construct drawings. Although you can plot within the Construct and View drawings, plotting from the Project Navigator provides additional flexibility. The features of the Sheets tab in the Project Navigator allow you to define a custom titleblock and border as a sheet template and print the entire sheet set using a page setup override.

NOTE

Refer to **e Tutorials** of Chapter 3 for download from the Student Companion Site of CengageBrain http://www.cengagebrain.com described in the Preface E tutorials include:
E Tutorial 3-1 Publishing Sheets to Drawing Web Format for a Project
E Tutorial 3-2 Setting the Sheet Template for a Project

TUTORIAL 3-5: IMPORTING AND CREATING WALL STYLES

1. Verify that the Accessing Tutor content for Chapter 3 of the CengageBrain has been downloaded to your Accessing Tutor folder on your computer as described in Organizing Tutorial Directories in the Preface.

2. Open AutoCAD Architecture 2012 and select the **Project Browser** from the Quick Access toolbar. Use the Project Selector drop-down list to navigate to your *Student\Ch3* directory. Select **New Project** to open the **Add Project** dialog box. Type **4** in the Project Number field and type **Ex 3-5** in the Project Name field. Check Create from template project, choose **Browse**, and edit the path to *Accessing Tutor\Imperial\Ch3\Ex 3-5\Ex 3-5.apj* or [*Accessing Tutor\Metric\Ch3\Ex 3-5\Ex 3-5.apj*] in the **Look in** field. Click **Open** to dismiss the **Select Project** dialog box. Click **OK** to dismiss the Add Project dialog box. Click **Close** to dismiss the Project Browser.

3. Select the **Constructs** tab of the **Project Navigator**. Double-click on **Floor 2** to open the Floor 2 drawing.

4. In this step, you will begin the process of importing a wall style using the Style Manager. Select the Manage tab, select Style & Display panel, and choose **Style Manager** from the ribbon. Choose the **Open Drawing** button from the Style Manager toolbar. Open **Lab 3-1** from the *Accessing Tutor\Imperial\Support\Lab 3-1.dwg* or [*Accessing Tutor\Metric\Support\Lab 3-1.dwg*].

5. Expand Architectural Objects of the **Lab 3-1.dwg** in the left pane. Choose **Wall Styles** category. Select the **Frame Brick** wall style, right-click, and choose **Copy** from the shortcut menu. Choose the current drawing **Floor 2** in the left pane, right-click, and choose Paste. The wall style has been transferred from Lab 3-1 to your current drawing. Click **OK** to dismiss the Style Manager dialog box.

6. Select the four exterior walls, right-click, and choose Properties. Edit the Style field in the Design tab of the Properties palette to **Frame Brick** and edit Justify to Baseline. Move the cursor from the Properties palette and choose ESC to clear the selection.

7. Choose **Zoom Window** from the **Zoom** flyout menu of the Navigation bar and respond to the workspace prompts as follows:

   Specify first corner: 100',35'[30500, 11000] ENTER
   Specify opposite corner: 110',45' [33500, 14000] ENTER
   (Notice that the interior walls do not merge as shown in Figure 3.83.)

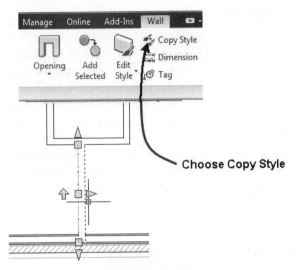

Choose Copy Style

FIGURE 3.83 *Merger of interior partitions*

8. Select the interior wall as shown in Figure 3.83 and choose **Copy Style** from the Wall tab, General panel of the ribbon to open the **Wall Style Properties-Standard (2)** dialog box.

9. Select the **General** tab, overtype the name **Interior** in the **Name** field to remove the "Standard (2)" name, and type **Interior partitions** in the **Description** field.

10. Select the **Components** tab, click in the Name field, and overtype **Wood**. Edit the Priority of the wall to 500, Width = **3 1/2 [89] (set variable value to 0)**, Edge Offset = **−3 1/2 [−89] (set variable value to 0)**, Function = **Structural**, and check positive and negative dimension checkboxes. Toggle ON **Interior** for the positive and negative direction of the wall style in the lower-left corner of the dialog box.

11. Click the **Add Component** button, click in the name column, and type **Gypsum** to specify the name. Edit the Priority = **1200**, Width = **1/2 [13]**, Edge Offset = **−4 [−102]**, Function = **Non-Structural**, and clear all **Dimension** checkboxes.

12. Click the **Add Component** button, click in the name column, and type **Gypsum** to specify the name. Edit the Priority = **1200**, Width = **−1/2 [−13]**, Edge Offset =**1/2 [13]**, Function = **Non-Structural**, and clear all **Dimension** checkboxes.

13. Click **OK** to dismiss the **Wall Style Properties** dialog box.

14. Select **Zoom Extents** from the **Zoom** flyout menu of the **Navigation** bar.

15. To change the interior walls to the **Interior** wall style, select all interior walls using a crossing selection; edit the Style = **Interior** and Justify = **Baseline** in the **Properties** palette. Move the cursor from the palette area, press ESC to end the selection, and edit.

16. Select **Zoom Window** from the **Zoom** flyout menu of the **Navigation** bar and respond to the workspace prompts as follows:

    Specify first corner: 100',35'[30500, 11000] ENTER
    Specify opposite corner: 110',45' [33500, 14000] ENTER
    (Notice that the interior walls merge after priority is set for each style.)

17. Select the Project tab, and choose Close Current Project from the toolbar of the Project tab. Choose Close all Project Files.

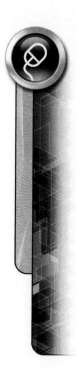

TUTORIAL 3-6: CREATING WALL MODIFIERS

1. Open AutoCAD Architecture 2012 and select **Project Browser** from the Quick Access toolbar. Use the Project Selector drop-down list to navigate to your *Student \Ch3* directory. Select **New Project** to open the **Add Project** dialog box. Type **5** in the **Project Number** field and type **Ex 3-6** in the **Project Name** field. Check **Create from template project**, choose **Browse**, and edit the path to *\Accessing Tutor\Imperial\ Ch3\Ex 3-6\Ex 3-6.apj* or *Accessing Tutor\Metric\Ch3\Ex 3-6\Ex 3-6.apj* in the **Look in** field. Click **Open** to dismiss the **Select Project** dialog box. Click **OK** to dismiss the **Add Project** dialog box. Click **Close** to dismiss the **Project Browser**.

2. Select the **Constructs** tab of the **Project Navigator**. Double-click **Floor 2** to open the Floor 2 drawing.

3. Select the View tab. Select the entities at p1, p3, and the vertical wall at p2 as shown in Figure 3.84. Select **Zoom Object** from the Zoom flyout menu of the **Navigation bar**.

4. Move the cursor over OSNAP on the status bar, right-click, and choose **Settings** from the shortcut menu. Clear all object snaps except **Node.** Choose the *Allow general object snap settings to act upon wall justification line* checkbox. Turn ON **Object Snaps**. Select the **Polar Tracking** tab and check **Polar Tracking** ON. Select the **Dynamic Input** tab, check **Enable Pointer Input**, **Enable Dimension Input where possible**, and **Show command prompting and command input near the crosshairs**. Click **OK** to dismiss the **Drafting Settings** dialog box.

5. To create a wall modifier style, select the vertical wall at p2 as shown in Figure 3.84 to display the Wall tab. Choose the Edit Wall Modifier Style from the Plan Modifier flyout to open the Style Manager.

6. Verify that the **Wall Modifier Styles** category is selected in the left pane, right-click, and select **New**. The Wall Modifier Style will be applied to a wall in Step 13.

7. Overtype **Bar_pilaster** in the left pane.

8. Select **Bar_pilaster** in the left pane, right-click, and select **Set From** from the shortcut menu. Select the polyline at **p1** as shown in Figure 3.84. Click **OK** to close the Style Manager.

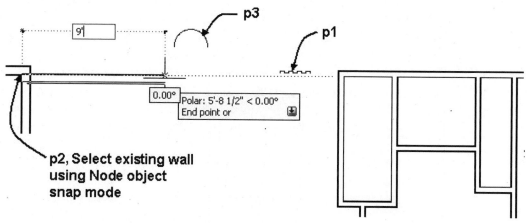

FIGURE 3.84 *Selection of polyline and wall start point*

9. Select an interior wall and choose Add Selected from the General panel of the Wall tab. Edit the Properties palette Base Height = 42" [1060]. Respond to the following workspace prompts to add the wall shown in Figure 3.84.

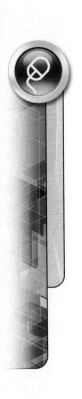

Start point or: *(Select the **Node** point of the existing wall at p2 shown in Figure 3.84.) (Move the cursor to the right with a polar angle of 0°.)*

Endpoint or: **9' [2750]** ENTER

Endpoint or: ENTER *(Press ENTER to end the command.)*

10. To remove the solution tip, select the wall segment created in Step 9 and choose **Wall Reverse Justification Line (WallReverseBaseline)** from the **Wall Reverse** fly-out of the **Modify** panel in the **Wall** contextual tab. The direction of the wall is reversed and the solution tip removed.

11. Move the cursor over **OSNAP** on the **Application** status bar, right-click, and choose **Settings** from the shortcut menu. Check **Endpoint** and retain **Node** object snap modes. Click **OK** to dismiss the **Drafting Settings** dialog box.

12. Select the wall at **p1** as shown in Figure 3.85, and choose **Add** from Plan Modifier fly-out of the Modify panel in the Wall tab. Respond to the workspace prompts as shown below and refer to Figure 3.85.

Select start point: *(Select the wall at p2 using the running Endpoint object snap as shown in Figure 3.85.)*

(Move the cursor left with a polar angle of 180°.)

Select endpoint: **8 [200]** ENTER

(Move the cursor up.)

Select the side to draw the modifier: *(Select a point near p3 as shown in Figure 3.85.)*

Enter wall modifier depth <10 1/4>: **1.5 [38]** ENTER *(Specify depth of wall modifier.)*

(Add Wall Modifier dialog box opens; refer to Figure 3.85 to edit the Wall Modifier Style = **Bar_Pilaster**, *Start Elevation = 0, End Elevation = 0, and set Wall Component = Gypsum. (There are two gypsum components listed; choose the top gypsum component.)*

(Click **OK** *to dismiss the Add Wall Modifier dialog box and create the wall modifier.)*

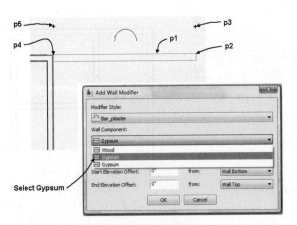

FIGURE 3.85 *Location of wall modifier*

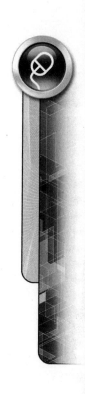

13. Retain the selection of the wall at p1 as shown in Figure 3.85 and choose Add from the Plan Modifier flyout of the Modify panel in the Wall tab. Respond to the workspace prompts as shown below.

> Select start point: *(Select the wall at p4 using the running Endpoint object snap as shown in Figure 3.85.)*
>
> *(Move the cursor right with a polar angle of 0°.)*
>
> Select endpoint: **8 [200]** ENTER
>
> *(Move the cursor up.)*
>
> Select the side to draw the modifier: *(Select a point near p5 as shown in Figure 3.85.)*
>
> Enter wall modifier depth <10 1/4>: **1.5 [38]** ENTER *(Specify depth of wall modifier.)*
>
> *(The* **Add Wall Modifier** *dialog box opens; edit the* Wall Modifier Style = **Bar_Pilaster**, Wall Component = **Gypsum**, Start Elevation = **0**, End Elevation = **0** *as shown in Figure 3.85. There are two gypsum components listed; choose the top gypsum component.)*
>
> *(Click* **OK** *to dismiss the Add Wall Modifier dialog box.)*
>
> *(Wall modifier created.)*

14. To create a wall modifier from existing geometry, choose Convert Polyline to Wall Modifiers of the Plan Modifier flyout of the Modify panel in the Wall tab as shown in Figure 3.86. Respond to the following workspace prompts:

> Select a polyline: *(Select the arc at p1 shown in Figure 3.86.)*
>
> Erase layout geometry? [Yes/No] <No> Y ENTER.

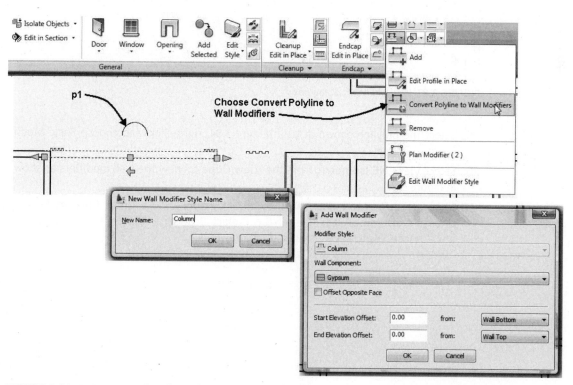

FIGURE 3.86 *Creating a Wall Modifier from Existing Geometry*

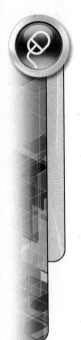

15. Type **Column** in the **New Name** field of the **New Wall Modifier Style Name** dialog box shown in Figure 3.86. Click **OK** to dismiss the New Wall Modifier Style Name dialog box.

16. Edit the **Add Wall Modifier** dialog box as shown in Figure 3.86 and click **OK** to dismiss the dialog box.

17. Select the wall modifier placed in the previous step and choose Plan Modifier (3) from the Plan Modifier flyout of the Modify panel to open the **Wall Modifiers** dialog box. The **Column** plan modifier will be moved in the next step by editing the settings in the Wall Modifiers dialog box.

18. Select the Column plan modifier and edit the Start Position Offset = **0 [0]** from **Midpoint of Wall.** Verify that the wall modifiers are positioned as shown in Figure 3.87. Click **OK** to dismiss the Wall Modifiers dialog box.

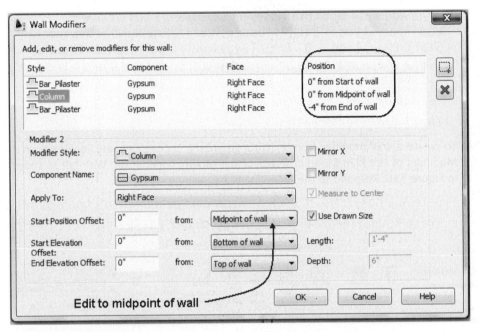

FIGURE 3.87 *Wall Modifiers dialog box*

19. Select the polyline shown at p1 in Figure 3.84, right-click, and choose Basic Modify Tools > Delete.

20. Select View, **NE Isometric** from the **View Cube** to view the wall modifiers as shown in Figure 3.88.

21. Choose the Project tab of the Project Navigator. Choose **Close Current Project.** Choose **Close** all project files of the Project Browser – Close Project Files dialog box. Choose Yes to save changes to all drawings.

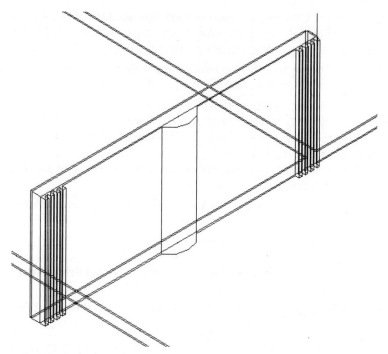

FIGURE 3.88 *Wall modifiers created for wall*

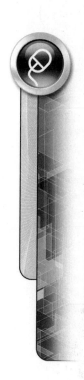

TUTORIAL 3-7: CREATING THE BASEMENT FLOOR PLAN

1. Verify that the Accessing Tutor content for Chapter 3 of the CengageBrain has been downloaded to your Accessing Tutor folder on your computer as described in Organizing Tutorial Directories in the Preface.

2. Open AutoCAD Architecture 2012 and choose **Project Browser** from the Quick Access toolbar. Use the **Project Selector** drop-down list to navigate to your *Student \Ch3* directory. Select **New Project** to open the **Add Project** dialog box. Type **6** in the Project Number field and type **Ex 3-7** in the Project Name field. Check **Create** from template project, choose **Browse**, and edit the path to *\Accessing Tutor\Imperial\Ch3\ EX 3-7\Ex 3-7.apj* or *\Accessing Tutor\Metric\Ch3\EX 3-7\Ex 3-7.apj* in the **Look in** field. Click **Open** to dismiss the **Select Project** dialog box. Click **OK** to dismiss the **Add Project** dialog box. Click **Close** to dismiss the **Project Browser**.

3. Select the **Constructs** tab of the **Project Navigator**. Double-click on **Floor 2** to open the Floor 2 drawing.

4. Choose **Top** from the of the **View Cube**.

5. In the Project Navigator, select Floor 2, right-click, and choose **Copy Construct to Levels** from the shortcut menu to open the **Copy Construct to Levels** dialog box.

6. Check **Level 1 [G]** for the new construct. Click **OK** to dismiss the **Copy Construct to Levels** dialog box.

7. Select the Floor 2 (1) or [Floor 2 (G)] construct, right-click, and select **Rename** from the shortcut menu. Type **Basement** as the name of the construct. Choose **Repath project now** from the Project Navigator – Repath Project dialog box. Double-click on **Basement** to open the file.

8. The basement wall style will be developed from some of the content of the Frame Brick wall style created in Ex 3-1. However, new material definitions are needed for the concrete footing and the concrete masonry units. Select the Render tab. Double-click the Render Tools button of the Tools panel to display Visualization palette set.

9. Select the Materials palette. Choose the **Concrete.Cast-In-Place.Flat Grey** tool and drag the pointer to the workspace. Press ESC when prompted to Select a component or object. The material definition is now imported into the drawing.

10. Choose the Home tab. Choose the Tools command from the Tools flyout at left. Choose the Content Browser from the Tools flyout of the Tools panel.

11. Open the Visualization Catalog. Navigate to AEC Material Tools > US Imperial > Masonry > Unit Masonry [AEC Material Tools > US Metric > Masonry > Unit Masonry]. Choose page 4. Select the i-drop for **Masonry.Unit.Masonry.CMU. Stretcher.Running.** Drag the i-drop to the workspace, and press ESC when prompted to Select a component or an object. The material definition is added to the drawing and can be applied to a wall style.

12. Select an exterior wall and choose Edit Style > Wall Styles from the General panel of the Wall tab. Expand Wall Styles under Architectural Objects. Select Frame Brick, right–click, and choose New Style. Overtype Basement as the name of the new wall style.

13. Choose the **General** tab, and type **BASEMENT** in the Name field and **Brick veneer basement wall** in the **Description** field.

14. Select the **Components** tab and enter the following components as shown in Table 3.27 and in Figure 3.89.

CAUTION When specifying width and edge offset, choose the drop-down and set the variable value to 0.

TABLE 3.27 *Wall components*

Index	Name	Priority	Width	Edge Offset	Function		Dimension	Bottom Elevation	Top Elevation
Offset	From	Offset	From						
1	8CMU [200CMU]	300	7 5/8 [194]	−7 1/8 [−181]	Structural	None	4'−0 [1219]	Baseline	0 [0]	Base Height
2	12 CMU [300CMU]	300	11 5/8 [295]	−7 1/8 [−181]	Structural	Pos. & Neg.	−8 [−203]	Baseline	4'−0 [1219]	Baseline
3	Brick	800	3 5/8 [92]	1½ [38]	Non-Structural	None	4'−0 [1219]	Baseline	0 [0]	Wall Top
4	FTG	200	2'−0 [610]	−1'−1 1/2 [−343]	Structural	None	−1'−4 [−406]	Baseline	−8 [−203]	Baseline

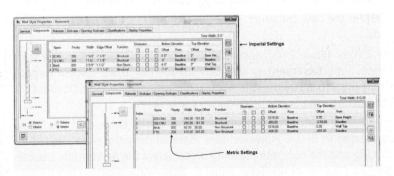

FIGURE 3.89 *Components tab for Basement wall style*

15. Select the Materials tab, click in the Material Definition column, and assign the material definition to the components as shown in Table 3.28.

TABLE 3.28 *Material Definition assignments*

Component	Material Definition
8CMU [200CMU]	Masonry, Unit Masonry, CMU, Stretcher, Running
12CMU [300 CMU]	Masonry, Unit Masonry, CMU, Stretcher, Running
FTG	Concrete, Cast-In Place Flat.Grey
Brick	Masonry.Unit Masonry.Brick.Modular.Running.Brown

16. Select the **Display Properties** tab and check the **Style Override** checkbox for the **Plan High Detail** display representation. The **Plan High Detail** display representation is selected since this drawing will be placed on a sheet at the 1/4" = 1'–0" [1:50] scale in Tutorial 3-9. Select the **Layer/Color/Linetype** tab, edit the **Below Cut Plane**, click in the **Linetype** column, and select **Hidden2** from the **Select Linetype** dialog box. Click **OK** to dismiss the Select Linetype dialog box.

17. Select the Cut Plane tab and verify the dialog box as follows: clear Override Display Configuration Cut Plane, check Automatically Choose Above and Below Cut Plane Heights, and clear the Manual Above and Below Cut Plane Heights checkbox as shown in Figure 3.90. Click OK to dismiss all dialog boxes.

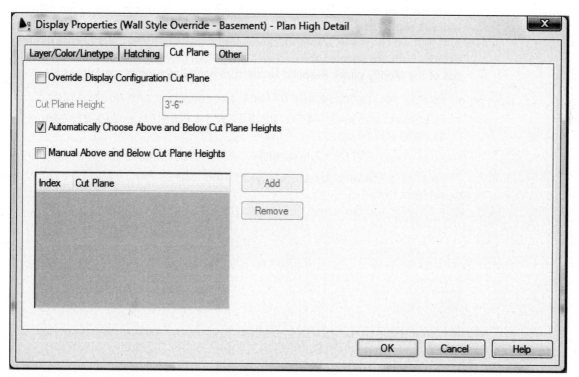

FIGURE 3.90 *Cut Plane tab of Basement wall style*

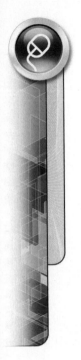

18. Verify that the **Model** tab is active in the **Application** status bar. Verify that Display Configuration is set to High Detail in the Drawing Window status bar. Select the **Cut Plane** height in the Drawing Window status bar to open the Global Cut Plane dialog box and edit the **Cut Height** to **5'-0"** **[1500]**. Click **OK** to dismiss the Global Cut Plane dialog box.

19. Click the **Work layout** in the **Application** status bar. Choose the View tab. Click in the left viewport and select **Zoom Extents** from the **Zoom** flyout menu of the **Navigation Bar**. Select **2D Wireframe** from the **Visual Styles** flyout menu of the Visual Styles panel in the View tab. Toggle OFF GRID on the status bar. Click in the right viewport and choose **Zoom Extents** from the Zoom flyout of the Navigation bar. Select **2D Wireframe** from the **Visual Styles** flyout menu of the Visual Styles panel in the View tab.

20. In this step, you will change the wall style of the exterior walls and use the **FloorLine** command to project the bottom of the wall down to wall bottom and display the footing. Click in the left viewport, select the four exterior walls, click in the **Style** field in the Design tab of the Properties palette, change the style to **Basement**, and verify that **Base** height is **8'** **[2500]**. Retain selection of the walls, and choose **Modify Floor Line** from the **Roof/Floor Line** flyout of the **Modify** panel. Respond to the command line prompts as follows:

    FloorLine *(Choose the* Offset *option from the Workspace.)*

    Enter offset <0>: **−16 [−406]** ENTER *(Extends floors down as shown in Figure 3.91.)*

    FloorLine: ENTER *(Press ENTER to end FloorLine command.)*

21. In this step, you will project the brick component up above the wall base height to cover the floor system. The total thickness of the floor system placed on the masonry wall will consist of a sill plate 1 1/2" [38], 2 × 10 floor joists 9 1/4" [235], and subfloor 3/4" [19] which creates a total thickness equal to 11 1/2" [292]. Verify that the four exterior walls are selected and choose Modify Roof Line from the Roof/Floor Line flyout of the Modify panel. Respond to the command line prompts as follows:

    RoofLine: *(Choose the* Offset *option from the Workspace.)*

    Enter offset <−1'−4">: **11.5 [292]** ENTER *(Specify distance to extend walls up.)*

    RoofLine: ENTER *(Press ENTER to end the command.)*

    (Walls extended as shown in Figure 3.91. Press ESC to clear selection.)

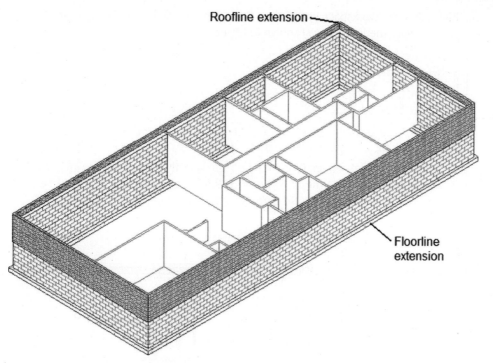

FIGURE 3.91 *Walls extended with FloorLine and RoofLine commands*

22. Click in the right viewport.

23. In this step, you will begin the edit of the interior walls to create rooms for the basement. Select the wall at **p1** as shown in Figure 3.92, and choose Intelligent Cleanup from the Modify panel of the Wall tab. Respond to the workspace prompts to extend the wall **p1** to wall **p2** as shown in Figure 3.92

> Select boundary wall for T-Cleanup or: Select wall at p2 as shown in Figure 3.92.
>
> *(Press* ESC *to clear selection.)*

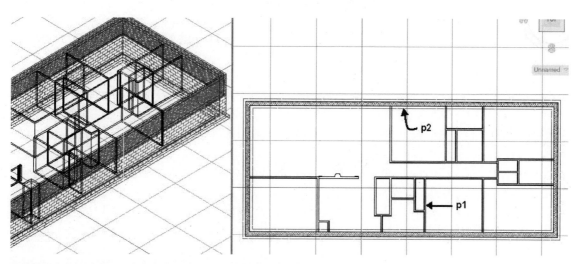

FIGURE 3.92 *Selection of walls for Intelligent Cleanup command*

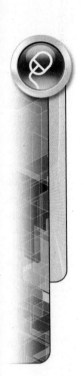

24. Right-click the workspace and choose Basic Modify Tools > Trim from the shortcut menu. Select the interior wall shown in Figure 3.93.

> Select objects or <select all>: 1 found *(Select wall at* **p1** *as shown in Figure 3.93.)*
>
> Select objects: ENTER (Press ENTER to end cutting edge selection.)
>
> Select object to trim or shift-select to extend or: *(Select wall at* **p2** *as shown in Figure 3.93.)*
>
> Select object to trim or shift-select to extend or: *(Select wall at* **p3** *as shown in Figure 3.93.) Select object to trim or shift-select to extend or: (Press ESC to end the command.)*

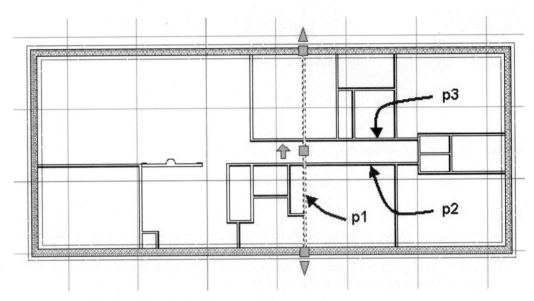

FIGURE 3.93 *Selection of walls for the Trim command*

25. Right-click the workspace and choose Basic Modify Tools > Delete from the shortcut menu. Select the interior walls shown in Figure 3.94 to erase the walls as shown in the following command line prompts:

> Select objects: Specify opposite corner: 10 found *(Create crossing selection, click near* **p1** *and then near* **p2** *as shown in Figure 3.94.)*
>
> Select objects: Specify opposite corner: 5 found, 15 total *(Create crossing selection, click near* **p3** *and then near* **p4** *as shown in Figure 3.94.)*
>
> Select objects: 1 found, 16 total *(Select wall at* **p5** *as shown in Figure 3.94.)*
>
> Select objects: 1 found, 17 total *(Select wall at* **p6** *as shown in Figure 3.94.)*
>
> Select objects: ENTER *(Press ENTER to end selection.)*

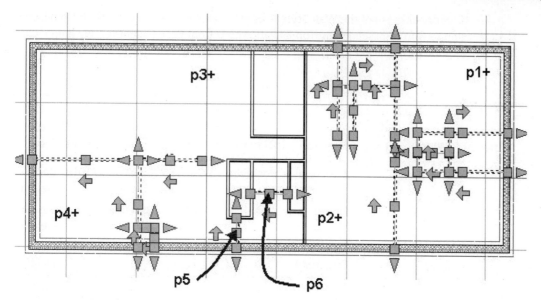

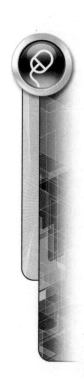

FIGURE 3.94 *Walls selected for the Erase command*

26. Select the wall at p1 as shown in Figure 3.95. Choose Intelligent Cleanup from the Modify panel of the Wall tab. Respond to the workspace prompts to extend the wall p2 to wall p1 as shown in Figure 3.95.

> Select boundary wall for T-Cleanup or: *(Select wall at p2 as shown in Figure 3.95. Press ESCAPE to clear selection of walls.)*

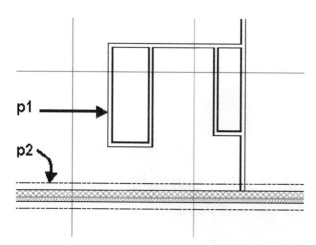

FIGURE 3.95 *Walls extended using the Intelligent Cleanup command*

27. Select an interior wall and choose Edit Style > Wall Styles to open the Wall Style Properties dialog box. Expand Wall Styles under Architectural Objects. Select Interior, right-click, and Rename. Overtype Basement_Int as the name of the wall style.

28. Select the **General** tab of the **Wall Style Properties** dialog box and overtype **Basement_Int** as the name of the style.

29. In this step, you will modify the Basement_Int wall style to start the bottom of the wall at the future concrete slab elevation and set the bottom elevation to 0. In addition, you will edit the bottom elevation and top elevation offsets. Select the

Components tab of the **Wall Style Properties** dialog box and edit the Bottom Elevation to 0 from Baseline and Top Elevation to 0 from Base Height for all components as shown in Table 3.29.

TABLE 3.29 *Component definition*

Index	Name	Priority	Width	Edge Offset	Function	Dimension	Bottom Elev. Offset	From	Top Elev. Offset	From
1	Wood	500	3 ½ [89]	−3 ½ [−89]	Structural	Positive	0 [0]	Baseline	0 [0]	Base Height
2	Gypsum	1200	1/2 [13]	−4 [−102]	Non-Structural	None	0 [0]	Baseline	0 [0]	Base Height
3	Gypsum	1200	−1/2 [−13]	½ [13]	Non-Structural	None	0 [0]	Baseline	0 [0]	Base Height

30. Click **OK** to dismiss the Wall Style Properties – Basement_Int dialog box.

31. Select all interior walls and change the **Style** to **Basement_Int** in the **Properties** palette. Move the cursor away from the **Properties** palette and press ESC to clear the selection.

32. Click in the left viewport. Select **Realistic** from the **Visual Styles** flyout menu of the Visual Styles panel in the **View** tab to view the interior walls shown in Figure 3.96.

FIGURE 3.96 *Pictorial view of basement walls*

33. Select the Project tab of the Project Navigator. Choose Close Current Project from the toolbar of the Project tab. Choose Close all project files of the Project Browser – Close Project Files dialog box. Choose Yes to save all drawings.

TUTORIAL 3-8: CREATING A CRAWL SPACE FOUNDATION PLAN

1. Verify that the Accessing Tutor content for Chapter 3 of the CengageBrain has been downloaded to your Accessing Tutor folder on your computer as described in Organizing Tutorial Directories in the Preface.

2. Open AutoCAD Architecture 2012 and select **Project Browser** from the Quick Access toolbar. Use the Project Selector drop-down list to navigate to your *Student \Ch3* directory. Select **New Project** to open the **Add Project** dialog box. Type **Ex 3-8** in the Project Number field and type **Ex 3-8** in the Project Name field. Check **Create from template** project, choose **Browse**, and edit the path to *\Accessing Tutor\ Imperial\Ch3\Ex 3-8\Ex 3-8.apj or \Accessing Tutor\Metric\Ch3\Ex 3-8\Ex 3-8.apj* in the **Look in** field. Click **Open** to dismiss the **Select Project** dialog box. Click **OK** to dismiss the **Add Project** dialog box. Click **Close** to dismiss the **Project Browser**.

3. In this step, you will revise the levels for the foundation and first-floor plan as shown in Figure 3.97. This project was previously created for a basement; in this step, you will revise the levels for a crawl space. Select the **Project** tab of the Project Navigator and click the **Edit Levels** button to open the **Levels** dialog box. Clear the **Auto-Adjust Elevation** checkbox. Edit Level 1 [G] as follows: Level 1 [G] Floor Elevation = **0**, Floor to Floor Height = **4'–0 [1200]**, Description = **Crawl**. Edit Level 2 [1] as follows: Floor Elevation = **4'–11.5" [1511]**, Floor to Floor Height = **9'–0" [2700]**. Click **OK** to dismiss the Levels dialog box. Click **Yes** to regenerate all views and dismiss the AutoCAD dialog box.

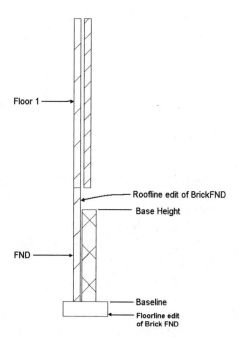

FIGURE 3.97 *Wall section of wall for each level*

4. Select the **Constructs** tab of the **Project Navigator**. Select the **Floor 2** construct, right-click, and choose **Copy Construct to Levels**. Check **Level 1 [G]** for the new construct. Click **OK** to dismiss the dialog box.

5. Select **Floor 2(1) [Floor 2(G)]**, right-click, and select **Rename**. Overtype **FND** as the name of the new construct. Choose *Repath project now* on the Project navigator – Repath Project dialog. Select **Floor 2**, right-click, select **Rename**, and overtype **Floor 1**. Choose *Repath project now* on the Project navigator – Repath Project dialog.

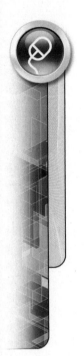

6. Double-click **FND** to open the drawing. Select the **Work** tab in the **Application** status bar, and click in the right viewport. Choose Zoom Extents from the Zoom flyout of the Navigation bar. Toggle OFF **Grid** in the **Application** status bar.

7. In this step, you will remove non-bearing interior walls and retain the load bearing walls. The load bearing walls will be developed into footing wall styles for the support of piers and load bearing walls of Floor 1. Right-click the workspace and choose **Basic Modify Tools > Delete**. Erase all the walls except the load bearing wall, as shown in the following workspace sequence:

Select objects: Specify opposite corner: (*Create a crossing selection and click at* **p1** *and then at* **p2** *as shown in Figure 3.98.*)

Select objects: (*Hold SHIFT and select load bearing wall* **p3** *to remove it from the selection set.*)

Select objects: (*Press ENTER to end the command.*)

(*The interior partition is retained to identify the centerline for the pier footing as shown in Figure 3.98.*)

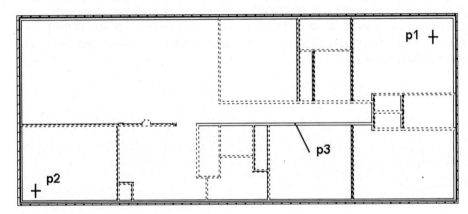

FIGURE 3.98 *Walls selected for Erase command*

8. Select an exterior wall to display the Wall tab. Choose the Wall Styles command from the Edit Style flyout of the General panel to open the Style Manager. Select **File > Close All** from the **Style Manager** menu to close unused files.

9. In this step, you will modify the wall style for exterior walls to include components designed to cover the floor framing system. Select **Open Drawing** from the **Style Manager** toolbar, edit the **Look in** directory to your *Accessing Tutor\Imperial \Support\Ex 3-7\Constructs\Basement.dwg* or *Accessing Tutor\Metric\Support\Ex 3-7\ Constructs\Basement.dwg*. Click **Open** to complete the selection.

10. Verify that the *Basement.dwg* and the *FND.dwg* drawings are displayed in the left pane. Select the **Wall Styles** category for the Basement drawing in the left pane. Select **Toggle View** on the **Style Manager** toolbar. Expand *Basement.dwg* and *FND.dwg* drawings in the Wall Style category as shown in Figure 3.99.

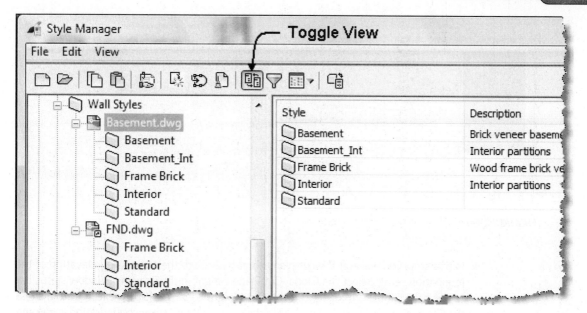

FIGURE 3.99 *Style Manager*

11. Click the **BASEMENT** wall style of the Basement drawing in the left pane, continue to hold down the left mouse button, drag the wall style over the *FND.dwg*, and release.

12. Select the **BASEMENT** wall style of FND in the left pane, right-click, and choose **Rename** from the shortcut menu. Overtype the name **BrickFND**.

13. Click **BrickFND** in the left pane, select the **Components** tab, delete the **12CMU [300CMU]** wall component, and edit the remaining components as shown in Table 3.30 and Figure 3.100. Toggle **Exterior** for the positive and **Interior** for the negative direction of the wall. Notice that the brick component extends above the CMU in the left view window of the **Components** tab. Click **OK** to dismiss the Style Manager. Press ESC to clear the selection of wall.

TABLE 3.30 *Wall components*

Index	Name	Priority	Width	Edge Offset	Function	Dimension	Bottom Elevation Offset	From	Top Elevation Offset	From
1	8CMU [200CMU]	300	7 5/8 [194]	−7 1/8 [−181]	Structural	Pos. & Neg.	0 [0]	Baseline	0 [0]	Base Height
2	Brick	800	3 5/8 [92]	1 ½ [38]	Non-Structural	None	0 [0]	Baseline	0 [0]	Wall Top
3	FTG	200	2'−0 [610]	−1'−1 1/2 [−343]	Structural	None	−8 [−203]	Baseline	0 [0]	Baseline

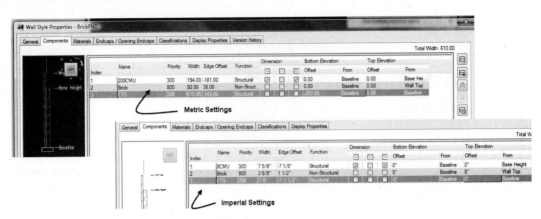

FIGURE 3.100 *Components of BrickFND wall style*

14. In this step, you will edit the cut plane to pass through the anticipated elevation of the foundation vents. Select the Global Cut Plane Height in the Drawing Window status bar to open the Global Cut Plane. Edit the **Cut Height** to **3'–0" [900]**. Click **OK** to dismiss the Global Cut Plane dialog box. Verify that High Detail display configuration is current.

15. Click in the left viewport and choose Zoom **Extents** from the **Zoom** flyout menu of the Navigation bar. Select the four exterior walls and edit the **Properties** palette as follows: Style = **BrickFND**, Height = **4'–0" [1219]**. Move the cursor from the Properties palette and press ESC to clear the selection.

16. In this step you will define the Wall Bottom as –8" below the baseline to display the footing of the wall as defined in the wall style (refer to Table 3.30). Press Ctl + A of the keyboard to select all the walls of the drawing, right-click, and choose **Roof/Floor Line > Modify Floor Line** from the shortcut menu. Respond to the command line prompts as follows:

    FloorLine : *(Choose the* Offset *option from the workspace.)*

    Enter offset <0>: **–8 [–200]** ENTER *(Extends walls down -8 below baseline to display footing.)*

    FloorLine: ENTER *(Press ENTER to end FloorLine command. Press ESC to clear selection of walls.)*

17. In this step you will define the wall top to 11.5" above the base height to extend the brick veneer above the CMU wall component. Select the four exterior walls, right-click, and select Roof/FloorLine > Modify Roof Line from the shortcut menu. Respond to the command line prompts as follows:

    RoofLine *(Choose the* Offset *option to extend walls up.)*

    Enter offset : **11 ½ [292]** ENTER *(Extends walls up.)*

    RoofLine: ENTER *(Press ENTER to extend the Brick component up to the top of wall above the base height. Press ESC to clear wall selection.)*

18. Click in the right viewport. Select the interior wall, and select **Copy Style** from the General panel of the Wall tab of the ribbon to open the Wall Style Properties dialog box.

19. Select the **General** tab and type **Pier_FND** in the **Name** field and **Pier continuous footing** in the **Description** field.

20. Select the **Components** tab and edit as shown in Table 3.31. Delete the Gypsum components and rename the index 1 component to FTG. Edit the positive and negative sides of the wall to Interior.

TABLE 3.31 *Wall component settings for the pier footing*

Index	Name	Priority	Width	Edge Offset	Function	Dimension	Bottom Elevation Offset	Bottom Elevation From	Top Elevation Offset	Top Elevation From
1	FTG	200	2'–0 [600]	–1'–0 [–300]	Structural	Center	–8 [–200]	Baseline	0 [0]	Baseline

21. Select the Materials tab and edit the FTG component Material Definition to Concrete. Cast-In-Place.Flat.Grey.

22. Select the **Display Properties** tab and check the **Style Override** checkbox for the current display representation (Plan High Detail). Select the **Layer/Color/Linetype** tab and edit as follows. Edit **Below Cut Plane**, click in the **Linetype** column, and select **Hidden2** from the **Select Linetype** dialog box. Click **OK** to dismiss the **Select Linetype** dialog box. Verify settings in the **Display Properties** dialog box as shown in Figure 3.101. Select the Cut Plane tab and check Automatically Choose Above and Below Cut Plane Heights checkbox.

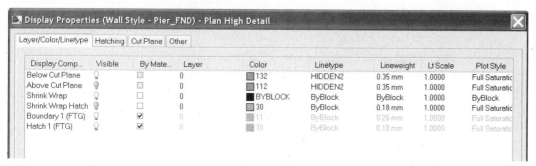

FIGURE 3.101 *Display Properties dialog box for Pier_FND wall style*

23. Click **OK** to dismiss the **Display Properties** dialog box.

24. Click **OK** to dismiss the **Wall Style Properties** dialog box. Retain the selection of the wall.

25. Verify that the wall at p1 is selected as shown in Figure 3.102, and choose Intelligent Cleanup from the Modify panel of the Wall tab. Respond to the workspace prompts as follows:

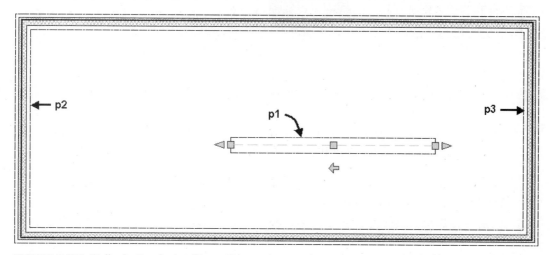

FIGURE 3.102 *Wall selection for Intelligent Cleanup command*

Select boundary wall for T-Cleanup or: *(Select wall at* p2 *as shown in Figure 3.102.)*

(Right-click choose Repeat WALLCLEANUPI *from the shortcut menu.)*

Select boundary wall for T-Cleanup or: *(Select wall at* p3 *as shown in Figure 3.102.)*

(Press ENTER *to end selection.)*

26. To check the alignment of the pier footing with the walls of the Floor 1 drawing, choose the Constructs tab. Choose Floor 1, right-click, and choose Xref Overlay.

    The Floor 1 is displayed in the left viewport and not in the right viewport. Verify that the right viewport is current. Choose the Cut Plane value in the Drawing Window status bar to open the Global Cut Plane dialog. Edit the cut height to 12'. Click OK to dismiss the dialog box.

    Choose the Insert tab, extend the Reference panel, and choose the Xref Fading slider to the right to 55. The first floor is displayed dim relative to the foundation as seen in each viewport.

27. Type Ctr + A to select all objects of the drawing; choose Edit in Section from the General panel of the Multiple Objects tab.

    Respond to the workspace prompts as follows.

    Specify first point of section line or Enter to change UCS: (Select a point near p1 as shown in Figure 3.101.)

    Specify next point of section line: (Select a point near p2 as shown in Figure 3.101.)

    Specify next point of section line or: (Press Enter to end section line.)

    Specify section extents: (Select a point near p3.)

    Click the Surface Hatch Toggle shown in 3.103 to view the FND drawing in section.

    Choose Exit Edit in Section of the Edit in View message box.

28. Choose the Project tab of the Project Navigator. Choose **Close Current Project** from the toolbar of the Project tab. Choose **Close all project files** of the Project Browser – Close Project Files. Choose Yes to save all drawings.

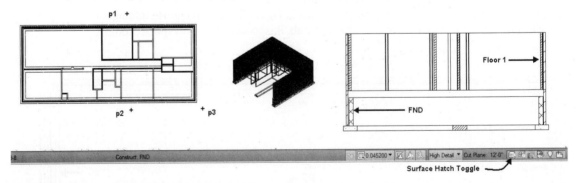

FIGURE 3.103 *Foundation walls complete*

TUTORIAL 3-9: PLOTTING FLOOR PLANS

1. Verify that the Accessing Tutor content for Chapter 3 of the CengageBrain has been downloaded to your Accessing Tutor folder on your computer as described in Organizing Tutorial Directories in the Preface.

2. Open AutoCAD Architecture 2012 and select **Project Browser** from the Quick Access toolbar. Use the Project Selector drop-down list to navigate to your *Student\ Ch 3* directory. Select **New Project** to open the **Add Project** dialog box. Type 7 in the Project Number field and type **Ex 3-9** in the **Project Name** field. Check **Create from template project**, choose **Browse**, and edit the path to *\Accessing Tutor\Imperial\ Ch3\ Ex 3-9\Ex 3-9.apj* or *\Accessing Tutor\Metric\Ch3\Ex 3-9\Ex 3-9.apj* in the **Look in** field. Click **Open** to dismiss the **Select Project** dialog box. Click **OK** to dismiss the **Add Project** dialog box. Click **Close** to dismiss the Project Browser.

3. Select the **Views** tab of the **Project Navigator**, select the **Views** folder, right-click, and choose **New View Dwg > General**. Type **Basement View** in the name field of the **Add General View** dialog box. Click **Next** to open the **Context** page and check Level **1 [G]** for Division 1. Click **Next** to verify that the Basement construct is checked in the **Content** page. Check the **Open in drawing editor** checkbox. Click **Finish** to dismiss the dialog box.

4. Choose ¼"=1'−0" [1:50] from the Annotation Scale flyout. Choose High Detail display configuration. Click the Cut Plane value to open the Global Cut Plane dialog box and edit the Cut Height to 5'−0" [1500]. Click OK to dismiss the dialog box.

5. Verify that the **Views** tab of the **Project Navigator** is current, select the **Views** category, right-click, and choose **Regenerate**.

6. In this step, you will copy a template file from the Accessing Tutor to your project folder for the storage of Page Setup Overrides. Right-click over **Start** on the Windows taskbar, and choose **Explore** to open the **Windows Explorer**. Navigate to the C:\Accessing Tutor\Imperial\Support [C:\Accessing Tutor\Metric\Support] folder. Select Accessing Page Setups (Imperial Stb).dwt or Accessing Page Setups (Metric Stb).dwt, right-click, and choose **Copy**. Navigate to your student folder\Ch3\Ex 3-9, right-click, and choose **Paste**. Close the **Windows Explorer**.

7. Select the Sheets tab. Verify that the Sheet Set view is displayed, select the Ex 3-9 sheet set title, right-click, and choose **Properties** to open the Sheet Set Properties dialog box as shown in Figure 3.104.

8. To specify the location of the template file for page setup overrides, click the right margin shown in Figure 3.104, **p1** of the **Page setup overrides file** field to display the Browse button. Click the Browse button to navigate to your student folder\Ch 3\ Ex 3-9 and select Accessing Page Setups (Imperial Stb).dwt or Accessing Page Setups (Metric Stb).dwt. Click **Open** to dismiss the dialog box.

9. In this step, you specify a default template for all sheets. Click the right margin of the **Sheet creation template** field shown in Figure 3.104, **p2** of the **Sheet Set Properties** dialog box to display the Browse button. Click the Browse button to open the **Select Layout as Sheet Template** dialog box. Verify that the sheet template is *Aec Sheet (Imperial stb).dwt /Aec Sheet (Metric stb).dwt*. If necessary, choose the Browse button at **p3** and navigate to *Vista - C:\ProgramData\Autodesk\ACA 2012\enu\Template folder*.

10. Select **Arch D (24 × 36) [ISO A1 (594 × 841)]** from the layout list as shown in Figure 3.104. Click **OK** to dismiss all dialog boxes. Click *Apply changes to all nested subsets* in the **Sheet Set – Confirm Changes** dialog box.

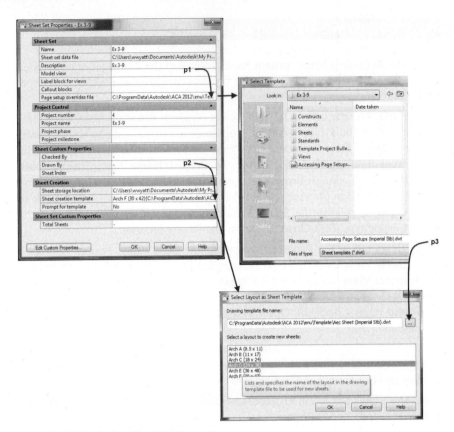

FIGURE 3.104 *Setting Sheet Set Properties*

11. Expand the Architectural category. Select the **Plans** subset of the Architectural category, right-click, and choose **New > Sheet** to open the **New Sheet** dialog box.

12. Type **A-3** in the **Number** field and **Floor Plan Basement Level** in the **Sheet Title** field of the **New Sheet** dialog box. Verify that the Sheet template is Arch D (24 × 36) [ISO A1 (594 × 841)]. Click the *Open in drawing editor* checkbox. Click **OK** to dismiss the dialog box.

13. Select the **Views** tab, select the **Basement View**, and drag the view into the A-3 sheet. Verify that the scale of the model space view during insertion is **1/4" = 1'–0"** [1:50] scale.

14. Verify that Sheet name **A-3 Floor Plan Basement Level** is displayed in the Titleblock. Double-click inside the viewport and verify the **Display Configuration High Detail** from the **Drawing Window** status bar. Select Cut plane value in the Drawing Window status bar to open the Global Cut Plane dialog box. Edit the **Cut Height** to **5'–0" [1500]**. Click **OK** to dismiss the Global Cut Plane dialog box.

15. Select the Sheets tab, select the **Plans** subset, right-click, and choose **New > Sheet** to open the New Sheet dialog. Type **A-4** in the Number field and **Floor Plan Level 2** in the Sheet title field. Verify that the Sheet template is Arch D (24 × 36) or [ISO A1 (594 × 841)]. Verify that *Open in drawing editor* is checked. Click **OK** to dismiss the dialog box.

16. Select the Views tab. Select Floor 2 View, right-click, and choose Properties. Click **Context** and verify that level 2 [1] is checked. Select **Content** and verify that Constructs and Floor 2 are checked. Click **OK** to dismiss the dialog box.

17. In the **Views** tab, select the **Floor 2 View** and drag the view onto the A-4 sheet. Verify that the scale during insertion is set to **1/4" = 1'–0" [1:50]** scale. Click to position the view on the sheet.

18. Verify that Sheet name **A-4** is displayed in the Titleblock, select the **Display Configuration** flyout menu, select the viewport, right-click, and choose **High Detail** from the **Drawing Window** status bar.

19. To create a page setup that can be applied in the sheet tab, select the Sheets tab. Select the sheet set title Ex 3-9, right-click, and choose **Publish > Manage Page Setups**. The Page Setup Manager opens. Click the **New** button of the **Page Setup Manager** to open the **New Page Setup** dialog box. Type **D Plot [A1 Plot]** in the **New Page Setup name** field. Select **<Default output device>** and click **OK** to define the settings for the page in the **New Page Setup** dialog box. Specify the plotter available to your computer that will plot D [A1] size paper and edit the paper size to **ARCH expand D (24.00 × 36.00 Inches) [ISO expand A1 (594.00 × 841.00 MM)]**, **What to plot** to **Layout** and **Scale** to **1:1**. Choose **Aec Standard.stb** from the Plot style table (pen assignments) flyout. Click **OK** to dismiss the dialog box. Click Close to dismiss the Page Setup Manager. The Page Setup is named in the Accessing Page Setup (Imperial Stb).dwt or Accessing Page Setup (Metric Stb).dwt file.

> If you do not have a D-size output device, choose the output device available.

20. Select the Plans category, right-click, and choose **Publish > Publish using Page Setup Override > D Plot[A1 Plot]**. All sheets specified in the Plans category will plot to the plotter specified in this Page Setup.

21. Select the Project tab. Choose Close Current Project from the toolbar of the Project tab. Choose Close all project files. Choose Yes to save each project file.

> Refer to **e Projects** of Chapter 3 for download from the Student Companion Site of CengageBrain http://www.cengagebrain.com described in the Preface, which includes Ex 3-10 Creating a Wall Style.

SUMMARY

- Wall styles are created in the Style Manager and are accessed using the Wall Style command (WallStyle).

- Wall components are created by the WallStyle command; they represent subassemblies of a wall and can be defined with unique widths and elevations.

- The edge offset distance of a wall component positions one edge of the wall component from the baseline.

- The width of a wall component establishes the distance between the surfaces of a wall component.

- The wall component priority number controls the cleanup of wall components when they intersect other walls.

- Wall endcap styles are applied to a wall component to close the end of a wall component at wall openings or wall ends.

- Wall styles include display properties control settings that control layer, visibility, color, and linetype of wall entities.

- Display control of a wall style can include the display of materials or hatching of wall components.

- Wall styles can be exported to resource files and imported from resource files to other files.

- Wall endcap styles can be edited in place.

- Applying wall modifier styles to a wall will create vertical projections along a wall.
- Body modifiers can add a mass element to a wall.
- Wall modifier styles are created from polylines.
- The RoofLine and FloorLine commands allow you to extend the wall above the base height and below the baseline.
- Copy the first-floor construct to other levels to create foundation plans and additional floor plans.
- Plot Sheet drawings from the Sheets tab of the Project Navigator. View drawings are attached as reference files to Sheet drawings.

NOTE

Refer to **Review Questions** folder, which includes review questions for Chapter 3 of Student Companion Site of **CengageBrain** http://www.cengagebrain.com described in the Preface.

Space Planning and Mass Modeling

INTRODUCTION

Space planning is a necessary component of the design of any building. One approach is to study the space arrangement of an existing building or existing floor plans to develop summaries of spaces allocated per function. Another approach is to sketch or lay out a space diagram to determine the quantity, location, and traffic between spaces. Walls can be added to enclose the spaces of the space diagram or you can develop mass elements for the study of a building shape that will contain the spaces. This chapter will also include the development of mass elements and groups to represent the shape of the building.

OBJECTIVES

Upon completion of this chapter, you will be able to

- Create and Modify 2D, Extrusion, and Freeform spaces using the Space Generate and SpaceAdd tools.

- Divide and combine spaces using the LineworkDivide and LineworkMerge commands.

- Summarize the spaces and zones of the drawing with the Zones and Space Evaluation commands.

- Create and edit mass elements and mass groups.

- Create a Drape mass element.

- Create parametric constraints of mass elements.

GEOMETRY FORMS OF SPACES

The content of Chapters 2 and 3 allowed you to develop skills in drawing walls and the details of wall styles. The first part of this chapter will apply space planning of such existing floor plans that have been developed. The **Space Generate** tool allows you to create spaces from existing walls of a floor plan. When you choose the Space Generate command, you are prompted to identify an internal point from which a boundary for the space is identified. The space is graphically represented by a hatch pattern in plan view. Spaces as shown in Figure 4.1 can be created in the following geometry types: 2D, Extrusion, or Freeform. The **SpaceAdd** and Space Generate tools can be used to create 2D, Extrusion, and Freeform spaces. The 2D space shown in Figure 4.1 is a simple two-dimensional graphic that represents the two dimensions of a room. The **Extrusion** space includes the third dimension with a floor boundary and ceiling boundary to represent the actual building components. Freeform spaces are created from mass elements or from walls, floor slabs, and roof slabs if they form a bounding box. The X and Y dimensions of wall objects define the size of the 2D and Extrusion spaces placed with the Space Generate tool. The Extrusion space shown in Figure 4.1 includes the representation of a ceiling boundary and floor boundary.

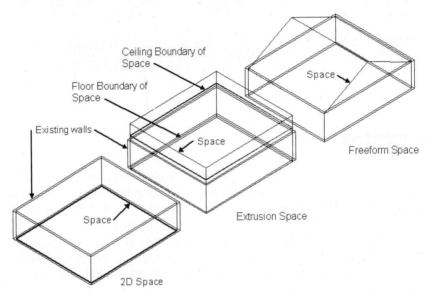

FIGURE 4.1 *Types of Space Geometries*

CREATING ASSOCIATIVE SPACES FROM GEOMETRY

Spaces can be defined as associative or nonassociative. If you develop spaces from a floor plan, the spaces can be defined to associate with the floor plan geometry. Associative spaces created in AutoCAD Architecture 2012 automatically update to changes in wall locations. The associative space links the building to the space, whereas the nonassociative space is not linked to existing walls. Nonassociative spaces can be used to develop space diagrams. Walls can be added to nonassociative spaces to create the floor plan. The **Space Generate** tool, shown in Figure 4.2, can be used to create associative or nonassociative spaces. The Space Generate tool creates spaces using the Standard space style. The functionality of the Space Generate tool can be applied to any of the SpaceAdd tools of the Spaces palette when you edit the **Create type** to **Generate.** Access the Space Generate tool as shown in Table 4.1.

TABLE 4.1 *Accessing Space Generate tool*

Command prompt	**SPACEADD**
Ribbon	Select the Home tab > Build panel. Choose the Generate Space from the Space flyout (refer to Figure 4.2).
Tool palette	Choose Space Generate from the Design palette of the Design palette group.

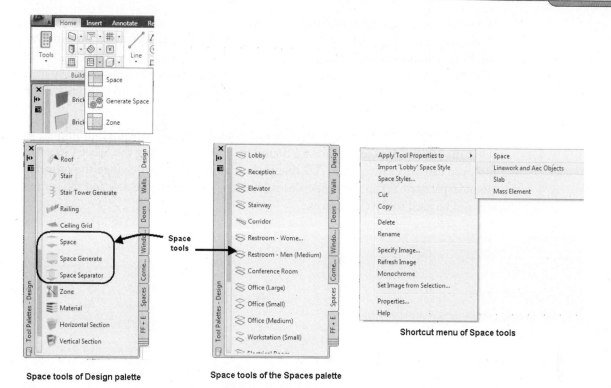

FIGURE 4.2 *Accessing the Space Generate and SpaceAdd tools*

When you choose the Space Generate tool, the Create type is preset to Generate as shown in Figure 4.3. When the Create type is preset to Generate, the **Properties palette** includes a special Generate Space section as shown in Figure 4.3. You can specify the Style and Geometry Type from drop-down list as shown in Figure 4.3. The Extrusion geometry type is shown in Figure 4.3. If AEC objects form a closed shape in the drawing, you will be prompted to pick internal point. When you move the cursor inside the closed shape, the boundary will be displayed with red lines; left-click inside the shape to create the space.

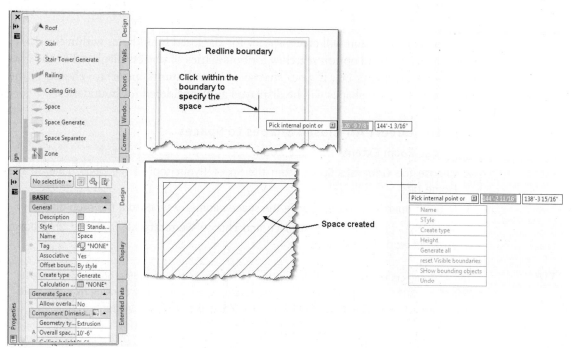

FIGURE 4.3 *Creating spaces from objects using Space Generate*

If, however, no Aec objects or AutoCAD entities such as lines, circles, polylines, arcs, blocks, and ellipses are defined to bound spaces in the Properties palette, the **Analyzing Potential Spaces** dialog box is displayed as shown in Figure 4.4. You can choose "Use all visible objects to bound spaces" to create spaces from all visible closed spaces. If you choose this option, the Bound Spaces property of all closed entities will be edited to Yes, allowing you to create a space.

If you choose the Select objects that should bound spaces, you will be prompted to select the entities or objects to form a boundary. Prior to selecting the Space Generate tool, you can select the AutoCAD linework of a floor plan and change the property Bound Spaces to Yes in the Properties palette as shown in Figure 4.4, and the geometry can be used to develop spaces.

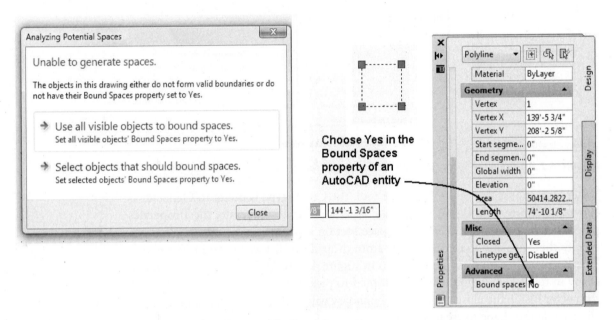

FIGURE 4.4 *Converting boundaries to bound spaces using Space Generate*

The Generate all option will create spaces for all closed boundaries within the drawing. The Generate all option searches for boundaries of objects displayed in the graphics window; therefore, if you choose **Zoom Extents** prior to choosing the command, all closed shapes will be displayed and considered for boundaries.

Steps to Convert All Closed Shapes to Spaces

1. Choose **Zoom Extents** from the **Zoom** flyout of the Navigation bar.
2. Choose the **Generate Space** from the Space flyout tool of the Build panel in the Home tab.
3. In the Properties palette, choose the space geometry type from the drop-down list.
4. Move the cursor to the workspace and choose **Generate all** from the drop-down arrow of workspace options.

| TIP | The AEC objects in attached reference files can serve as the bounding objects for a space. Therefore, in the Project environment, you develop spaces in a View drawing from the walls of a construct floor plan. Refer to Project Ex 4-9 at the end of the chapter. |

The unique options of Properties palette for creating spaces when Create Type is set to **Generate** are described below. Additional properties of a space are described later in the chapter.

General

Style—The Style list allows you to select the space styles included in the drawing. Unless space styles have been imported into the drawing, only the Standard style is listed.

Name—The Name drop-down list includes names defined for the spaces used in the drawing. The names are derived from List Definitions. List Definitions are defined and viewed in the Multi-Purpose Objects > List Definitions category of the Style Manager.

Tag—The Tag field includes a list of multi-view blocks designed for placing tags on the space. The Aec8_Room tag [M_Aec8_Room tag] is included in the AEC Model template drawing. Additional tags can be imported from the Schedule Tags > Room and Finish Tags categories of Documentation Tool Catalog – Imperial or Documentation Tool Catalog – Metric.

Associative—The Yes/No options of associative allow you to specify the associative property of spaces during creation and editing of existing spaces. Spaces inserted using the **Generate** Create Type are associative since they are restrained by a boundary.

Offset boundaries—The offset boundaries drop-down list allows you to choose Manual, By Standard, or By Style.

Create Type—The **Create Type** specifies the method of creating spaces. When Space Generate is selected, the Create Type is preset to Generate to create associative or nonassociative spaces. The Insert, Rectangular, and Polygon options allow you to create freestanding nonassociative spaces based on geometric shapes.

Calculation Modifiers—The **Calculation Modifiers** are formulas defined in a calculation modifier style, which is created in the Style Manager.

Generate Space

Allow overlapping spaces—The Yes option of Allow overlapping spaces allows you to create overlapping spaces within a boundary each time the Space Generate command is applied.

Component Dimensions

Geometry Type—The Type drop-down list includes 2D, Extrusion, and Freeform geometry types.

USING THE SPACE ADD TOOL TO CREATE SPACES FROM A FLOOR PLAN

The **Space Generate** tool located on the Design tool palette creates spaces using the Standard space style or space styles that have been imported into the drawing. The **SpaceAdd** tools of the Spaces palette have unique space styles assigned to each tool. When you choose the SpaceAdd tool, you can edit the Create type using Generate, Rectangular, Polygon, or Insert tools. When the Create type is set to Generate, each space tool provides the same options as the Space Generate command. Additional space styles can be imported into the drawing in the Style Manager or from Design Tool Catalog – Imperial and Design Tool Catalog – Metric of the Content Browser. The shortcut menu of the SpaceAdd tools includes commands that will convert existing spaces and other objects to the space style associated with the selected SpaceAdd tool. The commands of the shortcut menu are shown in Table 4.2.

TABLE 4.2 *Commands of the SpaceAdd shortcut menu*

Right-click on a SpaceAdd tool and choose one of the following:	Purpose
Apply Tool Properties to > Space	Converts selected space to the space style of the tool.
Apply Tool Properties to > Linework and Aec Objects	Converts selected linework or Aec Object to a space with the same style of the space tool.
Apply Tool Properties to > Slab	Converts selected slab to a space with the same style of the space tool.
Apply Tool Properties to > Mass Elements	Converts the selected mass element to a space with the same geometric form and size of the mass element.

When you choose **Apply Tool Properties to > Linework and AEC Objects,** you are prompted in the workspace to select objects to convert. You can use this command to convert sketches into space diagrams. The Apply Tool Properties to > Linework and AEC Objects can be used to convert the rectangle shown in Figure 4.5 to the Reception space style. If the objects selected are closed, the **Convert to Space** dialog box opens allowing you to specify the cut plane height and choose to erase layout geometry. The geometry of the sketch is converted to a nonassociative space.

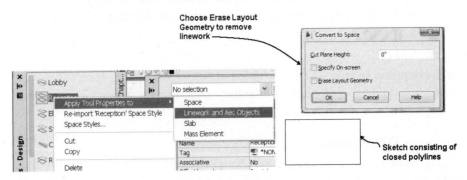

FIGURE 4.5 *Creating a space from closed polylines*

TAGGING SPACES

Tags can be placed on each space that is inserted into the drawing. The Aec8_ Room_ Tag [M_Aec8_Room_Tag] definition is a tag included in the Aec Model imperial and metric templates. The SpaceObjects and RoomFinishObjects property sets are available for the tag. When the Aec8_Room_Tag [M_Aec8_Room_Tag] is inserted using the SpaceObjects property set, an incremental room number and a name can be placed. The default name for the tag is SPACE; however, it can be changed in the Name field in the Design tab of the Properties palette for the space. A list of names appropriate to a space style is imported with the space. Additional tags can be imported from the Documentation Tool Catalog (Imperial) > Schedule Tags > Room & Finish Tag or Documentation Tool Catalog (Metric) > Schedule Tags > Room & Finish Tag. If you choose the i-drop for the Room Tag with Dimensions and tag an existing space or press ESC when prompted to select a space, the tag will be available as a tag in the **Design** tab of the **Properties** palette. The following steps outline the procedure for placing the Aec8_Space_Tag [M_Aec8_Space_Tag] with spaces created by the Space Generate tool. The Aec8_Space_Tag [M_Aec8_Space_Tag] updates to reflect changes in the area.

Steps for Placing Tags with Spaces

1. Choose **Annotate tab.** Choose the **Annotation Tools** command from the **Tools** panel to display the Document palette set.

2. Choose the Tags palette. Choose More Tag Tools to open **Documentation Tool Catalog – Imperial > Schedule Tags > Room & Finish Tags** or **Documentation Tool Catalog – Metric > Schedule Tags > Room & Finish Tags**.

3. Click and drag the i-drop for a tag—for example, Room Tag (w/Dimensions)—into the workspace; you will be prompted to select a space. Press ESC to end the command and import the property into the drawing. The **Room Tag (w/Dimensions)** tool imports the Aec8_Space_Tag [M_Aec8_Space_Tag] multi-view block as shown in Figure 4.6.

4. Choose a space tool from the **Spaces** palette. Edit the General section in the Design tab of the Properties palette as follows: Tag = Aec8_Space_Tag [M_Aec8_Space_Tag], Create type = **Generate**. Edit the **Generate Space** section of the **Properties** palette as follows: Allow overlapping Space = **No.** In the **Component Dimensions** section, edit Geometry Type = **Extrusion**.

5. Click inside a room defined by walls to specify the space. Space is created and tagged as shown in Figure 4.6.

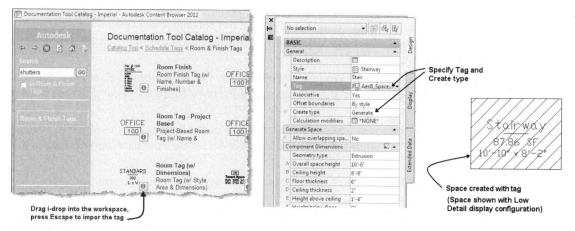

FIGURE 4.6 *Importing tags for Space Generate*

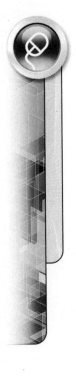

TUTORIAL 4-0: ADDING SPACES TO A FLOOR PLAN

1. **Download Chapter 4 files from the Accessing Tutor\Imperial or Accessing Tutor\ Metric category of the Student Companion site of CengageBrain** http://www. cengagebrain.com described in the Preface to your Accessing Tutor folder.

2. Select Open from the Quick Access toolbar.

3. Select *Accessing Tutor\Imperial\Ch4\Ex 4-0.dwg* or *Accessing Tutor\Metric\Ch 4\ Ex 4-0.dwg*.

4. Choose **SaveAs > AutoCAD Drawing** from the Application menu and save the drawing as **Lab 4-0.dwg** in your student directory.

5. Since annotation is added to the drawing in this tutorial, edit the Annotation Scale = 1/4" = 1'-0" [1:50] and Display Configuration = High Detail.

6. Double-click the **Annotation Tools** from the Annotate tab of the ribbon to display the Document palette set.

7. Select the Tags palette. Choose More Tag Tools to open the **Content Browser** to the **Documentation Tool Catalog – Imperial > Schedule Tags** or **Documentation Tool Catalog – Metric > Schedule Tags**. Open **Room & Finish Tags**.

8. Click and drag the i-drop for **Room Tag (w/Dimensions)** tag into the workspace; you will be prompted to Select object to tag. Press ESC to end the command. The tag properties will be imported into the drawing, and the tag will be available in the Properties palette.

9. Click the **Home** tab of the ribbon. Click the **Tools** button of the **Build** panel. Select the Spaces palette of the Design palette group. Click the **Lobby** space of the **Spaces** palette.

10. Edit the Properties palette as follows: Tag = **Aec8_Space_Tag** [M_Aec8_Space_Tag], Create type = **Generate**, Associative = **Yes,** Geometry type = **Extrusion**.

11. Move the cursor to point **p1** as shown in Figure 4.7. When the red polygon is displayed for the boundary, click to specify the boundary.

12. Click **Office (Medium)** space of the **Spaces** palette.

13. Verify the Properties palette as follows: Tag = **Aec8_Space_Tag [M_Aec8_Space_ Tag]**, Create type = **Generate,** Geometry type = **Extrusion.** Move the cursor to point **p2** as shown in Figure 4.7. When the red polygon is displayed for the boundary, click to specify the boundary. Continue to click inside the rooms at **p3, p4, p5, p6, p7, p8,** and **p9.**

14. Click the **Restroom – Women (Medium)** space of the Spaces palette. Move the cursor to point **p10**. When the red polygon is displayed for the boundary, click to specify the boundary.

15. Click the **Restroom – Men (Medium)** space of the **Spaces** palette. Move the cursor to point **p11**. When the red polygon is displayed for the boundary, click to specify the boundary.

16. Click the **Closet** space of the **Spaces** palette. Move the cursor to point **p12**. When the red polygon is displayed for the boundary, click to specify the boundary.

17. Click the **Corridor** space of the **Spaces** palette. Move the cursor to point **p13**. When the red polygon is displayed for the boundary, click to specify the boundary.

18. Choose the **Layer List** from the Layer panel of the Home tab. Click the light-bulb of the drop-down list to turn OFF the visibility of **A-Area-Spce-Patt** as shown in Figure 4.8. Save and close the drawing.

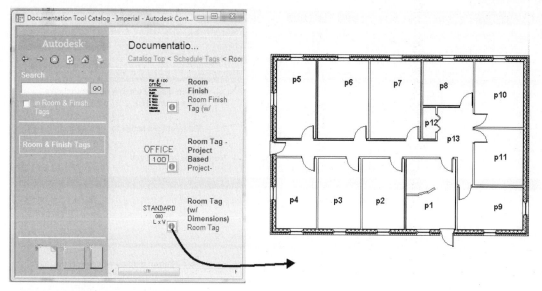

FIGURE 4.7 *Room geometry for spaces*

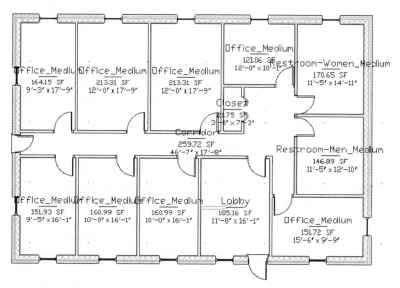

FIGURE 4.8 *Display of space objects turned OFF*

EDITING SPACES

The commands to modify an existing space are included on the contextual Space tab, which is displayed when you select a space. These commands are also listed on the shortcut menu of a space. The ribbon display of a nonassociative space is shown at top in Figure 4.9. When an associative space is selected, the Space contextual tab includes commands as shown at bottom in Figure 4.9. An associative space boundary will update itself when walls are moved or new walls that cross the space are added. The Update command of the Modify panel allows you to update a space to any boundary changes. Grips are not displayed for the associative space as shown at the bottom in Figure 4.9, whereas the nonassociative space shown at the top includes grips for editing its boundary. When a nonassociative space is changed to associative, the space will change to fit the boundary. AEC Modify Tools are also included in the shortcut menu of a nonassociative space.

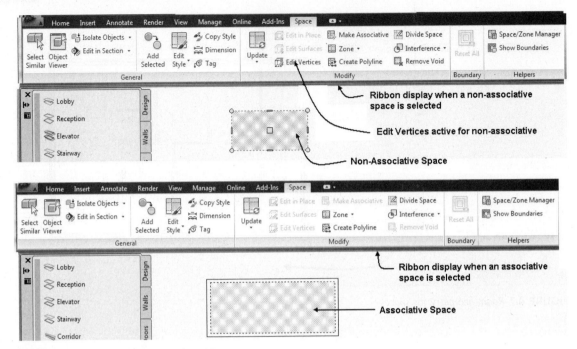

FIGURE 4.9 *Space contextual tab for editing spaces*

Updating an Associative Space

If walls are moved after creating the space, the associative space will update to reflect the change. The automatic update of spaces is a feature controlled by the *Automatically update associative spaces* checkbox as shown in the AEC Object Settings tab of the Options dialog box shown in Figure 4.10. Associative spaces can be created in Construct or View drawings of a project. When you create spaces in a Construct drawing, the modifications of the wall will be updated for the space. The drawing, created in a view, of a project from boundaries defined by walls in the reference construct drawing automatically updates the View drawings when the Construct drawing changes. Nonassociative spaces when converted to associative can be updated to a boundary by selecting the **Selected Space** option from the flyout of **Update** in the ribbon (refer to Figure 4.9). You can update all associative spaces of the drawing if you choose **All Associative Spaces** from the flyout of the **Update** button in the Ribbon. The Update tools can also be used to update a selected space to its boundary of previous release drawings since the automatic update of spaces was not included in all previous releases. Access the Update command as shown in Table 4.3.

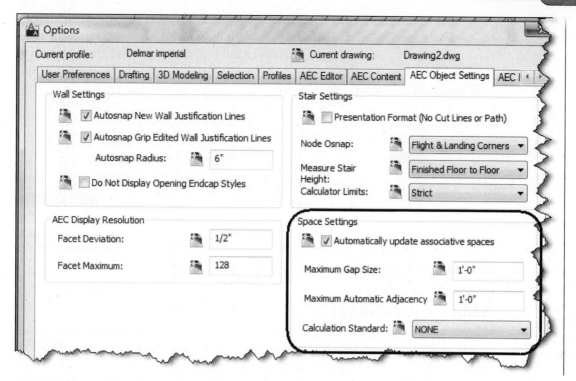

FIGURE 4.10 *Automatically update associative spaces setting in Options*

TABLE 4.3 *Accessing the SPACEUPDATESELECTEDSPACES command*

Command prompt	SPACEUPDATESELECTEDSPACES
Ribbon	Select an associative space to display the Space contextual tab. Choose Selected Space from the Update flyout in the Modify panel.
Shortcut menu	Choose an associative space, right-click, and choose Update Space Geometry > Selected Spaces or Update Space Geometry > All Associative Spaces.

If you edit drawings from releases prior to 2007, the Automatically update associative spaces checkbox is clear by default. Therefore, check the Automatically update associative spaces checkbox to update spaces when walls are modified.	**TIP**

If you Freeze or turn OFF the A-Area-Spce-Patt [Aec-Area-Spce-Patt] layer, the space will automatically update to boundary changes when you can grip edit walls or change other Aec Objects. If you Freeze the A-Area-Spce [Aec Space] layer, the space will not update when editing the walls. The SPACEUPDATESELECTEDSPACES command can be used to update spaces that were frozen.	**TIP**

When you create a space using the **Generate** option of **Create type,** a space is created with an associative relation to the bounding walls. When you select the space, there are no grips displayed to alter the space. However, you can toggle the associative relation to No in the Properties palette to temporarily cut the link to the bounding walls and thus the grips of the nonassociative space are available. The grips of a nonassociative space are displayed as shown at the right in Figure 4.11. If you select the nonassociative space, choose Make Associative from the Modify panel of the Space tab or choose Make Associative from the shortcut menu; the space becomes associative and will change to fit any boundary changes.

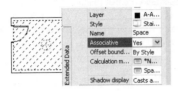

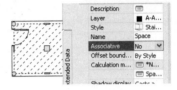

Associative Space Selected   Associativity Off-Grips Displayed   Choose Make Associative to Set Associativity to Yes

FIGURE 4.11 *Grips of Associative and Nonassociative Spaces*

Editing Associative Spaces

You can make revisions to existing spaces to subtract the areas of such objects as structural columns or multi-view blocks by editing the Bound Space property of the AEC object to Yes in the Properties palettes. Pipe chase symbols are multi-view blocks located in the Chases category of the Documentation Tool Catalog – Imperial or Documentation Tool Catalog – Metric. The **SpaceInterferenceAdd** command can be used to subtract the area of stairs from a space. Objects with bounding boxes such as columns and stairs can be subtracted from the space using the interference tool. Access the **SpaceInterferenceAdd** command as shown in Table 4.4. Table 4.5 provides a list of objects and entities and the method of subtraction.

TABLE 4.4 *Accessing the SpaceInterferenceAdd command*

Command prompt	SPACEINTERFERENCEADD
Ribbon	Select a space to display the Space contextual tab and choose Interference from the Modify panel.
Shortcut menu	Select the space, right-click, and choose Interference Condition > Add.

TABLE 4.5 *Revising spaces for symbols and objects*

Component	Method of Subtraction
Structural Columns	Select the column; edit Bound Spaces to Yes in the Properties palette.
AutoCAD Entity	Select the entity; edit Bound Spaces to Yes in the Advanced section of the Properties palette.
Multi-view blocks developed from closed polylines	Select the multi-view block; edit Bound Spaces to Yes in the Properties palette.
Stairs	Select the space, choose Interference > Add from the Modify panel of the Space tab, and then select the stair. Choose the Reflected display configuration to view the interference geometry of the stair.

If you add a wall that crosses a space as shown in Figure 4.12, the wall can function as a divider of the space. However, some walls that should not divide a space are added to a plan. You can control the function of the wall by the **Bound Spaces** property of the wall. If you set Bound Spaces to **Yes,** the wall will serve as a boundary for the spaces. When Bound Spaces is set to **No** and a wall is added that crosses the space, the space does not divide. The **By Style** option refers to the checkbox *Objects of this style may act as a boundary for associative spaces* of the General tab in the wall style properties dialog box. The space shown in Figure 4.12 includes a wall without the Bound Spaces property that penetrates the space. The space is not divided by the wall.

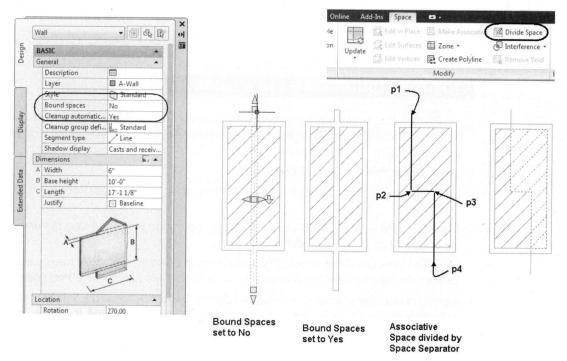

FIGURE 4.12 *Modifying an associative space*

Dividing Associative Space Using SpaceSeparator

You can divide an associative space by specifying two or more points. The first and last segments must cross the boundary of the associative space. Access the **SpaceSeparator** command as shown in Table 4.6.

TABLE 4.6 *Accessing the SpaceSeparator command*

Command prompt	SPACESEPARATOR
Ribbon	Select an associative space to display the Space contextual tab and choose Divide Space from the Modify panel.
Palette	Select the **SpaceSeparator** command of the Design palette in the Design palette group.

The SpaceSeparator command can be used to divide an associative space as shown in Figure 4.12.

> (Select an associative space and choose Divide Space from the Modify panel of the Space contextual tab in the ribbon.)
>
> Select first point: *(Select a point near p1 shown in Figure 4.12.)*
>
> Specify next point: *(Select a point near p2 shown in Figure 4.12.)*
>
> Specify next point: *(Select a point near p3 shown in Figure 4.12.)*
>
> Specify next point: *(Select a point near p4 shown in Figure 4.12.)*

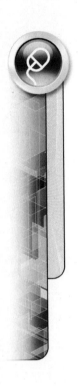

TUTORIAL 4-1: EDITING SPACES TO A FLOOR PLAN

1. Verify that the Accessing Tutor\Imperial or Accessing Tutor\Metric content for Chapter 4 of the CengageBrain has been downloaded to your Accessing Tutor folder on your computer as described in Organizing Tutorial Directories in the Preface.

2. Select **Open** from the **Quick Access** toolbar.

3. Select Accessing Tutor\Imperial\Ch4\Ex 4-1.dwg or Accessing Tutor\Metric\Ch 4\ Ex 4-1.dwg.

4. Choose SaveAs > AutoCAD Drawing from the Application menu and save the drawing as Lab 4-1.dwg in your student directory.

5. Click the Home tab of the ribbon. Choose layer drop-down list from the Layers panel. Turn on the light-bulb for the A-Area-Spce [Aec-Space] and A-Area-Spce-Patt layers.

6. Right-click over the command window and choose Options. Choose the Selection tab and verify in the Selection Preview section that **When a command is active** and **When no command is active** are checked. Click OK to dismiss the Options dialog box.

7. Select the polyline at p1 as shown in Figure 4.13. Verify that the Polyline entity is selected in the Properties palette. Expand the Advanced category. Scroll down within the Properties palette to display Bound Spaces. Edit Bound Spaces to Yes. Press ESC to clear the selection and view revised space geometry.

8. Select the chase at p2 as shown in Figure 4.13. Edit Bound Spaces to Yes in the Basic section of the Properties palette. Space is revised to exclude the chase. Press ESC to end the command.

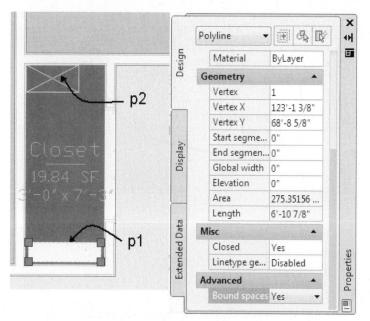

FIGURE 4.13 *Interference condition created with space*

9. Choose **Zoom Window** from the Zoom flyout of the **Navigation bar**. Respond to the workspace prompts as shown below:

   Specify first corner: **98',91' [30000,27750]**

   Specify opposite corner: **114',65' [35000,20000]**

10. Choose the Layer drop-down list of the Layers panel. Select the light-bulb to turn off the A-Area-Spce [Aec-Space] and A-Area-Spce-Patt layers.

11. Toggle Polar Tracking ON in the Application status bar.

12. Right-click over the command window and choose Options from the shortcut menu. Choose the Aec Object Settings tab. Verify that Automatically update associative spaces is toggled ON. Click OK to dismiss the Options dialog box.

13. Select the wall at **p1** as shown in Figure 4.14 and verify that **Bound Spaces** is set to By Style **(Yes)** in the **Properties** palette. Choose the location grip at the midpoint of the wall and move the cursor right. Type **1' [300]** in the dynamic input workspace prompt. Press ESC to clear the selection.

14. Choose the **Layer** drop-down list of the Layers panel. Select the light-bulb to turn on the A-Area-Spce-Patt and A-Area-Spce [Aec-Space] layers. The space is updated to fit the new boundary as shown in Figure 4.14.

15. Save and close the drawing.

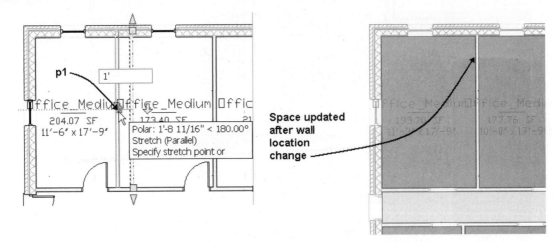

FIGURE 4.14 *Spaces updated to fit wall location*

INSERTING A SPACE

A nonassociative space is created when you choose the **Space (SpaceAdd)** tool from the Design palette or the Space tool from the Space flyout menu in the Build panel of the Home tab. You can set the Create type to Insert, Rectangle, or Polygon in the Properties palette. Only nonassociative 2D and Extrusion Geometry types can be created with the Insert, Rectangle, or Polygon Create types. The Generate option of Create type can be used to create spaces of all geometry types that are associative or nonassociative spaces. Access the **SpaceAdd** command as shown in Table 4.7.

TABLE 4.7 *Accessing SpaceAdd tool*

Command prompt	SPACEADD
Ribbon	Choose Space from the Space flyout in the Build panel of the Home tab.
Tool palette	Choose Space from the Design palette or a space tool from the Spaces palette.

When you choose the Create type option, the Component Dimensions and Actual Dimensions sections of the Properties palette change to permit you to specify the space. When you choose the **SpaceAdd** command, the Properties palette opens, allowing you to choose 2D and Extrusion types as shown in Figure 4.15. The Properties palette for placing an Extrusion space using the Rectangular Create type is shown in Figure 4.15. The space style defines the name of the space and parameters such as minimum and maximum dimensions and area of the space. The active values allow you to control the size of the space based upon the Create type selected. You can edit the ceiling and floor boundary locations of the Extrusion space in the Properties palette. The Component Dimensions of the Extrusion space can be modified as shown in Figure 4.15. When you place a space, the area of the floor and the area of the ceiling are extracted and will be displayed in the Properties palette of an existing space.

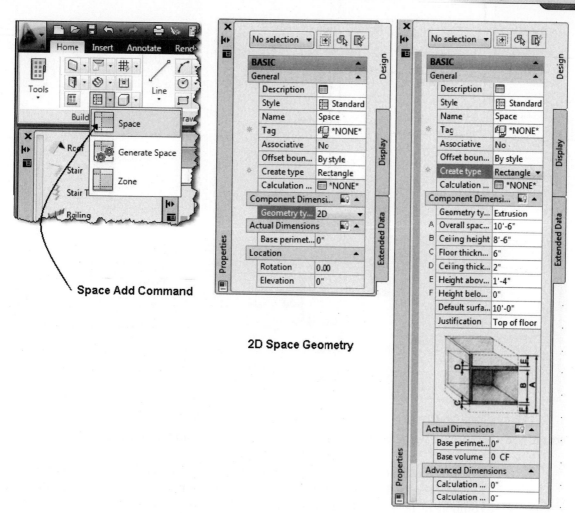

Space Add Command

2D Space Geometry

Extrusion Space Geometry

FIGURE 4.15 *Properties of a space*

If you select a space that has been inserted, the Properties palette provides detailed information regarding the actual dimensions of the space as shown in Figure 4.16. You can toggle the Geometry Type from **Freeform** or **Extrusion** to the **2D geometry** type in the Properties palette. The Freeform space that is created using Space Generate extracts ceiling, wall, and floor properties from size and shape of the bounding objects. If you select a 2D geometry space and toggle it to Extrusion or Freeform, default properties for the ceiling and floor heights are assigned since the space was not developed from three-dimensional boundaries.

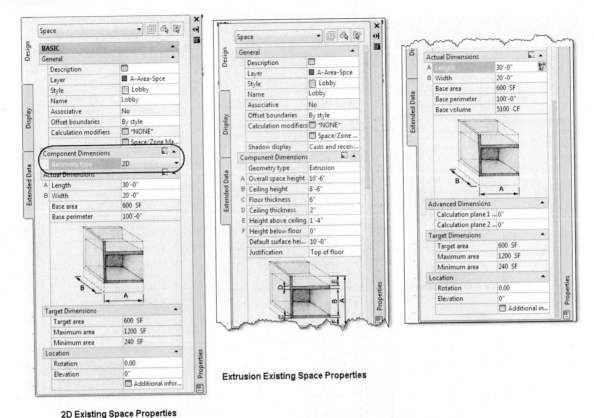

2D Existing Space Properties

Extrusion Existing Space Properties

FIGURE 4.16 *Content in the Design tab of the Properties palette for a 2D and Extrusion existing space*

The **Display** tab of the Properties palette, shown at the left in Figure 4.17, allows you to control space display by object, style, or drawing default. If you click in the Display component field, the flyout lists all the display components of a space. If a display component is selected, you can edit such options as display, color, linetype, and lineweight.

The **Extended Data** tab of the Properties palette shown at the right in Figure 4.17 lists additional properties of the space that can be developed into a schedule. The additional properties can be added to property sets in the Style Manager. The Base-AreaMinusInterference reflects the resultant area of a space that includes interference subtractions for columns or chases. The method to add property sets and schedules to a drawing is presented in Chapter 11.

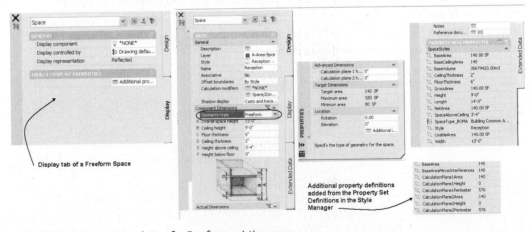

FIGURE 4.17 *Properties palette of a Freeform existing space*

Setting Properties of a Space

The contents of the Properties palette when inserting a space with Create type set to **Insert** are described in the following sections.

Design Tab—Basic

General

Description—The Description button allows you to add text to describe the space.

Style—The Style list includes the predefined space styles. The Standard space style is the default space style in drawings.

Name—The Name drop-down list includes names defined for the style created within the List Definitions category of the Style Manager.

Tag—The Tag drop-down lists Aec8_Room_Tag [M_Aec8_Room_Tag] multi-view block tags. If you select a tag from the drop-down list, you will be prompted to specify the insertion point for the room tag when the space is created. You can import additional tags from the Schedule Tags > Room & Finish Tags of the Documentation Tool Catalog (Imperial) and Documentation Tool Catalog (Metric).

Associative—Spaces inserted with the Insert Create Type are nonassociative. Existing spaces can be specified as Associative: Yes or No.

Offset Boundaries—The Offset Boundaries options include By Style or Manual. The offset distance for Base, Usable, Gross, and Net boundaries can be specified in the space style or by selecting the grips to manually adjust the offsets. Boundary offsets are specified in floor measurement standards.

Create Type—The Create Type is preset to Insert; additional options include: Rectangle, Polygon, or Generate.

Calculation Modifiers—The Calculation Modifiers field lists calculation modifiers defined in the Style Manager. Calculation Modifiers can be created to reduce the area of a space to account for the tiling finish in a toilet.

Component Dimensions

Geometry Type—The Geometry Type option includes 2D and Extrusion when Insert is the Create type.

A-Overall space height—The Overall space height specifies the distance between the bottom of the space below the floor and the height above the ceiling for existing spaces.

B-Ceiling height—The Ceiling height is the distance between the top of the floor component and the bottom of the ceiling component for existing spaces.

C-Floor thickness—The Floor thickness allows you to specify the thickness boundary that simulates the floor of the space.

D-Ceiling thickness—The Ceiling thickness allows you to specify the thickness of the boundary that simulates the ceiling of the space for existing spaces.

E-Height above ceiling—The Height above ceiling field defines the distance between the top of the ceiling and the top of the Overall space height for existing spaces.

F-Height below floor—The Height below floor consists of the distance of space that projects below the floor boundary for existing spaces.

Default surface height—The Default surface height is the overall height between the top of the Height above ceiling and the bottom edge of the Height below floor for existing spaces.

Justification—The Justification drop-down list consists of Top of floor, Bottom of floor, or Bottom of space below floor options for existing spaces.

Actual Dimensions

Specify on Screen—The Specify on Screen option is displayed for the Insert space type. A Yes response allows you to select locations in the workspace to define the length and width of the space. A No response activates the A-Length and B-Width fields.

Constrain—The Constrain field, displayed for the Insert space type, lists the following options: Area, Length, Width, and None. If Constrain is set to Area, the dimensions specified are restricted to area limits of the style. If the Area field is specified, the length and width dimensions of the space will vary as necessary to provide the area. The Length and Width constraints will restrict the length and width as specified. The None constraint allows you to specify the size of the space without limitations of area, length, or width.

A-Length—When the Insert space type is specified and Constrain is set to None, Area, or Width, the A-Length field specifies the length dimension of the space.

B-Width—The B-Width field is active if Constrain is set to None, Area, or Length. The B-Width field specifies the width dimension of the space.

Base area—The Base area option displayed for the Insert space type lists the default area of the space defined by the space style.

Base perimeter—The Base perimeter option displayed for the Insert space type lists the default perimeter of the space defined by the space style.

Base volume—The Base volume option displayed for the Insert space type lists the default volume of the space defined by the space style.

Advanced Dimensions

Calculation plane 1 height—The Calculation plane 1 height allows you to specify a height elevation of a plane used to determine the area of a freeform space at the specified height.

Calculation plane 2 height—The Calculation plane 2 height allows you to specify an alternate height elevation of a plane used to determine the area of a freeform space at the specified height.

Target Dimensions

Target area—The Target area is listed for the space as specified in the space style.

Maximum area—The Maximum area is listed for the space as specified in the space style.

Minimum area—The Minimum area is listed as specified in the space style.

Location

> **Rotation**—The Rotation value listed includes the angle of rotation for the nonassociative space.

> **Elevation**—The Elevation value listed specifies the elevation height of the space.

The height dimension of a space is inherited from the last insertion. The SpaceAdd command does not extract the height dimension from walls or other geometry when the Extrusion space is created.

NOTE

When you choose the SpaceAdd command, edit the **Create Type** field to specify how to insert a space. The Generate option discussed earlier in the chapter places an associative space bound by entities and Aec Objects. The Insert, Rectangle, or Polygon options are provided for creating nonassociative spaces. The options of the **Create Type** field control the workspace prompts and the content in the Design tab of the Properties palette. If you choose **Insert** option, you will be prompted in the workspace to specify the insertion point and rotation as shown in Figure 4.18. The actual dimensions can be specified in the workspace or in the properties palette, depending upon the settings of **Specify on screen**. Specify on screen is set to No as shown in Figure 4.18; therefore, you enter the length and width in the Properties palette. The **Rectangle** create type sizes a rectangle when you specify the start and endpoints of a diagonal of the rectangle, whereas the **Polygon** create type allows you to specify the vertices of the polygon as shown in Figure 4.18.

Insert

Specify insertion point and rotation Actual Dimensions can be specified on screen. Constrain can be based upon Length, Width, and Area

Rectangle

Specify Start corner and End corner to locate the diagonals of the space.

Polygon

Specify the location of vertices of a polygon

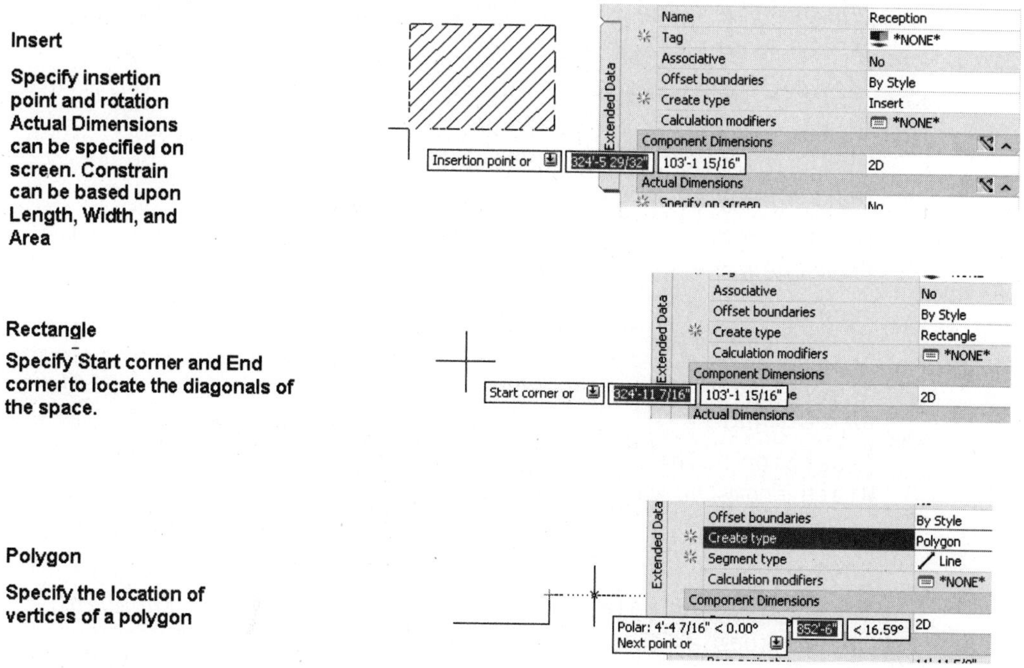

FIGURE 4.18 *Creating Type settings of a new space*

DRAGGING THE INSERTION POINT

The default location for the insertion point of a space placed with the Insert option is located at the lower-left corner of the space. You can toggle the insertion point in a counterclockwise direction to the other corners of the space by selecting the **Drag**

point option in the workspace options. Each successive selection of the Drag point option continues to move the insertion point around the space in a counterclockwise direction as shown in Figure 4.19.

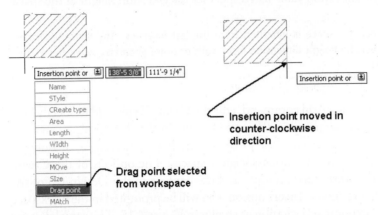

FIGURE 4.19 *Insertion point moved using Drag point option*

Setting **Yes** to **Specify on Screen** and **Constrain** to **None** allows you to size the length or width in the workspace. If you are placing the Standard style, the size of the space can be specified in the workspace to create a space up to 10,000 square feet. The following workspace prompts sized the space to fit the length and width dimensions as shown in Figure 4.20.

(Choose the Space *(SpaceAdd) tool on the Design palette. Set create type to Insert in the Properties palette.*

Set Specify on Screen *to Yes and* Constrain *to None. Move the cursor from the Properties palette into the workspace.)*

Insertion point or: SI *(Press the down arrow, and choose Size from the workspace options.)*

Insertion point or : L *(Press the down arrow, and choose* L *from the workspace options.)*

Length <9'-9 9/16>: *(Choose point* p1 *using the Endpoint object snap.)*

Specify second point: *(Choose point* p2 *using the Endpoint object snap.)*

Insertion point or : WI *(Press the down arrow, and choose* **Width** *from the workspace options.)*

Width <9'-9 9/16>: *(Choose point* p2 *using the Endpoint object snap.)*

Specify second point: *(Choose point* p3 *using the Endpoint object snap.)*

Insertion point or : *(Choose point* p4 *using the Endpoint object snap.)*

Rotation or <0.00>: *(Press* ENTER *to accept 0 rotation angle.)*

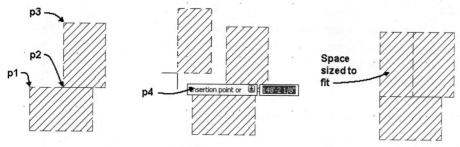

FIGURE 4.20 *Creating spaces with Specify on screen option*

MODIFYING NONASSOCIATIVE SPACES

The AEC Modify commands located on the shortcut menu of a space allow you to trim, divide, subtract, merge, crop, array, reposition from, space evenly, and center the space. The AEC Modify tools can be applied to other AEC objects such as doors, windows, walls, and detail components. The following description demonstrates a sample of the uses of the AEC Modify tools to modify the spaces. Access the **Trim** tool as shown in Table 4.8.

TABLE 4.8 *Accessing the LineworkTrim command*

Command Prompt	LINEWORKTRIM
Shortcut menu	Select a nonassociative space, right-click, and choose AEC Modify Tools > Trim.

(*Select a nonassociative space, right-click, and choose* AEC Modify > Trim.)

Select the first point of the trim line or ENTER to pick on screen: (*Select point* **p1** *as shown in Figure 4.21.*)

Select the second point of the trim line: (*Select point* **p2** *as shown in Figure 4.21.*)

Select the side to trim: (*Select a point near* **p3** *as shown in Figure 4.21.*)

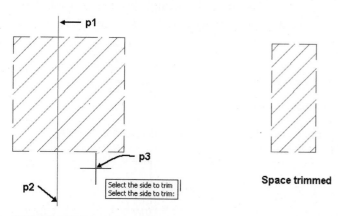

Space trimmed

FIGURE 4.21 *Trimming a space*

Divide

The Divide command **LineworkDivide** will divide the space into two components. Access the LineworkDivide command as shown in Table 4.9. The dividing lines are specified in the workspace prompts as shown below. The spaces are retained, allowing you to resize the spaces.

TABLE 4.9 *Accessing the LineworkDivide command*

Command prompt	LINEWORKDIVIDE
Shortcut menu	Select a nonassociative space, right-click, and choose AEC Modify > Divide.

(*Select a nonassociative space, right-click, and choose* AEC Modify > Divide.)

Select the first point of the dividing line or ENTER to pick on screen: (*Select a point near* p1 *as shown in Figure 4.22.*)

Select the second point of the dividing line: (*Select a point near* p2 *as shown in Figure 4.22.*)

(*Space is divided as shown at the right in Figure 4.22.*)

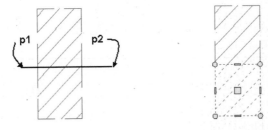

Space Divided

FIGURE 4.22 *Using LineworkDivide command to divide a space*

Subtract

The **Subtract (LineworkSubtract)** command allows you to subtract the area enclosed by a closed shape from a space. Access the **Subtract** command as shown in Table 4.10. The area of the rectangle is subtracted from the space as shown in Figure 4.23 and the following workspace prompts.

TABLE 4.10 *Accessing the LineworkSubtract command*

Command prompt	LINEWORKSUBTRACT
Shortcut menu	Select a nonassociative space, right-click, and choose AEC Modify > Subtract.

Select object(s) to subtract or NONE to pick rectangle: (*Select the rectangle at* p1 *as shown in Figure 4.23.*) 1 found

Select object(s) to subtract or NONE to pick rectangle: ENTER (*Press* ENTER *to end selection.*)

Erase selected linework? [Yes/No] <No>: Y *(Choose* Yes *to erase the space inside the rectangle.)*

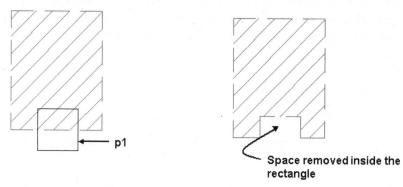

FIGURE 4.23 *Modifying a space using LineworkSubtract*

Merge

The **Merge (LineworkMerge)** command can be used to merge two spaces that overlap into one space. Access the Merge command as shown in Table 4.11. The spaces shown in Figure 4.24 were merged as shown in the following workspace prompts.

TABLE 4.11 *Accessing LineworkMerge command*

Command prompt	**LINEWORKMERGE**
Shortcut menu	Select the space, right-click, and choose AEC Modify > Merge.

(Select a space, right-click, and choose AEC Modify > Merge.*)*

Select object(s) to merge or NONE to pick rectangle: 1 found *(Select the space at* **p1** *as shown in Figure 4.24.)*

Select object(s) to merge or NONE to pick rectangle: *(Press* ENTER *to end selection.)*

Erase selected linework? [Yes/No] <No>: Y *(Choose* Yes *to erase boundary.)*

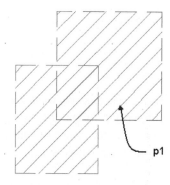 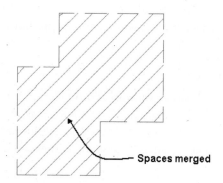

FIGURE 4.24 *Merging spaces*

Crop

If you sketch a shape using linework within a space, you can use the **Crop (Linework-Crop)** command to convert the shape into a space. If you are editing a space with offset boundaries set to manual, select all trigger grips to turn on the display of all boundaries prior to executing the command as shown in Figure 4.25. Access the Crop command as shown in Table 4.12.

TABLE 4.12 *Accessing the LineworkCrop command*

Command prompt	**LINEWORKCROP**
Shortcut menu	Select the space, right-click, and choose AEC Modify > Crop.

(*Select a space, right-click, and choose* AEC Modify > Crop.)

Select object(s) to form crop boundary or NONE to pick rectangle: 1 found (*Select the line at* **p1** *as shown in Figure 4.25.*)

Select object(s) to form crop boundary or NONE to pick rectangle: (*Press* ENTER *to end selection.*)

Erase selected boundary? [Yes/No] <No>: Y (*Choose* Yes *to erase all boundaries.*)

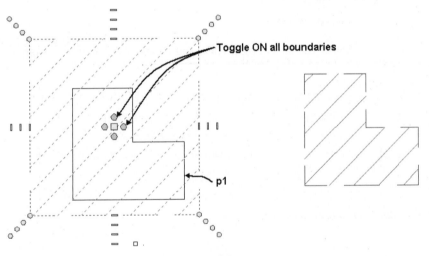

FIGURE 4.25 *Defining a space using LineworkCrop*

Array

The **Array (LineworkArray)** command allows you to array a space, a specified distance, or a number. The **Pick array distance** option is described in the following workspace prompts; however, additional options allow you to enter an array count or clear distance. Access the **LineworkArray** command as shown in Table 4.13.

TABLE 4.13 *Accessing the LineworkArray command*

Command prompt	**LINEWORKARRAY**
Shortcut menu	Select the space, right-click, and choose AEC Modify > Array.

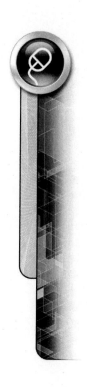

(*Select the space, right-click, and choose* `AEC Modify > Array.`)

`Select an edge to array from or ENTER to pick two points:` (*Select a point at* **p1** *as shown in Figure 4.26.*)

`Drag out array elements or: p` (*Choose the down arrow and choose* `Pick array distance/Enter count.`)

`Specify new array distance <3'-0">: _endp of` (*Choose the endpoint at* **p2** *as shown in Figure 4.26.*)

`Specify second point: _endp of` (*Choose the endpoint at* **p3** *as shown in Figure 4.26.*)

`Drag out array elements or:` (*Move cursor right to array space; to specify array count click near* **p4** *as shown in Figure 4.26.*)

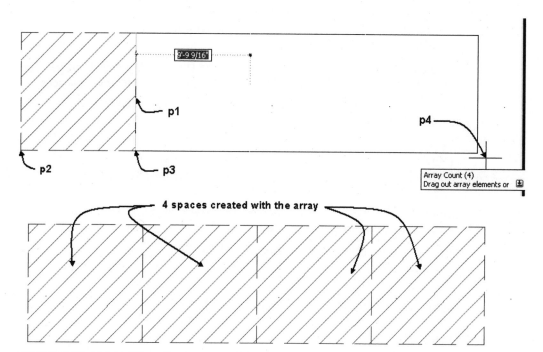

FIGURE 4.26 *AEC Modify tools to Array a Space*

TUTORIAL 4-2: CREATING A SPACE DIAGRAM

1. Select **New** from the Quick Access toolbar.

2. Choose **SaveAs > AutoCAD Drawing** from the Application menu. Save the drawing as **Lab 4-2** in your student directory.

3. Choose the Home tab. If the Design palette group is not displayed, click the Tools button of the Build panel to display the Design palette group.

4. Verify that only Polar Tracking, Object Snap, Object Snap Tracking, Dynamic Input, and Quick Properties are toggled ON on the status bar. Right-click over the Dynamic Input toggle and choose **Settings**. Verify that **Enable Pointer Input**, **Enable Dimensional Input where possible**, and **Show command prompting and command input near the crosshairs** are checked. Click **OK** to dismiss the dialog box.

5. Move the cursor over the **Object Snap** button on the status bar, right-click, and choose **Settings**. Edit the **Drafting Settings**, **Object Snap** tab as follows: clear all Osnap modes except **Node**. Check Allow general object snap settings to act upon wall justification line. Click the **Options** button on the **Object Snap** tab, toggle ON the **Display polar tracking vector**, **Display full-screen tracking vector**, **Display Auto Track tooltip, Marker, Magnet**, and **Display AutoSnap tooltip** options. Click **OK** to dismiss the **Options** dialog box and click **OK** to dismiss the Drafting Settings dialog box.

6. Choose the **Spaces** palette of the **Design** palette set.

7. Choose the **Office (Small)** space from the **Spaces** palette. Verify the following in the **Properties** palette: Offset boundaries = **By Style**, Create type = **Insert**, Specify on screen = **No**, and Constrain = **None**. Move the cursor from the Properties palette to the workspace and respond to the workspace prompts as follows:

   > Insertion point or: (Choose a point in the middle of the workspace.)
   >
   > Rotation or: ENTER (Press ENTER to accept 0 rotation. Press ESC to end the command.)

8. Choose **Low Detail** display configuration from the Drawing Window status bar. The hatch pattern of Low Detail is User Single, which allows easy selection of spaces.

9. Select the Office (Small) space, right-click, and choose AEC Modify Tools > Array. Respond to the workspace prompts as follows:

   > Select an edge to array from or ENTER to pick two points: *(Move the cursor to the edge at* **p1** *as shown in Figure 4.27. When the cyan color line is displayed along the edge, click to select the edge.)*
   >
   > Drag out array elements or: **p** *(Choose the down arrow and choose* **Pick array distance**. *)*
   >
   > Specify new array distance <3'-0" [1000]>: _endp of *(Shift+right-click, choose endpoint object snap, and select* **p2** *as shown in Figure 4.27.)*
   >
   > Specify second point: _endp of *(Shift+right-click, choose the endpoint object snap, and choose* **p3** *as shown in Figure 4.27.)*
   >
   > Drag out array elements or: *(Move cursor right to array space. Click when the Array Count =* **4** *in the tooltip.)*

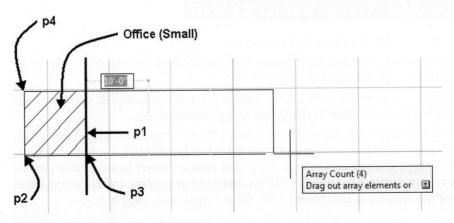

FIGURE 4.27 *Arraying of an office space*

10. Choose the Corridor space from the Spaces palette. Verify the following in the Properties palette: Offset boundaries = By Style, Create type = Insert, Specify on screen = No, Constrain = None, Length = 36' [11000], and Width = 5' [1500]. Move the cursor from the Properties palette to the workspace and respond to the workspace prompts as follows:

> Insertion point or: *(Select the Node of the* Office(Small) *space or* **p4** *as shown in Figure 4.27.)*
>
> Rotation or: ENTER *(Press ENTER to accept 0 rotation. Press ESC to end the command.)*

11. Select the Office (Small) space at p1 as shown in Figure 4.28, right-click, and choose Select Similar. Right-click and choose Basic Modify Tools > Copy. Respond to the workspace prompts as follows:

> Specify base point or: *(Select a point using the* **Node** *object snap at* **p2** *as shown in Figure 4.28.)*
>
> Specify second point or [Array] <use first point as displacement>: *(Select a point using the* **Node** *object snap at* **p3** *as shown in Figure 4.28.)*
>
> Specify second point or [Exit/Undo] <Exit>: *(Press ESC to end the command.)*

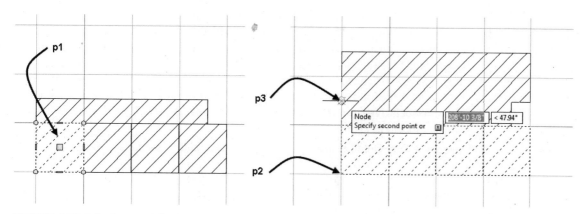

FIGURE 4.28 *Selecting space for copy*

12. Choose the Lobby space from the Spaces palette. Verify the following in the Properties palette: Offset boundaries = By Style, Create type = Insert, Specify on screen = No, Constrain = None, Length = 24' [7300], and Width = 24' [7300]. Move the cursor from the Properties palette to the workspace and respond to the workspace prompts as follows:

> Insertion point or: *(Choose the down arrow of the workspace and choose* **Drag point** *to toggle the insertion point to the lower-right corner of the space. Click to insert the space at* **p2** *as shown in Figure 4.28.)*
>
> Rotation or: ENTER *(Press* ENTER *to accept 0 rotation. Press* ESC *to end the command.)*

13. In this step, you will add walls to enclose the spaces. Choose the Wall tool from the Build panel in the Home tab. Edit the Properties palette as follows: Bound space = Yes, Cleanup automatically = Yes, Width = 6 [150], Base height = 8'-0" [2500],

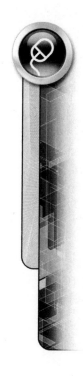

Justify = Baseline. Draw the walls using the Node object snap as shown in Figure 4.29. Extend the walls beyond the intersecting walls as shown.

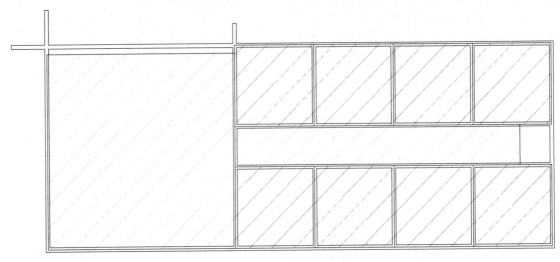

FIGURE 4.29 *Walls added to define the space*

14. Move the cursor over the perimeter of a space to display the rollover highlight of the space, click and verify that a Space object has been selected in the Properties palette, and choose Quick Select in the Design tab of the Properties palette to open the Quick Select dialog box. Edit Apply to = Entire drawing, Object type = Space, Operator = Select All; check Include in new selection set and Append to current selection set. Click Ok to dismiss the dialog box and display the spaces. Right-click and choose **Make Associative** from the shortcut menu. The spaces are adjusted to the wall locations as shown in Figure 4.30.

15. Select Medium Detail display configuration. Save and close the drawing.

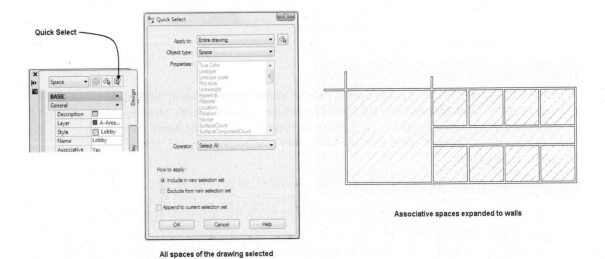

All spaces of the drawing selected

Associative spaces expanded to walls

FIGURE 4.30 *Spaces resized to fit walls*

CREATING A SPACE STYLE

Space styles from commercial, education, medical, and residential palettes can be added using More Space Tools command of the Spaces palette. The More Space Tools command opens the Content Brower to Design Tool Catalog – Imperial or the Design Tool Catalog – Metric. Additional space styles can be created and edited in the **Style Manager**. Space styles are created and edited in the *Architectural Objects\ Space Styles* folder of the **Style Manager** for each drawing. If you choose **Open** in the Style Manager, the Open Drawing dialog box opens; choose the **Content** folder of **Places** in the file dialog box, and you can import space styles from metric and imperial folders of *{Vista – C:\ProgramData\Autodesk\ACA 2012\enu\Styles}* directory.

Applying CopyAndAssignStyle to Space

The **CopyAndAssignStyle** command (see Table 4.14) allows you to create a space style from an existing style. The command opens the **Space Style Properties** dialog box, which copies the space and allows you to rename and edit the properties without opening the Style Manager.

TABLE 4.14 *CopyAndAssignStyle command access*

Command prompt	COPYANDASSIGNSTYLE
Ribbon	Select a Space and choose Copy Style from the General panel of the Space contextual tab.
Shortcut menu	Select a Space, right-click, and choose Copy Space Style and Assign Style.

Space Style Properties

When you insert a space, the name of the space is associated with the function. Each space is defined by a style definition that can be edited in the Style Manager or its **Space Style Properties** dialog box. You can select a space and choose Edit Style from the General panel to open the Space Style Properties dialog box. The tabs of the Space Style Properties dialog box are: **General, Design Rules, Materials, Classifications, Display Properties,** and **Version History**. The **Design Rules** tab of the Space Style Properties dialog box is shown in Figure 4.31. This tab allows you to define a target length, width, and area for the space. These dimensions form the upper and lower limits for the dynamic sizing of the space. The **Materials** tab allows you to define materials to the ceiling and floor components of the space. The Design Rules tab, unique to space styles, is described as follows:

Area—The target area, minimum area, and maximum area for a space are defined in the **Dimensions** tab.

Length—The target, minimum, and maximum length values allow you to define the length of the space.

Width—The target, minimum, and maximum width values allow you to define the width of the space.

Net Offset—The **Net offset** value defined in the style determines the offset distance from the base boundary to locate the Net boundary.

Gross—The gross distance defined in the style creates an offset from the base boundary to locate the Gross boundary.

Usable—The usable distance defined in the style creates an offset from the base boundary to locate the Usable boundary.

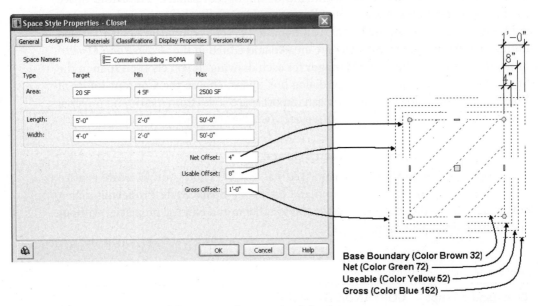

FIGURE 4.31 *Design Rules tab of the Space Style Properties dialog box*

Modifying the Boundaries of a Space

Four unique boundaries are inserted when a space is placed: **Base, Net, Usable**, and **Gross**. The boundaries may be coincident to each other or may be offset. The offset of the boundaries can be defined in the style definition. If boundary offset is set to By Style in the Properties palette, the boundaries will be displayed according to the values specified in the Design Rules tab in Figure 4.31. The purpose of the boundaries is to establish areas defined in standards of measurement for leasing or appraising space. The Standard Method for Measuring Floor Area In Office Buildings - BOMA (Building Owners and Managers Association) is a standard that specifies the location of the boundary. A general description of the boundaries follows:

Base boundary—The base boundary is a measure of the actual room floor space appropriate for maintenance and operations of a building.

Net boundary—The net boundary is an area less than the base boundary that has been adjusted for reduced ceiling heights as occurs in attic rooms.

Usable boundary—The usable boundary is the area adjusted for renting, pricing, and tax base by standards. The usable boundary can extend from the inside of exterior walls to the middle of an interior wall width.

Gross boundary—The gross boundary is an area adjusted in standards for cost and price purposes. The floor area is adjusted according to the exterior and interior wall functions.

Grips of the Nonassociative Space with Manual Boundary Offsets

The grips of the nonassociative space with manual offset boundaries are shown in Figure 4.32. The boundary offsets can be defined by **By Style** or **Manual** in the Properties palette. The space shown in Figure 4.32 is set to Manual; therefore, the grips are provided for the adjustment of each vertex and edge of the space. The round trigger grips allow you to choose which boundary you are editing. A dark gray trigger grip indicates that the edit is off and the light gray trigger grip indicates that the boundary edit is on. Click a dark gray trigger grip to toggle ON the boundary. Only the grips of the Base Boundary shown at the left in Figure 4.32 can be edited, whereas all the grips shown at the right are displayed since all trigger grips are toggled ON. The boundary edits can be removed by choosing **Reset All** from the Boundary panel of the ribbon.

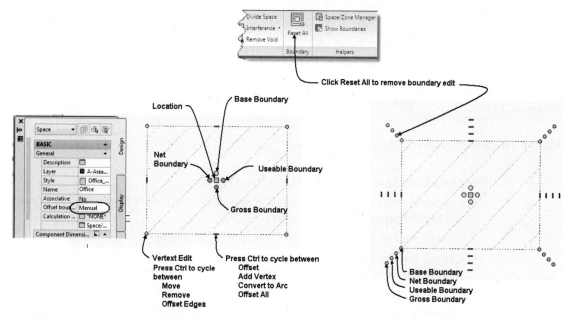

FIGURE 4.32 *Grips of a Nonassociative Space*

> **NOTE**
>
> Access Instructional Video 4.1 Creating Spaces from the in the Instructional Video category of the Student Companion site of CengageBrain http://www.cengagebrain.com described in the Preface.

EVALUATING SPACES FOR SPACE PLANNING

Summaries of spaces can be displayed using Zones, Space Zone Manager, and Space Evaluation. A List Definition can be created to enhance the naming system for spaces. The purpose of list definitions is to create a preset list for spaces, zones, or manual property definitions. Access List Definitions as shown in Table 4.15.

TABLE 4.15 *Accessing List Definition command*

Ribbon	Choose Manage tab. Choose Style Manager from the Style & Display panel. Expand Multi-Purpose Objects. Choose List Definitions.

The list is specified for the space or zone template in their respective style definitions. Therefore, when you place a space, its associated list definition includes the preset names from the list definition. Spaces inserted from the Spaces palette include names from the Commercial Building – BOMA list definition style. The names are displayed in the drop-down list of names in the Properties palette as shown in Figure 4.33. The following steps outline the procedure for creating the name list.

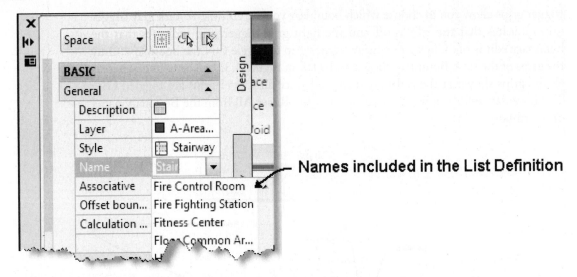

FIGURE 4.33 *Accessing Names in the Properties palette*

Steps to Create a List Definition for a Space

1. Choose Manage tab. Choose Style Manager from the Style & Display panel. Expand the **Multi-Purpose Objects** category.

2. Choose the **List Definitions**, right-click, and choose **New**.

3. Choose the **General** tab of the right pane as shown in Figure 4.34 and overtype the name of the new list in the Name field.

4. Choose the **Applies To** tab of the right pane and check the Space Names checkbox.

5. Choose the **Items** tab of the right pane. Click the **Add** button as shown at the right in Figure 4.34. Edit the name column of each name for the list.

6. Continue within the Style Manager and expand the **Architectural Objects > Space Styles**.

7. Choose a space style in the left pane, right-click, and choose **Edit**. Choose the **Design Rules** tab of the right pane.

8. Edit the drop-down list to specify the name list for the space style as shown in Figure 4.35. When this space is inserted, the list definition linked to the space style will be displayed in the **Name** field in the Design tab of the Properties palette.

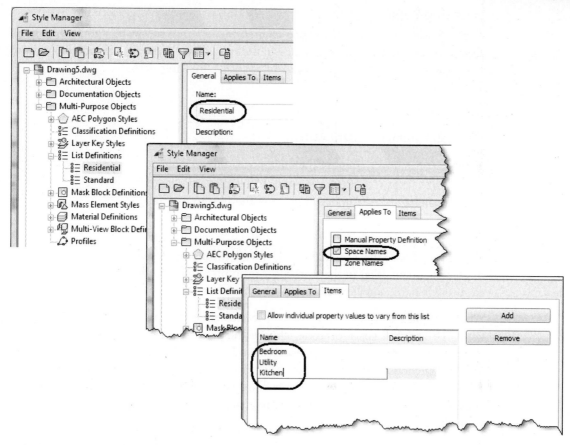

FIGURE 4.34 *Creating list definitions*

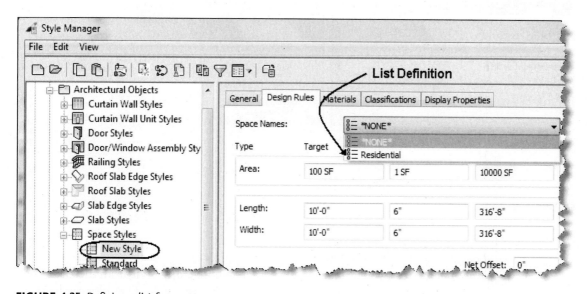

FIGURE 4.35 *Defining a list for a space*

Inserting Zones

Zones can be inserted in a drawing to link one or more spaces to a zone. The zone allows you to sum the spaces included in the zone. Zone styles that group one or more zones can be created. Access **ZoneAdd** as shown in Table 4.16.

TABLE 4.16 *Accessing the ZoneAdd tool*

Command prompt	ZONEADD
Ribbon	Choose the Build panel of the Home tab. Choose Zone from the Space command flyout (refer to Figure 4.36).
Design palette	Choose Zone from the Design palette.

When you choose the **ZoneAdd** tool, you are prompted to specify the insertion point to place a zone marker in the workspace prompts. You can edit the Properties palette to specify a name for the zone.

After you place a zone, one or more spaces or zones can be attached to the selected zone. To attach a space or zone, choose the zone marker, right-click, and choose **Attach Space/Zones** from the shortcut menu and select spaces in the workspace. When you choose the zone marker, the grips are displayed as shown in Figure 4.36. Choose the "+" grip to convert your cursor to a pick box, then choose spaces to add to the zone. If spaces have been attached to a zone, choose the "−" grip to detach spaces from the zone.

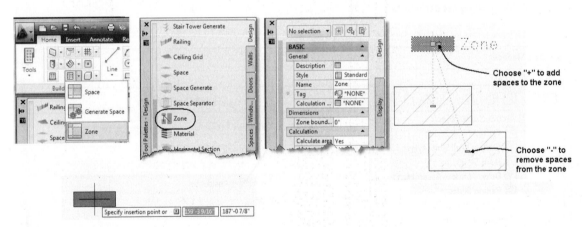

FIGURE 4.36 *Grips of a Zone*

The color and other properties of spaces assigned to a zone are set in the style of a zone. Therefore, if you insert a zone and then choose **Copy Style** from the Zone contextual tab of the ribbon, you can name and edit the zone style. The color can be set in the **Layer/Color/Linetype** tab of the **Display Properties** tab of a zone style as shown in Figure 4.37.

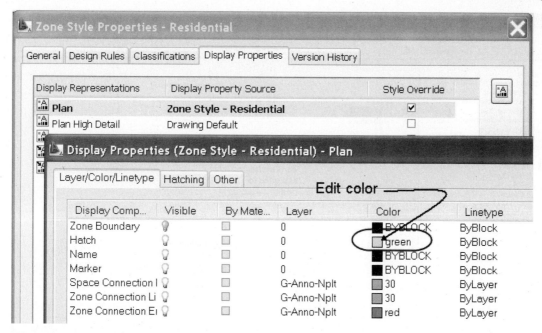

FIGURE 4.37 *Modifying the zone display*

Creating Zone Templates

A zone template allows you to link zone styles together into a group. When a zone template is placed, the zone styles defined for the template are inserted into the drawing. A zone template could be created for the living, sleeping, and service zones of a residence.

Steps to Creating a Zone Style and Zone Template

1. Choose the Manage tab from the ribbon. Choose the Style Manager from the Style & Display panel.

2. Expand the Documentation Objects category.

3. Choose the **Zone Styles** category, right-click, and choose **New. Overtype a new name** to create new zone styles for assignment to the zone template.

4. To create a new zone template, choose the **Zone Templates** category, right-click, and choose **New**. Overtype the name of zone template.

5. Choose the **General** tab of the right pane and overtype the name of the zone template in the Name field.

6. Choose the **Content** tab. Choose the **Unnamed** category, right-click, and choose **Rename**. Type the name of the category.

7. To create subcategories, choose a category, right-click, and choose **New**. Overtype the name of the zone category. Choose each zone category and edit the zone style assigned to the node in the **General** section shown at the right in Figure 4.38.

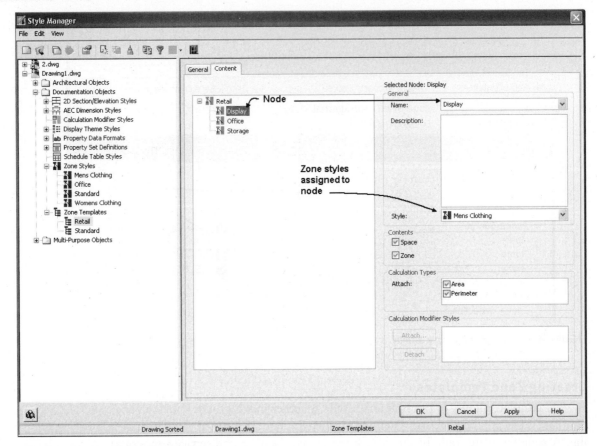

FIGURE 4.38 *Creating zone categories*

Placing a Zone Template

After you create the zone template in the Style Manager, choose the BOMA Useable **Zone Template (ZoneCreateFromTemplate)** tool from the Spaces palette to place the new zone template as shown in Table 4.17 and Figure 4.39. When you choose the command, the style is preset to the BOMA Useable Zone style and you are prompted to specify the insertion point. However, you can edit the Style name in the Properties palette to insert other zone templates that you have created or imported. The options of the command allow you to specify the distance between the zone styles of the template.

TABLE 4.17 *Accessing the ZoneCreateFromTemplate command*

Command prompt	**ZONECREATEFROMTEMPLATE**
Tool palette	Choose BOMA Useable Zone Template from the Spaces palette. Edit the Zone Template in the Properties palette.

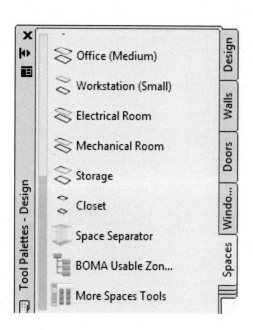

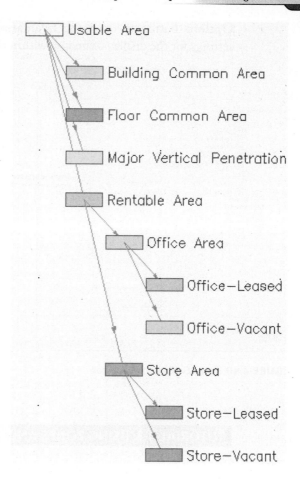

FIGURE 4.39 *Accessing the Zone Template tool*

Summarizing Spaces Using Space Evaluation Tools

You can tag a space object to identify the room number or space name. The tools for placing tags are presented in Chapter 11. The results of the **SpaceEvaluation** tool consist of the area of the floor, ceiling, and walls of each space. The results can be exported to an Excel file. The Zone and Zone Template provide one or more categories to link spaces. A summary of the spaces per zone category can be accessed from the Space Evaluation tool. Access the SpaceEvaluation command as shown in Table 4.18.

TABLE 4.18 *Accessing the Space Evaluation tool*

Command prompt	SPACEEVALUATION
Ribbon	Choose the Annotate tab. Choose the Evaluate Space from the Scheduling panel drop-down.
Tool palette	Choose Space Evaluation from the Scheduling palette of the Document palette set.

When you choose **Space Evaluation** from the Scheduling palette of the Document palette set, the **Space Evaluation** dialog box opens as shown in Figure 4.40. The left pane lists the zone categories and spaces of open drawings. The right pane lists the **Calculation Types** and the **Evaluation Options** button. Click the **Evaluation**

Options button to open the **Evaluation Properties** dialog box, which includes settings for the display of images within the Excel or Text file.

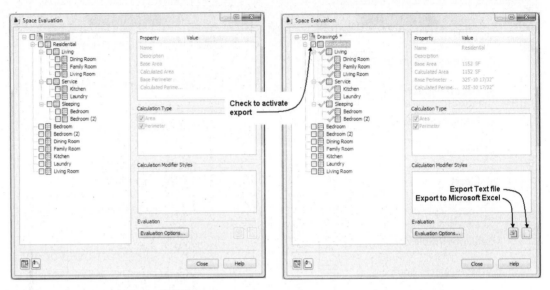

FIGURE 4.40 *Space Evaluation dialog box*

TUTORIAL 4-3: USING ZONES AND SPACE EVALUATION TOOLS

1. Verify that the Accessing Tutor\Imperial or Accessing Tutor\Metric content for Chapter 4 of the CengageBrain has been downloaded to your Accessing Tutor folder on your computer as described in Organizing Tutorial Directories in the Preface.

2. Open Accessing Tutor\Imperial\Ch4\Ex 4-3.dwg or Accessing Tutor\Metric\Ch 4\ Ex 4-3.dwg. Save the file as Lab 4-3.dwg to your student folder.

3. Edit Annotation Scale = ¼" = 1'-0" [1:50] and make sure Display Configuration is set to High Detail in the Drawing Window Status Bar.

4. Choose **Zone (ZoneAdd)** from the **Space** flyout menu of the Build panel in the Home tab. Insert 4 zones above the floor plan.

5. Select **Zone 4** to display the Zone contextual tab of the ribbon. Edit the **Name** in the **Properties** palette to **Sleeping**. Retain selection of Zone 4 and choose **Copy Style** from the Zone contextual tab to open the **Zone Style Properties** dialog box.

6. Choose the **General** tab and edit the **Name** to **Sleeping**.

7. Choose the **Display Properties** tab. Check **Style Override** of the **Plan High Detail** display representation. Choose the **Layer/Color/Linetype** tab. Turn **ON Visibility** for the **Zone Boundary**. Click in the color column for **Zone Boundary** to open the **Color** dialog box. Choose the **Blue (color 5)** color. Click **OK** to dismiss the **Select Color** dialog box.

8. Choose the **Hatch** display component. Click in the color column to open the **Select Color** dialog box. Choose the **Blue (color 5)** color. Click **OK** to dismiss all dialog boxes. Press ESC to clear selection.

9. Select **Zone 3** to display the Zone contextual tab of the ribbon. Edit the **Name** in the Properties palette to **Living**. Retain selection of Zone 3 and choose **Copy Style** from the Zone contextual tab to open the **Zone Style Properties** dialog box. Choose the **General** tab and edit the **Name** to **Living**.

10. Choose the **Display Properties** tab. Check **Style Override** of the **Plan High Detail** display representation. Choose the **Layer/Color/Linetype** tab. Turn **ON Visibility** for the **Zone Boundary**. Click in the color column for Zone Boundary to open the **Select Color** dialog box. Choose the **Green (color 3)** color. Click **OK** to dismiss the **Select Color** dialog box.

11. Choose the **Hatch** display component. Click in the color column to open the **Color** dialog box. Choose the **Green (color 3)** color. Click **OK** to dismiss all dialog boxes. Press ESC to clear selection.

12. Choose **Zone 2** to display the Zone contextual tab of the ribbon. Edit the **Name** in the Properties palette to **Service**. Retain selection of Zone 2 and choose **Copy Style** from the Zone contextual tab to open the **Zone Style Properties** dialog box. Choose the **General** tab and edit the **Name** to **Service**.

13. Choose the **Display Properties** tab. Check **Style Override** of the **Plan High Detail** display representation. Choose the **Layer/Color/Linetype** tab. Turn **ON Visibility** for the **Zone Boundary**. Click in the color column for **Zone Boundary** to open the **Color** dialog box. Choose the **Yellow (color 2)** color. Click **OK** to dismiss the **Select Color** dialog box.

14. Choose the **Hatch** display component. Click in the color column to open the **Color** dialog box. Choose the **Yellow (color 2)** color. Click **OK** to dismiss all dialog boxes. Press ESC to clear selection.

15. Choose **Zone** in the workspace. Edit the **Name** property to **Residence** in the Properties palette. Retain the selection of the Zone, choose the "+" grip, and select the **Sleeping**, **Service**, and **Living** zones as shown in Figure 4.41. Press ENTER and then ESC to end the selection.

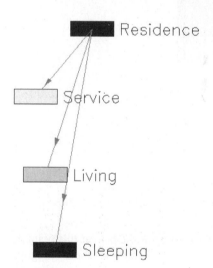

FIGURE 4.41 *Zones of the Residence*

16. In the next three steps, you will select spaces to assign to the zones. Choose Low Detail display configuration to view the spaces with User Single hatch pattern and allow for easy selection of the space.

17. Choose the **Sleeping** zone to display the "+" sign. Choose the "+" and select the four bedroom spaces and the adjoining closet [cupboard] as shown in Figure 4.42. Press ENTER to end the selection and press ESC to end the command.

18. Choose the **Service** zone to display the "+" sign. Choose the **Kitchen**, **Hallway**, **Laundry**, and **Garage 2 Car [Garage_Double]** spaces. Press ENTER to end the selection and press ESC to end the command.

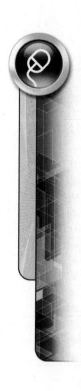

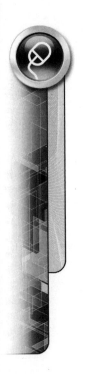

19. Choose the **Living** zone to display the "+" sign. Choose the **Living Room [Living Room-Large]**, **Dining Room**, and **Entry-Foyer [Entrance-Hall]** spaces. Press ENTER to end the selection and press ESC to end the command.

20. Choose the High Detail display configuration to view the spaces with the zone color.

21. Choose the **Service** zone, right-click, and choose **Properties**. The number of spaces and the total area for the Service zone are displayed in the Properties palette as shown in Figure 4.42.

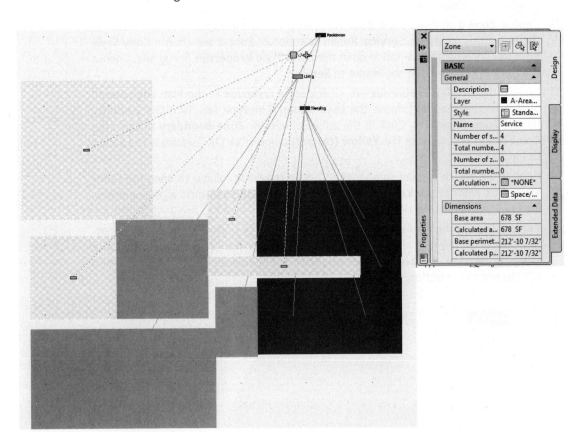

FIGURE 4.42 *Attaching spaces to zones*

22. Choose the **Annotate** tab of the ribbon. Extend the **Scheduling panel**. Choose the **Evaluate Space** tool.

23. Select the **Residence** category to determine the total area of the plan at this point in the design. You can choose each category to determine the space total per category as shown in Figure 4.43.

24. Save and close the drawing.

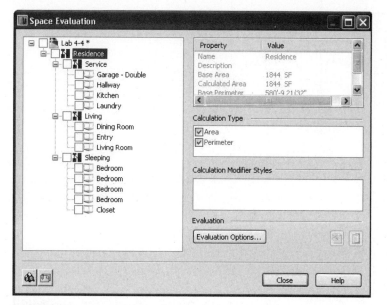

FIGURE 4.43 *Space Evaluation dialog box*

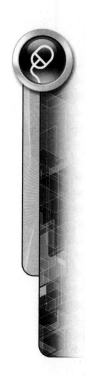

CREATING FREEFORM SPACES

Freeform spaces are unique since the space is three-dimensional and identical in form to the shape and size of the bounding elements of the model. Freeform spaces can be created from mass elements or from bounding objects. The walls, roof slabs, and floor slabs must be connected without gaps. The components must overlap to create intersections. The procedure to create and view the properties of a freeform space is shown below.

Steps to Create a Freeform Space

1. View the building in plan view. All wall enclosing elements must be displayed in the workspace.

2. Choose **Space Generate** from the **Design** palette to open the **Properties** palette.

3. Edit **Geometry Type** to **Freeform**.

4. Move the cursor to the workspace and left-click inside the building. Press ESC to end the command.

Information regarding the space, including doors and windows, is displayed in the properties of the space of the **Space/Zone Manager** dialog box as shown in Figure 4.44. The surface information can be useful for analysis within the Autodesk AutoCAD MEP software. If you click the Space/Zone Manager button located in the Properties palette, the Space/Zone Manager dialog box opens as shown in Figure 4.44. Check the **Show All Zones and Spaces** and **Show Space Surfaces** checkboxes to display the properties of the space in the left pane.

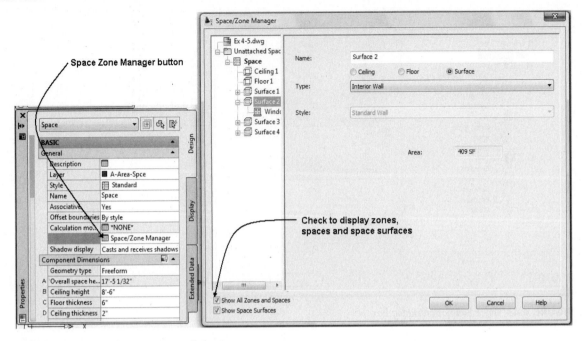

FIGURE 4.44 *Space/Zone Manager dialog box*

If the freeform space is viewed in isometric, you can choose each surface in the Space/Zone Manager, and its image will display in red in the pictorial view as shown in Figure 4.45.

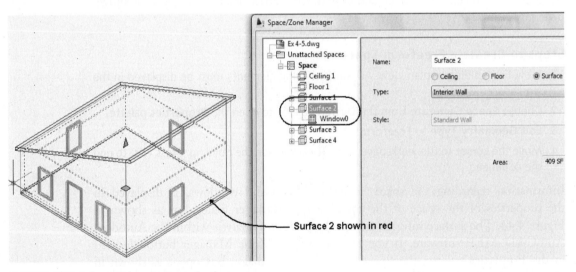

FIGURE 4.45 *Information regarding a freeform space*

Surface information for doors and windows can be added to Extrusion and Freeform spaces; however, the information is not linked to the actual door and window objects that have been placed in the walls. If you choose an Extrusion space, the grips are shown at the left in Figure 4.46. You can add doors and windows to the selected space if you choose **Edit Surfaces** from the shortcut menu to display the "+" grips, as shown in Figure 4.46. If you choose the "+" grip, you can add a door or window to the surface. A negative grip is displayed for the openings; choose the "−" grip to remove a door or window.

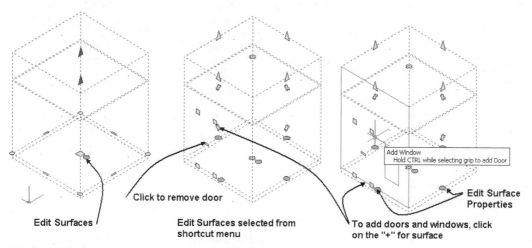

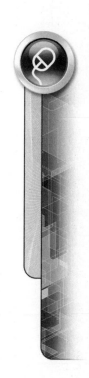

FIGURE 4.46 *Inserting doors or windows in a freeform space*

TUTORIAL 4-4: CREATING A FREEFORM SPACE

1. Verify that the Accessing Tutor\Imperial or Accessing Tutor\Metric content for Chapter 4 of the CengageBrain has been downloaded to your Accessing Tutor folder on your computer as described in Organizing Tutorial Directories in the Preface.

2. Open Accessing Tutor\Imperial\Ch4\Ex 4-4.dwg or Accessing Tutor\Metric\Ch 4\ Ex 4-4.dwg. Save the file as **Lab 4-4** in your student folder.

3. Choose the Home tab. Choose **Space Generate** from the Space flyout of the Build panel.

4. Edit Geometry Type to Freeform in the Properties palette.

5. Move the cursor to the workspace and left-click inside the building. Press ESC to end the command.

6. Choose **SE Isometric** from the View Cube.

7. Select the space at p1 as shown in Figure 4.47, right-click, and choose **Space/Zone Manager** from the shortcut menu.

8. Verify that the **Show All Zones and Spaces** and **Show Space Surfaces** checkboxes are checked in the lower-left corner of the Space/Zone Manager. Reposition the Space/Zone Manager, if necessary, to view the freeform space and the Space/Zone Manager. Expand Space and then expand **Surface 4** to view its contents as shown in Figure 4.47. Click OK to dismiss the dialog box.

9. Save and close the drawing.

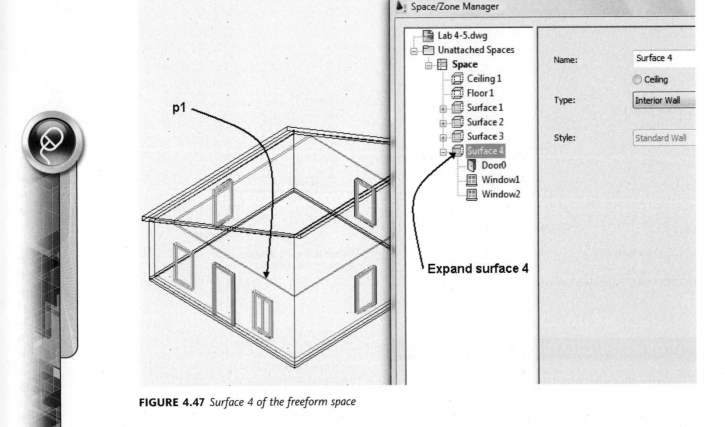

FIGURE 4.47 *Surface 4 of the freeform space*

CREATING MASS MODELS

Designers can create physical models to refine their design and to communicate to clients. The mass elements accessed from the **Massing** tool palette shown in Figure 4.48 can be used to create computer models to serve the same purpose as physical models of the building or of building components. Mass models can be developed to create a shape that will enclose the spaces of a space diagram. Computer models are created by joining together a combination of mass elements in the form of twelve possible prism shapes, extrusions of profiles, drapes, or revolutions of profiles. The mass elements can be joined together through Boolean operations of addition, subtraction, or intersection. The mass elements can also be used to simulate such building components as columns, floors, or fireplaces in other drawings. Once a mass model is created, you can slice the model to create polylines that can be used for the development of floor plans. Polylines can be converted to a space, and the space can be divided by walls.

Inserting Mass Elements

When a mass element tool is selected, the Properties palette opens, which displays the dimensions and other properties of the element as shown in Figure 4.48. The **Shape** drop-down list in the Design tab of the Properties palette includes the following shapes: Arch, Barrel Vault, Box, Cone, Cylinder, Dome, Doric, Drape, Gable, Pyramid, Sphere, Isosceles Triangle, Right Triangle, Extrusion, Revolution, and Freeform. Custom mass elements can also be created as extrusions or revolutions from a closed polyline. The prisms of the Shape list are shown in Figure 4.48. You can select a shape from the drop-down list in the Design tab of the Properties palette or select the tool from the Massing tool palette. Mass elements are created on the A-Area-Mass [Aec-Mass-Elements] layer, which has color 70.

The dimensions of the mass element can be defined on screen or in the Properties palette. The **MassElementAdd** command (refer to Table 4.19) is used to add the mass elements shown on the Massing tool palette.

TABLE 4.19 *MassElementAdd command access*

Command prompt	**MASSELEMENTADD**
Ribbon	Choose Box flyout menu of the Build panel in the Home tab.
Tool palette	Select any of the following from the Massing tool palette: Arch, Barrel Vault, Box, Pyramid, Isosceles Triangle, Right Triangle, Cone, Cylinder, Dome, Sphere, Gable, Drape, Extrusion, or Revolution.

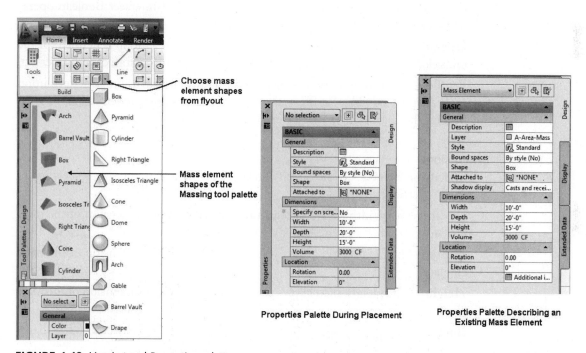

FIGURE 4.48 *Massing and Properties palette*

When a mass element tool is selected, the Properties palette opens, allowing you to define the dimensions of the element. The options in the Design tab of the Properties palette vary according to the mass element you are adding. The options in the Design tab of the Properties palettes of a mass element shown in Figure 4.48 are described below.

Design Tab—Basic
General

Description—The Description button opens the Description dialog box, which allows you to add a description of the mass element.

Style—The Style fields lists the element styles included in the drawing. Element styles allow you to attach materials to the element.

Layer—The Layer field displays the name of the layer of existing mass elements. Mass elements are placed on the layer defined by the current layer key style.

Bound spaces—The Bound spaces field allows you to toggle Yes or No for the element to function as a boundary for a space.

Shape—The Shape list includes each of the following prisms: Arch, Barrel Vault, Box, Cone, Cylinder, Dome, Doric, Gable, Pyramid, Sphere, Isosceles Triangle, Right Triangle, Extrusion, Revolution, and Freeform.

Attached to—The Attached to drop-down list displays the mass groups. If an element is attached to a group, the add, subtract, and intersect Boolean operations can be applied to members of the group.

Operation—The Operation drop-down list is active if a group is defined for the mass element. The Operation field has the options of Add, Subtract, and Intersect.

Profile—The Profile option is displayed only when Extrusion and Revolution tools are selected. The Profile option allows you to select predefined AEC Profiles for the Extrusion or Revolution to create a shape.

Dimensions

Specify on Screen—If the Specify on Screen is set to No, the size is specified in the Properties palette.

Width—The Width field allows you to insert the width dimension of the mass element.

Depth—The Depth field allows you to insert the depth dimension of the mass element.

Height—The Height field allows you to insert the height dimension of the mass element.

Radius—The Radius field is active if the mass element includes a radial shape. The radius of the radial shape is defined in this field.

Rise—The Rise field allows you to specify the rise of the Gable mass element.

Location

Rotation—The Rotation field displays the angle of rotation.

Elevation—The Elevation field lists the vertical distance from $Z = 0$.

Add Information—The Add Information button opens the Location dialog box to describe an existing mass element.

Each mass shape has a specified insertion point or handle located in the center of the bottom plane. The insertion point is coincident with the Location grip as shown in Figure 4.49. The Doric column is not a mass element on the Massing tool palette; however, it can be selected from the **Shape** drop-down list in the Design tab of the Properties palette. To create a Doric column, select any mass element from the Massing palette and choose the Doric shape from the drop-down list in the Properties palette. Revolution and Extrusions are methods in the Shape list in which mass elements are created from profiles. Any closed polyline can be converted to an AEC Profile in the Style Manager and extruded or revolved to create a mass element. If other mass elements are placed on top of an existing element, the top node of a mass element can be snapped to with the Node object snap to align mass elements together.

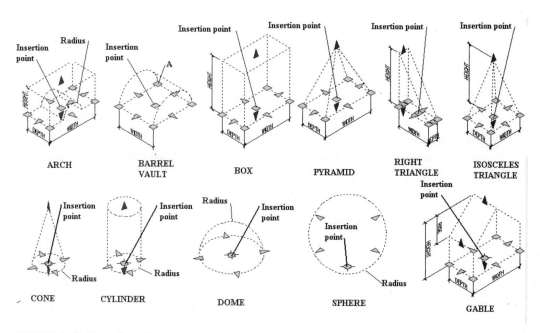

FIGURE 4.49 *Mass element prisms*

Steps to Insert a Mass Element (Gable)

1. Select the **Work** layout in the **Application** status bar. Click in the left viewport. Select **SW Isometric** from the **View Cube**.
2. Select the **Gable** mass element of the **Massing** tool palette.
3. Move the cursor to the workspace; the Gable mass element is phantom-displayed with the cursor. Dimensions of the mass element can be specified on screen or in the Properties palette.
4. Edit the **Properties** palette as follows: Dimensions—Specify on screen = **No**, Width = **10'** **[3000]**, Depth = **20'** **[6000]**, Height = **15'** **[4500]**, and Rise = **5'** **[1500]**.
5. Move the cursor to the workspace and specify the insert point and rotation as shown in the workspace of Figure 4.50.

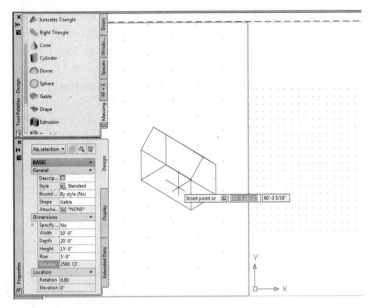

FIGURE 4.50 *Creating a mass element*

The top view of the mass element is hatched as shown in Figure 4.51; therefore, place mass elements in a pictorial view since the hatch is not displayed in model view.

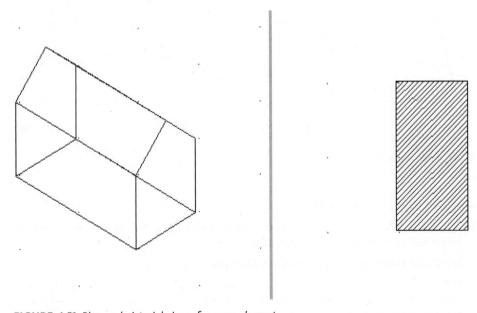

FIGURE 4.51 *Plan and pictorial view of a mass element*

Creating Mass Elements Using Extrusion and Revolution

In addition to the twelve mass element prisms, you can create a mass element from an extrusion or revolution of a profile. The **ExtrudeLinework** command (see Table 4.20) can be used to create an extrusion from a closed polyline. After creating the mass element using extrusion, you can change the shape to any of the shapes listed on the Massing tool palette. The outline of a building component or the entire building can be drawn using a polyline. Select a polyline, right-click, and choose **Convert** to > **Mass element** (**ExtrudeLinework**) to create a mass element from the shape of the polyline. The height of the extrusion is specified in the command line as follows:

> *(Select a closed polyline, right-click, and choose* Convert to > Mass Element. *)*
>
> Erase selected linework? [Yes/No] <No>: Y *(Specify* Yes *to erase polyline.)*
>
> Specify extrusion height <20'-0">: *(Specify height of mass element.)*

TABLE 4.20 *ExtrudeLinework command access*

Command prompt	**EXTRUDELINEWORK**
Shortcut menu	Select a closed polyline, right-click, and choose Convert to > Mass Element.

Extrusions and revolutions can be created by selecting the **Extrusion** or **Revolution** tool from the **Massing** tool palette and then specifying height and/or radius and the name of a profile in the Properties palette. Profiles are created in the Style Manager as presented in Chapter 3. The revolution or extrusion is placed by specifying the insert point and rotation.

TUTORIAL 4-5: CREATING MASS ELEMENT COMPONENTS FOR A FIREPLACE

1. Verify that the Accessing Tutor\Imperial or Accessing Tutor\Metric content for Chapter 4 of the CengageBrain has been downloaded to your Accessing Tutor folder on your computer as described in Organizing Tutorial Directories in the Preface.

2. Open Accessing Tutor\Imperial\Ch 4\Ex 4-5.dwg or Accessing Tutor\Metric\Ch 4\ Ex 4-5.dwg.

3. Choose **Save As > AutoCAD Drawing** from the Application menu and save the drawing as **Lab 4-5** in your student directory.

4. If the Design palettes are not displayed, select **Home** tab and choose Tools from the Build panel.

5. Select the **Gable** tool from the **Box** flyout menu of the Build panel.

6. Edit the **Properties** palette as follows: Style = **Standard**, Shape = **Gable**, Attach to = **None**, Specify on Screen = **No**, Width = **4'-4" [1300]**, Depth = **3'-0" [900]**, Height = **8' [2400]**, and Rise = **8 [200]**. Move the cursor to the workspace and specify the **Insert** point and **Rotation** as shown in the following workspace sequence:

   > Insert point or: 15',20' [4500,6000] ENTER *(Specify the insert point for the mass element.)*
   >
   > Rotation or <0.00>: 0 ENTER *(Specify the rotation of the mass element. Do not exit the command. Continue to the next step.)*

7. Edit the Properties palette as follows: Style = Standard, Shape = Box, Attach to = None, Specify on Screen = No, Width = 5'-0" [1500], Depth = 1'-6" [450], Height = 1'-2" [350]. Move the cursor from the Properties palette to the workspace and specify the Insert point and Rotation as shown in the following workspace prompts:

   > Insert point or: 24',10' [7000,3000] ENTER *(Specify the insert point for the mass element.)*

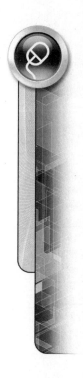

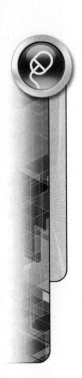

Rotation or <0.00>: **0** ENTER *(Specify the rotation of the mass element. Do not exit the command. Continue to the next step.)*

8. Edit the Properties palette as follows: Style = Standard, Shape = Box, Attach to = None, Specify on Screen = No, Width = 2'-0" [600], Depth = 2'-0" [600], and Height = 24' [7300]. Move the cursor to the workspace and specify the Insert point and Rotation as shown in the following workspace sequence:

   Insert point or : **6', 20' [1800,6000]** ENTER *(Specify the insert point for the mass element.)*

   Rotation or <0.00>: **0** ENTER *(Specify the rotation of the mass element.)*

   (Press ENTER *to end the command.)*

9. Select the polyline at p1 as shown in Figure 4.52, right-click, and choose Convert To > Mass Element. Respond to the following workspace prompts to specify the mass element:

   Erase selected linework? [Yes/No] <No>: **N** *(Select* **No** *to retain the polyline.)*

   Specify extrusion height <1'-0">: **2'-5" [750]** ENTER

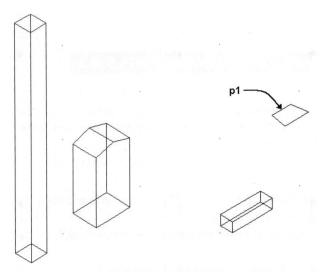

FIGURE 4.52 *Selection of a polyline to create a Profile*

10. Select the extruded mass element and verify properties in the Properties palette as follows: Style = **Standard**, Profile = **EMBEDDED**, Attach to = **None**, Width = **3' [900]**, Depth = **1'-11" [600]**, and Height = **2'-5" [750]**. Mass elements are developed as shown in Figure 4.53.

11. Choose Zoom Extents from the Zoom flyout of the Navigation bar. Save and close the drawing.

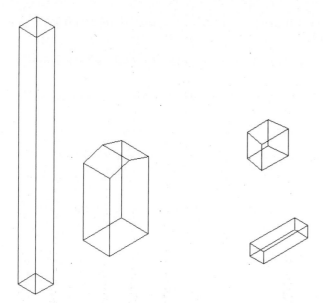

FIGURE 4.53 *Mass elements created*

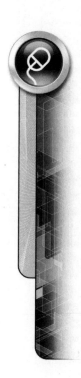

Creating Mass Elements Using Drape

The **Drape** mass element is used to create irregular 3D surfaces that drape polylines. The mesh density specified in the command specifies the number of faces created to define the surface. The polylines create the elevation points for the mass elements. The **Drape** command is very useful in creating simulation of land contours. The polylines used in the Drape command should vary in the Z coordinate direction according to the elevation of the land. Access the **Drape** tool from the Massing tool palette as shown in Table 4.21.

TABLE 4.21 *Drape command access*

Command prompt	DRAPE
Ribbon	Choose Drape from the Box flyout menu of the Build panel in the Home tab.
Tool palette	Select Drape from the Massing tool palette.

The **Drape** command is applied to the polylines shown in Figure 4.54 and is used in the following command sequence:

```
Select objects representing contours: (Select the contour
lines using a crossing window, select p1 as shown in
Figure 4.54.)

Specify opposite corner: (Select a point at p2 as shown in
Figure 4.54. 13 found

Select objects representing contours: ENTER (Press ENTER to
end the command.)

Erase selected contours [Yes/No] <No>: ENTER

Generate regular mesh [Yes/No] <Yes>: ENTER
```

Select rectangular mesh corner: *(Select a point near* p3 *as shown in Figure 4.54.)*

Select opposite mesh corner: *(Select a point near* p4 *as shown in Figure 4.54.)*

Mesh subdivision along X direction <30>: ENTER *(Press ENTER to accept the default mesh size.)*

Mesh subdivision along Y direction <30>: ENTER

Enter base thickness <10'-0>: 6 [150] ENTER *(Type 6 [150] to specify the base thickness.)* The topography has been extrapolated.

(Click the Model *tab in the* Application *status bar and select* SW Isometric *from the* View *Cube. Select the View tab. Select* Realistic *from the* Visual Styles *flyout menu of the* Visual Styles *panel to view the drape as shown at the right in Figure 4.54.)*

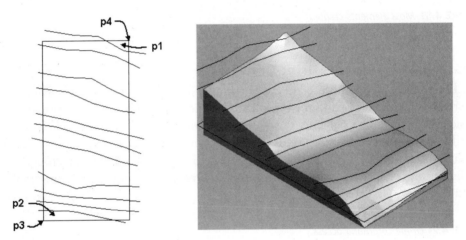

FIGURE 4.54 *Selection of points for Drape*

Modifying Mass Elements

To modify mass elements, you can edit the grips of the mass element or edit the dimensions in the Properties palette. The functions of the grips of the Gable mass element are described in Figure 4.55. Selecting a wedge-shaped grip changes one dimension of the mass element, whereas the square grips change two dimensions or the location. Using the stretch operation with grips allows you to modify mass elements to fit within existing mass elements. When the grips of the mass element shown at the right in Figure 4.55 are edited, you can press the TAB key to edit the dynamic dimensions. When you insert a mass element, it is inserted at Z = 0; however, the height grip located on the base can be selected and the elevation of the base changed.

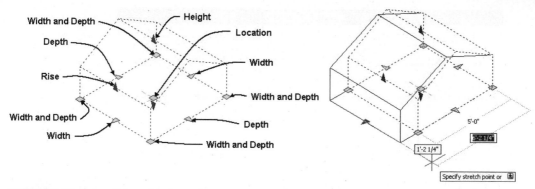

FIGURE 4.55 *Grips of the gable mass element*

Trimming Mass Elements

Mass elements can be trimmed to create custom mass elements. The command **Trim by Plane** (**MassElementTrim**) creates a trim plane that cuts across the element. You can then select which side of the trim plane to retain. Access the **Trim by Plane** command from the shortcut menu of the mass element as shown in Table 4.22.

TABLE 4.22 *Mass Element Trim by Plane command access*

Command prompt	MASSELEMENTTRIM
Ribbon	Select a mass element to display the Mass Element contextual tab. Choose Trim by Plane from the Modify panel.
Shortcut menu	Select a mass element, right-click, and choose Trim by Plane.

The Trim by Plane command is used in the following command line sequence to trim the Doric column.

> Specify trim plane start point or: *(Select* p1 *using* Endpoint *object snap as shown in Figure 4.56.)*
>
> Specify trim plane endpoint: *(Select* p2 *using* Endpoint *object snap as shown in Figure 4.56.)*
>
> Select mass element side to remove: *(Select* p3 *using* Endpoint *object snap as shown in Figure 4.56.)*

The plane defined to trim the mass element is extruded from a line specified by two points or from a plane defined by three points. After you specify the first point, the trim plane is dynamically shown in red as you move your cursor to create an additional point that crosses the mass element and defines a line on the plane. The points selected for the trim plane can be located on or off the mass element.

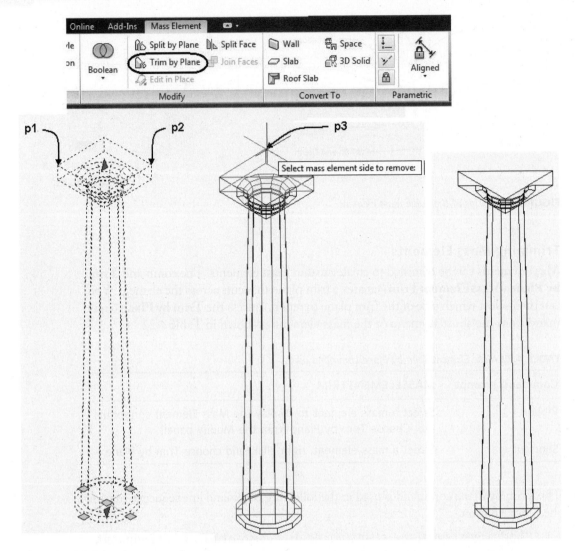

FIGURE 4.56 *Trimming a mass element*

Splitting Mass Elements

Mass elements can be divided by a cutting plane to create two mass elements. The **Split by Plane** command (**MassElementDivide**) creates a cutting plane that cuts the mass element. Access the **Split by Plane** command as shown in Table 4.23. The Split by Plane command is used in the following workspace sequence to divide the Gable mass element. The **Move** command can be used after the split to separate the mass elements.

> Specify divide plane start point or: _mid of *(Select point* p1 *using the* Midpoint *object snap as shown in Figure 4.57.)*
>
> Specify divide plane endpoint: _mid of *(Select point* p2 *using the* Midpoint *object snap as shown in Figure 4.57.)*

TABLE 4.23 *Command access for the Split by Plane command*

Command prompt	MASSELEMENTDIVIDE
Ribbon	Select a mass element to display the Mass Element contextual tab and choose Split by Plane from the Modify panel (refer to Figure 4.57).
Shortcut menu	Select the mass element, right-click, and choose Split by Plane.

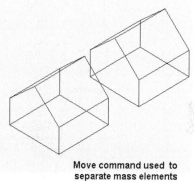

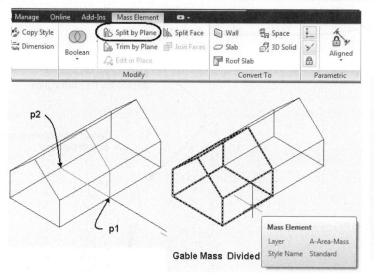

FIGURE 4.57 *Splitting a mass element*

Editing Faces

The mass element can be modified by editing its faces. When you split one of the faces of the mass element, you can stretch a face to a new position. The **Split Face(MassElementFaceDivide)** command (refer to Table 4.24) creates a freeform mass element; therefore, the split face and the remaining faces of the mass element can be edited. After you split the face of the mass element, additional grips are displayed on each face as shown in Figure 4.58. Select the grips and drag the face to a new position.

TABLE 4.24 *Split Face command access*

Command prompt	MASSELEMENTFACEDIVIDE
Ribbon	Select a mass element to display the Mass Element contextual tab and choose Split Face from the Modify panel (refer to Figure 4.58).
Shortcut menu	Select a mass element, right-click, and choose Split Face.

The Split Face command is applied to the gable shown in Figure 4.58 in the following command line sequence:

```
(Select the mass element and choose Split Face from the
Modify panel in the Mass Element tab.)
```

Select first point on face: _mid of *(Select point* p1 *using the* Midpoint *object snap as shown in Figure 4.58.)*

Select second point on the same face: _mid of *(Select point* p2 *using the* Midpoint *object snap as shown in Figure 4.58.)*

Gable converted to Free Form.

(Press ENTER to end the command.)

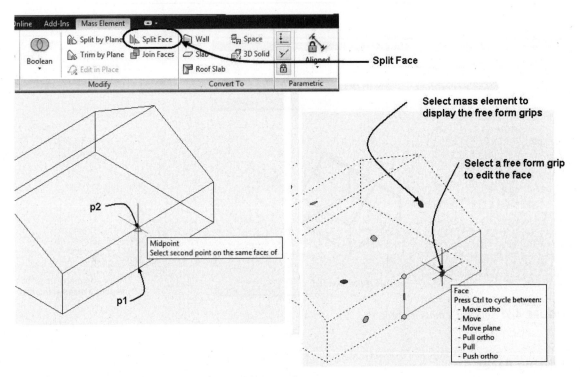

FIGURE 4.58 *Split Face command applied to Gable mass element*

After you split the face, the mass element is in free form, and if you select the mass element, the free form grips are displayed as shown in Figure 4.59. You can cycle through the grip options when you press CTRL prior to moving the cursor. After a face is split, the face can be rejoined, if it remains **coplanar**, by selecting the **Join Faces(MassElementFaceJoin)** command described in Table 4.25.

TABLE 4.25 *Join Faces command access*

Command prompt	MASSELEMENTFACEJOIN
Ribbon	Select a mass element with previously split faces to display the Mass Element contextual tab and choose Join Faces from the Modify panel.
Shortcut menu	Select a mass element with previously split faces, right-click, and choose Join Faces.

(Select a coplanar face and choose Join Faces *from the* Modify *panel of the Mass Element contextual tab.)*

```
Select edges of coplanar faces: 1 found (Select dividing face
line at p1 [shown as red] as shown in Figure 4.59)

Select edges of coplanar faces: ENTER (Press ENTER to end
selection.) Free Form converted to Gable.
```

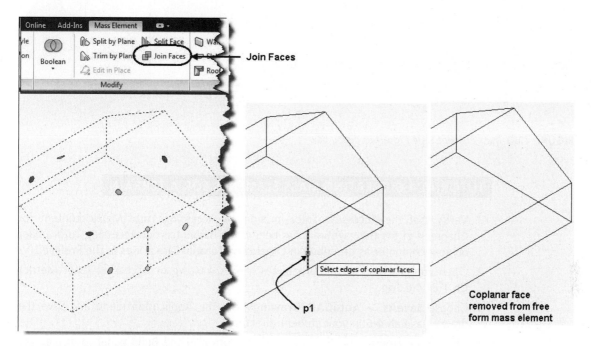

FIGURE 4.59 *Faces joined with Join Faces*

Adding Materials to Mass Elements

A style can be created for mass elements. A benefit of creating a mass element style is that you can assign materials to the style. The material definition is applied to the Body display component of the mass element. Mass elements with different styles can be assembled to represent building components or terrain. Masonry fireplaces and masonry wall projections can be modeled with mass elements and assigned brick or concrete masonry unit materials. Mass element styles can be created and edited in the *Multi-Purpose Objects\Mass Element Styles* category of the drawing in the Style Manager. However, to create a new style and assign the style to an existing mass element, select a mass element and choose Copy Style from the General panel of the Mass Element contextual tab.

Steps to Apply a Material to a Mass Element Style

1. Choose the **Render** tab. Choose **Render Tools** of the Tool panel to display the **Visualization** palette. Choose a sample material tool from the Material palette.

2. Respond to the command prompt; select a component or an object and select the mass element as shown in Figure 4.60. Press ENTER to apply the material to the style of the mass element.

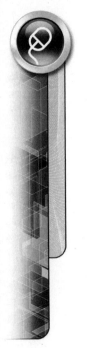

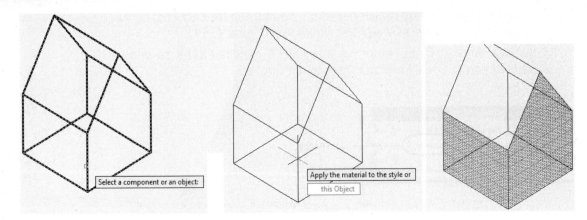

FIGURE 4.60 *Mass Element Style Properties dialog box*

TUTORIAL 4-6: CREATING MASS ELEMENTS FOR A TERRAIN

1. Verify that the Accessing Tutor\Imperial or Accessing Tutor\Metric content for Chapter 4 of the CengageBrain has been downloaded to your Accessing Tutor folder on your computer as described in Organizing Tutorial Directories in the Preface.

2. Open your Accessing Tutor\Imperial\Ch 4\Ex 4-6.dwg or Accessing Tutor\Metric\ Ch 4\Ex 4-6.dwg.

3. Choose **SaveAs > AutoCAD Drawing** from the Application menu and save the drawing as **Lab 4-6** in your student directory.

4. Choose the **Home** tab. Choose the **Tools** button of the Build panel to display the **Design** palette set. Toggle OFF Object Snap in the status bar.

5. Verify that the right viewport is current and select **Drape** from the **Massing** tool palette. Respond to the command line prompts as shown below:

> Select objects representing contours: *(Select a point near* **p1** *as shown in Figure 4.61.)*
>
> Specify opposite corner: *(Select a point near* **p2** *as shown in Figure 4.61.)* 11 found
>
> Select objects representing contours: ENTER *(Press ENTER to end selection.)*
>
> Erase selected contours [Yes/No] <No>: N ENTER
>
> Generate regular mesh [Yes/No] <Yes>: Y ENTER
>
> Select rectangular mesh corner: *(Use* SHIFT *and right-click to select* **Endpoint** *and select a point near* **p3** *as shown in Figure 4.61.)*
>
> Select opposite mesh corner: *(Use* SHIFT *and right-click to select* **Endpoint** *and select a point near* **p4** *as shown in Figure 4.61.)*
>
> Mesh subdivision along X direction <30>: ENTER *(Specify mesh size equal to* **30**.*)*
>
> Mesh subdivision along Y direction <30>: ENTER
>
> Enter base thickness <6>: 6 [150] ENTER *(Specify base thickness.)* The topography has been extrapolated. The resulting mass element includes a cyan hatch pattern shown at the z location of the cut plane height for the display configuration.

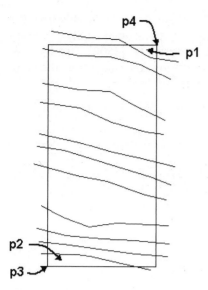

FIGURE 4.61 *Selection of contour lines for Drape mass element*

6. To create a mass element style for the drape, select the mass element and choose **Copy Style** from the General panel of the Mass Element contextual tab. Choose the **General** tab of the Mass Element Style Properties dialog box and overtype **Terrain** as the name of the mass element. Click **OK** to dismiss the **Mass Element Style Properties-Terrain** dialog box.

7. Click in the left viewport. Select the Render tab.

8. Select the Render Tools of the Tools panel to display the **Materials** tool palette from the **Visualization** palette set.

9. Choose **Site Construction. Planting.Groundcover.Grass.Thick** material from the Materials palette. Move the cursor to the workspace and select the mass element as shown in Figure 4.62. Press ENTER to apply the material to the style of the drape.

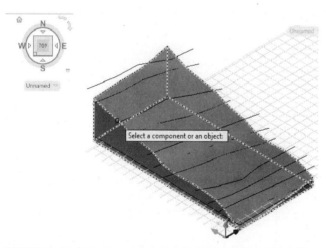

FIGURE 4.62 *Assigning material to the Body component of the mass element*

10. Choose Realistic visual style from the Visual Styles flyout of the Visual Styles panel of the View tab.

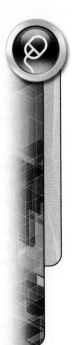

11. Choose the Home tab. Verify the left viewport is current, and select **Gable** from the **Box** flyout of the Build panel. Edit the **Properties** palette as follows: Style = **Standard**, Shape = **Gable**, Specify on Screen = **No**, Width = **28'** **[8500]**, Depth = **68'** **[20700]**, Height = **22'** **[6700]**, and Rise = **5'** **[1500]**. Move the cursor from the Properties palette and respond to the command prompts as follows:

Insert point or: 80',70',8'[28000,12000,3000] ENTER

(Specify the insert point for the mass element.) Rotation or <0.00>: 90 ENTER *(Specify the rotation of the mass element.)*

(Press ENTER to end the command.)

(A view of the Gable mass element and drape is located in the left viewport as shown in Figure 4.63.)

12. Save and close the drawing.

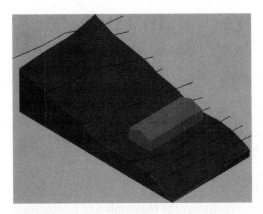

FIGURE 4.63 *View of building placed in drape mass element*

Creating Groups for Mass Elements

Complex models are created by defining a group and attaching elements to the group as shown in Figure 4.64. The **Mass Group** tool of the Massing tool palette (refer to Table 4.26) allows you to create a group. The Mass Group tool inserts a group marker, which represents the group. Mass elements or AutoCAD solids can be attached to the group.

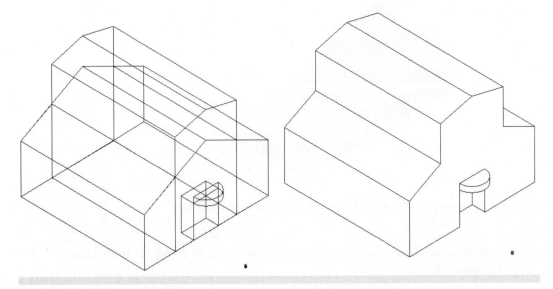

FIGURE 4.64 *Mass element and mass groups used to create mass model*

TABLE 4.26 *Accessing the Mass Group tool*

Command prompt	MASSGROUPADD
Tool palette	Select Mass Group from the Massing tool palette.

When you select the **Mass Group** tool, you are prompted to select elements to attach and a location as shown in the following workspace prompts. Grips of the mass group marker are displayed when an element of the group is selected as shown in Figure 4.65. The mass group marker is used to graphically display the presence of a group, and you can select the "+" to add elements to the group or select the "−" to remove elements from the group. A name for the mass group is assigned when the marker is placed. If you select a mass element that does not belong to a mass group, the computer-generated names of all mass groups will be listed in the **Attached to** drop-down list in the Properties palette; you can then add the element to a group.

```
(Select Mass Group from the Massing palette.)
Select elements to attach: 1 found (Select a mass element.)
Select elements to attach: ENTER (Press ENTER to end
selection.)
Location: (Select a location in the workspace.)
Attaching elements... 1 element(s) attached.
(The grips of the mass group marker are displayed when
an element attached to the group is selected as shown in
Figure 4.65.)
```

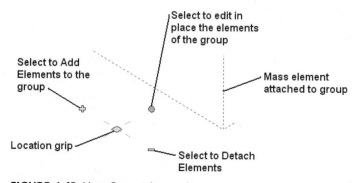

FIGURE 4.65 *Mass Group trigger grips*

The mass group marker and members of the group are placed on the A-Area-Mass-Grp layer [Aec-Mass-Groups] (color 13). The command sequence above does not prompt you to assign a name for the group. AutoCAD Architecture 2012 will assign a name to the mass group, such as 4A8. The assigned name is the AutoCAD handle name. AutoCAD handle names exist for all entities created in AutoCAD; they are not intended to describe the entity.

Adding Mass Elements to a Group

You can assign mass elements to a group when they are created, or you can assign their group definition after insertion by modifying the properties of the mass element. If a mass group has been established in the drawing, when you select a mass element tool from the Massing tool palette, you can edit the Attached to field to assign a group upon insertion. The Boolean Add operation is assigned by default to mass elements defined to a group upon insertion.

Attaching Elements to a Group

If a mass group exists, you can assign mass elements to a group by selecting the group to display the group marker shown in Figure 4.65. You can add elements to the group if you choose the "+" sign grip of the mass group marker or right-click and choose **Attach Elements** (refer to Table 4.27). The mass elements combine to form one object, the mass group, by adding their volume and geometry to the mass of the group. Mass elements can also be defined to subtract or intersect their mass with other mass elements of a group. These Boolean operations can be changed as the model develops. Several mass groups can be defined for the drawing according to the needs of the model. All mass elements that are attached to a group are highlighted if you select any member of the group as shown at right in Figure 4.66.

TABLE 4.27 *Attach Elements to group command access*

Command prompt	**MASSGROUPATTACH**
Ribbon	Select any member of the group to display the Mass Group contextual tab and choose Attach Elements from the Modify panel.
Shortcut menu	Select any member of the group to display the mass group marker, right-click, and choose Attach Elements.
Properties palette	Select the element, right-click, and choose Properties; if a mass group exists in the drawing, select the group from the flyout of the Attached to field.

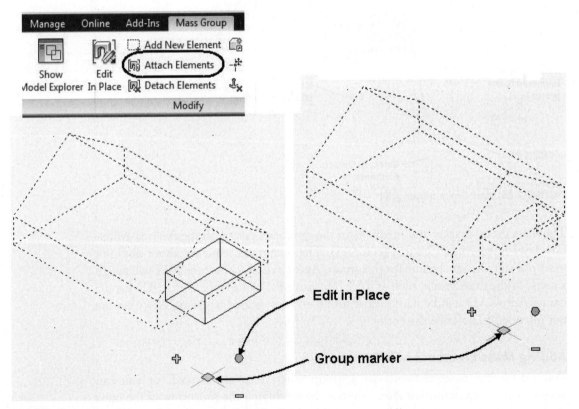

FIGURE 4.66 *Mass elements attached to mass group*

Detaching Elements from a Mass Group

Elements that have been attached to a group can be detached from the group. Detached mass elements are not erased from the drawing. The mass element no longer functions in the group to add, subtract, or intersect its mass with other elements of the group. To detach an element from a group, select the group to display the group marker, choose the "−" grip of the marker, or right-click and choose Detach Elements from the shortcut menu (refer to Table 4.28). The members of the mass group are displayed in sepia (color 13), whereas detached mass elements are displayed in green (color 70).

TABLE 4.28 *MassGroupDetach command access*

Command prompt	MASSGROUPDETACH
Ribbon	Select a mass group to display the Mass Group contextual tab and choose Detach Elements from the Modify panel.
Shortcut menu	Select the mass group, right-click, and choose Detach Elements.

Boolean Operations with Mass Elements

Mass elements can be defined to add, subtract, or intersect their mass with the other elements. You can select a mass element or a mass group and perform union, subtract, or intersect Boolean operations on the mass elements. Mass elements not attached to a group are combined when you select the options of the Boolean flyout as shown in Figure 4.67.

If a mass group exists in the drawing, when new elements are added, the mass group and operation can be specified in the Properties palette. After a mass element becomes a member of the group, when you select the mass element all elements of the group and the group marker are displayed. To isolate and edit a mass element that belongs to the group, select a member to display the Mass Group contextual tab and choose Edit In Place from the ribbon or choose the **Edit in Place** grip as shown in Figure 4.67. The **Edit in Place** option turns off the display of the group and allows you to select the mass elements for edit. After editing the mass elements, select **Finish** from the ribbon. When the element is isolated, the operation of a mass element within a group can be edited in the Properties palette or by selecting the Mass Group Operation from the ribbon shown at right in Figure 4.67.

Steps to Edit In Place Add Operation of a Mass Group

1. Select the mass group to display the Mass Group contextual tab.
2. Choose **Edit In Place** from the Modify panel of the ribbon.
3. Select the mass element at p1 as shown in Figure. 4.67. Choose Additive from the **Mass Group Operation** flyout.
4. Choose Finish from the Edits panel of the Edit In Place: Mass Group tab.

The Additive operation can be assigned to an element using the **MassElementOpAdd** command; refer to Table 4.29.

TABLE 4.29 *Mass Group Operation > Additive command access*

Command prompt	MASSELEMENTOPADD
Ribbon	Select a mass group to display the Mass Group contextual tab and choose Edit In Place. Choose Additive from the Mass Group Operation flyout of the Modify panel.
Shortcut menu	Select the Edit In Place grip, select the mass element, right-click, and choose Mass Group Operation > Additive.
Properties palette	Select the Edit In Place grip, select the mass element, and choose Add from the Operation drop-down list in the Design tab of the Properties palette.

Addition is similar to performing a union of the mass elements. Figure 4.67 shows two mass elements shown as **p1** and **p2** that are attached to a group using the additive operation. The Barrel Vault and the Gable are displayed at right as a group.

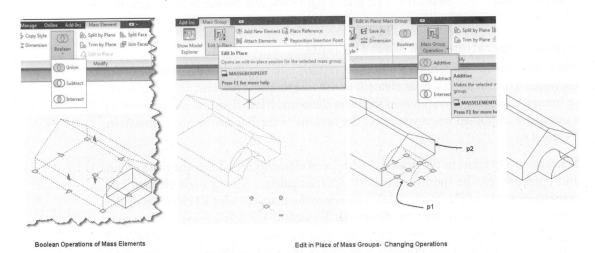

Boolean Operations of Mass Elements

Edit in Place of Mass Groups- Changing Operations

FIGURE 4.67 *Barrel Vault added to the Gable mass element*

Subtracting Mass Elements

When a mass element is defined to subtract, its volume is cut from the group as shown in Figure 4.68. The **MassElementOpSubtract** command (refer to Table 4.30) is used to define the Subtract operation of a mass element. The element selected for the subtraction must be attached to the group after other elements from which it is subtracting its mass.

Steps to Edit In Place Subtract Operation of a Mass Group

1. Select the mass group to display the Mass Group contextual tab.
2. Choose **Edit In Place** from the Modify panel of the ribbon.
3. Select the mass element at p1 as shown in Figure 4.68. Choose Subtractive from the Mass Group Operation flyout.
4. Choose Finish from the Edits panel of the Edit in Place: Mass Group tab.

TABLE 4.30 *Accessing the Subtract operation from the Mass Group command*

Command prompt	MASSELEMENTOPSUBTRACT
Ribbon	Select a mass group to display the Mass Group contextual tab and choose Edit In Place. Choose Subtractive from the Mass Group Operation flyout of the Modify panel.
Shortcut menu	Select the Edit In Place grip, select the mass element, right-click, and choose Mass Group Operation > Subtractive.
Properties palette	Select the Edit In Place grip, select the mass element, right-click, and choose Subtract from the Operation drop-down list in the Design tab of the Properties palette.

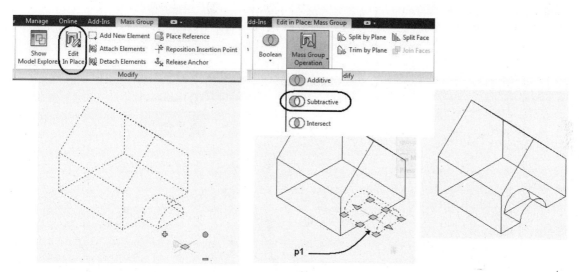

FIGURE 4.68 *Barrel Vault subtracted from the Gable mass element*

Using the Intersect Operations

The **MassElementOpIntersect** command (refer to Table 4.31) joins two mass elements and creates a mass group, which consists of the mass or volume that is common to the two mass elements. The following command sequence modifies the mass element to intersect.

Steps to Edit In Place Intersect Operation of a Mass Group

1. Select the mass group to display the Mass Group contextual tab.
2. Choose **Edit In Place** from the Modify panel of the ribbon.
3. Select the mass element at p1 as shown in Figure 4.69. Choose Intersect from the Mass Group Operation flyout.
4. Choose Finish from the Edits panel of the Edit In Place: Mass Group tab.

TABLE 4.31 *Mass Element Intersection command access*

Command prompt	MASSELEMENTOPINTERSECT
Ribbon	Select a mass group to display the Mass Group contextual tab and choose Edit In Place. Choose Intersect from the Mass Group Operation flyout of the Modify panel.
Shortcut menu	Select the Edit In Place grip, select the mass element, right-click, and choose Mass Group Operation > Intersect.
Properties palette	Select the Edit In Place grip, select the mass element, right-click, and choose Intersect from the Operation drop-down list of Properties palette.

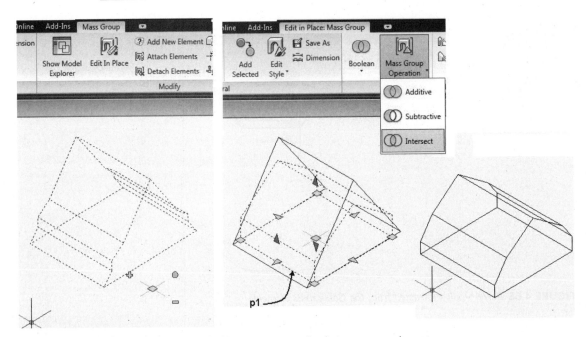

FIGURE 4.69 *Gambrel roof created with the intersect operation between mass elements*

The Intersect operation allows you to create custom-shaped mass elements that combine the two mass elements and retain the volume and shape that is common. The mass elements shown at the left in Figure 4.69 consist of a Gable and an Isosceles Triangle. The mass elements are combined in the mass group. The Isosceles Triangle is defined to intersect with the Gable to form a gambrel roof shape, as the mass group shows on the right in Figure. 4.69.

TUTORIAL 4-7: USING MASS GROUPS TO CREATE THE FIREPLACE

1. Verify that the Accessing Tutor\Imperial or Accessing Tutor\Metric content for Chapter 4 of the CengageBrain has been downloaded to your Accessing Tutor folder on your computer as described in Organizing Tutorial Directories in the Preface.

2. Open your Accessing Tutor\Imperial\Ch4\Ex 4-7.dwg or Accessing Tutor\Metric\ Ch 4\Ex 4-7.dwg.

3. Choose **SaveAs > AutoCAD Drawing** from the Application menu bar and save the drawing as **Lab 4-7** in your student directory.

4. Right-click over the **Object Snap** toggle in the status bar and select **Settings,** check only the **Midpoint** object snap, and toggle **ON Object Snap On (F3)**. Click **OK** to dismiss the Drafting Settings dialog box. Toggle ON Dynamic Input in the application status bar.

5. Select the Home tab. Choose Move from the Modify panel. Move the mass elements by responding to the following prompts as shown in Figure 4.70.

> Select objects: *(Select the mass element at* **p1** *as shown in Figure 4.70.)*
>
> Select objects: ENTER *(Press ENTER to end selection.)*
>
> Specify base point or: *(Select the line at* **p1** *using the* **Midpoint** *object snap.)*
>
> Specify second point of displacement or <use first point as displacement>: *(Select the line at* **p2** *using the* **Midpoint** *object snap.)*
>
> *(Right-click and choose* **Repeat Move** *from the shortcut menu.)*
>
> Select objects: *(Select the box mass element at* **p3** *as shown in Figure 4.70.)*
>
> Select objects: ENTER *(Press ENTER to end selection.)*
>
> Specify base point or: *(Select the line at* **p3** *using the* **Midpoint** *object snap.)*
>
> Specify second point of displacement or <use first point as displacement>: *(Select the line at* **p4** *using the* **Midpoint** *object snap.)*
>
> *(Right-click and choose* **Repeat Move** *from the shortcut menu.)*
>
> Select objects: *(Select the mass element at* **p5** *as shown in Figure 4.70.)*
>
> Select objects: ENTER *(Press ENTER to end selection.)*
>
> Specify base point or: *(Select the line at* **p5** *using the* **Midpoint** *object snap.)*
>
> Specify second point of displacement or <use first point as displacement>: *(Select the line at* **p6** *using the* **Midpoint** *object snap.)*

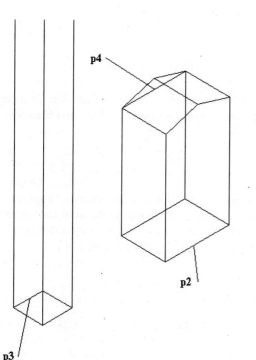
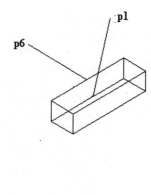
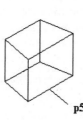

FIGURE 4.70 *Moving mass elements to create a fireplace mass group*

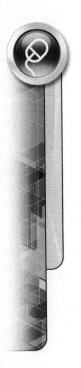

6. Select Massing palette.

7. Select **Mass Group** from the **Massing** tool palette. Respond to the following work-space prompts to place the marker in the drawing:

> Select elements to attach: *(Select a point near* p1 *as shown in Figure 4.71.)*
>
> Specify opposite corner: *(Select a point near* p2 *as shown in Figure 4.71.)* 5 found
>
> Select elements to attach: ENTER *(Press* ENTER *to end selection.)*
>
> Location: *(Select a point near* p3 *as shown in Figure 4.71.)*

8. Select the **Mass Group** at p4 in Figure 4.71 to display the Mass Group tab. Choose **Edit In Place** from the Modify panel.

9. Select the hearth shown at p5 in Figure 4.71 to display its grips. Choose the **Subtractive** option from the Mass Group Operation flyout of the Modify panel in the Edit In Place: Mass Group tab. Choose Finish from the Edits panel to save the changes.

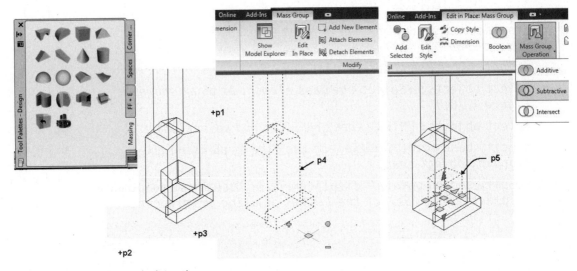

FIGURE 4.71 *Creating and editing the mass group*

10. Choose the Render tab. Double-click the Render Tools button of the Tools palette to display the Visualization palette set. Choose the Materials palette of the Visualization palette set. Choose the **Masonry.Unit Masonry.Brick. Modular** tool from the **Materials** palette. Select a mass group when prompted to select a component or an object.

11. In this step you will modify the material definition to display the brick hatch on the top and bottom planes of an object. Select the mass group, right-click, and choose **Properties.** Choose the **Display** tab. Edit the Display tab as follows: **Display component** = Entity. Verify **By material** = Yes. Click the **Additional Properties** button of the **Advanced** section to open the **Display Properties (Material Definition Override-Masonry.Unit Masonry.Brick.Modular)** dialog box. Choose the **Other** tab, select **Top** and **Bottom** checkboxes, and verify that **Left, Right, Front,** and **Back** boxes are checked for the **Surface Hatch Placement**. The material will be displayed on the top surfaces of the fireplace. Click **OK** to dismiss the dialog box.

12. Choose the View tab. Choose **Realistic** from the **Visual** Styles flyout of the Visual Styles panel to view the mass group as shown in Figure 4.72.

13. Save changes and close the drawing.

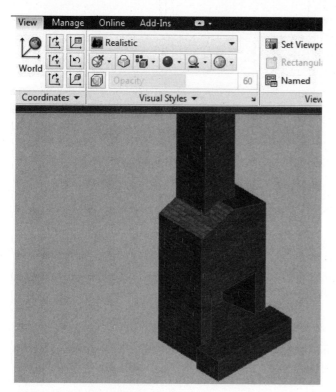

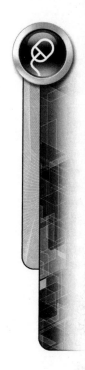

FIGURE 4.72 *Choose Realistic from the Visual Styles panel*

Creating Parametric Constraints for Mass Elements

Parametric constraints can be applied to mass elements to position mass elements relative to other mass elements or walls, columns, beams, and curtain walls. The constraints are two-dimensional. The Coincident, Collinear, Fix, Aligned, and Angular constraints are available for mass elements. All the constraints were presented in Chapter 2 except the Coincident constraint. The Coincident constraint will constrain a point to lie anywhere on mass element. Access the Coincident constraint as described in Table 4.32.

TABLE 4.32 *Slice command access*

Command prompt	GCCOINCIDENT
Tool palette	Select a mass element > Choose Coincident from the Parametric panel of the Wall contextual panel.

A coincident constraint is placed in the following command responses to position a point on a wall to a point on the mass element.

> (Select a mass element, choose Coincident from the Parametric panel of the Wall contextual tab.)
>
> Select first point or: (Select the mass element at p1 as shown in Figure 4.73.)
>
> Select second point or: (Select the wall at p2 as shown in Figure 4.73. Wall is positioned to the first point selected on the mass element.)

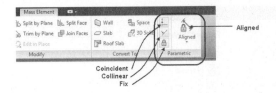

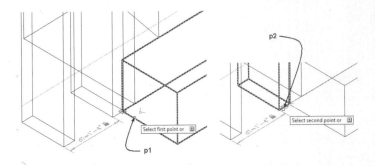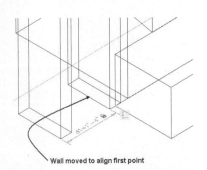

FIGURE 4.73 *Placing the coincident constraint*

NOTE

Refer to **e Tutorials** of Chapter 4 for downloading from the Student Companion Site of CengageBrain http://www.cengagebrain.com described in the Preface. **The E Tutorial 4-8 Using Parametric Constraints to Place Mass Elements is provided.**

Creating a Floorplate from Models

After you develop a mass model of a building, you can use the Slice tool to create a horizontal slice of the model. The **Slice** (**AECSLICECREATE**) tool allows you to slice through the model at elevations equivalent to floor elevations. The slice can be converted to a polyline for the development of floorplates.

The **Slice** (**AecSliceCreate**) tool (refer to Table 4.33) is located on the Massing tool palette. The Slice tool creates slice markers. Mass elements or mass groups are attached to slice markers to create mass elements, and a multistory building would be sliced at the elevations of each finish floor. The exact elevation of finish floors may have to be adjusted later in the design, and therefore, the slice elevation can be edited and the cut will move on the mass elements. The floorplate becomes a horizontal section from which floor plans can be developed, or the geometry can be used to develop space plans.

TABLE 4.33 *Slice command access*

Command prompt	AECSLICECREATE
Tool palette	Select Slice from the Massing tool palette.

When you select the **Slice** command, you are prompted to specify the number of slices, the location for the slice marker, rotation, the elevation of the first slice, and the distance between slices. The slice marker is a graphic marker indicating the slice; the actual slice is executed when the slice marker is attached to the mass elements or mass group. The slice marker and the geometry of the slice are created on the A-Area-Mass-Slce [Aec-Mass-Slices] layer, which has the color 51 (a hue of yellow) and Dashed2 linetype. Shown below is the command sequence used for creating the slice shown in Figure 4.74.

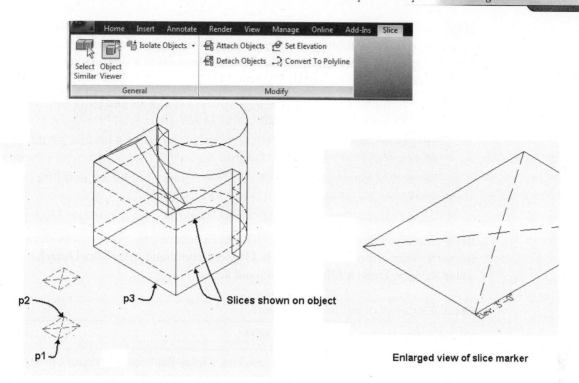

FIGURE 4.74 *Placing a slice*

(Select Slice *from the Massing tool palette.)*

Number of slices <1>: **2** ENTER

First corner: *(Select a point near point* p1 *in Figure 4.74.)*

Second Corner or: *(Select a point near point* p2 *in Figure 4.74.)*

Rotation <0.00>: ENTER *(Press ENTER to accept the 0° rotation.)*

Starting height <0>: **3' [900]** ENTER

Distance between slices <1'-0>: **9' [2700]** ENTER

(Slice markers are created for each of the slices as shown at the right in Figure 4.74.)

The slice marker that is created by the Slice command includes the text describing the elevation of each slice as shown at the top in Figure 4.74. The location, size, and rotation of the marker can be anywhere in the drawing. After placing the marker, select the marker, right-click, and choose Attach Objects from the shortcut menu. The **Attach Objects** command (**AecSliceAttach**) is used to assign the objects to be sliced. See Table 4.34 for Attach Objects command access.

TABLE 4.34 *Attach Objects to Slice command access*

Command prompt	**AECSLICEATTACH**
Ribbon	Select a slice marker to display the Slice contextual tab and choose Attach Objects from the Modify panel of the Ribbon (see Figure 4.74).
Shortcut menu	Select a marker, right-click, and choose Attach Objects.

When you choose the **Attach Objects** command (**AecSliceAttach**) from the short-cut menu, you are prompted to select elements or mass groups to attach for the slice. The workspace prompts for attaching the mass elements to a slice marker are shown in the workspace sequence and in Figure 4.74.

> *(Select the slice markers, choose* `Attach Objects` *from the* `Modify` *panel of the* `Slice` *contextual tab.)*
>
> `Select elements to attach:` *(Select the mass element at* `p3` *as shown in Figure 4.74.)* `1 found`
>
> `Select elements to attach:` `ENTER` *(Press* `ENTER` *to end the selection of mass elements or mass groups.)*
>
> *(Slices are created as shown at the left in Figure 4.74.)*

Mass elements and mass groups that have been attached to a slice can be detached from the slice through the **Detach Objects** command (**AecSliceDetach**). See Table 4.35 for **Detach Objects** command access.

TABLE 4.35 *Detach Objects command access*

Command prompt	AECSLICEDETACH
Ribbon	Select a slice marker to display the Slice contextual tab and choose Detach Objects from the Modify panel of the Ribbon (see Figure 4.74).
Shortcut menu	Select a slice, right-click, and choose Detach Objects.

When you select a slice or slice marker, right-click, and choose **Detach Objects,** you are prompted in the workspace to specify the elements. You can detach elements or a group from the slice. The workspace prompts for detaching a mass element from a slice are shown below.

> *(Select a slice, right-click, and choose* `Detach Objects`*.)*
>
> `Select elements to detach:` `1 found` *(Select the mass element to detach.)*
>
> `Select elements to detach:` *(Press* `ENTER` *to end the selection of mass elements or mass groups.)*

The elevation of the slice can be revised after the **Slice** command has been executed. Changing the elevation of the slice allows you to adjust the finish floor elevation as the design changes. To change the elevation of the slice, select the slice or the slice marker and choose the **Set Elevation** command (**AecSliceElevation**) from the Modify panel of the Slice contextual tab. See Table 4.36 for Set Elevation command access.

TABLE 4.36 *Set Elevation command access*

Command prompt	AECSLICEELEVATION
Ribbon	Select a slice marker to display the Slice contextual tab and choose Set Elevation from the Modify panel of the Ribbon (see Figure 4.74).
Shortcut menu	Select a slice marker or slice, right-click, and choose Set Elevation.

When you select the **Set Elevation** command, you are prompted in the workspace to specify a new elevation. If you type a different elevation in response to the command prompt, the slice will move to the specified elevation. The geometry of the slice cannot be edited. The AutoCAD **Move** command can be used to move the slice marker in the drawing. When the slice marker is moved in the z direction, the slice location on the mass element moves.

Converting the Slice to a Polyline

The slice can be converted to a polyline for the development of walls. To convert a slice to a polyline, select the slice, right-click, and choose the **Convert to Polyline** command (**AecSliceToPline**). See Table 4.37 for **Convert to Polyline** command access.

TABLE 4.37 *Convert to Polyline command access*

Command prompt	AECSLICETOPLINE
Ribbon	Select a slice marker to display the Slice contextual tab and choose Convert to Polyline from the Modify panel of the Ribbon (see Figure 4.74).
Shortcut menu	Select a slice, right-click, and choose Convert to Polyline.

When you select the **AecSliceToPline** command, the slice geometry shown is converted to a polyline, which can be edited according to the desired shape. The polyline is created on the current layer.

Creating Walls from the Polyline. The polyline created from the slice can be used as the base for the development of walls. If you right-click on a wall tool and choose **Apply Tool Properties to > Linework (WallToolToLineWork)**, the polyline will be converted to the wall style of the selected wall tool.

PROJECT TUTORIAL 4-9: CREATING SPACES FROM A FLOOR PLAN

1. Open AutoCAD Architecture 2012 and select **Project Browser** from the Quick Access toolbar. Use the **Project Selector** drop-down list to navigate to your *Student\Ch4* directory. Select **New Project** to open the **Add Project** dialog box. Type **8** in the **Project Number** field and type **Ex 4-9** in the **Project Name** field. Check **Create** from template project, choose **Browse**, and edit the path to *\Accessing Tutor\ Imperial\Ch4\EX 4-9\Ex 4-9.apj* or *\Accessing Tutor\Metric\Ch4\EX 4-9\Ex 4-9.apj* in the **Look in** field. Click **Open** to dismiss the **Select Project** dialog box. Click **OK** to dismiss the **Add Project** dialog box. Click **Close** to dismiss the **Project Browser**.

2. Select the **Views** tab of the **Project Navigator**. Choose the **Views** category and choose **New View Dwg > General** to create a new drawing named Space Study 2 for context = level 2[1] and content = Floor 2. Check Open in drawing editor. Choose Finish to open the drawing.

3. Set the annotation scale to **1/4″ = 1′-0″ [1:50]** and **High Detail** display configuration. Import Room Tag w/ Dimensions tag from the Documentation Tool Catalog.

4. Create a Residential Spaces tool palette. Drag the following spaces from the **Design Tool Catalog** to a new palette: Bedroom, Bath (Small) [Bathroom (Small)], Closet [Cupboard], Hallway, Kitchen, Family Room [Lounge], Laundry and Living Room (Small), and Stairway.

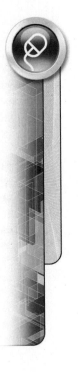

5. Draw a line from p1 to p2 using the node object snap. Draw a line from p3 to p4 using the node object snap. Select each line and edit the Bound Spaces property to Yes in the Properties palette.

6. Create 2D associative spaces as shown in Figure. 4.75. Apply the Room Tag w/Dimensions to each space.

7. Create Sleeping, Service, and Living Zones. Attach the following spaces to each zone:

- Sleeping zone: Bedroom, Closet [Cupboard] adjacent to bedrooms
- Service zone: Bath (Small) [Bathroom (Small)], Hallway, Kitchen, Laundry, Stairway, and Living Room Closet [Cupboard]
- Living zone: Living Room (Small), Family Room [Lounge]

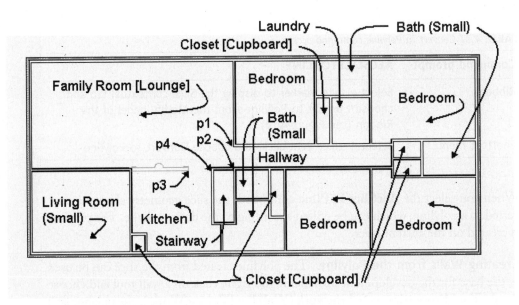

FIGURE 4.75 *Space locations for project*

SUMMARY

- Spaces can be created with 2D, Extrusion, and Freeform geometry.
- Associative spaces can be updated to the boundary formed by AEC objects.
- The Space Generate tool can be used to create spaces from an existing floor plan.
- Space styles can be inserted from the Spaces palette to create nonassociative spaces for a space diagram.
- Create Space Styles in the **Style Manager** to include length, width, and area parameters as well as material definitions for the ceiling and floor.
- To convert polylines to spaces, select **Spaces** from the Design palette, right-click, and choose **Apply Tool Properties to > Linework and Aec Objects**.
- Access editing tools from the AEC Modify shortcut menu to divide, subtract, merge, crop, and array spaces.
- The SpaceEvaluation command allows you to determine the areas of the spaces in a drawing.
- Mass models are created using mass groups, which include mass elements modeled from regular prisms, AutoCAD solids, extrusions, revolutions, and drapes.

- Select the mass group marker tool from the Massing tool palette and attach mass elements to create a mass model.
- Select the **Slice** tool from the Massing tool palette to create a slice through a mass model. AEC slices can be converted to polylines, which can be developed into walls.

Refer to e Review Questions **folder, which includes review questions for Chapter 4 of Student Companion Site of** CengageBrain **http://www.cengagebrain.com described in the Preface.**

NOTE

CHAPTER 5

Placing Doors and Windows

INTRODUCTION

Doors, windows, corner windows, and openings are placed in the drawing as objects with three-dimensional properties. When doors, windows, and openings are placed in a wall, the wall is automatically trimmed and edited. The swing, hinge location, and positions within and along the wall can be edited in AutoCAD Architecture. Door and window styles can be created with material assignments. These styles can also be imported from resource files and can be exported to resource files to improve design.

OBJECTIVES

After completing this chapter, you will be able to

- Insert doors, windows, corner windows with precision using dynamic dimensions, cycle measure to, reference point on and relative to grid options.

- Change the size and location of doors, windows, and openings in the Properties palette.

- Create door and window styles.

- Shift doors and windows within and along walls using the **RepositionWithin** and **RepositionAlong** commands.

- Edit the display representation of windows to include muntins and shutters.

- Place and modify openings in walls using the **OpeningAdd** command.

PLACING DOORS IN WALLS

Doors are placed in a drawing by selecting the **Door** tool from Door flyout menu in the Build panel of the Home tab as shown in Figure 5.1. The Door command can be selected from the Design tool palette of the Design palette set. Each tool allows you to specify the properties of the door, window, or opening in the Design tab of the Properties palette. Additional door styles can be selected from the Doors tool palette.

Doors placed from the Build panel of the Home tab or the Design palette are the Standard style. Access the **DoorAdd** command as shown in Table 5.1.

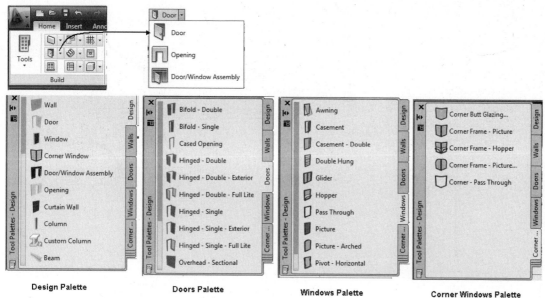

Design Palette **Doors Palette** **Windows Palette** **Corner Windows Palette**

FIGURE 5.1 *Design, Doors, Windows, and Corner Windows tool palettes*

TABLE 5.1 *DoorAdd command access*

Command prompt	DOORADD
Ribbon	Choose the Home tab. Choose Door from the Door flyout menu in the Build panel as shown in Figure 5.1.
Tool palette	Select Door from the Design palette as shown in Figure 5.1.
Tool palette	Select a door style from the Doors tool palette as shown in Figure 5.1.

To add a door, click **Door** on the Build panel or the Design palette, edit the Design tab of the Properties palette, and then select a wall or grid assembly to insert the door. The workspace prompts are shown below.

```
Select wall, grid assembly or ENTER (Select a wall.)
Insert point or: (Select an insert point for the door.)
```

When you select this command, your cursor changes to a pick box to enable you to select the wall or grid assembly. Selecting the wall or grid assembly locks the door to the anchor of the wall. The door object rubber-bands along the wall or grid assembly when you move your cursor. You anchor the door to a location along the wall or grid assembly by selecting a point with the cursor when prompted for the insert point.

You can edit the Design tab of the Properties palette to specify the dimensions and location of the door and return to the workspace, and the DoorAdd command remains active. The Width, Height, and Swing angle are specified in the Dimensions section as shown in Figure 5.2.

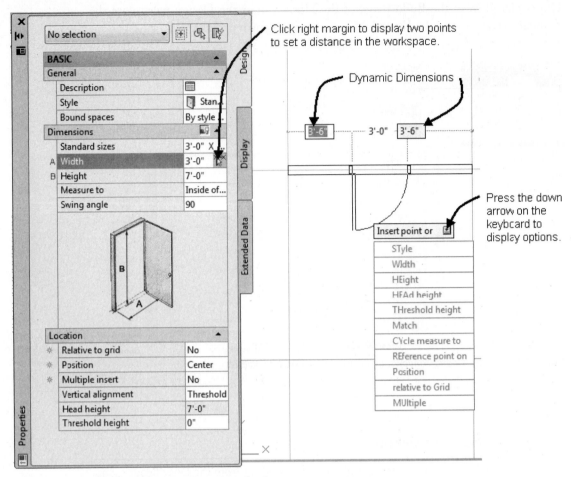

FIGURE 5.2 *Design tab of the Properties palette for doors*

The Location section of the Design tab of the Properties palette provides three key options for placing the door. The *Relative to grid* option allows you to position the door relative to a wall or a structural column grid. The relative to grid is set to No in Figure 5.2; therefore the door is centered in the wall segment. After specifying the *Relative to grid* option, the options for *Position* allow placement using the Center, Offset, and Unconstrained options. The Offset option allows the placement of the door a specified offset distance from the ends of the wall as shown in Figure 5.3, whereas the Unconstrained option allows you to position the door at any point along the wall or grid assembly. The Unconstrained option allows the position of door by typing values in the dynamic dimension or specifying a location with the mouse.

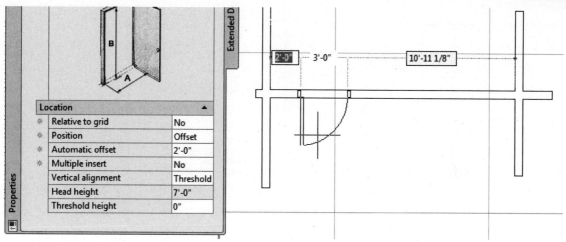

FIGURE 5.3 *Door placed with Position set to Offset*

If **Relative to grid** is set to **Yes**, the door may be centered between column grids. The Center, Offset, and Unconstrained Position options may be chosen relative to the grid. When Multiple Insert is set to Yes, the Number to insert and Spacing options are added to the Properties palette. You can specify the quantity of doors for insert and the spacing between doors. The two doors are inserted relative to the column grid as shown Figure 5.4. The Spacing may be set to equally or a specific offset distance value.

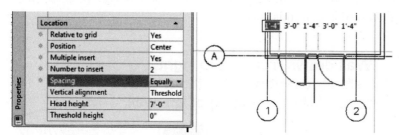

FIGURE 5.4 *Inserting doors relative to a grid*

Each time the Properties palette is edited, a phantom image of the revised door is displayed about the cursor. Toggling between the drawing area and the Properties palette allows you to place doors of various properties in one or more walls.

After you select the wall, dynamic dimensions are displayed, indicating the location of the door as shown in Figure 5.3. Dynamic dimensions are displayed if Dynamic Input is toggled ON in the Application status bar.

The door can be positioned by specifying a head or threshold height in the properties palette. The door location is specified with a -10" threshold, which lowers the threshold of the door down 10" from the bottom of the wall as shown in Figure 5.5.

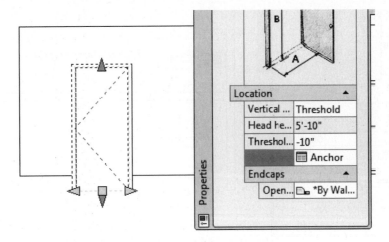

Modify Threshold Value to Specify Location

FIGURE 5.5 *Edit of threshold to position door in wall*

When you place a door with the Position set to Unconstrained, you cannot set the hinge point prior to insertion. The swing can be specified by moving the cursor to the other side of the wall prior to specifying the insert point. When the insert point is selected, this finalizes the position for the door swing. The default location of the hinge point of an unconstrained door is toward the start point of the wall. In Figure 5.6 the walls were drawn from left to right. When unconstrained, door placement allows you only to move the cursor location above and below the wall to set the swing as shown in Figure 5.6. The hinge point and swing of doors can be changed by grips, which are discussed later in this chapter.

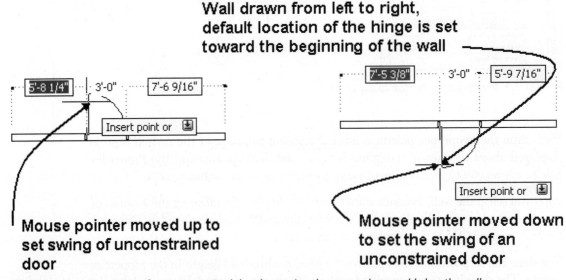

FIGURE 5.6 *Setting swing of an unconstrained door by moving the cursor above and below the wall*

Locating the Insertion Point of the Door

Although the insertion point of the door can be precisely placed using the Position constraint, dynamic input allows you to position the door by editing the dynamic dimension during placement. Options within the DoorAdd command allow you to toggle the door insert point to the center or edge of the door by cycling through the insertion point. In addition, the **Reference point on** option allows you to select points in the drawing to position the door.

Dynamic Dimensions

A door can be placed in a wall with precision using dynamic dimensions. Dynamic dimensions will be displayed when you select a wall or grid assembly. The dynamic dimensions shown in Figure 5.7 are displayed from the endpoints of the wall to the door. The active dynamic dimension is shown with a blue background. You can press TAB to toggle the highlight for each dimension. In Figure 5.7 TAB has been pressed to highlight the right dimension. When a dimension is highlighted, you can type a number that will display in the workspace to set the dimension as shown in Figure 5.7.

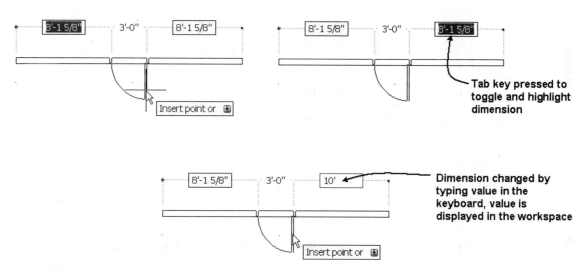

FIGURE 5.7 *Specifying door placement using dynamic dimensions*

Using the Cycle Measure to Option

The insertion point shown on the left in Figure 5.8 is locating the dynamic dimensions to the center of the door. The dynamic dimension can be changed to place the handle for insertion to center, left, and right positions by right-clicking and choosing **Cycle measure to**. Each time you choose Cycle measure to, the handle is toggled to either center, left, or right as shown in Figure 5.8. Cycle measure to is an option for doors, windows, and openings.

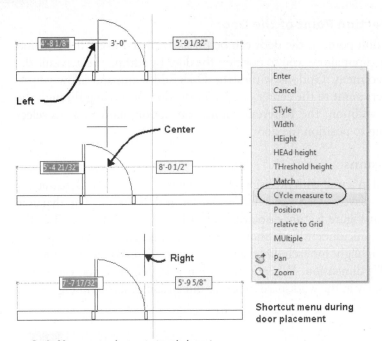

Cycle Measure to chosen to toggle insert
point from center, left, and right locations

FIGURE 5.8 *Cycle measure to option used to change dynamic dimension*

The **Reference point on** option is displayed in the shortcut menu during door, window, and opening placement when the Relative to grid is set to No and the Position is set to Unconstrained. The Reference point on option allows you to measure from features of a wall or other objects to establish the location of the door. The Reference point on option requires you to select a point in the drawing to reference the dynamic dimension. After you select the reference point, move the cursor to set the direction for the measurement and type the correct distance in the workspace. The door shown in Figure 5.9 was placed in the wall using Reference point on as shown in the following command prompts.

> Select wall, grid assembly or ENTER: (Select the wall at **p1** as shown in the upper-left corner of Figure 5.9)
>
> Insert point or: (Right-click and choose **Reference point on** from the shortcut menu to select option.)
>
> Pick the start point of the dimension: **_endpoint of** (SHIFT + right-click, choose **End** object snap, and select point **p2** as shown in the upper-right corner of Figure 5.9) (Move the cursor to the right.)
>
> Insert point or: **5' [1500]** ENTER (Type **5' [1500]** in the workspace to position the door as shown in the lower-left corner of Figure 5.9.)
>
> Insert point or: ENTER (Press ENTER to end the command.)

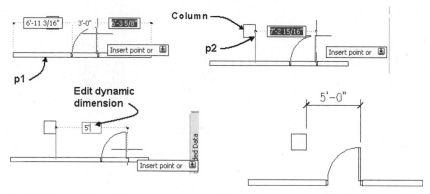

FIGURE 5.9 *Reference point on option used to locate the door*

Using Grips to Modify a Door

The grips of a door can be used to move the door along the wall, change the hinge location, change the swing, and change the width of the door. Selecting the center grip allows you to stretch the door along the wall to a new location. Editing with this grip is shown in Figure 5.10; the grip is selected and the cursor moved to the right. When you stretch a grip, you can type the distance to move the door in command line. The door can also be shifted along the wall with the **RepositionAlong** command discussed later in this chapter.

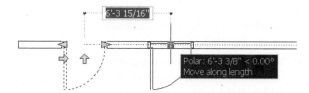

FIGURE 5.10 *Using grips to move the door along the wall*

Selecting the grip when the door is viewed in pictorial allows you to edit the door vertically with grips as shown in Figure 5.11. After you have selected the center grip, each time you press CTRL the movement of the cursor is restricted to shifting the door vertically, within the wall, or along the wall. The door shown in Figure 5.11 at the left is stretched down using the **Move vertically** option of grips. Shifting the door vertically can be useful when a door from the one level must extend down to the lower level. Garage doors of the first floor often extend down into the foundation wall.

Doors can also be stretched within the wall to position the jamb to the edge of the wall. Pressing CTRL while editing the center grip allows you to shift the door within the wall. The door in Figure 5.11 is shown stretched out of the wall. Doors that have been repositioned within the wall remain anchored to the wall. The shifting of the door within the wall can also be done using the **RepositionWithin** command discussed later in this chapter.

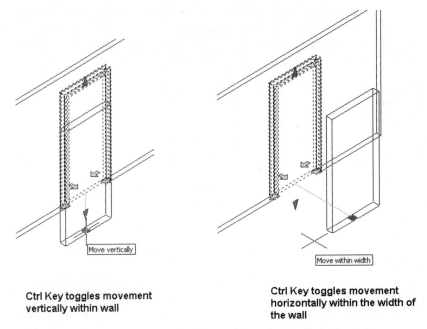

Ctrl Key toggles movement vertically within wall

Ctrl Key toggles movement horizontally within the width of the wall

FIGURE 5.11 *Using grips to move the door within and vertically in the wall*

The grips located at the jamb are arrows indicating the direction of edit. After you select a door to display the grips, if you move the cursor over the frame grip, a dynamic dimension will display the current width of the door. Select the frame grip, and dynamic dimensions are activated. When the dynamic dimension is displayed, movement of the cursor will display two dimensions: the new width of the door and the change in door width. You can stretch the grip to the gray marks shown in Figure 5.12. The gray marks are graphical representations of widths defined in the standard sizes of the door style. The DoorAdd tools with preset styles located on the Doors palette include standard sizes. The new door size can then be typed in the workspace when the dynamic dimension is highlighted as shown in Figure 5.12. This option of grip editing, shown in Figure 5.12, allows you to stretch the doorjamb to the edge of the wall by tabbing through dynamic dimensions. The left frame location remains anchored when you edit the right frame grip as shown in Figure 5.12.

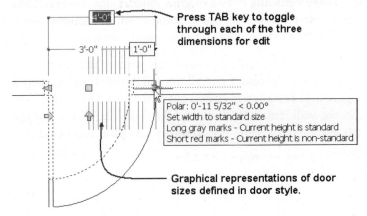

Press TAB key to toggle through each of the three dimensions for edit

Polar: 0'-11 5/32" < 0.00°
Set width to standard size
Long gray marks - Current height is standard
Short red marks - Current height is non-standard

Graphical representations of door sizes defined in door style.

FIGURE 5.12 *Editing the door frame grip*

The **Flip** trigger grip shown at the right in Figure 5.13 if selected will flip the swing of the door to the opposite side of the wall. The hinge can be flipped by selecting the **Flip** trigger grip shown at the left in Figure 5.13. The hinge and swing of the doors shown in Figure 5.13 were changed using the Flip trigger grips. After editing the door using grips, press ESC to clear the grip display.

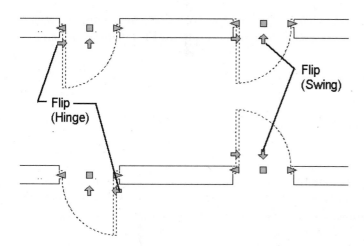

FIGURE 5.13 *Changing hinge and swing of a door with trigger grips*

The hinge location can also be changed with the **OpeningFlipHinge** command, which flips the location of the hinge.

Access the **OpeningFlipHinge** command as shown in Table 5.2.

TABLE 5.2 *OpeningFlipHinge command access*

Command prompt	OPENINGFLIPHINGE

When you type this command in the command line, you are prompted to select doors, windows, or openings. The hinge location will flip for all selected doors, windows, or openings. This command allows you to flip the swing of all selected doors without specifying the location of the hinge. The **OpeningFlipHinge** command was applied to the left door to flip the hinge to the right, as shown in the following command sequence and in Figure 5.14.

```
Command: OpeningFlipHinge
Select doors, windows, or openings: (Select the door near p1
as shown in Figure 5.14.) 1 found
Select doors, windows, or openings: ENTER (Press ENTER to end
the selection.)
(Hinge of door flipped as shown at the right in Figure 5.14.)
```

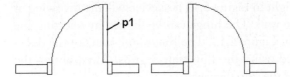

FIGURE 5.14 *Editing the hinge location with the OpeningFlipHinge command*

The swing location can also be changed by the **OpeningFlipSwing** command, which flips the swing of one or more doors. Access the **OpeningFlipSwing** command as shown in Table 5.3.

TABLE 5.3 *OpeningFlipSwing command access*

Command prompt	OPENINGFLIPSWING

When you type this command in the command line, you are prompted to select doors or windows. The swing location will flip for all selected doors and windows. The OpeningFlipSwing command was applied to the door on the left to flip the swing as shown in the following command sequence and in Figure 5.15.

```
Command: OpeningFlipSwing

Select doors or windows: (Select the door at p1 as shown on
the left in Figure 5.15.) 1 found

Select doors or windows: ENTER (Select ENTER to end door
selection.)

(Door swing is flipped down as shown on the right in
Figure 5.15.)
```

FIGURE 5.15 *Editing door swing with OpeningFlipSwing command*

Defining Door Properties

The properties of doors can be changed by editing the Properties palette. When you select an existing door, the Properties palette changes to include additional information that is not included when you add a door. Additional fields are added to allow you to change features of the inserted door. Therefore, to change one or more doors, select the doors and then edit the Properties palette.

The Display tab of the Properties palette allows you to modify the display of the door. Display can be controlled by This object, Door Style, and Drawing default settings. The Display tab of a door is shown in Figure 5.16. The Display representation is set to Plan since the Medium Display Configuration is current. Choose the Display component from the drop-down list to view the settings of the display components of the Display controlled by setting.

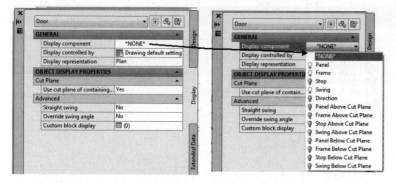

FIGURE 5.16 *Editing door display in the Properties palette*

CREATING AND USING DOOR STYLES

Sample door styles, which provide customized frame and leaf properties, are included in the Doors palette. When a door is selected from the Doors palette, the DoorAdd command is selected with style preset according to the tool selected. Additional styles can be imported into the drawing from other drawings or created in the **Style Manager.** Additional door styles are also located in the Design Tool Catalog – Imperial and Design Tool Catalog – Metric of the Content Browser. Choose **More Door Tools** of the Doors palette to open the tool catalogs to door styles.

Access the **Door Styles** command as shown in Table 5.4.

TABLE 5.4 *Door Styles of the Style Manager command access*

Command prompt	**DOORSTYLE**
Command prompt	**AECSTYLEMANAGER**
Ribbon	Select a door to display the Door contextual tab. Choose Edit Style from the General panel to open the Door Style Properties dialog.
Ribbon	Select a door to display the Door contextual tab. Choose Door Styles from the Edit Style flyout menu of the General panel to open the Style Manager.
Ribbon	Choose Manage tab. Choose Style & Display panel. Expand Architectural Objects, choose Door Styles in the Style Manager. Choose Style Manager.
Shortcut menu	Select a door in the workspace, right-click, and choose Edit Door Style.
Shortcut menu	Select a door style from the Doors palette, right-click, and choose Door Styles.

When the **Style Manager** command is executed, the Style Manager opens as shown in Figure 5.17. The Door Styles are listed in the Architectural Objects category of the current drawing. The commands of the Style Manager toolbar are described in "Creating and Editing Wall Styles" in Chapter 3.

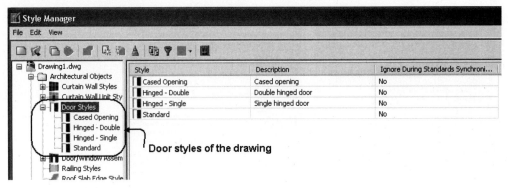

FIGURE 5.17 *Door Styles in the Style Manager*

Creating a New Door Style

Create a new door style by selecting the Door Styles folder in the left pane of the **Style Manager**, right-clicking, and choosing **New Style** from the shortcut menu. Overtype a name for the new style in the left pane. The properties of the door style are defined by editing the right pane of the Style Manager.

TUTORIAL 5-0: EXPLORING A DOOR STYLE

1. Download Chapter 5 files from the Accessing Tutor\Imperial or Accessing Tutor\ Metric category of the Student Companion site of CengageBrain **http://www. cengagebrain.com** described in the Preface to your Accessing Tutor folder.

2. Open drawing Ex 5-0 in your Accessing Tutor\Imperial\Ch 5 or Accessing Tutor\ Metric\Ch 5 folder.

3. Choose **Save As** > **AutoCAD Drawing** from the Application menu, and save the drawing as **Lab 5-0** in your student directory.

4. Select the door to display Door contextual tab. Choose Edit Style from the General panel of the Door contextual tab.

5. Choose the **General** tab of the Door Style Properties dialog box. Verify that the check box for *Objects of this style may act as a boundary for associative spaces* is checked. This checkbox specifies that doors of this style will serve as part of the boundary for the creation of spaces for rooms.

6. Choose the **Dimensions** tab of the Door Style Properties dialog box. The **Dimensions** tab of the **Door Style Properties** dialog box allows you to set dimensions of the frame, stop, and door thickness, as shown in Figure 5.18.

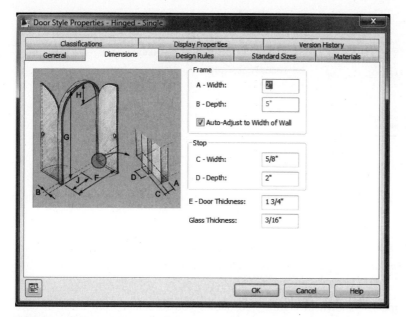

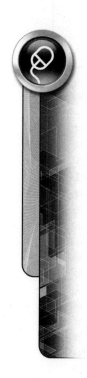

FIGURE 5.18 *Dimensions tab of the Door Style Properties dialog box*

7. The Auto-Adjust to Width of Wall checkbox of the Dimensions tab determines whether the depth is fixed or varies according to the width of the wall. When this checkbox is selected, the frame depth adjusts to the wall width, as shown at the right in Figure 5.19. If Auto-Adjust to Width of Wall is not checked, the frame depth is fixed to that specified in the B-Depth field. The door on the left in Figure 5.19 has a fixed door frame depth.

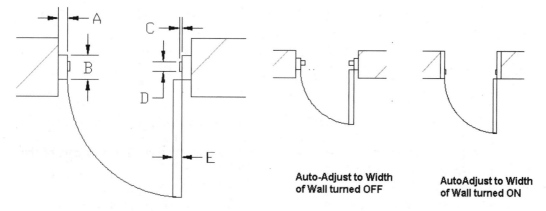

FIGURE 5.19 *Frame dimensions of a door style*

8. Choose the **Design Rules** tab as shown in Figure 5.20. The Design Rules tab allows you to choose from the following list of Predefined shapes: rectangular, half-round, quarter-round, arch, gothic, and peak pentagon. This list is extensive and satisfactory for most design applications. The Use Profile radio button is active if an AEC Profile is defined in the drawing. The AEC Profile is converted from a closed polyline and applied to define the shape of a door.

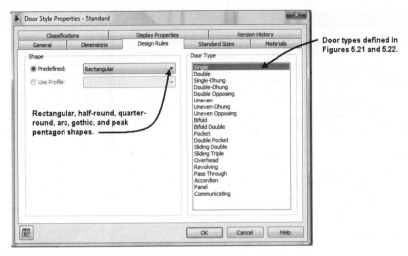

FIGURE 5.20 *Design Rules tab of the Door Style Properties dialog box*

9. The **Door Type** list shown in Figure 5.20 allows you to select from a list of door types. The types of doors that can be created are shown in Figures 5.21 and 5.22. The name assigned to each type is shown below the figure. Note that the garage door is named Overhead and is shown in Figure 5.22.

Door Type

Single   Double   Single-Dhung   Double-Dhung   Double Opposing   Uneven

Uneven-Dhung   Uneven Opposing   Bifold   Bifold Double   Pocket   Double Pocket

FIGURE 5.21 *Door types available in AutoCAD Architecture*

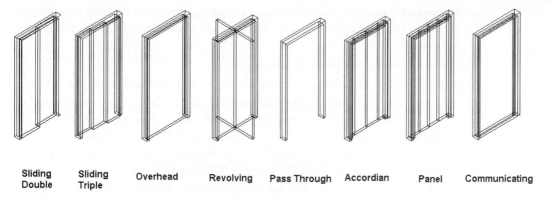

Sliding Double   Sliding Triple   Overhead   Revolving   Pass Through   Accordian   Panel   Communicating

FIGURE 5.22 *Door types including the Overhead garage door*

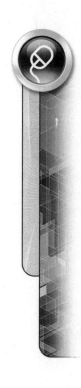

10. Choose the **Standard Sizes** tab of the Door Style Properties dialog box. The sizes listed in the Standard Sizes table include **Description, F-Width, G-Height, H-Rise**, and **J-Leaf.** These dimensions are illustrated on the **Dimensions** tab in Figure 5.23. The H-Rise is inactive unless the door shape is Gothic, Arch, or Peak Pentagon. The J-Leaf column is inactive unless the door type is set to Uneven.

11. Choose the Add New Standard Size button to create additional standard sizes. Selecting the Add New Standard Size button opens the **Add Standard Size** dialog box as shown in Figure 5.23. You can create sizes by editing the **Width, Height, Rise**, and **Leaf** fields of the **Add Standard Size** dialog box. You can choose the Edit Standard Sizes or Remove Standard Sizes to modify the list.

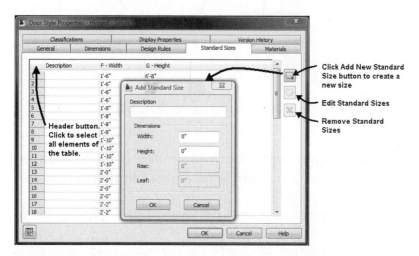

FIGURE 5.23 *Standard Sizes tab of the Door Style Properties dialog*

12. Choose the **Materials** tab. Materials have been defined for each component of the Door. The current material for the Stop component is Doors & Windows.Metal Doors & Frames.Aluminum Frame.Anodized.Dark Bronze.Satin. Left-click in the Material Definition column for the Stop component to display the material definitions of the drawing in the flyout list. You could choose Doors & Windows.Wood Doors.Ash to change the material definition.

13. Verify that the current display configuration is **Medium Detail** in the Drawing Window status bar. Choose the **Display Properties** tab of the Door Style Properties dialog box. Choose the **Plan** display representation. Choose the **Edit Display**

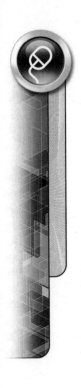

Properties button to open the Display Properties (Drawing Default)-Plan. Choose the Layer/Color/Linetype tab. The Stop component is turned Off for this display representation.

14. Choose OK to dismiss all dialog boxes.

15. Choose the **High Detail** display configuration in the Drawing Window status bar.

16. Select the door, and choose **Edit Style** from the Door contextual tab. Choose the Display Properties tab. Select the **Plan High Detail** display representation. Choose Edit Display Properties to top open the Display Properties (Drawing Default) - Plan High Detail. Choose the Layer/Color/Linetype tab. Imperial users only, click the light bulb to turn on the display of the Stop as shown in Figure 5.24.

17. Click the Frame Display tab. This tab is present only for the Plan High Detail display representation. The frame type can be selected from the Type list.

18. Edit the Frame Display. Type to U-Shaped as shown in Figure 5.24. Click OK to dismiss all dialog boxes and view the U-shaped frame.

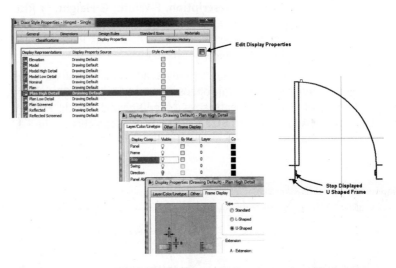

FIGURE 5.24 *Editing Display Properties to display stop and modify frame type*

19. Choose **Medium Detail** display configuration from the Drawing Window status bar.

20. Select the door to display the Door contextual tab. Choose **Edit Style** from the General panel of the Door contextual tab.

21. Choose the **Display Properties** tab. Choose the Threshold Plan display representation. Choose the **Edit Display Properties** button at right.

22. Choose the Layer/Color/Linetype tab. Toggle ON the lightbulb for Threshold B. Choose the Other tab. Edit the D - Depth value to 2" [50]. Choose OK to dismiss all dialog boxes and display a threshold for the door as shown in Figure 5.25.

23. Save and close the drawing.

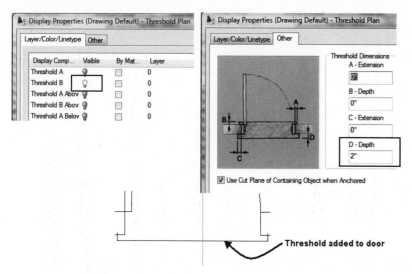

FIGURE 5.25 *Editing Display Properties to add a threshold*

Access Instructional Video 5.1 Placing Doors from the Instructional Video category of the Student Companion site of CengageBrain **http://www.cengagebrain.com** described in the Preface.

NOTE

TUTORIAL 5-1: CREATING A DOOR STYLE USING AN AEC PROFILE

1. Verify that the Accessing Tutor Imperial or Accessing Tutor Metric content for Chapter 5 of the CengageBrain has been downloaded to your Accessing Tutor folder on your computer as described in Organizing Tutorial Directories in the Preface.

2. Open drawing Ex 5-1 in your Accessing Tutor\Imperial\Ch 5 or Accessing Tutor\Metric\Ch 5 directory.

3. Choose **Save As > AutoCAD Drawing** from the Application menu, and save the drawing as **Lab 5-1** in your student directory.

4. Select the polyline at **p1** as shown in Figure 5.26, right-click, and choose Convert To > Profile Definition. Respond to the workspace prompts as follows:

   Insertion Point or: (Press down arrow on the keyboard; choose the Add Ring option.)

   Select a closed polyline, spline, ellipse, or circle for an additional ring: (Select the line at **p2** as shown in Figure 5.26.)

   Insertion Point or: _endpoint of (Press Shift, right-click, and choose the Endpoint object snap; then select the end of the line at **p3** as shown in Figure 5.26.)

   Type Chamfer in the New Profile Definition dialog box. Click OK to dismiss the dialog box.

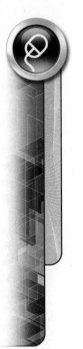

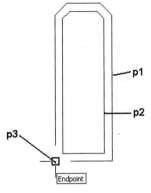

FIGURE 5.26 *Selected polyline for profile*

5. Select the **Doors** tool palette, right-click a door style, and choose **Door Styles** from the shortcut menu.

6. Select the **Door Styles** category, right-click, and choose **New** from the shortcut menu. Overtype the name **Chamfer** in the left pane to name the door style.

7. Verify that the **Chamfer** style is current, and choose the **Design Rules** tab.

8. To assign the profile to the new style, select the **Use Profile** radio button and select the **Chamfer** Profile. Verify that Door Type is set to **Single.**

9. Select the **Inline Edit Toggle** button of the Style Manager toolbar as shown in Figure 5.27.

10. Select the **Viewer** tab in the right pane; select **SW Isometric** view, **3D Hidden** Visual Styles, and **Medium Detail** display as shown in Figure 5.27.

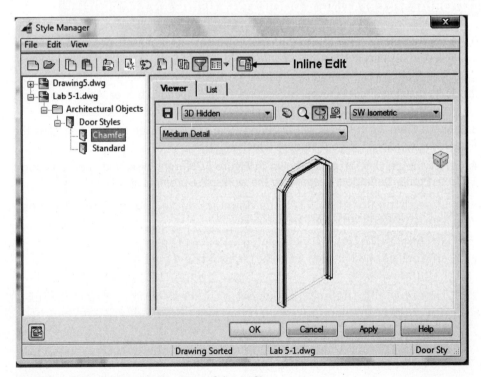

FIGURE 5.27 *Door created in the shape of the profile*

11. Click **OK** to dismiss the **Style Manager.**

12. Select **Door** from the **Design** tool palette. Edit the **Properties** palette as follows: **Style = Chamfer, Width = 3' [900], Height = 6'-8" [2000], Measure to = Inside of frame, Swing angle = 45, Relative to grid = No, Position = Center, Multiple insert = No.** Place the door in the center of the wall as shown in the following workspace sequence:

> Select wall, grid assembly or ENTER: (Select the wall at **p4** as identified in the drawing.).
>
> Insert point or: (Select the point at **p5** as identified in the drawing.)
>
> Insert point or: (Press ESC to end the command.)

13. Choose **SW Isometric** from the **View Cube.**

14. Toggle ON Polar Tracking on the status bar.

15. Select the door, and choose **Edit In Place** from the Profile panel of the contextual Door tab. Select **Yes** to convert the profile to its actual size, and dismiss the AutoCAD Architecture warning dialog box. Select the middle grip at the bottom edge of the inner panel at **p1** as shown in Figure 5.28, move the cursor up (polar angle = 90), and type **6 [150]** in the dynamic dimension to stretch the segment up. Select **Finish** on the Edits panel of the **Edit In Place: Door** contextual tab to save the profile change.

16. The profile of the door is modified as shown at the right in Figure 5.28.

17. Save and close the drawing.

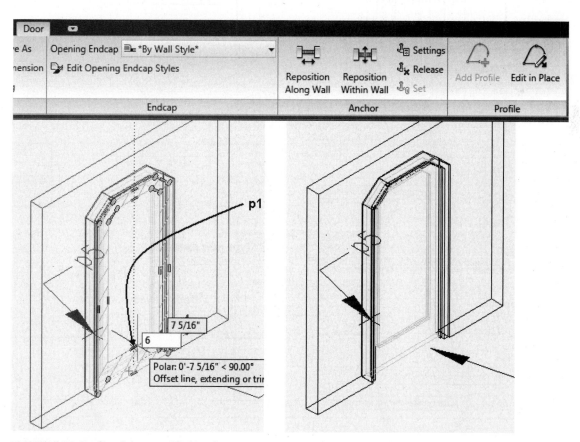

FIGURE 5.28 *Profile of door modified in place*

Shifting a Door within the Wall

A door frame can be shifted within a wall using grips or the **RepositionWithin** command. The prompts of this command allow you to measure the distance from a point on the frame to a point on the wall. The distance between the two points can then be edited with dynamic dimensions to adjust the offset. Access **RepositionWithin** as shown in Table 5.5.

TABLE 5.5 *RepositionWithin command access*

Command prompt	REPOSITIONWITHIN
Ribbon	Select a door to display the Door contextual tab. Choose Reposition Within Wall on the Anchor panel.
Shortcut menu	Select a door, right-click, and choose Reposition Within Wall.

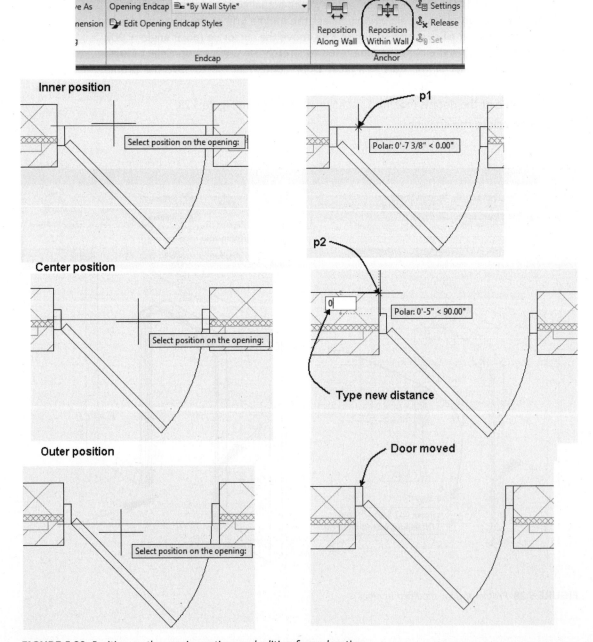

FIGURE 5.29 *Position on the opening options and editing frame location*

When the **Reposition Within Wall** command is selected from the contextual tab of a selected door, you are prompted to specify a position on the opening. When you move the cursor over the frame, the center, inner, and outer edges of the frame are highlighted with a red line to help you select the position on the opening as shown at the left in Figure 5.29.

The next prompt is to select a reference point. The reference point should be the point within the wall where you want to move the "position on the opening." After you select these two points, a dynamic dimension is displayed that allows you to edit the distance between the points. The following workspace sequence moved the edge of the frame (**p1**) to the face of the wall (**p2**) as shown in Figure 5.29.

> Select position on the opening: (Select a point on the frame **p1** as shown at the right in Figure 5.29.)
>
> Select a reference point (Select a point on a feature of the wall **p2** as shown at the right in Figure 5.29.)
>
> Enter the new distance between the selected points <0>: **0** (Overtype a distance in the dynamic dimension.)

Shifting the Door along a Wall

The **RepositionAlong** command will shift a door along a wall a specified distance. Access **RepositionAlong** as shown in Table 5.6.

TABLE 5.6 *RepositionAlong command access*

Command prompt	REPOSITIONALONG
Ribbon	Select a door to display the Door contextual tab. Choose Reposition Along Wall on the Anchor panel.
Shortcut menu	Select a door, right-click, and choose Reposition Along Wall.

When you select a door, right-click, and choose **Reposition Along Wall** from Door contextual tab, you are prompted to specify a position on the opening, which will serve as a handle to position the door. The position on the opening, shown as a red line, can be the center or either edge of the frame as shown in Figure 5.30. The next command line prompt is to select a reference point, which can be any point on the wall. After the two points are selected, a dynamic dimension is displayed, allowing you to edit the distance between the two points.

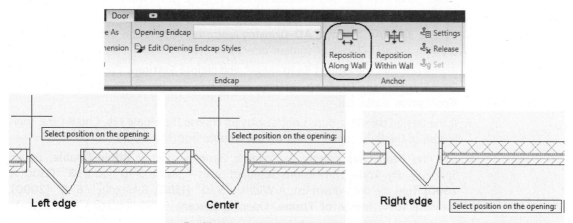

FIGURE 5.30 *Center and Edge options to position frame*

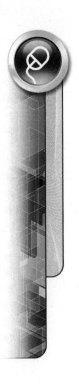

The **RepositionAlong** command was used in the following command line sequence to move the door 4' to the end of the wall.

> Select position on the opening: *(Select center of frame at p1 as shown in Figure 5.31.)*
>
> Select a reference point *(Select node at end of wall near p2 as shown in Figure 5.31.)*
>
> Enter the new distance between the selected points <5'-0">: 4' [<1500.00>: 1200] *(Type 4' [1200] in the dynamic dimension.)*

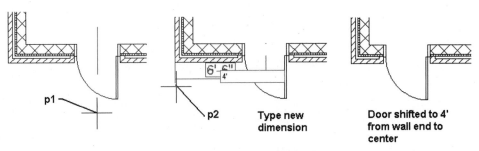

p1   p2   Type new dimension   Door shifted to 4' from wall end to center

FIGURE 5.31 *Door adjusted with RepositionAlong command*

The **RepositionAlong** command can be used to edit doors inserted with the incorrect Offset value. The offset value of the RepositionAlong command positions an existing door in a manner similar to placing a new door with the offset position option of the DoorAdd command.

NOTE

The RepositionWithin and RepositionAlong commands can also be used to shift windows and openings in a wall.

TUTORIAL 5-2: POSITIONING DOORS RELATIVE TO WALL AND CUSTOM COLUMN GRID

1. Verify that the Accessing Tutor\Imperial or Accessing Tutor\Metric content for Chapter 5 of the CengageBrain has been downloaded to your Accessing Tutor folder on your computer as described in Organizing Tutorial Directories in the Preface.

2. Select Open from the Quick Access toolbar. Choose Ex 5-2 from the Accessing Tutor\ Imperial\Ch 5 or *Accessing Tutor\Metric\Ch 5* folders.

3. Choose **Save As > AutoCAD Drawing** from the Application menu, and save the drawing as **Lab 5-2** in your student directory.

4. Right-click over the **Object Snap** toggle, and choose **Settings.** Clear all modes. Click **OK** to dismiss the **Drafting Settings** dialog box. Toggle ON Dynamic Input in the **Application** status bar.

5. If the Design palette group is not displayed, choose the Home tab. Choose the Tools button of the Build panel. Select **BiFold Double** from the **Doors** tool palette.

6. Edit the **Properties** palette as follows: General-Style = **Bifold-Double**, Bound Spaces = **By Style**, Dimensions-Standard Sizes—choose **5'-0" × 6'-8" [1500 × 2000]** from the drop-down list, A-Width = **5'-0" [1500]**, B-Height = **6'-8" [2000]**, Measure to = **Inside of Frame**, Opening percent = **50**, Location - Relative to

grid = **No**, Position = **Offset**, Automatic offset = **2' [600]**, Multiple Insert = **No**, Vertical Alignment = **Threshold,** and Threshold height = **0**.

7. Place the door by responding to the workspace prompts as follows:

   Select wall, grid assembly or ENTER: *(Select the wall at* **p1** *as shown in Figure 5.32.)*

   Insert point or: *(Select a point below the selected wall near* **p2** *to specify the swing as shown in Figure 5.32.)*

   Insert point or: *(To insert an additional door select a point near* **p3** *to specify the swing and location as shown in Figure 5.32.)*

   (Right-click and choose ENTER *from the shortcut menu.)*

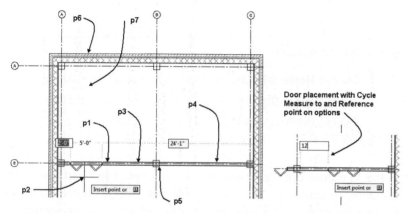

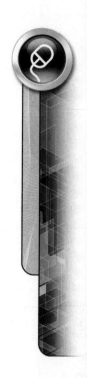

FIGURE 5.32 *Placing doors using Reference point from and Cycle Measure to options*

8. Select **Bifold-Double** from the **Doors** tool palette. Verify the **Properties** palette as follows: General-Style = **Bifold-Double**, Bound Spaces = **By Style**, Dimensions: A-Width = **5'-0" [1500]**, B-Height = **6'-8" [2000]**, Measure to = **Inside of Frame**, Opening percent = **50**. Edit the following in the Properties palette: Location-Relative to grid = **No**, Position = **Unconstrained**, Multiple insert = **No**, Vertical Alignment = **Threshold**, and Threshold height = **0**.

9. Place the door by responding to the workspace prompts as follows:

   Select wall, grid assembly or ENTER: *(Select the wall at* **p4** *as shown in Figure 5.32.)*

   (Right-click and choose Reference point on *from the shortcut menu.)*

   Pick the start point of the dimension: *(Hold the Shift key down, right-click, and choose the Endpoint object snap; select point* **p5** *as shown in Figure 5.32. Move the cursor right.)*

   Insert point or: 12 [300] ENTER *(Type distance in the workspace.)*

   Insert point or: *(Press* ESC *to end the command.)*

10. Select **Hinged Single** from the **Doors** tool palette. Edit the **Properties** palette as follows: General-Style = **Hinged - Single**, Bound Spaces = **By Style**, Dimensions: A-Width = **3'-0" [900]**, B-Height = **6'-8" [2000]**, Measure to = **Inside of Frame**, Swing angle = **90**, Location-Relative to grid = **Yes**, Position = **Center**, Multiple insert = **No**, Vertical Alignment = **Threshold**, and Threshold height = **0**.

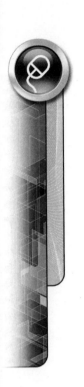

11. Place the door by responding to the workspace prompts as follows:

    `Select wall, grid assembly or ENTER:` *(Select the wall at* **p6** *as shown in Figure 5.32.)*

    `Insert point or:` *(Select a point near p7 as shown in Figure 5.32.)*

    (Right-click and choose ENTER *from the shortcut menu.)*

12. Save and close the drawing.

PLACING WINDOWS IN WALLS

The **WindowAdd** command is used to place windows in a drawing. Access the **WindowAdd** command as shown in Table 5.7.

TABLE 5.7 *WindowAdd command access*

Command prompt	WINDOWADD
Ribbon	Choose Home tab. Choose Window from the Build panel.
Design palette	Select Window from the Design palette as shown in Figure 5.33.
Tool palette	Select a window style from the Windows tool palette.

When you select the **WindowAdd** command, you are prompted to select a wall or grid assembly. Windows are inserted in the drawing with precision using the dynamic dimensions **Reference point on** and **Cycle measure to** techniques as presented in the DoorAdd command. The description, style, sizes, and Position and Relative to grid are similar to the options of the door. The content in the Design tab of the Properties palette is displayed when you insert windows as shown in Figure 5.33.

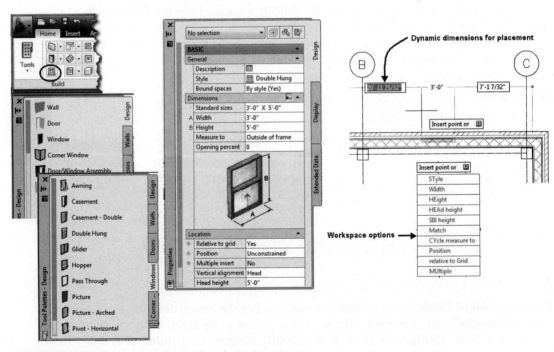

FIGURE 5.33 *Properties palette of a window during insertion*

The swing angle is specified in the Dimensions section for hinged windows such as casements. The Opening percent option is included when placing windows with slide components such as double-hung windows. The Rise option will be active on window styles created from the arch, gothic, peak pentagon, and trapezoid shapes. The rise is the distance from the top of the window in which the arch begins. A one-foot rise was used to create the arch, gothic, peak pentagon, and trapezoid windows shown in Figure 5.34.

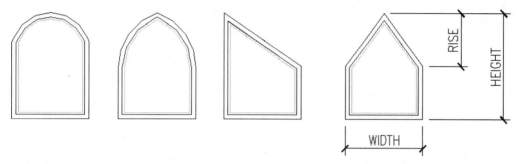

FIGURE 5.34 *Rise of windows*

The Head Height is the distance from the baseline of the wall to the inside of the frame or outside of the frame at the top of the window. The Sill Height is the distance from the baseline of the wall to the bottom of the window frame.

After you select the wall during the insertion of the window, the window will dynamically slide along the wall until an insertion point is defined by a left-click of the mouse. The Properties palette can be edited before or after an insertion point for the window is specified. After selecting the wall, press the down arrow on the keyboard to display the options as shown in Figure 5.33. The workspace options can be accessed from the shortcut menu during window placement. All workspace options except Match, Undo, Cycle measure to, and Reference point on can be set in the Properties palette.

Defining Window Properties

When you select existing windows, the Properties palette includes additional information such as the layer and endcap settings. The Properties palette can be used to edit the properties of one or more windows. The display tab allows you to edit how the window is displayed. The Display component, Display controlled by, and the current Display Representation are shown in the General section. The Display component drop-down lists all display components of the window. When you choose a display component from the drop-down list, you can specify how it is displayed. The Display controlled by options consist of This object, Window Style, and Drawing default setting. The Display controlled by option determines the content in the Design tab of the Properties palette for the selected component. The Display representation lists the display representations available for the current display configuration as shown in Figure 5.35. If a display component is selected in the General section, display properties such as layer, color, and linetype can be edited for the display component.

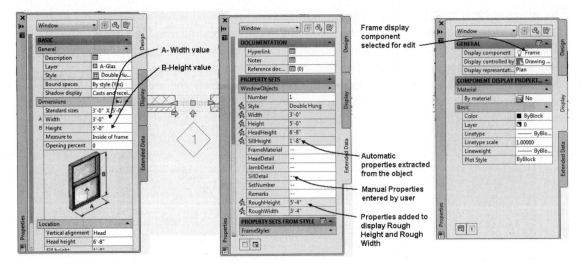

FIGURE 5.35 *Properties palette of an existing window*

The Extended Data Tab includes a **Property Sets** list as shown in Figure 5.35. The data for the properties are inserted in the drawing when window tags are placed. The **A-Width** and **B-Height** values are properties extracted from the drawing and used in the sample Window Schedule table. Therefore, if you enter the rough or masonry opening values in the **Width** and **Height** fields in the Design tab of the Properties palette, these values will automatically be displayed in the Window Schedule. Additional properties can be defined to customize the window schedule to include sash size and model numbers. (The development of schedules is presented in Chapter 11.) The **Property Sets** list for a window is shown at the right in Figure 5.35. The **RoughHeight** and **RoughWidth** properties have been added to the WindowObjects properties shown in Figure 5.35. Since the Measure to Inside of Frame is specified in the Dimensions section of the Design tab, the frame width is added to the **A-Width** and **B-Height** values. The **RoughHeight** and **RoughWidth** values can be included in a window schedule.

When the **Measure to Outside of frame** option is selected, the width and height measurements are applied to the window frame rather than the sash opening size. The **Measure to Outside of frame** dimension would be appropriate if you want to specify the masonry opening or rough opening. **Measure to Inside of frame** will apply the width and height dimensions to specify the window sash opening size. The width of frame dimension specified in the window style is added to the width and height of the sash opening size to determine the overall **RoughHeight** and **RoughWidth** properties.

NOTE When you use the Measure to Outside of frame option, the A-Width and B-Height values are equal to the RoughWidth and RoughHeight property values. The RoughWidth and RoughHeight properties are inserted in the drawing when the window tag is placed. The default window schedule table automatically extracts the A-Width and B-Height property values regardless of "Measure to" options.

When you use **Measure to Inside of frame**, the frame width is added to the A-Width and B-Height values to define the RoughWidth and RoughHeight property values. The default window schedule table can be modified to include the RoughWidth and RoughHeight property values.

Editing a Window with Grips

The grips of a window can be used to edit the swing, hinge point, size, head height, sill height, and location within and location along a wall described for doors earlier in this chapter. The grip locations of a window are shown in Figure 5.36.

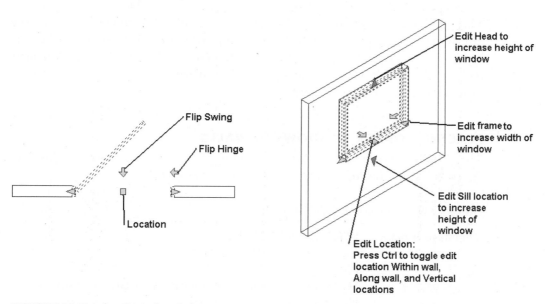

FIGURE 5.36 *Grip locations of a window*

THE WINDOWS PALETTE

Windows can also be inserted by selecting a window style from the Windows palette as shown in Figure 5.33. There are ten styles listed on the Windows palette. When you select a tool from the palette, the **WindowAdd** command is executed with the style preset. Additional styles can be imported from other drawings or created in the Style Manager. Additional styles are located in the Imperial and Metric folders of *{Vista - C:\ProgramData\Autodesk\ACA 2012\enu\Styles}* directory. Choose More Window Tools from the Windows palette to open the Design Tool Catalog – Imperial or Design Tool Catalog – Metric. Bay and Bow Window styles are included in the Design Tool Catalog-Imperial; however, Bay and Bow window styles are not included in the Design Tool Catalog-Metric.

CREATING A NEW WINDOW STYLE

Window styles can be created that include a saved list of standard sizes, frame, and sash parameters. Using styles of windows improves uniformity in placing windows. You can create and edit window styles using the **Style Manager** or the **WindowStyle** command. Access **WindowStyle** as shown in Table 5.8.

TABLE 5.8 *WindowStyle command access*

Command prompt	WINDOWSTYLE
Ribbon	Select a window to display the Window tab. Choose Window Styles to open the Style Manager to the window objects category.
Ribbon	Select a window to display the Window tab. Choose Edit Style to open the Window Style Properties dialog.
Shortcut menu	Select Window in the Design palette or Windows palette, right-click, and choose Window Styles.
Shortcut menu	Select a window, right-click, and choose Edit Window Style.

Selecting the **Window Styles** command from the Edit Style flyout menu in the Build panel opens the Style Manager to the *Architectural Objects\Window Styles* category. This command will display the Style Manager with only the Window Styles category displayed in the left pane. Because the Window Styles folder is preselected when the Style Manager opens, to create a new style, select **New Style** from the **Style Manager** toolbar. The default name for the new style is displayed in the left pane. Overtype a name for the new style in the left pane.

PLACING CORNER WINDOWS IN WALLS

The **WindowCornerAdd** command is used to place corner windows in a drawing. Access the **WindowCornerAdd** command as shown in Table 5.9.

TABLE 5.9 *WindowCornerAdd command access*

Command prompt	WINDOWCORNERADD
Ribbon	Choose Home tab. Choose CornerWindow from the Build panel.
Design palette	Select CornerWindow from the Design palette as shown in Figure 5.37.
Tool palette	Select a corner window style from the Corner Window tool palette.

When you select the **CornerWindowAdd** command, you are prompted in the workspace as follows:

```
Select wall to insert corner window: (Select the wall at p1 as
shown in Figure 5.37.) Insert point or: (Select a point near
p2 as shown in Figure 5.37 to specify the corner. Press Esc to
end the command.)
```

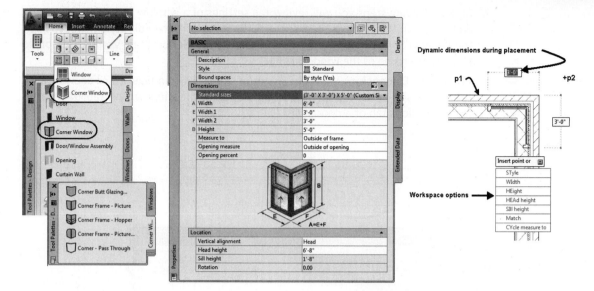

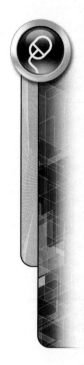

FIGURE 5.37 *Accessing the corner window*

The Properties palette shown in Figure 5.37 allows you to specify the width of each window for the height specified. The Corner Windows tool palette shown in Figure 5.37 provides a sample of corner window styles.

TUTORIAL 5-3: PLACING A WINDOW AND CORNER WINDOW - EXPLORING THEIR STYLES

1. Verify that the Accessing Tutor\Imperial or Accessing Tutor\ Metric content for Chapter 5 of the CengageBrain has been downloaded to your Accessing Tutor folder on your computer as described in Organizing Tutorial Directories in the Preface.

2. Choose **Open** from the **Quick Access** toolbar. Choose **Ex 5-3** from the *Accessing Tutor\Imperial\Ch 5* or Accessing Tutor\ Metric\Ch 5 folder.

3. Choose **Save As > AutoCAD Drawing** from the Application menu, and save the drawing as **Lab 5-3** in your student directory.

4. If the Design palette group is not displayed, choose the **Home** tab. Choose the **Tools** button of the **Build** panel. Choose the Windows palette. Select **Double Hung** from the Windows palette.

5. Edit the Properties palette as follows: General-Style = **Double Hung**, Bound spaces = **By style (Yes)**, Dimensions-Width = **3' - 0" [900]**, B-Height = **5'-0" [1500]**, Measure to = **Outside of Frame**, Location-Relative to grid = **No**, Position = **Center**, Multiple Insert = **No**, Vertical Alignment = **Head**, and Head height = **6'-8" [2000].** Respond to the workspace prompts as shown below:

       Select wall, grid assembly or ENTER: *(Select the wall at p1 as shown in Figure 5.38.)*

       Insert point or: *(Select a point near p2 as shown in Figure 5.38.)*

       (Press ESC to end the command.)

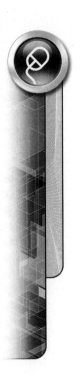

6. In the following steps you will review the settings of a window style. Select the window placed in the previous step, choose **Edit Style** from the Edit Style flyout to open the Window Style Properties dialog box. Choose the **General** tab. Verify that Double Hung is the name of the window style.

7. Choose the **Dimensions** tab. The Dimensions tab shown in Figure 5.38 allows you to specify the size of the frame, sash, and glass size of the window. This tab is similar to the Dimensions tab of a door style.

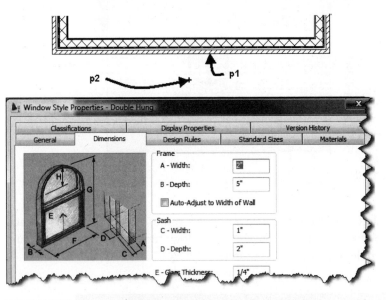

FIGURE 5.38 *Dimensions tab of a window style in the Style Manager*

8. Choosing the **Design Rules** tab allows you to specify the shape and window type of the style. The Design Rules tab is shown in Figure 5.39. You can create an AEC Profile from a closed polyline, and apply the AEC Profile to define the shape of the window in the **Shape** options. The **Window Type** options define the operation of the sash for the window.

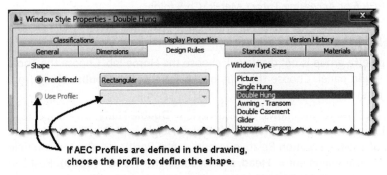

FIGURE 5.39 *Design Rules tab of a window style in the Style Manager*

9. Verify that the Predefined radio button is toggled ON. Choose the drop-down list of predefined shapes. The predefined list includes the following: rectangle, round, half-round, quarter-round, oval, arch, gothic, Isosceles triangle, right triangle, peak pentagon, octagon, hexagon, and trapezoid. Each of these shapes can be applied to a window type to create a window style. The Use Profile option allows you to create an irregular-shaped polyline that when converted to an AEC Profile becomes the shape of the window.

10. The Window Type list box includes various types of windows that are manufactured, such as awning, casement, double hung, and glider. The Double-Hung window is shown bold to define the window type for this window style.

11. Choose the **Standard Sizes** tab to view a table of available sizes for a window style. The Standard Sizes tab is shown in Figure 5.40. The following dimensions are defined in a Standard Size: F-Width, G-Height, H-Rise, and J-Leaf. These dimensions are illustrated on the **Dimensions** tab of the **Window Style Properties** dialog box. The H-Rise is inactive unless the window shape is Arch, Gothic, Peak, or Trapezoid. The J-Leaf column is inactive unless you specify an "Uneven" window type.

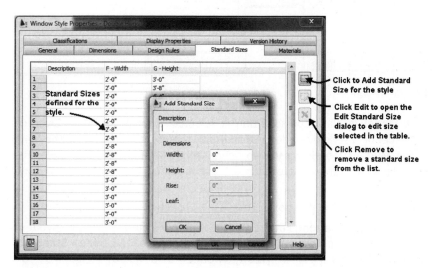
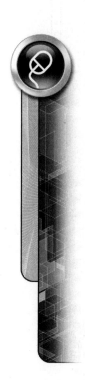

FIGURE 5.40 *Standard Sizes tab of a window style in the Window Style Properties dialog box*

12. Choose the **Materials** tab. This tab lists the material definitions for each of the window components.

13. Choose the **Classification** tab. The Classification tab is empty since no classifications are defined in the drawing.

14. Choose the **Display Properties** tab as shown at right in Figure 5.41. Choose the Plan display representation. Choose the Edit Display Properties button to open the Display Properties (Drawing Default) - Plan. The Frame, Sash, and Glass components are turned on for this display representation. Choose OK to dismiss the Display Properties (Drawing Default) - Plan dialog box.

15. Choose the **Model** display representation. Choose the Edit Display Properties button to open the Display Properties (Drawing Default) - Model dialog box as shown at right in Figure 5.41. The Model display representation consists only of three display components: Frame, Sash, and Glass. Choose OK to dismiss the Display Properties (Drawing Default) - Model dialog box.

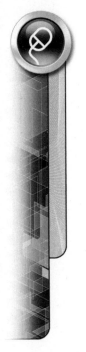

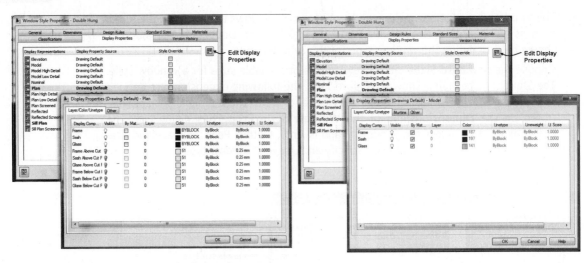

FIGURE 5.41 *Display Properties tab of a window*

16. Verify that **Model** display representation is selected. Choose the Style Override check box to define a style override for this display representation. Choose the Muntins tab. The Muntins tab allows you to define muntins for the window style. The model and elevation display representations are modified to include muntins and shutters.

17. Choose OK to dismiss all dialog boxes.

18. Verify the **Home** tab is current; choose **Corner Window** from the Window flyout of the Build panel. Edit as necessary the Properties palette as follows: Style = **Standard**, A-Width = **7'-0"**, E Width 1 = **4'-0" [1200]**, Width 2 = **3'-0" [900]**, Height = **5' - 0" [1500]**, Location-Vertical alignment = **Head**, and Head height = **6'-8" [2000]**. Move the cursor to the workspace and respond as follows:

```
Insert point or: (Select a point near the corner at p1 as
shown in Figure 5.42.)
```

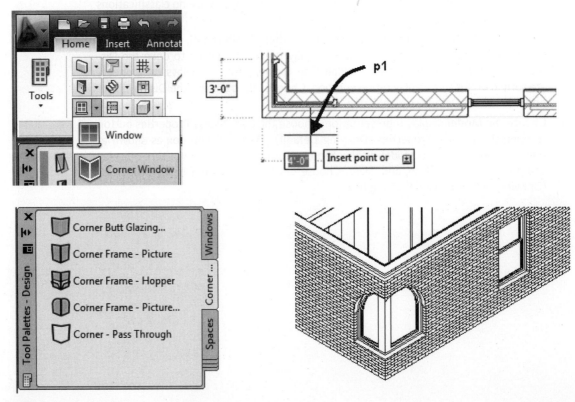

FIGURE 5.42 *Placing a corner window*

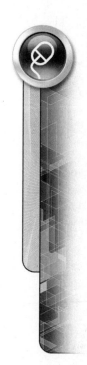

19. In this step you will change the style of the corner to a style located on the Corner Window tool palette. Choose the Corner Window palette of the Design palette group. Choose the Corner Frame - Picture Half Round of the Corner Windows tool palette, right-click, and choose Apply Tool Properties to > Window. Select the window placed in the previous step to modify its style.

20. Choose Southwest Isometric from the View Cube to view the corner window as shown at right in Figure 5.42.

21. Save and Close the drawing.

CREATING MUNTINS FOR WINDOWS

The **Display Properties** dialog box for the Elevation and Model display representations includes a **Muntins** tab as shown in Figure 5.41. **Muntins** can be defined in the window style or in the object override display. Muntins divide the glass area of the window to make the window attractive. The muntin patterns can be Rectangular, Diamond, Prairie 9 Lights, Prairie 12 Lights, Starburst, Sunburst, or Gothic as shown in Figure 5.43.

The starburst, sunburst, and gothic patterns are restricted to certain types and shapes of windows. Table 5.10 lists the window types and window shapes in which the Starburst, Sunburst, and Gothic muntins can be used. The window shape and type are defined in the window style. The Rectangular and Diamond muntin patterns can be used on any window type or shape.

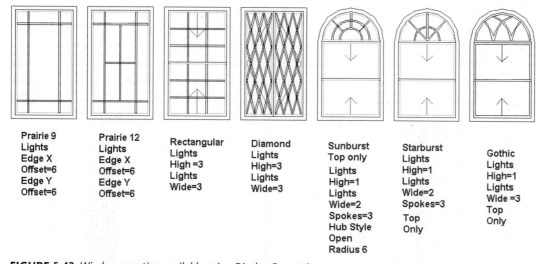

Prairie 9
Lights
Edge X
Offset=6
Edge Y
Offset=6

Prairie 12
Lights
Edge X
Offset=6
Edge Y
Offset=6

Rectangular
Lights
High =3
Lights
Wide=3

Diamond
Lights
High=3
Lights
Wide=3

Sunburst
Top only
Lights
High=1
Lights
Wide=2
Spokes=3
Hub Style
Open
Radius 6

Starburst
Lights
High=1
Lights
Wide=2
Spokes=3
Top
Only

Gothic
Lights
High=1
Lights
Wide =3
Top
Only

FIGURE 5.43 *Window muntins available using Display Properties*

TABLE 5.10 *Muntin options for window types*

Window Types for Starburst Sunburst - Gothic Muntins	Window Shapes for Starburst	Window Shapes for Sunburst	Window Shapes for Gothic
Awning	Round	Round	Round
Single Hopper	Half Round	Half Round	Half Round
Single Transom	Quarter Round Top	Quarter Round Top	Gothic
Vertical Pivot			Peak Pentagon
Horizontal Pivot			Arch
Double Hung			
Glider			
Single Hung			
Single Casement			
Picture			

Muntins are defined by editing the Elevation or Model display representations. You can access the Muntins tab from Display Properties dialog box as shown at the right in Figure 5.41. The elevation or model display representation is set current when the window is viewed in isometric or elevation view. You can also create muntins defined within a window style by the Display tab of the Properties palette. To create muntins from the Properties palette, select the window when viewed in isometric or elevation, right-click, and choose Properties. Choose the Display tab of Properties; edit **Display controlled by** to **Window Style.** When you specify Display controlled by Window Style, the Add Style Override dialog box opens as shown in Figure 5.44. Click the **Muntins** button to open the Muntins Block Display dialog box. When you click the **Add** button of the Muntins Block Display dialog box, the **Muntins Block** dialog box opens as shown in Figure 5.45.

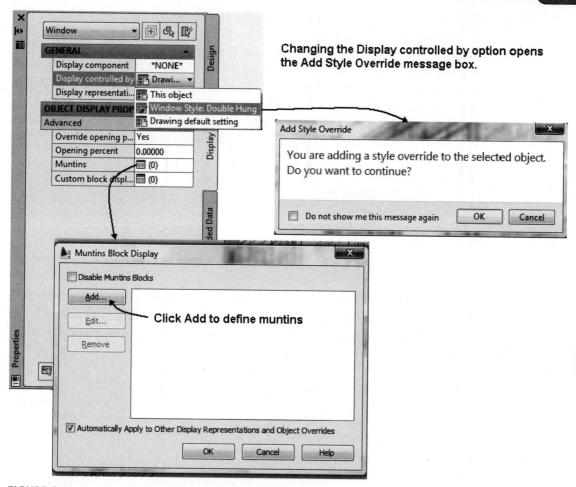

Changing the Display controlled by option opens the Add Style Override message box.

FIGURE 5.44 *Creating Muntins from the Display tab of Properties*

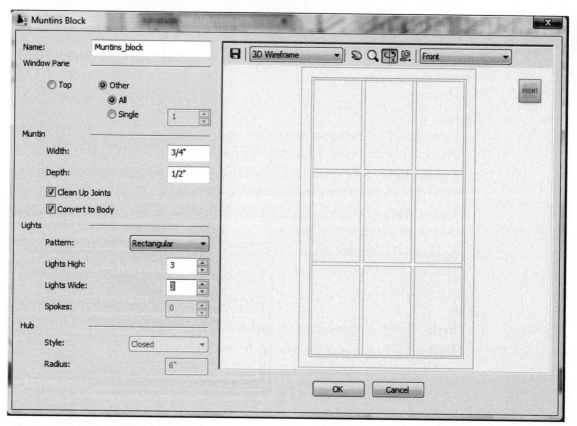

FIGURE 5.45 *Muntins Block dialog box*

The definition of the muntin block is created by editing the fields shown on the left; the resulting muntin pattern is displayed on the right. The radio buttons located in the Window Pane section allow you to specify the muntins for either the top sash or all sashes of the window. The Lights section allows you to specify the number of divisions within the sash and the pattern. The rectangular pattern is shown in Figure 5.46; however, the drop-down list includes the following patterns: Rectangular, Diamond, Prairie 9 Lights, Prairie 12 Lights, Starburst, Sunburst, and Gothic. The options of the Muntins Block dialog box are described below.

Name
The Name field is defined by AutoCAD, or you can type a specific name of the muntin pattern.

Window Pane
Top—Select the Top radio button to turn on the display of muntins only in the top sash area.

Other—Select the Other radio button to specify muntins in all sash areas or a single sash specified by its index number. If **Other** and **Single** are selected, you specify the sash number to the right of the **Single** radio button. The index number is a number identifying the sash location within the window. The number system starts numbering the sash in the lower-left corner and increases the number assigned as you move in a counterclockwise direction from the lower-left sash location. The index numbers of the two sashes of a double hung are 1 for the lower window and 2 for the upper sash.

Muntin
Width—The Width field specifies the width dimension of the muntins.

Depth—The Depth field specifies the depth dimension of the muntins.

Clean Up Joints—Check ON **Clean Up Joints** to trim the intersection of muntins lines as shown in Figure 5.46.

Convert to Body—The **Convert to Body** checkbox creates a 3D body to represent the muntins.

Lights
Patterns—The **Patterns** drop-down list consists of Rectangular, Diamond, Prairie 9 Lights, Prairie 12 Lights, Starburst, Sunburst, and Gothic patterns.

Lights High—Specify in the Lights High field the number of divisions in each sash vertically by clicking on the arrow to the right.

Lights Wide—Specify in the Lights Wide field the number of divisions to divide the sash in a horizontal direction.

Spokes—Specify the number of spokes to create in the sash if the starburst or sunburst patterns are selected.

Hub
Style—The hub style can be set to closed or open for the sunburst style.

Radius—The radius of the hub can be specified.

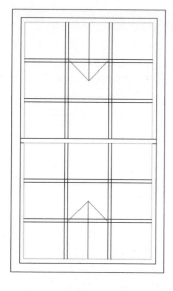 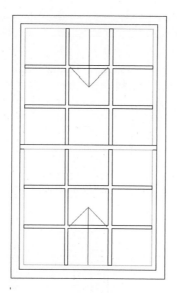

Cleanup Joints OFF   Cleanup Joints ON

FIGURE 5.46 *Display of muntins using Clean Up Joints toggle*

ADDING SHUTTERS TO A WINDOW STYLE

Shutters are created in a window by adding a custom block that consists of the shutter graphics. The shutter graphics block is included in the Shutters-Dynamic or Shutters-12 window styles located in the Design Tool Catalog – Imperial catalog of the Content Browser. Shutters are not included in the metric styles.

The Shutters-Dynamic style adjusts the shutter width according to the width of the window. The Shutters-12 style includes a window with a fixed 12 wide shutter. Each style has added the shutters custom block to a window with rectangular shape using the picture window type. Therefore, you can insert the Shutters-Dynamic or Shutters-12 window styles into a drawing. The Shutters-Dynamic or Shutters-12 window styles become the base for the style; therefore, to create a double-hung window with shutters, copy the style and change the window type to Double Hung in the **Design Rules** tab of the **Window Style Properties** dialog box.

The shutters created in the Shutters-Dynamic and Shutters-12 styles are custom blocks applied to the Elevation and Model display representations. Therefore, if the shutter styles have been used in a drawing, the custom blocks are imported into the drawing. You can add the custom blocks to other windows in the drawing. The custom blocks are added by clicking the **Add** button on the **Other** tab of the **Display Properties – Window Elevation Display Representation** or the **Display Properties – Window Model Display Representation** dialog box. Clicking the **Add** button opens the **Custom Block** dialog box as shown in Figure 5.47. Click the **Select Block** button to open the **Select A Block** dialog box. The shutters are then selected as custom blocks from the block list and attached to the left and right sides of the window as shown in Figure 5.47.

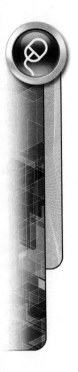

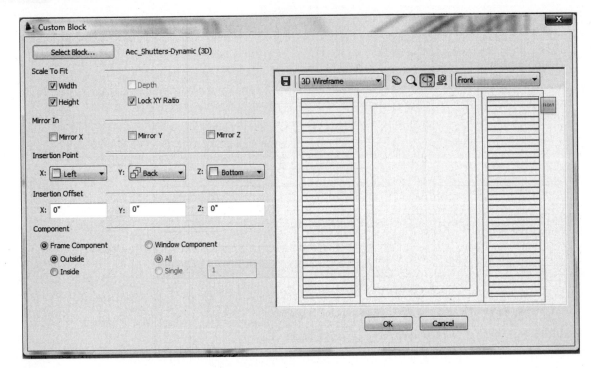

FIGURE 5.47 *Adding custom blocks to the window*

TUTORIAL 5-4: CREATING WINDOWS WITH SHUTTERS

1. This tutorial is imperial-only since shutters are not included in the Design Tool Catalog – Metric catalog.

2. Verify that the Accessing Tutor\Imperial content for Chapter 5 of the CengageBrain has been downloaded to your Accessing Tutor folder on your computer as described in Organizing Tutorial Directories in the Preface.

3. To perform the tutorial using imperial tools, choose **Open** from the **Quick Access** toolbar. Choose **Ex 5-4** from the *Accessing Tutor\Imperial\Ch 5* folder.

4. Choose **Save As > AutoCAD Drawing** from the Application menu, and save the drawing as **Lab 5-4** in your student directory.

5. Right-click over the **Object Snap** toggle, and choose **Settings.** Clear all modes. Click **OK** to dismiss the **Drafting Settings** dialog box. Toggle ON Dynamic Input in the **Application** status bar.

6. Verify that the Design palette set is displayed. Right-click over the title bar of the tool palettes, and choose **New Palette** to create a new palette. Overtype **Shutters** as the name of the new palette.

7. Choose the Home tab. Choose **Content Browser** from the Tools flyout menu of the Build panel. Open the Design Tool Catalog – Imperial.

8. Type **Shutters** in the **Search** field of the catalog. Click the **GO** button as shown in Figure 5.48 to display the shutters tools. Drag the i-drop for the **Shutters - 12** onto the new shutters tool palette. Close the Content Browser.

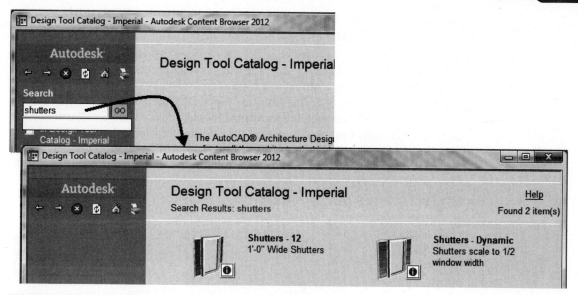

FIGURE 5.48 *Inserting the Shutters 12 tool from the Content Browser*

9. Choose the **Shutters 12** tool. Edit the Design tab of the **Properties** palette as follows: General-Style = **Shutters 12**, Bound spaces = **By Style**, Dimensions-A-Width = **3'–0"**, B-Height = **5'–0"**, Measure to = **Outside of Frame**, Opening percent = **0**, Location-Relative to grid = No, Position = Offset, Automatic offset = **3'–0"**, Vertical Alignment = **Head**, and Head height = **6'–8"**. Respond to the workspace prompts as shown below.

> Select wall, grid assembly or ENTER: *(Select the wall at* **p1** *as shown in Figure 5.49.)*
>
> Insert point or: *(Select a point near* **p2** *as shown in Figure 5.49.)*
>
> *(Press ESC to end the command.)*

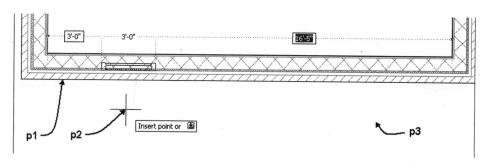

FIGURE 5.49 *Locations for window*

10. In this step you will create a new window style that includes the shutter block with the double-hung window style. Select the window, and choose **Copy Style** from the General panel of the Window contextual menu. Choose the **General** tab. Type **Double-Hung-Shutters** in the **Name** field. Choose the **Design Rules** tab. Edit the **Window Type** to **Double Hung.** Select OK to dismiss the dialog box.

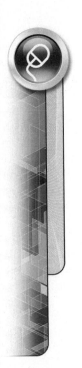

11. Choose the View **SW Isometric** from the **View Cube**. Choose **Realistic** from the Visual Styles flyout menu of the Visual Styles panel in the View tab.

12. To create muntins for the window, select the window, right-click, and choose Properties. Choose the Display tab of the Properties palette. In the Display tab verify that Display Representation = **Model** and Display controlled by = Window Style. Choose the Muntins button to open the **Muntins Block Display** box. Click the **Add** button shown in Figure 5.50 to open the **Muntins Block** dialog box. Edit the Muntins Block as follows: Window Pane, and toggle on the radio buttons for **Other** and **All**; Muntin Width = **3/4** and Depth = **1/2**; Lights Pattern = **Rectangular**; Lights High = **3**; **Lights Wide** = **3**. Click **OK** to dismiss all dialog boxes.

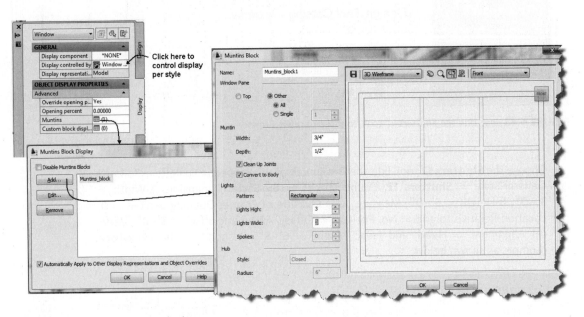

FIGURE 5.50 *Creating muntin display*

13. Select the window, right-click, and choose **Add Selected** from the shortcut menu. Respond to the workspace prompts as shown below.

    ```
    Insert point or: (Select a point near p3 as shown in
    Figure 5.49. Press ESC to end the command.)
    ```

14. In this step you will create a tool for the tool palette from the new style. Select the window, move the cursor over the window, left-click on the image of the window, continue to hold the left mouse button down, and release over the new tool palette as shown in Figure 5.51.

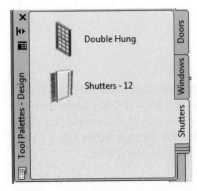

FIGURE 5.51 *Creating a shutters window tool*

15. Verify that the Shutters palette is current (Figure 5.51). Right-click over the shutters palette, and choose Delete Palette.

16. Save and close the drawing.

CREATING OPENINGS

You can create openings in walls without inserting doors or windows by using the **OpeningAdd** command. Access the **OpeningAdd** command as shown in Table 5.11.

TABLE 5.11 *OpeningAdd command access*

Command prompt	OPENINGADD
Ribbon	Choose the Home tab. Choose Opening from the Door flyout of the Build panel.
Design palette	Select Opening from the Design palette.

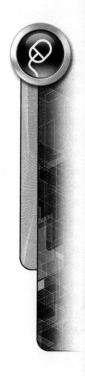

Openings are placed in a wall by setting the shape, width, and height in the Properties palette in a manner similar to placing doors or windows. The opening does not include a frame. The features of the opening are specified in the Properties palette as shown in Figure 5.52. The features are described below.

You may click in the Shape option of the Properties palette to specify the shape for the opening. The shape list includes the following: rectangular, round, half-round, quarter-round, oval, arch, trapezoid, gothic, isosceles triangle, right triangle, peak pentagon, octagon, hexagon, or Custom. Selecting the Custom option allows you to specify an AEC Profile to define the shape. The Profile feature is added to the Properties palette if **Custom** is selected from the **Shape** drop-down list. The **Profile** lists the AEC Profiles in the drawing.

The opening can be placed using Unconstrained, Center, and Offset of the Position options in the Location section of the Properties palette. The opening is placed in a similar manner as doors and windows.

Defining Opening Properties

Editing the Properties palette can change the properties of openings. When you select an existing opening, the Properties palette is expanded to include the following additional features:

Layer—The Layer field displays the name of the layer of the insertion point of the opening.

Anchor—The Anchor button opens the **Anchor** dialog box, which lists the position of the opening relative to the justification line.

Endcaps—The opening endcaps of the wall for the opening can be edited by selecting the opening endcaps from the drop-down list.

Adding an Opening with Precision

The opening can be added by editing the dynamic dimensions or by selecting the **Reference point on** or **Cycle measure to** options to insert the opening with precision. After you select the wall, you can edit the Properties palette or choose the options from the dynamic input prompts of the workspace as shown in Figure 5.52.

Modifying Openings

Openings can be modified by selecting the opening and editing its features in the Properties palette. The grips of the opening allow you to change its location, width, height, sill height, and head height as shown in Figure 5.53. Openings can be edited by selecting the grip and editing the dynamic dimensions in the workspace.

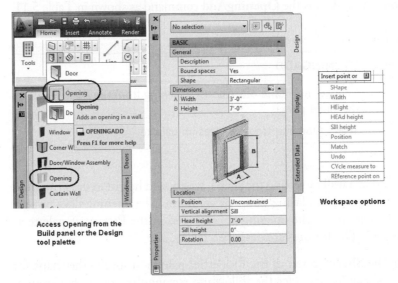

FIGURE 5.52 *Properties palette and options of an opening*

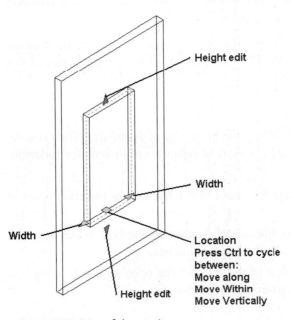

FIGURE 5.53 *Grips of the opening*

ADDING AND EDITING THE PROFILE OF DOORS, WINDOWS, AND OPENINGS

When you place doors and windows, the shape is defined in the style definition. When you place openings, the shape is defined in the Properties palette. The shape can be a predefined shape or Custom. The Custom option allows you to select a profile defined in the drawing, which defines the opening shape. The **Opening AddProfile** command allows you to create or assign a profile to an existing door, window, or opening created without the use of a profile. Access the **OpeningAddProfile** command as shown in Table 5.12.

TABLE 5.12 *OpeningAddProfile command access*

Command	OPENINGADDPROFILE
Ribbon	Select a door, window, opening, or door/window assembly; and choose Add Profile from the Profile panel of the contextual tab.
Shortcut menu	Select a door, window, opening, or door/window assembly; right-click; and choose Add Profile.

If you select a door, window, opening, or door/window assembly (created without a shape profile); right-click; and choose **Add Profile**, a dialog box opens that allows you to create a new profile to define the shape of the object. If you choose an opening, to display the contextual tab, then choose Add Profile; the Add Opening Profile dialog box opens as shown in Figure 5.54. The Add Opening Profile dialog box includes a **Profile Definition** field that allows you to add the profile from scratch or select an existing profile from the drop-down list. The **New Profile Name** field allows you to type a new name for the profile. When you click **OK**, the Add Opening Profile dialog box closes, and the grips of the profile are displayed. You can select the grips as shown in Figure 5.54 to edit the profile. You can select the negative grip symbol at the midpoint of the segment and stretch the segment to a new location, or select a vertex grip to edit the location of a vertex of the profile.

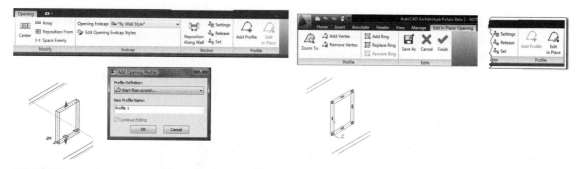

FIGURE 5.54 *Changing the profile with Edit In Place*

The **Edit In Place: Opening** tab is displayed in the ribbon to assist you in editing the profile. You can select the grip to modify the shape, or you can right-click and choose from the shortcut menu to modify the shape. The options of the shortcut menu and Edit in Place tab are presented in Chapter 3; refer to Using Edit in Place with Wall

Sweep. After editing the profile, select Save As, Cancel, or Finish from the Edit in Place tab. When the changes are saved by choosing Save As or Finish, the grips are cleared.

After creating and modifying the profile, if you choose the opening, the Edit In Place button is activated from the Profile panel of the Opening tab as shown in Figure 5.54. The **OpeningAddProfile** command adds a profile to the door, window, or opening and can be used to modify the profile. Access the **OpeningProfileEdit** command as shown in Table 5.13. The OpeningProfileEdit command displays the grips of the profile and the **Edit In-Place Opening** tab.

TABLE 5.13 *OpeningProfileEdit command access*

Command	OPENINGPROFILEEDIT
Ribbon	Select an opening (defined by a profile), and choose Edit in Place from the Profile panel of the Opening tab.
Shortcut menu	Select an opening (defined by a profile), right-click, and choose Edit Profile In Place.

USING AEC MODIFY TOOLS TO MOVE DOORS, WINDOWS, AND OPENINGS

The **AEC Modify Tools** for doors, windows, and openings can be used to adjust the location of the objects in the wall. The AEC Modify Tools for doors and windows are accessed as shown in Table 5.14.

TABLE 5.14 *Linework Array Tool for Doors, Windows, and Openings*

Command Prompt	LINEWORKARRAY
Shortcut menu	Select a door, window, or opening; right-click; and choose AEC Modify > Array.
Ribbon	Select a door, window, or opening; and choose Array from the Modify panel of the contextual tab

The LineWorkArray command was presented in Chapter 2 to array walls within a floor plan. The Array command allows you to array a door, window, or opening in a wall. The following workspace sequence created an array of windows in a wall.

```
Select a window, choose Array from the Modify panel of the
Window contextual tab shown in Figure 5.55.

Select an edge to array from or Enter to pick two points:
(Move the cursor to display the axis; choose the axis at p1 as
shown in Figure 5.55.)

Drag out array elements or [Offset/Clear distance/Pick array
distance/Enter count]:c (Type C to specify Clear Distance)

Drag out array elements or [Offset/Clear distance/Pick array
distance/Enter count]:4'[1200] (Type 4'[1200] in the dynamic
dimension to specify the clear distance.)

Drag out array elements or [Offset/Clear distance/Pick array
distance/Enter count]: (Move the cursor to a location at p2 to
display 4 windows as shown in Figure 5.55.)
```

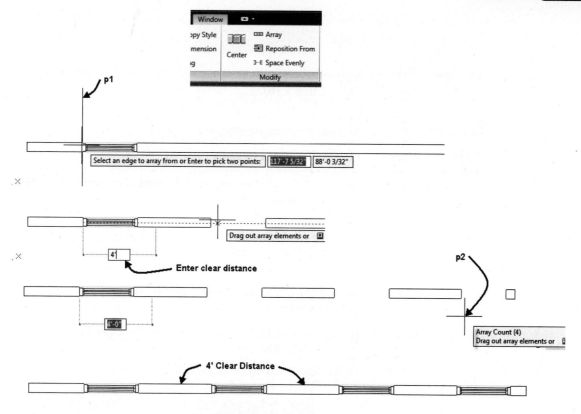

FIGURE 5.55 *Using the Linework Array command to add windows*

The RepositionFrom command allows you to move a door relative to a location on the door, window, or opening a specified distance from a point (Table 5.15).

TABLE 5.15 *Accessing RepositionFrom Command for Doors, Windows, and Openings*

Command Prompt	AECREPOSITIONFROM
Shortcut menu	Select a door, window, or opening; right-click; and choose AEC Modify > Reposition from.
Ribbon	Select a door, window, or opening; and choose Reposition from of the Modify panel of the Window contextual tab.

An application of the RepositionFrom command is shown in the following work-space sequence:

Select a window, choose **Reposition From** of the Modify panel in the contextual tab.

Select an edge to reposition from or ENTER to pick two points: *(Move the cursor to display the axis; choose the axis at* **p1** *as shown in Figure 5.56)*

Specify a point to reference the new dimension from: *(Select the window at* **p2** *as shown in Figure 5.56)*

```
Specify the new distance between the selected edge and the
selected point: 12" [300] (Type the new distance in the
workspace. Press Escape to end the command.)
```

The **Space Evenly** command allows you to space the selected windows evenly between two specified points (Table 5.16).

TABLE 5.16 *Accessing Space Evenly for Doors, Windows, and Openings*

Command Prompt	**LINEWORKSPACEEVENLY**
Shortcut menu	Select a door, window, or opening; right-click; and choose AEC Modify > Space Evenly.
Ribbon	Select a door, window, or opening; and choose Space Evenly from the Modify panel of the Window contextual tab.

The two windows shown in Figure 5.56 were spaced evenly within the wall as shown in the following workspace sequence.

```
Select the windows at p3 and p4, choose Space Evenly from the
Modify panel of the Window contextual tab.

Select an axis to space evenly on or ENTER to specify with two
points: (Press ENTER to specify points.)

Select the first point: (Select the end of wall at p5 as shown
in Figure 5.56.)

Select the second point: (Select the end of wall at p6 as
shown in Figure 5.56.)

(Windows repositioned as shown at the right in Figure 5.56.)
```

The LineWorkCenter command allows you to center a door, window, or opening between two points (Table 5.17).

TABLE 5.17 *Accessing LINEWORKCENTER for Doors, Windows, and Openings*

Command Prompt	**LINEWORKCENTER**
Shortcut menu	Select a door, window, or opening; right-click; and choose AEC Modify > Center.
Ribbon	Select a door, window, or opening; and choose Center from the Modify panel of the Window contextual tab.

The LineWorkCenter command was used to center the window in the wall in the following workspace sequence:

```
Select the window shown at p7, choose Center from the Modify
panel shown in Figure 5.56.

Select an axis to center on or ENTER to specify with two
points: (Press ENTER to specify points.)

Select first point: (Select the end of wall at p8 as shown in
Figure 5.56.)

Select second point: (Select the end of wall at p9 as shown in
Figure 5.56.)
```

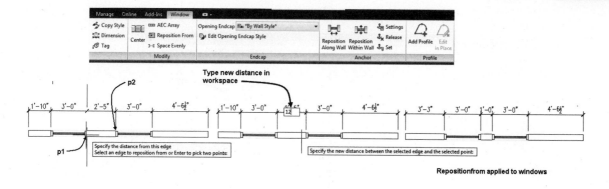

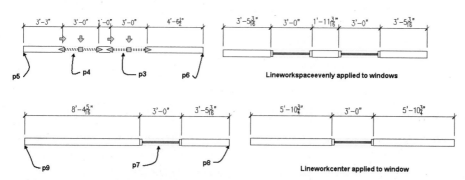

FIGURE 5.56 *Applications of AEC Modify Tools*

APPLYING TOOL PROPERTIES

The Existing doors, windows, and openings can be changed to the settings of the current tool with the **Apply Tool Properties to** command. When a tool of the tool palette is highlighted, right-click and choose **Apply Tool Properties to > Door** as shown in Figure 5.57. The **Apply Tool Properties to** command will prompt you to select objects for edit. The objects selected are changed to the current tool; you can edit the Properties palette for additional size editing. An application of the **Apply Tool Properties to** command is shown in the following steps. The Apply Tool Properties option of the Opening tool allows you to convert doors, door/window assembly, and windows to an opening. The Apply Tool Properties command allows you to change doors to windows.

Steps to Apply Tool Properties to Doors

1. Select a tool, such as **Bifold-Double** tool, from the **Doors** tool palette, right-click, and choose **Apply Tool Properties to > Door** from the shortcut menu as shown in Figure 5.57. The **Apply Tool Properties to Door** command allows you to change doors to windows.

2. Select doors in the drawings for edit as shown in the following workspace sequence.

   Select Door(s): 1 found *(Select a door at* **p1** *as shown in Figure 5.57.)*

   Select Door(s): 1 found, 2 total *(Select a door at* **p2** *as shown in Figure 5.57.)*

   Select Door(s): ENTER

3. Doors remain selected; edit the Properties palette to complete the edit.

4. Press ESC to end the **Apply Tool Properties to** command.

   Doors are changed to Bi-fold type.

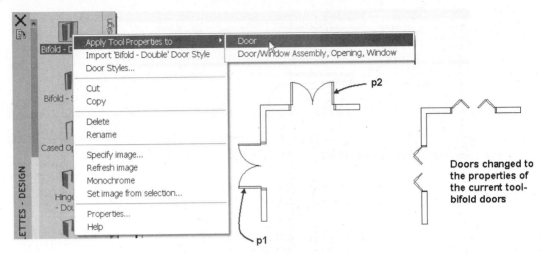

FIGURE 5.57 *Apply Tool Properties to command*

IMPORTING AND EXPORTING DOOR AND WINDOW STYLES

Door and window styles can be imported into a drawing from the Design Tool Catalog – Imperial and Design Tool Catalog – Metric of the Content Browser. Once styles are i-dropped into the drawing, they are listed in the Style Manager. The files included in the style directory are shown in Table 5.18.

TABLE 5.18 *Door and Window Styles of Autodesk AutoCAD Architecture*

Imperial	Description
Door Styles	Fifty-one styles including panel, arched, half-round, and over-head garage doors with windows
Window Styles	Seventy-two styles including jalousie, bay, bow, shutters, and arched windows
Metric	
Door Styles	Fifty-one styles including six panel, arched, glazed, and louvered
Window Styles	Sixty-three styles including louver and shapes such as arched, Gothic, half-round, oval, and trapezoid

Styles can be exported to a styles directory. The styles can then be imported into new drawings from the styles directory as you develop them. The resource files then become a depository of frequently used styles that can be imported into future drawings. The steps for importing door styles are shown below.

Steps to Import Door Styles from Source Drawings

1. Open the drawing targeted to receive the door styles.
2. Select **Door** from the **Design** tool palette, right-click, and choose **Door Styles** to open the Style Manager.
3. Select **Open Drawing** from the **Style Manager** toolbar. Navigate to a style resource drawing in the Open Drawing dialog box. Click the **Open** button to select the drawing, and dismiss the **Open Drawing** dialog box.

4. Expand the **Architectural Objects** > **Door Styles** category of the style resource drawing file in the left pane.

5. Select a style, right-click, and choose **Copy** from the shortcut menu.

6. Select the current drawing file name in the left pane, right-click, and choose **Paste.**

7. The imported door style is displayed in the *Architectural Objects\Door Styles* folder of the current file.

PROJECT TUTORIAL 5-5: INSERTING DOORS

1. Verify that the Accessing Tutor Imperial or Accessing Tutor Metric content for Chapter 5 of the CengageBrain has been downloaded to your Accessing Tutor folder on your computer as described in Organizing Tutorial Directories in the Preface.

2. Choose **Project Browser** from the Quick Access toolbar. Use the Project Selector drop-down list to navigate to your *Student\Ch 5* student directory. Select **New Project** to open the **Add Project** dialog box. Type **9** in the **Project Number** field and type **Ex 5-5** in the **Project Name** field. Check **Create** from template project, choose **Browse**, and edit the path to *\Accessing Tutor\Imperial\Ch5\Ex 5-5\Ex 5-5.apj* or *\ Accessing Tutor\Metric\Ch5\Ex 5-5\Ex 5-5.apj* in the **Look in** field. Click **Open** to dismiss the **Select Project** dialog box. Click **OK** to dismiss the **Add Project** dialog box. Click **Close** to dismiss the Project Browser.

3. Select the **Constructs** tab of the **Project Navigator.** Double-click on **Floor 2** to open the Floor 2 drawing. Choose Top from the View Cube.

4. If the tool palettes are not displayed, choose the Tools button of the Home tab. Choose the **Design** palette set.

5. Verify that Dynamic Input is ON in the Application status bar, select Dynamic Input, right-click, and choose Settings; then check the Enable Pointer Input, Enable Dimension Input where possible, and Show command prompting and command input near the crosshairs checkboxes. Click OK to dismiss all dialog boxes. Turn Off Object Snap on the Application status bar.

6. Select **Zoom Window** from the Zoom flyout menu of the Navigation bar and respond to the workspace prompts as follows:

```
Specify first corner: 100',58' [31500, 14000] ENTER
Specify opposite corner: 119',45' [37500, 18000] ENTER
```

7. Select **Door** from the Door flyout menu in the Build panel of the Home tab.

8. Edit the **Design** tab of the **Properties** palette as follows: General-Style = **Standard**, Bound spaces = **By style**, Dimensions-A-Width = **2'-0" [600]**, B-Height = **6'-8" [2000]**, Measure to = **Inside of Frame**, Swing angle = **45**, Location-Relative to Grid = **No**, Position = **Offset**, Automatic offset = **6" [150]**, Multiple Insert = **No**, Vertical Alignment = **Head**, and Head height = **6'-8" [2000]**.

9. Place the door by responding to the workspace prompts as follows:

```
Select wall, grid assembly, or ENTER: (Select the wall at p1
as shown in Figure 5.58 to place the 2'-0" [600] door.)

Insert point or: (Select a point below the selected wall near
p2 to specify the swing as shown in Figure 5.58.)

(Click in the Properties palette and edit the A-Width to 2'-6"
[750].)

Insert point or: (Select a point near p3 as shown in
Figure 5.58.)
```

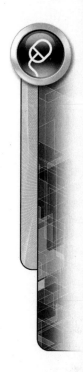

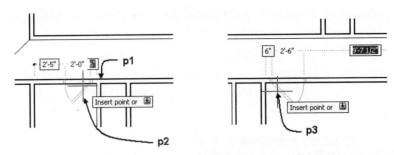

FIGURE 5.58 *Selecting wall for placing door*

> Insert point or: (Select a point near **p1** as shown in Figure 5.59.)
>
> Insert point or: (Press ESC to end the command.)

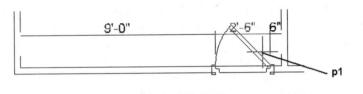

FIGURE 5.59 *Placing an additional door*

10. Select **Zoom Window** from the Zoom flyout menu of the **Navigation** bar, and respond to the command prompts as follows:

    > Specify first corner: 72',72' [22000,22000] ENTER
    >
    > Specify opposite corner: 105',60' [32000,18000] ENTER

11. Select the Doors tool palette, and select **Hinged-Single-Exterior** from the tool palette.

12. Edit the **Properties** palette as follows: A-Width = **3'-0"** [900]. Verify the remaining settings: General-Style = **Hinged-Single-Exterior**, Dimensions-B-Height = **6'-8"** [2000], Measure to = **Inside of Frame**, Swing angle = **45**, Location-Relative to Grid = **No**, Position = Unconstrained, Multiple Insert = **No**, Vertical Alignment = **Head**, and Head height = **6'-8"** [2000].

13. Select the wall at **p1**, move the cursor down to set the door swing inside as shown in Figure 5.60, press TAB until the left dimension is highlighted as shown, and then type **17'** [5000] ENTER in the field of the workspace. Press ESC to end the command.

14. Select the door to display the trigger grips, and choose the **Flip** grip for the hinge to move the hinge to the right.

15. Select the door at **p2** as shown in Figure 5.60, right-click, and choose **Edit Style** from the General panel of the Door contextual tab.

16. Select the **Dimensions** tab of the **Door Style Properties** dialog box. Edit the Frame: A-Width = **2 1/4"** [55], B-Depth = **5 9/16"** [140], Stop: C-Width = **3/8"** [10],

D-Depth = **2″** **[50]**, E-Door Thickness = **1 3/4″** **[40]**, and Glass Thickness = **3/16″** **[4]** as shown in Figure 5.61. Click the **OK** button to accept the changes and dismiss the Door Style Properties dialog box.

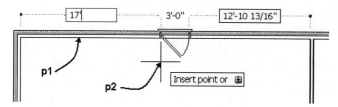

FIGURE 5.60 *Editing dynamic dimension to place a window*

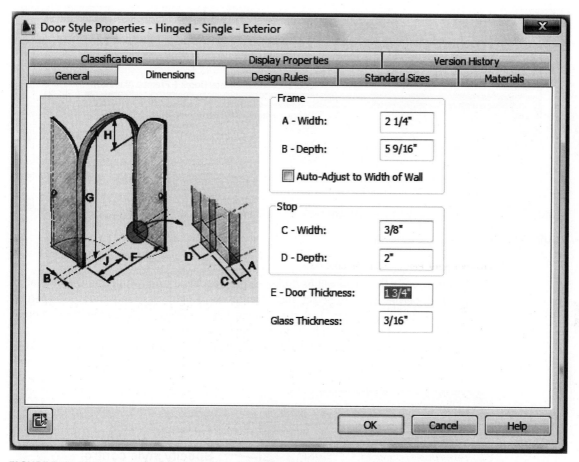

FIGURE 5.61 *Dimensions tab of the Door Style Properties dialog box*

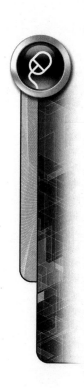

17. In this step you will apply a material definition to a display component of the door.

18. Choose the Content Browser from the Tools flyout menu of the Build panel of the Home tab. Choose the Visualization Catalog. Navigate to AEC Material Tools > US Imperial > Doors and Windows > Wood Doors [AEC Material Tools > US Metric > Doors and Windows > Wood Doors]. Drag the i-drop for the **Doors & Windows.Wood Doors.Ash** material into the workspace. Respond to the workspace prompts as follows:

```
Select a component or an object: (Select the panel of the
exterior door.)
```

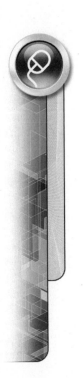

```
Apply the material to the style or [this Object]: ENTER
(Press ENTER to apply the change to the style).
```

19. Select the door placed near **p2** as shown in Figure 5.60, and choose **Edit Style** from the General panel of the Door contextual tab. Select the **Materials** tab of the **Door Style Properties-Hinged Single Exterior** dialog box. Verify that the material of the panel is **Doors & Windows.Wood Doors.Ash** from the **Material Definition** list.

20. To edit the material of the Stop, click in the **Material Definition** column of the Stop, and select **Doors & Windows.Wood Doors.Ash** from the **Material Definition** list as shown in Figure 5.62. Click OK to dismiss the Door Style Properties dialog box.

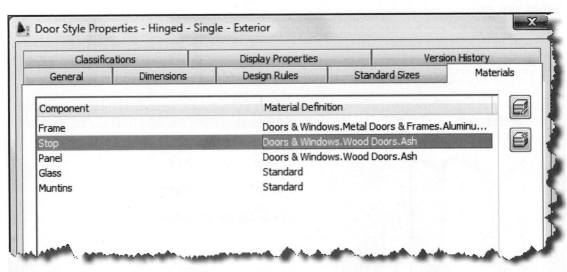

FIGURE 5.62 *Materials tab of a door style*

21. In this step a threshold will be added to the door style from within the **Display** tab of the Properties palette. Select the exterior door, right-click, and choose Properties. Choose the Display tab. Edit the Display tab as follows: Display representation = Threshold Plan, Display controlled by = Door Style: Hinged–Single Exterior, and Display component = None. Choose the Threshold dimension button as shown in Figure 5.63.

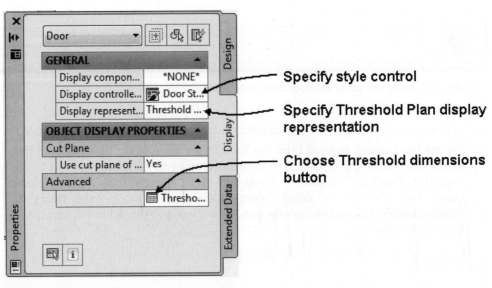

FIGURE 5.63 *Display tab of the Properties palette*

22. Edit the Threshold Dimension dialog box as follows: C-Extension = **2″** **[50]** and D-Depth = **1″** **[25]** as shown in Figure 5.64. Click OK to dismiss the dialog box.

23. In the Display tab of the Properties palette, click the Display component field to Threshold A, and toggle the light bulb ON to display **Threshold A.** Click OK to dismiss the Modify Display Component at Style Override Level dialog box.

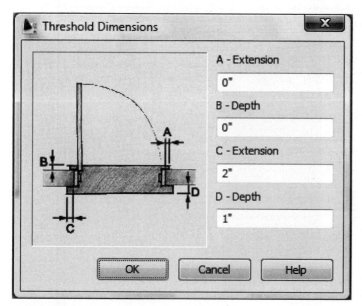

FIGURE 5.64 *Threshold Dimensions dialog box*

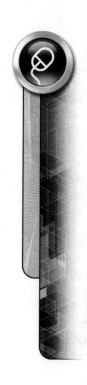

24. Select the exterior door at **p2** as shown in Figure 5.60, right-click, and choose **Object Viewer** from the shortcut menu. Select **SE Isometric** from the View flyout menu. Edit the display representation to **High Detail**, and select **Realistic** visual style to view the door with materials applied as shown in Figure 5.65. Press ESC to close the Object Viewer.

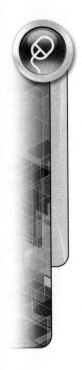

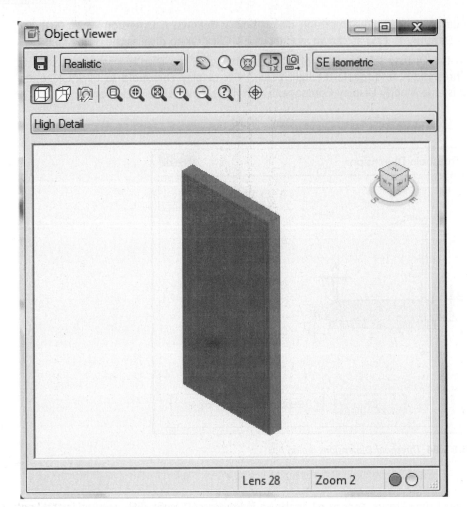

FIGURE 5.65 *Materials display in Object Viewer*

25. Select **Zoom Window** from the **Zoom** flyout menu of the Navigation bar, and respond to the command line prompts as shown below.

    Specify first corner: 90',70' [27400, 21300] ENTER

    Specify opposite corner: 100',65' [30500, 20000] ENTER

26. Select the exterior door, and choose **Reposition Within Wall** from the Anchor panel of the Door contextual tab. Respond to the command prompts as shown below to move the jamb flush with the interior wall.

    Command: RepositionWithin

    Select position on the opening: *(Select a point near* p1 *as shown in Figure 5.66.)*

    Select a reference point: _endpoint of *(SHIFT and right-click, choose* **Endpoint** *from the shortcut menu, and select the endpoint of the wall at* p2 *as shown in Figure 5.66.)*

    Enter the new distance between the selected points <0>: 0 ENTER *(Type the desired distance between the two points to reposition the door as shown in Figure 5.66.)*

27. Save the drawing. Choose Close Current Project from the Project Navigator. Choose Close all project files from the Project Browser – Close project files dialog.

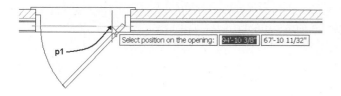

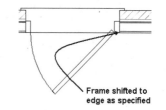

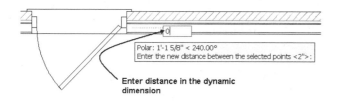

Enter distance in the dynamic dimension

Frame shifted to edge as specified

FIGURE 5.66 *Moving the door with Reposition Within*

PROJECT TUTORIAL 5-6: INSERTING WINDOWS AND CREATING WINDOW STYLES

1. Verify that the Accessing Tutor\Imperial or Accessing Tutor\Metric content for Chapter 5 of the CengageBrain has been downloaded to your Accessing Tutor folder on your computer as described in Organizing Tutorial Directories in the Preface.

2. Open Autodesk AutoCAD Architecture 2012, and select **Project Browser** from the Quick Access toolbar. Use the Project Selector drop-down list to navigate to your student directory. Select **New Project** to open the **Add Project** dialog box. Type **10** in the **Project Number** field, and type **Ex 5-6** in the **Project Name** field. Check **Create** from template project, choose **Browse**, and edit the path to \Accessing TutorImperial\Ch5\Ex5-6\Ex5-6.apj or Accessing Tutor\Metric\Ch5\Ex 5-6\Ex 5-6.apj in the **Look in** field. Click **Open** to dismiss the **Select Project** dialog box. Click **OK** to dismiss the **Add Project** dialog box. Click **Close** to dismiss the Project Browser.

3. Select the **Constructs** tab of the **Project Navigator**. Double-click on **Floor 2** to open the Floor 2 drawing.

4. Select **Zoom Window** from the **Zoom** flyout of the Navigation bar, and respond to the following workspace prompts:

   ```
   Specify first corner: 72',72' [22000,22000] ENTER
   Specify opposite corner: 110',64' [33500,20000] ENTER
   ```

5. Verify that **Dynamic Input** is **ON** and **Object Snap** is **OFF.** Choose the Home tab. If the Design palette set is not displayed, choose the Tools button of the Build panel.

6. Choose the **Windows** tool palette, and choose **Casement-Double** tool.

7. Edit the Design tab of the **Properties** palette as follows: General-Style = **Casement-Double**, Bound spaces = **By Style**, Dimensions-A-Width = **4'–6" [1370]**, B-Height = **3'–6" [1070]**, Measure to = **Outside of Frame**, Swing angle = **45**, Location-Relative to grid = No, Position = **Unconstrained**, Multiple Insert = **No,** Vertical Alignment = **Head**, and Head height = **6'–8" [2000].** Respond to the command prompts as shown below:

   ```
   Select wall, grid assembly or ENTER: (Select the wall at p1 as
   shown in Figure 5.67.)

   Insert point or: (Select a point near p2 as shown in Figure
   5.67 to display the 6'-3" dimension.)
   ```

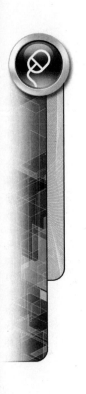

(Move the cursor to the right beyond the door, right-click, and choose the **Cycle measure to** *and Verify the insertion point is set to Center of the window options.)*

(Press TAB to highlight the dimension on the right as shown in Figure 5.68.)

Insert point or: **5' [1500]** ENTER *(Type the desired distance in the dynamic dimension.)*

Insert point or: ESC *(Press ESC to end the command.)*

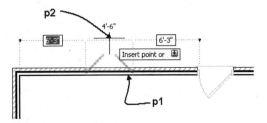

FIGURE 5.67 *Insertion point for window*

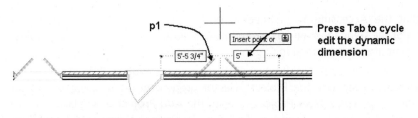

FIGURE 5.68 *Editing the dynamic dimension of the window*

8. Select **More Window Tools** from the Windows palette to open the Design Tool Catalog – Imperial [Design Tool Catalog – Metric]. Select page 2 of the Windows category.

9. Select the i-drop for the **Casement-Double Arched** window, and drag the i-drop into the drawing as shown in Figure 5.69. Press ESC to end the **WindowAdd** command before selecting a wall. Close the **Content Browser.**

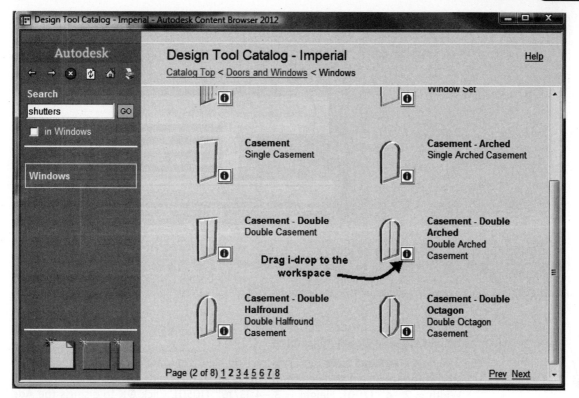

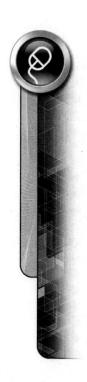

FIGURE 5.69 *Casement Double Arch window in the Content Browser*

10. Select the casement window at the right at **p1** as shown in Figure 5.68, and edit the style to **Casement – Double Arched** in the **Properties** palette.

11. Select the left grip of the Casement – Double Arched window, move the cursor left, and type **5' [1500]** in the dynamic dimension as shown in Figure 5.70. Edit the **Height** in the **Properties** palette to **5' [1500]**. Move the cursor from the Properties palette, and press ESC to clear the grips selection.

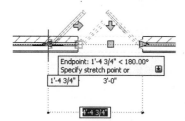

FIGURE 5.70 *Editing window width using dynamic dimensions*

12. Select the window placed in step 10, and choose **Window Styles** from the Edit Style flyout of the General panel in the Window contextual tab to open the Style Manager.

13. Verify that the Window Styles category of Floor 2.dwg is selected in the left pane, right-click, and choose **New** from the shortcut menu.

14. Overtype **Andersen_Casement** in the left pane as the name of the new style.

15. Verify that the **Andersen_Casement** name is selected in the left pane.

16. Select the **General** tab in the right pane, and type **Double Casement Windows** in the **Description** field. Select the **Dimensions** tab in the right pane, and edit the Frame section: A-Width = **2" [50]**, B-Depth = **5 9/16" [140]**, Sash C-Width = **1/2" [12]**, D-Depth = **2" [50]**, and E-Glass Thickness = **3/16" [5]** as shown in Figure 5.71.

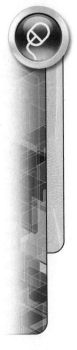

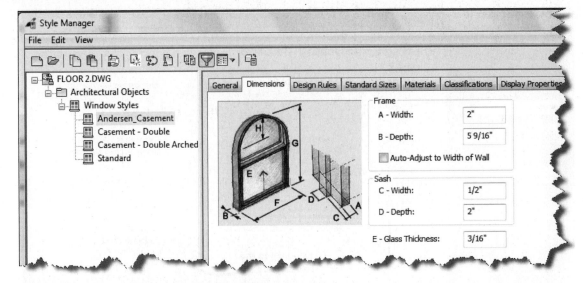

FIGURE 5.71 *Dimensions tab of the window style in the Style Manager*

17. Select the **Design Rules** tab, and edit the Shape to **Predefined: Rectangular** and Window Type to **Double Casement.**

18. Select the **Standard Sizes** tab, click the **Add New Standard Size** button, and edit the **Add Standard Size** dialog box shown in Figure 5.72 as follows: Description = **CW235**, Width = **2'–4" [700]**, Height = **3'–4 13/16" [1050]**. Click **OK** to dismiss the **Add Standard Size** dialog box.

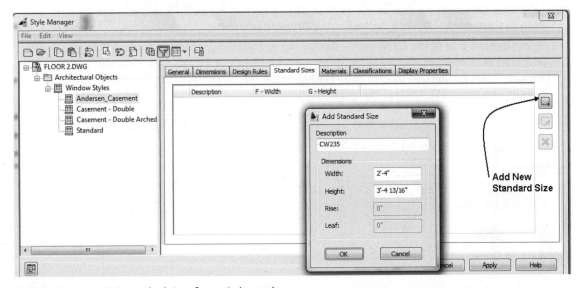

FIGURE 5.72 *Creating standard sizes for a window style*

19. Click the **Add New Standard Size** button, and edit the **Add Standard Size** dialog box as follows: Description = **CW135**, Width = **4'-8 1/2" [1350]**, Height = **3'-4 13/16" [1050]**. Click **OK** to dismiss the Add Standard Size dialog box.

20. Select the **Materials** tab and edit the Frame, Sash, and Muntins material definition to **Doors & Windows.Metal Doors & Frames.Aluminum Windows.Painted.White** and Glass to **Doors & Windows.Glazing.Glass.Clear**. The material definitions were imported with the Casement-Double Arched window from the **Content Browser**.

21. Click **OK** to dismiss the **Style Manager**.

22. Select the casement window at **p1** as shown in Figure 5.73. Click in the **Style** field of the **Properties** palette, select **Andersen_Casement**, and select the **4'-8 1/2" × 3'-4 13/16" [1350 × 1050]** standard size.

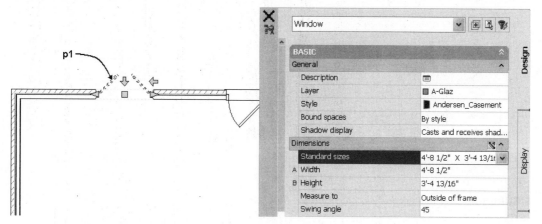

FIGURE 5.73 *Editing existing windows*

23. Move the cursor from the **Properties** palette, and press ESC to clear the grips selection.

24. Select **NE Isometric** from the **View Cube**.

25. Save and close the drawing.

26. Click Close Current Project from the toolbar of the Project tab on the Project Navigator.

PROJECT 5-7: INSERTING DOORS IN A FLOOR PLAN

Verify that the Accessing Tutor\Imperial or Accessing Tutor\Metric content for Chapter 5 of the CengageBrain has been downloaded to your Accessing Tutor folder on your computer as described in Organizing Tutorial Directories in the Preface.

Create project number 11 named **Ex 5-7** from template *\Accessing Tutor\Imperial\ Ch5\ Ex 5-7\Ex 5-7.apj or \Accessing Tutor\Metric\Ch5\Ex5-7\Ex 5-7.apj.* Open the Floor 2 Construct drawing, and insert additional doors as shown in Figure 5.74 and Table 5.19. Use Offset set to 3" [75] when placing the doors. View the building from **SW Isometric** when complete.

TABLE 5.19 *Door sizes for project*

DOOR MARK	SIZE	STYLE	Note
A	2'-6" × 6'-8" [750 × 2000]	Hinged Single	
B	2'-0" × 6'-8" [600 × 2000]	Hinged Single	
C	3'-0" × 6'-8" [900 × 2000]	Hinged Single Exterior	
D	6'-0" × 6'-8" [1800 × 2000]	Bifold Double	
E	3'-6" × 6'-8" [1000 × 2000]	Bifold Double	
F	4'-0" × 6'-8" [1200 × 2000]	Hinged Double-6 Panel Half Lite	I-drop from the Design Tool Catalog > Doors and Windows > Doors

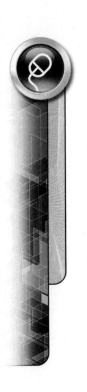

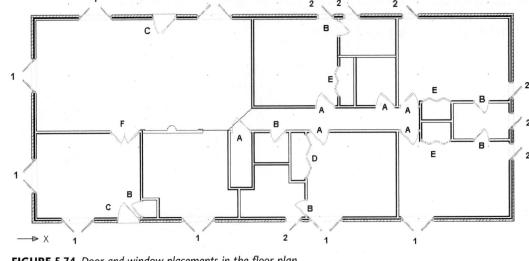

FIGURE 5.74 *Door and window placements in the floor plan*

PROJECT 5-8: INSERTING WINDOWS IN A FLOOR PLAN

Create a new project number 12 named **Ex 5–8** from the template *Accessing Tutor\ Imperial\Ch5\Ex 5-8\Ex 5-8.apj* or *Accessing Tutor\Metric\Ch5\Ex 5-8\Ex 5-8.apj*. Open the Floor 2 Construct drawing.

Create a window style named **Pella_Case_Sgl**, set the window type to **Single Casement**, and set the frame to A-Width = **2"** **[50]**, B-Depth = **5 9/16"** **[140]**, Sash C-Width = **1-3/8"** **[35]**, D-Depth = **2"** **[50]**, and E-Glass Thickness = **3/16"** **[5]**. Define the following material definitions: **Doors & Windows.MetalDoors & Frames.AluminumWindows. Painted.White** material definition to the Frame, Sash, and Muntins and **Doors & Windows. Glazing.Glass.Clear** the material definition to the Glass component.

Create one standard size for the style that is 2'−4" × 3'−4 13/16 [711 × 1037], and enter a description **PW2434**.

Insert the following windows, as shown in Figure 5.74 and Table 5.20. Use Offset/ Center with 6" [150] offset when placing windows. Reposition all exterior windows to be 3" [75] from the exterior brick surface of the wall. Edit the front door, and reposition the exterior door frame to the interior surface of the wall.

TABLE 5.20 *Window sizes*

WINDOW MARK	SIZE	STYLE
1	4'-8 1/2" × 3'-4 13/16" [1435 × 1037]	Anderson_Casement
2	2'-4" × 3'-4 13/16" [711 ×1037]	Pella_Case_Sgl

SUMMARY

- Doors are inserted in a drawing with the DoorAdd command.
- Door styles are created in the Style Manager.
- The insertion handle for doors and windows is located on the end of the unit near the start of the wall.
- The insertion point for doors, windows, and openings can be toggled from each edge of the frame and centered by selecting the Cycle measure to option.
- Change the hinge and swing of doors and windows with Flip Hinge and Flip Swing.
- Use grips to edit the location along a wall, within a wall, and the vertical position of doors and windows.
- Dynamic dimensions can be toggled to edit the size and location of doors and windows.
- Shutters can be added as custom blocks to windows, and shutter windows styles can be imported into a drawing.
- Openings can be created with a specified width and height.
- Doors, windows, and openings can be created in the shape of an AEC Profile.
- Door and window styles can be imported from and exported to resource files.

Refer to **e Projects** of Chaper 5 for download from the Student Companion Site of Cengage-Brain **http://www.cengagebrain.com** described in the Preface. The folder includes the Project Ex 5-9 - Inserting Doors and Windows in a Basement Plan.

Refer to **e Review Questions** folder for download of review questions for Chapter 5 of Student Companion Site of **CengageBrain http://www.cengagebrain.com** described in the Preface

<blockquote>
CHAPTER

6

Door/Window Assemblies and Curtain Walls
</blockquote>

INTRODUCTION

Door/window assemblies and curtain walls allow you to create a unit with multiple window, door, and panel components. The door/window assembly and curtain walls are created based upon a grid. The grid allows you to divide the horizontal and vertical dimensions of the unit into a number of equal divisions or manually specify grid locations.

OBJECTIVES

After completing this chapter, you will be able to

- Insert door/window assemblies using the DoorWinAssemblyAdd command and curtain walls using the CurtainWallAdd command.

- Edit the properties of the door/window assembly and curtain wall.

- Create and edit divisions, infills, frames, and mullions.

- Define fixed cell dimension, fixed number of cells, and manual location of grid.

- Assign door and window styles as infills.

- Add or modify the profile used for frame and mullion components.

- Apply interference to a door/window assembly to trim the AEC object.

ADDING DOOR/WINDOW ASSEMBLIES

The style of a door/window assembly defines the content of the grid, which includes the divisions, frames, cell infills, and mullions. There are 13 pages of door/window assembly styles included in the Door and Window Assemblies category of the Design Tool Catalog - Imperial [Design Tool Catalog - Metric] shown in Figure 6.1. Included are styles with arched, peaked, and trapezoid shapes that can include transoms and skylights.

FIGURE 6.1 *Door and Window Assemblies of the Content Browser*

When you select **Door/Window Assembly** from the Design palette, refer to Table 6.1, you are prompted to select a wall or grid assembly. Unless styles have been imported, the Standard style is the only option and is inserted. The door/window assembly is anchored to the wall. The width, height, and other features of the Door/Window Assembly can be specified in the Properties palette as shown in Figure 6.2. The options specified in the Properties palette for a door/window assembly are described below.

TABLE 6.1 *Door/Window Assembly (DoorWinAssemblyAdd) command access*

Command prompt	DOORWINASSEMBLYADD
Ribbon	Choose Home tab. Choose Door/Window Assembly from the Door flyout of the Build panel.
Design palette	Select Door/Window Assembly from the Design palette shown in Figure 6.2.

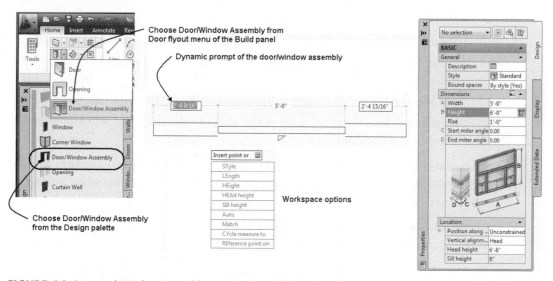

FIGURE 6.2 *Door and Window Assembly command and Properties palette*

Design Tab—Basic
General
Description—The Description button opens a **Description** dialog box in which to add a description.

Style—The Style displays the current window and door assembly; the drop-down list includes all styles loaded in the drawing.

Bound Spaces—The Bound Spaces option allows you to specify that the assembly contains spaces.

Dimensions
A-Length—The Length field allows you to specify the length of the assembly.

B-Height—The Height field allows you to type the distance from the baseline to the top of the unit. The Height value includes the rise.

Rise—The Rise field allows you to type the vertical dimension of the assembly from the base height to the top of the window assembly. This dimension applies to arched, trapezoid, and peaked shapes.

C-Start Miter—The **Start Miter** is the angle of the frame, mullion, and infill at the start of the assembly.

D-End Miter—The **End Miter** is the angle of the frame, mullion, and infill at the end of the assembly.

Location
Position along wall—The options of Position along wall are **Offset/Center** or **Unconstrained**. The **Offset/Center** option will restrict placement to the center of the wall segment or a specified distance from an intersecting wall. The **Unconstrained** option allows placement at any location along the wall or space boundary.

Offset Distance—The Offset Distance option is available when the Offset/Center constraint is selected. The offset distance is the distance between the unit and the intersecting wall.

Vertical Alignment—Vertical alignment can be specified to the head or sill.

Head Height—The Head Height is the distance from the head of the assembly to the baseline of the wall.

Sill Height—The Sill Height is the height of the sill from the baseline.

Rotation—The Rotation angle is the angle of the door and window assembly from the 3 o'clock position.

When you select the **Door/Window Assembly** tool from the Design palette, you are prompted to select a wall or space boundary as shown in the following workspace sequence:

```
Select wall, grid assembly or ENTER: (Select a wall.)
Insert point or: (Click to specify the insert point of the
assembly.)
```

The workspace options shown in Figure 6.2 are identical to those of a door or window. The options of the command can be selected by selecting from the drop down of the dynamic prompt, or, if you right-click, you can choose them from the shortcut menu. The dynamic dimensions can be edited to insert the assembly with precision in the same manner as discussed for doors and windows. The style definition of the assembly controls what is placed in the cells and the layout of the assembly. The Standard style for a door/window assembly is shown in Figure 6.3.

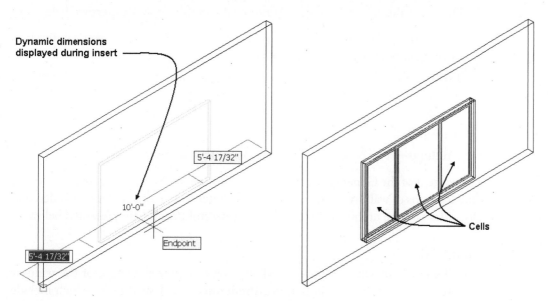

FIGURE 6.3 *Specifying the insert point of the Door/Window Assembly*

This assembly consists of three cells. The cells could include doors or windows. Access the **DoorWinAssemblyStyle** command as shown in Table 6.2 to open the **Door/Window Assembly Style Properties** dialog box. The content of the style is defined in the tabs of the right pane as shown in Figure 6.4.

TABLE 6.2 *Edit Door/Window Assembly Style command access*

Command prompt	DOORWINASSEMBLYSTYLE
Ribbon	Select a door/window assembly, and choose Edit Style to open the Door/Window Assembly Style Properties dialog box.
Ribbon	Select a door/window assembly, and choose Door/Window Assembly Styles to open the Style Manager.
Shortcut menu	Select a door/window assembly, right-click, and choose Edit Door/Window Assembly Style.
Shortcut menu	Select the Door/Window Assembly tool from the Design palette, right-click, and choose Door/Window Assembly Styles.

COMPONENTS OF A DOOR/WINDOW ASSEMBLY STYLE

The style of the door/window assembly is based upon a grid. The length and height dimensions specified when adding the assembly become the horizontal and vertical dimensions of the grid. The style defines the size and number of divisions in the grid. The intersection of the divisions creates a cell, which can be filled with a window, door, or panel unit. Each grid consists of the following elements as defined in the **Design Rules** tab of the **Door/Window Assembly Style Properties** dialog box:

Divisions—The Divisions of the grid specify the horizontal or vertical divisions of the grid to create the number of cells.

Infills—The Cell Infills define the components for each cell. The cell infill can be a panel, door, or window.

Frames—The Frame is the component that surrounds the outer edge of the primary grid.

Mullions—The Mullion is the component that separates cells.

The content of the General, Materials, Classification, and Display Properties tabs are similar to tabs of wall styles as presented in Chapter 3. The Shape, Design Rules, and Overrides tabs are unique to Door/Window Assemblies and are presented below.

Shape Tab

The **Shape** tab shown in Figure 6.4 allows you to define the shape of the window assembly. There are two radio buttons, **Predefined** and **Use Profile**, which include drop-down lists of shapes. The options of the **Shape** tab are described below.

Predefined—Predefined shapes are specific shapes listed in the drop-down list such as rectangular, round, half-round, quarter-round, oval, arch, gothic, isosceles triangle, right triangle, peak pentagon, octagon, hexagon, or trapezoid.

Use Profile—The Use Profile option allows you to select from a list of profiles. Therefore, you can create a window assembly based on any customized shape that has been defined as a profile. If there are no profiles defined in the drawing, this option is inactive.

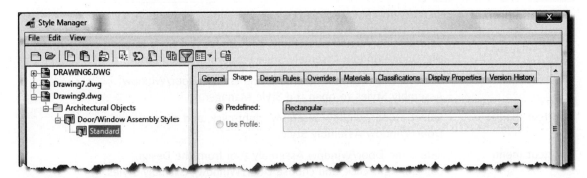

FIGURE 6.4 *Shape tab of the Door/Window Assembly Style*

Design Rules Tab

The majority of the components defined in the door/window assembly are specified in the **Design Rules** tab as shown in Figure 6.5. This tab includes a toolbar for creating new cell assignments, frames, or mullions, or deleting components. The commands of the toolbar are shown at the left in Figure 6.5.

Overrides Tab

The **Overrides** tab of the **Door/Window Assembly Style Properties** dialog box lists any overrides that have been defined as exceptions to the style as shown in Figure 6.6. Exceptions to a style are defined when you transfer design to the object using the **GridAssemblyCopyFromStyle** command and save back changes to the style. This tab includes information only if a specific style includes exceptions defined in the style.

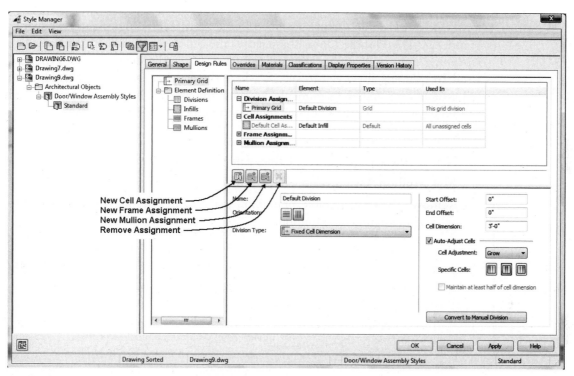

FIGURE 6.5 *Design Rules tab of the Door/Window Assembly Style Properties dialog box*

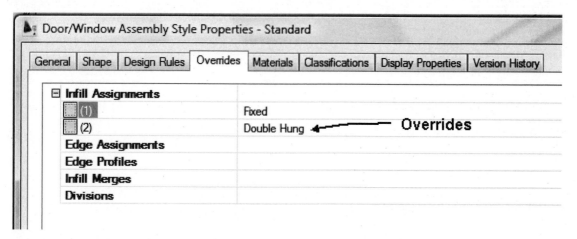

FIGURE 6.6 *Overrides tab of the Door/Window Assembly Style Properties dialog box*

Defining the Components Using Design Rules

The **Design Rules** tab consists of a tree on the left, which consists of the Primary Grid and its Element Definitions. If you select an element in the tree of the left pane, its properties will be displayed in the upper-right pane and can be edited in the lower-right pane. As shown in Figure 6.7, the primary grid has a vertical orientation. Selecting the Orientation buttons in the lower-right pane will change the direction in which the grid is subdivided.

If you select the **Primary Grid** in the left pane, you can choose a component field in the upper-right pane of the dialog box and edit its components in the lower-right pane. You can click in the **Element**, **Type**, and **Used In** columns shown in Figure 6.7 to select from the drop-down list. In Figure 6.7, the **New Cell Assignment** field was selected to expand the lower portion of the dialog box and display the settings for the cell. A **Double Door** is assigned to the **New Cell Assignment** as shown in Figure 6.7. The **Double Door** is applied to the cell at the middle of the unit. The remaining units are assigned the **sidelight** since it is the default. The **Type** and **Used In** columns allow you to specify where the element will be used. Additional cell assignments, frame assignments, and mullion assignments can be created and deleted when you select the buttons shown in Figure 6.7 as follows:

New Cell Assignment—Creates a new cell assignment definition.

New Frame Assignment—Defines a new frame assignment.

New Mullion Assignment—Defines a new mullion assignment.

Remove—Select cell, frame, or mullion assignment, and then choose **Remove** to delete the assignment definition.

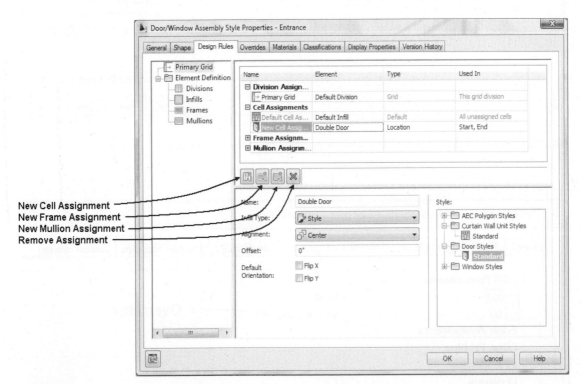

FIGURE 6.7 *Editing the Design Rules tab*

Tutorial 6-0 provides an opportunity for you to explore a sample of the options available in the **Design Rules** tab of the Door/Window Assembly Style Properties dialog box.

TUTORIAL 6-0: EXPLORING THE DESIGN RULES TAB OF A DOOR/WINDOW ASSEMBLY

1. Download Chapter 6 files from the Accessing Tutor\Imperial or Accessing Tutor\ Metric category of the Student Companion site of CengageBrain **http://www. cengagebrain.com** described in the Preface to your Accessing Tutor folder.

2. Open **Ex 6-0.dwg** from the Accessing Tutor\Imperial\Ch6 or Accessing Tutor\Metric\ Ch6 directory.

3. Choose **Save As > AutoCAD Drawing** from the Application menu, and save the drawing as **Lab 6-0** in your student directory.

4. Select the Door/Window Assembly object in the drawing, and choose Edit Style from the Edit Style flyout of the General panel in the Door/Window Assembly tab.

5. Choose the Design Rules tab. If you select an element in the tree of the left pane, its properties will be displayed in the upper-right pane and can be edited in the lower-right pane. Choose Primary Grid in the left pane as shown in Figure 6.8. The door/ window assembly is developed from a grid; the primary grid has a vertical orientation as shown in lower pane.

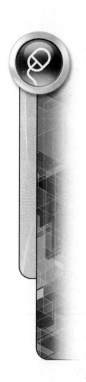

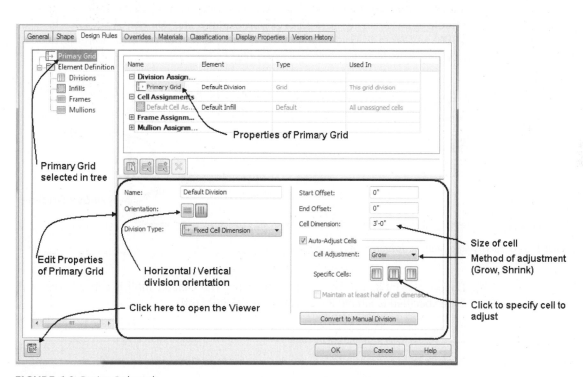

FIGURE 6.8 *Design Rules tab*

6. The categories listed in the Element Definition of the left pane are used to define the content of the door/window assembly. Choose each of the Element Definitions to review the settings for the component in the lower-right pane.

7. Choose the **Divisions** category of the left pane; the upper-right pane lists the Default Division whereas the lower-right pane lists the division type as Fixed Cell Dimension

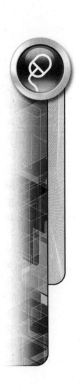

as shown in Figure 6.8. The Orientation toggles shown in Figure 6.8 allow you to change the grid orientation to horizontal or vertical. The settings for the Default Division with the Fixed Cell Dimension provide a grid 3'-0" [900] cells, allowing the middle grid division to adjust.

8. Options for the Division Type are Fixed Cell Dimension, Fixed Number of Cells, or Manual. Choose the Division Type flyout to specify **Fixed Number of Cells** as shown in Figure 6.9. Choose the Viewer in the lower-left corner to view the door/window assembly after the edit. The number of cells is specified to 2; therefore there are only two cells.

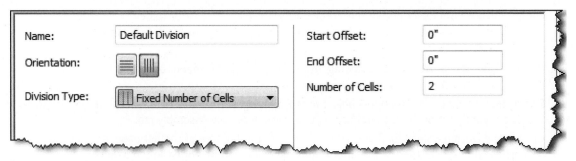

FIGURE 6.9 *Specifying Fixed Number of Cells*

9. Choose the Division Type flyout to specify Manual. Click Add Gridline as shown in Figure 6.10 to create gridline. Edit the Offset value to **3'-0" [900]** and From to **Grid Middle** as shown in Figure 6.10. Click Add Gridline, and edit the Offset value to -3'-0" [-900] and From to Grid Middle. Choose the Viewer button in the lower-left corner to view the manual door/window assembly.

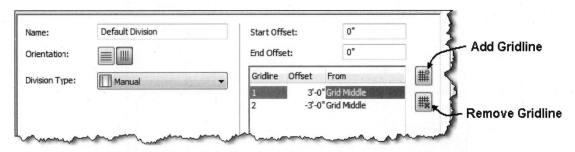

FIGURE 6.10 *Specifying a Manual division type*

10. Choose the Infills category in the left pane. The Default Infill is current. The Infill Type specified is Simple Panel as shown in Figure 6.11.

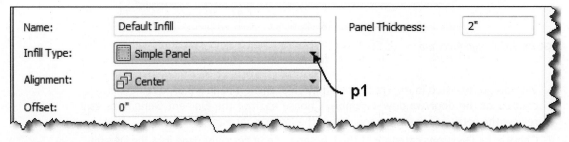

FIGURE 6.11 *Defining the simple panel infills*

11. The Infill Type can be specified as Style. Click the Infill Type flyout, and choose Style. The following styles are listed in the right pane: polygons, curtain wall unit, doors, or window styles as shown in Figure 6.12.

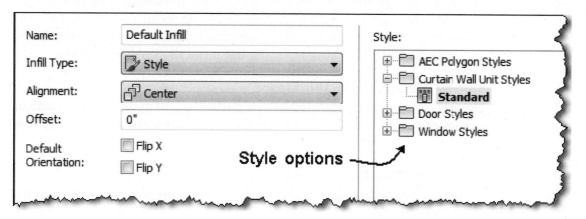

FIGURE 6.12 *Defining a style infill type*

12. Choose the Frames category in the left pane. The Default Frame settings are shown in the lower-right pane and in Figure 6.13. You can modify the width and depth values or specify an AEC Profile to define the frame.

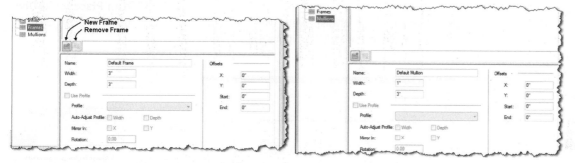

FIGURE 6.13 *Setting the size of the frame and mullion*

13. Choose the Mullions category in the left pane. The Width and Depth values of the mullion can be edited in the lower-right pane as shown at right in Figure 6.13.

14. Save and close the drawing.

MODIFYING DOOR/WINDOW ASSEMBLIES

Door/window assemblies can be edited by selecting the door/window assembly and then editing the features in the Design tab of the Properties palette. The layer, style, length, height, rise, start miter angle, end miter angle, vertical alignment, head height, sill height, and rotation can be changed on the **Design tab** of the Properties palette. In addition, you can select and edit each component of the assembly. Therefore, if the assembly consists of two double-hung windows and a picture window, you can select the picture window and edit it in the Properties palette. Additional details regarding the door/window assembly style can be changed by selecting options from the shortcut menu.

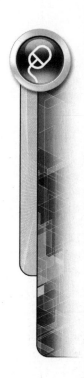

Changing the Door/Window Assembly Style in the Drawing

The components of the door/window assemblies that have been placed in the drawing can be changed in the workspace, allowing you to see the changes, rather than editing the **Design Rules** tab and returning to the drawing to view the changes. These changes can be saved to the style or remain as an override to the style. A door/window assembly is changed by selecting the grips and editing them in the workspace. Commands located on the shortcut menu allow additional editing.

Editing the Grid in the Workspace

The grid definition of a door/window style can be changed in the workspace with or without editing the style. You must view the door/window assembly in elevation or pictorial view to display the grips.

Steps to In-Place Edit of Door/Window Assembly

1. Select a door/window assembly; the grips of the unit are displayed as shown at the left in Figure 6.14.
2. Select the Edit Grid trigger grip (gray circle grip).
3. When you select the Edit Grid trigger grip, you are prompted to select one of the following options: Division in place, Cell, or Frame and mullion assignment. Choose the *Division in place* option. You are then prompted to select an edge to edit. Move the cursor over the edge of the door/window assembly to display the hatch pattern as shown in Figure 6.14. Click to select the grid edge.
4. When In-Place Editing begins, additional grips are displayed on the assembly to help you edit the grid as shown at the right in Figure 6.14. The **Edit in Place: Grid Division** contextual tab is displayed, which includes the Finish command to save the edits as shown at the right in Figure 6.14.

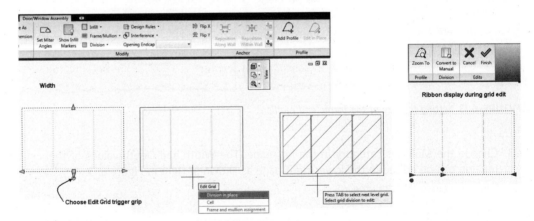

FIGURE 6.14 *Editing grips of the door/window assembly*

The grips for editing a door/window assembly vary based upon the design rules defining the unit. The door/window assembly shown in Figure 6.15 is defined with the Fixed Cell Dimension division type.

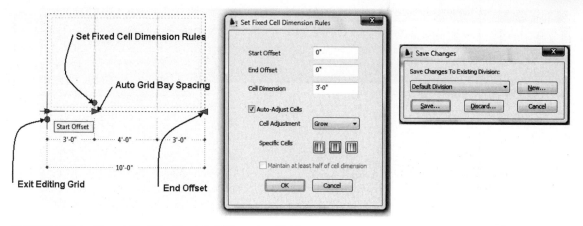

FIGURE 6.15 *In-Place Edit of the Fixed Cell Dimension Division*

The grips shown at the left allow you to change the fixed cell dimension and to change the spacing and the offsets at the start and end. The grid can be edited by selecting the grips or selecting the Set Fixed Cell Dimension Rules grip. Selecting the **Set Fixed Cell Dimension Rules** grip shown in Figure 6.15 will open the **Set Fixed Cell Dimension Rules** dialog box as shown in Figure 6.15, allowing you to specify a new set of design rules. When you click **OK** to dismiss the **Set Fixed Cell Dimension Rules** dialog box, the **In-Place Edit** is still active. Selecting **Finish** on the Edits panel of the **Edit in Place: Grid Dimension** tab of the ribbon will open the **Save Changes** dialog box as shown in Figure 6.15. The changes can be saved to the current division name, or you can specify a new division name. If you save the change to the current division name, all door/window assemblies that use that division name will change. If you click the **New** button and specify a new division name, the change will be saved only to the current door/window assembly and other assemblies of the same style will not change. Clicking the **Discard** button of the **Save Changes** dialog box ends In-Place Editing and discards all changes. Clicking the **Save** button shown in Figure 6.15 will save the changes to the door/window assembly as exceptions to the style. The Exit Editing Grid trigger grip can also be selected during In-Place Edit to open the **Save Changes** dialog box.

If the Door/Window Assembly design rules are set to **Fixed Number of Cells,** the In-Place edit of the division includes positive and negative signs on the grid as shown in Figure 6.16. When you move your cursor over the positive sign, dynamic dimensions will be displayed to allow you to add a bay as shown at right in Figure 6.16. If you move the cursor over the negative sign, you can click to decrease the number of bays. Upon completion of the grip edit, choose Finish on the Edit in Place: Grid Division contextual tab to open the Save Changes dialog box. The Save Changes dialog box allows you to specify the name for the division edit.

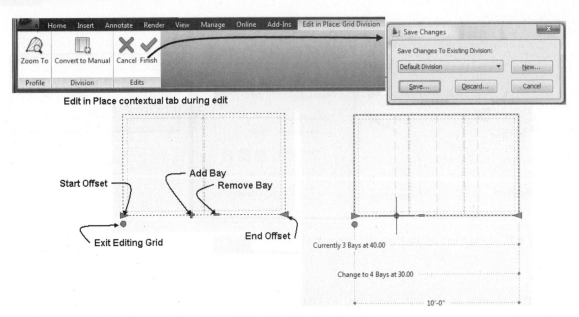

FIGURE 6.16 *In-Place Edit of Fixed Number of Cells Door/Window Assembly*

The editing of a **Manual** division type will allow you to click on the arrow grip to display dynamic dimensions. You can click on a grid division to shift its position as shown in Figure 6.17. A plus grip is displayed at the midpoint of the cell; when this grip is selected, you can add a manual location at the midpoint of the cell. The Remove manual grid grip can be selected to remove a manual location for the grid. Upon completion of the grip edit, choose Finish from the Edits panel in the Edit In Place contextual tab to open the Save Changes dialog box. The Save Changes dialog box allows you to specify a name for division edit.

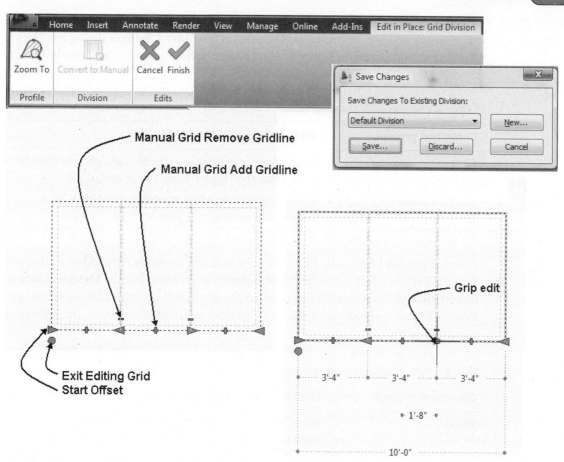

FIGURE 6.17 *In-Place Edit with Manual Design Rules of Door/Window Assembly*

Saving In-Place Edit Changes to the Style

After you save the changes with Edit In-Place Edit to a **New** division name, the changes can be saved to the style by transferring control to the object. The **Transfer to Object (GridAssemblyCopyFromStyle)** option can be selected before or after an In-Place Edit. This option toggles control to the workspace for either editing the style definition of the selected door/window assembly or reverting a door/window assembly to its original definition. Access **GridAssemblyCopyFromStyle** as shown in Table 6.3.

TABLE 6.3 *Transfer to Object command access*

Command prompt	GRIDASSEMBLYCOPYFROMSTYLE
Ribbon	Select a door/window assembly, and choose Transfer to Object from the Design Rules flyout of the Modify panel.
Shortcut menu	Select a door/window assembly, right-click, and choose Design Rules > Transfer to Object.

When you select a door/window assembly and choose the **Design Rules >Transfer to Object** option from the Door/Window Assembly tab, you can continue to edit the grid of the door/window assembly in the workspace. When editing is complete, you can save the changes reflected in the door/window assembly back to the style by selecting **Save to Style** from the Design Rules flyout menu (see Table 6.4 for command access). When you select this command, the **Save Changes** dialog box opens, allowing you to save the changes to the current style or to click the **New** button to create a new style name to save the changes.

TABLE 6.4 *Save to Style command access*

Command prompt	GRIDASSEMBLYSAVECHANGES
Ribbon	Select the door/window assembly, and choose Save to Style from the Design Rules flyout menu.
Shortcut menu	Select the door/window assembly, right-click, and choose Design Rules > Save to Style.

The changes can be discarded and the door/window assembly returned to its original definition if you select a door/window assembly and then select the **Design Rules > Revert to Style** command (see Table 6.5 for command access). This command removes all edits and redisplays the door/window assembly based upon the last saved definition of the style.

TABLE 6.5 *Revert to Style command access*

Command prompt	GRIDASSEMBLYMAKESTYLEBASED
Ribbon	Select the door/window assembly, and choose Revert to Style Design Rules from the Design Rule flyout menu.
Shortcut menu	Select the door/window assembly, right-click, and choose Design Rules > Revert to Style Design Rules.

Modifying the Infill

The infill can be edited by merging the cells or overriding the infill assignment to a cell. Prior to editing the infill, the cell markers must be turned on. Turning ON the cell markers allows you to select a cell for editing. **Cell Markers** are displayed in the Elevation or Model view in the center of the cell as shown in Figure 6.18. Access the **Infill > Show Markers (GridAssemblySetEditDepthAll)** command to turn on cell markers as shown in Table 6.6.

TABLE 6.6 *Show Markers command access*

Command prompt	GRIDASSEMBLYSETEDITDEPTHALL
Ribbon	Select the door/window assembly, and choose Show Infill Markers (see Figure 6.18) from the Modify panel of the Door/Window Assembly tab.
Shortcut menu	Select the door/window assembly, right-click, and choose Infill > Show Markers.

The infill of a cell can be merged with another cell; this allows you to convert a 4-division door/window assembly to a 3-division assembly as shown at the right in Figure 6.18. Access the **Infill > Merge (GridAssemblyMergeCells)** command to merge cells as shown in Table 6.7.

TABLE 6.7 *Merge command access*

Command prompt	GRIDASSEMBLYMERGECELLS
Ribbon	Select the door/window assembly, and choose Merge from the Infill flyout menu shown in Figure 6.18.
Shortcut menu	Select the door/window assembly, right-click, and choose Infill > Merge.

The command requires you to select the cell marker for each cell as shown in the following workspace sequence:

```
Select the door/window assembly, choose Merge from the
Infill flyout of the Modify panel.

Select cell A: (Select cell marker at p1 as shown at the left
in Figure 6.18.)

Select cell B: (Select cell marker at p2 as shown at the left
in Figure 6.18.)

(Cells merged as shown at the right in Figure 6.18.)
```

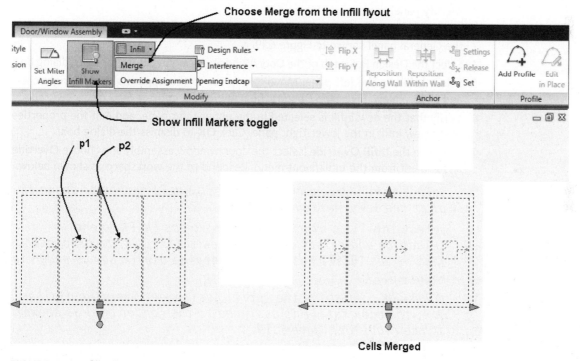

FIGURE 6.18 *Infill cells merged*

Overriding the Infill

Overriding Infill Assignments allows you to define a different infill for a cell. This allows you to change one cell without changing all cells defined in the style. The infill override definition must first be defined as an infill within the style before you can apply it as an override. Cell markers must be turned ON to allow you to select the cell. Access the **Override Assignment (GridAssemblyAddCellOverride)** command as shown in Table 6.8.

TABLE 6.8 *Override Assignment command access*

Command prompt	GRIDASSEMBLYADDCELLOVERRIDE
Ribbon	Select the door/window assembly, and choose Override Assignment from the Infill flyout menu of the Modify panel.
Shortcut menu	Select the door/window assembly, right-click, and choose Infill > Override Assignment.

With the cell markers turned ON, select the door/window assembly, right-click, and choose **Infill > Override Assignment** from the shortcut menu. You are prompted to select a cell to override; select the cell marker of the cell to edit. The **Infill Assignment Override** dialog box opens, allowing you to select a different infill element. The following steps outline the procedures for creating an infill and applying the new infill as an **Override Assignment**.

Steps to Create an Override Assignment

1. View the door/window assembly in Elevation or Pictorial view.
2. Select the door/window assembly, and choose **Infill > Show Infill Markers** from the **Modify** panel as shown in Figure 6.19.
3. To create a new infill, select the door/window assembly, and choose **Edit Style** from the General panel shown in Figure 6.19.
4. Select the Design Rules tab of the Door/Window Assembly Style Properties dialog box.
5. Select **Infills** in the left pane. Move the cursor to the upper-right pane, right-click, and choose **New**. Overtype Door as the name of the infill.
6. Verify that the new infill is selected in the upper-right pane, and edit the properties of the new infill in the lower-right pane. Click **OK** to dismiss the dialog box.
7. To create the **Infill Override**, select the door/window assembly and choose Override Assignment from the Infill flyout menu. Respond to the workspace as shown below.

> Select infill to override: *(Select a cell marker as shown at p1 at the left in Figure 6.19.)*
>
> Select infill to override: ENTER *(Press ENTER to end selection and select* door, *the new infill, in the drop-down list of the* Infill Assignment Override *dialog box as shown in Figure 6.19.)*
>
> *(Click* OK *to dismiss the* Infill Assignment Override *dialog box; the door infill is assigned to the center cell as shown at the right in Figure 6.19.)*

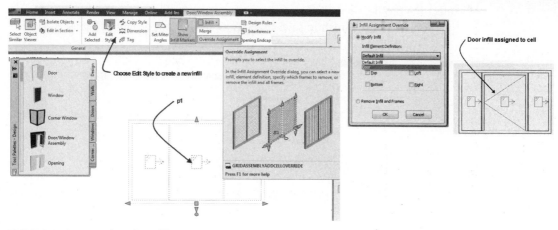

FIGURE 6.19 *Overriding the Infill Assignment*

Editing the Profile of the Frame or Mullion

The profile of the frame or mullion can be edited in the workspace. The profile can be edited with the following three commands: **Add Profile**, **Edit Profile**, and **Override Assignment**. The profile can be saved as an override to the style, or an additional profile can be created and saved in the style. Access the **Frame/Mullion > Add Profile** (**GridAssemblyAddProfileOverride**) command as shown in Table 6.9.

TABLE 6.9 *Frame/Mullion > Add Profile command access*

Command prompt	GRIDASSEMBLYADDPROFILEOVERRIDE
Ribbon	Select the door/window assembly, and choose Add Profile from the Frame/Mullion flyout menu as shown in Figure 6.20.
Shortcut menu	Select the door/window assembly, right-click, and choose Frame/Mullion > Add Profile.

When you select the **GridAssemblyAddProfileOverride** command, you are prompted to select a frame or mullion. The **Add Frame Profile** dialog box opens if you are editing a frame, which allows you to select a profile from the drop-down list. The drop-down list includes **Start from Scratch** and other profiles that have been created in the drawing. The **Start from Scratch** option creates a new profile, numbered as shown in Figure 6.20. The profile can be applied as a shared frame element definition or as a frame profile override. The **Add Profile** command was used to change the frame as shown in the following steps.

Steps to Add a Profile Override

1. Select a door/window assembly, and choose Add Profile from the Frame/Mullion flyout menu of the Modify panel in the Door/Window Assembly tab.

2. Select a frame or mullion as shown in Figure 6.20. The **Add Frame Profile** dialog box opens as shown in Figure 6.20. Specify the name of the new profile or use the **Start from scratch** default name. Click **OK** to dismiss the **Add Frame Profile** dialog box and begin in-place editing.

3. Select **Zoom To** from the **Edit in Place: Grid Assembly Profile** contextual tab. An enlarged view of the frame or mullion is displayed.

4. Select the profile, right-click, and choose from the Profile panel of the ribbon to edit the profile. Choose **Finish** to save changes as shown at the right in Figure 6.20.

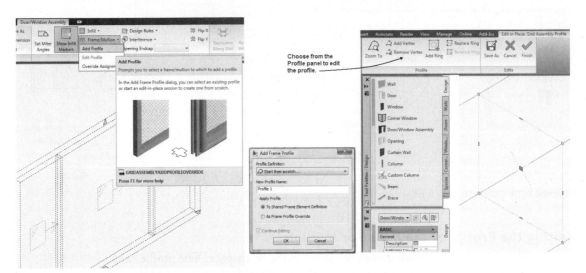

FIGURE 6.20 *Editing a frame profile*

Editing a Profile In-Place

The **Edit Profile In-Place** command is a shortcut menu option (see Table 6.10) that allows you to edit a profile that has been added or edited to define a frame or mullion. This command applies the tools of **Edit In-Place** to edit the vertices of the modified profile.

TABLE 6.10 *Edit Profile In-Place command access*

Command prompt	GRIDASSEMBLYPROFILEEDIT
Ribbon	Select the door/window assembly, and choose Edit Profile from the Frame/Mullion flyout menu.
Shortcut menu	Select the door/window assembly, right-click, and choose Frame/Mullion > Edit Profile In-Place.

Start this command by selecting a door/window assembly that includes a profile that has been previously edited. Choose **Edit Profile** from the flyout menu of Frame/ Mullion. When you select a frame that has a profile, the **Edit in Place: Grid Assembly Profile** tab will be displayed as shown in Figure 6.21. The **Zoom To** option will enlarge the view of the frame or mullion, allowing you to easily edit the profile as shown at the right in Figure 6.21. Select the profile to display its grips. The profile can then be modified by grip editing, or you can choose from the Profile panel of the ribbon. The procedure to edit a profile is shown in the following steps:

Steps to Edit a Modified Profile Using In-Place Edit

1. Select a previously modified door/window assembly, right-click, and choose **Edit Profile of the Frame/Mullion flyout menu**.

2. Select a frame or mullion of the grid assembly to edit. The **Edit in Place: Grid Assembly Profile** tab is displayed, and grips are displayed on the profile as shown at the left in Figure 6.21.

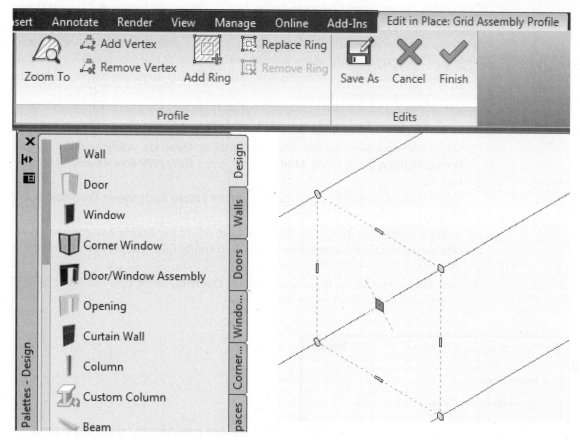

FIGURE 6.21 *Editing a modified frame profile*

3. Select **Zoom To** from the **Edit in Place: Grid Assembly Profile** tab.

4. Select the profile to display its grips, and choose commands from the Profile panel to edit the profile.

5. Select **Finish** from the Edits panel of the Edit in Place: Grid Assembly Profile to save the changes.

Overriding the Frame or Mullion Assignment

The **Override Assignment** (**GridAssemblyAddEdgeOverride**) command allows you to define a different frame for the top, bottom, or sides of the door/window assembly. Access the **Override Assignment** command as shown in Table 6.11.

TABLE 6.11 *Override Assignment command access*

Command prompt	GRIDASSEMBLYADDEDGEOVERRIDE
Ribbon	Select a door/window assembly, and choose Override Assignment from the Frame/Mullion flyout menu of the Modify panel.
Shortcut menu	Select a door/window assembly, right-click, and choose Frame/Mullion > Override Assignment.

When you execute this command, you are prompted to select a frame, and the **Frame Assignment Override** dialog box opens, which allows you to change the frame.

The Frame Assignment Override dialog box is shown in Figure 6.22. The drop-down list of frames is limited to frame definitions defined in the style. The Frame Assignment Override dialog box also allows you to delete the frame at the selected location. This option would allow you to set the frame at the top or header position wider than the other frame components. The following steps list the procedure to create a frame assignment override.

Steps to Create a Frame Override to the Edge Assignment

1. Select the door/window assembly, and choose **Override Assignment** from the **Frame/Mullion** flyout of the **Modify** panel in the **Door/Window Assembly** contextual tab.
2. Select the header frame or an edge to open the **Frame Assignment Override** dialog box.
3. Select a frame definition from the drop-down list of the **Frame Assignment Override** dialog box. The frame element shown in Figure 6.22 was previously defined in the style.
4. Click **OK** to dismiss the **Frame Assignment Override** dialog box. The frame header is increased as shown in Figure 6.23.

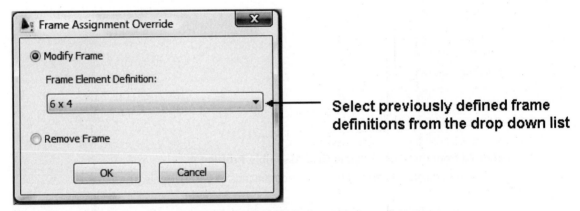

FIGURE 6.22 *Selecting the frame element definition of the Frame Assignment Override*

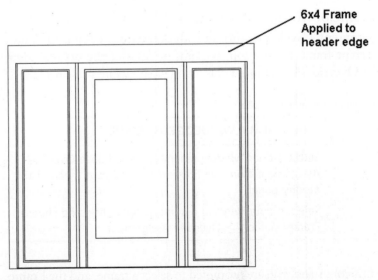

FIGURE 6.23 *Frame adjusted using Override Assignment*

Creating an Override of a Division Assignment

The division definition of the door/window assembly can be overridden. The division definition can be changed when you select the **Division > Override Assignment** command from the shortcut menu as shown in Table 6.12.

TABLE 6.12 *Division > Override Assignment command access*

Command prompt	**GRIDASSEMBLYADDDIVISIONOVERRIDE**
Ribbon	Select the door/window assembly, and choose Override Assignment from the Division flyout of the Modify panel.
Shortcut menu	Select the door/window assembly, right-click, and choose Division > Override Assignment.

This allows you to change how the unit is divided without editing the style definition. In Figure 6.24 the door/window assembly on the left was edited to create four equal divisions. The division definition must exist in the style prior to selecting the Override Assignment. The division definitions are included in the drop-down list shown in Figure 6.25.

Steps to Create an Override of a Division Assignment

1. Select the door/window assembly, and choose **Override Assignment** from the **Division** flyout menu of the **Modify** panel.
2. Select an edge of the door/window assembly.
3. Select a division definition from the drop-down list of the **Division Assignment Override** dialog box as shown in Figure 6.25. The Four Equal division element definition was previously defined in the style.
4. Click **OK** to dismiss the **Division Assignment Override** dialog box. Divisions of the unit are changed to four equal divisions as shown in Figure 6.24.

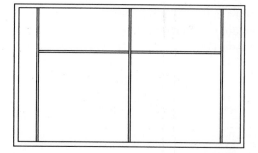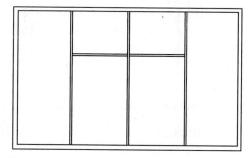

FIGURE 6.24 *Overriding a door/window assembly division definition*

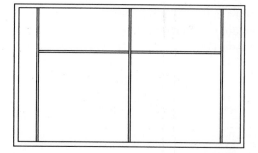

FIGURE 6.25 *Editing the division of the door/window assembly*

TUTORIAL 6-1: CREATING AN ENTRANCE DOOR/WINDOW ASSEMBLY

1. Verify that the Accessing Tutor\Imperial or Accessing Tutor\Metric content for Chapter 6 of the CengageBrain has been downloaded to your Accessing Tutor folder on your computer as described in Organizing Tutorial Directories in the Preface.

2. Open **Ex 6-1.dwg** from the Accessing Tutor\Imperial\Ch6 or Accessing Tutor\Metric\ Ch6 directory.

3. Choose **Save As** > **AutoCAD Drawing** from the Application menu, and save the drawing as **Lab 6-1** in your student directory.

4. In the **Design** tool palette, right-click on a Door/Window Assembly tool and choose **Door/Window Assembly Styles** to open the **Style Manager**.

5. Verify that the Door/Window Assembly Styles category is selected in the left pane, right-click, and choose **New** from the shortcut menu.

6. Overtype **Entrance** as the name for the new style in the left pane.

7. Click **Entrance** in the left pane to display the edit tabs in the right pane.

8. Select the General tab, and type Entrance door with side lights in the Description field.

9. Select the **Select Keynote** button, and verify that the keynote database is *{Vista: C:\ ProgramData\Autodesk\ACA 2012\enu\Details\Details (US)\AecKeynotes (US).mdb [C:\ ProgramData\Autodesk\ACA 2012\enu\Details\Details (UK)\AecKeynotes (UK NBS).mdb]}*.

10. In the Select Keynote dialog box click the "+" to expand Division 08-Openings. Select the 08 43 00 Storefronts. In the *AEC Keynotes-Assemblies (Metric).mdb* database expand B-Shell > B 20 Exterior Enclosure and select B2020 Exterior Windows. Click **OK** to dismiss the Select Keynote dialog box.

11. Select the **Design Rules** tab, select **Divisions** in the left pane, and edit the lower-right pane as follows: Name = **Entrance**, Orientation = **Vertical**, Division Type = **Manual**, Start Offset = **0**, End Offset = **0**. Select **Add Gridline**, and create the following grid lines, as shown in Table 6.13 and in Figure 6.26.

TABLE 6.13 *New gridline settings*

Gridline	Offset	From
1	-1'-8 1/2" [-525]	Grid Middle
2	1'-8 1/2" [525]	Grid Middle

12. Select **Infills** in the left pane, and type **Sidelight** in the name field of the lower-right pane.

13. Edit the **Infill Type** in the lower-right pane to **Style**, expand the **Window Style** folder in the style window, and select **Standard** as shown in Figure 6.27.

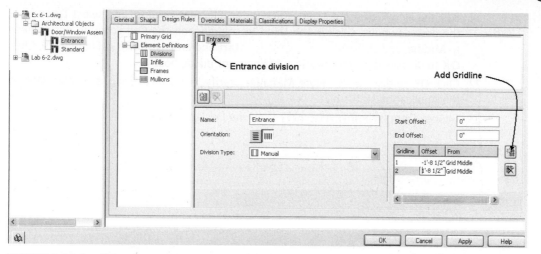

FIGURE 6.26 *Specifying the Divisions in the Design Rules*

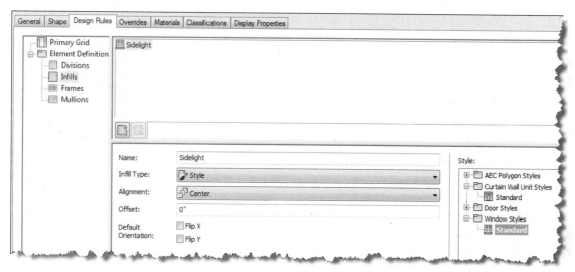

FIGURE 6.27 *Specifying the window style for the infill*

14. Edit the Alignment to **Center**.

15. Right-click over the upper-right pane, and choose **New**. Overtype **Door** in the name field of the upper-right pane.

16. Edit the **Infill Type** in the lower-right pane to **Style**, expand the **Door Styles** in the Style window, and select **Standard**.

17. Edit the Alignment to **Center**.

18. Select **Frames** in the left pane, and edit the frame dimension in the lower-right pane as follows: Name = **Default Frame**, Width = **2" [50]**, Depth = **4-9/16" [120]**, Offset X = **0**, Y = **0**, Start = **0**, and End = **0** as shown in Figure 6.28.

19. Select **Mullions** in the left pane, and edit the mullion dimension in the lower-right pane as follows: Name = **Default Mullion**, Width = **1" [25]**, Depth = **3" [75]**, Offset X = **0**, Y = **0**, Start = **0**, and End = **0**.

20. Select **Primary Grid** in the left pane, select **Cell Assignments** in the upper-right pane, and choose the **New Cell Assignment** button in the right pane as shown in Figure 6.29.

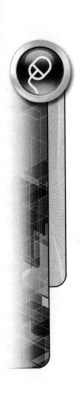

21. Edit the New Cell Assignment. Click in the **Element** column, and select **Door** from the drop-down list. Click in the **Used In** column, edit the Cell Location Assignment to **Middle**, and clear the checkboxes for **start** and **end** as shown in Figure 6.29. Click **OK** to dismiss the **Cell Location Assignment** dialog box.

22. Click **OK** to dismiss the **Style Manager**. Verify that Object Snap is toggled **OFF** in the Application status bar.

23. Choose **Door/Window Assembly** from the Design palette.

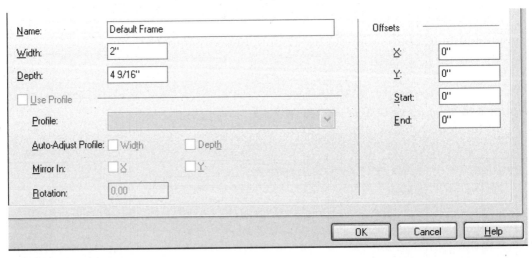

FIGURE 6.28 *Specifying the size of the default frame*

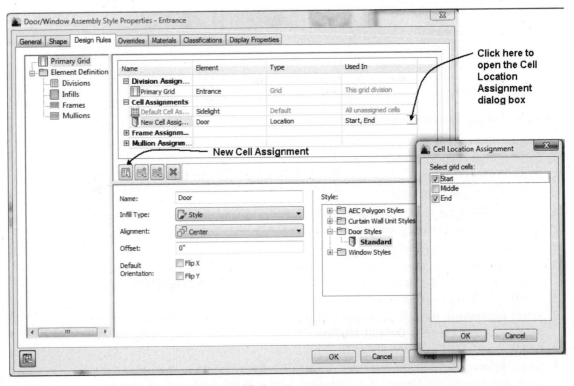

FIGURE 6.29 *Specifying the door unit for the middle location*

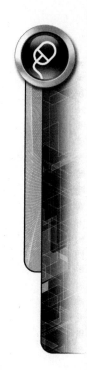

24. Edit the Properties palette as follows: Style = Entrance, A-Width = 7'-6" [2300], B-Height = 6'-8" [2000], Start Miter = 0, End Miter = 0, Position along wall = Center, Vertical Alignment = Head, and Head Height = 6'-8" [2000].

25. Specify the Insert point as shown in the following workspace sequence:

> Select wall, grid assembly, or ENTER: *(Select the wall in the drawing.)*
>
> Insert point or: *(Select a point above the wall near its middle.)*
>
> Insert point or: ENTER *(Press ENTER to end the command.)*

26. Select **SW Isometric** from the View Cube.

27. Select the left-window unit as shown in Figure 6.30, right-click, and choose Properties. Select the Display tab of the Properties palette.

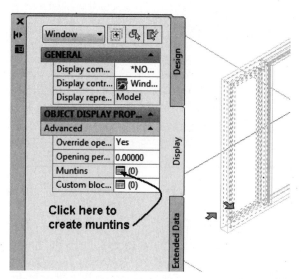

FIGURE 6.30 *Selecting the window unit of the door/window assembly*

28. In this step muntins will be added to the display of the window style. Edit Display controlled by = Window Style: Standard. Select OK to Add Style Override dialog box to create an override to the style definition. Choose the Muntins button as shown in Figure 6.30, and click the Add button to open the Muntins Block Display dialog box shown in Figure 6.31. In the Window Pane section, click the Other button and click All. In the Muntin section, set Width = 3/4" [19], Depth = 1/2" [13], and check the Clean Up Joints and Convert to Body checkboxes. In the Lights section, set Pattern = Rectangular, Lights High = 6, and Lights Wide = 2.

29. Click OK to dismiss the Muntins Block dialog box.

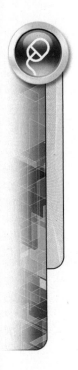

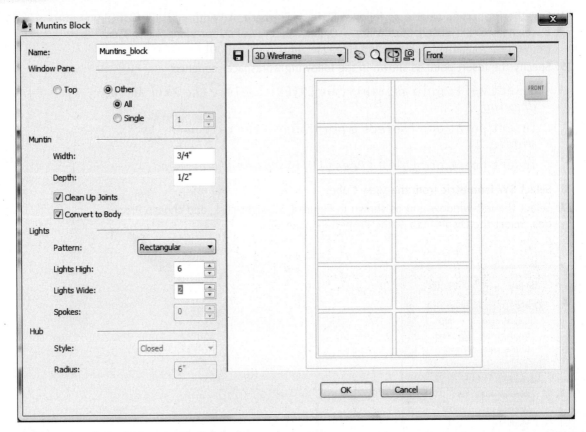

FIGURE 6.31 *Editing the Muntins Block dialog box*

30. Verify that the Automatically Apply to Other Display Representations and Object Overrides checkbox is checked. Click OK to dismiss the dialog box. Press Escape to end selection.

31. Select the door of the door/window assembly, right-click, and choose Edit Door Style.

32. Select the Dimensions tab, and set the Frame-A-Width = 2″ [50], B-Depth = 4 9/16″ [120], Stop-C-Stop Width = 2″ [50], D-Depth = 1 1/2″ **[38]**, E-Door Thickness = **1 3/4″ [44]**, and Glass Thickness = **1/4″ [6]**. Click **OK** to dismiss the Door Style Properties dialog box.

33. Select Front from the View Cube.

34. Select the frame of the door/window assembly, right-click, and choose Edit Door/Window Assembly Style.

35. In this step you will create a custom frame that will be applied as an override in future steps. Select the Design Rules tab, select Frames in the left pane, move the cursor over the upper-right pane, right-click, and choose New from the shortcut menu. Edit the lower-right pane as follows: Name = Threshold, Width = 5/8″ [15], Depth = 6″ [150]. Check the Use Profile checkbox, and select the Threshold profile from the drop-down list. (This profile was created in the current file from the geometry at the right prior to your starting this tutorial.) For Auto-Adjust Profile, check Width and Depth. Click **OK** to dismiss the Door/Window Assembly Style Properties dialog box.

36. Select the frame of the door/window assembly, and choose Frame/Mullion > Override Assignment from flyout menu of the Modify panel in the Door/Window Assembly tab. Respond to the workspace sequence as shown below.

Select an edge: *(Select the sill of the frame.)*

37. Edit the Frame Assignment Override dialog box: select the Modify Frame radio button and select Threshold from the drop-down list as shown in Figure 6.32. Click OK to dismiss the Frame Assignment Override dialog box.

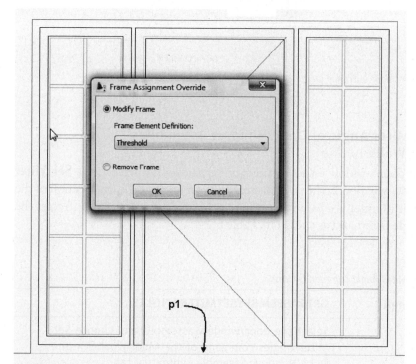

FIGURE 6.32 *Overriding the frame assignment*

38. Choose **Top** from the View Cube. Choose **Edit in Section** from the Edit in Section flyout of the General panel in the Door/Window Assembly tab. Respond to the work-space prompts as follows:

    Specify first point of section line or Enter to change UCS: *(Select a location near **p1** as shown in* Figure 6.33.)

    Specify next point of section line: *(Select a location near **p2** as shown in* Figure 6.33.)

    Specify next point of section line or [Break]: *(Press Enter to section line location.)*

    Specify section extents: (Select a point near **p3** as shown in Figure 6.33. The threshold is shown at right in Figure 6.33.)

39. Choose Exit from the Edit in View message box to end the edit.

40. Save and close the drawing.

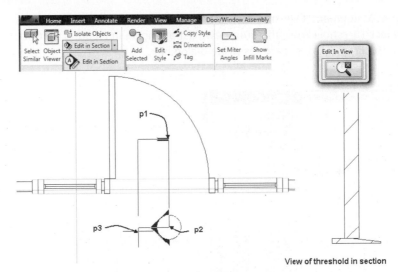

View of threshold in section

FIGURE 6.33 *Front door entrance created as a door/window assembly with threshold.*

Setting the Miter Angle of Door/Window Assemblies and Curtain Walls

When ends of door/window assembly units or curtain walls intersect, the **Set Miter Angle** command will modify the ends to create a miter angle at the intersection of the units. The angle of miter is created according to the angle of intersection. Access the **Set Miter Angle** command as shown in Table 6.14.

TABLE 6.14 *Set Miter Angle command access*

Command prompt	GRIDASSEMBLYSETMITERANGLES
Ribbon	Select the door/window assembly, and choose Set Miter Angles command of the Modify panel in the Door/Window Assembly contextual tab.
Shortcut menu	Select the door/window assembly, right-click, and choose Set Miter Angle.

Steps to Create a Miter

1. Select the door/window assembly at **p1** as shown in Figure 6.34, and choose **Set Miter Angle**.
2. Select the second grid assembly at **p2** as shown in Figure 6.34. Units are mitered as shown in Figure 6.34.
3. The miter is displayed only in pictorial view as shown in Figure 6.34.

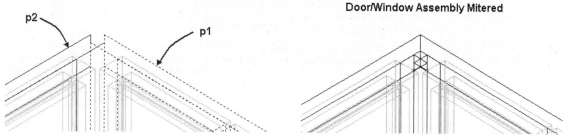

FIGURE 6.34 *Miter start and end of door/window assembly*

Applying Interference Add to the Assembly

If a door/window assembly or curtain walls intersect with a column, the objects can be trimmed to the intersection of the AEC object with the **Interference > Add** command. Access the **Interference > Add** command as shown in Table 6.15.

TABLE 6.15 *Interference > Add command access*

Command prompt	GRIDASSEMBLYINTERFERENCEADD
Ribbon	Select a door/window assembly, and choose Add from the Interference flyout menu of the Modify panel in the Door/Window Assembly tab.
Shortcut menu	Select door/window assembly, right-click, and choose Interference > Add.

The **Interference > Add** command allows you to trim the assembly by an AEC object. The procedure for trimming an object by the door/window assembly is shown below.

Steps to Use Interference Add to Trim the Assembly

1. Select a door/window assembly at **p1** as shown in Figure 6.35, and choose Add from the **Interference** flyout menu in the Modify panel.
2. Respond to the workspace prompts as shown below:

    Select AEC objects to add: 1 found *(Select a column at* p2*.)*

    Select AEC objects to add: ENTER *(Press* ENTER *to end selection.)*

    Apply to infill? [Yes/No] <Yes>: ENTER *(Press* ENTER *to apply intersection to the infill.)*

    Apply to frames? [Yes/No] <Yes>: ENTER *(Press* ENTER *to apply intersection to the frames.)*

    Apply to mullions? [Yes/No] <Yes>: ENTER *(Press* ENTER *to apply intersection to the mullions.)*

```
1 object(s) added as interference [1193]
```
(Interference applied as shown in Figure 6.35.)

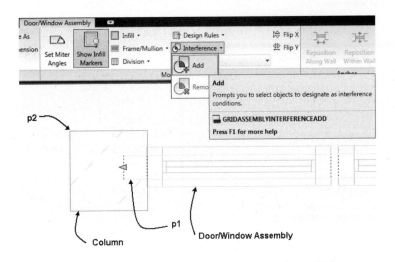
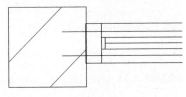

FIGURE 6.35 *Applying interference to the door/window assembly unit*

Removing Interference with AEC Objects

The **Interference > Remove** command on the shortcut menu will remove the interference between AEC objects. When the interference is removed, the assembly or curtain wall will overlap the column as shown in Figure 6.35. Access the **Interference > Remove** command as shown in Table 6.16.

TABLE 6.16 *Interference > Remove command access*

Command prompt	**GRIDASSEMBLYINTERFERENCEREMOVE**
Ribbon	Select the door/window assembly with interference property, and choose Remove from the Interference flyout menu of the Modify panel in the Door/Window Assembly contextual tab.
Shortcut menu	Select the door/window assembly, right-click, and choose Interference > Remove.

CREATING CURTAIN WALLS

Curtain walls are added to the drawing to provide a representation of non-bearing walls, which can include doors, windows, or panels. The curtain wall style is developed about a grid, which is divided in horizontal and vertical divisions. Windows, doors, or panels can be assigned to the cells formed by the horizontal and vertical divisions in the same manner as Door/Window Assemblies. Unlike the Door/Window Assembly, the floor line or roof line can be adjusted to extend above the base height or below the baseline of the curtain wall. Curtain walls are created on the **A-Glaz-Curt layer.** Curtain wall units are mini curtain walls that have properties similar to curtain walls. The curtain wall unit can be inserted as an infill into a curtain wall. The curtain wall unit is a style-based object that has frames, mullions, and divisions that are independent of the curtain wall.

Using the Curtain Wall Tool

The **CurtainWallAdd** command is used to place the curtain wall. See Table 6.17 for command access.

TABLE 6.17 *CurtainWallAdd command access*

Command prompt	CURTAINWALLADD
Ribbon	Choose Curtain Wall from the Wall flyout menu of the Build panel of the Home tab.
Design palette	Select Curtain Wall from the Design palette.

When you select the **Curtain Wall** tool from the Design tool palette, the Properties palette opens, describing the current dimensions and attributes of the curtain wall. The properties of curtain walls are similar to window/door assembly styles. The options in the Design tab of the Properties palette for the **Curtain Wall** command are described below and shown in Figure 6.36.

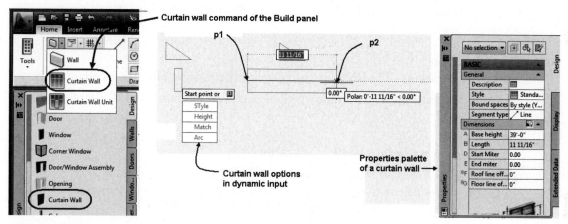

FIGURE 6.36 *Properties palette of the CurtainWallAdd command*

Design Tab—Basic
General

> **Description**—The Description button opens a Description dialog box. Type to add a description.
>
> **Style**—The Style lists the current curtain wall style; the drop-down list includes all styles loaded in the drawing.
>
> **Layer**—The Layer field lists the layer name of existing curtain walls.
>
> **Bound spaces**—The Bound spaces options include By Style, Yes, and No. This option controls the function of the curtain wall to serve as a boundary for a space.
>
> **Segment type**—The segment types for curtain walls are Line and Arc.

Dimensions

A-Base height—The Base height field allows you to type the distance from the baseline to the top of the unit.

B-Length—The Length field dynamically displays the length of the curtain wall. The length of existing curtain walls is displayed in this field.

D-Start Miter—The Start Miter is the angle of the frame, mullion, and infill at the start point of the wall.

E-End Miter—The End Miter is the angle of the frame, mullion, and infill at the end of the curtain wall.

F-Roof line offset from base height—The Roof line offset from base height extends the top of the curtain wall above the base height. A negative roof line offset value will shorten the curtain wall below the base height.

G-Floor line offset from baseline—The Floor line offset from baseline extends the bottom of the curtain wall below the baseline of the curtain wall. A negative floor offset value will shorten the curtain wall above the baseline.

Design Tab—Advanced

Roof/floor line—Select the **Roof/floor line** button to open the **Roof and Floor Line** dialog box. The contents of the Roof and Floor Line dialog box were presented in Chapter 2.

Location

Rotation—The Rotation field displays the rotation of existing curtain walls.

Elevation—The Elevation field displays the elevation above $Z = 0$ of the curtain wall.

Placing a Curtain Wall

To place a curtain wall, select **Curtain Wall** from the Design tool palette. Edit the Properties palette to specify the style, height, and segment type, and select the start and endpoints of the curtain wall, as shown in Figure 6.36.

```
Start point or: (Select the start point at p1 as shown in
Figure 6.36.)

Endpoint or: (Select the endpoint at p2 as shown in
Figure 6.36.)

Endpoint or: ENTER (Press ENTER to end the command.)
```

CONVERTING WALLS TO CURTAIN WALLS

The **Apply Tool Properties to > Walls (CurtainWallToolToWalls)** command will change an existing wall to a curtain wall with the same length and height as the wall. The command allows you to specify the style of the curtain wall. Access the **CurtainWallToolToWalls** command as shown in Table 6.18.

TABLE 6.18 *Apply tool properties to > Walls command access*

Shortcut menu	Select the Curtain Wall tool, right-click, and choose Apply Tool Properties to > Walls.

When you select the **Curtain Wall** tool, right-click, and choose **Apply Tool Properties to > Walls,** you are prompted to select walls in the command line and specify the justification. The justification can be left, right, or baseline. The justification will shift the curtain wall baseline in alignment with the left wall component, right wall component, or the baseline component of the selected wall. The command line sequence shown below converted the walls to curtain walls.

```
Command: CurtainWallToolToWalls
Select walls: 1 found (Select the wall at point p1 as shown in
Figure 6.37.)
Select walls: 1 found (Select the wall at point p2 as shown in
Figure 6.37.)
Select walls: ENTER (Press ENTER to end selection of walls.)
Curtain wall baseline alignment <Baseline>: (Choose the
Right justification option from the workspace option list.)
Erase layout geometry? [Yes/No] <N>: ENTER (Press ENTER to
retain the wall geometry.)
2 new Curtain Wall(s) created.
(Press ESC to end the command. Isometric view of walls
converted to curtain walls shown in Figure 6.37.)
```

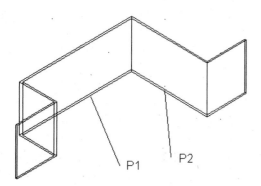 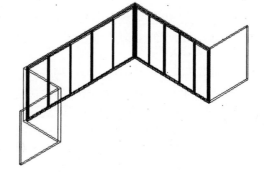

FIGURE 6.37 *Walls converted to curtain walls*

CURTAIN WALL STYLES

Curtain wall styles allow you to preset the number of horizontal and vertical divisions of the curtain wall and the type of unit inserted within the divisions. The style definition controls the appearance of the curtain wall. If you change the style after the curtain wall has been inserted, the change will be reflected in all curtain walls that exist with the same style name. You can override the style definition to change one or more units within the curtain wall style as presented for door/window Assemblies. Curtain wall styles are created and edited in the **Style Manager** (refer to Table 6.19). The content of the right pane of the Style Manager for a style is similar to the tabs of a door/window assembly style presented earlier in this chapter.

TABLE 6.19 *Style Manager command access*

Command prompt	AECCURTAINWALLSTYLE
Ribbon	Select a curtain wall, and choose Curtain Wall Styles from the Edit Style flyout in General panel of the Curtain Wall contextual tab.
Shortcut	Select a curtain wall, right-click, and choose Edit Curtain Wall Style from the shortcut menu.

When the **Style Manager** opens, the Curtain Wall Styles category is selected. Select **New Style** from the Style Manager toolbar, and overtype the name of the new curtain wall style in the left pane. The grid and content of the curtain wall is specified in the **Design Rules** tab as shown in Figure 6.38. The **Primary Grid** shown in Figure 6.38 is defined to create horizontal divisions. The cell dimension of 6' is created with the remaining cell defined to shrink the bottom, whereas the Secondary grid creates vertical divisions within the primary grid. If the Secondary Grid were removed, the curtain wall would consist only of the top and bottom cells. Unlike the Infill types for door/window assemblies, the Infill type of a curtain wall can include the following: AEC Polygon Styles, Curtain Wall Unit Styles, Door Styles, Window Styles, and Door/Window Assembly Styles.

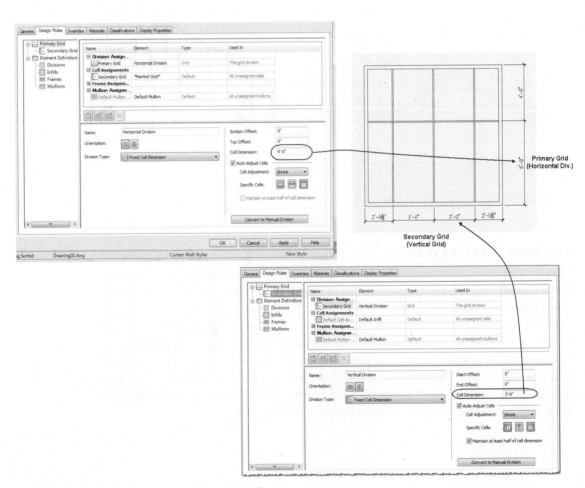

FIGURE 6.38 *Division assignments of a curtain wall*

TUTORIAL 6-2: CREATING A CURTAIN WALL STYLE

1. Verify that the Accessing Tutor\Imperial or Accessing Tutor\Metric content for Chapter 6 of the CengageBrain has been downloaded to your Accessing Tutor folder on your computer as described in Organizing Tutorial Directories in the Preface.

2. Open **Ex 6-2.dwg** from the *Accessing Tutor\Imperial\Ch6* or *Accessing Tutor\Metric\ Ch6* directory. Save the drawing to your Chapter 6 student folder as *Lab 6-2.dwg*.

3. Right-click on the **Curtain Wall** tool of the Design palette and choose **Apply tool properties to > Walls**. Respond to the workspace prompts as follows:

   Select walls: (*Select the walls using a crossing selection from* p1 *to* p2 *as shown in Figure 6.39.*)

   Select walls: (*Press* ENTER *to end selection.*)

   Curtain wall baseline alignment <Baseline>: (*Choose* Baseline *from the drop-down options.*)

   Erase layout geometry? [Yes/No] <No>: Y (*Choose the* Yes *option to erase the walls.*)

4. Select **SW Isometric** from the **View Cube**.

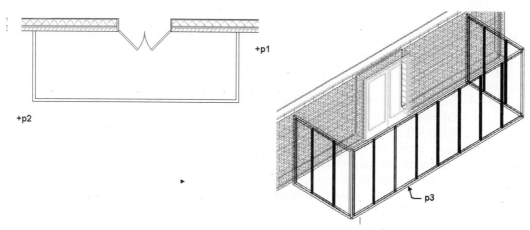
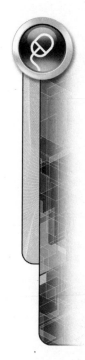

FIGURE 6.39 *Curtain wall created from existing walls*

5. Select the curtain wall at **p3** as shown in Figure 6.39, and choose Copy Style from the General panel to open the Curtain Wall Style Properties dialog box. Choose the **General** tab. Overtype **Sun Room** as the name of the curtain wall. Choose the **Design Rules** tab. In this step you will adjust the default settings to create horizontal divisions for the primary grid. Choose **Primary Grid** in the left pane, as shown in Figure 6.40. Edit the lower-right pane as follows: Name = **Horizontal Division**, Orientation = **Horizontal**, Division Type = **Fixed Cell Dimension**, Bottom Offset = **0**, Top Offset = **0**, Cell Dimension = **6' [1800]**. Check the Auto-Adjust Cells checkbox and set Cell Adjustment = **Shrink** and Specific cells = **Bottom** only. Click **OK** to dismiss the Curtain Wall Style Properties – Sun Room dialog box.

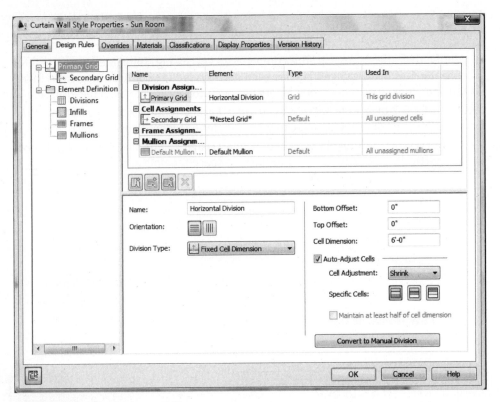

FIGURE 6.40 *Edit of Primary Grid*

6. Select the curtain wall at **p3** as shown in Figure 6.39, and choose **Edit Style** from the General panel of the Curtain Wall contextual **tab.** Choose the Floating **Viewer** in the lower-left corner to preview the changes. Select **SW Isometric** from the View drop-down menu and **3D Hidden** from the Visual Styles drop-down menu of the Viewer. To rename the Secondary Grid, choose the **Secondary Grid** in the left pane at **p1** in Figure 6.41, double-click, and overtype the name **Top Grid**. Edit the lower-right pane as follows: Name = **Vertical Division**, Orientation = Vertical, Division Type = Fixed Cell Dimension, Start Offset = **0**, End Offset = **0**, Cell Dimension = 3' [900]. Check the Auto-Adjust Cells checkbox and set Cell Adjustment = Shrink and Specific Cells = Start and End as shown in Figure 6.41. Click OK to dismiss the dialog box.

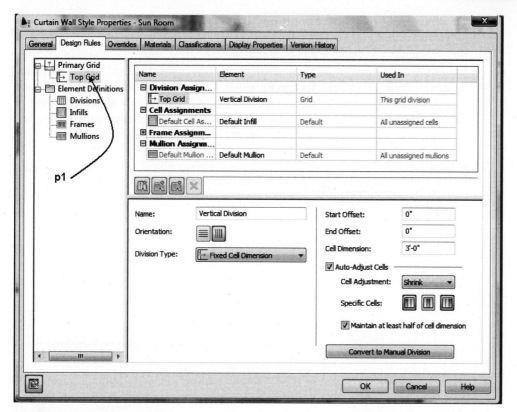

FIGURE 6.41 *Edit of Top Grid*

7. Select the curtain wall at **p3** as shown in Figure 6.39, and choose **Edit Style** from the General panel of the Curtain Wall contextual **tab.** Choose the **Primary Grid** in the left pane. Choose the **New Cell Assignment** button of the right pane. Edit the new grid Element type to **Nested Grid**. Double-click the **New Cell Assignment** name in the left pane, and overtype the **Bottom Grid**. Select the Primary Grid in the left pane to display the Top Grid and Bottom Grid as shown in Figure 6.42. Close the Viewer. Click **OK** to dismiss the Style Manager.

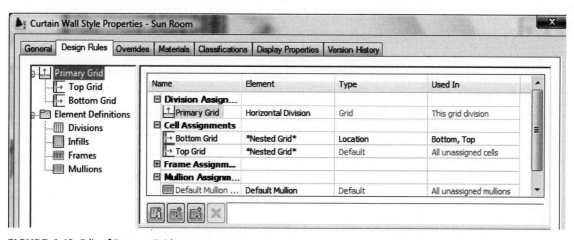

FIGURE 6.42 *Edit of Bottom Grid*

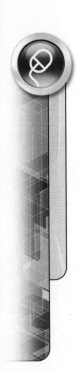

8. To import the Awning window style into the drawing, choose the **Awning** window style from the **Windows** tool palette. Drag the tool into workspace, and press ESC when prompted to select a wall or grid assembly.

9. Select the curtain wall at **p3** as shown in Figure 6.39, right-click, and choose **Edit Curtain Wall Style**. Choose the **Design Rules** tab. Choose the **Infills** in the left pane, and edit the Default Infill as follows: Name = **Awning**, Infill Type = **Style**, Alignment = **Center**, Offset = **0**. Expand **Window Styles** and choose the **Awning** window style from the drop-down list of Window Styles at the right as shown in Figure 6.43.

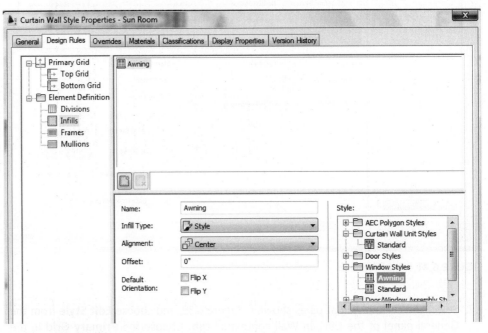

FIGURE 6.43 *Specifying the Awning infill*

10. Verify that Infills category is selected in the left pane, and choose the **New** button in the right pane. Edit the properties of the new infill as follows: Name = **Panel**, Infill Type = **Simple Panel**, Alignment = **Center**, and Panel Thickness = **1 [25]** as shown in Figure 6.44.

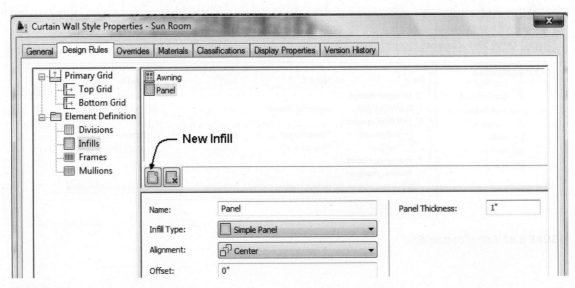

FIGURE 6.44 *Creating the panel infill*

11. Choose the **Primary Grid** in the left pane. To define where to apply the **Bottom Grid**, click in the right margin of the **Used In** field of the Bottom Grid as shown in Figure 6.45. Edit the **Cell Location Assignment** to only the **Bottom** cells. Click **OK** to dismiss the Cell Location Assignment dialog box.

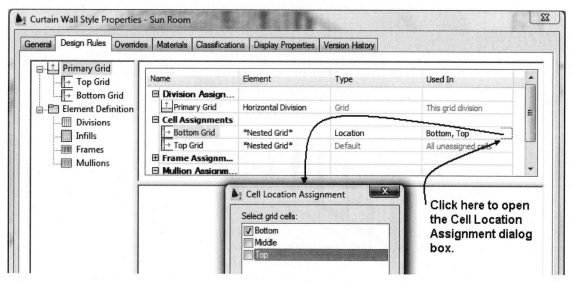

FIGURE 6.45 *Cell assignments of grids*

12. To define the element for the cells of the bottom grid, choose **Bottom Grid** in the left pane. Click in the **Element** column, and choose **Panel** for the **Default Cell Assignment** as shown in Figure 6.46. Click **OK** to dismiss the Curtain Wall Style Properties - Sun Room dialog box.

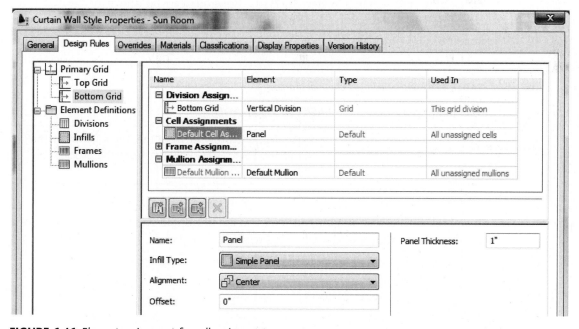

FIGURE 6.46 *Element assignment for cell assignment*

13. Select the curtain walls at **p1** and **p2** as shown in Figure 6.47, right-click select Properties, and choose **Sun Room** from the Style drop-down list of the Design tab of Properties.

14. Save and close the drawing.

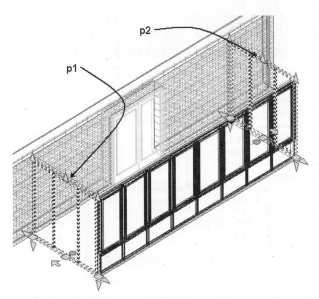

FIGURE 6.47 *Editing curtain wall style*

AEC POLYGONS FOR CURTAIN WALL STYLES

AEC Polygons can be created and inserted as an infill to curtain walls. The AEC Polygon can be hatched, providing color and texture to the curtain wall infill to simulate the actual building materials. AEC Polygons can be drawn or created from closed polylines. Access the **AecPolygonAdd** command from the **Content Browser** as shown in Table 6.20.

TABLE 6.20 *AecPolygonAdd command access*

Command prompt	**AECPOLYGONADD**
Content Browser	Choose Content Browser from the Tools flyout of the Build panel. Select AEC Polygon from the Stock Tool Catalog > Helper Tools.

If you select the **AecPolygonAdd** tool, you are prompted to specify the vertices of a closed shape as shown in the following workspace sequence:

```
Specify start point or: (Select point p1 as shown in
Figure 6.48.)

Specify next point or: (Select point p2 as shown in
Figure 6.48.)

Specify next point or: (Select point p3 as shown in
Figure 6.48.)

Specify next point or: (Select point p4 as shown in
Figure 6.48.)
```

Specify next point or: *(Choose* Close *from the dynamic prompts of workspace.)*

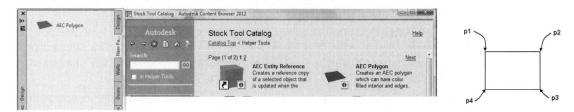

FIGURE 6.48 *Points selected to create polygon*

An existing closed polyline can be converted to an AEC Polygon using the **AecPolygonToolToPline** command. Access the **AecPolylineToolToPline** command as shown in Table 6.21.

TABLE 6.21 *Convert Closed Polyline to AEC Polygon command access*

Command prompt	**AECPOLYGONTOOLTOPLINE**
Palette shortcut	Drag the AEC Polygon i-drop from the Stock Tool catalog to a tool palette. Right-click the AecPolygon tool, and choose Apply Tool Properties to > Closed Polyline.

When you choose the **Apply Tool Properties to > Closed Polyline (AECPolygonToolToPline)** tool, you are prompted in the workspace as follows:

(Right-click on the AecPolygon *tool and choose* Apply Tool Properties to > Closed Polyline.*)*

Select polylines: *(Select a polyline.)*

Erase layout geometry? [Yes/No] <No>: *(Select the* Yes *option from the workspace prompts.)*

After creating an AEC Polygon, you can add hatching to the interior of the polygon. The hatching pattern can simulate materials applied as infills to the cell assignments when the AEC Polygon is defined as an infill for a curtain wall. The following steps outline the procedure to define the properties of an AEC Polygon style.

Steps to Adding a Hatch Pattern to an AEC Polygon

1. Select an AEC Polygon, and choose **Copy Style** from the General panel of the AEC Polygon contextual tab to open the **AEC Polygon Style Properties – style name** dialog box. Choose the **General** tab, and type a name (for example, **Precast Concrete**) in the **Name** field. Click **OK** to dismiss the AEC Polygon Style Properties – Precast Concrete dialog box.

2. Select the AEC Polygon right-click, and choose Properties.

3. Select the **Display** tab and edit Display controlled by = AEC Polygon Style. The Add Style Override dialog box opens; click **OK** to edit the style and dismiss the warning box.

4. Click in the Display component field, and choose Interior Hatch. Toggle the light bulb ON for Interior hatch. Set the **Interior Hatch** component current as shown in Figure 6.49. Click **OK** to dismiss the Modify Display Component at Style Override Level.

5. Choose the Select component button (p1 in Figure 6.49) of the Properties palette, and select the hatch at p2 as shown in Figure 6.49. Click the **Pattern** button (p3 in Figure 6.49) in the Hatching section of the Display tab as shown in Figure 6.49 to open the Hatch Pattern dialog box.

6. Edit the **Hatch Pattern** dialog box: select the **Predefined** pattern type. Select a pattern name, such as **AR-CONC**, from the pattern drop-down list. Click **OK** to dismiss the Hatch Pattern dialog box. Click **OK** to dismiss the Modify Display Component at Style Override Level.

7. Click in the **Scale/Spacing** column of the Interior Hatch display component, and edit the scale factor. The scale factor is set to **1 [1]** as shown in Figure 6.49.

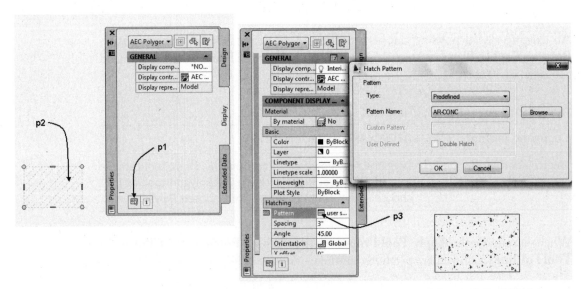

Pattern applied to AEC Polygon

FIGURE 6.49 *AEC Polygon Style Properties dialog box*

AEC Polygons can be applied as an infill of a curtain wall. The infill will assume the display properties of the AEC Polygon wall. The following steps outline the procedure to assign the AEC Polygon to an infill.

Steps to Assigning an AEC Polygon to an Infill

1. Select an existing curtain wall, right-click, and choose **Edit Curtain Wall Style** to open the **Style Manager** dialog box.

2. Select the **Design Rules** tab of the right pane of the Style Manager.

3. Select **Infills** in the left pane and select the **New Infill** button. Type a name, such as **Precast**, to name the new infill in the bottom pane. Edit the Infill type to **Style,** and expand the **AEC Polygon Styles** in the Style window. Select the name of an AEC Polygon, such as **Precast Concrete**, as shown in Figure 6.50.

4. Select **Primary Grid** in the left pane, and select **Cell Assignment** in the right pane. Select the **New Cell Assignment** button, click in the **Element** field of the New Cell Assignment, and specify an infill name, such as **Precast Concrete**. Note that the AEC Polygon infill can be assigned to various cells in the Cell Location Assignment dialog box or as an Infill Override Assignment.

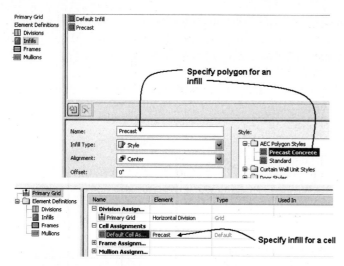

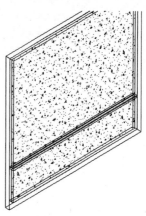

Polygon applied to the curtain wall

FIGURE 6.50 *AEC Polygon assigned as an infill style*

5. Click **OK** to dismiss the Curtain Wall Style Properties – style name dialog box. Select **SE Isometric** to view the AEC Polygon assigned to the cells of the curtain wall, as shown at the right in Figure 6.50.

CONVERTING SKETCHES TO CURTAIN WALLS

Curtain walls and curtain wall units can be created from Sketches. The tools to convert a sketch to a curtain wall or curtain wall unit are located on the shortcut menu of the Curtain Wall tool when placed on a tool palette as shown in Tables 6.22 and 6.23.

TABLE 6.22 *Curtain Wall > Apply Tool Properties to > Elevation Sketch tool access*

Shortcut menu	Select the Curtain Wall tool from the Design palette, right-click, and choose Apply Tool Properties to > Elevation Sketch.

TABLE 6.23 *Curtain Wall Unit > Apply Tool Properties to > Elevation Sketch tool access*

Shortcut	Select Curtain Wall Unit tool from a tool palette, right-click, and choose Apply Tool Properties to > Elevation Sketch.

The sketch shown at the left in Figure 6.51 consists of lines and arcs. The sketch was converted to a curtain wall in the following workspace sequence.

> (*Right-click on the* Curtain Wall Unit *tool and choose* Apply Tool Properties to > Elevation Sketch.*)
>
> Select elevation linework: (*Create a crossing window: select a point near* **p1** *as shown in Figure 6.51.*)
>
> Specify opposite corner: (*Select a point near* **p2** *as shown in Figure 6.51.*) 3 found
>
> Select elevation linework: ENTER (*Press ENTER to end selection.*)

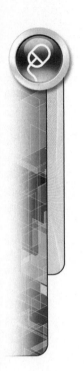

Select baseline or ENTER for default: ENTER *(Press* ENTER *to accept default.)*

Erase layout geometry? [Yes/No] <No>: ENTER *(Press* ENTER *to retain geometry.)*

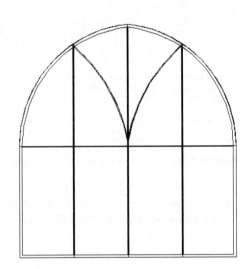
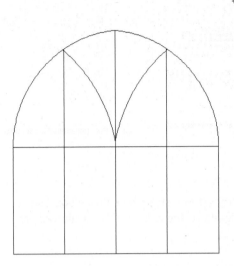

FIGURE 6.51 *Converting a sketch to a curtain wall*

PROJECT EXERCISE 6-3: CREATING A DOOR/WINDOW ASSEMBLY

1. Verify that the Accessing Tutor\Imperial or Accessing Tutor\Metric content for Chapter 6 of the CengageBrain has been downloaded to your Accessing Tutor folder on your computer as described in Organizing Tutorial Directories in the Preface.

2. Open *Ex 6-3* from the *Accessing Tutor\Imperial\Ch6* or *Accessing Tutor\Metric\Ch 6* directory. Save the drawing as **Lab 6-3** in your student directory. To insert the style of a double-hung window, right-click the **Double Hung** tool of the Windows tool palette and choose **Import 'Double Hung' Window Style**.

3. Create a new door/window assembly style named **DH-Fixed assembly** that has a vertical primary grid. Set the cell to **fixed** cell dimension and the cell dimension to **24" [600]**. Use the **Auto-Adjust** feature in the **Design Rules** to grow the middle cell. Create two new infills (see Table 6.24).

TABLE 6.24 *Infill type and style*

Name	Infill Type	Alignment	Style
DH	Style	Center	Double-hung window style
Fixed	Style	Center	Standard window style

4. Set the default cell assignment to the **DH** infill. The **Fixed** infill should not be used in any location. Close the Style Manager.

5. Insert a door/window assembly using the **DH-Fixed** style, A-Width = **8'-1"** **[2400]**, and B-Height = **5'** **[1500]** with the **Position along wall** option set to **Center** constrained. Place the assembly in the middle of the wall. View the wall in the SW Isometric view.

6. Turn on cell markers, and merge the inner two cells. Create an override of the cell infill for the inner cells, and set the inner cell to the **Fixed** cell infill as shown in Figure 6.52.

7. Save and close the drawing.

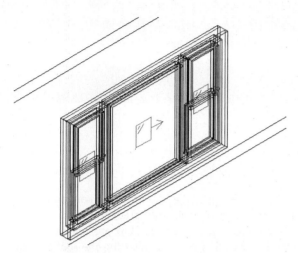

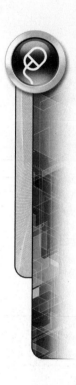

FIGURE 6.52 *DH-Fixed door/window assembly style*

SUMMARY

- Door/window assemblies are created based upon a grid.
- The primary grid can be horizontal or vertical and defined as a fixed cell dimension, fixed number of cells, or manual.
- Components of each grid intersection can be a simple panel, style, or nested grid.
- A nested grid defined for a cell can create additional divisions of a cell.
- Frame and mullions can be created from profiles.
- The display of cell markers allows you to select cells using the **GridAssemblyMergeCells** command.
- Profiles used for frames and mullions can be edited in place.
- The divisions, cells, frames, and mullions assignments can be overridden for a style.
- The **GridAssemblyAddInterference** command allows you to trim the assembly by an AEC object such as a column.

Refer to e **Review Questions** folder which includes review questions for Chapter 6 of Student Companion Site of **CengageBrain http://www.cengagebrain.com** described in the Preface.

NOTE

Creating Roofs and Roof Slabs

INTRODUCTION

Simple or complex roofs can be placed on a building through the **RoofAdd** command. The roof can be applied to the walls of the building or created from a polyline. This chapter includes tutorials that develop the following roofs: gable, gambrel, hip, and dormer. Additional detail and flexibility in editing the roof is obtained by converting the roof to roof slabs. The use of roof slabs to develop overhang features will also be presented in this chapter. A roof slab can be extended, trimmed, or mitered to create a complex roof. The commands for creating roofs and roof slabs are included on the Design tool palette.

OBJECTIVES

After completing this chapter, you will be able to

- Use RoofAdd to create hip, gable, shed, and gambrel roofs.

- Edit roof plate height, roof slope, thickness, overhang, and fascia angle using the Properties palette.

- Use grips to change a hip roof to a gable roof.

- Convert walls and linework to a roof.

- Convert roofs, walls, and linework to roof slabs.

- Create roof slabs and dormer roofs that intersect the main roof.

- Extend walls to the roof line.

CREATING A ROOF WITH ROOFADD

The commands for creating and editing roofs are located on the Design tool palette and Home tab as shown in Figure 7.1. The **Roof** tool (**RoofAdd**) is used to place a roof. New and existing roofs are changed by editing the Properties palette. Existing roofs are edited by their grips and by choosing commands from the shortcut menu of the selected roof. Existing closed walls, lines, and arcs can be converted to a roof by using the **Apply Tool Properties to** shortcut option of the **RoofAdd** command located on the Design tool palette.

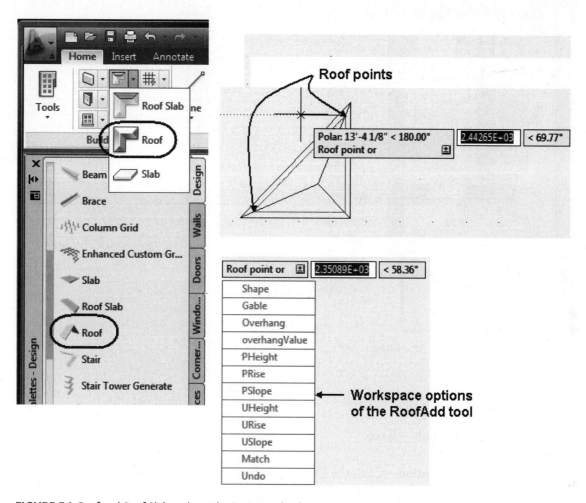

FIGURE 7.1 *Roof and Roof Slab tools on the Design tool palette*

Use the Roof tool (RoofAdd command) to create roofs by selecting the walls at the corners to specify the roof. See Table 7.1 for command access.

TABLE 7.1 *RoofAdd command access*

Command prompt	**ROOFADD**
Ribbon	Select Roof from the Roof Slab flyout menu of the Build panel.
Design palette	Select Roof from the Design palette.

Selecting the **Roof** tool from the Design tool palette opens the Properties palette as shown in Figure 7.2. The roof is created by editing the values in the Properties palette and selecting points **p1, p2, p3,** and **p4.** The completed roof is shown in the lower-left corner of Figure 7.2.

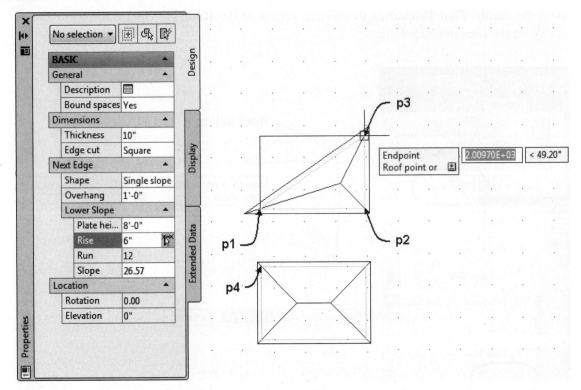

FIGURE 7.2 *Properties palette for creating a roof*

The features specified in the Properties palette are described below.

Design Tab—Basic
General

> **Description**—Click the Description button to open the **Description** dialog box to type a description.

> **Bound Spaces**—The Bound Spaces option allows you to specify if the roof will function as a boundary for spaces.

Dimensions

> **Thickness**—The Thickness field is the distance between the faces of the roof. The default value is 10 [200]. Roof thickness should be set according to the dimensions of the rafters.

> **Edge cut**—The options for the Edge cut are **Square** and **Plumb**. The **Square** option creates a fascia perpendicular to the face of the roof. The **Plumb** option creates a vertical fascia.

Next Edge

> **Shape**—There are three Shape options: **Single slope, Double slope**, and **Gable**. A single slope roof extends a roof plane at an angle from the **Plate Height** (see Figure 7.3). The single slope shape also creates a hip roof since each edge is sloped. A double slope roof includes a single slope and adds another slope, which begins at the intersection of the first slope and the height specified for the second slope. The last-used single or double shape serves as the default shape. The Gable roof has a vertical roof edge.

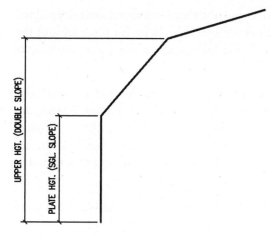

FIGURE 7.3 *Plate heights for single and double slopes*

Overhang—The Overhang is the horizontal distance the roof plane is extended down from the plate height at the angle of the roof to create the overhang.

Lower Slope—The Lower Slope is the angle of the roof plane from horizontal. Slope angles can be entered directly or derived from the rise value. If slope angles are entered, the equivalent rise value is displayed in the **Rise** field.

Plate Height—The Plate Height field allows you to specify the height of the top plate from which the roof plane is projected. The height is relative to the XY plane, which has a Z coordinate of zero. The plate height can vary for one or more planes of a roof.

Rise—The Rise is the slope value entered to obtain the slope angle of a single slope roof. The rise to run ratio value specifies the angle of the roof based upon a run value of 12 (Imperial). The run value for metric pitch is based upon 100. As shown in Figure 7.2, a rise of 6 [50] creates a 6/12 [50/100] roof, which forms a slope angle of 26.57 degrees. The slope [pitch] angle formed is the arc tangent of the rise value divided by 12 [100].

Run—The Run is the base value of 12 [100].

Upper Slope—The upper slope fields are displayed only for the double-slope-shaped roofs.

Upper Height—The Upper Height field is the height defined to begin the second slope of a double slope roof. If **Shape** is toggled to **Double slope**, the **Upper Height** and upper **Rise** fields are displayed in the Properties palette. The second slope begins at the **Upper Height** distance defined from the XY plane.

Rise—The Rise value determines the slope angle of the second slope. The Rise value for the upper slope is based upon a run of 12 [100].

Run—The Run has a base value of 12 [100].

Slope—The Slope is the angle of the roof plane from horizontal. Slope angles can be entered directly or derived from the rise value. If slope angles are entered, the equivalent rise value is displayed in the **Rise** field.

Location

Rotation—The Rotation field displays the rotation angle of the roof.

Elevation—The **Elevation** field displays the elevation of the roof relative to insert points.

The Properties palette consists of fields to define the shape, overhang, and slope of the roof. The creation of the roof is based upon a specified slope for the roof and the height of the plate. Some of the options in the Design tab of the Properties palette are **display only**.

The options of the command can be selected from the Properties palette or the dynamic prompt of the workspace as shown in Figure 7.1. The command options are as follows: Shape, Gable, Overhang, OverhangValue, PHeight, PRise, PSlope, UHeight, URise, USlope, and Match. All of the options except Match can be selected in the Properties palette. The Match option allows you to select a roof to match some or all of its properties.

When the **RoofAdd** command is executed, the Properties palette opens, and you are prompted to select roof points for each corner of the roof. The points selected for the roof edge are located at the bottom of the wall with a Z coordinate of zero. Selecting the points using the **Intersection** object snap locates the edge of the roof plane on either the inner or outer wall surface of a corner. The **Node** object snap locates the roof plane on the justification line of the wall. Selecting the outer wall surface for the roof points generates a roof construction similar to the truss roof construction shown in Figure 7.4. Note that the lower plane of the roof intersects the outer wall surface.

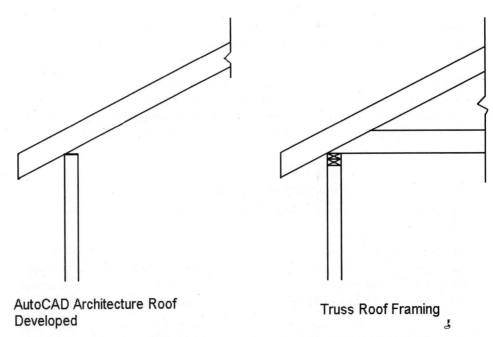

AutoCAD Architecture Roof
Developed

Truss Roof Framing

FIGURE 7.4 *Roof developed by the selection of outer lines of the wall to define the roof*

If the inner wall surface is selected to determine the roof plane, the construction is similar to that of conventional framed rafters. Using the **Intersection** object snap mode to select the inner roof plane results in the lower plane of the roof being generated from the top of the inner wall, as shown in Figure 7.5.

As points are selected, the roof is dynamically constructed. If errors are made in selecting points, right-click and choose **Undo** from the shortcut menu or type **U** in the command line. Roof planes will be created at the slope angle specified in the

Properties palette from the edge specified by each pair of points. Because a sloped roof plane is generated from each edge, a **hip** roof is created by default when the **Single slope** or **Double slope** shape is specified.

The plate height for a roof should be set equal to the wall height. The plate height is measured relative to the XY plane with a zero Z coordinate. The plate height and slope can be changed as you select vertices of the roof.

TIP

After placing the roof, you can select the roof and walls and choose **Edit in Section** from the General panel of the Roof contextual tab to view and adjust the plate height as necessary to match actual framing. The AecEditInSection command can be used to view and edit the roof and wall in section. The plate height or elevation of the roof can be edited in the Properties palette in order to move the roof plane and simulate actual construction.

NOTE

You can select Zoom *Window* from the *Zoom* flyout of the Navigation bar to zoom a window around desired roof points while creating a roof. *A* Zoom command selected from the Zoom flyout of the Navigation bar is a transparent command and does not cancel the active command.

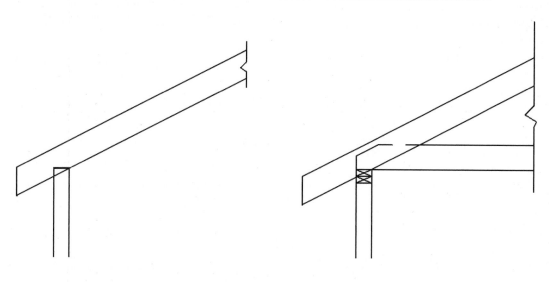

AutoCAD Architecture
Roof Developed

Conventional Roof Framing

FIGURE 7.5 *Roof developed by the selection of inner lines of the walls to define the roof*

Creating a Hip Roof

To create a hip roof, set the **Shape** to **Single slope** or **Double slope** and set the **Rise** for the roof planes. Selecting the corners of the building identifies each edge of the roof. The slope of each edge can be set independently of other roof edges and should be defined prior to selecting the second point that defines that edge of the roof. The roof is created dynamically as the corners of the building are selected; therefore, the **Undo** button can be clicked to deselect a corner of the roof as it is developed. Tutorial 7-4, "Creating a Hip Roof," is a project-based tutorial located at the end of the chapter.

DEFINING ROOF PROPERTIES

The properties of a roof can be set prior to the creation of the roof or after the roof has been developed. Double-clicking to select a roof opens the Properties palette, which allows you to edit the roof properties. Changes made in the Properties palette are reflected immediately in the roof. Additional fields are displayed in the Properties palette as shown in Figure 7.6. The fields of the **Design, Display,** and **Extended Data** tabs of three sample roofs are described in Tutorial 7-0.

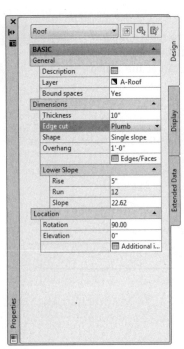

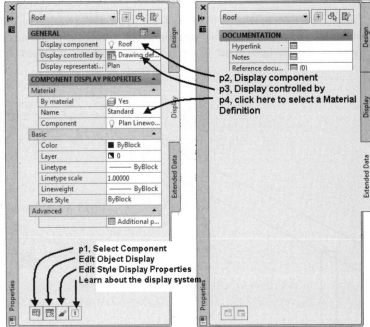

FIGURE 7.6 *Properties palette of an existing roof*

TUTORIAL 7-0: EXPLORING ROOF PROPERTIES

1. Download Chapter 7 files from the Accessing Tutor\Imperial or Accessing Tutor\ Metric category of the Student Companion site of CengageBrain http://www. cengagebrain.com described in the Preface to your Accessing Tutor folder.

2. In this tutorial you will review the settings of the Roof Properties palette for three roofs. Open Accessing Tutor\Imperial\Ch7\Ex 7-0.dwg or Accessing Tutor\Metric\ Ch 7\Ex 7-0.dwg and save the drawing as **Lab 7-0** in your Student directory.

3. Double-click to select the roof of building 1 as shown in Figure 7.7. The Shape of this roof is defined as **Single Slope** in the Properties palette. Each edge of the roof slopes to create a Hip Roof. Press **Escape** to end selection.

4. Select the roof of building 2 as shown in Figure 7.7. The Shape of this roof is defined as **Double Slope** in Shape field of the Properties palette. Each edge of the roof

consists of two faces. The plate height and rise of each face are described in the Lower Slope and Upper Slope sections of the Design tab. Press Escape to end selection.

5. Select the roof of building 3 as shown in Figure 7.7; this roof was created by using the **Gable** and **Single Slope** shapes. Press Escape to end selection.

6. Select the roof for buildings 1 and 2 as shown in Figure 7.7. Choose **Zoom Object** from the Zoom flyout of the Navigation bar.

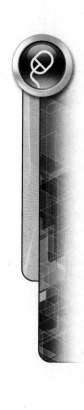

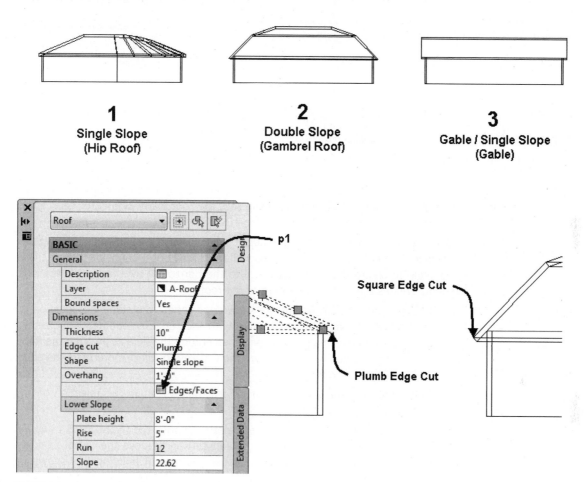

1
Single Slope
(Hip Roof)

2
Double Slope
(Gambrel Roof)

3
Gable / Single Slope
(Gable)

FIGURE 7.7 *Setting the Edge cut of a roof*

7. Select the roof for building 1 shown at left. The Edge cut of this roof is **Plumb** which provides a vertical fascia as shown in Figure 7.7. Press Escape to end selection.

8. Select the roof for building 2 shown at right. The Edge cut of this roof is **Square** which provides a fascia perpendicular to the roof face.

9. Choose Top from the View Cube. Select the roof for building 1 which includes a curved edge. Choose the **Edges/Faces** button at **p1** as shown in Figure 7.8 to open the Roof Edges and Faces dialog box. The Roof Edges and Faces dialog box shown in Figure 7.8 includes a list of roof planes that were created. Edge number 0 is the first edge created when the roof was placed. An edge is specified by selecting the first two points to create the first roof segment. The remaining edges are listed in ascending order in the same sequence as they were created. The properties of each edge are described in columns of the list box.

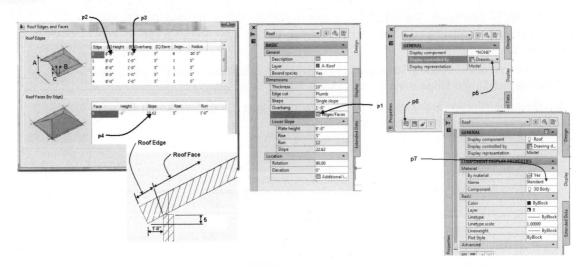

FIGURE 7.8 *Edit of Roof Edges and Faces dialog box*

10. You may click the field at **p2** of Figure 7.8 to adjust the (A) Height which is the plate height or the distance from the XY plane that the roof plane begins for each edge. The roof is generated at the specified slope from this point.

11. Click in the field at **p3** in Figure 7.8 to adjust the (B) Overhang shown as the horizontal distance the edge projects from the wall line. The roof extends at the specified slope angle to create the overhang equal to the horizontal overhang distance. The overhang creates the soffit component of the roof.

12. The (C) Eave, Segment, and Radius are shown and reflect the slope and location of the roof. The slope of each Face associated with each edge is modified by selecting an Edge in the Roof Edges pane at top and then editing the field at **p4** from the Roof Faces (by Edge) pane. You may hold down the Ctrl key to select more than one edge from the Roof Edges pane for edit.

13. Choose OK to dismiss the Roof Edges and Faces dialog. Retain the selection of the roof for building 1.

14. Choose the **Display** tab of the Properties palette. Choose the flyout at **p5** to verify Display controlled by is set to Drawing default setting. Choose the Select Component button shown at **p6** in Figure 7.8. Choose the roof of building 1 as shown in Figure 7.7.

15. The settings for the Roof display component are shown at right in Figure 7.8.

16. You may modify the material specified for the selected component if you click at p7 as shown in Figure 7.8 to select a different material definition of the drawing.

17. If you choose a different material definition, the Modify Display Component at Drawing Default Level dialog box opens. Click OK to assign this material to the roof at the drawing default level.

18. Save and close the drawing.

CREATING A GABLE ROOF

A gable roof is created in a procedure similar to that for creating the hip roof. Define the gable ends of the roof by selecting **Gable** in the Properties palette when the vertical edges of the roof plane are defined. When a gable roof edge is created, the overhang on that edge can be changed in the Properties palette. The steps for creating a gable roof are described below and in Figure 7.9.

Steps to Create a Gable Roof

1. Select the **Roof** tool from the Design palette.

2. Edit the Properties palette: set Shape = **Single slope**, Overhang = **16" [400]**, and select point **p1** as shown at the left in Figure 7.9.

3. Retain settings in the Properties palette and select point **p2**.

4. Edit the Properties palette: set Shape = **Gable**, Overhang = **3" [75]**, and select point **p3**.

5. Edit the Properties palette: set Shape = **Single slope**, Overhang = **16" [400]**, and select point **p4**.

6. Edit the Properties palette: set Shape = **Gable**, Overhang = **3" [75]**, and select point **p5**.

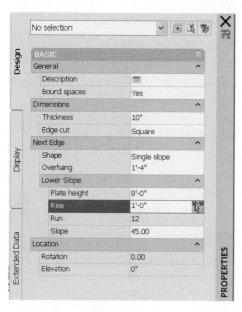

Properties palette
settings for placing
points p1, p2, p4

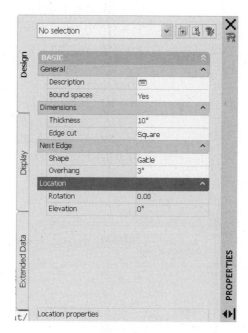

Properties palette
settings for placing
points p3 & p5

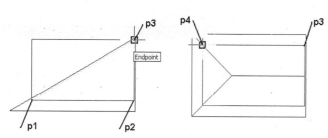
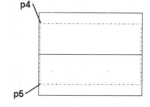

FIGURE 7.9 *Selection of points to create a gable roof*

TUTORIAL 7-1: CREATING A GABLE ROOF

1. Verify that the Accessing Tutor\Imperial or Accessing Tutor\Metric content for Chapter 7 of the CengageBrain has been downloaded to your Accessing Tutor folder on your computer as described in Organizing Tutorial Directories in the Preface.

2. Open Accessing Tutor\Imperial\Ch7\Ex 7-1.dwg or Accessing Tutor\Metric\Ch 7\ Ex 7-1.dwg and save the drawing as **Lab 7-1** in your Student directory.

3. Move the cursor to the Object Snap toggle of the status bar and right-click; then choose **Settings** from the shortcut menu. Click the **Clear All** button, check ON the **Intersection** mode of object, Check ON **Object Snap On (F3)** at the top of the **Object Snap** tab, choose the **Dynamic Input** tab, and check the **Enable Pointer Input**, **Enable Dimension Input where possible**, and **Show command prompting and command input near the crosshairs** checkboxes. Click **OK** to dismiss the Drafting Settings dialog box.

4. Select the **Roof** command from the **Design** palette. Edit the **Properties** palette as follows: Basic-Dimensions: Thickness = **6″ [150]**, Edge cut = **Plumb**, Next Edge: Shape = **Gable**, Overhang = **6″ [150]**, as shown in Figure 7.10. Respond to the following workspace prompts by selecting roof points as specified below.

> Roof point or: *(Select the wall using the intersection of the outer lines of the corner at* p1 *as shown in Figure 7.10.)*
>
> Roof point or: *(Select the wall using the intersection of the outer lines of the corner at* p2 *as shown in Figure 7.10.)*
>
> *(Edit the Properties palette as follows: Shape =* Single slope, *Overhang =* 20 [500], *Plate height =* 8'-0″ [2400], *and Rise =* 8 [67].*)*

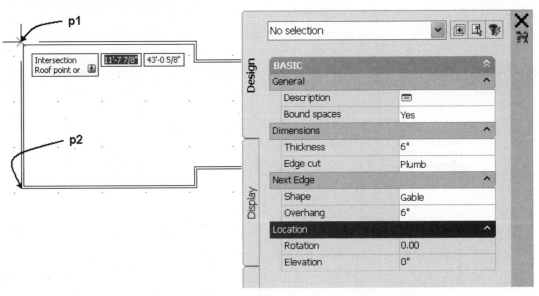

FIGURE 7.10 *Editing the Properties palette for roof edge 1*

> Roof point or: *(Select the wall using the intersection of the outer lines of the corner at* p3 *as shown in Figure 7.11)*
>
> *(Edit the Properties palette as follows: Shape =* Gable, *Overhang =* 6 [150].*)*

Roof point or: *(Select the wall using the intersection of the outer lines of the corner at* **p4** *as shown in Figure 7.11.)*

(Edit the Properties palette as follows: Shape = `Single slope`, *Overhang =* `20 [500]`, *Plate height =* `8'-0" [2400]`, *and Rise =* `8 [67]`.*)*

Roof point or: *(Select the wall using the intersection of the outer lines of the corner at* **p5** *as shown in Figure 7.11.)*

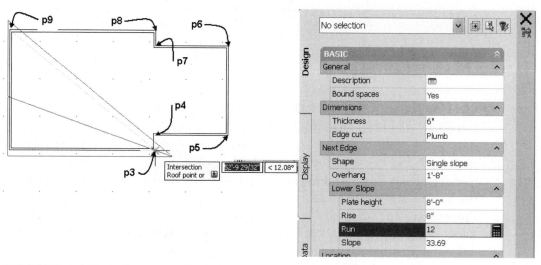
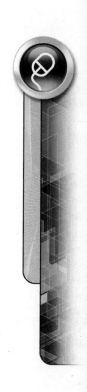

FIGURE 7.11 *Editing the Properties palette for point 3*

(Edit the Properties palette as follows: Shape = `Gable`, *Overhang =* `6 [150]`.*)*

Roof point or: *(Select the wall using the intersection of the outer lines of the corner at* **p6** *as shown in Figure 7.11.)*

(Edit the Properties palette as follows: Shape = `Single slope`, *Overhang =* `20 [500]`, *Plate height =* `8'-0" [2400]`, *and Rise =* `8 [67]`.*)*

Roof point or: *(Select the wall using the intersection of the outer lines of the corner at* **p7** *as shown in Figure 7.11.)*

(Edit the Properties palette as follows: Shape = `Gable`, *Overhang =* `6 [150]`.*)*

Roof point or: *(Select the wall using the intersection of the outer lines of the corner at* **p8** *as shown in Figure 7.11)*

(Edit the Properties palette as follows: Shape = `Single slope`, *Overhang =* `20 [500]`, *Plate height =* `8'-0" [2400]`, *and Rise =* `8 [67]`.*)*

Roof point or: *(Select the wall using the intersection of the outer lines of the corner at* **p9** *as shown in Figure 7.11.)*

Roof point or: ENTER

5. Select **SE Isometric** from the View Cube of the Navigation bar.

6. Select the View tab. Choose **3D Hidden** from the Visual Styles flyout of the Visual Styles panel; the final roof should appear as shown in Figure 7.12.

7. Save and close the drawing.

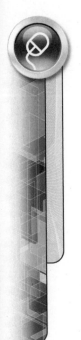

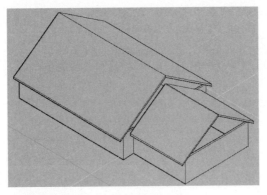

FIGURE 7.12 *Completed gable roof*

EDITING AN EXISTING ROOF TO CREATE GABLES

It is often easier to create roof gables by developing a roof for the entire structure and then editing selected roof planes to create the gable. A roof for an entire structure can be developed with a single slope. The single slope roof creates a hip roof. The **Roof Edges and Faces** dialog box can be used to edit the roof planes. If you select a roof and then select the **Edges/Faces** button in the Design tab of the Properties palette, all roof edges and faces will be displayed in the **Roof Edges and Faces** dialog box. You could select the edge for edit in this dialog box; however, it is difficult to identify which edge of the roof you are editing. The roof edges are listed in the Roof Edges and Faces dialog box in the same order as the edges were created when the roof was developed. This approach is appropriate if you need to edit a feature for all edges.

Using Edit Edges/Faces to Edit Roof Planes

Each plane of a roof can be edited with the **Edit Edges/Faces** command (**RoofEditEdges**). This command allows you to select a plane of the roof and edit its properties. See Table 7.2 for command access.

TABLE 7.2 *Edit Edges/Faces command access*

Command prompt	ROOFEDITEDGES
Ribbon	Select the roof and choose Edit Edges from the Modify panel of the Roof contextual tab.
Shortcut menu	Select the roof, right-click, and choose Edit Edges/Faces.

When the **Edit Edges/Faces** command (**RoofEditEdges**) is selected from the Modify panel of the ribbon as shown in Figure 7.13, you are prompted to "Select a roof-edge" in the workspace. When you select an edge, the **Roof Edges and Faces** dialog box opens as shown in Figure 7.13. In contrast to clicking the **Edges/Faces** button in the Properties palette, this command limits the display to only information regarding the plane selected. The Edit Edges/Faces command allows you to select one or more roof edges for edit and change the edge in the Roof Edges and Faces dialog box. When the slope is changed, the adjoining roof planes extend or adjust to the different slope. The steps to editing an edge to create a gable are described below.

Steps to Edit Roof Edges and Faces

1. Select the roof and choose **Edit Edges** from the Modify panel in the Roof tab as shown in Figure 7.13. Respond to the workspace prompts as shown below.

   > Select a roofedge: *(Select roof edge at p1 as shown in Figure 7.13.)*
   >
   > Select a roofedge: ENTER *(Press ENTER to end selection.)*

2. Edit the **Roof Edges and Faces** dialog box: click in the **Overhang** column of the top **Roof Edges** section and overtype an overhang value. Then click in the **Roof Faces (by Edge)** section, click in the **Slope** column, and overtype the slope value to **90** as shown in Figure 7.13. Click **OK** to dismiss the Roof Edges and Faces dialog box. The slope of the roof has changed as shown at the right in Figure 7.13.

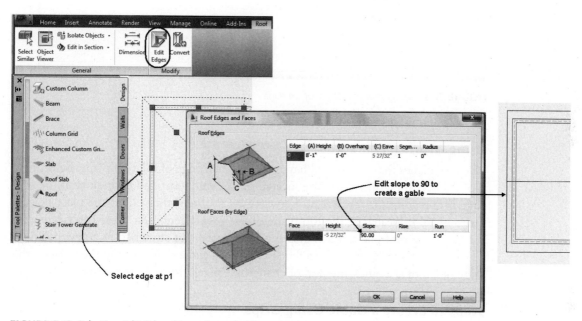

FIGURE 7.13 *Selecting Edit Edges/Faces from the shortcut menu*

If more than one edge is selected, the edges will be listed in the Roof Edges section of the Roof Edges and Faces dialog box. To select more than one edge, hold down the CTRL key to add to the selection as shown in Figure 7.14. After selecting multiple edges, click in the column for overhang or slope, and the change will be applied to all selected edges.

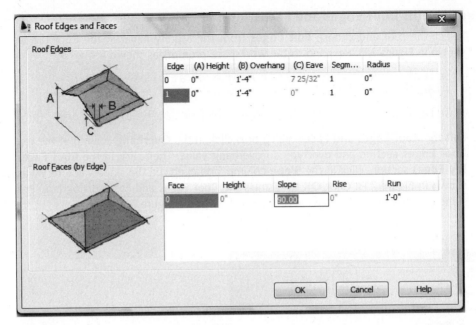

FIGURE 7.14 *Editing multiple edges of an edge*

Using Grips to Edit a Roof

The roof planes that are created have grips on each edge of the roof plane. Selecting a roof displays the grips. Grips are located on the midpoint and endpoints of the roof edge at the point of intersection between the roof and the wall at the plate height. Grips are also located at the top plane of the roof where each plane intersects with the adjoining roof plane. Each grip can be selected and stretched to a new position. The grips of the ridge can be stretched to create a gable roof. Editing the grips on the ridge of a roof is a quick method of changing a plane from hip to a gable.

Steps to Create a Gable Using Grips

1. Select a roof to display its grips.
2. Select the ridge grip at **p1** as shown in part A of Figure 7.15.
3. Stretch the hot grip to the left and specify the new location by selecting a point to the left near **p2** as shown in part B of Figure 7.15.

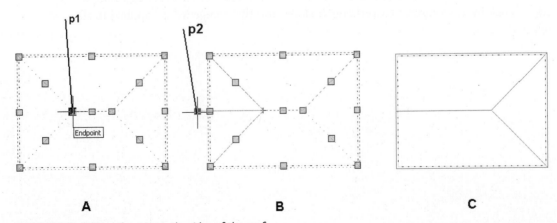

FIGURE 7.15 *Editing the grip on the ridge of the roof*

4. Press ESC to remove the display of grips. The hip roof plane is converted to a gable as shown in part C of Figure 7.15.

Grips allow each plane to be stretched to a new position, and the adjoining planes adjust to the new position. Each ridge grip can be stretched to any point near or outside the building to convert the hip roof to a gable.

CREATING A SHED ROOF

Create a shed roof by selecting the **Gable** shape in the Properties palette for three of the four sides of a roof. The slope of the non-gable side of the roof creates the shed. The **Gable** shape cannot be selected for more than three sides. Each time it is used, the roof for that edge of the building has a vertical slope. Shed roofs are often used in combination with other roof types.

CREATING A GAMBREL ROOF USING A DOUBLE-SLOPED ROOF

A double-sloped roof allows you to create a gambrel roof, which consists of two roof planes, with the first beginning at the top plate height. This first roof plane shown in Figure 7.16 is projected from the top plate at the lower-slope angle of 30/12 until it intersects with the height of the upper slope. The second roof plane shown in Figure 7.16 begins at 14' from the intersection with the lower slope and continues to the ridge at a slope of 6/12. Walls are not necessary for specifying the position of the upper slope. The gambrel roof is created when you use the Double slope shape for two sloped edges and the Gable shape to create the gable ends of the building.

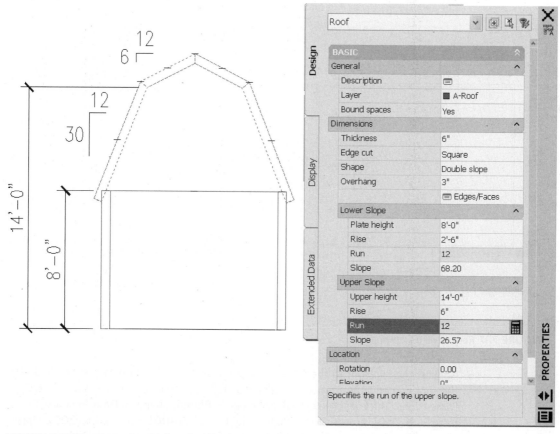

FIGURE 7.16 *Creating a Gambrel roof with double slope*

TUTORIAL 7-2: CREATING AND EDITING A GAMBREL ROOF

1. Verify that the Accessing Tutor\Imperial or Accessing Tutor\Metric content for Chapter 7 of the CengageBrain has been downloaded to your Accessing Tutor folder on your computer as described in Organizing Tutorial Directories in the Preface.

2. Open *Accessing Tutor\Imperial\Ch7\Ex 7-2.dwg* or *Accessing Tutor\Metric\Ch7\Ex 7-2.dwg* and save the drawing as **Lab 7-2** in your student directory.

3. Set the object snap by moving the cursor to the Object Snap toggle of the status bar and right-clicking; choose **Settings** from the shortcut menu. Select the **Intersection** mode and clear all other object snap modes. Verify that **Object Snap On (F3)** is checked. Select the **Dynamic Input** tab; verify that the **Enable Pointer Input, Enable Dimension Input**, and **Show command prompting and command input near the crosshairs where possible** checkboxes are checked. Select **OK** to dismiss the Drafting Settings dialog box.

4. Clear all toggles except Object Snap, Dynamic Input, and Quick Properties on the status bar.

5. Right-click the **Roof** tool of the **Design** palette and choose **Apply Tool Properties to > Linework and Walls**. Respond to the workspace prompts as shown below:

   Select objects: *(Create a crossing selection by selecting a point near* **p1** *as shown in Figure 7.17.)*

   Specify opposite corner: *(Move the cursor left, and select a point near* **p2** *as shown in Figure 7.17.)*

   Select objects: ENTER *(Press* ENTER *to end selection.)*

   Erase layout geometry? [Yes/No] <No>: ENTER *(Press* ENTER *to retain the walls in the drawing.)*

   (Roof is created as shown at the right in Figure 7.17.)

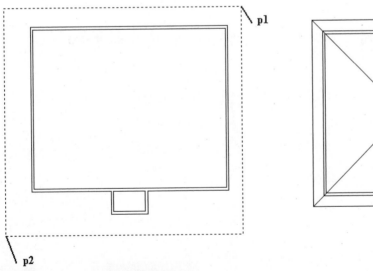
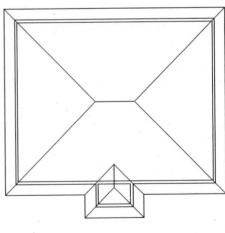

FIGURE 7.17 *Roof created from walls*

6. Convert the Hip roof to the Gambrel shape by editing the Properties palette. Select the roof and edit the Properties palette as follows: Basic: Layer = **A-Roof [A270G]**, Dimensions: Thickness = **6 [150]**, Edge cut = **Plumb**, Shape = **Double slope**, Overhang = **6 [150]**, Lower Slope: Plate height = **8' [2400]**, Rise = **24 [200]**, Upper Slope: Upper height = **16' [4800]**, Rise = **5 [40]**, as shown in Figure 7.18.

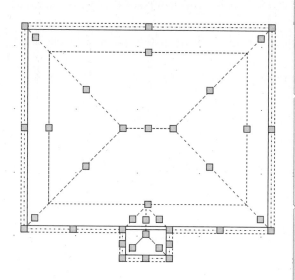

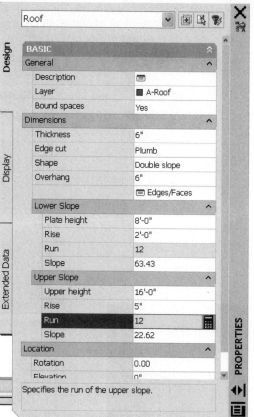

FIGURE 7.18 *Properties of roof defined*

7. Move the cursor from the Properties palette, press ESC, and toggle **ON ORTHO mode** on the status bar.

8. Create a gable using grips, select the roof, and verify that grips are displayed. Select the ridge grip of the upper slope at **p1** as shown in Figure 7.19. Stretch the grip to a point near **p2** as shown at the right in Figure 7.19.

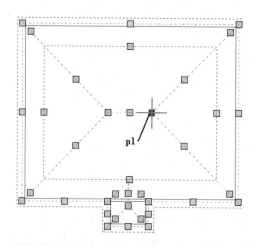

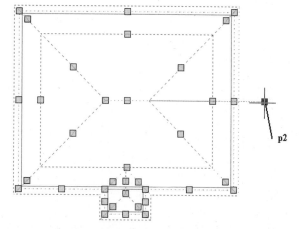

FIGURE 7.19 *Upper slope ridge grip edited*

9. Create a gable for the remaining slope using grips by selecting the grip of the lower slope at **p1** as shown in Figure 7.20. Stretch the grip to a point near **p2** as shown at the right in Figure 7.20.

10. Drag grips at **p3** and **p4** as shown at the left in Figure 7.20 to **p5** by repeating steps 7 and 8 above.

11. Press ESC and select the **SE Isometric** from the View Cube to view the roof edit.

12. Select **Top** from the View Cube.

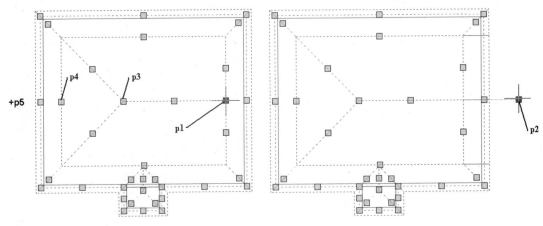

FIGURE 7.20 *Lower slope grip edited*

13. In this step and the following one, a gable is created using the Edit Edges/Faces command. Select the roof and choose **Edit Edges** from the Modify panel of Roof contextual tab. Respond to the command prompts as shown below:

    Select a roof edge: *(Select edge at p1 as shown in Figure 7.21.)*

    Select a roof edge: ENTER *(Press ENTER to end the selection and open the Roof Edges and Faces dialog box.)*

14. Select **Edge 0** in the **Roof Edges** section of the Roof Edges and Faces dialog box. Click in the **Slope** column of Face 0 and change the slope to **90**. Click in the **Slope** column of Face 1 and change the slope to **90**. Click **OK** to dismiss the Roof Edges and Faces dialog box.

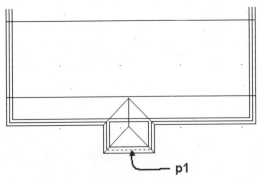

FIGURE 7.21 *Creating gable using RoofEditEdges command*

15. In this and the following step the slope of each face of the edges is modified to create a single slope shape. Select the roof, right-click, and choose **Edit Edges/Faces** from the shortcut menu. Select edges at **p1** and **p2** as shown in Figure 7.22. Press ENTER to open the Roof Edges and Faces dialog box.

16. Select **Edge 0** in the **Roof Edges** section of the Roof Edges and Faces dialog box. Click in the **Slope** column of Face 0 and change the slope to **22.62 [21.80]**. Select **Edge 1** in the **Roof Edges** section and change the slope of Face 0 to have a slope of **22.62 [21.80]**, as shown in Figure 7.22. Click **OK** to dismiss the Roof Edges and Faces dialog box. Press Escape to end selection.

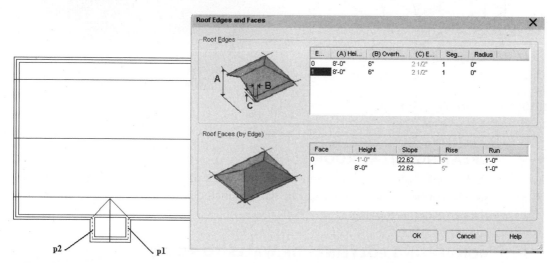

FIGURE 7.22 *Creating equal slope for upper and lower faces*

17. Select the **SE Isometric** from the **View** Cube.

18. In this step, the walls are projected to the roof. Select the walls at **p1, p2,** and **p3** as shown at the left in Figure 7.23. Choose **Modify Roof Line** from the Roof/Floor Line flyout of the Modify panel in the Wall contextual menu. Respond to the command line prompts as shown below:

> RoofLine: *(Choose* Auto project *option from the drop-down dynamic prompt.)*
>
> Select objects: 1 found *(Select the gambrel roof.)*
>
> Select objects: ENTER *(Press ENTER to end the selection.)*

19. Select the View tab. Choose **Conceptual** from the **Visual Styles** flyout menu of the **Visual Styles** panel to view the walls extended as shown at the right in Figure 7.23.

20. Save and close the drawing.

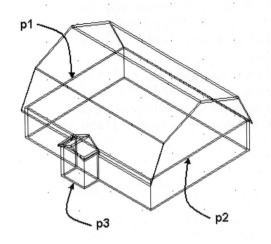

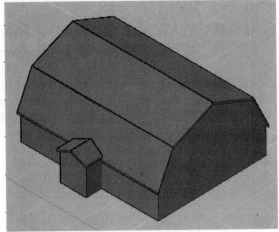

Walls projected to roof

FIGURE 7.23 *Extending walls to roof*

CREATING A FLAT ROOF

A flat roof is often used in commercial construction. Create a flat roof using a slab, which is presented in Chapter 8. The slab can include the cant strip edge style around the perimeter of the slab.

CONVERTING POLYLINES OR WALLS TO ROOFS

The **Apply Tool Properties to > Linework and Walls (RoofToolToLinework)** command will create a roof from a closed polyline or a closed series of walls without specifying each corner of the roof. The walls or polyline can be selected using a crossing or window selection method or they can be selected individually. See Table 7.3 for **Apply Tool Properties to > Linework and Walls** command access.

TABLE 7.3 *Apply Tool Properties to > Linework and Walls to create a Roof command access*

Shortcut menu	Select the Roof tool from the Design tool palette, right-click, and choose Apply Tool Properties to > Linework and Walls.

When the **Apply Tool Properties to > Linework and Walls** command is executed, you are prompted to select walls or polylines as shown below. In response to this prompt, select all walls or a closed polyline to create a roof over the walls or polyline. After you select the walls or polyline, you can specify the properties in the Properties palette. The **RoofToolToLinework** command is a quick method of creating a roof over curved walls.

(Select the Roof *tool on the Design palette, right-click, and choose* Apply Tool Properties to > Linework and Walls *as shown in Figure 7.24.)*

Choose walls or polylines to create roof profile.

Select objects: *(Select polyline at* **p1** *as shown in Figure 7.24 to convert to a roof.)*

Select objects: ENTER *(Press ENTER to end selection.)*

Erase layout geometry? [Yes/No] <No>: ENTER *(Choose No from the dynamic input options.)*

(The Roof is created as shown at the right in Figure 7.24.)

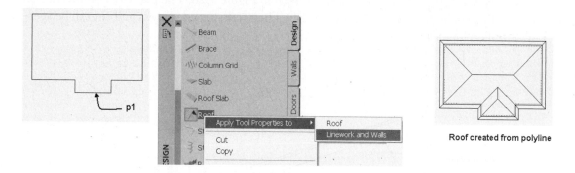

Roof Tool To Linework command

FIGURE 7.24 *Creating a roof from the polyline geometry*

CREATING ROOF SLABS

Roof slabs include features such as fascia, soffit, and frieze components. Unlike the roof discussed earlier, each roof slab is a single object that can be edited independent of adjoining roof slabs. If you edit the slope of an edge of a roof, the adjoining edges will adjust to continue the roof. Therefore, the initial roof should be developed using the **Roof** command and then the roof object converted to roof slabs to customize the edges and individual planes of the roof. Roof slabs created from the roof object are not linked; therefore, any changes to the roof object are not reflected in the roof slab. Roof slabs can be extended, trimmed, and mitered to create custom roof shapes. Mass elements and polygons can be used as a cutting edge to cut holes in a roof slab. Roof slabs are also used to trim a roof for a dormer. The treatment of the edges of a roof slab, as shown in Figure 7.25 can include fascia, soffit, and frieze components, which are saved in roof edge style.

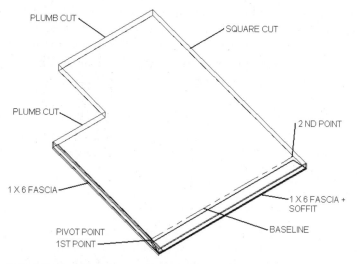

FIGURE 7.25 *Roof slab edge style applied to a roof slab*

There are fourteen roof slab styles located in the Roof Slabs and Slabs > Roof Slabs category of Design Tool Catalog—Imperial or Design Tool Catalog—Metric. The roof slab is created by specifying a series of vertices to define a closed shape. The first point specified becomes the pivot point. A pivot point marker can be turned on in the object display. The pivot point is shown in Figure 7.26. The first two points establish the baseline to hinge or pivot the slab. The overhang, edge style, and cut can be specified for each edge of the slab.

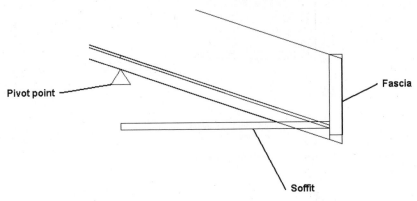

FIGURE 7.26 *Pivot point of a roof slab*

Roof slabs are created by selecting **Roof Slab** (**RoofSlabAdd** command) from the Design palette as shown in Table 7.4.

TABLE 7.4 *RoofSlabAdd command access*

Command prompt	ROOFSLABADD
Ribbon	Choose Roof Slab from the Build panel shown in Figure 7.27
Design palette	Select Roof Slab from the Design palette.

When you select the **RoofSlabAdd** command, the Properties palette displays the default values of the roof slab as shown in Figure 7.27. You are prompted to specify the start point. The start point becomes the default slab insertion point and the pivot point. The next point specified creates the baseline, which is the hinge line for the slope of the slab. The roof slab must consist of at least three vertices and a slope angle.

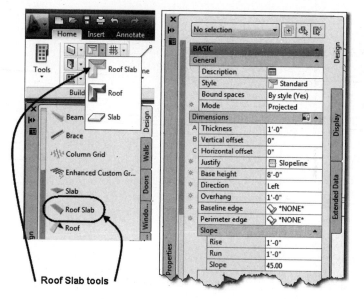

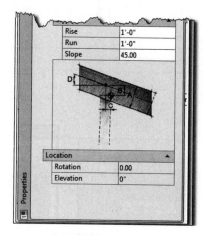

FIGURE 7.27 *Accessing the Roof Slab and Properties palette of a Roof Slab*

The Properties palette allows you to set the style, mode, thickness, base height, overhang, justify, slope, and direction of the roof slab. The options in the Design tab of the Properties palette are described below.

Design Tab—Basic
General

Description—The Description button opens a **Description** dialog box in which you can type a description of the roof slab.

Style—The Style drop-down list consists of the **Standard** style. Additional styles can be accessed from the Roof Slabs and Slabs > Roof Slabs category of the Design Tool Catalog—Imperial or Design Tool Catalog—Metric. The roof slab styles can include fascia, frieze, and soffit components.

Bound Spaces—The Bound Spaces option allows you to define the roof slab to function as a boundary of spaces.

Mode—The **Mode** options include **Direct** or **Projected**. The **Direct** mode creates the roof plane from the vertices specified. The **Projected** mode projects from the specified vertices up the distance as specified in the base height from the selected points.

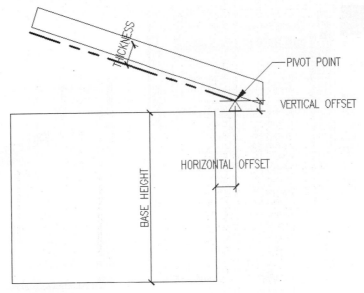

FIGURE 7.28 *Horizontal and Vertical Offset of a roof slab*

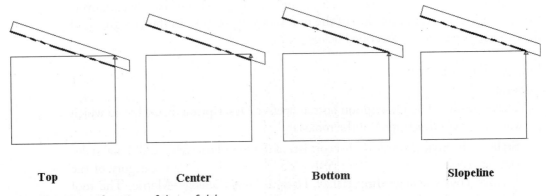

Top   Center   Bottom   Slopeline

FIGURE 7.29 *Justify options of the roof slab*

Dimensions

A-Thickness—The Thickness specifies the thickness or depth of the roof slab object.

B-Vertical offset—The Vertical offset specifies the vertical distance from the top of the wall to the roof slab baseline. The top of the wall is established by the base height property for a projected slab. A direct slab is developed with a 0 base height.

C-Horizontal offset—The Horizontal offset distance shifts the slab horizontally relative to the baseline. A positive horizontal offset shifts the slab to the right of the baseline as shown in Figure 7.28. A negative horizontal offset shifts the roof slab to the left of the baseline.

Justify—The Justify option specifies the position of the slab component defined in the Standard style. The **Top, Center, Bottom,** and **Slopeline** justifications of the Standard roof slab style are shown in Figure 7.29. The location of components is defined in the roof slab style relative to the baseline.

Base height—The Base height is the distance the slab is created from the $Z = 0$ plane. The base height is applied to the slab created in the projected mode.

Direction—The Direction toggle allows you to create a roof slab to the left or right of first two selected points when you use **Ortho close**.

Overhang—The Overhang field specifies the horizontal distance from the pivot point to the fascia line.

Baseline edge—The Baseline edge field allows you to specify roof slab edge style for the edge of the roof along the baseline edge.

Perimeter edge—The Perimeter edge field allows you to specify a roof slab edge style for the perimeter edges of the roof slab, excluding the baseline edge. **SLOPE:**

Rise—The Rise specifies the vertical displacement per horizontal run unit. You can edit the Rise field to change the angle of existing roof slabs.

Run—The Run specifies the horizontal component of the slope. This value is usually 12 [100] as base value. You can edit the **Run** value to change the angle of existing roof slabs.

Slope—The **Slope** displays the corresponding angle as a result of the rise and run values. The angle will be displayed resulting from the rise and run, or you can type a value for the angle.

The options of the command can be selected from the Properties palette, command line, or workspace prompts. The following options are displayed in the command line: Style, Mode, Height, Thickness, Slope, Overhang, Justify, Match. The Match option is available only from the command line and prompts you to select other roof slabs and match their properties. However, you can drag a roof slab to a tool palette and create a tool from the roof slab. If you right-click the new tool and choose **Apply Tool Properties to > Roof Slab**, you can convert other roof slabs to the style property of the roof slab tool.

Using the Direction Property to Create Roof Slabs

Roof slabs can be created by selecting a series of points to define the vertices of the roof slab. The **Direction** property specified in the Properties palette allows you to create a roof slab to the left or right of the first roof edge line. The roof slab is created by selecting two points to specify the pivot point and baseline, and then selecting **Ortho close** from the shortcut menu. Selecting **Ortho close** from the shortcut menu generated the roof to the left of the baseline as shown in part **A** in Figure 7.30 since Direction was set to Left. The roof is generated to the left or right of the first edge segment according to the **Direction** specified in the Properties palette.

The **Ortho close** option can also be used if additional points are specified as shown in part **B** in Figure 7.30. The slab is specified by selecting points **p1**, **p2**, **p3**, **p4**, and **p5**, and then right-clicking and choosing **Ortho close** from the shortcut to close the roof to point **p1**. The roof shape is completed as necessary to form a right angle between the last roof edge segment and the first roof edge segment.

The **Close** option of the shortcut menu can be used to close the roof shape from the last specified point to the first specified point as shown in part **C** in Figure 7.30. The **Close** option of the shortcut menu will close the last segment to the first edge segment without regard to the angle formed.

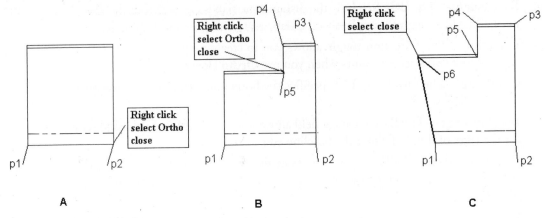

FIGURE 7.30 *Creating a slab using Close and Ortho close options*

Creating the Roof Slab from a Wall, Roof, or Polyline

A roof slab can be created from existing walls, roof, or a closed polyline. The shortcut menu of the **Roof Slab** tool of the Design tool palette includes the option **Apply Tool Properties to > Linework, Walls, and Roof (RoofSlabToolToLinework** command) as shown in Table 7.5. When you select this command, you can convert walls, roofs, or linework to a roof slab.

TABLE 7.5 *Apply Roof Slab Properties to > Linework, Walls, and Roof command access*

Design palette	Select Roof Slab from the Design palette, right-click, and choose Apply Tool Properties to > Linework, Walls, and Roof.

When you apply the Apply Tool Properties to > Linework, Walls, and Roof command to a polyline, you are prompted in the workspace to select the polyline and define the pivot point, mode, height, and justification as shown in the following workspace prompts.

```
Select roof, walls, or polylines: 1 found (Select polyline at
p1 as shown in Figure 7.31.)

Select roof, walls, or polylines: ENTER (Press ENTER to end
selection.)

Erase layout geometry? [Yes/No] <No>: ENTER (Choose No
option from the dynamic input prompt.)

Creation mode [Direct/Projected] <Projected>: ENTER (Choose
the Projected option from the dynamic input prompt.)

Specify base height <0>: 8' [2400] ENTER (Type the distance
value to project the roof.)

Specify slab justification [Top/Center/Bottom/Slopeline]
<Bottom>: ENTER (Choose the Bottom option from the dynamic
input prompts.)

Select the Pivot edge for the slab: (Move the cursor over an
edge as shown in Figure 7.31 and click when the red line is
coincident with the desired pivot edge.)
```

You can apply the Apply Tool Properties to > Linework, Walls, and Roof tool to two-dimensional roof plans to create a roof slab for each closed polyline.

TIP

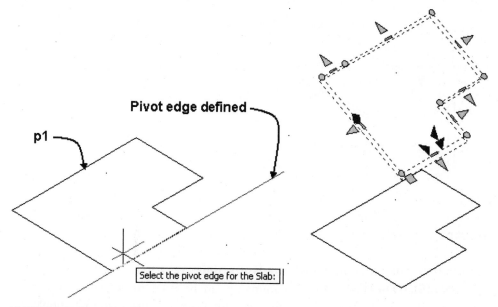

FIGURE 7.31 *Converting the closed polyline to a roof slab*

Creating the Roof Slab from a Wall

The **RoofSlabToolToLinework** command can also be used to generate the roof slab from a wall. The wall base height is used as the project base height distance, and the wall length establishes the length and width of the roof slab. The roof slab created in Figure 7.32 was created based upon the wall shown at **p1**. The roof slab is created with equal length and width. The default slope of 45° and thickness of the roof can be edited in the Properties palette. Right-click the Roof Slab tool in the Design palette; choose Apply Tool Properties to Linework, Walls, and Roof. Respond to the workspace prompts as follows.

> Select roof, walls, or polylines: 1 found *(Select the wall at* **p1** *as shown in Figure 7.32.)*
>
> Select roof, walls, or polylines: ENTER *(Press ENTER to end selection.)*
>
> Erase layout geometry? [Yes/No] <No>: ENTER *(Press ENTER to retain the wall.)*
>
> Specify slab justification [Top/Center/Bottom/Slopeline] <Bottom>: ENTER *(Press ENTER to specify a bottom justification.)*
>
> Specify wall justification for edge alignment [Left/Center/ Right/ Baseline] <Baseline>: r *(Choose* **Right** *justification from the dynamic input options.)*
>
> Specify slope Direction [Left/Right] <Left>: ENTER *(Choose* **Left** *from the dynamic input options.)*

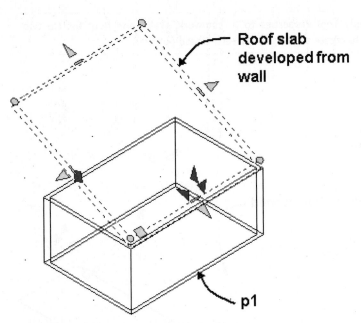

FIGURE 7.32 *Creating a roof slab based upon a wall*

Converting an Existing Roof to a Roof Slab

The **RoofSlabToolToLinework** command can be used to convert a roof to a roof slab. The roof slab will remain separate from the original roof; however, the slope and thickness of the roof slab will be equal to the roof as shown on the left in Figure 7.33. If you retain the roof when creating the roof slab, you can turn OFF the **A-Roof [A270G]** layer to hide the display of the roof. The roof slab is created on layer **A-Roof-Slab [A370G]**.

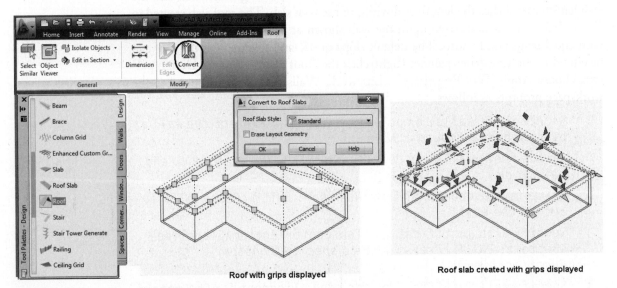

Roof with grips displayed

Roof slab created with grips displayed

FIGURE 7.33 *Creating a roof slab from a roof*

In addition, you can select a roof, right-click, and select **Convert to Roof Slabs** from the shortcut menu to convert a roof to a roof slab. Access **RoofSlabConvert** as shown in Table 7.6.

TABLE 7.6 *Converting a roof to a roof slab*

Command prompt	**ROOFSLABCONVERT**
Ribbon	Select a roof and choose Convert from Modify panel of the Roof contextual tab.
Shortcut menu	Select a roof, right-click, and select Convert to Roof Slabs.

When you select a roof, choose **Convert** from the Modify panel of the Roof tab to open the Convert to Roof Slabs dialog box as shown in Figure 7.34. Specify the roof slab style and respond to the Erase Layout Geometry checkbox. Click **OK** to dismiss the Convert to Roof Slabs dialog box.

Since the dimensions of a roof and a roof slab could be identical, it may be difficult to distinguish between the two objects during selection for edit. However, you can cycle through the selection to select the roof slab if you move the cursor over the objects to display the highlight and hold down SHIFT + SPACEBAR. Press and release the spacebar to toggle the highlight to the roof slab; then click to select. The grips of the roof slab are shown at the right in Figure 7.34. After i-dropping roof slab styles from the Content Brower onto to a tool palette, you can convert a roof to a roof slab. The roof slab style associated with a roof slab tool is applied to a roof as shown in the following workspace sequence.

```
(Right-click a roof slab tool located on a palette, choose
Apply Tool Properties to > Linework, Walls, and Roof.)
Select roof, walls, or polylines: 1 found (Select the roof
object.)
Select roof, walls, or polylines: ENTER (Press ENTER to end
the selection.)
Erase layout geometry? [Yes/No] <No>: ENTER (Press ENTER to
retain the roof object.)
```

MODIFYING ROOF SLABS

A roof slab can be modified by editing its properties in the Properties palette or editing the grips. In addition, the shortcut menu and Roof Slab contextual tab include commands to trim, extend, or miter roof slabs. If you select an existing roof slab, additional fields are added to the roof slab Properties palette as shown in Figure 7.34. The additional fields in the Design tab of the Properties palette shown in Figure7.34 include the **Edges** button, which opens the **Roof Slab Edges** dialog box. The Roof Slab Edges dialog box allows you to assign roof edge styles to the edges of the roof slab. The following **additional** fields in the Properties palette for existing roof slabs can be modified to change a roof slab.

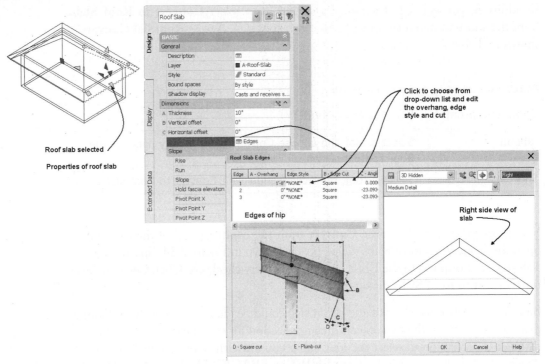

FIGURE 7.34 *Using the Properties palette to modify a roof slab*

Design Tab—Basic

General

Layer—The Layer field displays the name of the layer of the insertion point of the roof slab. Roof slabs are placed on the **A-Roof-Slab** [A370G] layer when the AIA (256 Color) [BS1192 Cisfb (256 Color)] Layer Key Style is used in the drawing.

Dimensions

Edges—The Edges button opens the **Roof Slab Edges** dialog box. The Roof Slab Edges dialog box shown in Figure 7.34 describes the hip plane shown at the right in the object viewer. The options of the dialog box allow you to adjust the overhang, edge style, edge cut, and angle. The Edge Style drop-down list shows the roof edge styles of the drawing. Roof edge styles can be imported in the Style Manager from the *Roof Slab & Roof Slab Edge Styles (Imperial).dwg* or *Roof Slab & Roof Slab Edge Styles (Metric).dwg* file of the Imperial and Metric folders in the **Content** of Places *{Vista—C:\ProgramData\Autodesk\ACA 2012\ enu\Styles} folder.*

SLOPE:

Hold fascia elevation—The Hold fascia elevation field includes the **No** option, which allows the fascia elevation to change as the slope of the roof is changed. The **By adjusting overhang** option increases or decreases the overhang to retain the same fascia height. The **By adjusting baseline height** option increases or decreases the baseline height to retain the overhang elevation.

Pivot Point X—The Pivot Point X value is the world coordinate X value of the pivot point.

Pivot Point Y—The Pivot Point Y value is the world coordinate Y value of the pivot point.

Pivot Point Z—The Pivot Point Z value is the world coordinate Z value of the pivot point.

Location

Elevation—The Elevation field displays the elevation of the insertion point of the roof slab.

Additional information—The Additional information button opens the **Location** dialog box, which lists the insertion point, normal, and rotation values of the roof slab.

Using Grips to Edit a Roof Slab

The grips of a roof slab, shown in Figure 7.35, allow you to change the properties and vertices of the roof slab. The *circle* grip markers allow you to change the *vertices* of the roof slab. The pivot point grip displays as a cyan-colored square when viewed in plan and a rhombus when viewed in pictorial. The *wedge*-shaped grip markers on each edge allow you to change the edge overhang. The *negative* symbol grip allows you to edit the location of the edge. If you hold down CTRL, you can toggle through the options to add an edge maintaining slope, add an edge changing slope, move the edge maintaining the slope, move the edge changing the slope, convert to arch, or offset all edges. The remaining wedge-shaped grips located on the face allow you to edit the thickness, horizontal offset, and vertical offset. The angle of the plane can be changed by selecting the **Angle** grip (green colored rhombus-shaped grip). The Angle grip is not displayed in plan view. When you select a grip, the dynamic dimensions are displayed; you can then press TAB to change the dimension that is highlighted. When a dynamic dimension is highlighted, you can overtype a new dimension in the workspace to change the size of the feature. The edge grip at the top edge of the roof has been selected for the roof slab shown at the left in Figure 7.36. The edge overhang grip at the right in Figure 7.36 is stretched to increase the overhang.

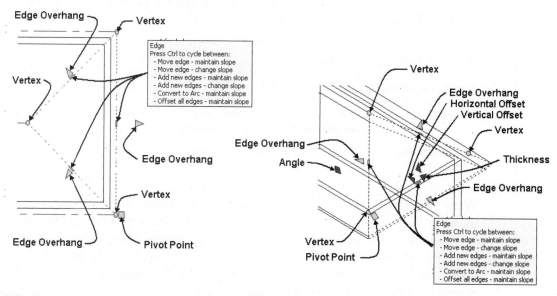

FIGURE 7.35 *Using grips to edit the roof slab properties*

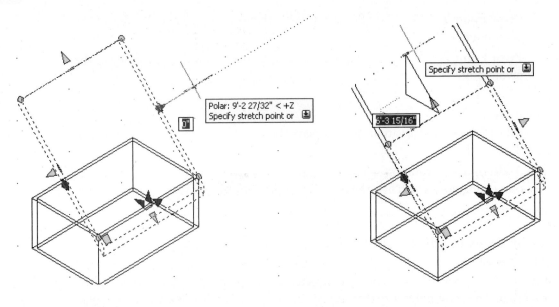

Editing Edge Overhang
                                                Editing Edge Location

FIGURE 7.36 *Editing roof slab edges*

TOOLS FOR EDITING ROOF SLABS

The commands to edit roof slabs are on the shortcut menu of a selected roof slab. Included on the shortcut menu are the commands to trim, extend, miter, cut, add vertex, remove vertex, roof dormer, and modify using Interference operations.

Trimming a Roof Slab

The **RoofSlabTrim** command can be used to trim a roof slab that intersects with a polyline, wall, or slab. The cutting object is projected to intersect in the current user coordinate system. Therefore, if the cutting object does not intersect the roof slab, it is projected up, based upon the current UCS to identify the cutting plane. Access the **RoofSlabTrim** command as shown in Table 7.7.

TABLE 7.7 *RoofSlabTrim command access*

Command prompt	ROOFSLABTRIM
Ribbon	Select the roof slab and choose Trim from the Modify panel of the Roof Slab contextual tab.
Shortcut menu	Select the roof slab, right-click, and choose Trim.

When you select a roof slab, right-click, and choose **Trim** from the shortcut menu shown in Figure 7.37, you are prompted to select the trimming object and to specify the side to remove when trimmed. In the following command line sequence, the roof slab is trimmed by the wall. The wall height is below the roof slab; however, because the slab and wall are viewed from the top view, the wall is projected based upon the current UCS to identify the cutting edge.

(Select the roof slab at **p1** *as shown at the left in Figure 7.37. Choose* **Trim** *from the Modify panel of the Roof Slab contextual tab. Respond to the following workspace prompts.)*

Select trimming object (a slab, wall, or polyline): *(Select the wall at* **p2** *as shown at the left in Figure 7.37.)*

Specify side to be trimmed: *(Select a point near* **p3** *as shown at the left in Figure 7.37. The Roof slab is trimmed as shown in Figure 7.37.)*

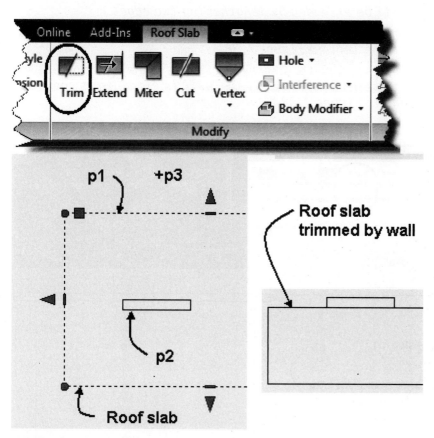

FIGURE 7.37 *Trim Roof Slab command used to trim the roof slab*

Extending a Roof Slab

The **RoofSlabExtend** command can be used to extend a roof slab to wall or slab. The cutting object is projected based upon the current user coordinate system. Therefore, if the object does not intersect the roof slab, it is projected up, based upon the current UCS to identify the extension. Access the **RoofSlabExtend** command as shown in Table 7.8.

TABLE 7.8 *RoofSlabExtend command access*

Command prompt	ROOFSLABEXTEND
Ribbon	Select a roof slab and choose Extend from the Modify panel of the Roof Slab contextual tab.
Shortcut menu	Select a roof slab, right-click, and choose Extend.

When you select a slab, right-click, and choose **Extend** from the shortcut menu, you are prompted to select the object to extend and the edges to lengthen. In the following command sequence, the roof slab was extended to the wall as shown in Figure 7.38.

(Select a roof slab at p1 *as shown in Figure 7.38 and choose* Extend *from the Modify panel of the Roof Slab contextual tab. Respond to the following command line prompts.)*

Select an object to extend to (a slab or wall): *(Select the wall at* p2 *as shown at the left in Figure 7.38.)*

(Slab will be extended by lengthening two edges.)

Select first edge to lengthen: *(Select the roof slab at* p3 *as shown at the left in Figure 7.38.)*

Select second edge to lengthen: *(Select the slab at* p4 *as shown at the left in Figure 7.38. Wall is extended as shown at the right in Figure 7.38.)*

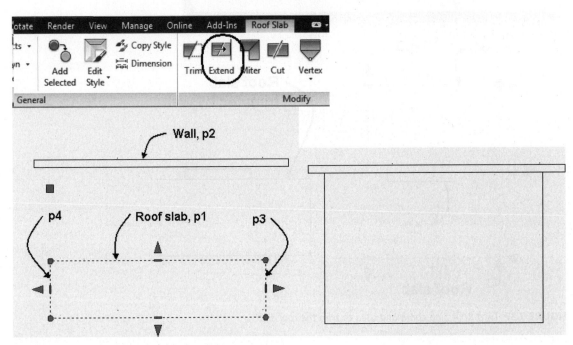

FIGURE 7.38 *Extend Roof Slab used to extend the roof slab*

Mitering Roof Slabs

The **RoofSlabMiter** command can be used to extend or trim roof slabs. The RoofSlabMiter command edits only one edge of each roof slab. Access the **Miter** command as shown in Table 7.9.

TABLE 7.9 *RoofSlabMiter command access*

Command prompt	**ROOFSLABMITER**
Ribbon	Select the roof slab and choose Miter from the Modify panel of the Roof Slab tab.
Shortcut menu	Select the roof slab, right-click, and choose Miter.

When you select the **RoofSlabMiter** command, you are prompted to miter either the Intersection or Edges of the roof slabs. If you miter by intersection, the slabs will be trimmed according to the intersection of the roof slabs. The miter by edges requires you to select the edges of the roof slabs. The two roof slabs shown in Figure 7.39 were mitered by intersection as described in the following workspace sequence:

(Select a roof slab and choose Miter *from the Modify panel of the Roof Slab contextual tab.)*

Miter by [Intersection/Edges] <Intersection>: ENTER *(Choose the* Intersection *option of the dynamic prompt.)*

Select first roof slab at the side to keep: *(Select the roof slab at* p1 *as shown at the left in Figure 7.39.)*

Select second roof slab at the side to keep: *(Select the roof slab at* p2 *as shown at the left in Figure 7.39.)*

(Roof slabs are mitered as shown at the right in Figure 7.39.)

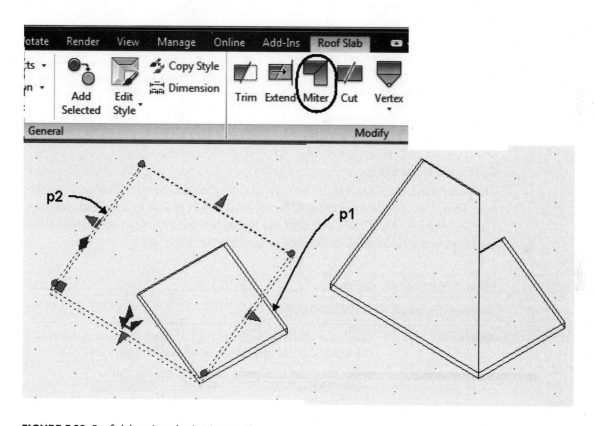

FIGURE 7.39 *Roof slabs mitered using Intersection*

The Edges miter method can be used to extend two roof slabs to create the miter. Although the roof slabs shown in Figure 7.40 do not intersect, edges **p1** and **p2**, if extended, would create a gable roof. The two roof slabs were mitered using the Edges option, as shown in the following command sequence:

(Select a roof slab and choose Miter *from the Modify panel of the Roof Slab contextual tab.)*

Miter by [Intersection/Edges] <Intersection>: *(Choose* Edge *from the dynamic drop-down options.)*

Select edge on first slab: *(Select the edge at* **p1***, as shown at the left in Figure 7.40.)*

Select edge on second slab: *(Select the edge at* **p2***, as shown at the left in Figure 7.40.)*

(Edges extended to form miter as shown at the right in Figure 7.40.)

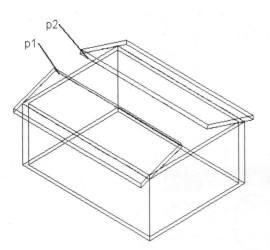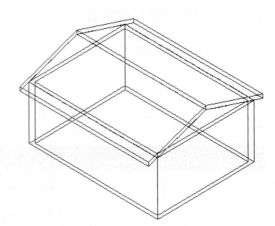

FIGURE 7.40 *Roof slabs mitered by Edges method*

Cutting a Roof Slab

The **RoofSlabCut** command can be used to cut a roof slab by any 3D object or polyline. Therefore, this command could be used to cut a roof slab by a wall or chimney. The intersecting object must intersect the perimeter line or consist of a closed polyline. Access the **RoofSlabCut** command as shown in Table 7.10.

TABLE 7.10 *RoofSlabCut command access*

Command prompt	ROOFSLABCUT
Ribbon	Select the roof slab and choose Cut from the Modify panel of the Roof Slab contextual tab as shown in Figure 7.41.
Shortcut menu	Select the roof slab, right-click, and choose Cut.

When you select a slab, right-click, and choose **Cut** from the shortcut menu, you are prompted to select the objects that intersect the roof slab. The **RoofSlabCut** command was used to cut the roof slab by the intersecting polyline shown in Figure 7.41 and the following workspace sequence:

(Select the roof slab at **p1** *as shown at the left in Figure 7.41 and choose* **Cut** *from the Modify panel of the Roof Slab contextual tab.)*

Select cutting objects (a polyline or connected solid objects): 1 found *(Select the polyline at* **p2** *as shown in Figure 7.41.)*

Select cutting objects (a polyline or connected solid objects) ENTER *(Press ENTER to end the selection.)*

Erase layout geometry [Yes/No] <No>: N *(Choose **No** option from the dynamic prompt options.)*

(Verify that the roof slab is cut by selecting the slab as shown at the right in Figure 7.41.)

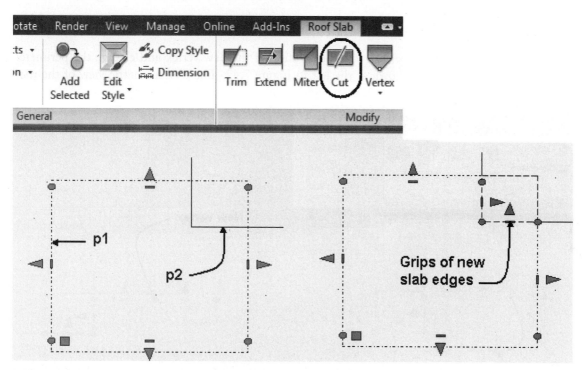

FIGURE 7.41 *Roof slab cut using the RoofSlabCut command*

Adding a Roof Slab Vertex

The **RoofSlabAddVertex** command can be used to add a vertex along the edge of a roof slab. The new vertex will have a grip that can be stretched to a location for editing the roof. The RoofSlabAddVertex command is used to modify the perimeter of the slab. Each edge of the perimeter consists of grips at the midpoint and ends of the roof edge. Access the **RoofSlabAddVertex** command as shown in Table 7.11.

TABLE 7.11 *RoofSlabAddVertex command access*

Command prompt	ROOFSLABADDVERTEX
Ribbon	Select the roof slab and choose Add from the Vertex flyout menu of the Modify panel in the Roof Slab tab.
Shortcut menu	Select the roof slab, right-click, and choose Add Vertex.

When you select the RoofSlabAddVertex command, you are prompted to select the roof slab and specify the location for the vertex. The RoofSlabAddVertex command was used to add a vertex to the roof slab at the intersection, as shown in Figure 7.42 and the following workspace sequence:

(Select a roof slab and choose Add *from the Vertex flyout menu of Modify panel in the Roof Slab tab.)*

Specify point for new vertex: _mid of *(Hold* SHIFT *down, right-click, and choose* Mid *from the shortcut menu; then select the edge near* p1 *as shown at the left in Figure 7.42.)*

(Select the slab to verify the new grip point as shown at the right in Figure 7.42.)

Grips can be used to edit the location of the new vertex and change the perimeter of the roof as shown at the left in Figure 7.42. The SW Isometric view of the roof slab is shown at the right in Figure 7.42.

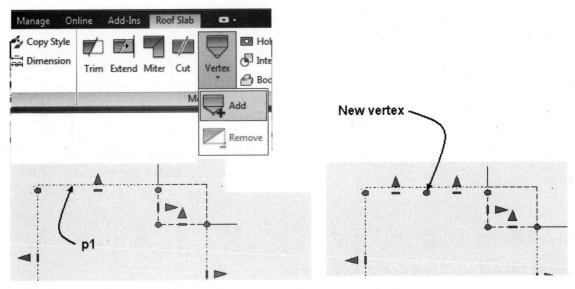

FIGURE 7.42 *Vertex added to roof slab*

Removing the Vertex of the Roof Slab

The **RoofSlabRemoveVertex** command can be used to remove a vertex from the perimeter of the roof. The RoofSlabRemoveVertex command is used to modify the perimeter of the slab. Access the **RoofSlabRemoveVertex** command as shown in Table 7.12.

TABLE 7.12 *RoofSlabRemoveVertex command access*

Command prompt	**ROOFSLABREMOVEVERTEX**
Ribbon	Select the roof slab and choose Remove from the Vertex flyout menu of the Modify panel in the Roof Slab tab.
Shortcut menu	Select the roof slab, right-click, and choose Remove Vertex.

When you select the roof slab and choose Remove Vertex, you are prompted to select a roof slab vertex as shown in the following workspace sequence.

(Select the roof slab and choose Remove *from the* Vertex *flyout menu of the Modify panel in the Roof Slab tab.)*

Select a vertex: *(Select the vertex at* **p1** *as shown at the left in Figure 7.43.)*

(Roof slab vertex removed as shown at the right in Figure 7.43.)

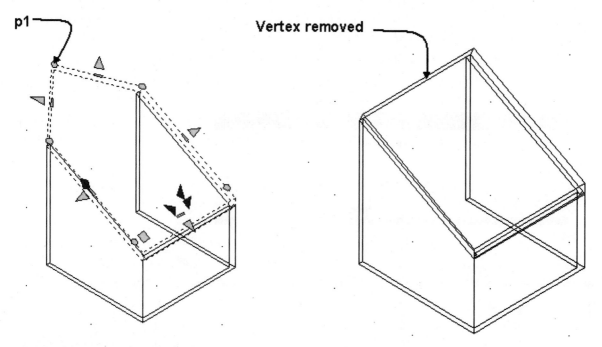

FIGURE 7.43 *Remove Vertex command used to remove midpoint vertex*

Creating Holes in a Roof

The **RoofSlabAddHole** command can be used to add a hole in a roof slab. The hole is projected from closed polylines or a 3D object. The **Hole** command on the shortcut menu can be used to create a hole or remove a hole from the roof slab. The projection of the polyline is based upon the current UCS. When 3D objects that consist of voids are used to cut the hole, you are prompted to select the inside or outside to cut the hole. Access the **RoofSlabAddHole** command as shown in Table 7.13.

TABLE 7.13 *RoofSlabAddHole command access*

Command prompt	ROOFSLABADDHOLE
Ribbon	Select the roof slab and choose Add from the Hole flyout of the Modify panel in the Roof Slab contextual tab.
Shortcut menu	Select the roof slab, right-click, and choose Hole > Add.

When you select a slab and choose **Add** from the Hole flyout menu, you are prompted to select a closed polyline or other connected solid objects. The objects will be projected to the roof plane to cut the hole. The **RoofSlabAddHole** command was used to cut a hole in the roof slab, as shown in Figure 7.44 and the following workspace sequence:

(Select a roof slab and choose **Add** *from the Hole flyout menu of the Modify panel in the Roof Slab contextual tab.)*

Select a closed polyline or connected solid objects to define a hole: 1 found *(Select the polyline at* **p1** *as shown in Figure 7.44.)*

Select a closed polyline or connected solid objects to define a hole: ENTER *(Press* ENTER *to end selection.)*

Erase layout geometry? [Yes/No] <No>: ENTER *(Choose* **No** *from the options in the dynamic prompt.)*

(Polyline is projected to the roof plane to cut the hole as shown at the right in Figure 7.44.)

(Roof slab is cut by the projection of the closed polyline as shown in Figure 7.44.)

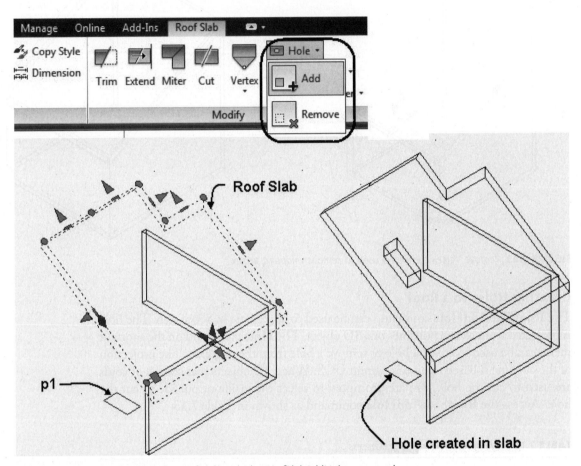

FIGURE 7.44 *Hole created in the roof slab with the RoofSlabAddHole command*

Removing Holes from a Roof

The **RoofSlabRemoveHole** command is used to remove holes that exist in slabs. Access the **RoofSlabRemoveHole** command as shown in Table 7.14.

TABLE 7.14 *Roof SlabRemoveHole command access*

Command prompt	**ROOFSLABREMOVEHOLE**
Ribbon	Select the roof slab and choose Remove from the Hole flyout menu of the Modify panel.
Shortcut menu	Select the roof slab, right-click, and choose Hole > Remove.

When you select a slab and choose **Remove** from the Hole flyout of the Modify panel, you are prompted to select a hole to remove. The hole is removed when you select it, as shown in the following workspace sequence and Figure 7.45.

> *(Select a roof slab and choose* Remove *from Hole flyout menu of the Modify panel.)*
>
> Select an edge of hole to remove: *(Select the hole at* p1 *as shown at the left in Figure 7.45.)*
>
> *(Hole is removed as shown at the right in Figure 7.45.)*

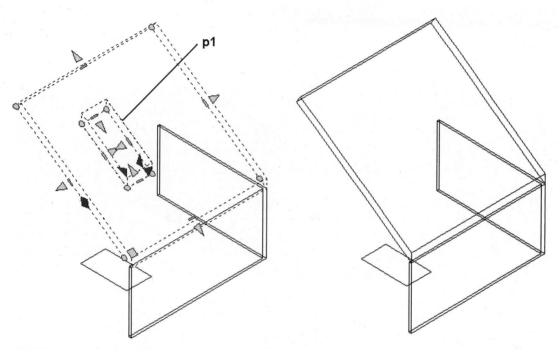

FIGURE 7.45 *Hole removed from slab using the RoofSlabRemoveHole command*

Using Interference to Combine Objects to Roof Slabs

The **RoofSlabInterferenceAdd** command allows you to **Add** or **Subtract** other AEC Objects to a roof slab. The **Add** option can be used to add a mass element to the roof slab as shown in Figure 7.46. The subtract option removes the mass element from the roof slab. These elements can be used to add or subtract their shape to or from a roof slab. The subtract option could subtract from the roof slab a mass element created to represent a chimney. If you move the mass element after an addition or subtraction, the roof slab will be modified based upon the new position of the objects. Access the **RoofSlabInterferenceAdd** command as shown in Table 7.15.

TABLE 7.15 *Interference Condition > Add command access*

Command prompt	ROOFSLABINTERFERENCEADD
Ribbon	Select the roof slab and choose Add from the Interference flyout of the Modify panel.
Shortcut menu	Select the roof slab, right-click, and choose Interference Condition > Add.

When you select a roof slab and choose **Add** from the **Interference** flyout menu, you are prompted to select an object. The AEC object can be a mass element that is added to the roof slab. The mass element in Figure 7.46 was added to the roof slab as shown in the following workspace sequence.

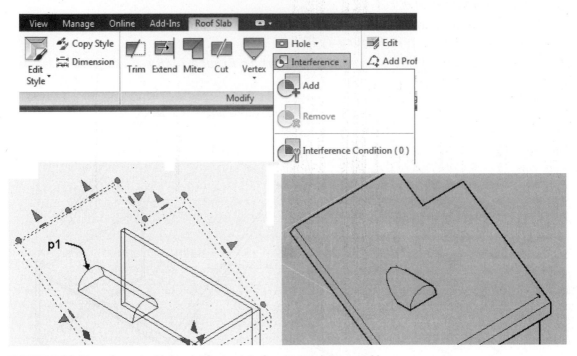

FIGURE 7.46 *Mass element added to roof using Interference Condition > Add*

(Select a roof slab and choose Add *from the Interference flyout menu of the Modify panel.)*

Select objects to add: 1 found *(Select the mass element at* p1 *as shown at the left in Figure 7.46.)*

Select objects to add: *(Press* ENTER *to end the selection.)*

Enter shrinkwrap effect [Additive/Subtractive]: A *(Choose* Additive *from the dynamic prompt.)*

1 object(s) attached.

(Select Hidden *from the flyout menu of the Visual Styles panel of the View tab to view the union as shown at the right in Figure 7.46.)*

The **RoofSlabInterferenceRemove** command can be used to remove an object from the interference condition with the roof slab. The remove returns the mass element as an independent object. Access the **RoofSlabInterferenceRemove** command as shown in Table 7.16.

TABLE 7.16 *Interference Condition > Remove command access*

Command prompt	**ROOFSLABINTERFERENCEREMOVE**
Ribbon	Select the roof slab and choose Remove from the Interference flyout menu of the Modify panel.
Shortcut menu	Select the roof slab, right-click, and choose Interference Condition > Remove.

Adding Body Modifiers to Roofslabs

The **RoofSlabBody** command (see Table 7.17) can be used to add a body modifier to a roof slab. When the body modifier is added to a roof slab, the mass element geometry can modify the slab by adding, subtracting, or replacing the roof slab. The **Replace** option allows you to create a roof slab from the geometry of the mass element.

TABLE 7.17 *Body Modifier commands accessed from the shortcut menu*

Command prompt	ROOFSLABBODY
Ribbon	Select the roof slab and choose Add from the Body Modifier flyout menu of the Modify panel.
Shortcut menu	Select the roof slab, right-click, and choose Body Modifiers > Add.
Command prompt	**ROOFSLABBODYEDIT**
Ribbon	Select the roof slab and choose Edit In Place from the Body Modifier flyout menu of the Modify panel.
Shortcut menu	Select the roof slab, right-click, and choose Body Modifier > Edit In Place.
Command prompt	**ROOFSLABBODYRESTORE**
Ribbon	Select the roof slab and choose Restore from the Body Modifier flyout menu of the Modify panel.
Shortcut menu	Select the roof slab, right-click, and choose Body Modifier > Restore.

The following workspace sequence converted the curved mass element to a roof slab. The curved mass element shown in Figure 7.47 was converted from a closed polyline that was drawn while viewing the slab from the front view.

(*Select the roof slab at* **p1** *as shown in Figure 7.47 and choose Add from the* **Body Modifier** *flyout menu.*)

Select objects to apply as body modifiers: 1 found (*Select the mass element at* **p2** *as shown in Figure 7.47.*)

Select objects to apply as body modifiers: (*Press* ENTER *to end selection.*)

(*Edit the* **Add Body Modifier** *dialog box. Choose* **Replace** *from the Operation drop down and* **Erase Selected Objects**. *Click* OK *to dismiss the Add Body Modifier dialog box.*)

(*Roof slab is created in the shape of the mass element as shown at the right in Figure 7.47.*)

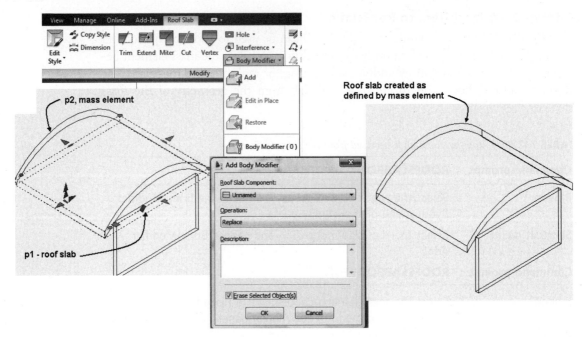

FIGURE 7.47 *Mass element subtracted from a roof slab*

If a body modifier has been added to a roof slab, the shortcut menu of the roof slab allows you to add or remove the body modifier and edit in place the mass element, as shown in Table 7.17.

Creating a Roof Dormer

The **Roof Dormer** (**RoofSlabDormer**) command allows you to create a dormer from walls that penetrate the roof. The walls may or may not be trimmed by the main roof. The dormer walls must include four walls, which penetrate the roof slabs. Access the **Roof Dormer** command as shown in Table 7.18:

TABLE 7.18 *Roof Dormer command access*

Command prompt	**ROOFSLABDORMER**
Ribbon	Select the roof slab and choose Add from the Dormer panel.
Shortcut menu	Select the roof slab, right-click, and choose Roof Dormer.

When you select the **Roof Dormer** command, you are prompted to select the roof slab and the other objects that form the dormer. The roof slab is sliced by the other elements to form the dormer, as shown in the following workspace sequence:

(*Select the roof slab at* **p1** *as shown at the left in Figure 7.48; choose* Add *from the* **Dormer** *panel.*)

Select objects that form the dormer: 1 found (*Select wall at* **p2** *as shown at the left in Figure 7.48.*)

Select objects that form the dormer: 1 found, 2 total (*Select wall at* **p3** *as shown at the left in Figure 7.48.*)

Select objects that form the dormer: 1 found, 3 total *(Select wall at* **p4** *as shown at the left in Figure 7.48.)*

Select objects that form the dormer: 1 found, 4 total *(Select wall at* **p5** *as shown at the left in Figure 7.48.)*

Select objects that form the dormer: 1 found, 5 total *(Select dormer roof slab at* **p6** *as shown at the left in Figure 7.48.)*

Select objects that form the dormer: 1 found, 6 total *(Select dormer roof slab at* **p7** *as shown at the left in Figure 7.48.)*

Select objects that form the dormer: enter *(Press ENTER to end selection.)*

Slice wall with roof slab [Yes/No] <Yes>: ENTER *(Choose the* **Yes** *option from the dynamic input options to slice the dormer wall to remove the lower portion of dormer wall.)*

(Dormer is created as shown at the right in Figure 7.48. To complete the operation, use the **Erase** *command to erase the wall at* **p8**.*)*

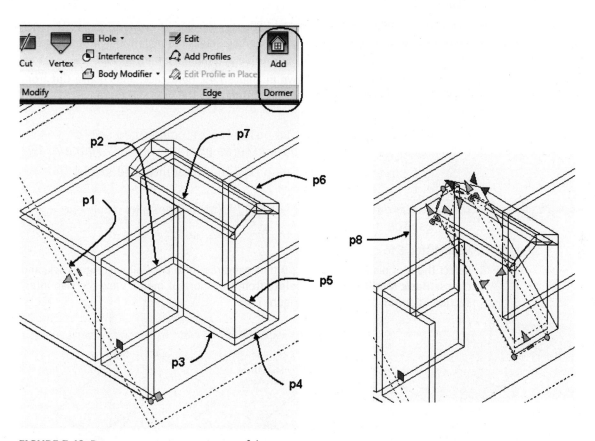

FIGURE 7.48 *Dormer geometry to create a roof dormer*

DETERMINING ROOF INTERSECTIONS

The **Roof Dormer** command determines the intersection of the dormer roof with the main roof. However, other roofs can intersect with the main roof without forming a dormer. When a roof slab intersects with another roof slab with different slope and plate height, the intersection is not automatically determined. To determine the valley intersections of multiple roof slabs:

1. Create a mass group and attach the intersecting roof slabs to the group to determine the intersection.
2. Draw a polyline that crosses an edge of the roof slab and traces the intersection; use this polyline as a cutting object with the RoofSlabCut command to cut each roof slab.
3. Remove the unnecessary extensions of each roof slab.

Steps to Cut and Edit Roof Slabs

1. Convert each roof to roof slabs.
2. Select **Mass Group** from the **Massing** palette and attach each roof slab **p1**, **p2**, and **p3** shown in Figure 7.49 to the group. Select any coordinate for the location of the mass group marker.
3. View the roofs from Top, select **Polyline** from the Line flyout of the Draw panel in the Roof Slab tab, and trace the intersection of the roof using the **Intersection** object snap to create a polyline. The isometric view of the polyline is shown at **p5** in Figure 7.49.
4. Select **SW Isometric** from the View Cube.
5. Select the group to display the group marker, select the negative sign of the group marker, and then select the roof slabs to detach them from the group.
6. Select the roof slab at **p6** as shown in Figure 7.49, right-click, and choose **Cut** from the Modify panel of the Roof Slab tab. Respond to the workspace prompt as shown below:

   ```
   Select cutting objects (a polyline or connected solid
   objects): (Select the polyline at p5 as shown in Figure 7.49.)

   Select cutting objects (a polyline or connected solid
   objects): ENTER

   Erase layout geometry [Yes/No] <No> (Choose No from the
   workspace options.)
   ```

7. Repeat step 6 to cut the plane **p7**.
8. Select the roof planes at **p8** and **p9** that penetrate into the building, right-click, and choose **Basic Modify Tools** > **Delete** from the shortcut menu. View the slab intersection as shown at the right in Figure 7.49.

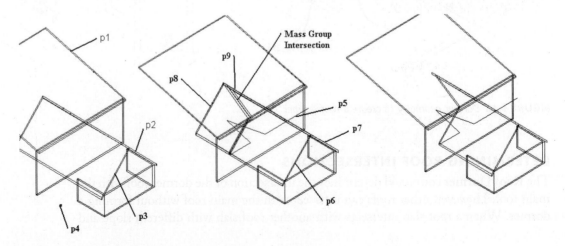

FIGURE 7.49 *Cutting a roof slab at the intersection of roof planes*

CREATING ROOF SLAB STYLES

Roof slab styles include preset values for thickness and predefined roof slab edge styles. Additional details such as fascia, soffit, and frieze components are added by assigning roof slab edge styles to the edge of the roof slab. Materials and components can be defined to the roof slab to represent the roofing and framing materials. The Standard roof slab style is the default when a roof slab is inserted from the Design palette. The Standard roof slab style does not include soffit, fascia, and frieze components. Additional roof slab styles included in the Design Tool Catalog—Imperial > Roof Slabs or Design Tool Catalog—Metric of the **Content Browser** can be i-dropped from the tool catalog into the drawing or onto a tool palette. If you place roof slab styles onto a tool palette, you can modify the style in the Style Manager.

Access the **Roof Slab Styles** command as shown in Table 7.19.

TABLE 7.19 *Roof Slab Styles command access*

Command Prompt	ROOFSLABSTYLEEDIT
Ribbon	Select Roof Slab and choose Roof Slab Styles from the Edit Style flyout of the General panel.
Shortcut menu	Select Roof Slab tool on the Design tool palette, right-click, and select Roof Slab Styles.

When you access this command by right-clicking the **Roof Slab** tool of the Design tool palette and choosing **Roof Slab Styles** from the shortcut menu, the Style Manager opens to the Roof Slab Styles folder of the Style Manager. You could create a new roof slab style or modify existing styles included in the drawing in the Style Manager. The procedure to import a roof slab is presented in the following steps.

Steps to Import a Roof Slab Style from the Content Browser to a Tool Palette

1. Right-click over the titlebar of the tool palettes and choose New Palette. Overtype the name Roof Slabs.

2. Choose the Content Browser from the Tools flyout of the Build panel in the Home tab. Open the Design Tool Catalog—Imperial or Design Tool Catalog—Metric of the **Content Browser**. Navigate to Roof Slabs and Slabs > Roof Slabs category. Drag the i-drop for 04 − 1 × 4 Fascia + Soffit [100 − 25 × 100 Fascia + Soffit] onto the new palette as shown in Figure 7.50.

3. Right-click the 04 − 1 × 4 Fascia + Soffit [100 − 25 × 100 Fascia + Soffit] roof slab tool located on a palette and choose Import 04 − 1 × 4 Fascia + Soffit [100 − 25 × 100 Fascia + Soffit].

4. Right-click the 04 − 1 × 4 Fascia + Soffit [100 − 25 × 100 Fascia + Soffit] roof slab tool of the palette and choose Roof Slab Styles.

The Style Manager opens, displaying the roof slab styles of the drawing.

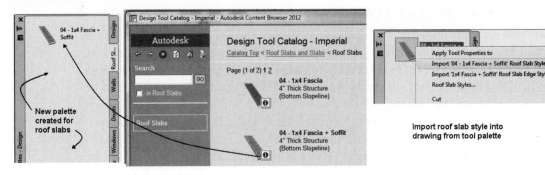

FIGURE 7.50 *Importing a roof slab style from the Content Browser*

The content of the roof slab style consists of the following tabs: General, Components, Materials, Classifications, Display Properties, and Version History. The content of the Components tab is unique and is presented below for the 04 − 1 × 4 Fascia + Soffit roof slab style. The components are developed from the Baseline of the roof slab as shown in Figure 7.51. The Thickness and Thickness Offset values are specified in the respective columns for each component. When you click in the right margin of the Thickness and Thickness Offset fields, they expand as shown in Figure. 7.51. The values can be specified fixed, independent of the roof slab thickness, if you choose 0 in the field as shown at the right in Figure 7.51. Dependent values defined by a formula using the Base Thickness of the slab and arithmetic operations can be specified as shown at the left in Figure 7.51. The Base Thickness value is specified in the Properties palette for the roof slab. The Thickness and Thickness Offset for the 04 − 1 × 4 Fascia + Soffit as shown in Figure 7.51 are specified as fixed values independent of the base thickness.

Components Tab

Index—The Index field displays the index number of the component.

Name—Click in the Name field to type the name of the component.

Thickness—Click in the right margin of the Thickness field. Edit the drop-down fields to specify the component thickness.

Thickness Offset—Click in the right margin of the Thickness Offset field. Edit the drop-down list to specify the distance from the baseline to begin the component.

Add Component—Select the Add Component button to create a new roof slab component.

Remove Component—Select a component in the list of components and choose the Remove Component button to delete the component.

Move Component Up In List—Select a component in the list and choose the Move Component Up In List button to move the component up in the list of components.

Move Component Down In List—Select a component in the list and choose the Move Component Down In List button to move the component down in the list.

Offset Increment—Edit the Offset Increment value and choose the Up or Down arrows at the left to adjust the offset of the component.

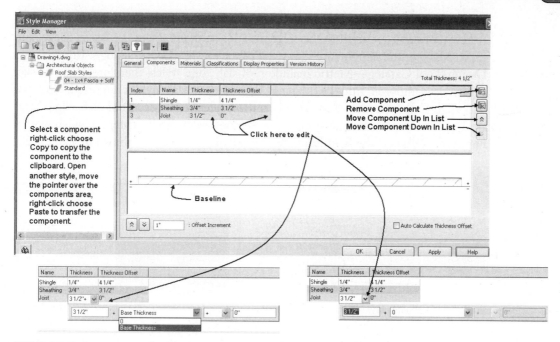

FIGURE 7.51 *Components tab of a roof slab style in the Style Manager*

You can import components from other roof slab styles in the Style Manager. If you choose a component in the list of components, right-click, and choose Copy, the component definition is copied to the clipboard. Open another roof slab style in the Style Manager, choose the Components tab, move the cursor over the tab, right-click, and choose Paste. The component is added to the style.

Materials for each component are assigned in the Materials tab in a similar manner as the wall style. The content of the **Display Properties** tab allows you to turn ON or OFF the visibility of the display components. The outline of the roof slab is displayed with hidden lines since the **Above Cut Plane Outline** display component linetype is set to HIDDEN2 as shown in Figure 7.52.

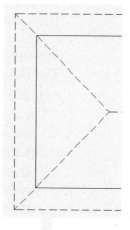

Roof outline shown with hidden lines

FIGURE 7.52 *Display Properties tab of a roof slab style*

Creating a New Roof Slab Edge Style

A new roof slab edge style can be created to define the shape of the fascia, frieze, and soffit. Many of the sample roof slab styles of the Design Tool Catalog—Imperial and Design Tool Catalog—Metric include roof slab edge styles; therefore, the roof slab edge style is imported into the drawing when the roof slab style is imported. The roof slab edge style is created and edited in the **Style Manager**. Roof slab edge styles are included in the Architectural Objects\Roof Edge Styles category as shown in Figure 7.53. Access the **RoofSlabEdgeStyleEdit** command as shown in Table 7.20.

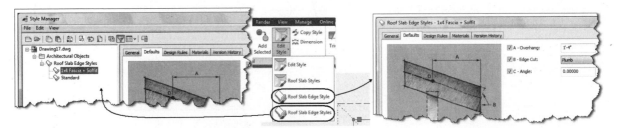

FIGURE 7.53 *Defaults tab of a roof slab edge style in the Style Manager*

TABLE 7.20 *RoofSlabEdgeStyleEdit command access*

Command prompt	**ROOFSLABEDGESTYLEEDIT**
Ribbon	Select a roof slab that includes a roof slab edge style and choose Roof Slab Edge Style from the Edge Style flyout of the General panel to open the Roof Slab Edge Styles dialog box shown in Figure 7.53.
Shortcut menu	Select a roof slab that includes a roof slab edge style, right-click, and choose Edit Roof Slab Edge Style.

Roof slab edge styles are listed in the right pane of the Style Manager. The styles can be copied for export or edited in the Style Manager. The left pane of the Style Manager shown in Figure 7.53 lists the Roof Slab Edge Styles included in the current drawing. The content of the roof slab edge style consists of five tabs: **General**, **Defaults**, **Design Rules**, **Materials**, and **Version History**. The Defaults and Design Rules tabs are unique to this style and are described below:

Defaults Tab

The **Defaults** tab shown in Figure 7.53 allows you to define the overhang distance and the angle of cut for the edge. A description of the options in the tab follows.

A-Overhang—The Overhang specifies the horizontal distance to extend the roof edge from the insertion point.

B-Edge Cut—The Edge Cut specifies the roof edge vertical plane as either plumb or square. A square orientation places the fascia plane perpendicular to the roof-decking plane. The plumb option orients the fascia plane perpendicular to the floor or a horizontal plane.

C-Angle—The Angle field specifies the angle of the fascia relative to the plane specified by the orientation. Negative angles are measured clockwise from the specified orientation plane, whereas positive angles are measured counterclockwise from the orientation plane.

Design Rules

The **Design Rules** tab shown in Figure 7.54 consists of fields for defining the profiles and location of fascia and soffit components. The shapes of the fascia and soffit components are developed from profiles. The profiles of the current drawing are listed in the drop-down list for the components. Custom profiles can be created from closed polylines in the Style Manager. A description of the options of the **Design Rules** tab follows.

Fascia

Fascia—The Fascia checkbox turns ON the display of a fascia. If it is toggled ON, the **Profile** drop-down list is active, and you can select a profile for the fascia.

A-Auto-Adjust to Edge Height—The Auto-Adjust to Edge Height checkbox adjusts the fascia automatically according to the roof slab height changes. If this is toggled OFF, the fascia dimensions are fixed according to the size defined in the profile.

Soffit

Soffit—The Soffit checkbox turns ON the display of the soffit. If it is toggled ON, you can select a profile for the soffit from the drop-down list.

B-Auto-Adjust to Overhang Depth—The Auto-Adjust to Overhang Depth checkbox allows the soffit to automatically scale according to changes in the roof slab. If this is toggled OFF, the soffit is fixed according to the dimensions defined in the profile.

C-Angle—The Angle specifies the soffit angle from horizontal. A zero angle places the soffit in a horizontal position, whereas a positive angle will tilt the soffit up from the insertion point at J as shown in Figure 7.54.

D-Horizontal Offset from Roof Slab Baseline—The Horizontal Offset from Roof Slab Baseline field specifies the horizontal distance from the soffit to the roof slab baseline. This field is grayed out if no profile is specified for the soffit.

E-"Y" Direction—The "Y" Direction field specifies the distance from the fascia insertion point to the soffit insertion point.

F-"X" Direction—The "X" Direction field specifies the distance in x direction from the fascia insertion point to the soffit insertion point.

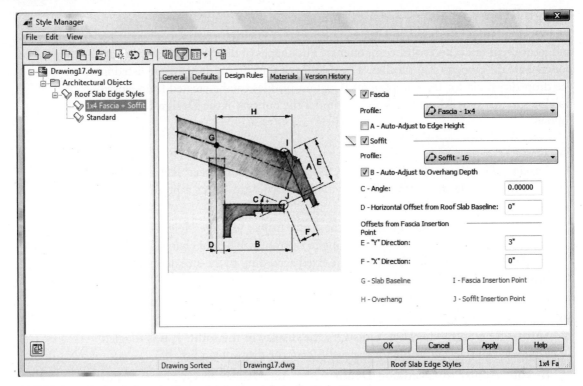

FIGURE 7.54 *Design Rules tab of a roof slab edge style in the Style Manager*

Assigning Roof Slab Edge Styles to Slab Edges

Roof slab edge styles can be assigned to a roof slab edge by selecting the **Edges** button in the Properties palette of an existing roof slab to open the **Roof Slab Edges** dialog box. The **Roof Slab Edges** dialog box lists all edges of the roof slab. Although the edges are numbered according to the sequence of development, it may be difficult to determine which edge of the slab you are editing in the dialog box, as shown in Figure 7.55.

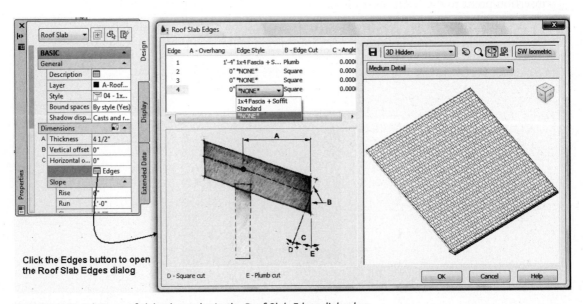

FIGURE 7.55 *Editing roof slab edge styles in the Roof Slab Edges dialog box*

However, the **Edit Roof Slab Edges** command allows you to select an edge for edit, and the **Edit Roof Slab Edges** dialog box will open with information regarding only the selected edge as shown in Figure 7.56. Access the **Edit Roof Slab Edges** command as shown in Table 7.21.

TABLE 7.21 *Edit Roof Slab Edges command access*

Command prompt	ROOFSLABEDGEEDIT
Ribbon	Select a roof slab and choose Edit from the Edge panel of the Roof Slab contextual tab.
Shortcut menu	Select a roof slab, right-click, and choose Edit Roof Slab Edges.

When you select a roof slab and choose **Edit** from the Edge panel, you are prompted to select an edge as shown in the following workspace sequence:

(Select a roof slab and choose Edit *from the Edge panel.)*

Select edges of one roof slab: *(Select an edge of a roof slab.)*

Select edges of one roof slab: ENTER *(Press ENTER to end selection and open the Edit Roof Slab Edges dialog box as shown in Figure 7.56.)*

The top-left section of the **Edit Roof Slab Edges** dialog box lists the edge number, and you can click in the columns for each edge and edit the **A-Overhang**, **Edge Style**, **B-Edge Cut**, and **C-Angle**. The roof slab edge styles included in the drawing are listed in the **Edge Style** drop-down list as shown in Figure 7.56.

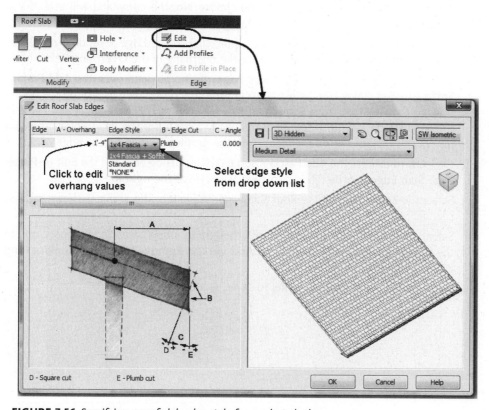

FIGURE 7.56 *Specifying a roof slab edge style for a selected edge*

USING ADD EDGE PROFILE IN THE WORKSPACE

A profile can be added to the drawing for the fascia without opening the **Style Manager** and creating a new profile. The **Add Edge Profiles** command is included in the shortcut menu of a selected roof slab. This command allows you to create a profile while remaining in the workspace. The edge selected for the command must have a defined edge style. Access the **Add Edge Profiles** command as shown in Table 7.22.

TABLE 7.22 *Add Edge Profiles command access*

Command prompt	**ROOFSLABADDEDGEPROFILES**
Ribbon	Select a roof slab and choose Add Profiles from the Edge panel.
Shortcut menu	Select a roof slab, right-click, and choose Add Edge Profiles.

When you select a roof slab and choose **Add Profiles** from the Edge panel, you are prompted to select a roof slab edge, which must include a defined roof slab edge style. The command allows you to specify a profile in the drawing for the selected edge and opens the profile for edit using Edit In Place. The Edit In Place: Roof Slab Edge Profile contextual tab is displayed, which provides the tools for edit. The profile can then be edited by adding or deleting vertices that define the profile. The procedure to create a simple profile for an edge is shown in the following steps.

Steps to Add an Edge Profile

1. Select a roof slab and choose **Add Profiles** from the Edge panel.

2. Select a roof slab edge. If no roof slab edge style and overhang are defined for the edge, you are prompted to **Set the edge style** in the **Add Profile – No Edge Style** dialog box shown in Figure 7.57. If you choose **No**, the command will end. If you choose **Yes**, the Edit Roof Slab Edges dialog box opens.

3. To edit the Edit Roof Slab Edges dialog box, click in the Edge Style drop-down list, select **Standard**, and type a distance for the overhang in the **Overhang** field.

4. Click **OK** to dismiss the Edit Roof Slab Edges dialog box and open the **Add Fascia/ Soffit Profiles** dialog box. Select a profile from the drop-down lists of profile definitions or choose **Start from Scratch** as shown in Figure 7.58. You can type a name in the **New Fascia Profile Name** field to specify a name for the new profile. The content of this dialog box is dependent upon the profiles defined for the edge.

5. Click **OK** to dismiss the Add Fascia / Soffit Profiles dialog box, and the Edit **In-Place: Roof Slab Edge Profile** tab is displayed as shown at the right in Figure 7.58.

6. Select the **Zoom To** from the Roof Slab Edge Profile tab. The profile is shown in cross section.

7. Select a grip on the profile and stretch the grip to a new location, or choose one of the following from the Profile panels: **Add Vertex**, **Remove Vertex**, or **Replace Ring** as shown in Figure 7.58.

8. Select **Finish** from the **Edits** panel to save the edit.

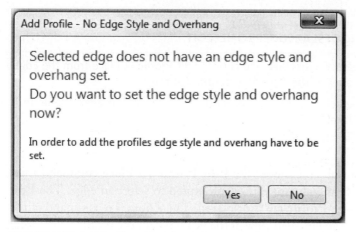

FIGURE 7.57 *Add Profile – No Edge Style message box*

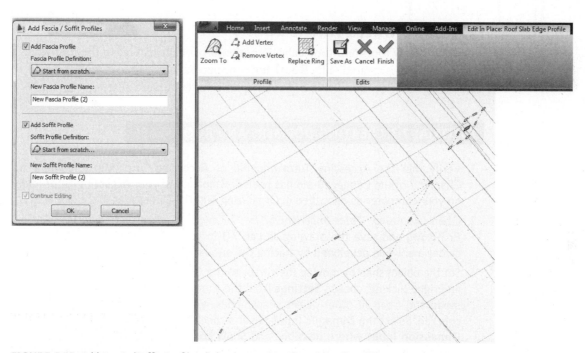

FIGURE 7.58 *Add Fascia/Soffit Profiles dialog box and In-Place Edit of profile*

Changing the Profile with Edit Edge Profile In Place

After a profile has been changed using the Add Edge Profiles command, additional editing can be performed with the Edit Profile In Place command. This command allows you to perform additional editing of the profile and to add or remove vertices from the profile in the workspace. Access the Edit Edge Profile In Place command as shown in Table 7.23.

TABLE 7.23 *Edit Edge Profile In Place command access*

Command prompt	ROOFSLABEDGEPROFILEEDIT
Ribbon	Select a roof slab and choose Edit Profile In Place from the Edge panel of the Roof Slab contextual tab.
Shortcut menu	Select a roof slab, right-click, and choose Edit Edge Profile In Place.

When you select a roof slab and choose the **Edit Profile In Place** command from the Edge panel of the Roof Slab tab, you are prompted to select an edge. The **Edit In-Place: Roof Slab Edge Profile** tab is displayed. The **Zoom To** command of the **Edit In-Place** tab creates an enlarged view of the profile. Edit the profile by selecting a grip, or choose **Add Vertex, Remove Vertex,** or **Replace Ring** from the **Profile** panel of the **Edit In Place: Roof Slab Edge Profile** tab. Choose **Finish** from the **Edit In Place: Roof Slab Edge Profile** tab to save changes and end the edit.

EXTENDING WALLS TO THE ROOF

The walls below roofs and roof slabs do not automatically extend to the roof when a roof or roof slab is created. The **RoofLine** command allows you to extend the walls to the roof or roof slab. See Table 3.7 of Chapter 3 for command access.

TUTORIAL 7-3: CREATING DORMERS AND ROOF SLABS

1. Verify that the Accessing Tutor\Imperial or Accessing Tutor\Metric content for Chapter 7 of the CengageBrain has been downloaded to your Accessing Tutor folder on your computer as described in Organizing Tutorial Directories in the Preface.

2. Open *Accessing Tutor\Imperial\Ch7\Ex 7-3.dwg* or *Accessing Tutor\Metric\Ch7\ Ex 7-3.dwg* and save the drawing as **Lab 7-3** in your student directory. This file includes a polyline developed by tracing the outline of the first floor plan.

3. Set the object snap by moving the cursor to the Object Snap toggle of the status bar and right-clicking; choose **Settings** from the shortcut menu. Select the **Intersection** mode and clear all other object snap modes. Verify that **Object Snap On (F3)** is checked. Select the **Dynamic Input** tab and check the **Enable Pointer Input, Enable Dimension Input where possible,** and **Show command prompting and command input near the crosshairs** checkboxes. Click **OK** to dismiss the Drafting Settings dialog box.

4. Clear all toggles except Object Snap, Dynamic Input, and Quick Properties on the status bar.

5. The main roof is developed from a polyline created from the geometry of the first floor plan. Select the **Roof** command from the Design palette. Edit the Properties palette as follows: Basic—Dimensions: Thickness = **6 [150]**, Edge cut = **Plumb,**

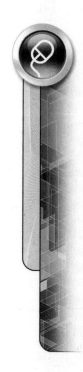

Next Edge: Shape = **Gable**, Overhang = **12 [300]**, as shown in Figure 7.59. Respond to the following workspace sequence by selecting roof points as specified below:

> Roof point or: (*Select the polyline using the* Intersection *object snap at the outer lines of the corner at* p1 *as shown in Figure 7.59.*)
>
> Roof point or: (*Select the polyline using the intersection of the outer lines of the corner at* p2 *as shown in Figure 7.59.*)
>
> (*Edit the Properties palette as follows: Shape =* Single slope, *Overhang =* 12 [300], *Plate height =* 0, *and Rise =* 12 [100].)
>
> Roof point or: (*Select the polyline using the intersection of the outer lines of the corner at* p3 *as shown in Figure 7.59.*)
>
> (*Edit the Properties palette as follows: Shape =* Gable, *Overhang =* 12 [300].)
>
> Roof point or: (*Select the polyline using the intersection of the outer lines of the corner at* p4 *as shown in Figure 7.59.*)
>
> (*Edit the Properties palette as follows: Shape =* Single slope, *Overhang =* 12 [300], *Plate height =* 0, *and Rise =* 12 [100].)
>
> Roof point or: (*Select the polyline using the intersection of the outer lines of the corner at* p1 *as shown in Figure 7.59.*)
>
> Roof point or: ENTER (*Press* ENTER *to end Roof command.*)

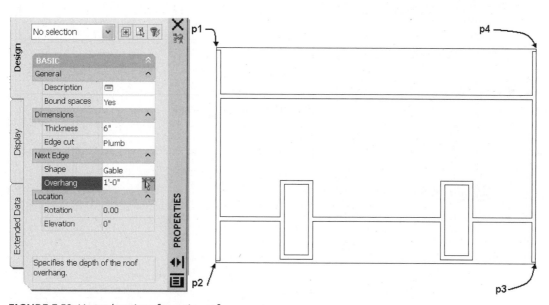

FIGURE 7.59 *Vertex locations for main roof*

6. Add the dormer roofs by selecting the **Roof** command from the Design palette. Edit the Properties palette as follows: Basic—Dimensions: Thickness = **6 [150]**, Edge cut = **Plumb**, Next Edge: Shape = **Single slope**, Overhang = **0**, Lower Slope-Plate Height = **8' [2400]**, Rise = **4 [33]**. Respond to the following command line prompts by selecting roof points as specified in Figure 7.60.

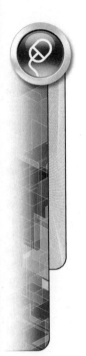

Roof point or: *(Select the wall using the intersection of the outer lines of the corner at* **p1** *as shown in Figure 7.60.)*

Roof point or: *(Select the wall using the intersection of the outer lines of the corner at* **p2** *as shown in Figure 7.60.)*

(Edit the Properties palette as follows: Shape = **Gable**, *Overhang =* **0**.*)*

Roof point or: *(Select the wall using the intersection of the outer lines of the corner at* **p3** *as shown in Figure 7.60.)*

(Edit the Properties palette as follows: Shape = **Single slope**, *Overhang =* **0**, *Lower Slope-Plate Height =* **8'** **[2400]**, *Rise =* **4** **[33]**.*)*

Roof point or: *(Select the wall using the intersection of the outer lines of the corner at* **p4** *as shown in Figure 7.61.)*

(Edit the Properties palette as follows: Shape = **Gable**, *Overhang =* **0**.*)*

Roof point or: *(Select the wall using the intersection of the outer lines of the corner at* **p1** *as shown in Figure 7.60.)*

Roof point or: ENTER *(Press* ENTER *to end the command.)*

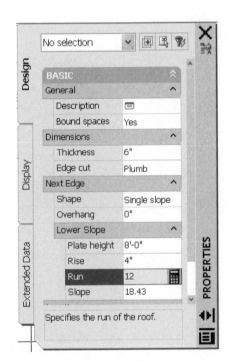

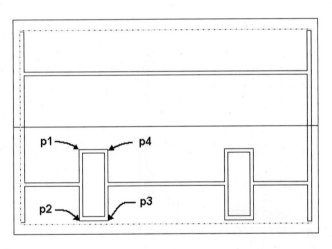

FIGURE 7.60 *Vertex locations for dormer roof*

7. Repeat step 5 to add a roof for the dormer at the right.

8. Select **SW Isometric** from the View Cube.

9. Select the **Roof Slab** command on the Design palette, right-click, and choose **Apply Tool Properties to > Linework, Walls, and Roof**. Respond to the workspace prompts as shown below:

   Select roof, walls, or polylines: 1 found *(Select main roof at* **p1** *as shown in Figure 7.61.)*

```
Select roof, walls, or polylines: 1 found, 2 total (Select
dormer roof at p2 as shown in Figure 7.61.)

Select roof, walls, or polylines: 1 found, 3 total (Select
dormer roof at p3 as shown in Figure 7.61.)

Select roof, walls, or polylines: ENTER (Press ENTER to end
selection.)

Erase layout geometry? [Yes/No] <No>: Y (Choose the Yes
option from the workspace options to remove the roof.)
```

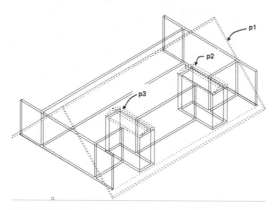

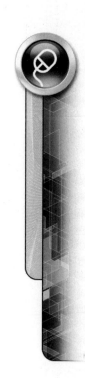

FIGURE 7.61 *Selection of existing roofs to convert to roof slabs*

10. Select **Top** from the View Cube.

11. The cut plane lines are displayed on the roof slab since the roof slab was developed from a roof with a plate height equal to zero. To remove the cut plane lines from the main roof slab, select the front roof slab at **p1** as shown in Figure 7.62, right-click, and choose **Properties**. Select the **Display** tab of the Properties palette. Set Display component = None and Display controlled by = Drawing default setting. Edit the Cut Plane section and set **Height** to **14' [4500]**. Move the cursor from the Properties palette and press ESC to end the edit.

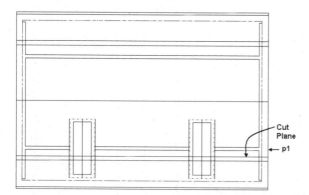

FIGURE 7.62 *Selection of roof to edit display of cut plane (metric roof slab similar)*

12. Select **SW Isometric** from the View Cube.

13. Select the front roof slab near **p1** as shown in Figure 7.63. Choose **Add** from the **Dormer** panel. Respond to the workspace prompts as shown below:

> Select objects that form the dormer: *(Select* **p2** *as shown at the left in Figure 7.63.)* 1 found
>
> Select objects that form the dormer: *(Select* **p3** *as shown at the left in Figure 7.63.)* 1 found, 2 total
>
> Select objects that form the dormer: *(Select* **p4** *as shown at the left in Figure 7.63.)* 1 found, 3 total
>
> Select objects that form the dormer: *(Select* **p5** *as shown at the left in Figure 7.63.)* 1 found, 4 total
>
> Select objects that form the dormer: *(Select* **p6** *as shown at the left in Figure 7.63.)* 1 found, 5 total
>
> Select objects that form the dormer: *(Select* **p7** *as shown at the left in Figure 7.63.)* 1 found, 6 total
>
> Select objects that form the dormer: ENTER *(Press ENTER to end selection.)*
>
> Slice wall with roof slab [Yes/No] <Yes>: ENTER *(Choose* Yes *from the workspace prompts.)*

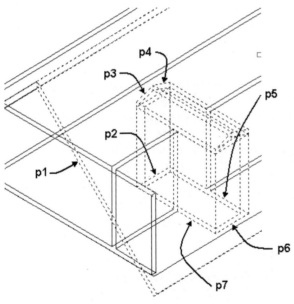
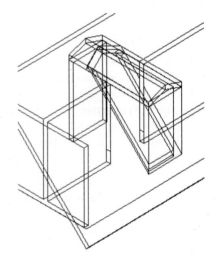

FIGURE 7.63 *Selection of components to create dormer*

14. Repeat step 12 to create the dormer at the right in Figure 7.64.

15. Select walls **p1**, **p2**, **p3**, **p4**, **p6**, and **p8** as shown in Figure 7.64. Choose **Modify Roof Line** from the Roof/Floor Line command flyout of the Modify panel in the Wall contextual tab. Respond to the workspace prompts as shown below:

> RoofLine: a *(Choose* Auto project *from the workspace options.)*
>
> Select objects: *(Select roof slab at* **p9** *as shown in Figure 7.64.)* 1 found
>
> Select objects: *(Select roof slab at* **p10** *as shown in Figure 7.64.)* 1 found, 2 total
>
> RoofLine: ENTER *(Right-click and choose ENTER from the shortcut menu to end the command.)*

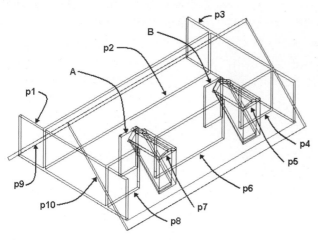
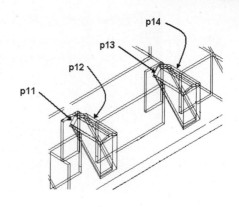
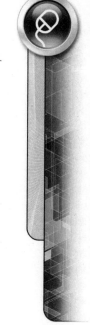

FIGURE 7.64 *Selection of walls for edit to roofline*

16. Select the walls at **p5** and **p7** as shown in Figure 7.64, right-click, and choose **Roof/Floor Line > Modify Roof Line** from the shortcut menu. Respond to the workspace prompts as shown below:

    RoofLine: a *(Choose* Auto project *from the workspace options.)*

    Select objects: *(Select roof slab at* p11 *as shown in Figure 7.64.)* 1 found

    Select objects: *(Select roof slab at* p12 *as shown in Figure 7.64.)* 1 found, 2 total

    Select objects: *(Select roof slab at* p13 *as shown in Figure 7.64.)* 1 found, 3 total

    Select objects: *(Select roof slab at* p14 *as shown in Figure 7.64.)* 1 found, 4 total

    Select objects: ENTER *(Press* ENTER *to end the selection.)*

    [2] Wall cut line(s) converted.

    RoofLine: ENTER *(Right-click and choose* ENTER *from the shortcut menu to end the command.)*

17. Select the Home tab. Select **Erase** from the **Modify** panel and respond to the workspace prompts as shown below:

    Select objects: 1 found *(Select wall* A *as shown in Figure 7.64.)*

    Select objects: 1 found, 2 total *(Select wall* B *as shown in Figure 7.64.)*

    Select objects: *(Right-click and choose* ENTER *to end the selection.)*

18. Select **SW Isometric** view from the View Cube.

19. Select the View tab; choose **3D Hidden** from the Visual Styles flyout menu of the Visual Styles panel of the View tab to view the roof as shown in Figure 7.65.

20. Save and close the drawing.

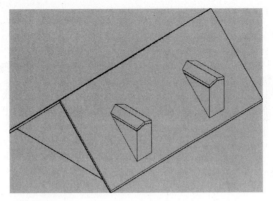

FIGURE 7.65 *Completed roof slab edit*

TUTORIAL 7-4: CREATING A HIP ROOF

1. Verify that the Accessing Tutor\Imperial or Accessing Tutor\Metric content for Chapter 7 of the CengageBrain has been downloaded to your Accessing Tutor folder on your computer as described in Organizing Tutorial Directories in the Preface.

2. Open Autodesk AutoCAD Architecture 2012 and select **Project Browser** from the Quick Access toolbar. Use the **Project Selector** drop-down list to navigate to your *Student\Ch7* student directory. Select **New Project** to open the **Add Project** dialog box. Type **14** in the Project Number field and type **Ex 7-4** in the Project Name field. Check **Create from template project**, choose **Browse**, and edit the path to *Accessing Tutor\Imperial\Ch7\Ex 7-4\Ex 7-4.apj* or *Accessing Tutor\Metric\Ch7\Ex 7-4\Ex 7-4.apj* in the Look in field. Click **Open** to dismiss the Select Project dialog box. Click **OK** to dismiss the Add Project dialog box. Click **Close** to dismiss the Project Browser.

3. Select the **Constructs** tab of the Project Navigator and double-click **Floor 2** of the Constructs tab.

4. Move the cursor to the Object Snap toggle on the status bar, right-click, and choose **Settings** from the shortcut menu. Edit the **Object Snap** tab of the Drafting Settings dialog box: clear all object snap modes, and check the **Intersection** object snap. Check the **Object Snap On (F3)** checkbox and clear the **Object Snap Tracking On (F11)** checkbox. Select the **Dynamic Input** tab, toggle ON **Enable Pointer Input, Enable Dimension Input where possible**, and **Show command prompting and command input near the crosshairs**. Click **OK** to dismiss the Drafting Settings dialog box.

5. Turn OFF the display of brick and brick hatching by selecting the exterior wall; right-click and choose **Properties** from the shortcut menu.

6. Select the **Display** tab of the **Properties** palette. Choose the Edit Style Display button as shown in Figure 7.66. Select the Display Properties tab of the Wall Style Properties dialog. Choose the checkbox for Style Override of the Plan High Detail display representation to open the Display Properties (Wall Style Override—Frame Brick)—Plan High Detail.

7. Choose the Layer/Color/Linetype tab toggle off the light bulb of the following components as shown in Figure 7.66: Shrink Wrap, Boundary 2 (Brick), Hatch 2 (Brick), Boundary 3 (Air Gap), Hatch 3 (Air Gap), Boundary 4 (Sheathing), Hatch 4 (Sheathing), Boundary 5 (Gypsum), Hatch 5 (Gypsum).

8. Click OK to dismiss all dialog boxes.

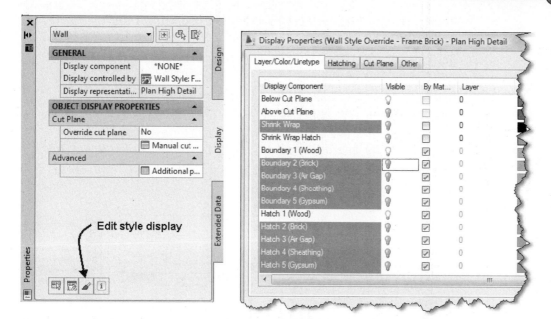

FIGURE 7.66 *Editing display component visibility from the Display Properties tab of a wall style*

9. Choose ESC to end selection of the wall.

10. Choose the **Home** tab. If the Design palette set is not displayed, choose Tools of the Build panel. Choose Design palette. Select the **Roof** command from the Design palette and edit the Design tab of the Properties palette as follows: Basic—Dimensions: Thickness = **6 [150]**, Edge cut = **Plumb**, Next Edge: Shape = **Single slope**, Overhang = **20 [500]**, Lower Slope: Plate height = **8' [2500]**, Rise = **6 [50]**, as shown in Figure 7.67. Respond to the following prompts by selecting roof points as specified below:

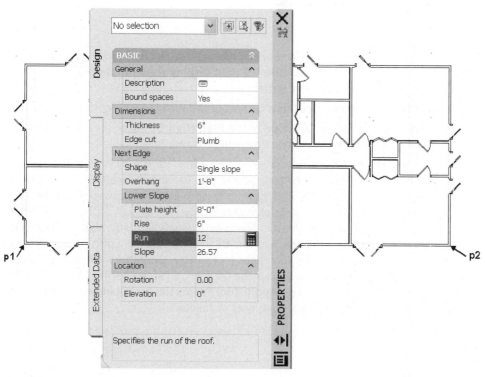

FIGURE 7.67 *Settings in the Properties palette*

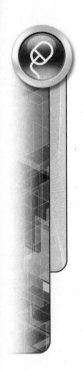

Roof point or: *(Select the intersection of lines representing the outside of stud, point* **p1** *in Figure 7.67.)*

Roof point or: *(Select the intersection of lines representing the outside of stud, point* **p2** *in Figure 7.67.)*

Roof point or: *(Select the intersection of lines representing the outside of stud, point* **p3** *in Figure 7.68.)*

Roof point or: *(Select the intersection of lines representing the outside of stud, point* **p4** *in Figure 7.68)*

Roof point or: ENTER *(Press* ENTER *to end the selection of points.)*

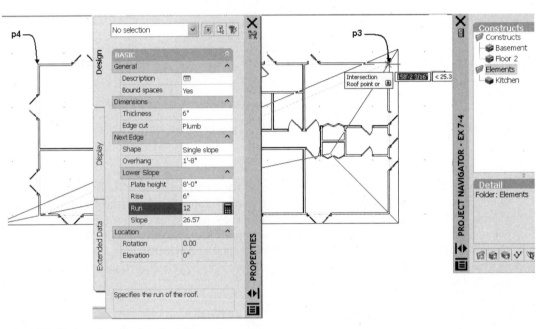

FIGURE 7.68 *Location of points for a hip roof*

11. Turn ON the display of brick and brick hatching by selecting an exterior wall, right-click, and choose Properties. Choose the **Edit display style properties** button of the **Display** tab of the Properties palette.

12. Verify the Display Properties tab is current of the Wall Style Properties dialog box.

13. Clear the check for **Style Override** of the **Plan High Detail** display representation. Click **OK** to dismiss the Wall Style Properties—Frame Brick dialog box.

14. In this step you will import the 04 − 1 × 6 Fascia + Soffit [100 − 25 × 150 Fascia + Soffit] tool. Right-click over the tool palette title bar and choose New Palette. Over-type Roof Slabs to name the new palette. Choose Content Browser from the Tools flyout menu of the Build panel. Double-click to open the Design Tool Catalog—Imperial or Design Tool Catalog—Metric. Navigate to Roof Slabs and Slabs > Roof Slabs category. Click and drag the i-drop for the 04 − 1 × 6 Fascia + Soffit [100 − 25 × 150 Fascia + Soffit] tool onto the Roof Slabs tool palette. Close the tool catalog. To convert the roof to a roof slab, right-click the 04 − 1 × 6 Fascia + Soffit [100 − 25 × 150 Fascia + Soffit] tool and choose **Apply Tool Properties to > Linework, Walls, and Roof**. Respond to the workspace prompts as follows:

Select roof, walls, or polylines: 1 found *(Select the roof.)*

Select roof, walls, or polylines: *(Right-click to end the selection.)*

Erase layout geometry? [Yes/No] <No>: Y *(Choose* **Yes** *from the workspace options.)*

(The baselines of the roof slabs are shown in plan view.)

15. In this step you will modify the roof slab style to display the outline of the roof slab using hidden lines. Choose the ridge of a roof slab and choose **Edit Style** of the General panel to open Roof Slab Styles—04 − 1 × 6 Fascia + Soffit [100 − 25 × 150 Fascia + Soffit]. Choose the **Display Properties** tab; the Plan High Detail display representation is selected. Check the **Style Override** checkbox to open the Display Properties (Roof Slab Style Override—04 − 1 × 6 Fascia + Soffit [100 − 25 × 150 Fascia + Soffit])—Plan High Detail dialog box. Turn on visibility for the **Above Cut Plane Outline** and turn OFF all other display components as shown in Figure 7.69. Repeat this step for the **Plan** display representation. Click **OK** to dismiss all dialog boxes.

16. Choose the **SW Isometric** from the View Cube. Select the View tab. Choose **Realistic** from the Visual Styles flyout menu of the Visual Styles panel to view the roof as shown Figure 7.70.

17. Save and close your drawing. Choose the Project tab. Choose the **Close Current Project** button in the Project tab toolbar located at the bottom of the Project tab.

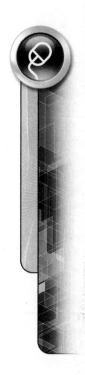

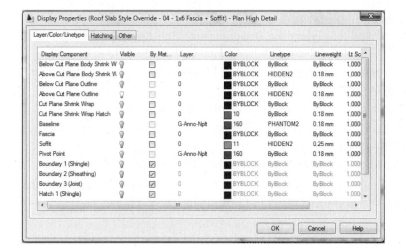

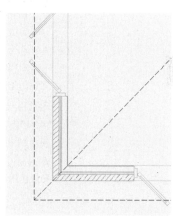

Roof outline shown with hidden lines

FIGURE 7.69 *Editing the display representation for the plan*

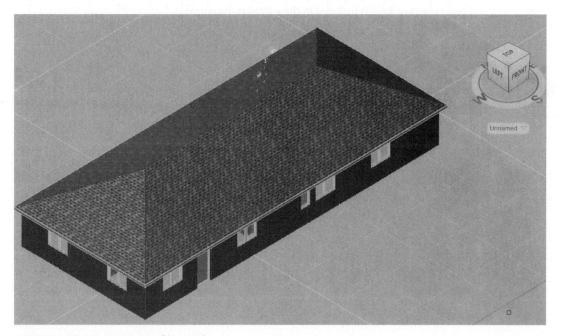

FIGURE 7.70 *Isometric view of hip roof*

SUMMARY

- Roofs are created with the **RoofAdd** command located on the Design tool palette and the Build panel of the Home tab.
- Using the **RoofAdd** command with a single slope shape will create a hip roof.
- The Properties palette allows you to define the thickness of the roof plane, slope, and overhang.
- Existing roofs can be edited by changing the slope, plate height, shape, and overhang in the Properties palette.
- The angle of the roof can be defined by the ratio of the rise to run or by a slope angle.
- Create a gable end of a gable roof by setting the roof edge slope to 90 degrees.
- Create a shed roof by selecting the **Gable** shape for three of the four roof edges.
- Create the gambrel roof by applying the double slope shape to two edges of the roof.
- The **Apply Tool Properties to > Roof** command is a shortcut menu option of the **Roof** command located on the Design palette. The **Apply Tool Properties to** command can be used to create a roof based upon linework or walls.
- A roof object can be converted to roof slabs by selecting the roof, right-clicking, and choosing **Convert to Roof Slabs**.
- Roof slabs can be trimmed, extended, and mitered to create complex roofs.
- Fascia, soffit, and frieze millwork components can be added to a roof slab by editing the roof slab edge style.
- A profile is used to create components for the roof slab edge style.
- A profile used in a roof slab edge style can be created or modified by the **Add Edge Profiles** or **Edit Edge Profile in Place** command of the roof slab shortcut menu.

NOTE Refer to **e Projects** of Chaper 7 for download from the Student Companion Site of CengageBrain http://www.cengagebrain.com described in the Preface. The Project Ex 7-5 Creating a Shed Roof is included in the e Project document.

NOTE Refer to **e Review Questions** folder which includes review questions for chapter 7 of Student Companion Site of **CengageBrain** http://www.cengagebrain.com described in the Preface.

CHAPTER
8

Creating Slabs for Floors and Ceilings

INTRODUCTION

Slabs are used to represent the floor or roof system of a building. The slab used in wood frame construction represents the thickness of the floor joist and flooring system. If a concrete slab is used, the slab thickness is set to the thickness of the concrete slab. Slabs can also be used to represent a flat or low-sloped roof in commercial construction. Sample slab styles located in the Design Tool Catalog—Imperial and Design Tool Catalog—Metric of the Content Browser include materials and components to represent the construction of floor systems.

OBJECTIVES

After completing this chapter, you will be able to:

- Use the SlabAdd command to add a slab.
- Modify a slab by editing the Properties palette and using grips.
- Create a slab that includes a slab edge style.
- Convert polylines and walls to a slab.
- Edit the slab edge style of a slab.

ADDING AND MODIFYING A FLOOR SLAB

A slab is added to the drawing by selecting the Slab tool (SlabAdd command) located on the Design tool palette as shown in Figure 8.1. Access the Slab tool as shown in Table 8.1. When you select the Slab tool, the Properties palette opens, displaying the properties of the floor slab. The slab has properties similar to the roof slab. The slab has thickness and can be sloped to provide drainage. The Properties palette and workspace options are shown in Figure 8.1 and described below.

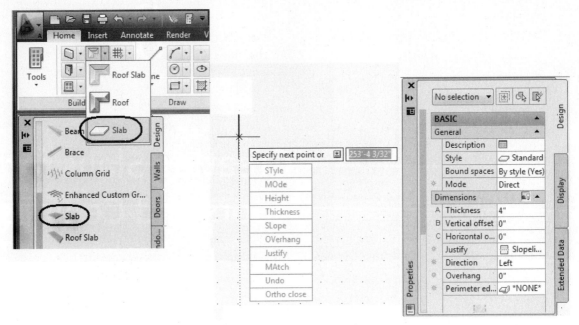

FIGURE 8.1 *Slab command of the Design tool palette and Properties palette*

TABLE 8.1 *SlabAdd command access*

Command prompt	SLABADD
Ribbon	Select the Slab tool from the Roof Slab flyout of the Build panel in the Home tab.
Palette	Select the Slab tool from the Design palette as shown in Figure 8.1.

Design Tab—Basic

General

Description—The Description button opens the **Description** dialog box, allowing you to type a description.

Style—The Style drop-down list displays the styles that have been created or imported into the drawing. Additional styles can be imported from the Design Tool Catalog—Imperial and Design Tool Catalog—Metric of the Content Browser.

Bound Spaces—The Bound Spaces options allow you to specify if the slab will function as a boundary for a space.

Mode—The Mode options include **Direct** and **Projected**. The **Direct** mode creates the slab at the XYZ coordinate of the vertices specified. The **Projected** mode projects from the specified vertices up from the distance specified in the **Base height**.

Dimensions

A-Thickness—The Thickness specifies the base thickness or depth of the slab.

B-Vertical offset—The Vertical offset specifies the vertical distance from the insert points to the slab baseline.

C-Horizontal offset—The Horizontal offset distance shifts the slab horizontally relative to the baseline.

Justify—The Justify options include **Top**, **Center**, **Bottom**, and **Slopeline**. Justification is a property used when placing slabs and is not retained in the Properties palette for existing slabs. The Justify options position the top, center, bottom, or slopeline of the slab relative to the pivot point. The slopeline is the position on the slab baseline relative to the bottom face.

Base height—The Base height is the distance the slab is created from the Z = 0 plane. The base height is applied to a slab created in the **Projected** mode. The **Base height** is not an option with a Direct Mode slab.

Direction—The Direction toggle allows you to create a slab to the left or right of first two selected points when you use **Ortho close**.

Overhang—The Overhang value specifies the horizontal distance from the pivot point to extend the slab.

Perimeter edge—The Perimeter edge field allows you to specify a slab edge style for the perimeter.

SLOPE:

Rise—The Rise specifies the vertical displacement per horizontal run unit of projected slabs. The Slope section is only displayed for Projected mode slabs.

Run—The Run specifies the horizontal component of the slope. The base value is usually 12[100].

Slope—The **Slope** displays the corresponding angle as a result of the rise and run values.

Location

Rotation—The Rotation field displays the rotation angle of the slab.

Elevation—The Elevation field displays the current elevation of the slab. The elevation of existing slabs can be edited.

When you select the **Slab** tool (**SlabAdd** command) from the Design palette, you are prompted to specify points for the vertices of the slab. The location of the slab in the Z direction is controlled by the Mode property of the slab. The Direct mode will place the slab vertices with the same Z component as the feature selected. The Projected mode will place the slab with a Z component as specified in the Base height field in the Design tab of the Properties palette. The thickness values specify the base thickness as shown in Figure 8.2.

The Standard slab style is applied when you choose the SlabAdd command. When the Standard slab style is used, the baseline is at the top of the slab and the thickness value projects down from the baseline as shown at the right in Figure 8.2. Therefore, if you are placing a slab of the Standard slab style on top of a wall, set the base height to the top of the slab. The Thickness and Base height is set dependent upon the component positions defined in the slab style. The Wood Floor System slab style located in the Roof Slabs and Slabs > Slabs category of the Design Tool Catalog—Imperial and Design Tool Catalog—Metric of the Content Browser includes the wood flooring, subfloor, and space for the joists. The imperial Wood Floor System slab style includes a profile defined for slab edge style to represent the rim joist for 2″ × 12″ joist. The metric Wood Floor System slab style does not include a profile defined for the slab edge. The thickness value is applied up in a positive direction from the baseline in the Wood Floor System slab style. The BeamAdd command will be used in Chapter 13 to add joists within the slab.

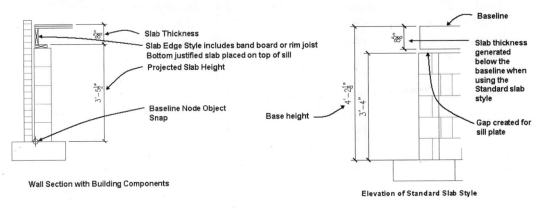

Wall Section with Building Components

Elevation of Standard Slab Style

FIGURE 8.2 *Determining slab thickness and base height from a section*

> **NOTE**
>
> Access Instructional Video **8.1 Creating a Slab** from the Instructional Video category of the Student Companion site of CengageBrain http://www.cengagebrain.com described in the Preface.

TUTORIAL 8-0: INTRODUCTION TO A PROJECTED SLAB

1. Download Chapter 8 files from the Accessing Tutor\Imperial or Accessing Tutor\ Metric category of the Student Companion site of CengageBrain http://www. cengagebrain.com described in the Preface to your Accessing Tutor folder.

2. Open Accessing *Tutor\Imperial\Ch8\EX 8-0.dwg* or *Accessing Tutor\Metric\Ch 8\ Ex 8-0.dwg*.

3. Choose **Save As** > AutoCAD Drawing from the Application menu and save the drawing as **Lab 8-0** in your student directory.

4. Move the cursor to the Object Snap toggle of the status bar, right-click, and choose **Settings** from the shortcut menu. Edit the **Object Snap** tab of the **Drafting Settings** dialog box: clear all object snap modes and check **Node** object snap. Check the **Object Snap ON (F3)** checkbox and check **Object Snap Tracking ON (F11)**. Select the **Polar Tracking** tab; check ON **Polar Tracking (F10)**. Select the **Dynamic Input** tab. Verify that the **Enable Pointer Input, Enable Dimension Input where possible,** and **Show command prompting and command input near the crosshairs** checkboxes are checked. Click **OK** to dismiss the Drafting Settings dialog box.

5. Choose the Home tab. Choose the Slab tool from the Roof Slab flyout of the Build panel. *Edit the Properties palette as follows: General: Style =* **Standard**, *Mode =* **Projected**, *Dimensions: Thickness =* **8 5/8" [219]**, *B-Vertical Offset =* **0**, *C-Horizontal =* **0**, *Justify =* **Top**, *Base height =* **4'-2-1/8" [1273]**, *Direction =* **Left**, *Overhang =* **0**, *Perimeter edge =* **None**, *Slope: Rise =* **0"**, *Run =* **12"[100]**. *Respond to workspace prompts as follows.)*

> Specify next point or: *(Use the Node object snap to select point* **p1** *as shown in Figure 8.3.)*
>
> Specify next point or: *(Use the Node object snap to select point* **p2** *as shown in Figure 8.3.)*

Specify next point or: *(Use the Node object snap to select point* **p3** *as shown in Figure 8.3.)*

Specify next point or: *(Use the Node object snap to select point* **p4** *as shown in Figure 8.3.)*

Specify next point or: *(Use the Node object snap to select point* **p1** *as shown in Figure 8.3.)*

Specify next point or: *(Press ENTER to create the slab.)*

Specify next point or: ESC *(Press ESCAPE to end the command.)*

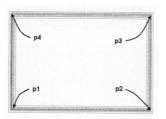
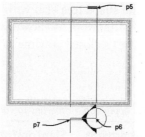
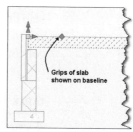

FIGURE 8.3 *Location of vertex points for the slab*

6. Press CTRL + A to select all objects of the drawing. Choose Edit in Section of the General panel of the Multiple Objects tab.

   Specify first point of section line or Enter to change UCS: (Select a point near **p5** as shown in Figure 8.3.)

   Specify next point of section line: (Select a point near **p6** as shown in Figure 8.3.)

   Specify next point of section line or [Break]: (Press Enter to end the section line.)

   Specify section extents: (Select a point near **p7** as shown in Figure 8.3.)

7. Select the slab to view the grips of the slab. The grips are located at the top of the slab as shown in Figure 8.3. The thickness of the slab is projected down from the baseline location.

8. Choose **Exit Edit In View** from the Edit in View message box.

9. Save and close the drawing.

The location of the vertices of slabs often may not align with wall geometry or with the available object snaps. You can view the walls in isometric and temporarily turn off the display of selected wall components in the style to view the concrete masonry unit. The vertices of the slab can be specified using the Direct or Projected modes and the endpoint object snap to choose the edges of the concrete masonry unit wall component. The Vertical Offset value in the Properties palette allows you to move the slab up above the specified points to provide a space for the sill. The slab for one level can be copied and inserted in other drawings. The slab can be developed on walls of the Foundation plan and exported to other levels using Windows clipboard operations. Select the slab, right-click, and choose Clipboard > Copy with Base Point. Open the new drawing right-click, and choose Clipboard > Paste to original coordinates. When the slab is inserted in the first floor plan, the pictorial view is realistic because a floor system is present.

TIP   When you select the slab location with the Node object snap of a wall, there are nodes at the top and bottom of the baseline. Therefore, view the wall in isometric to verify location of points.

In addition, you can draw a polyline that represents the outline of the slab and convert this polyline shape to a slab with the **SlabToolToLinework** command. Access the SlabToolToLinework as shown in the Table 8.2. The SlabToolToLinework allows you to convert a closed polyline to a slab. The slab can be direct or projected from the polyline.

TABLE 8.2 *Apply Tool Properties to command access*

Command prompt	SlabToolToLinework
Shortcut menu	Select the Slab tool on the Design palette, right-click, and choose Apply Tool Properties to > Linework and Walls.

When you select the Slab tool on the Design palette, right-click, and choose Apply Tool Properties to, the cascade menu includes the following options: Linework and Walls, Slab and Space. The Linework and Walls option allows you to select a polyline and create the slab as shown in the following workspace sequence.

> *(Select the* Slab *tool from the Design palette, right-click, and choose* Apply Tool Properties to > Linework and Walls *from the shortcut menu.)*
>
> Select walls or polylines: *(Select the polyline at* p1 *as shown in Figure 8.4.)*
>
> Select walls or polylines: ENTER *(Press ENTER to end selection.)*
>
> Erase layout geometry? [Yes/No] <No>: *(Choose* No *from the dynamic input options to retain the polyline.)*
>
> Creation mode [Direct/Projected] <Projected>: *(Choose* Projected *from the dynamic prompts to project the slab in the z direction.)*
>
> Specify base height <0>: 36 [915] ENTER *(Specify distance to project slab in Z direction.)*
>
> Specify slab justification [Top/Center/Bottom/Slopeline] <Bottom>: *(Choose the* Bottom *justification option from the workspace prompts.)*
>
> Specify the pivot point. *(Move the cursor over an edge of the slab to display a red line, click when the red line aligns with necessary pivot point, as shown in Figure 8.4.)*

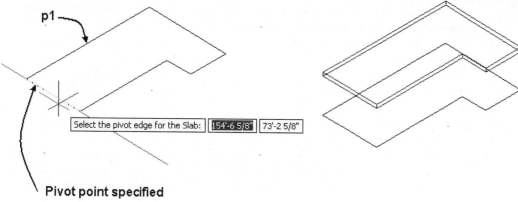

FIGURE 8.4 *Slab created from polyline shape*

The Apply Tool Properties to > Linework and Walls command can also be applied to the walls of a foundation. The slab can be positioned as defined by the baseline of the walls. The wall base height is used as the slab base height. The slab height can be adjusted in the Elevation field of the Design tab of Properties. In the following workspace sequence, a slab is placed in alignment with the baseline of the foundation wall as shown in Figure 8.5. The slab baseline is then located at the specified wall justification.

(Select the Slab *tool on the Design tool palette, right-click, and choose* Apply Tool Properties to > Linework and Walls *from the shortcut menu.)*

Select walls or polylines: *(Select the wall at* p1 *as shown in Figure 8.5.)*

Select walls or polylines: *(Select the wall at* p2 *as shown in Figure 8.5.)*

Select walls or polylines: *(Select the wall at* p3 *as shown in Figure 8.5.)*

Select walls or polylines: *(Select the wall at* p4 *as shown in Figure 8.5.)*

Select walls or polylines: ENTER *(Press ENTER to end selection.)*

Erase layout geometry? [Yes/No] <No>: *(Choose the* No *option from the dynamic prompt to retain the geometry.)*

Specify slab justification [Top/Center/Bottom/Slopeline] <Bottom>: *(Choose* Bottom *from the dynamic prompt options.)*

Specify wall justification for edge alignment [Left/Center/Right/ Baseline] <Baseline>: *(Choose the* Baseline *justification to position the slab edge on the baseline of the wall.)*

Select the pivot edge for the Slab: *(Select a point near* p3 *as shown in Figure 8.5.)*

The Standard slab is placed. Adjusting the Thickness to 10" [250] and adding 8" [250] to the Elevation value in the Properties palette, the slab is positioned as shown in the section at the right in Figure 8.5.

TIP	You can select the wall and display the grips to identify the start point and justification of the wall prior to applying slab properties to the wall.

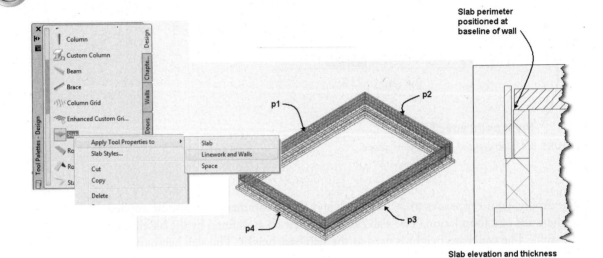

FIGURE 8.5 *Slab created based upon a wall*

Converting a Space to Slab

Extrusion and Freeform spaces consist of ceiling and floor components to represent the architectural features. The ceiling and floor space components can be converted to slabs. The floor component of a space could represent a concrete slab for a space. Material definitions can be assigned to the ceiling and floor slabs to represent the components. The **SlabConvertSpace** command allows you to convert a slab from a space. Refer to Table 8.3 for SlabConvertSpace.

TABLE 8.3 *Accessing the SlabConvertSpace command*

Command prompt	**SLABCONVERTSPACE**
Tool palette	Select the Slab tool, right-click, and choose Apply Tool Properties to > Space.

A slab is created from a space as outlined in the following workspace sequence:

(Choose the Slab *tool of the Design palette, right-click, and choose* Apply Tool Properties to > Space.*)*

Select spaces to convert: *(Select a space at* p1 *as shown in Figure 8.6.)* 1 found

Select spaces to convert: ENTER *(Press* ENTER *to end selection.)*

(Edit the Convert Space to Slab dialog box; check Convert Ceiling to Slab *and* Convert Floor to Slab, *as shown in Figure 8.6. Click* OK *to dismiss the dialog box.)*

(The slabs are created based upon the sizes of the space components; however, they can be modified in the Properties palette.)

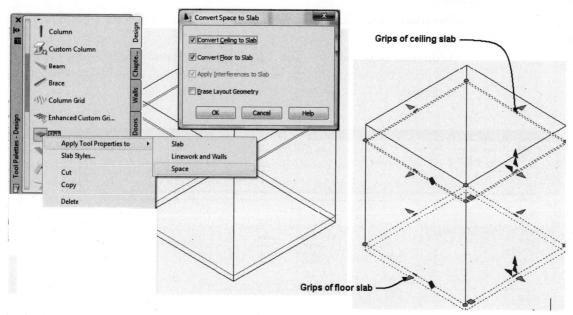

FIGURE 8.6 *Slab created from space*

MODIFYING THE SLAB USING PROPERTIES AND GRIPS

A slab can be edited by changing its properties in the Properties palette or editing the grips. Slabs can be trimmed, extended, or mitered in a manner similar to roof slabs. If you select an existing slab, additional fields are added to the slab Properties palette as shown in Figure 8.7. The additional fields described below include the Layer field, dimensions, and the Edges button. The Edges button allows you to assign slab edge styles to the edges of the slab.

Design Tab—Basic
General
> **Layer**—The Layer field displays the name of the layer of the slab. Slabs are placed on the **A-Slab** [S232G] layer when the AIA (256 Color) Layer Key Style is used in the drawing.

Dimensions
> **Edges**—The Edges button opens the **Slab Edges** dialog box, which is identical to the **Slab Edges** dialog box of roof slabs discussed in Chapter 7. This dialog box allows you to specify an overhang, edge style, edge cut, and angle.
> **SLOPE:**
> **Hold fascia elevation**—The Hold fascia elevation field includes the following options: **No, By adjusting overhang**, and **By adjusting baseline height**, as discussed in Chapter 7.
> **Pivot Point** X—The Pivot Point X value is the world coordinate X value of the pivot point.
> **Pivot Point** Y—The Pivot Point Y value is the world coordinate Y value of the pivot point.
> **Pivot Point** Z—The Pivot Point Z value is the world coordinate Z value of the pivot point.

Location

Elevation—The Elevation field allows you to adjust the vertical position of the slab.

Additional information—The Additional information button opens the Location dialog box, which lists the insertion point, normal, and rotation descriptions of the slab.

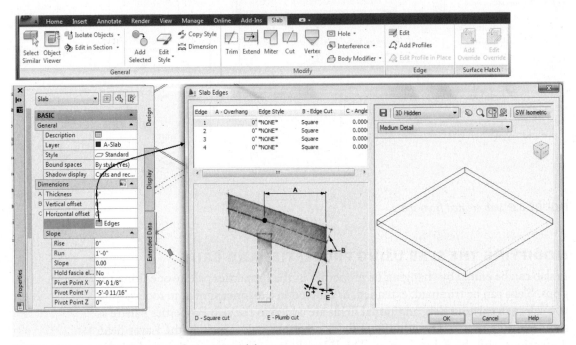

FIGURE 8.7 *Properties dialog box of an existing slab*

Using Grips to Edit Slabs

The grips of a slab allow you to change the pivot point, thickness, edge location, angle, and vertices. The grips of a slab are identical to those of the roof slab; therefore, refer to "Using Grips to Edit a Roof Slab" in Chapter 7. The Slab contextual tab shown in Figure 8.7 includes the commands listed in Table 8.4. The commands are similar to the commands used to edit roof slabs. Holes can be added to a slab using the **SlabAddHole** command to represent framing openings for stairs, skylight shafts, and masonry fireplaces. Holes are created by cutting the hole in the shape of a closed polyline. The closed polyline is drawn in the shape of the framing opening. Refer to "Tools for Editing Roof Slabs" in Chapter 7 to review the commands used to edit slabs.

TABLE 8.4 *Slab Edit commands of the shortcut menu*

Command Prompt	Modify Panel of the Slab Contextual Tab
SLABTRIM	Select a slab and choose Trim.
SLABEXTEND	Select a slab and choose Extend.
SLABMITER	Select a slab and choose Miter.
SLABCUT	Select a slab and choose Cut.
SLABADDVERTEX	Select a slab and choose Add from the Vertex flyout menu.
SLABREMOVEVERTEX	Select a slab and choose Remove Vertex from the Vertex flyout menu.

Continued

TABLE 8.4 *Continued*

Command Prompt	Modify Panel of the Slab Contextual Tab
SLABADDHOLE	Select a slab and choose Add from the Hole flyout menu.
SLABREMOVEHOLE	Select a slab and choose Remove from the Hole flyout menu.
SLABBODY	Select a slab and choose Add from the Body Modifiers flyout menu.
SLABBODYRESTORE	Select a slab and choose Restore from the Body Modifiers flyout menu.
SLABINTERFERENCEADD	Select a slab and choose Add from the Interference Condition flyout.
SLABINTERFERENCE REMOVE	Select a slab and choose Remove from the Interference Condition flyout.

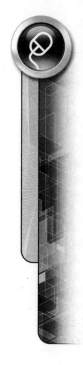

TUTORIAL 8-1: CREATING A CATHEDRAL CEILING

1. Verify that the Accessing Tutor\Imperial or Accessing Tutor\Metric content for Chapter 8 of the CengageBrain http://www.cengagebrain.com has been downloaded to your Accessing Tutor folder on your computer as described in Organizing Tutorial Directories in the Preface.

2. Open Accessing Tutor\Imperial\Ch8\EX 8-1.dwg or Accessing Tutor\Metric\Ch 8\ Ex 8-1.dwg.

3. Choose **Save As** > AutoCAD Drawing from the Application menu and save the drawing as **Lab 8-1** in your student directory.

4. Move the cursor to the Object Snap toggle of the status bar, right-click, and choose **Settings** from the shortcut menu. Edit the **Object Snap** tab of the **Drafting Settings** dialog box: clear all object snap modes and check **Node** object snap. Check the **Object Snap ON (F3)** checkbox and check **Object Snap Tracking ON (F11)**. Select the **Polar Tracking** tab; check ON **Polar Tracking (F10)**. Select the **Dynamic Input** tab. Verify that the **Enable Pointer Input**, **Enable Dimension Input where possible**, and **Show command prompting and command input near the crosshairs** checkboxes are checked. Click **OK** to dismiss the Drafting Settings dialog box.

5. Toggle OFF Snap Mode and Grid Display.

6. Select **Slab** from the **Design** palette and edit the **Properties** palette as follows: Basic-General: Style = **Standard**, Bound Spaces = **By Style**, Mode = **Projected**, Dimensions: Thickness = **6″ [150]**, Vertical Offset = **0**, Horizontal Offset = **0**, Justify = **Top**, Base height = **8′-6″ [2650]**, Direction = **Left**, Overhang = **0**, Perimeter edge = **None**, Slope: Rise = **3″ [25]** and Run = **12″ [100]**. Then respond to the following workspace prompts to place the slab:

   > Specify start point or: *(Select a point near* p1 *using the Node object snap as shown in Figure 8.8.)*
   >
   > Specify next point or: *(Select a point near* p2 *using the Node object snap as shown in Figure 8.8.)*
   >
   > *(Move the cursor up to set the polar tracking angle vector to 90.)*
   >
   > Specify next point or: 6′ [1800] ENTER *(Type* 6′ [1800] *distance to create a vertical edge of the slab 6′ from the front wall.)*

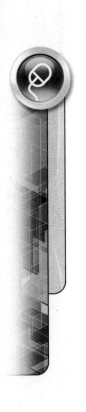

(Move the cursor left to display a polar angle of 180°, right-click, and choose Ortho close *from the shortcut menu to close the slab as shown in Figure 8.9.)*

(Press ESC *to end the command.)*

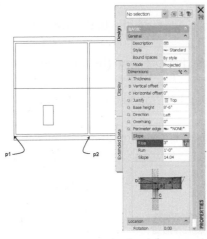

FIGURE 8.8 *Insert point for slab*

7. Select **Slab** from the Design palette and edit the Properties palette as follows: Basic-General: Style = **Standard**, Bound Spaces = **By Style**, Mode = **Projected**; Dimensions: Thickness = **6″ [150]**, Vertical Offset = **0**, Horizontal Offset = **0**, Justify = **Top**, Base height = **8′-6″ [2650]**, Direction = **Right**, Overhang = **0**, Perimeter edge = **None**, Slope: Rise = **3″ [25]**, and Run = **12″ [100]**. Then respond to the following command prompts to place the slab:

Specify start point or: *(Select a point near* p1 *using the Node object snap as shown in Figure 8.9.)*

Specify next point or: *(Select a point near* p2 *using the Node object snap as shown in Figure 8.9.)*

(Move the cursor down to set the polar tracking angle vector to 270.)

Specify next point or: 6′ [1800] ENTER *(Type* 6′ [1800] *distance to create a vertical edge of the slab 6′ below the wall.)*

(Move the cursor to the right, set the polar angle = 0, right-click, and choose Ortho close *from the shortcut menu to close the slab.)*

(Press ESC *to end the command.)*

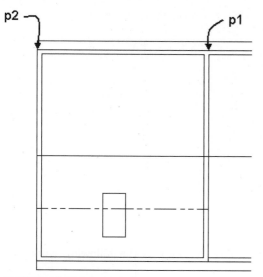

FIGURE 8.9 *Slab insert points*

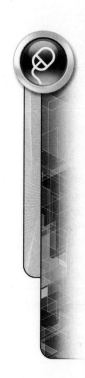

8. In this step you will miter the two slabs: respond to the workspace prompts as shown below.

> *(Select slab at* **p1** *as shown in Figure 8.10 and choose* `Miter` *from the Modify panel of the Slab contextual tab.)*
>
> `Miter by [Intersection/Edges] <Intersection>: E` *(Choose the* `Edges` *option at the dynamic prompts of the workspace.)*
>
> `Select edge on first slab:` *(Select slab at* **p1** *as shown in Figure 8.10.)*
>
> `Select edge on second slab:` *(Select slab at* **p2** *as shown in Figure 8.10.)*

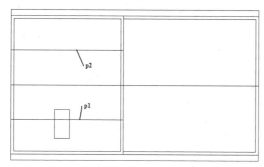

FIGURE 8.10 *Selected slabs for Miter*

9. Select **SW Isometric** from the View Cube to view the slabs mitered as shown in Figure 8.11.

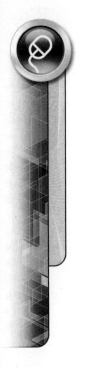

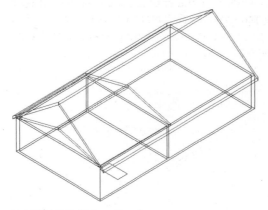

FIGURE 8.11 *Slabs mitered*

10. In this step you cut a hole in the ceiling: respond to the following workspace prompts.

> *(Select slab at* **p1** *as shown in Figure 8.12 and choose* **Add** *from the Hole flyout menu of the Slab contextual tab.)*

> Select closed polylines or connected solid objects to define holes: 1 found *(Select the polyline shown at* **p2** *in Figure 8.12.)*

> Select closed polylines or connected solid objects to define holes: ENTER *(Press* ENTER *to end selection.)*

> Erase layout geometry? [Yes/No] <No>: ENTER *(Press* ENTER *to retain the polyline.)*

> *Press Esc to clear object selection.*

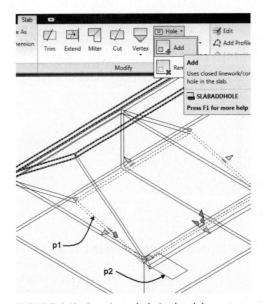

FIGURE 8.12 *Creating a hole in the slab*

11. Select the roof slab at **p1** shown in Figure 8.13 and choose **Add** from Hole flyout menu of the Modify panel in the Slab contextual tab. Create a hole in the roof slab by selecting the polyline at **p2** as shown in Figure 8.13. Retain the polyline geometry. Press Esc to clear object selection.

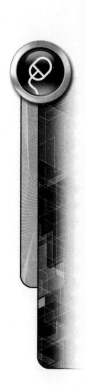

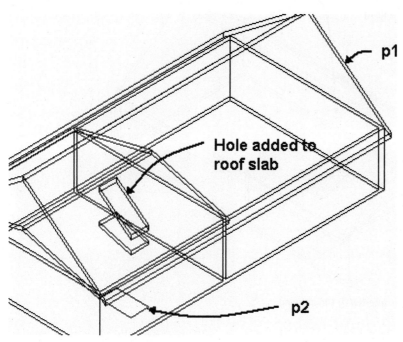

FIGURE 8.13 *Hole added to slab*

12. Select **Slab** from the Design palette and edit the Properties palette as follows: Basic-General: Style = **Standard**, Bound Spaces = **By Style**, Mode = **Projected**, Dimensions: Thickness = **6 [150]**, Vertical Offset = **0**, Horizontal Offset = **0**, Justify = **Top**, Base height = **8'-6 [2650]**, Direction = **Left**, Overhang = **0**, Perimeter edge = **None**, and Rise = **0**. Select the following vertices at **p1, p2, p3, p4**, and **p1** as shown in Figure 8.14.

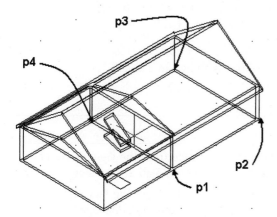

FIGURE 8.14 *Slab vertex points*

13. Slabs representing cathedral and horizontal ceilings are shown in Figure 8.15. Save and close the drawing.

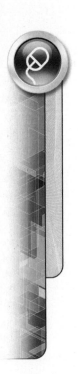

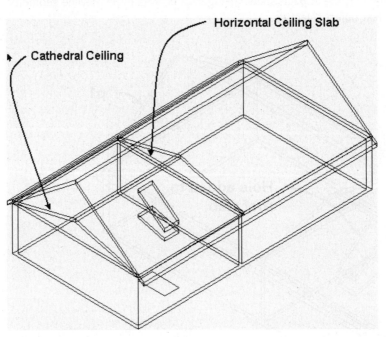

FIGURE 8.15 *Slabs inserted in the drawing*

ACCESSING SLAB STYLES

The Standard slab style is used when you choose SlabAdd from the Design palette. Additional slab styles are located in the Design Tool Catalog—Imperial > Roof Slabs and Slabs > Slabs or Design Tool Catalog—Metric > Roof Slabs and Slabs > Slabs of the Content Browser shown in Figure 8.16. The styles can be accessed in the Style Manager when you choose Content in Places (*Vista—C:\ProgramData\ Autodesk\ACA 2012\enu*) and open the Slab & Slab Edge Styles (Imperial) or Slab & Slab Edge Styles (Metric) drawing located in the Imperial and Metric folders of *Styles* directory. The slab styles include preset thickness, slab edge styles, and material definitions. The features of a slab style are created to represent ceilings with surface finishes, floors with surface finishes, grass, and landscaping surfaces.

The slab styles can be edited in the Style Manager or you can select a slab style and choose Edit Style from the Slab contextual tab as shown in Table 8.5.

TABLE 8.5 *Edit Slab Style command access*

Command prompt	**SLABSTYLEEDIT**
Ribbon	Select a slab and choose Edit Style from the General panel of the Slab contextual tab to open the Slab Style dialog box.
Ribbon	Select a slab and choose Slab Styles from the Edit Style flyout menu of the General tab to open the Style Manager.
Shortcut menu	Select a slab, right-click, and select Edit Slab Style.

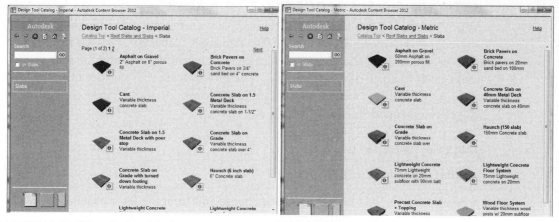

FIGURE 8.16 *Slab styles listed in the Design Tool Catalog of the Content Browser*

The tabs of the Slab Styles dialog box include the same content as the Roof Slab Styles dialog box; therefore, refer to "Roof Slab Styles" in Chapter 7 for a review of the tab contents. The Components tab of the Wood Floor System slab style shown in the Style Manager in Figure 8.17 allows you to specify the slab thickness and thickness offset. The Components tab for the Wood Floor System includes three components: Wood Floor Framing, Subfloor, and Wood Flooring. The Base Thickness value that is entered in the Properties palette should be equal to the vertical dimension of the floor joists. The subfloor and wood flooring components are defined in the style with offset from the base thickness. Therefore, when you place a slab with this style, set the Thickness value in the Properties palette equal to floor joist size. This slab style can be used for any size floor joist with 3/4" [19] subfloor and 3/4" [19] finish flooring. The Materials tab lists the material definitions used in the style.

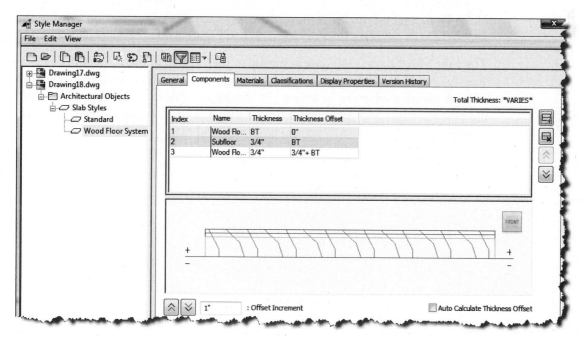

FIGURE 8.17 *Slab Styles dialog box*

ATTACHING A SLAB EDGE STYLE TO A SLAB

When you place a slab using one of the sample slab styles from the Content folder, a slab edge style may be defined within the slab style. Slab edge styles are used in roof slabs to define the fascia and soffit geometry. The geometry for the rim joist or cant strips can be defined in slab edge styles. The geometry for slab edge styles are defined from AEC Profiles. Slab edge styles are created from profiles. Access the Edit Slab Edges command as shown in Table 8.6.

TABLE 8.6 *SlabEdgeEdit command access*

Command prompt	**SLABEDGEEDIT**
Ribbon	Select a slab and choose Edit from the Edge panel of the Slab contextual tab.
Shortcut menu	Select a slab, right-click, and choose Edit Slab Edges.

When you select *Edit* from the Edges panel of the Slab contextual tab, you are prompted to select an edge as shown in the following workspace sequence.

```
Select edges of one slab: (Select a slab.)
Select edges of one slab: ENTER (Press ENTER when finished
selecting edges.)
(The Edit Slab Edges dialog box opens as shown in Figure 8.18.)
```

When you select one or more slabs, the Edit Slab Edges dialog box opens as shown in Figure 8.18. The upper-left window of the dialog box allows you to select a slab edge style from the Edge Style drop-down list, which displays the slab edge styles included in the drawing. Many of the slab styles included in the Design Tool Catalogs (accessible through the Content Browser) include slab edge styles within their slab style. Therefore, if you import a slab style from the Slab & Slab Edge Style (Imperial) [(Metric)] drawing, the associated slab edge style is listed in the Perimeter edge field in the Design tab of the Properties palette. Selected slab edge styles included in the sample slabs of the Slabs category are shown in Figure 8.19.

The Edit Slab Edges dialog box is similar to the Edit Roof Slab Edges dialog box. The A-Overhang, B-Edge Cut, and C-Angle can be edited to modify the slab edge. A negative value can be entered in the overhang field to shrink the edge from the baseline edge. Slab edge styles can be assigned to the edge in the Edge Style column.

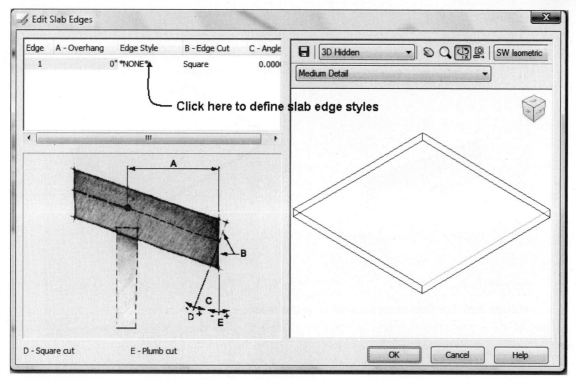

FIGURE 8.18 *Specifying a slab edge style for the edge of a slab*

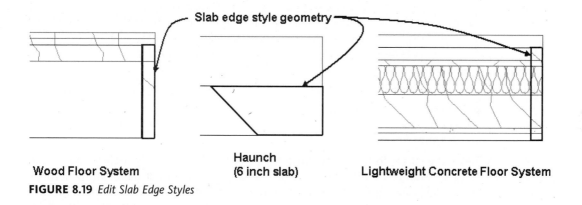

FIGURE 8.19 *Edit Slab Edge Styles*

Slab edge styles can also be assigned to a slab edge by selecting the Edges button in the Properties palette of a selected slab. Selecting the Edges button opens the Slab Edges dialog box, which lists all edges of the slab as shown in Figure 8.20. The edges are numbered according to the sequence of development; however, it may be difficult to determine which edge of the slab you are editing in the dialog box. Therefore, use the Edit Slab Edge command to select an edge and edit the Edit Slab Edges dialog box.

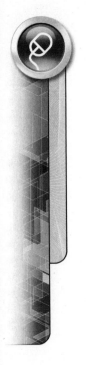

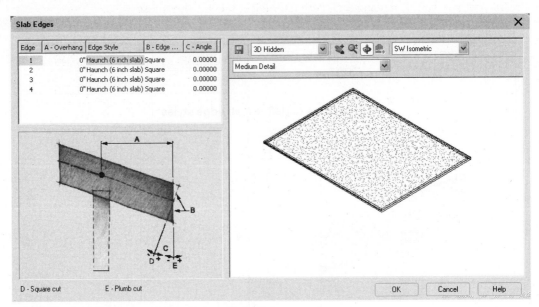

FIGURE 8.20 *Slab Edges dialog box displays all slab edges*

TUTORIAL 8-2: CREATING A FLAT ROOF WITH A SLAB

1. Verify that the Accessing Tutor\Imperial or Accessing Tutor\Metric content for Chapter 8 of the CengageBrain http://www.cengagebrain.com has been downloaded to your Accessing Tutor folder on your computer as described in Organizing Tutorial Directories in the Preface.

2. Open Accessing Tutor\Imperial\Ch8\EX 8-2.dwg or Accessing Tutor\Metric\Ch8\EX 8-2.dwg.

3. Choose Save As > AutoCAD Drawing from the Application menu and save the drawing as Lab 8-2 in your student directory.

4. Move the cursor to the Object Snap toggle of the status bar, right-click, and choose Settings from the shortcut menu. Edit the Object Snap tab of the Drafting Settings dialog box: clear all object snap modes and check Node object snap. Check the Object Snap On (F3) checkbox and check Object Snap Tracking On (F11). Select the Polar Tracking tab; check Polar Tracking (F10). Select the Dynamic Input tab. Verify that the Enable Pointer Input, Enable Dimension Input where possible, and Show command prompting and command input near the crosshairs checkboxes are checked. Click OK to dismiss the Drafting Settings dialog box.

5. Turn OFF Snap Mode and Grid Display on the Application status bar.

6. Right-click over the tool palette titlebar and choose New Palette. Overtype Slabs to name the new palette. Choose Content Browser from the Tools flyout of the Build panel in the Home tab. Open the Design Tool Catalog–Imperial or Design Tool Catalog–Metric. Navigate to the Roof Slabs and Slabs > Slabs category. Choose the i-drop for the Cant slab and drag to the new Slab palette. Minimize the Content Browser.

7. Choose the Cant tool from the Slabs tool palette. To place a slab with a Cant strip edge, edit the Properties dialog box as follows: Basic-General: Style = Cant, Bound Spaces = By Style, Mode = Projected, Dimensions: Thickness = 6" [150], Vertical Offset = 0, Horizontal Offset = 0, Justify = Bottom, Base height = 8' [2438],

Direction = Left, Overhang = 0, Perimeter edge = Cant, Slope: Rise = -.02" [−.17], Run = 12" [100], and Slope = 359.90. Then respond to the following workspace prompts to place the slab:

Specify start point or: *(Select a point near* **p1** *using the Node object snap as shown in Figure 8.21.)*

Specify next point or: *(Select a point near* **p2** *using the Node object snap as shown in Figure 8.21.)*

Specify next point or: *(Select a point near* **p3** *using the Node object snap as shown in Figure 8.21.)*

Specify next point or: *(Select a point near* **p4** *using the Node object snap as shown in Figure 8.21.)*

Specify next point or: *(Select a point near* **p1** *using the Node object snap as shown in Figure 8.21.)*

Specify next point or: ENTER *(Press ENTER to create the slab.)*

Specify next point or: ESCAPE *(Press ESCAPE to end the command.)*

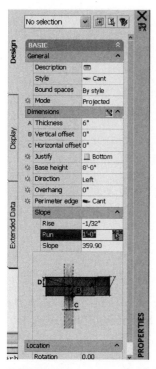
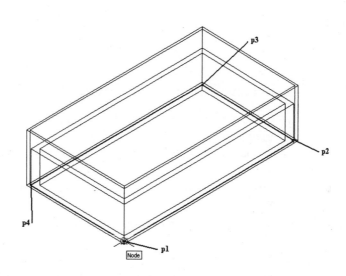

FIGURE 8.21 *Insert points of the slab*

8. Select **Front** from the View Cube.

9. Select **Zoom Window** from the Zoom flyout menu of the Navigation bar and respond to the following command prompts:

Specify first corner: 128',72', 10' [27500,15500, 3000] ENTER *(Specify the corner of the zoom window.)*

Specify opposite corner: 133',72',5' [28500,15500,2000] ENTER *(Specify the second corner of the zoom window. Cant strip added to slab as shown in Figure 8.22.)*

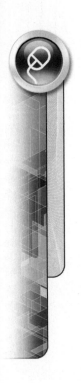

10. Select **Top** from the View Cube. Choose **SW Isometric** from the View Cube to view the flat roof as shown in Figure 8.23.

11. Save and close the drawing.

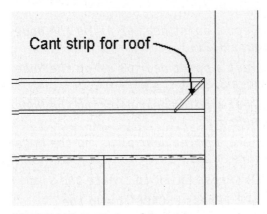

FIGURE 8.22 *Front view of Cant strip*

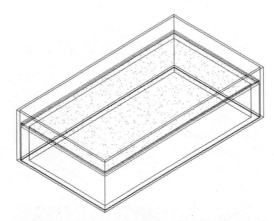

FIGURE 8.23 *SW Isometric view of roof (Imperial shown)*

Editing the Slab Edge Styles of a Slab

The style of the slab edge can be edited by choosing Edit Slab Edge Style from the shortcut menu of a selected slab. Refer to Table 8.7 for **SlabEdgeStyleEdit** command access.

TABLE 8.7 *SlabEdgeStyleEdit command access*

Command prompt	SLABEDGESTYLEEDIT
Ribbon	Select a slab that includes a slab edge style and choose Slab Edge Style from the Edit Style flyout menu of the General panel in the Slab tab.
Shortcut menu	Select a slab that includes a slab edge style, right-click, and choose Edit Slab Edge Style.

When you select a slab and choose **Slab Edge Style** from the Edit Style flyout menu, you are prompted to select a slab edge. If the selected edge does not include an edge style, the command ends. If the slab edge includes a slab edge style, the Slab Edge Styles dialog box opens, displaying the properties of the slab edge style. The Slab Edge Styles dialog box is similar to the Roof Slab Edge Styles dialog box. The Design Rules tab of this dialog box, shown in Figure 8.24, allows you to specify the profile for the edges of the slab.

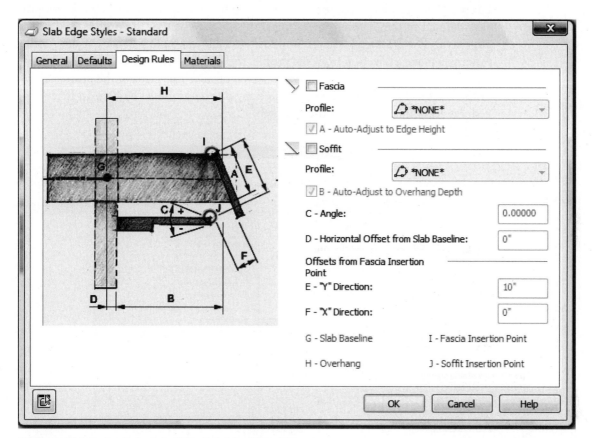

FIGURE 8.24 *Design Rules tab of the Slab Edge Styles dialog box*

Adding an Edge Profile in the Workspace

A profile can be added to the drawing for the edge of the slab while in the workspace without opening the Style Manager and creating a new profile. The Add Edge Profiles command is included in the shortcut menu of a selected slab. This command allows you to create a profile while remaining in the workspace. The edge selected for the command must have a defined edge style. Access the Add Edge Profiles command as shown in Table 8.8.

TABLE 8.8 *Add Edge Profiles command access*

Command prompt	SLABADDEDGEPROFILES
Ribbon	Select a slab and choose Add Profiles from the Edge panel of the Slab contextual tab.
Shortcut menu	Select a slab, right-click, and choose Add Edge Profiles.

When you select a slab and choose Add Profiles from the Edge panel, you are prompted to select a slab edge. The slab edge must include a defined roof slab edge style. If the slab edge includes an overhang, you can add a profile for the soffit and fascia. If no overhang is included, a message box opens and asks if you want to set an overhang value and edge style. The overhang value is specified in the Edit Slab Edges dialog box. When the Edit Slab Edges dialog box is dismissed, the Add Fascia/Soffit Profiles dialog box opens as shown in Figure 8.25. If no overhang is specified and the Standard edge style is specified, only a profile for a fascia can be created. The Add Fascia/Soffit Profiles dialog box includes a drop-down list of profiles defined in the drawing. Selecting from the list changes the profile used in the edge style. The Start from scratch profile option allows you to edit the current profile. After you make a selection from the profile, the Edit In Place: Slab Edge Profile contextual tab opens, allowing you to zoom to the profile. The grips of the profile shown at the right in Figure 8.25 can be selected to edit the profile. The Edit In Place procedures are similar to editing the roof slab.

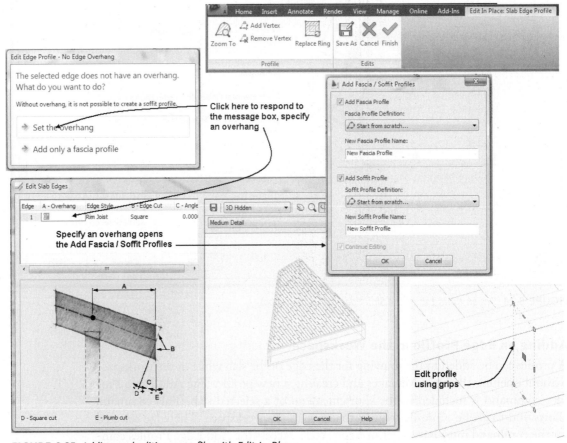

FIGURE 8.25 *Adding and editing a profile with Edit In Place*

EDITING AN EDGE PROFILE IN PLACE

After a profile has been changed using the Add Edge Profiles command, additional editing can be performed with the **Edit Edge Profile In Place (SlabEdgeProfile-Edit)** command. This command allows you to perform additional editing of the profile. The Edit Edge Profile In Place command allows you to add or remove vertices from the profile in the workspace. Access the Edit Edge Profile In Place command as shown in Table 8.9.

TABLE 8.9 *Edit Edge Profile In Place command access*

Command prompt	**SLABEDGEPROFILEEDIT**
Ribbon	Select a slab and choose Edit Profile In Place.
Shortcut menu	Select a slab, right-click, and choose Edit Edge Profile In Place.

When you select a roof slab and choose the Edit Profile In Place command from the Slab contextual tab, the Edit In Place: Slab Edge Profile tab is displayed. The commands of the Edit In Place: Slab Edge Profile tab allow you to edit slabs in a manner similar to profiles for roof slab edges.

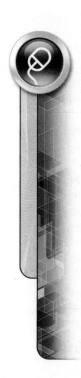

TUTORIAL 8-3: CREATING A SLAB FOR A CRAWL SPACE

1. Verify that the Accessing Tutor\Imperial or Accessing Tutor\Metric content for Chapter 8 of the CengageBrain http://www.cengagebrain.com has been downloaded to your Accessing Tutor folder on your computer as described in Organizing Tutorial Directories in the Preface.

2. Open Accessing Tutor\Imperial\Ch8\EX 8-3.dwg or Tutor\Metric\Ch8\Ex 8-3.dwg. In this tutorial you will modify a Rim Joist slab edge style and add an insulation component to the slab style.

3. Choose Save As > AutoCAD Drawing from the Application menu and save the drawing as Lab 8-3 in your student directory.

4. Move the cursor to the Object Snap toggle of the status bar, right-click, and choose Settings from the shortcut menu. Edit the Object Snap tab of the Drafting Settings dialog box: clear all object snap modes and check Node object snap. Check the Object Snap On (F3) checkbox and check Object Snap Tracking On (F11). Select the Polar Tracking tab; check Polar Tracking (F10). Select the Dynamic Input tab. Verify that the Enable Pointer Input, or Accessing Enable Dimension Input where possible, and Show command prompting and command input near the crosshairs checkboxes are checked. Click OK to dismiss the Drafting Settings dialog box.

5. Turn OFF Snap Mode on the Application status bar. Choose the Toggle off Surface Hatch Toggle in the Drawing Window status bar. Verify Allow/Disallow Dynamic UCS, Dynamic Input, and Quick Properties are toggled ON in the Application status bar.

6. Metric only: The rim joist slab edge style is not included in the Wood Floor System metric slab style. To import a rim joist profile for the perimeter edge style, choose the Manage tab. Choose Style Manager from the Style & Display panel. Then choose Open from the toolbar of the Style Manager. Choose Content from the Places panel. Choose Styles > Metric > Slab & Slab Edge Styles (Metric).dwg. Click Open to dismiss the Open Drawing dialog box. In the left pane of the Style Manager, expand the Slab & Slab Edge Styles (Metric).dwg, to display **Architectural Objects > Slab Edge Styles**. Select the **Rim Joist** slab edge style, then right-click and choose **Copy**. Select the current drawing, **Lab 8-3,** in the left pane of the Style Manager and right-click **Paste**. Click **OK** to dismiss the Style Manager and end the Metric step.

7. Choose the **Slabs** palette that you created in Tutorial 8-2. Choose the Home tab. Choose Content Browser from the Tools flyout of the Build panel. Open the Design Tool Catalog—Imperial or Design Tool Catalog—Metric. Navigate to the Roof Slabs and Slabs > Slabs category. Choose the i-drop for the Wood Floor System and drag it to the Slab palette. Close the Content Browser.

8. Choose **Wood Floor System**, right-click, and choose **Apply Tool Properties to** > **Linework and Walls**. Respond to the workspace prompts as follows:

   Select walls or polylines: *(Select all the walls using a crossing window from point* **p1** *to* **p2** *as shown in Figure 8.26.)*

   Select walls or polylines: ENTER *(Press ENTER to end the selection.)*

   Erase layout geometry? *(Choose* **No** *from the workspace prompts.)*

   Specify slab justification: *(Choose* **Bottom** *from the workspace prompts.)*

   Specify wall justification for edge alignment: *(Choose* **Baseline** *justification.)*

   Select the pivot edge for the Slab: *(Choose the wall edge at* **p3** *as shown in Figure 8.26.)*

9. Retain selection of the slab and edit the Properties palette as follows: Style = **Wood Floor System**, Bound spaces = **By Style**, Thickness = **9.25"** [235], Vertical Offset = **1.5"** [38], and Horizontal Offset = **0**. Click the **Edges** button within the Properties palette and verify **or** set the **Edge Style** to **Rim Joist** for each of the edges. Click **OK** to dismiss the Slab Edges dialog box. Press ESC to clear the selection.

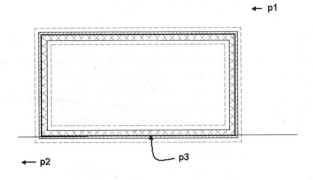

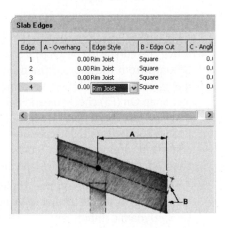

FIGURE 8.26 *Location of slab for a crawl space*

10. Select a point near **p1** as shown in Figure 8.27 and move the cursor to point p2; click to select the walls and slab. Choose **Edit in Section** from the General panel of the Multiple Objects tab. Respond to the workspace prompts as follows:

    Specify first point of section line or ENTER to change UCS: *(Select a point near* **p3** *as shown in Figure 8.27.)*

    Specify next point of section line: *(Select a point near* **p4** *as shown in Figure 8.27.)*

    Specify next point of section line or [Break]: *(Press ENTER to end selection.)*

    Specify section extents: *(Select a point near* **p5** *as shown in Figure 8.27.)*

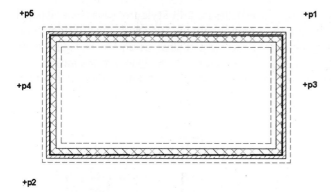

FIGURE 8.27 *Selection of object for Edit in Section view*

11. In this step you will edit the size of the rim joist. Select the slab as shown in the section at **p1** as shown in Figure 8.28 and choose **Edit Profile In Place** from the Edge panel of the Slab contextual tab. Select the Rim Joist at **p1** to display the grips as shown at the left. In this step the AEC Profile for the rim joist will be modified. Verify that Polar Tracking is On in the Application status bar, select the Edge grip shown at **p2** in Figure 8.28, move the cursor up to display the tracking vector, and type **2 [65] ENTER** in the workspace. Choose **Finish** in the Edits panel of the Edit In Place: Slab Edge Profile tab. The sill plate detail component can be added into the gap between the slab and foundation when the section is developed.

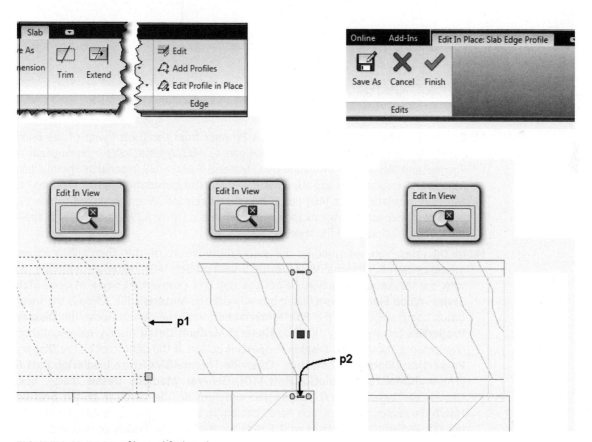

FIGURE 8.28 *AEC Profile modified in place*

12. Remain in Edit in View, choose the slab, and choose **Edit Style** from the General panel of the Slab contextual tab. Choose the **Components** tab and select the **Subfloor** component. Choose **Add Component** from the right toolbar as shown in Figure 8.29. Choose the **Index 3** component, click in the **Name** column, overtype **Insulation**, and click in the **Thickness** column and overtype **–6[–150]**. Click **OK** to dismiss the **Slab Styles-Wood Floor System** dialog box.

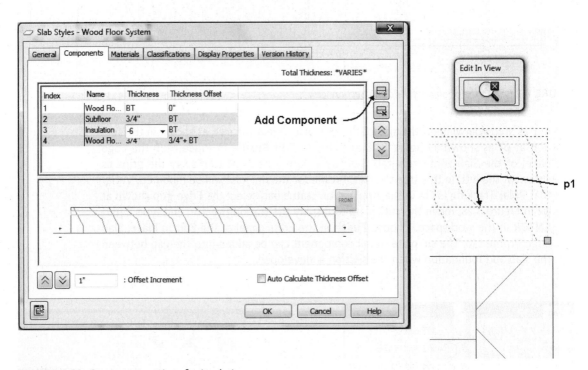

FIGURE 8.29 *Component settings for insulation*

13. Choose the Home tab. Choose Content Browser from the Tools flyout of the Build panel. Open the Visualization Catalog, navigate to AEC Material Tools > US Imperial > Thermal and Moisture > Insulation [AEC Material Tools > US Imperial > Thermal and Moisture > Insulation]. Drag the **Thermal & Moisture.Insulation.Batt.6 [Thermal & Moisture.Insulation.Batt.140]** i-drop into the workspace. When prompted to *Select a component or object*, click the pickbox at **p1** as shown in Figure 8.29. Press ENTER to apply the material definition to the style.

14. In this step you will modify the display properties of the Thermal&Moisture. Insulation.Batt.6 [Thermal & Moisture.Insulation.Batt.140] material definition to view the insulation in section. Select the slab and choose **Edit Style** to open **Slab Styles–Wood Floor System** dialog box. Choose the **Materials** tab. Choose the **Insulation** component. Click the **Edit Material** button at the right. Choose the **Display Properties** tab and verify that the **General Medium Detail** display representation is selected. Click the **Edit Display Properties** button at the right to open the **Display Properties (Material Definition Override-Thermal&Moisture.Insulation.Batt.6 [Thermal&Moisture.Insulation.Batt.140])-General Medium Detail** dialog box. Choose the **Layer/Color/Linetype** tab and turn on the visibility of the **Section Hatch**. To center the insulation hatch pattern in the joist cavity, choose the **Hatching** tab and edit the Y offset to **2.5 [64]** for the **Section Hatch** component. Click **OK** to dismiss all dialog boxes and view the insulation as shown at the right in Figure 8.30.

Accessing AutoCAD® Architecture 2012

Accessing AutoCAD® Architecture 2012

autodesk Press

WILLIAM G. WYATT

For product information and technology assistance, contact us at
Cengage Learning Customer & Sales Support, 1-800-354-9706

For permission to use material from this text or product,
submit all requests online at **www.cengage.com/permissions**.
Further permissions questions can be e-mailed to
permissionrequest@cengage.com

CHAPTER 4 SPACE PLANNING AND MASS MODELING 235

Introduction 235 • Objectives 235 • Geometry Forms of Spaces 235 • Creating Associative Spaces from Geometry 236 • Using the Space Add Tool to Create Spaces from a Floor Plan 239 • Tagging Spaces 240 • Editing Spaces 243 • Inserting a Space 250 • Dragging the Insertion Point 255 • Modifying Nonassociative Spaces 257 • Creating a Space Style 265 • Evaluating Spaces for Space Planning 267 • Creating Freeform Spaces 277 • Creating Mass Models 280

CHAPTER 5 PLACING DOORS AND WINDOWS 312

Introduction 312 • Objectives 312 • Placing Doors in Walls 312 • Creating and Using Door Styles 323 • Placing Windows in Walls 336 • The Windows Palette 339 • Creating a New Window Style 339 • Placing Corner Windows in Walls 340 • Creating Muntins for Windows 345 • Adding Shutters to a Window Style 349 • Creating Openings 353 • Adding and Editing the Profile of Doors, Windows, and Openings 355 • Using AEC Modify Tools to Move Doors, Windows, and Openings 356 • Applying Tool Properties 359 • Importing and Exporting Door and Window Styles 360

CHAPTER 6 DOOR/WINDOW ASSEMBLIES AND CURTAIN WALLS 374

Introduction 374 • Objectives 374 • Adding Door/Window Assemblies 374 • Components of a Door/Window Assembly Style 378 • Modifying Door/Window Assemblies 383 • Creating Curtain Walls 404 • Converting Walls to Curtain Walls 406 • Curtain Wall Styles 407 • AEC Polygons for Curtain Wall Styles 414 • Converting Sketches to Curtain Walls 417

CHAPTER 7 CREATING ROOFS AND ROOF SLABS 420

Introduction 420 • Objectives 420 • Creating a Roof with RoofAdd 420 • Defining Roof Properties 426 • Creating a Gable Roof 428 • Editing an Existing Roof to Create Gables 432 • Creating a Shed Roof 435 • Creating a Gambrel Roof Using a Double-Sloped Roof 435 • Creating a Flat Roof 440 • Converting Polylines or Walls to Roofs 440 • Creating Roof Slabs 441 • Modifying Roof Slabs 449 • Tools for Editing Roof Slabs 452 • Determining Roof Intersections 465 • Creating Roof Slab Styles 467 • Using Add Edge Profile in the Workspace 474 • Extending Walls to the Roof 476

CHAPTER 8 CREATING SLABS FOR FLOORS AND CEILINGS 487

Introduction 487 • Objectives 487 • Adding and Modifying a Floor Slab 487 • Modifying the Slab Using Properties and Grips 495 •

Accessing Slab Styles 502 • Attaching a Slab Edge Style to a Slab 504 • Editing an Edge Profile in Place 510

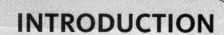

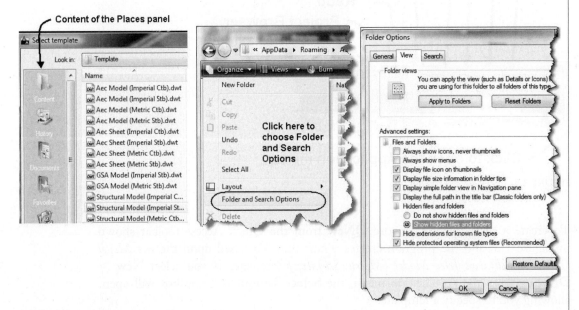

Path to style for the tool

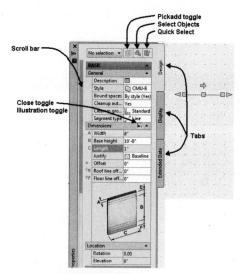

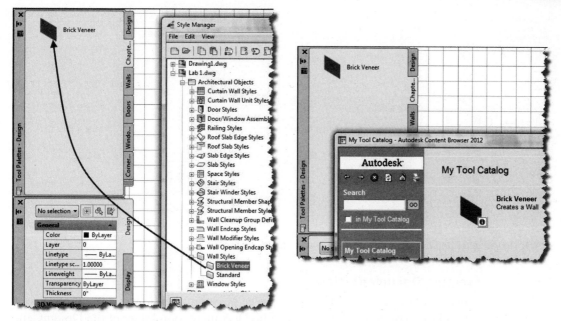

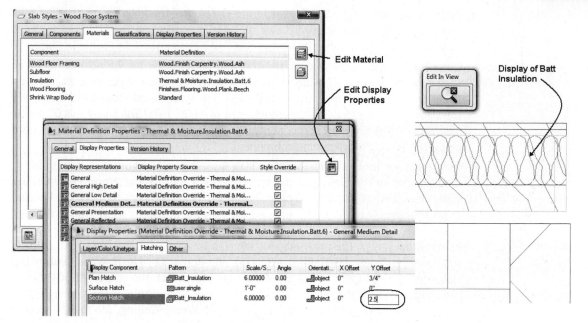

FIGURE 8.30 *Display settings for insulation material*

15. Choose the **Exit Edit in View** button of the Edit in View message box to close the section.

16. Choose **SW Isometric** from the View Cube. Select a wall and choose **Edit Style** from the General panel of the Wall contextual tab. Choose the **Display Properties** tab and clear the **Style Override** checkbox for the **Model** display representation. Click **OK** to dismiss the **Wall Style Properties—FND [BrickFND]** dialog box. Choose **Realistic** from the Visual Styles flyout menu of the Visual Styles panel in the View tab.

17. Save and close the drawing.

PROJECT 8-4: CREATING SLABS FOR A BASEMENT

1. Verify that the Accessing Tutor\Imperial or Accessing Tutor\Metric content for Chapter 8 of the CengageBrain http://www.cengagebrain.com has been downloaded to your Accessing Tutor folder on your computer as described in Organizing Tutorial Directories in the Preface.

2. Open AutoCAD Architecture 2012 and select File > Project Browser from the menu bar. Use the Project Selector drop-down list to navigate to your Accessing Tutor\Ch8 directory. Select New Project to open the Add Project dialog box. Type **15** in the Project Number field and type **Ex 8-4** in the Project Name field. Click Create from template project, choose Browse, and edit the path to *Accessing Tutor\Imperial\Ch8\ EX 8-4\Ex 8-4.apj* or *Accessing Tutor\Metric\Ch8\EX 8-4\Ex 8-4.apj* in the Look in field. Click Open to close the Select Project dialog box. Click OK to dismiss the Add Project dialog box. Click Close to dismiss the Project Browser.

3. Select the **Constructs** tab of the **Project Navigator** and open the **Basement** file.

4. Create slab **A** as shown in Figure 8.31. Import the **Wood Floor System** from the Slabs category of the Roof Slabs and Slabs category in the Design Tool Catalog—Imperial or Design Tool Catalog—Metric of the Content Browser into your drawing. Choose **Apply Tool properties to > Linework and Walls** to convert the exterior basement walls to the Wood Floor System. When you create the slab, specify Retain the layout geometry, **Bottom** justified, and baseline justified. Assign any of the four slab edges as the pivot edge for the slab. Select the slab, edit the Properties palette Thickness = **9.25″ [235]**, and adjust the Vertical Offset to **1 1/2″ [38]**.

5. Apply Isolate Objects > Edit In Section to slab A. [Metric only-import the Rim Joist slab edge style from the Lab 8-3.dwg of Tutorial 8-3. Define the Rim Joist for each edge of Slab A]. Adjust the rim joist to 9.25" [235] using the Edit Profile In Place.

6. Cut a hole in slab A for a proposed stairwell represented by the polyline at **p6** as shown in the drawing file. Choose slab **A** and choose **Copy Style** from the General panel of the Slab contextual tab. Name the slab style **Wood-Floor-Ceiling**. Modify the slab style and add a Gypsum component of **–1/2" [–13]** Thickness and **0"** Thickness Offset. The Gypsum component will represent the ceiling gypsum wall board below the Wood Floor Framing component.

7. Create a Slabs tool palette. To create slab **B** as shown in Figure 8.31, select slab A, drag the slab to the Slabs tool palette to create a tool for the **Wood-Floor-Ceiling slab**. View the building in isometric. Toggle Off Surface hatch to view the Polyline at **p6** as shown in the drawing. Convert the Polyline at p6, (as shown in the drawing file) to the Wood-Floor-Ceiling slab style. Project the slab 8'-0" [2438] using Bottom justification. Assign the pivot edge of the slab to align with the front wall. Edit the slab Thickness to **6" [150]**, Rise = **9" [75]**, and Run = **12" [100]**. After creating the slab, use grips to retain the slope of the slab and stretch the slab **5' [1500]** toward the inside of the house. Select slab **B** and choose **Copy Style** of the General panel in the Slab contextual tab. Name the slab style **Wood–Ceiling**. Modify the slab to remove the subfloor and wood flooring.

8. Choose the **Material** tool from the Design palette and assign **Finishes.Plaster and Gypsum Board.Gypsum Wallboard.Painted.White** material to the gypsum component for each slab.

9. Import the **Concrete Slab on Grade** tool from the Roof Slabs and Slab > Slabs category in the Design Tool Catalog—Imperial or Design Tool Catalog—Metric of the Content Browser. Create a slab for the garage area. Convert the slab to a polyline, as shown at **p5** in the drawing. The polyline is drawn on the layer Slab Geometry. Set the Mode to Direct, Top justified and assign the baseline pivot edge between points **p1** to **p2** to slope the slab down toward the right wall. Verify Thickness = **4" [102]**, adjust the Slope: Rise = **–.125" [–.5]** and Run = **12" [100].**

10. Create a slab for the living space using the **Concrete Slab on Grade** tool that passes through **p1, p2, p3**, and **p4**. Create the slab Justify = **Top**, Thickness = **4" [102]**, and Elevation = **0**. Use AecEditInSection to view and edit the slab in section.

11. Select **Realistic** from the Visual Styles flyout menu of the Visual Styles panel in the View tab.

12. Select **SW Isometric** from the View Cube.

13. Turn off all layers except the **A-Slab [S232G]** layer to view the slabs as shown in Figure 8.31. Save the file and close the project upon completion.

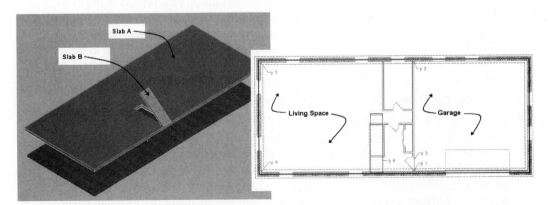

FIGURE 8.31 *Slab added to the basement floor plan*

SUMMARY

- Slabs are added with the **Slab** command located on the Design tool palette.
- A closed polyline can be converted to a slab.
- The Direct Mode slab is created at the same elevation as the points selected.
- A Projected Mode slab is created at a base height elevation from the selected points.
- The shortcut menu of a selected slab allows you to trim, extend, miter, cut, and add holes to the slab.
- Slab edge styles attached to a slab edge customize the edge of the slab.
- Slab edge styles and slab styles are defined in the **Style Manager**.
- Profiles of the slab edge can be edited in place with the **Add Edge Profiles** and **Edit Edge Profile in Place** commands.

Refer to e **Review Questions** folder which includes review questions for Chapter 8 of Student Companion Site of **CengageBrain http://www.cengagebrain.com** described in the Preface.

NOTE

CHAPTER 9

Stairs and Railings

INTRODUCTION

During the development of a floor plan, stairs must be inserted to determine the required space for the stairwell. This chapter will explain how to insert straight, U-shaped, multi-landing, and spiral stairs. When you place a stair, the design rules of a stair style automatically calculate the riser dimension, number of risers, size of tread, and the straight length. The style of the stair can control the landing extensions and landing size. The application of railings to the stairs and platforms will be presented in this chapter. Railing styles allow you to create guardrails appropriate for decks or railings, with custom shapes for interior stair applications.

OBJECTIVES

After completing this chapter, you will be able to

- Use the **Stair** command (**StairAdd**) to create Straight, Multi-landing, U-Shaped, U-Shaped Winder, and Spiral stairs.

- Create a stair tower in a project and a custom stair.

- Constrain the number of risers, tread size, riser size, and run of a stair using Calculation Rules.

- Edit the style, shape, turn type, width, height, and justification to create stairs.

- Identify the purpose of grips located on various stair shapes.

- Create, edit, import, and export stair and railing styles using the Style Manager or Content Browser.

- Control the visibility of the lower or upper portions of stairs using the display properties.

- Modify the stair edge with the following Customize Edge options: Offset, Project, Remove Customization, and Generate Polyline.

- Anchor an isolated stair to a stair landing and anchor a railing to an AEC object.

- Use the Railing command (RailingAdd) to create railings that are freestanding, attached to stairs, and attached to stair flights.

- Create railing styles.

CREATING THE STAIR CONSTRUCT

Stairs can be placed in a drawing without the use of a project; however, the tools of the **Project Navigator** assist in the development of the stair. A stair is created as a construct that spans multiple levels. The stair construct is assigned to each level the stair serves. Therefore, the stair would be assigned to levels 1 and 2 of a two-story house. When a View drawing is developed for each floor plan, the floor plan construct and stair construct for that level come together in the View drawing.

The stair construct can also be included in the model View drawing to visualize how the stair fits throughout the various levels. To create a stair construct, create a new Construct drawing and attach the construct floor plan for level 1 as a reference file. This allows you to determine an appropriate start and end point for the flight based upon existing walls of the floor plan. After creating the stair construct, you can detach the floor plan construct to remove the duplication of walls. The View drawing for a level displays the stair construct with the upper or lower flight components displayed appropriate to the level, as shown at the right in Figure 9.1.

Steps to Create a Stair Construct

1. Open the **Project Navigator** and create two levels. Select the **Constructs** tab of the project and create two constructs assigned to each level for the floor plan content of the level.

2. Select the Construct category, right-click, and choose **New Construct**. Type **Stair**, the name of the construct, and assign levels 1 and 2 as shown at the left in Figure 9.1. Check the Open in drawing editor checkbox.

3. Open the **Project Navigator** and select the **Construct** tab. Select the construct for level 1, Floor 1, right-click, and choose **Xref Overlay** from the shortcut menu to view the walls of the first floor.

4. Create the stair.

5. Select **Manage Xrefs** in the Drawing Window status bar to open the External References palette. Select Floor 1, right-click, and choose **Detach** to remove the display of level 1 walls.

6. Select the **Views** tab and open the View drawing for each level in the **Project Navigator** to view the stair shown at the right in Figure 9.1.

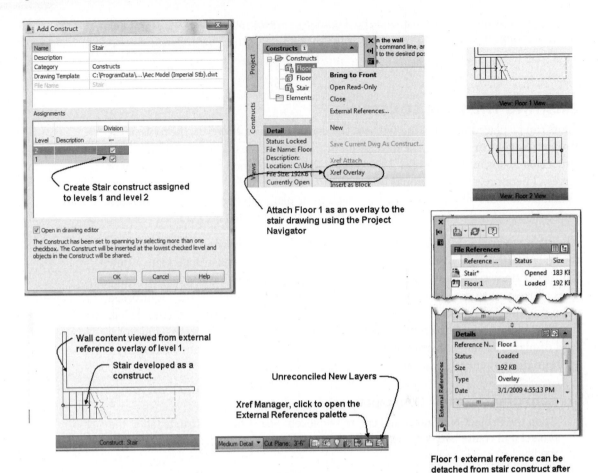

FIGURE 9.1 *Creating a Stair Construct drawing*

USING THE STAIRADD COMMAND

Stairs are placed in a drawing with the **Stair** (**StairAdd**) command selected from the Design palette or the Build panel of the Home tab as shown in Figure 9.2. Access the **Stair** command as shown in Table 9.1. The available shapes of stairs include U-Shaped, Multi-landing, Spiral, and Straight. The available space for the stair is a major consideration for the selection of the shape. The Properties palette for stairs shown in Figure 9.2 allows you to set the height and run of the stair based upon the **Calculation Rules**. The **Calculation Rules** button opens the **Calculation Rules** dialog box shown in Figure 9.5, which allows you to define the relationship of tread size, riser size, straight length, and riser count. Stair styles that include Cantilever, Concrete, Half Wall Rail, Ramp Concrete, Ramp Concrete Curb, Ramp Steel, Standard, Steel, Wood-Housed, and Wood Saddle can be imported into the drawing. Each of these styles changes the appearance of the stair and the means of support for the stair. Prior to placing the stair, you should first review the content in the Design tab of the Properties palette for stairs.

TABLE 9.1 *StairAdd command access*

Command prompt	STAIRADD
Ribbon	Select Stair from the Build panel of the Home tab.
Design palette	Select Stair from the Design palette.

Selecting the **Stair** command opens the Properties palette, and you are prompted in the workspace to specify the start point of the flight. The start flight and end flight points of a multi-landing stair are shown in Figure 9.2. The workspace sequence for the stair created in Figure 9.2 is shown below. The polar tracking tip indicates the distance between flight start and flight end. The number of risers is dynamically displayed in fraction format as you move the mouse to the flight end. The numerator of the fraction is the riser count currently displayed, and the denominator is the total risers required.

```
Flight start point or: (Select flight start for the multi-
landing stair at p1 as shown in Figure 9.2.)

Flight endpoint or: (Select flight end at p2 as shown in
Figure 9.2.)
```

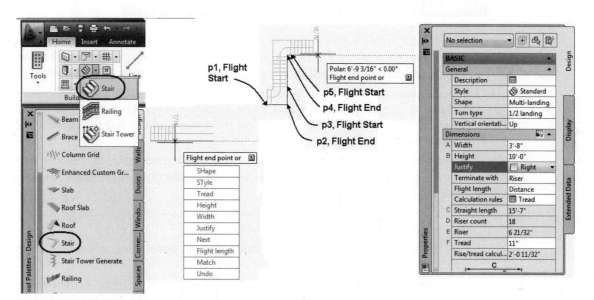

FIGURE 9.2 *Accessing the Stair command and specifying the location of flight*

```
Flight start point or: (Select flight start point at p3 as
shown in Figure 9.2.)

Flight endpoint or: (Select flight start point at p4 as shown
in Figure 9.2.)

Flight start point or: (Select flight start at p5 as shown in
Figure 9.2.)

Flight endpoint or: (Dynamic display of current cursor
location as shown in Figure 9.2.)
```

Properties of a Stair

The options of the StairAdd command are set in the Properties palette shown at the right in Figure 9.2. The following options can be selected from the workspace: Shape, Style, Tread, Height, Width, Justify, Next, Flight Length, Match, and Undo. Additional options can be selected from the Properties palette as described below. When you choose a Shape option, the content in the Design tab of the Properties palette changes according to the needs of the shape. The shape and turn options are shown in Figure 9.3. The **Calculation Rules** button opens the **Calculation Rules** dialog

box shown in Figure 9.5. This dialog box allows you to select among the following control options: A-Straight Length, B-Riser Count, C-Riser, and D-Tread. If Automatic is toggled ON for all fields, the stair dimensions are calculated based upon the **Design Rules** tab of the Stair Styles dialog box. Selecting a control option such as **Riser Count** will calculate the stair dimensions based upon the **Design Rules** tab of the **Stair Styles** dialog box, and the results will be displayed in the **Calculation Rules** dialog box. The tread dimension can be overridden in the Calculation Rules dialog box as shown in Figure 9.5. The name of the control option selected is then displayed in the Calculation Rules field. If you insert values in the Calculation Rules dialog box that violate the parameters specified in the Design Rules tab for the stair style, a warning dialog box is displayed and the solution tip is displayed on the stair, as shown at the right in Figure 9.5.

TUTORIAL 9-0: EXPLORING THE PROPERTIES OF STAIRS

1. Download Chapter 9 files from the Accessing Tutor\Imperial or Accessing Tutor\Metric category of the Student Companion site of CengageBrain http://www.cengagebrain.com described in the Preface to your Accessing Tutor folder.

2. Open Autodesk AutoCAD Architecture 2012. Open *Accessing Tutor\Imperial\Ch9\Ex 9-0.dwg* or *Accessing Tutor\Metric\Ch9\Ex 9-0.dwg*. Choose **Save As > AutoCAD Drawing** from the Application menu. Save the drawing as **Lab 9-0** to Ch9 folder of your student folder.

3. Right-click over the Object Snap toggle of the Application status bar; choose Settings. Clear all object snap modes. Check the Node object snap. Choose **OK** to dismiss the Drafting Settings dialog box.

4. The drawing file consists of the following four stairs of the following shapes: U-Shaped, Multi-Landing, Spiral, and Straight.

5. Select each stair and verify Style = **Standard** in the General section of the Design tab in the Properties palette.

6. Select the stair A as shown in Figure 9.3. Verify the Shape is U-Shaped in the General section of the Design tab in the Properties palette. Select the remaining stairs and verify the shapes as defined in Figure 9.3.

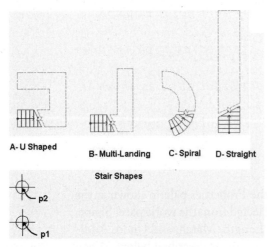

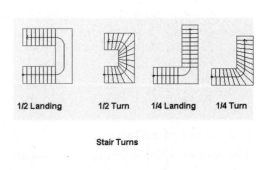

FIGURE 9.3 *Stair shape and turn options*

7. Choose the Stair tool from the Design palette. Choose the U-Shaped from the Shape flyout list in the Properties palette. Turn types may be specified for U-Shaped and

Multi-landing stairs. The 1/2 landing turn type was specified for the stair A as shown in Figure 9.3.

*(Move the cursor to Turn type field in the General section, choose the flyout and select **1/2 turn**. The Winder Style field is added to the General section. Winder Styles may be imported in the Style Manager.)*

8. Vertical Orientation includes Up and Down options. This option is available for all stair shapes and allows the stair to be developed Up or Down from the start point.

   *(Move the cursor to the Vertical Orientation field. Choose the **Up** option from the flyout menu for the Vertical Orientation field.)*

9. The Horizontal orientation field is displayed for Spiral and U-Shaped stairs. Clockwise or Counterclockwise horizontal options are available.

   *(Move the cursor to the Horizontal Orientation field; choose **Counterclockwise** orientation.)*

10. Respond to the workspace prompts as follows to create a stair.

    Flight start point or: *(Choose a point at* **p1** *to start the stair as shown in Figure 9.3.)*

    Flight end point or: *(Choose a point at* **p2** *to specify the end of the stair.)*

    Flight start point or: ESC*(Choose* **ESCAPE** *to exit the stair command.)*

11. The Dimensions section of the Design tab may be edited during or after placement of a stair. Select the Straight stair at D as shown in Figure 9.3. The **A-Width** field of the stair is 3'-8"[1000]. Edit the width to **4'-0"[1200]**. Retain the selection of the stair.

12. The B Height field of the Dimensions section allows you to enter the finish floor to finish floor design dimension for the stair. The Height value of the stair is 10'-0"[3000]. Edit the B Height to **16'-0" [4800]**.

13. Retain the selection of the stair. The Justify option of the straight stair is Center. Click the Justify option to view Left, Center, and Right justification options which are available for the Straight and Multi-landing stairs. Inside, Center, and Outside justify options are available for U-shaped and Spiral stair shapes. When placing a stair, the justify option provides the handle to specify the start and end of flights.

14. Retain the selection of the straight stair at D. The Terminate with field is available during and after placement of all stair shapes. The Terminate with options are riser, tread, or landing as shown in Figure 9.4.

    *(Move the cursor to the Terminate with field and choose **Landing**. The landing is added to the stair at the end of flight.)*

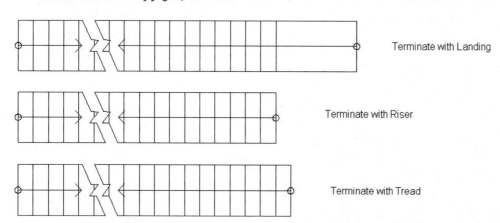

Terminate with Landing

Terminate with Riser

Terminate with Tread

FIGURE 9.4 *"Terminate with" options*

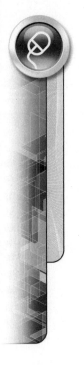

15. Choose the button in the Calculation rules field of the Design tab to open the Calculation Rules dialog box as shown in Figure 9.5. This dialog box includes the following four fields to control the design of the stair: A-Straight Length, B-Riser Count, C-Riser, and D-Tread. The fields of this dialog box may be set to Automatic (lighting bolt icon) or Manual (plus sign). Values entered must comply with the rules of the Design Rules tab of the style. Violations to the Design Rules will display the Stair Solutions tip as shown at right in Figure 9.5.

16. Edit the D-Tread value to 10" [225] and press Enter. Since this value violates the Design Rules, the Warning dialog box is displayed as shown in Figure 9.5.

17. Click OK to dismiss the Warning dialog box. Choose the D-Tread toggle to Automatic which will set all fields to Automatic; therefore the design of the stair is now controlled solely by the settings of the Design Rules tab of the stair style. Choose OK to dismiss the Calculation Rules dialog box.

18. The C-Straight length, D-Riser count, E-Riser, and F-Tread values listed in the Calculation rules dialog box are shown in the Properties palette. Choose OK to dismiss the Calculation Rules dialog box. Since Automatic was chosen for all fields in the Calculation Rules dialog box, these fields are not editable in the Properties palette.

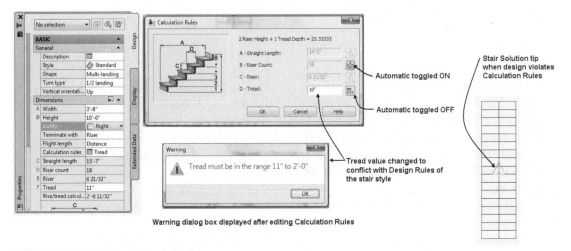

FIGURE 9.5 *Calculation Rules dialog box*

19. The Floor Settings of the stair may be adjusted in the Advanced section of the Properties palette. The A-Top Offset value within the Advanced section allows you to enter a value that represents the thickness of flooring at the top of the flight as shown in Figure 9.6. The C-Bottom Offset value represents the thickness of the flooring material at the bottom of the stair.

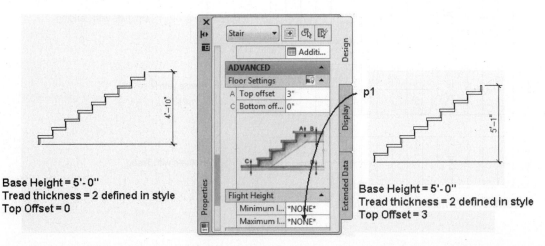

FIGURE 9.6 *Stair adjustment for Top Offset*

20. Select the D straight stair as shown in Figure 9.3. Choose **Isolate Objects** from the General panel. Choose **East** from the View Cube.

21. The landing at the flight end specified earlier in this tutorial is now displayed in elevation view. The **Flight Height** section within the Advanced section as shown at **p1** in Figure 9.6 allows you to control the location of intermediate landings based upon minimum or maximum limits. The Minimum limit type options include None, Risers, or Height. You can specify a minimum height or number of risers for the stair.

22. The Maximum Limit Type options include None, Risers, or Height. If a Risers or Height limit is selected, the Properties palette is expanded to include the Maximum Risers and Maximum Height options. Setting the maximum height limits allows the stair to be developed with a landing automatically placed in compliance with building code landing requirements.

23. Select from the Maximum limit type = Height option as shown in Figure 9.7. Edit the Maximum height to **12'[3000],** to add a landing to the stair in compliance with the maximum height.

24. Choose **SW Isometric** from the View Cube to view the intermediate landing as shown in Figure 9.7.

25. Choose the **Home** tab. Choose Layer Properties of the Layers panel to display the Layer Properties Manager. Thaw the **A-Slab** layer. Close the Layer Properties manager.

26. Select the stair; the Interference section within the Advanced category allows you to define the interference dimensions of the stair with walls and slabs. The Verify the Headroom height = 7'-0"[2100]. The slab will be cut to provide the 7'-0" [2000] headroom.

27. Press Escape. Choose the slab at p1 as shown in Figure 9.7. Choose Add from the Interference Condition flyout of the Modify panel in the Slab tab. Respond to the workspace prompts as follows:

```
Select objects to add: (Select the stair at p2 as shown in
Figure 9.7.) 1 found

Select objects to add: (Press Enter to end selection.)

Enter shrinkwrap effect [Additive/Subtractive]: s (Enter S
to create Subtractive Interference.)

1 object(s) attached.

Press Escape to end selection and view the slab modification
as shown at right in Figure 9.7.
```

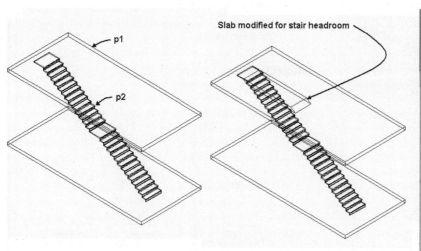
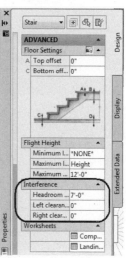

Slab modified for stair headroom

FIGURE 9.7 *Slab and wall cut using Headroom and Side Clearance*

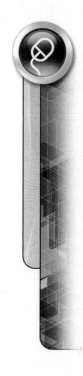

28. Select the straight stair D. Scroll if necessary to the Worksheets section of the Advanced section. The Worksheets sections provide the edit of Stair Components and Landing Extensions as shown in Figures 9.8 and 9.9. If Allow Each Stair to Vary are toggled ON in the Components tab of the Stair Style dialog box, the worksheets can be edited.

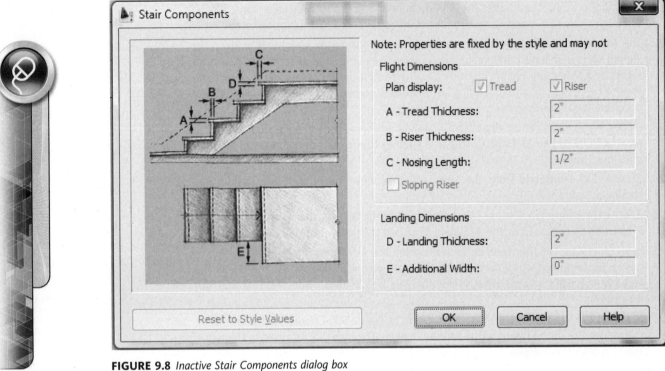

FIGURE 9.8 *Inactive Stair Components dialog box*

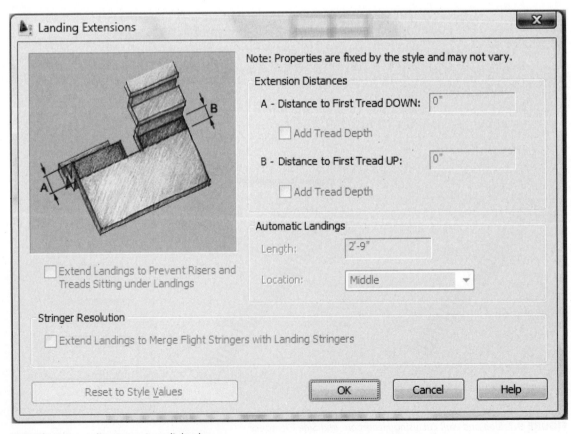

FIGURE 9.9 *Landing Extensions dialog box*

29. Save and close the drawing.

Using Properties to Modify an Existing Stair

A stair is modified by editing the Properties palette. Additional properties can be assigned to the stair in the stair style. After the stair is placed, handrails, balusters, and guardrails can be added through the **Railing** command. Existing railings are modified in the Properties palette. The following additional fields are displayed on the Design tab to edit an existing stair in the Properties palette.

Design Tab—Basic

General

> **Layer**—The name of the layer of the stair is displayed in the Layer field. The stair is placed on the **A-Flor-Strs [A240G]** layer when the AIA (256 Color) Layer Key Style is used in the drawing.

Location

> **Rotation**—The Rotation field displays the rotation of the stair in the XY plane.
>
> **Elevation**—The Elevation field displays the elevation of the insertion point of the stair.
>
> **Additional Information**—The Additional Information button opens the **Location** dialog box, which lists the insertion point, normal, and rotation description of the stair.

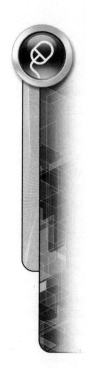

SPECIFYING FLIGHT POINTS TO CREATE STAIR SHAPES

The straight, multi-landing, U-shaped, and spiral stairs are created by selecting the **Shape** field in the Properties palette. The shape presets the sequence of prompts to create the stair. It is helpful to turn on polar tracking when placing straight, multi-landing, and U-shaped stairs. The polar tracking tip displays the distance you have moved the cursor from the last specified point. The straight stair can be created by selecting the start point of the stair at **p1** as shown in Figure 9.10, and the end of flight can be selected along the stair or beyond the dynamic display of the stair. The workspace prompts for the straight stair created in Figure 9.10 are shown below:

> Flight start point or: *(Select start of flight at* **p1** *as shown in Figure 9.10.)*
>
> Flight endpoint or: *(Select end of flight at* **p2** *as shown in Figure 9.10.)*
>
> Flight start point or: ESC *(Press ESC to end the command.)*

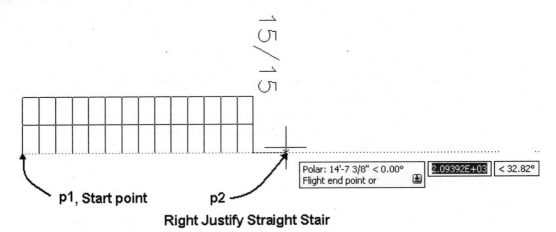

Right Justify Straight Stair

FIGURE 9.10 *Creating a straight stair*

Creating a Multi-Landing Stair

The multi-landing stair can include one or more landings and stair flights. The second stair flight can be straight or at an angle to the first flight. If the multi-landing stair turns at an angle, you can specify the type of turn in the **Turn type** field in the Design tab of the Properties palette. The turn types are shown in Figure 9.3. The **1/2 Landing** turn type and Right justify was selected to place the stair shown in Figure 9.11 for the workspace prompts shown below. The **1/2 Landing** is created by starting the second flight at the end of the landing. The display of the riser count is dynamically displayed as you place the endpoint of the flight.

Flight start point or: *(Using Endpoint object snap, select point* **p1** *as shown at the left in Figure 9.11.)*

Flight endpoint or: *(Using Endpoint object snap, select point* **p2** *as shown at the left in Figure 9.11.)*

Flight start point or: *(Move the cursor to the left until polar track tip distance equals at least 3', and then left-click to specify point* **p3** *as shown at the left in Figure 9.11.)*

Flight endpoint or: *(Move the cursor to the left, and select a point beyond the dynamic display of the stair near* **p4** *as shown at the right in Figure 9.11.)*

Flight start point or: ESC *(Press ESC to end the command.)*

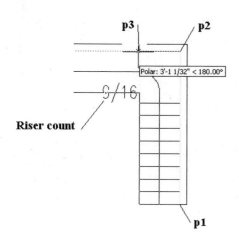

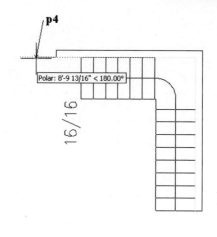

Specifying Start Points of Multilanding Stair

Specifying End of Multilanding Stair

FIGURE 9.11 *Start and end flight points of a multi-landing stair*

Access Instructional Video 9.1 Inserting a Multi-landing Stair from the Instructional Video category of the Student Companion site of CengageBrain http://www.cengagebrain.com described in the Preface.

NOTE

Creating a U-Shaped Stair

U-shaped stairs divide the total run into two equal flights. The stair can include a flat landing or a winder landing between the two runs. The turn of the landing is created by specifying the **1/2 Turn** or the **1/2 Landing**. The stair placed in Figure 9.12 includes 1/2 landing and requires the justification of the stair be set to the right to place the stair edge at **p1**. The stair can be generated with clockwise or counterclockwise Horizontal Orientation from the start point as specified in the Properties palette. The stair shown in Figure 9.12 was created with counterclockwise horizontal orientation. When placing a U-shaped stair with counterclockwise orientation, move the cursor to the left of the start point and specify a point greater than the width of the stair flights at **p2** as shown in Figure 9.12. The polar tracking tooltip displays the distance the cursor is moved from the start point of the flight.

Flight start point or: *(Select the end of the wall at* **p1** *as shown in Figure 9.12.)*

Flight endpoint or: *(Select a point near* **p2** *as shown in Figure 9.12.)*

Flight start point or: ESC *(Press* ESC *to end the command.)*

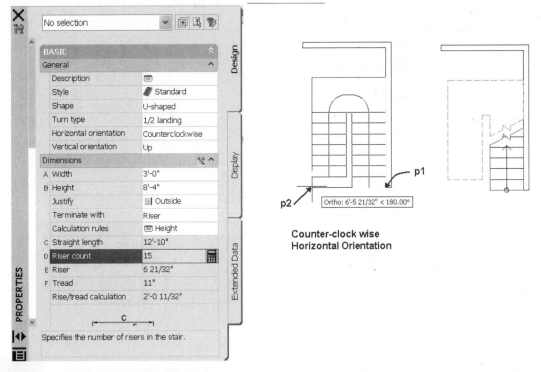

FIGURE 9.12 *U-shaped stair with landing*

After the stair is created, you can use the **Move** command to move it to fit existing walls. A U-shaped winder stair can be created if you set the Turn type = **1/2 turn** as shown in Figure 9.13. When you specify a 1/2 turn, the **Winder** property is added to the Properties palette. The **Balanced** winder is the default winder; however, additional winders can be imported in Style Manager from the Architectural Objects > Stair Winder Styles category of the Stair Styles (Imperial).dwg or Stair Styles (Metric).dwg located in Content (Styles > Imperial and Styles > Metric folders). The U-shaped winder stair is placed in a similar manner as the 1/2 landing U-shaped stair.

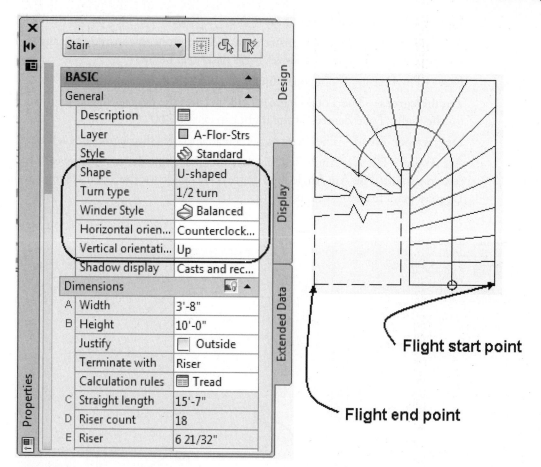

FIGURE 9.13 *Winder U-shaped stairs*

Creating Spiral Stairs

The **Spiral** shape is used to create curved or spiral stairs. A spiral stair is created by specifying the center and radius of the spiral stair; the stair **Calculation Rules** determine the number of risers and angle of the sweep. The angle of the arc can be specified as constrained or free in the **Arc Constraint** field in the Design tab of the Properties palette. The justification of the stair places the handle for insertion at the left, center, or right.

To place a spiral stair to fit an existing curved wall, select the Center object snap when specifying the center of the stair. The center of the spiral stair can be stretched by selecting the grip of the spiral stair. Editing the grips allows you to change the radius and location of the start point. The grips of the spiral stair are shown in Figure 9.19. The **Radius** grip of the spiral stair can be stretched to change the radius; therefore, the start point for the flight remains stationary. When you select the **Spiral** shape, the Properties palette expands to include the following additional fields to place the stair.

> **Horizontal Orientation**—The Horizontal Orientation field allows the stair to be developed from the start point clockwise or counterclockwise.
>
> **Specify on Screen**—The Specify on Screen drop-down list includes **Yes** or **No** options. Select the **Yes** option and select two locations in the workspace to specify the radius in the workspace.

Radius—If Specify on Screen is set to **No**, the **Radius** can be typed in this field.

Arc Constraint—The Arc Constraint includes options to constrain the angle of the arc of the stair. The drop-down list is described below.

Free—The Free arc is not constrained.

Total Degrees—The Total Degrees field specifies the degrees the arc can sweep for the flight. Total degrees of the flight can exceed 360°.

Degrees per Tread—The Degrees per Tread specifies the degree for each tread.

Arc Angle—The Arc Angle field displays the total angle the stair will sweep with free constraint. The Arc Angle field displays the angle for the Total Degrees and Degrees per Tread constraints.

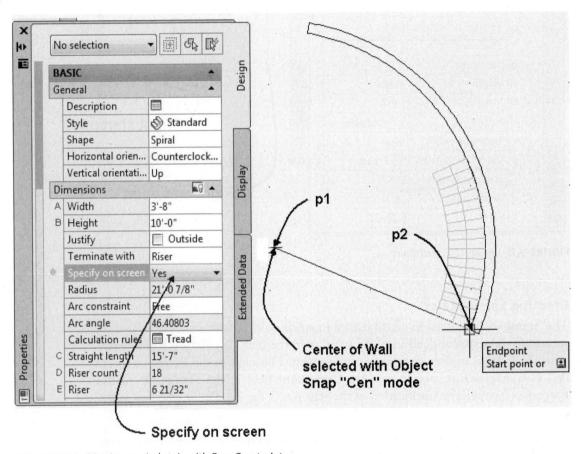

FIGURE 9.14 *Creating a spiral stair with Free Constraint*

The **Specify on Screen** option allows you to select the center of a wall using the Center object snap and then specify the start point to set the radius of the stair. The stair shown in Figure 9.14 was placed by setting the following options in the Properties palette and responding to the workspace sequence: Style = **Standard**, Shape = **Spiral**, Horizontal Orientation = **Counterclockwise**, Vertical Orientation = **Up**, Width = **3'-8**, Height = **10'**, Justify = **Outside**, Terminate with = **Riser**, Specify on Screen = **Yes**, Arc Constraint = **Free**.

Center of spiral stair or: _cen of (Press SHIFT + right-click, select Center from the Object Snap option, move the cursor over the wall, left-click when the Center object snap marker is displayed, and left-click to specify point **p1** *as shown in Figure 9.14.)*

Start point or: (Select the start point **p2** *as shown in Figure 9.14.)*

Start point or: ESC (Press ESC to end the command.)

In addition to the riser and tread code requirements, most building codes require that the tread depth 12" [300] from the narrow edge be at least 7 1/2" [190] to provide an adequate tread surface. The Properties palette does not include a code check for this tread dimension. To create a spiral stair that complies, set the **A-Width** to 24" [600], and then use the **StairOffset** command discussed in "Customizing the Edge of a Stair" in this chapter to offset the edge of the stair to the total desired stair width. You can check the minimum tread distance by drawing an arc with a radius equal to the spiral stair, offsetting the arc 12" [300] and then measuring the tread at this location. Measuring the tread along the arc that is 12" [300] from the narrow edge allows you to check the spiral stair for code compliance, as shown in Figure 9.15.

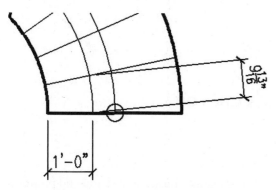

FIGURE 9.15 *Checking the minimum tread 12" from the narrow edge of the spiral stair*

Editing a Stair Using Grips

Each stair shape can be edited by its grips. Edge grips can be toggled OFF by selecting the trigger grip, and the grips can be used to edit the location, flight start, and flight end as shown at the top in Figure 9.16. When **Edit Edges** is toggled ON, you can adjust the width of the stair as shown at the bottom in Figure 9.16.

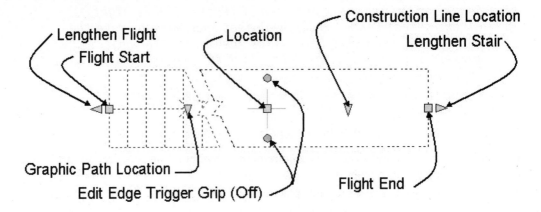

Edit Edge Grips OFF

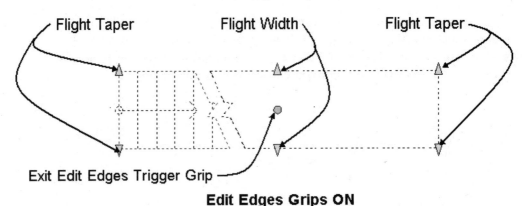

Edit Edges Grips ON

FIGURE 9.16 *Grips of the straight stair*

Stretching the grips of the stair allows you to quickly adjust the start and end points of the stair flight. If a multi-landing stair is edited by grips, it is shortened or lengthened a distance equal to the tread size. The **Flight Taper** and **Flight Width** grip allows you to stretch the edge of the stair. The **Location** grip moves the stair without stretching its components. The **Lengthen Stair** grip stretches the straight length of a Straight stair if **Straight Length** is **not** set to **Automatic** in the Calculation Rules. **Automatic** can be toggled OFF in Calculation Rules for an existing stair. **Flight Start** and **Flight End** grips adjust the location of the insertion point of the start and end of the flight. The **Construction Line** and **Graphic Path** grips edit the location of the path arrows. Movement of grip points to locations that violate the Design Rules will cause the display of the stair warning symbol shown in Figure 9.5.

Grips of the Multi-Landing Stair

The grips of the multi-landing stair consist of grips for each of the flights and the landing as shown in Figure 9.17. The trigger grips to **Edit Edges,** shown at the right in Figure 9.17, turn on the display of grips to edit the edges as shown at the left in Figure 9.17. The **Flight Taper** grip allows you to taper each flight of the multi-landing flight.

The **Move Flight** and **Turn Points** grips move the landing and the flight. The **Edge** and **Landing Width** grips change the shape of the landing without stretching the flight.

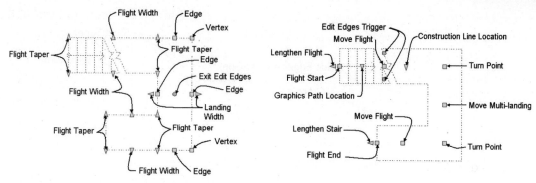

FIGURE 9.17 *Grips of the multi-landing stair*

Grips of the U-Shaped Stair

The grips of the U-shaped stair, shown in Figure 9.18, are similar to those of the multi-landing stair. The U-shaped stair consists of two flights that are fixed in length; therefore, selecting the **Flight Start** and **Flight End** grips will **not** stretch the flight. Selecting the **Flight Start** or the **Flight End** will **rotate** the unit rather than stretch the flight. The **Flip** grips shown in Figure 9.18 can be used to mirror flip the stair horizontally or vertically.

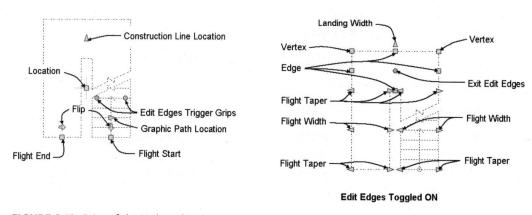

FIGURE 9.18 *Grips of the U-shaped stair*

Grips of the Spiral Stair

The grips of the spiral stair are shown in Figure 9.19. The **Flight Taper** and **Flight Width** grips are accessed when **Edit Edges** is toggled ON. The entire stair can be moved with the **Location** grip. The **Radius** grip allows you to drag the grip relative to the **Location** grip to change the radius.

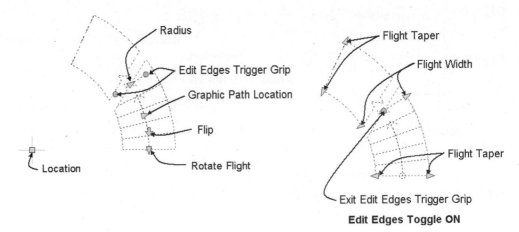

FIGURE 9.19 *Grips of the spiral stair*

CUSTOMIZING THE EDGE OF A STAIR

The edge of the stair can be customized to increase or decrease its width by a specific value or extend the stair to a wall or polyline. Previous edits of the stair can be removed with the customize edge commands. The **Customize Edge** shortcut menu option includes the following commands: **Edit, Offset, Project, Remove Customization**, and **Generate Polyline**. The **Edit** option toggles ON Edit Edges grips. The **Offset** option (**StairOffset** command) allows you to extend the width of the stair by a specified distance. Access **Offset** (**StairOffset** command) as shown in Table 9.2.

TABLE 9.2 *StairOffset command access*

Command prompt	STAIROFFSET
Ribbon	Select the stair and choose Offset from the Customize Edge flyout menu of the Modify panel in the Stair contextual tab.
Shortcut menu	Select the stair, right-click, and choose Customize Edge > Offset.

When you select the **Customize Edge > Offset** command from the shortcut menu for a stair, you are prompted in the workspace to type a positive distance to move the stair edge out or a negative distance to move the edge toward the inside of the stair. This command is used to adjust the width of the stair to fit the structure. The command allows you to control which edge of the stair to increase or decrease in width. The following workspace sequence was used to increase the width of the stair as shown in Figure 9.20.

> *(Select a stair and choose* Offset *from the* Customize Edge *flyout menu of the Modify pane in the Stair contextual tab.)*
>
> Select an edge of a stair: *(Select the stair at* **p1** *as shown in Figure 9.20.)*
>
> Enter distance to offset (positive = out, negative = in): 12 [300] ENTER *(Type* 12 [300], *a positive distance to offset the stair toward the outside of the stair as shown in Figure 9.20.)*

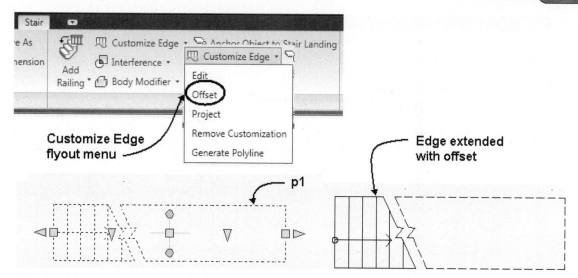

FIGURE 9.20 *Stair edge moved with Stair Offset command*

Projecting Stairs to Walls

The **Project** option of the **Customize Edge** shortcut menu (**StairProject** command) allows you to extend the stair to a wall or polyline. This option can be used to trim or extend a stair to a wall or polyline. Access the **StairProject** command as shown in Table 9.3.

TABLE 9.3 *StairProject command access*

Command prompt	STAIRPROJECT
Ribbon	Select the stair and choose Project from the Customize Edge flyout menu in the Modify panel of the Stair contextual tab.
Shortcut menu	Select the stair, right-click, and choose Customize Edge > Project.

The following workspace sequence was used to extend the stair shown in Figure 9.21 to the wall.

> *(Select a stair and choose* Project *from the Customize Edge flyout menu in the Modify panel of the Stair contextual tab.)*
>
> Select an edge of a stair: *(Select the stair at* p1 *as shown in Figure 9.21.)*
>
> Select a polyline or connected AEC objects to project to: 1 found *(Select the wall at* p2 *as shown in Figure 9.21.)*
>
> Select a polyline or connected AEC objects to project to: ESC *(Press* ESCAPE *to end selection.)*

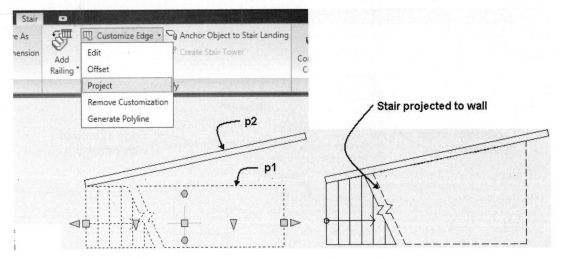

FIGURE 9.21 *Stair projected to a wall*

Removing Stair Customization

Previous customizations of the stair using the **Offset** and **Project** options can be removed with the **Remove Customization** option of the **Customize Edge** shortcut menu. The **Remove Customization** option (**StairRemoveCustomization**) removes all previous edits to the edge of the stair. Access the **StairRemoveCustomization** command as shown in Table 9.4.

TABLE 9.4 *StairRemoveCustomization command access*

Command prompt	STAIRREMOVECUSTOMIZATION
Ribbon	Select the stair and choose Remove Customization from the Customize Edge flyout menu of the Modify panel in the Stair contextual tab.
Shortcut menu	Select the stair, right-click, and choose Customize Edge > Remove Customization.

The customization of the stair edge was removed in the following workspace sequence as shown in Figure 9.22.

(Select a stair and choose Remove Customization *from the Customize Edge flyout menu in the Modify panel of the Stair contextual tab.)*

Select an edge of a stair: *(Select the edge of the stair at* p1 *to remove changes to the edge.)*

(Stair changed to original state as shown in Figure 9.22.)

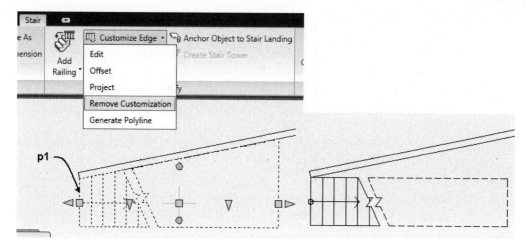

FIGURE 9.22 *Customization removed from a stair*

> The **Generate Polyline** option of the **Customize Edge** shortcut menu is used to create a polyline from the edge of the stair. The polyline can be edited to develop walls or other stair edges in the drawing.

TIP

Creating a Custom Stair from Linework

Custom stairs with irregular shapes can be developed from linework. The custom stair can be defined by a left side, left stringer path, center stringer path, right stringer path, and right side. Arcs, lines, and light-weight polylines can be used to define the geometry. Geometry may be specified to define stringer paths or the sides of the stair may serve as the path when the stair is generated. The first tread is developed from the current level. Access the **StairToolToLinework** command from the right-click menu of the **Stair** tool as shown in Table 9.5. The linework shown in Figure 9.23 was developed into a stair in the following workspace sequence. The **Treadprofile** option of the StairToolToLinework command allows you to create a stair with custom tread designs. The Design Rules and Calculation Rules do not apply to Custom stairs. Custom stairs cannot be converted to ramps. The Properties palette allows you to specify the riser count and height, as shown in Figure 9.23.

TABLE 9.5 *StairToolToLinework command access*

Command prompt	**STAIRTOOLTOLINEWORK**
Shortcut menu	Right-click a stair tool and choose Apply Tool Properties to > Linework.

(Right-click a stair tool and choose **Apply Tool Properties to** > **Linework.** *The tread geometry of the stair must intersect with the left and right side linework.)*

Pick left and right sides [Treadprofile]: *(Choose arc at* p1 *as shown in Figure 9.23.)* 1 found

Pick left and right sides [Treadprofile]: *(Choose arc at* p2 *as shown in Figure 9.23.)* 1 found, 2 total

Pick left and right sides [Treadprofile]: *(Press* ENTER *to end selection.)*

Select stair path [Automatic/Undo]<Automatic>: *(Press* ENTER *to choose* **Automatic** *path.)*

Select user defined left stringer path [RIght/Center/Undo] <use left side>: *(Press* ENTER *to choose the default left stringer path.)*

Select user defined right stringer path [LEft/Center/Undo] <use right side>: *(Press* ENTER *to choose the default right stringer path.)*

Select user defined center stringer path [RIght/LEft/Undo] <use stair path>: *(Press* ENTER *to accept the default stringer path.)*

Select first tread at current level [Undo]: *(Select the line at* **p3** *as shown in Figure 9.23.)* 1 found

Select remaining treads [Undo]: *(Select the line at* **p4** *as shown in Figure 9.23.)* 1 found, 2 total

Select remaining treads [Undo]: *(Select the line at* **p5** *as shown in Figure 9.23.)* 1 found, 3 total

Select remaining treads [Undo]: *(Select the line at* **p6** *as shown in Figure 9.23.)* 1 found, 4 total

Select remaining treads [Undo]: *(Select the line at* **p7** *as shown in Figure 9.23.)* 1 found, 5 total

Select remaining treads [Undo]: *(Select the line at* **p8** *as shown in Figure 9.23.)* 1 found, 6 total

Select remaining treads [Undo]: *(Select the line at* **p9** *as shown in Figure 9.23.)* 1 found, 7 total

Select remaining treads [Undo]: *(Select the line at* **p10** *as shown in Figure 9.23.)* 1 found, 8 total

Select remaining treads [Undo]: *(Select the line at* **p11** *as shown in Figure 9.23.)* 1 found, 9 total

Select remaining treads [Undo]: ENTER *(Press* ENTER *ends tread selection.)*

(Edit the Convert to Stair dialog box as shown in Figure 9.23. Check the **Erase Layout Geometry** *checkbox.)*

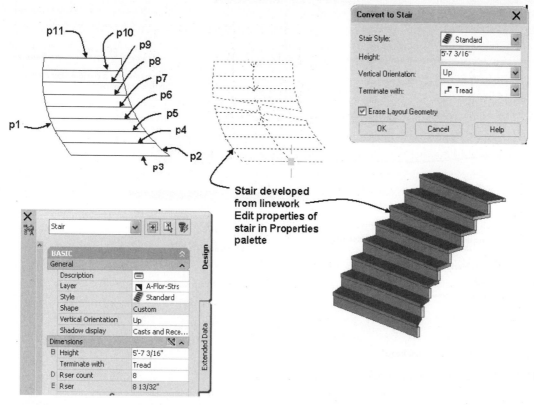

FIGURE 9.23 *Converting linework to stairs*

Converting Stairs to Custom Stairs

The resulting stair provides extensive component editing. The StairConvertCustom can be used to convert any stair to a custom stair and edit the stair components. Access the StairConvertCustom tool as shown in Table 9.6.

TABLE 9.6 *StairConvertCustom command access*

Command prompt	STAIRCONVERTCUSTOM
Ribbon	Select a stair and choose Convert to Custom from the Custom Stair panel in the Stair contextual tab.
Shortcut menu	Select a stair, right-click, and choose Convert to Custom Stair.

When you convert a stair object to a custom stair, custom stair only includes one grip as shown in Figure 9.24. You can select a custom stair and choose the following four tools on the Custom Stair panel in the Stair tab: Edit Tread/Riser, Match Tread/Riser, Set to Tread/Landing, or Replace Stringer Path as shown in Figure 9.24.

The shortcut menu of a custom stair, shown in Figure 9.24, allows you to edit the treads, riser, landing, or stringer stair components. You can select the grip for a tread/riser.

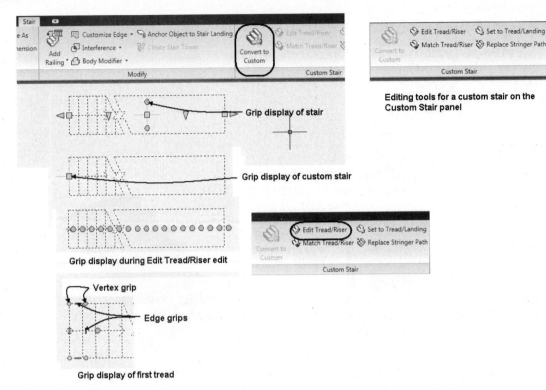

Editing tools for a custom stair on the Custom Stair panel

FIGURE 9.24 *Grips of a Custom Stair*

Modifying a Custom Stair

The Custom Stair allows you to modify the shape of any tread in the stair. Most often designs require the first or last tread shape be modified. The StairEditTread&Riser command (see Table 9.7) allows you to modify the tread or riser.

TABLE 9.7 *StairEditTread&Riser command access*

Command prompt	**STAIREDITTREAD&RISER**
Ribbon	Select a custom stair and choose Edit Tread/Riser of the Custom Stair panel in the Stair contextual tab.
Shortcut menu	Select a stair, right-click, and choose Modify Custom Stair > Edit Tread/Riser.

When you select a custom stair and choose Edit Tread/Riser from the Custom Stair panel, a grip is displayed on each tread and riser of the custom stair as shown in Figure 9.24. If you select a tread or riser grip, vertex and edge grips are displayed for the tread or riser as shown in Figure 9.25. You can toggle functions of the edge grip when you hold the Ctrl key down as you select the edge grip. The shape of the tread is modified as shown in Figure 9.25.

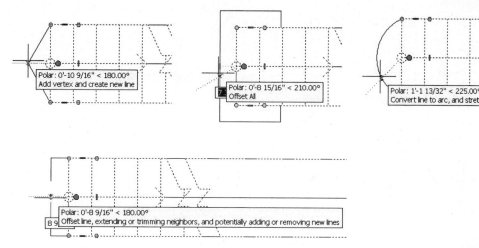

FIGURE 9.25 *Modifying the Edit Tread Grip*

The StairMatchTread command allows you to match an existing tread or riser of a custom stair and apply the shape for another custom stair. Access the StairMatch Tread command as shown in Table 9.8.

TABLE 9.8 *StairMatchTread command access*

Command prompt	**STAIRMATCHTREAD**
Ribbon	Select a custom stair and choose Match Tread/Riser.
Shortcut menu	Select a custom stair, right-click, and choose Modify Custom Stair > Match Tread/Riser.

The custom tread shown at the right in Figure 9.26 was applied to the left stair in the following workspace prompts.

> *(Choose a custom stair at* **p1** *in Figure 9.26 and choose Match Tread/Riser from the Custom Stair panel of the Stair contextual tab.)*
>
> *Command: StairMatchTread*
>
> *Match [Tread/Riser/Both] <Tread>:* T *(Type T to choose the Tread option.)*
>
> *Select source tread to match: (Select the tread at* **p2** *as shown in Figure 9.26.)*
>
> *Select destination tread(s) to replace: (Select tread at* **p3** *to customize stair.)*
>
> *Select destination tread(s) to replace [Undo]: ENTER (Press ENTER to end command and display stair as shown at the right in Figure 9.26.)*

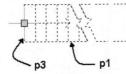
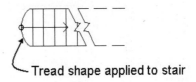

— Tread shape applied to stair

p3   p1

Applying StairMatchTread to stair

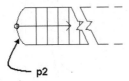

p2

Source stair with custom tread

FIGURE 9.26 *Converting linework to stairs*

The StairToggleTread command allows you to convert a tread to a landing. Access the StairToggleTread command as shown in Table 9.9.

TABLE 9.9 *StairToggletread command access*

Command prompt	**STAIRTOGGLETREAD**
Ribbon	Select a custom stair and choose Set to Tread/Landing from the Custom Stair panel.
Shortcut menu	Select a custom stair, right-click, and choose Modify Custom Stair > Toggle Tread/Landing.

When you select a stair and choose Set to Tread/Landing, you are prompted to select a tread and convert it to a landing. You can also convert a landing to a tread.

The geometry defining the path for a custom stair can be replaced to modify the stair. Access the StairCustomStringerPathReplace command as shown in Table 9.10.

TABLE 9.10 *StairEditTread&Riser command access*

Command prompt	**STAIRCUSTOMSTRINGERPATHREPLACE**
Ribbon	Select a custom stair and choose Replace Stringer Path from the Custom Stair panel.
Shortcut menu	Select a custom stair, right-click, and choose Modify Custom Stair > Replace Stringer Path.

When you access the StairCustomStringerPathReplace, you are prompted to define the path and the geometry. In the following workspace prompts the Center path was replaced.

```
(Select a custom stair that includes stringers, right-click,
and choose Modify Custom Stair > Replace Stringer Path.)
```

Replace stringer path [Left/Right/Center/Undo/eXit]
<eXit>: c

Select linework for stringer path: (*Select custom linework at* **p1** *as shown in Figure 9.27.*)

Replace stringer path [Left/Right/Center/Undo/eXit]
<eXit>: ENTER (*Press* ENTER *to end the command.*)

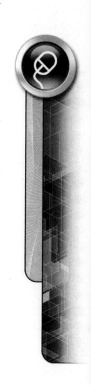

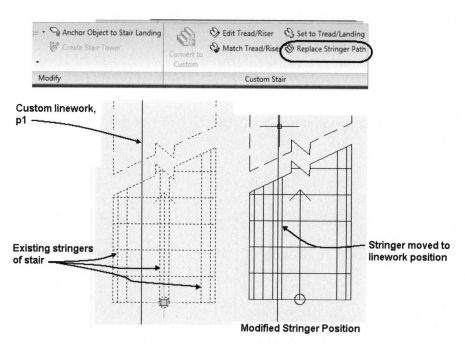

FIGURE 9.27 *Converting linework Command: StairCustomStringerPathReplace*

TUTORIAL 9-1: CREATING MULTI-LANDING STAIRS

1. Verify that the Accessing Tutor\Imperial or Accessing Tutor\Metric content for Chapter 9 of the CengageBrain http://www.cengagebrain.com has been downloaded to your Accessing Tutor folder on your computer as described in Organizing Tutorial Directories in the Preface.

2. Open Autodesk AutoCAD Architecture 2012. Open *Accessing Tutor\Imperial\Ch9\Ex 9-1.dwg* or *Accessing Tutor\Metric\Ch9\Ex 9-1.dwg*. Choose **SaveAs > AutoCAD Drawing** from the Application menu. Save the drawing as **Lab 9-1** to Ch9 folder of your student folder.

3. Right-click over Object Snap in the **Application** status bar and choose **Settings**. Choose Clear All, check the **Endpoint** object snap checkbox, and verify that **Object Snap** and **Object Snap tracking** are checked. Click **OK** to dismiss the Drafting Settings dialog box.

4. Select the **Stair** tool from the Design tool palette. Edit the Properties palette as follows: Style = **Standard**, Shape = **Multi-landing**, Turn type = **1/2 landing**, Vertical Orientation = **Up**, Width = **3'-0" [914]**, Height = **8'-4" [2540]**, Justify = **Left**, Terminate with = **Tread**, and Flight Length = **Distance**. Click the **Calculation Rules** button to open the Calculation Rules dialog box. Edit all parameters to **Automatic**. When parameters are automatic, the field is inactive and a lightening bolt button is

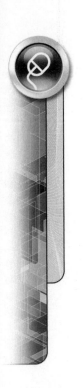

displayed at the right. Click **OK** to dismiss the Calculation Rules dialog box. Respond to the workspace prompts as follows:

> Flight start point or: *(Choose the wall at* **p1** *as shown in Figure 9.28 using the Endpoint object snap.)*
>
> Flight endpoint or: *(Choose the wall at* **p2** *as shown in Figure 9.28 using the Endpoint object snap. Toggle OFF Object Snap in the Application status bar.)*
>
> Flight start point or: *(Move the cursor up from* **p2**, *type* 1 [1] ENTER *in the dynamic input with the polar tracking vector angle of 90° to specify* **p3** *as shown in Figure 9.28.)*
>
> Flight endpoint or: *(Toggle ON Object Snap in the Application status bar and choose the wall at* **p4** *as shown in Figure 9.28 using the Endpoint object snap.)*
>
> Flight start point or: *(Press ESC to end the command.)*

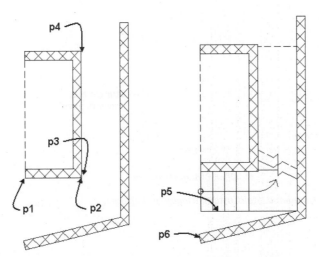

FIGURE 9.28 *Creating and modifying a multi-landing stair*

5. Select the stair and choose **Project** from the **Customize Edge** flyout menu in the Modify panel of the Stair contextual tab. Refer to Figure 9.28 and respond to the workspace prompts as follows:

> Select an edge of a stair: *(Select the stair at* **p5**.*)*
>
> Select a polyline or connected AEC objects to project to: *(Select the wall at* **p6**.*)* 1 found
>
> Select a polyline or connected AEC objects to project to: ENTER *(Press ENTER to end the command.)*

6. Select the stair and choose **Convert to Custom Stair** from the Custom Stair panel of the Stair contextual tab.

7. Retain selection of the stair and choose Edit Tread/Riser from the Custom Stair panel of the Stair contextual tab. A grip is displayed on each tread as shown in Figure 9.29. Select the grip at **p7** to display edge and vertex grips as shown at the right in Figure 9.29.

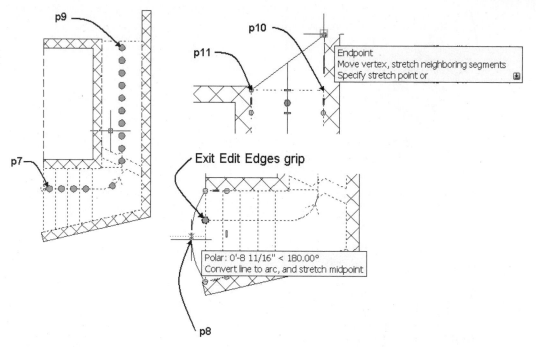

FIGURE 9.29 *Grip edit of tread*

8. Choose the edge grip at **p8** as shown in Figure 9.29. Press the CTRL key as necessary to display the Convert to Arc tip. Move the cursor left and type 4 [100] ENTER in the keyboard. Select the Exit Edit Edges grip. Press **ESC** to end the edit.

9. Select the stair and choose Edit Tread/Riser from the Custom Stair panel in the Stair contextual tab. A grip is displayed on each tread as shown in Figure 9.29. Select the grip at **p9** to display edge and vertex grips. Select the vertex grip at **p10** and drag the grip to the end of the wall as shown in Figure 9.29. [**Metric** only: select the vertex grip at p11, as shown in Figure 9.29, and drag the grip to the end of the wall at the corner.] Press ESC to end the stair edit.

10. Select the stair and choose **Isolate Objects** from the General panel of the Stair contextual tab. Choose **SW Isometric** from the View Cube. Choose **2D Wireframe** from the Visual Styles flyout menu of the Visual Styles panel in the View tab. The modified custom stair is shown in Figure 9.30.

11. Save and close the drawing.

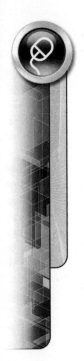

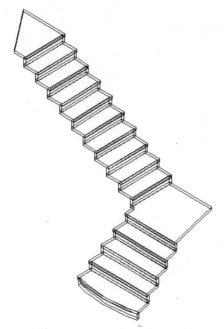

FIGURE 9.30 *Modified custom stair (Metric similar)*

CREATING STAIR STYLES

The **Standard** stair style is inserted in drawings created from the AEC Model templates. If other stair styles have been imported or created in the drawing, the style and other settings of the last stair inserted become the defaults for the current insertion. The Standard style is the simplest because it includes only risers and treads, with no stringer. The Standard stair style provides a simple representation of a stair. However, other stair styles add more detail regarding the stringer position and better represent the complexity of the stair. You can i-drop stair styles from the Design Tool Catalog—Imperial or Design Tool Catalog—Metric of the **Stairs and Railing >Stairs** category of the Content Browser. Stair styles can be imported in the Style Manager from the *Stairs Styles (Imperial)* or *Stairs Styles (Metric)* file located in the Imperial and Metric folders of the Styles folder accessed from Content of the Places panel *(Vista—C:\ProgramData\Autodesk\ACA 2012\enu\)*. The *Stair Styles (Imperial)* file includes the following styles: **Cantilever, Concrete, Half Wall Rail, Ramp Concrete, Ramp Concrete Curb, Ramp Steel, Standard, Steel, Wood-Housed,** and **Wood Saddle** as shown in Figure 9.31.

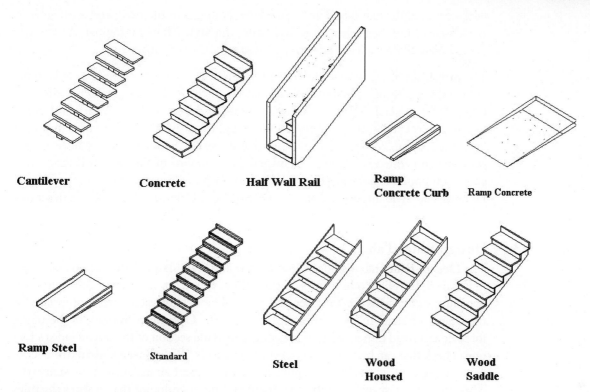

Cantilever   Concrete   Half Wall Rail   Ramp Concrete Curb   Ramp Concrete

Ramp Steel   Standard   Steel   Wood Housed   Wood Saddle

FIGURE 9.31 *Stair styles of the Stair Styles (Imperial) drawing*

Using the Style Manager to Create Stair Styles

Stair styles can be created, edited, imported, and exported in the Style Manager. The style of an existing stair can be edited by selecting the **Edit Stair Style** command. Access the **Edit Stair Style** (**StairStyleEdit**) command as shown in Table 9.11. The Stair Style definition includes control over the riser and tread dimensions and the insertion of stringers. The specific location and size of the stringer can be established through the **Stair Styles** command.

TABLE 9.11 *Edit Stair Style command access*

Command prompt	STAIRSTYLEEDIT
Ribbon	Select the stair and choose Edit Style from the Edit Style flyout menu to open the Stair Style Properties dialog box.
Shortcut menu	Select the stair, right-click, and choose Edit Stair Style.
Ribbon	Select the stair and choose Stair Styles from the Edit Style flyout menu in the General panel to open the Style Manager.
Tool palette	Right-click over the Stair tool and choose Stair Styles.

A new stair style can be created from an existing stair if you select a stair and choose Copy Style in the General panel. The **Copy Style** (CopyAndAssignStyle) command opens the Stair Styles dialog box; next, choose the General tab and overtype a new name in the General tab. To create a stair style in the Style Manager, select a stair and then choose **Stair Styles** from the Edit Style flyout menu of the General panel in the Stair contextual tab. The Style Manager opens to the Stair Styles category in the left pane, and then select **New Style** from the **Style Manager** toolbar. A new style named **New Style** is created; overtype a name for the new style.

The properties of a new style are defined by selecting the name of the new style in the left pane and editing the tabs of the right pane of the Style Manager. Since the contents of the General, Materials, Classifications, and Version History tabs are similar to other object styles, they will not be presented in this section. Stair styles imported from the Design Tool Catalog—Imperial and Design Tool Catalog—Metric can be reviewed and edited in the Style Manager. The contents of the Design Rules, Stringers, Components, Landing Extensions, and Display Properties tabs for the Wood-Housed stair style imported from the Design Tool Catalog—Imperial are described below.

Design Rules Tab

The **Design Rules** tab includes **Code Limits** and **Calculator Rule** sections as shown in Figure 9.32. The **Code Limits** section allows you to set maximum, minimum, and optimum dimensions for the riser and tread. The default values are appropriate for the design of stairs in a commercial building. An **Optimum Slope** sets the target for riser and tread dimensions. The **Calculator Rule** section of the tab includes a formula check that can be turned on or off using the **Use Rule Based Calculator** checkbox. The settings of this tab are applied when you insert an instance of the stair style. Therefore, if you specify a flight start point and flight endpoint that violates the rules of the tab, the stair solution tip will be displayed.

Code Limits

A-Maximum Slope—The Maximum Slope fields allow you to type the tallest riser and shortest tread that is acceptable in the building code.

B-Optimum Slope—The Optimum Slope represents the target values for the riser and tread when applied to the stair height and number of risers.

C-Minimum Slope—The Minimum Slope fields allow you to type the shortest riser and longest tread that is acceptable.

Calculator Rule

Use Rule Based Calculator—Selecting the Use Rule Based Calculator checkbox toggles ON/OFF the test of the riser and tread dimensions to comply with the formula.

Minimum-Maximum Limits—The Calculator Rule specifies the maximum and minimum value of the sum of two risers plus a tread. The formula rule shown in Figure 9.32 requires that two riser heights plus one tread depth must be no more than $2'-1''$. The **Calculator Rule** defines a range of dimensions acceptable for the riser and tread.

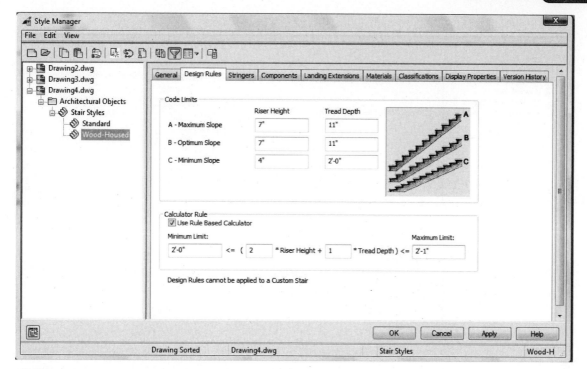

FIGURE 9.32 *Design Rules tab of the Wood-Housed stair style*

Stringers Tab

The **Stringers** tab is shown in Figure 9.33. It allows you to specify the dimensions for stair stringers.

Stringer—Click in the Stringer column to edit the name of a stringer.

Type—Click in the Type column to display the drop-down list, which includes the following types of stringers: Housed, Saddled, Slab, and Ramp.

Alignment—Click in the Alignment column to display the Alignment drop-down list that includes Align Left, Align Right, Center, or Full Width.

A-Width—The Width is the actual width of the stringer. Click in this column and overtype the width distance.

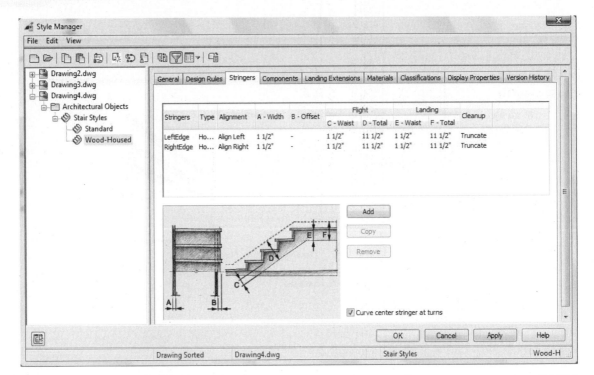

FIGURE 9.33 *Stringers tab of the stair style*

B-Offset—The Offset locates the distance from the stringer to the edge of the stair.

FLIGHT:

C-Waist—The Waist is the minimum width between the bottom surface of the stringer and the tread location on the stringer.

D-Total—The Total represents the dimension of the stringer inclusive of the stringer material.

LANDING:

E-Waist—The Waist represents the thickness of the flooring system of the landing.

F-Total—The Total represents the total thickness of the slab and stringer of the landing.

Cleanup—The Cleanup drop-down list allows you to choose **Cleanup** or **Truncate** for the end of the stringer. **Truncate** stops the stringer at the landing. **Cleanup** allows the stringer to continue under the landing.

Curve Center Stringer at Turns—The Curve Center Stringer at Turns checkbox converts the center stringer to a curve.

Add—Click the Add button to add a new stringer.

Copy—Select a stringer in the list window and then click the Copy button to create a copy of a stringer.

Remove—Select a stringer in the list window and then click the Remove button to delete the stringer.

Components Tab

The **Components** tab allows you to specify the dimensions of the various components of the stair. This tab, shown in Figure 9.34, consists of **Flight Dimensions** and **Landing Dimensions** sections. The riser, tread, and nosing dimensions are set in the **Flight Dimensions** section. The **Landing Dimensions** section allows you to specify the thickness and width of the landing. A description of the options of this tab follows.

> **Allow Each Stair to Vary**—The Allow Each Stair to Vary checkbox, if checked, allows the dimensions in the Components tab to be edited in the Properties palette when the stair is placed. If this box is cleared, component sizes are controlled in the stair style.

Flight Dimensions

> **Display Tread**—The Display Tread checkbox turns ON/OFF the display of the tread.
>
> **Display Riser**—The Display Riser checkbox turns ON/OFF the display of the riser. Turn OFF the riser to create an open riser stair.
>
> **A-Tread Thickness**—The Tread Thickness represents the thickness of the material used for the tread.
>
> **B-Riser Thickness**—The Riser Thickness represents the thickness of the material used for the riser.
>
> **C-Nosing Length**—The Nosing Length represents the distance the nosing overhangs the riser.
>
> **Sloping Riser**—The Sloping Riser checkbox, if checked, creates a riser tilted from vertical.

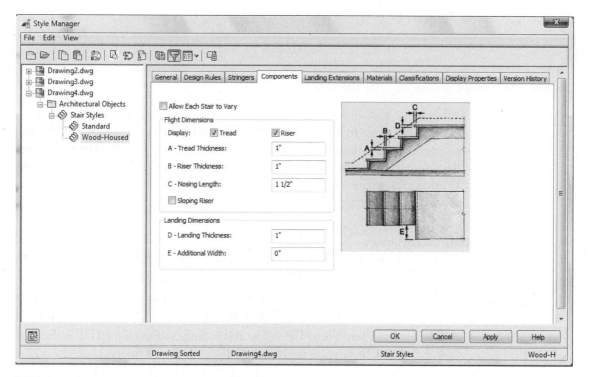

FIGURE 9.34 *Components tab of the stair style*

Landing Dimensions

D-Landing Thickness—The Landing Thickness represents the material thickness of the landing.

E-Additional Width—The Additional Width represents the additional width of the landing beyond the stair justification line.

Landing Extensions Tab

The landing extensions can be adjusted in the **Landing Extensions** tab as shown in Figure 9.35. The contents of the Landing Extensions tab are not applied to custom stairs.

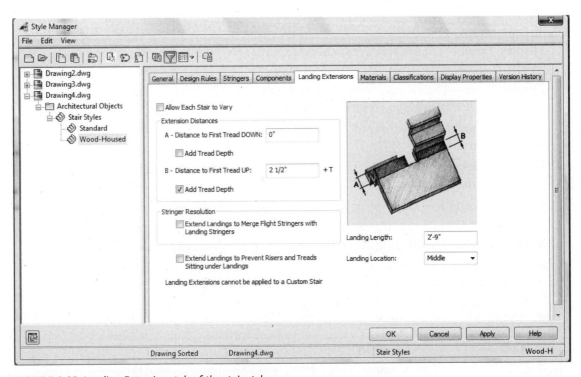

FIGURE 9.35 *Landing Extensions tab of the stair style*

Allow Each Stair to Vary—The Allow Each Stair to Vary checkbox, if checked, allows the dimensions of the Landing Extensions tab to be edited in the Properties palette by selecting the Landing Extensions button located in the **Advanced-Worksheets** section. If this box is clear, the landing extension components are controlled in the stair style definition.

Extension Distances

A-Distance to First Tread DOWN—The Distance to First Tread DOWN is the horizontal distance the landing is extended before the first step down.

Add Tread Depth—The Add Tread Depth checkbox, if checked, extends the landing before the down step by one tread depth.

B-Distance to First Tread UP—The B-Distance to First Tread UP is the horizontal distance the landing is extended before the first step up.

Add Tread Depth—The Add Tread Depth checkbox, if checked, extends the landing before the up step by one tread depth.

Stringer Resolution

Extend Landings to Merge Flight Stringers with Landing Stringers—This checkbox, if checked, extends the landing stringer to merge with flight stringers.

Extend Landing to Prevent Risers and Treads Sitting Under Landings— This checkbox, if checked, adjusts the landing position to create flush rectangular landings.

Landing Length—The Landing Length is the length of the landings placed with automatic landing placement. Automatic landings are placed when you set a minimum and maximum **Flight Height** in the **Advanced** section in the Design tab of the Properties palette.

Landing Location—The Landing Location drop-down list includes top, middle, and bottom locations. When **Automatic** landings are placed based upon the minimum and maximum **Flight Height**, the landing can be placed at the top, middle, and bottom locations to comply with the flight height restrictions.

Display Properties Tab

The **Display Properties** tab, shown in Figure 9.36, allows you to set the display of components of the stair in the stair style. The drawing default settings for layer and color are shown in Figure 9.36 for the **Plan** display representation. The Plan display representation is the display representation assigned for the **Medium Detail** display configuration. The Plan display representation turns on or off the riser and stringer above the cut plane. The stair in a first level View drawing displays the up components of the stair and the outline above the cut plane, whereas the stair in the second level View drawing includes the down components and the outline below the cut plane. This display system functions when the **Override Display Configuration** checkbox is cleared on the **Other** tab in the **Display Properties** dialog box shown in Figure 9.37. The Override Display Configuration checkbox is turned OFF for the Standard style and for the Imperial stair styles of the Design Tool Content catalog. In the Metric Design Tool catalog, on the other hand, the **Override Display Configuration** checkbox is checked for the standard stair style. Stair styles created with the Override Display Configuration checked require the user to edit the visibility of each stair display component to turn ON or OFF the up or down components in the **Layer/Color/Linetype** tab of the **Display Properties** tab.

The Plan Intermediate Level display representation is intended for the display of stair towers for levels between the first and top levels. The Plan Top Level display representation displays the down stair components of the top level, while the Plan Overlapping display representation can be applied to customize the display of components above, below, or overlapping.

TIP

Layer/Color/Linetype

Layer/Color/Linetype—The Layer/Color/Linetype tab lists the display components and the settings for visibility, by material, layer, color, linetype, lineweight, linetype scale, and plot style. The list of components varies according to the display representation.

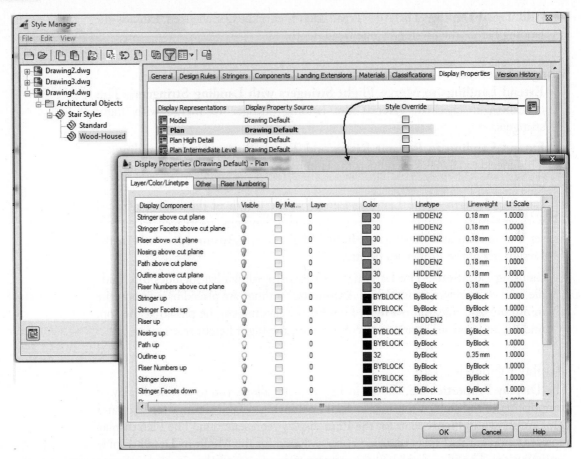

FIGURE 9.36 *Display Properties tab and the Display Properties Plan*

Other—The Other tab allows you to set the height of the cut plane, directional arrow, stair line, and break mark as shown in Figure 9.37.

CUT PLANE:

Override Display Configuration Cut Plane—If the Override Display Configuration Cut Plane checkbox is checked, the stair is displayed independent of the display configuration cut plane. Note that this checkbox was toggled ON as a default in releases prior to Autodesk AutoCAD Architecture 2008.

Show entire stair down—The Show entire stair down checkbox toggles ON the display of all stair components of stairs with down vertical orientation.

Height—The Height is the vertical distance from the $Z = 0$ coordinate of the cutting plane for creating the location of the lower half and the upper half of the stairs.

Cut Line Distance—The Cut Line Distance is the horizontal distance separating the break lines of the upper and lower stair sections.

Cut Line Angle—The Cut Line Angle is the angle formed relative to the treads for the break line symbol.

ARROW:

Size—The Size field allows you to define the dimensions of the directional arrow for the stair.

Offset—The Offset is the distance from the break line to the head of the directional arrow.

Dim Style—The **Dim Style** field lists the dimensioning style used to define the style of the directional arrow.

Arrow Size from Dimension Style Only—When a dimension style other than Standard is used, the size of the arrow can be controlled in the style.

STAIR LINE:

Shape—The Shape drop-down list includes curved or straight options.

Apply to—The Apply to drop-down list includes Entire Stair, Cut Plane Parallel, and Cut Plane Opposite.

Display Path—The Display Path drop-down list includes Graphics path and Construction line options for the direction of stair.

BREAK MARK:

Type—Type lists the break mark symbol options to represent the break, including None, Curved, Zigzag, and Custom shapes. If **Custom shape** is selected, the blocks of the drawing can be selected from the **Block** list. The break mark selected is displayed in the window below the **Block** field.

Block—The Block list displays blocks of the drawing that can be used for a custom break mark.

Riser Numbering

Riser Numbering—The Riser Numbering tab allows you to set the style and location of the text used to number the risers. You can edit the digit used as the first riser number as shown in Figure 9.37.

Style—The Style section allows you to specify the style, alignment, orientation, and height of text used for the riser numbers.

Location—The Location section allows you to specify the justification and x and y offsets of the text used for the riser numbers.

First Riser Number—The First Riser Number field allows you to specify the number for beginning the riser numbering.

Number Final Riser—The Number Final Riser checkbox turns ON/OFF the display of the final riser number.

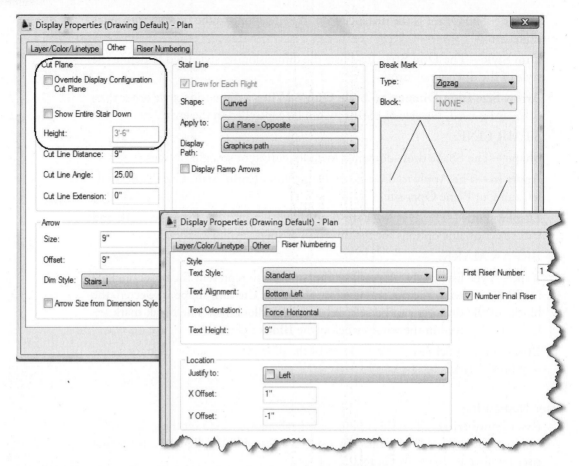

FIGURE 9.37 *Other and Riser Numbering tabs of the Display Properties dialog box*

TUTORIAL 9-2: CREATING A STAIR STYLE

1. Choose **New** from the Quick Access toolbar. Save the drawing to your student directory as Lab 9-2.dwg.

2. Toggle ON Ortho Mode and Dynamic Input in the **Application** status bar. Toggle OFF all other status bar buttons.

3. Select the **Stair** tool from the **Design** tool palette. Edit the **Properties** palette as follows: Shape = **Straight,** Width = **36"[914],** and Height = **10' [3050]**. Respond to the workspace prompts as follows:

   > Flight start point or: *(Select a point in the middle of the workspace.)*
   >
   > Flight endpoint or: *(Move the cursor right and choose a point beyond the dynamic display of the stairs.)*
   >
   > *(Press* ESC *to end the command.)*

4. Select the stair and choose **Copy Style** from the **General** panel to open the **Stair Styles—Standard (2)** dialog box. Choose the **General** tab and edit the name to **Residential**.

5. Select the **Design Rules** tab, as shown in Figure 9.38, and edit as follows: A-Maximum Slope: Riser Height = **7 3/4" [197]** and Tread Depth = **10" [254];**

B-Optimum Slope: Riser Height = **7 3/4"** **[197]** and Tread Depth = **10 1/2"** **[267]**; C-Minimum Slope: Riser Height = **6"** **[152]** and Tread Depth = **11"** **[279]**. Clear the **Use Rule Based Calculator** checkbox.

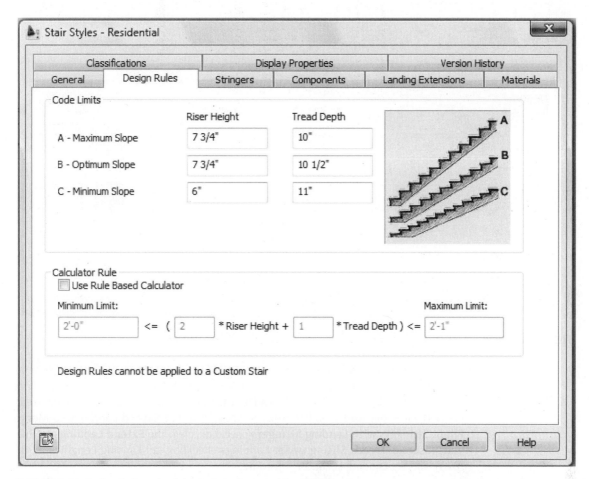

FIGURE 9.38 *Design Rules tab of the Stair Styles – Residential dialog box*

6. Refer to Figure 9.39 and select the **Stringers** tab, then edit as follows: click the **Add** button, select the **Unnamed** stringer in the **Stringers** column, and overtype **Left** to name the new stringer. Click in the **Type** column and choose **Housed**. Click in the **Alignment** column of the new stringer and select **Align Left** from the drop-down list. Click the **Add** button, select the **Unnamed** stringer in the **Stringers** column, and overtype **Right** to name the new stringer. Click in the **Type** column and choose **Housed**. Click in the **Alignment** column of the new stringer and select **Align Right** from the drop-down list. Select the **Add** button, select the **Unnamed** stringer in the **Stringers** column, and overtype **Middle** to name the new stringer. Click in the **Type** column and choose **Housed**. Click in the **Alignment** column of the new stringer and select **Center** from the drop-down list. Edit A-Width = **1 1/2"** **[38]**, C-Waist = **4"** **[100]**, and E-Waist = **4"** [100] for all stringers.

7. Refer to Figure 9.39 and select the **Components** tab, then edit as follows: verify that the **Allow Each Stair to Vary** checkbox is clear and that the **Display Tread** and **Riser** checkboxes are checked. Set A-Tread Thickness = **1"** **[25]**, B-Riser Thickness = **3/4"** **[19]**, and C-Nosing Length = **1"** **[25]**. Clear the **Sloping Riser** checkbox and set D-Landing Thickness = **1"** **[25]** and E-Additional Width = **0**.

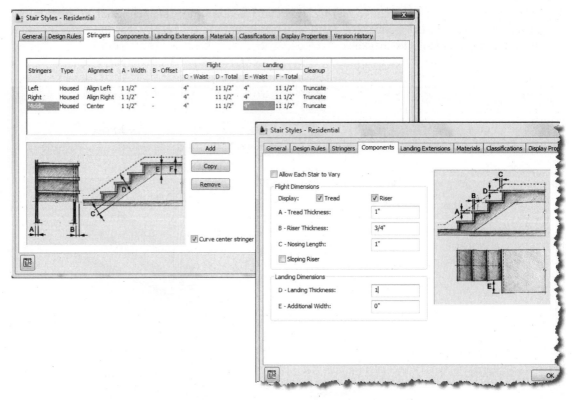

FIGURE 9.39 *Components and Stringers tabs of the Stair Styles – Residential dialog box*

8. Refer to Figure 9.40 and select the **Landing Extensions** tab, then edit as follows: clear the **Allow Each Stair to Vary** checkbox, set A-Distance to First Tread Down = **0**, clear the **Add Tread Depth** checkbox, set B-Distance to First Tread UP = **0**, clear the **Add Tread Depth** checkbox, clear the **Extend Landings to Merge Flight Stringers with Landing Stringers** checkbox, clear the **Extend Landings to Prevent Risers & Treads Sitting under Landings** checkbox, set Landing Length = **2'-9"** **[850]**, and set Landing Location = **Middle**.

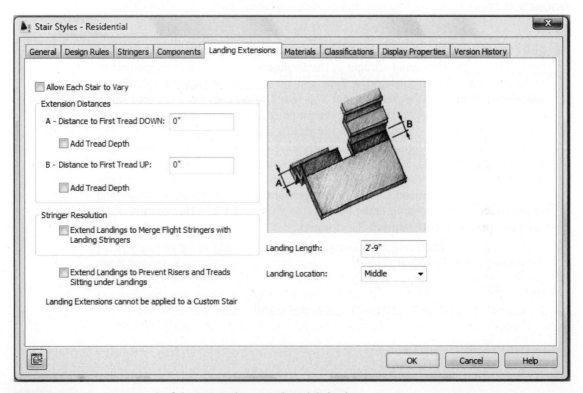

FIGURE 9.40 *Components tab of the Stair Styles – Residential dialog box*

9. Click **OK** to dismiss the **Stair Style – Residential** dialog box.

10. Select the Stair, right-click, and choose **Properties**. Click the **Calculation Rules** button in the Design tab of the Properties palette. Verify that all fields are set to **Automatic**. Verify that the Straight Length is **13'-1 1/2" [4005]**. Set the Tread toggle to **Manual**. Edit the Tread to **10" [254]** and the Straight Length changes to **12'-6" [3810]**. Click **OK** to dismiss the Calculator Rules dialog box. Press ESC to end selection of the stair.

11. Right-click over the Command line and choose **Options**. Choose the **AEC Object Settings** tab. Edit the Calculator Limits to **Relaxed**. Click **OK** to dismiss the Options dialog box.

12. Choose the **View** tab. Choose Regenerate Drawing from the Regenerate Drawing flyout of the Appearance panel.

13. In this step you will import a material definition into the drawing. Choose **Content Browser** from the Quick Access toolbar. Open the Visualization Catalog. Navigate to AEC Material Tools > US Imperial < Woods and Plastics < Finish Carpentry [Material Tools > US Metric < Woods and Plastics < Finish Carpentry]. Drag the i-drop for **Wood.Finish Carpentry.Wood.Ash** material definition into the workspace. Press ESC to import the material definition.

14. Select the stair, right-click, and choose Edit Stair Style from the shortcut menu. Select the Materials tab. Choose **Wood.Finish Carpentry.Wood.Ash** from the Material Definition drop-down list for the Tread and Riser components. Click **OK** to dismiss the dialog box.

15. Choose **SW Isometric** from the **View Cube**. Choose **Realistic** from the **Visual Styles** flyout menu of the **Visual Styles** panel in the View tab. Save and Close the drawing.

ANCHORING A STAIR TO A LANDING

The **AnchorToStairLanding** command allows you to create two or more flights of stairs that share a single landing or link an AEC object to a landing. Objects can be anchored to landings created using the multi-landing or U-shaped stairs that include a landing. The anchor links the flights such that changes in one flight will cause the AEC object to change accordingly. Access the **AnchorToStairLanding** command as shown in Table 9.12.

TABLE 9.12 *AnchorToStairLanding command access*

Command prompt	ANCHORTOSTAIRLANDING
Ribbon	Select a multi-landing stair and choose Anchor Object to Stair Landing from the Modify panel of the Stair contextual tab.
Shortcut menu	Select a multi-landing stair, right-click, and choose Stair Landing Anchor > Anchor Object.

When you select a multi-landing stair and choose **Anchor Object to Stair Landing**, you are prompted as follows in the workspace:

> *(Select the multi-landing stair at* p1 *as shown in Figure 9.41.)*
>
> Select AEC object to anchor to stair: *(Select the stair at* p2 *as shown in Figure 9.41.)*
>
> Select stair landing: *(Select the landing at* p3 *as shown in Figure 9.41. Stair is anchored to the location specified on the landing.)*

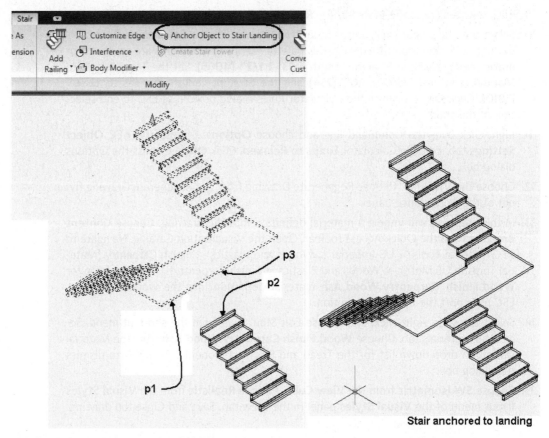

FIGURE 9.41 *Stair anchored to multi-landing stair*

The selected landing and the AEC object are then anchored together. The **Anchor-Release** command removes the anchor between the AEC object and the stair. Access the **AnchorRelease** command as shown in Table 9.13.

TABLE 9.13 *AnchorRelease command access*

Command prompt	**ANCHORRELEASE**
Shortcut menu	Select an anchored object, right-click, and choose Stair Landing Anchor > Release.

CREATING A STAIR TOWER

The **StairTowerGenerate** command is used to generate multi-landing or U-shaped stairs within a project. See Table 9.14 and Figure 9.42 for **StairTowerGenerate** command access.

TABLE 9.14 *StairTowerGenerate command access*

Command prompt	**STAIRTOWERGENERATE**
Tool palette	Select Stair Tower from the Design tool palette.
Ribbon	Choose Stair Tower from the Stair flyout menu of the Build panel in the Home tab.
Shortcut menu	Select a stair, right-click, and choose Stair Tower Generate.

The stair must be generated within a project from a stair Construct drawing assigned to each of the levels of a project. The stair should have its display controlled by the display configuration. (The stair display control is set by clearing the **Override Display Configuration Cut Plane** checkbox of the **Other** tab in the **Display Properties** dialog box. The **Display Properties** dialog box is accessed from the **Edit Display Properties** button in the **Display Properties** tab of the **Stair Style** dialog box.) The **StairTower-Generate** command creates additional stairs, slabs, and railings above the selected stair the distance specified in the floor-to-floor height for the levels of the project. The interference subtraction is also applied to each new stair and slab. Each stair can be edited independently of the stair for the first level. The stair shown at the right in Figure 9.42 was generated to the next level as shown in the following workspace sequence.

> *(Select the stair at* p1 *and choose* Stair Tower Generate from the Design *palette.)*
>
> Select railings and slabs: 1 found *(Select slab at* p2 *as shown in Figure 9.42.)*
>
> Select railings and slabs: ENTER *(Press ENTER to end selection.)*
>
> *(The* Select Levels *dialog box opens, which specifies the floor-to-floor height. Click* OK *to accept the settings and dismiss the Select Levels dialog box. The stair tower is generated as shown at the right in Figure 9.42.)*

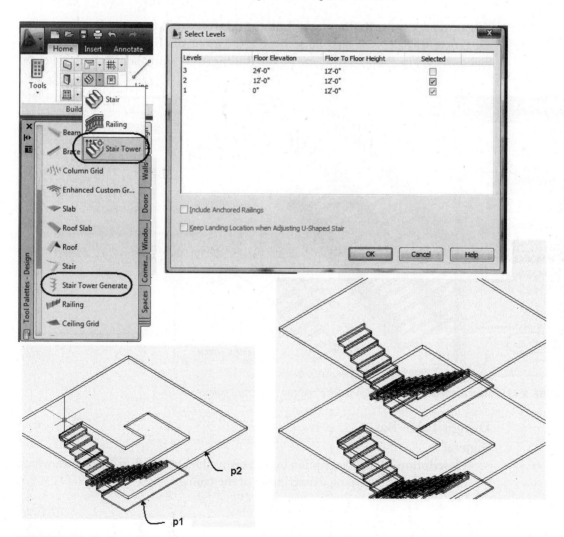

FIGURE 9.42 *Creating a stair tower*

If the floor-to-floor height of the project is changed after the stair tower is generated, select the stair and reapply Stair Tower Generate to update to the new elevations. When the stair Construct drawing is attached to the model view and to floor plan views, you can create the appropriate plan view of the stair when you edit the display configuration to Medium Detail, Medium Detail Intermediate Level, or Medium Detail Top Level in the View drawing.

CREATING RAILINGS

The **RailingAdd** command is used to add a railing to each side or the center of the stair. Railings can consist of posts, balusters, guardrails, bottom rails, and handrails. Railings can be added to a stair, stair flight, or be freestanding. See Table 9.15 for **RailingAdd** command access.

TABLE 9.15 *RailingAdd command access*

Command prompt	RAILINGADD
Tool palette	Select Railing from the Design tool palette.
Ribbon	Select Railing from the Stair flyout of the Build panel in the Home tab.
Shortcut menu	Select a stair, right-click, and choose Add Railing.

When you select the **Railing** tool (**RailingAdd** command) from the Design palette, the Properties palette opens, which displays the properties of the railing as shown in Figure 9.43. The Properties palette for the **RailingAdd** command is described below.

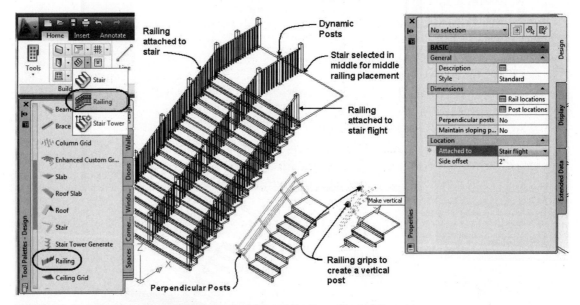

FIGURE 9.43 *Railing command of the Design palette and the Properties palette*

Design Tab—Basic
General
    Description—The Description button opens the **Description** dialog box, which includes fields for typing a description of the railing.

Layer—The Layer field is displayed when you select existing railings; it lists the name of the layer for the railing. The railing is placed on the A-FlorHral [A340G] layer if the AIA (256 Color) [BS1192 Cisfb (256 Color)] layer key style is used.

Style—The **Style** drop-down list displays the railing styles of the drawing.

Dimensions

Rail locations—The Rail locations button opens the **Rail Locations** dialog box. The Rail Locations dialog box is inactive for the Standard railing style because the features are controlled per style in the **Rail Locations** tab of the railing style.

Post locations—The Post locations button opens the **Post Locations** dialog box. The Post Locations dialog box is inactive for the Standard railing style because the features are controlled per style in the **Post Locations** tab of the **Railing Styles** dialog box.

Railing extensions—The Railing extensions button opens the Railing Extensions dialog box. The Railing extensions button is only displayed for railings when **Attach to** is set to Stair. The dialog box is inactive if the features are set in the style and active if the features are not set within the style.

Perpendicular Posts—The Perpendicular Posts toggle, if set to Yes, will place the corner post perpendicular to the slope line of the railing.

Maintain Sloping Posts at Landings—The Maintain Sloping Posts at Landings, if set to Yes, will retain the slope of the railing during the landing.

Location

Attached to—The Attached to drop-down list includes the **Stair, Stair Flight,** and **None** options. If Attached to is set to **Stair,** the railing is created as an attachment to the entire stair. The **Stair** option expands the **Properties** palette to include the **Automatic Placement** option discussed below. The **Stair Flight** option allows the railing to be attached to a flight segment of the multi-landing stair. The **None** option allows you to place a freestanding railing or guardrail.

Automatic Placement—The Automatic Placement options include **Yes** and **No**. If **Yes** is selected, the railing will be attached to the stair at the selected stair edge or the center of the stair. If Automatic Placement is set to **No**, the railing will be placed on the edge or center selected, and you will be prompted to select the start and end points of the railing.

Side offset—The Side offset is the distance from the centerline of the railing to the edge of the stair.

ADDING A RAILING

When you select **Railing** from the Design palette, the **Attach to** option in the Properties palette governs the workspace prompts. When **Attach to** is set to **Stair Flight**, the cursor changes to a pick box, and you are prompted to select the stair flight for the railing. When you select the stair, you are also specifying the location for the railing. Therefore, if you select the stair in the middle of the tread width, the railing is applied in the center as shown at **A** in Figure 9.44.

If the **Railing** command is selected and **Attach to** is set to **Stair**, then **Automatic Placement** is added to the Properties palette. When **Automatic Placement** is set to **Yes**, selecting the stair adds the railing to the entire stair edge as shown at **B** in Figure 9.44. If **Automatic Placement** is set to **No**, you are prompted to select the start and end points of the railing as shown in the following workspace sequence:

Select a stair or: *(Select the stair at* p1 *to determine a stair edge for the railing as shown at C in Figure 9.44.)*

Railing start point or: *(Select a point* p2 *to begin the railing as shown at C in Figure 9.44.)*

Railing endpoint or: *(Select a point* p3 *to end the railing as shown at C in Figure 9.44.)*

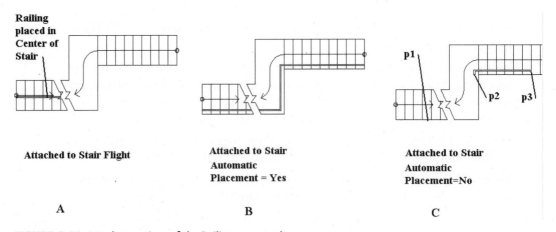

FIGURE 9.44 *Attach to options of the Railing command*

Using the Attached to: None Option

When the **Railing** command is selected and **Attach to** is set to **None**, you are prompted to specify the start and end points of the railing. The railing is not anchored to an AEC object; therefore, when you move an adjoining object, the railing does not move. This option can be used for placing railings on decks and balconies.

DEFINING RAILING STYLES

The design of the Standard railing is simple; therefore, railing styles can be created to represent typical interior and exterior railings. The style can specify the size, shape, and placement of handrails, guardrails, posts, and balusters as shown in Figure 9.45. Additional styles can be i-dropped into the drawing from the Stairs and Railings > Railings folder of the Design Tool Catalog (Imperial) or Design Tool Catalog (Metric) of the Content Browser.

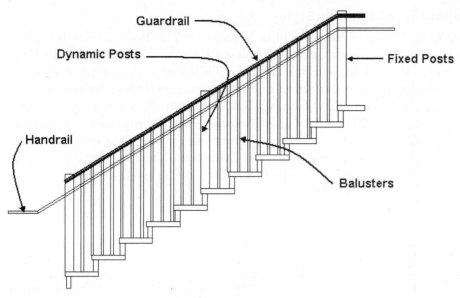

FIGURE 9.45 *Components of a railing*

Railing styles are created in the Style Manager. The railing style can be edited in the Style Manager or with the **Edit Railing Style** command; see Table 9.16 for command access.

TABLE 9.16 *Style Manager and Edit Railing Style command access*

Command prompt	RAILINGSTYLEEDIT
Ribbon	Select the railing and choose Edit Style from the Edit Style fly-out menu of the General panel in the Railing contextual tab to open the Railing Styles properties dialog box.
Ribbon	Select the railing and choose Railing Styles from the Edit Style flyout menu of the General panel in the Railing contextual tab to open the Style Manager.
Shortcut menu	Select the railing, right-click, and choose Edit Railing Style.
Tool palette	Right-click over the Railing tool and choose Railing Styles.

Steps to Import and Edit a Railing Style

1. Select **Content Browser** from the Tools flyout menu of the Build panel in the Home tab.
2. Open the Design Tool Catalog—Imperial or Design Tool Catalog—Metric.
3. Navigate to Stairs and Railings > Railings.
4. Drag the i-drop of a railing style into the workspace and press ESC.
5. Right-click the Railing tool of the Design palette and choose Railing Styles to open the Style Manager.
6. Select the imported railing style listed in the *Railing Styles* folder of the left pane; edit the properties of the style in the right pane.

Components of a Railing Style

A railing style can be edited in the Style Manager or in the Railing Styles dialog box. If a railing has been inserted, select the railing and choose **Edit Style** from the General panel of the Railing contextual tab. The Edit Railing Style command opens the **Railing Styles** dialog box as shown in Figure 9.46. The **Railing Styles** dialog box consists of the following tabs: **General**, **Rail Locations**, **Post Locations**, **Components**, **Extensions**, **Materials**, **Classifications**, **Display Properties**, and **Version History**. The content of the following tabs is unique for the railing style: **Rail Locations**, **Post Locations**, **Components**, and **Extensions**. The contents of the Display Properties tab include options to turn ON or OFF the up and down display components of the railing.

Rail Locations Tab

The **Rail Locations** tab shown in Figure 9.46 allows you to define the horizontal and vertical components of the railing.

> **Allow Each Railing to Vary**—When this checkbox is toggled ON, you can set values for the rail locations of the style in the Properties palette for each insertion in the drawing. If this checkbox is clear, the settings of this tab are specified only in the style and cannot be edited in the Properties palette.

Upper Rails

> **Guardrail**—Check the Guardrail checkbox to display a guardrail for the style. If the Guardrail checkbox is clear, the dimension fields are inactive.

> **Handrail**—Check the Handrail checkbox to display a handrail for the style. If the Handrail checkbox is clear, the dimension fields are inactive.

> **Horizontal Height**—The Horizontal Height is the height of the guardrail, handrail, or bottomrail in the Z direction.

> **Sloping Height**—The Sloping Height is the height of the guardrail, handrail, and bottomrail as they are placed on the stair parallel to the stair.

> **Offset from Post**—The Offset from Post is the distance from the guardrail, handrail, or bottomrail positioned from the centerline of the vertical posts.

> **Side for Offset**—The Side for Offset specifies the direction for placing the handrail, guardrail, and bottomrail relative to the posts. A positive offset distance places the railing the specified distance toward the center of the stair. Side of offset options in the drop-down list include right, left, center, and auto.

Bottomrail

> **Bottomrail**—The Bottomrail checkbox, if checked, will display a bottomrail.

> **Horizontal Height**—The Horizontal Height sets the vertical distance from $Z = 0$ to the bottomrail when placed in a horizontal position.

> **Sloping Height**—The Sloping Height sets the distance between stair and bottomrail as the railing is developed along the stair.

> **Number of Rails**—The Number of Rails specifies the number of bottomrails. If more than one bottomrail is specified, you can edit the **Spacing of Rails** in inches.

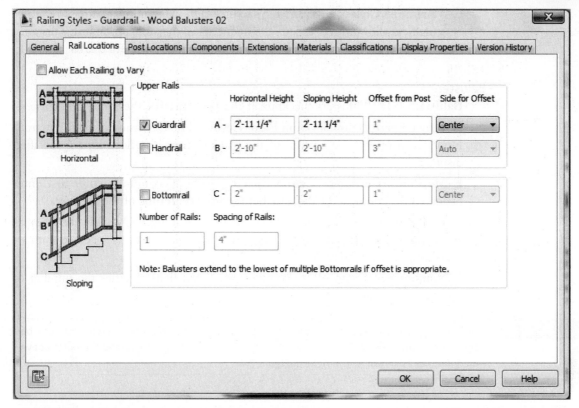

FIGURE 9.46 *Rail Locations tab of the Railing Styles dialog box*

Post Locations Tab

The **Post Locations** tab allows you set the frequency and position of posts placed along the rail. The following locations are specified in the **Post Locations** tab shown in Figure 9.47.

Fixed Posts

Fixed Posts—The Fixed Posts checkbox allows you to turn ON or OFF the display of fixed posts located at the beginning and end of the stair flight.

A-Extension of ALL Posts from Top Railing—The A-Extension of ALL Posts from Top Railing specifies the distance to extend the post above the top railing.

B-Extension of ALL Posts from Floor Level—The B-Extension of ALL Posts from Floor Level extends the bottom of the post up the specified distance from the floor level.

Fixed posts at Railing Corners—The Fixed posts at Railing Corners checkbox, when checked, forces the development of a post at the corners or change of direction of the railing.

Dynamic Posts

Dynamic Posts—The Dynamic Posts checkbox allows you to turn ON or OFF the display of posts between the fixed posts located at the beginning and end of the stair.

C-Maximum Center to Center Spacing—The Maximum Center to Center Spacing field allows you to specify the frequency of dynamic posts. If this distance is set to 4', dynamic posts will be placed along the rail no more than 4' apart.

Balusters

Balusters—The Balusters checkbox, if checked, specifies the display of balusters.

D-Extension of Balusters from Floor Level—The Extension of Balusters from Floor Level offsets the bottom of the baluster from the floor level a specified distance.

E-Maximum Center to Center Spacing—The Maximum Center to Center Spacing field specifies the distance between the centerline of the balusters.

Stair Tread Length Override—The Stair Tread Length Override checkbox, if checked, will override the distance between balusters to place a specified number of balusters on a stair tread.

F-Number per Tread—The Number per Tread field specifies the number of balusters per tread.

NOTE Deck railings can be created by specifying guardrail and bottom rails. Porch posts can be created by editing the A-Extension of ALL Posts from Top Railing to extend the posts above the guardrail as shown at the right in Figure 9.47.

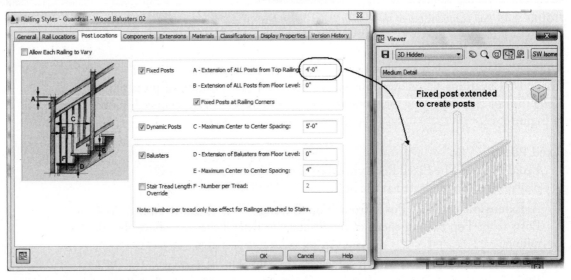

FIGURE 9.47 *Post Locations tab of the Railing Styles dialog box*

Components Tab

The **Components** tab allows you to specify the profile or shape used in extruding railing components. You can specify an AEC profile for a component by clicking in the **profile name** column shown in Figure 9.48. The **Profile Name** field becomes an active list, which displays circular, rectangular, or other available AEC profiles of the drawing.

> **Component**—The Components listed in Figure 9.48 are components of the railing developed from Profiles. Guardrail, handrail, posts, and balusters can be created by extruding profiles to create custom shapes. Components that are circular or rectangular can be created.

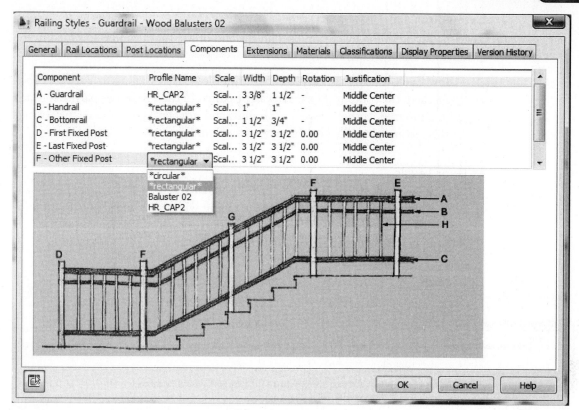

FIGURE 9.48 *Components tab including profiles used as baluster shapes*

Profile Name—The Profile Name drop-down list shows the AEC Profiles loaded in the drawing. A profile can be defined for the component by clicking in the Profile Name column of the component and selecting from the drop-down list. Figure 9.49 includes examples of polylines used to form the profile of components.

Scale—The Scale drop-down list allows you to scale the profile to Scale to Fit, Scale to Fit Width, Scale to Fit Depth, or No Scale.

Width—The Width dimension of the railing component.

Depth—The Depth dimension of the railing component.

Rotation—The Rotation dimension of the railing component in a counterclockwise direction.

Justification—The Justification option allows the profile to be positioned relative to the centerline of the railing using the following justifications: Top Left, Top Center, Top Right, Middle Left, Middle Center, Middle Right, Bottom Left, Bottom Center, Bottom Right, and Insertion point.

Guardrail Return 1.5 Plan   Handrail Return 1.5 Escutcheon Plan   Handrail Return 1.5 Plan   HR_Cap2

FIGURE 9.49 *Polylines of profiles used as handrail shapes*

Extensions Tab

The amount of railing extension beyond the end of the flight can be specified in the **Extensions** tab as shown in Figure 9.50.

> **Allow Each Railing to Vary**—When the Allow Each Railing to Vary checkbox is checked, the contents of the tab can be edited in the Properties palette for each railing inserted. If the Allow Each Railing to Vary checkbox is clear, the settings in this tab are controlled by the style definition.

At Floor Levels

> **Use Stair Landing Extension**—If the Use Stair Landing Extension checkbox is ON at floor levels, the extension of the railing will match the stair extension for the stairs and stair flights. When this checkbox is clear, the extension of the railing can be defined in this tab.

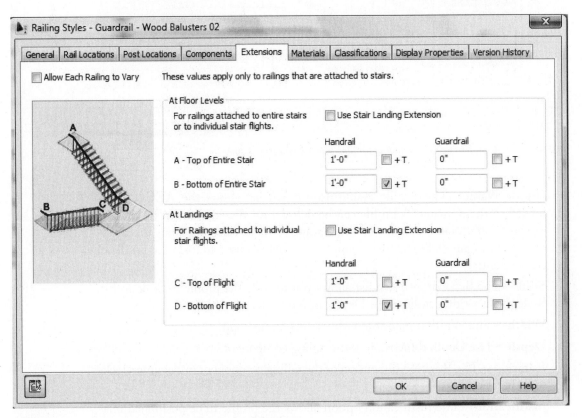

FIGURE 9.50 *Extensions tab of the Railing Styles dialog box*

> **Use Stair Landing Extension**—If Use Stair Landing Extension is toggled ON at floor levels, the extension of the railing will match the stair extension for the stairs and stair flights. When this toggle is clear, the extension of the railing can be defined in this tab.

> **A-Top of Entire Stair**—The Top of Entire Stair fields define the distance to extend the railing beyond the stair for the handrail and guardrail. The distance can be defined as a distance plus the depth of the tread to obtain a total distance.

B-Bottom of Entire Stair—The Bottom of Entire Stair fields define the distance to extend the railing beyond the stair at the bottom of the stair for the guardrail and handrail. The distance can be defined as a distance plus the depth of the tread to obtain a total distance.

At Landings

Use Stair Landing Extension—If the Use Stair Landing **Extension** checkbox is ON, the extension of the railing will match the stair flights at the landing. When this toggle is clear, the extension of the railing can be defined in this tab.

C-Top of Flight—The Top of Flight field defines the distance to extend the railing beyond the stair for the handrail and guardrail at the landing for the top flight. The distance can be defined as a distance plus the depth of the tread to obtain a total distance.

D-Bottom of Flight—The Bottom of Flight defines the distance to extend the railing beyond the stair for the handrail and guardrail at the landing for the bottom flight. The distance can be defined as a distance plus the depth of the tread to obtain a total distance.

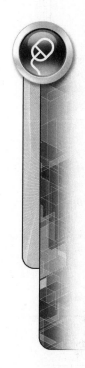

TUTORIAL 9-3: CREATING A RAILING STYLE

1. Choose New from the Quick Access toolbar. Save the drawing to your student directory as Lab 9-3.dwg.

2. This step imports a railing style into the drawing from the Content Browser. Select the Content Browser from the Tools flyout of the Build panel in the Home tab. Open the Design Tool Catalog – Imperial > Stairs and Railings > Railings [Design Tool Catalog – Metric > Stairs and Railings > Railings]. Go to page 2 of the catalog, select the i-drop of the Guardrail-Wood Balusters 01 style, drag it to the workspace, and press ESC to terminate the command without selecting a stair. The Guardrail-Wood Balusters 01 style is imported into the drawing.

3. Right-click the Railings tool of the Design tool palette and choose Railing Styles. Choose the Guardrail-Wood Balusters 01 in the left pane, right-click, and choose Rename. Overtype the name Handrail.

   Click the Handrail railing style in the left pane to display the edit tabs of the right pane. To create a simple handrail, refer to Figure 9.51 and select the Rail Locations tab, then edit as follows: clear the Allow Each Railing to Vary checkbox; in the Upper Rails section, check Guardrail; edit the A-Horizontal Height = 38" [965], Sloping Height = 38" [965], and Side for Offset = Center. Clear the Handrail and Bottomrail checkboxes.

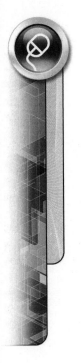

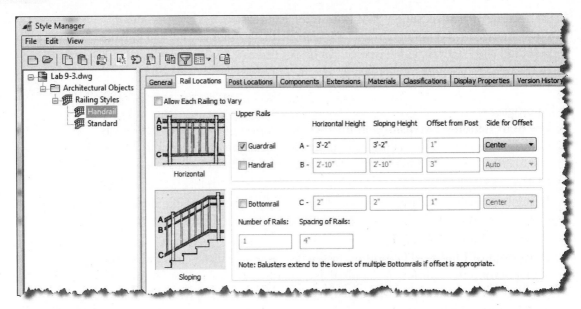

FIGURE 9.51 *Rail Locations tab of the railing style in the Style Manager*

4. Select the Post Locations tab and clear the Allow Each Railing to Vary, Fixed Posts, and Balusters checkboxes.

Refer to Figure 9.52 and select the **Extensions** tab, then clear the **Allow Each Railing to Vary** checkbox, check the **Use Stair Landing Extension** checkbox for **At Floor Levels**, and check the **Use Stair Landing Extension** checkbox for **At Landings**.

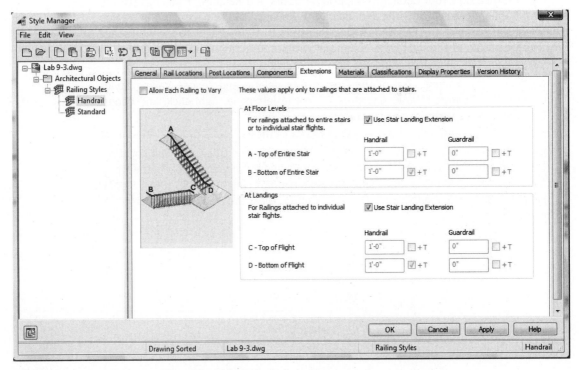

FIGURE 9.52 *Extensions tab of the railing style in the Style Manager*

5. To create a guardrail railing style that includes a handrail and balusters, select the **Handrail** style in the left pane, right-click, and choose **Copy** from the shortcut menu. Select the **Railing Styles** folder in the left pane, right-click, and choose **Paste**. Select **Handrail (2)** in the left pane, right-click, and choose **Rename**. Overtype **Guardrail** to rename the new style. Click the **Guardrail** style in the left pane to display the railing edit tabs of the right pane. Select the **General** tab and choose the **Floating Viewer** in the lower-left corner of the Style Manager. Resize the Viewer as shown in Figure 9.53. Choose **SW Isometric** from the View flyout menu in the Viewer.

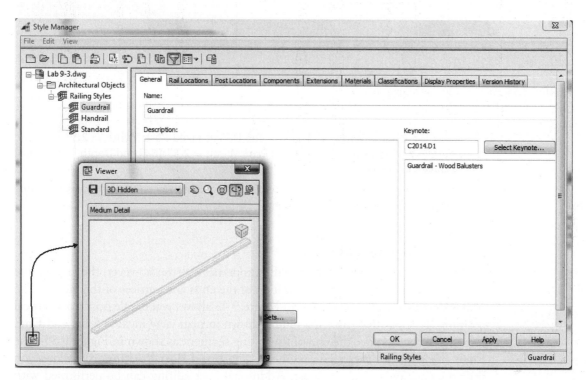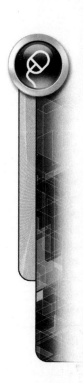

FIGURE 9.53 *Railing display in Viewer*

6. Select the **Post Locations** tab (see Figure 9.54) and edit as follows: clear the **Allow Each Railing to Vary** checkbox, check **Fixed Posts**, set A-Extension of All Posts from Top Railing = **0** and B-Extension of All Posts from Floor Level = **0**, check **Fixed Posts at Railing Corners**, clear **Dynamic Posts**, check **Balusters** ON and set D-Extension of Balusters from Floor Level = **0,** set E-Maximum Center to Center Spacing = **4″ [100]**, and clear Stair Tread Length Override.

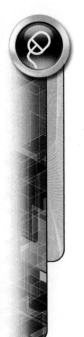

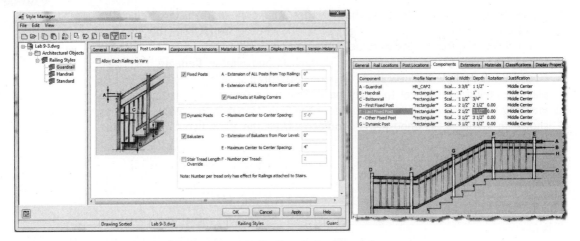

FIGURE 9.54 *Post Locations and Components tabs for Guardrail railing style*

7. Select the **Components** tab and edit the D-First Fixed Post: Width = **2.5″ [64]** and Depth = **2.5″ [64]**. Edit E-Last Fixed Post: Width = **2.5″ [64]** and Depth = **2.5″ [64]** as shown in Figure 9.54.

8. Choose the Floating Viewer to view the railing style. Click **OK** to dismiss the Style Manager. Save and close the drawing.

MODIFYING THE RAILING

Existing railings can be modified in the Properties palette; however, the grips of the railing can be used to change the location of the ends and vertices of the railing. The **Make Vertical** toggle grip shown in Figure 9.43 allows you toggle posts to a vertical position. Whereas the grips of a railing shown in plan view include **Start** and **End**, grips allow you to move the end of the railing segment as shown in Figure 9.55. The **Edge** grip stretches the railing segment. The **Fixed Post Position** grip allows you shift the position of the end post. Additional post editing can be performed with the **Post Location** option of the shortcut menu.

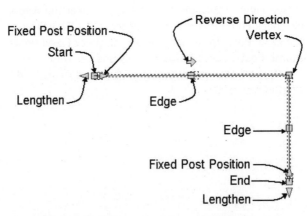

FIGURE 9.55 *Grip editing of railings not attached to stairs or stair flights*

Editing Post Locations

The Railing contextual tab of a selected railing includes commands to **Add**, **Remove**, **Hide**, **Show**, and **Redistribute** the posts. The **Add** option selects the **Railing-PostAdd** command as shown in Table 9.17.

TABLE 9.17 *RailingPostAdd command access*

Command prompt	**RAILINGPOSTADD**
Ribbon	Select a railing and choose Add from the Posts panel of the Railing contextual tab.
Shortcut menu	Select a railing, right-click, and choose Post Placement > Add.

When you select this option, you are prompted in the workspace to select a location for the additional posts, as shown in the following command sequence. Object snap and object tracking can be used to precisely place posts. The added posts become fixed post positions that can be removed.

> Specify position for Post: *(Select a location on the railing at* **p1** *as shown at the left in Figure 9.56.)*
>
> Specify position for Post: ENTER *(Press* ENTER *to end the command.)*
>
> *(Post is added as shown at the right in Figure 9.56.)*

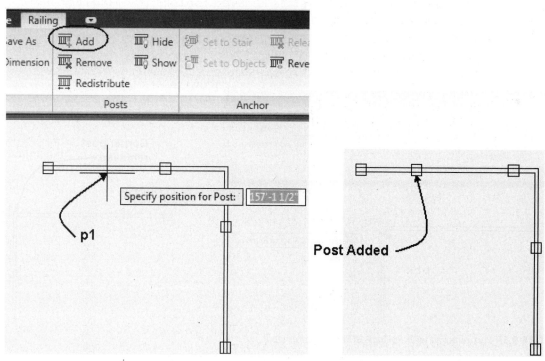

FIGURE 9.56 *Post added with Post Placement > Add*

Removing Posts

Posts that have been added with the **RailingPostAdd** command can be removed from the railing using the **RailingPostRemove** command. You cannot remove dynamic posts; however, when you remove fixed posts, the dynamic posts convert to fixed and are removed. Removing an added post redistributes the posts to be evenly placed along the railing. Access the **RailingPostRemove** command as shown in Table 9.18.

TABLE 9.18 *RailingPostRemove command access*

Command prompt	**RAILINGPOSTREMOVE**
Ribbon	Select a railing and choose Remove from the Posts panel of the Railing tab.
Shortcut menu	Select a railing, right-click, and choose Post Placement > Remove.

The following command sequence was used to remove the added post and fixed post as shown at the right in Figure 9.57. The dynamic posts were redefined as fixed posts when the fixed post was removed. The dynamic posts could not be removed independent of the fixed post removal.

(Select the railing and choose **Remove** *from the Posts panel of the Railing contextual tab.)*

Select a Railing Post to Remove: *(Select the manually added post at* **p1** *as shown in Figure 9.57.)*

Select a Railing Post to Remove: ENTER *(Press ENTER to end selection.)*

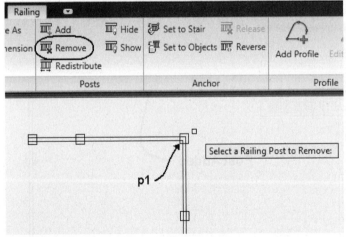
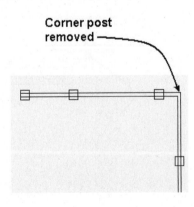
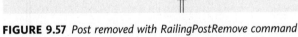

FIGURE 9.57 *Post removed with RailingPostRemove command*

Hiding Posts

Posts that have been added to a railing or located on the end of the railing can be hidden.

Posts placed dynamically between the fixed post position grips *cannot* be hidden.

The **RailingPostHide** command is used to hide posts. Access the **RailingPostHide** command as shown in Table 9.19.

TABLE 9.19 *RailingPostHide command access*

Command prompt	RAILINGPOSTHIDE
Ribbon	Select a railing and choose Hide from the Posts panel of the Railing contextual tab.
Shortcut menu	Select a railing, right-click, and choose Post Placement > Hide.

The **RailingPostHide** command was used in the following command sequence to hide the posts as shown at the right in Figure 9.58.

> *(Select a railing and choose* Hide *from the Posts panel of the Railing contextual tab.)*
>
> Select a Railing Post to Hide: *(Select posts at* p1 *and* p2 *as shown at the left in Figure 9.58.)*
>
> *(Posts are hidden as shown at the right in Figure 9.58.)*

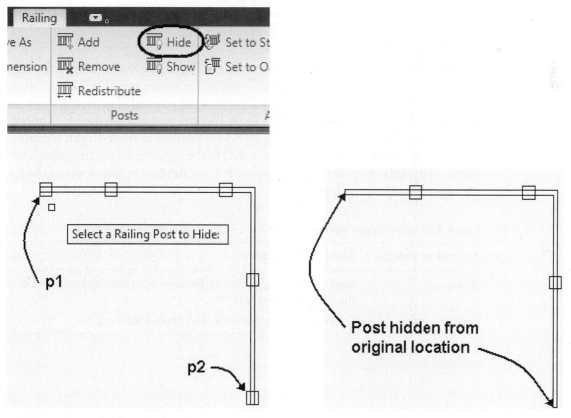

FIGURE 9.58 *Post hidden with Post Placement > Hide*

Showing Posts

Posts that have been hidden can be displayed for the railing when you select **Show** from the **Post** panel of the **Railing** contextual tab. The **RailingPostShow** command displays all hidden posts of the railing. Access the **RailingPostShow** command as shown in Table 9.20.

TABLE 9.20 *RailingPostShow command access*

Command prompt	**RAILINGPOSTSHOW**
Ribbon	Select a railing and choose Show from the Posts panel of the Railing contextual tab.
Shortcut menu	Select a railing, right-click, and choose Post Placement > Show.

Redistributing Posts

Posts that have been added to or removed from a railing create post locations not uniformly placed along the railing. The start and end posts can be stretched to a different location using grips. The **Redistribute** option resets the post distribution along the railing. Access **Redistribute** as shown in Table 9.21.

TABLE 9.21 *RailingRedistributePosts command access*

Command prompt	**RAILINGREDISTRIBUTEPOSTS**
Ribbon	Select a railing and choose Redistribute from the Posts panel of the Railing contextual tab.
Shortcut menu	Select a railing, right-click, and choose Post Placement > Redistribute.

Reversing the Railing

The **Reverse** option of the shortcut menu of a railing allows you to reverse the start and end of a railing. Access the **Reverse** command as shown in Table 9.22. A railing that has been edited to include additional posts can be reversed. The railing shown at the right in Figure 9.59 was reversed from its orientation shown on the left. The baluster design of the railing shown in Figure 9.59 is justified to **Back-Right**; therefore, when the railing is reversed, the balusters shift to the opposite side of the railing centerline. If the baluster position is center justified, the **Reverse** option does not change the appearance of the railing.

TABLE 9.22 *RailingReverse command access*

Command prompt	**RAILINGREVERSE**
Ribbon	Select a railing and choose Reverse from the Anchor panel of the Railing contextual tab.
Shortcut menu	Select a railing, right-click, and choose Reverse.

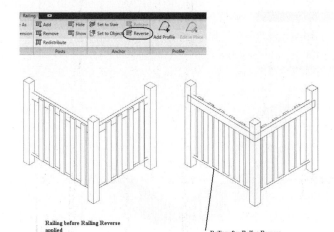

Railing before Railing Reverse applied

Railing after Railing Reverse applied

FIGURE 9.59 *Changing the start and end of the railing with RailingReverse command*

Anchoring a Railing

The **Anchor Railing (RailingAnchorToObjects)** command can be used to link a railing to an AEC object. Access the **Anchor Railing** command as shown in Table 9.23. The railing shown at the left has been anchored to the ramp shown in Figure 9.60.

TABLE 9.23 *RailingAnchorToObjects command access*

Command prompt	RAILINGANCHORTOOBJECTS
Ribbon	Select a railing and choose Set to Objects from the Anchor panel of the Railing contextual tab.
Shortcut menu	Select a railing, right-click, and choose Railing Anchor > Anchor to Object.

(Select the Railing and choose Set to Object *from the Anchor panel of the Railing contextual tab.)*

Select AEC Objects: 1 found *(Select the ramp at* p1 *as shown at the left in Figure 9.60.)*

Select AEC Objects: ENTER *(Press* ENTER *to end selection.)*

Automatic cleanup [Yes/No] <Yes>: y ENTER

Calculate height [Follow surface/At post locations only] <At post locations only>: ENTER

(The railing is anchored to the ramp as shown at the right in Figure 9.60.)

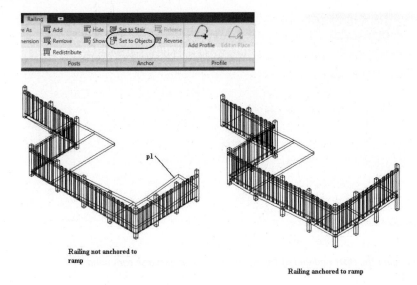

FIGURE 9.60 *Railing anchored to ramp*

The railing can be released from the AEC object by selecting the **AnchorRelease** command as shown in Table 9.24. When you select a railing that has been anchored to an object, choose **Release** from the Anchor panel. The railing remains in its position; however, future changes to the AEC object are independent of the railing.

TABLE 9.24 *Railing Anchor > Release command access*

Command prompt	**ANCHORRELEASE**
Ribbon	Select a railing that has been anchored to an object and choose Release from the Anchor panel of the Railing contextual tab.
Shortcut menu	Select a railing, right-click, and choose Railing Anchor > Release.

ADDING A PROFILE FOR RAILING COMPONENTS

A profile can be created in the workspace for the components of the railing. The **Add Profile** (**RailingAddComponentProfile**) command can be used to create a custom profile in the workspace for a selected railing component. Access the **Add Profile** command as shown in Table 9.25.

TABLE 9.25 *RailingAddComponentProfile command access*

Command prompt	**RAILINGADDCOMPONENTPROFILE**
Ribbon	Select a railing and choose Add Profile from the Profile panel of the Railing contextual tab.
Shortcut menu	Select a railing, right-click, and choose Add Profile.

Prior to selecting the **Add Profile** command, view the railing in an isometric view as shown in Figure 9.61. When you select the railing and choose **Add Profile**, you are prompted to select a railing component. If you select a post component, the **Add Post Profile** dialog box opens, allowing you to define the name for the new profile. When you click **OK** to dismiss the dialog box, the grips of a profile on the selected component are displayed and the **Edit In Place: Railing Profile** tab opens as shown in Figure 9.61. The profile can then be edited by modifying the grips or selecting options from the Edit In Place: Railing Profile contextual tab as discussed in the "Adding an Edge Profile in the Workspace" topic of Chapter 7.

After a profile has been changed, choose Finish from the Edits panel of the Edit In Place: Railing Profile. The **RailingAddComponentProfile** command can be repeated to edit the profile. When the Add Post Profile dialog box opens, choose the name of the existing profile and check Continue Editing to perform the edit of the profile.

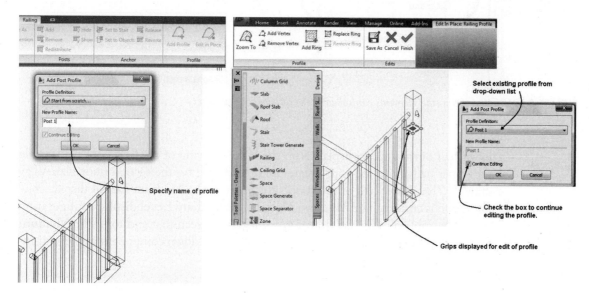

FIGURE 9.61 *Editing the profile of a railing*

DISPLAYING THE STAIR IN MULTIPLE LEVELS

Stairs are Construct drawings assigned to levels in a project. The upper or lower components of stairs are displayed according to the level of the View drawing. The stairs are displayed in this manner when no overrides to the display configuration are defined in the stair style. To identify whether or not display configuration overrides are defined in a stair style, choose the **Display Properties** tab of the Stair Styles dialog box. Select the **Style Override** checkbox for the display representation applied in the View drawing to open the **Display Properties (Stair Style Override)** dialog box. Choose the **Other** tab and verify that the **Override Display Configuration Cut Plane** checkbox is clear as shown in Figure 9.62. When this checkbox is clear, the

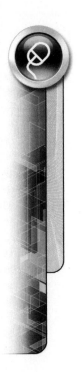

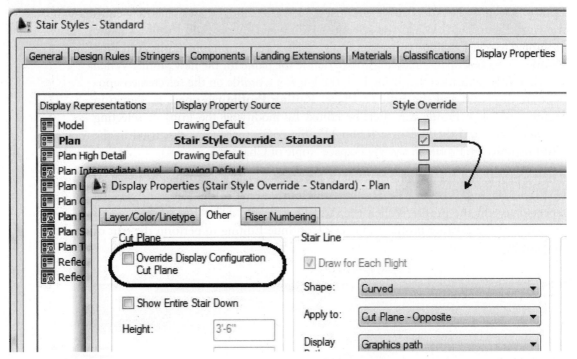

FIGURE 9.62 *Display of the stair elements controlled in the display configuration*

lower half of the stair is displayed on the View drawing of the first floor plan, while the upper half above the stair break line is displayed on the second floor plan View drawing. The lower half of the stair is displayed if the **Medium Detail** display configuration is current. The **Medium Detail Intermediate Level** display configuration should be applied in project drawings for levels between first and top. The **Medium Detail Top Level** display configuration displays the down component of the stair for the top level.

RECONCILING LAYERS

AutoCAD Architecture records a baseline list of layers used in a drawing when it is saved. When a stair drawing is created and attached to other drawings, additional layers will be added to the baseline of layers. This feature has been turned off (Layer-Eval = 0) in the project tutorials of this chapter. If this feature is turned on, the Unreconciled New layers message will display from the Unreconciled New Layers icon located in the lower-right corner of the Drawing Window status bar. Therefore, if the Unreconciled New Layers message is displayed in your future projects, select the hyperlink to open the Layer Manager. Select all layers, right-click, and choose Reconcile Layer from the shortcut menu.

PROJECT TUTORIAL 9-4: CREATING STRAIGHT STAIRS

1. Verify that the Accessing Tutor\Imperial or Accessing Tutor\Metric content for Chapter 9 of the CengageBrain http://www.cengagebrain.com has been downloaded to your Accessing Tutor folder on your computer as described in Organizing Tutorial Directories in the Preface.

2. Open Autodesk AutoCAD Architecture 2012 and select Project Browser from the Quick Access toolbar. Use the Project Selector drop-down list to navigate to your Student\Ch9 directory. Select New Project to open the Add Project dialog box. Type **EX 9-4** in the Project Number field and type **Ex 9-4** in the Project Name field. Check the Create from template project checkbox, choose Browse, and edit the path to \Accessing Tutor\Imperial\Ch9\Ex 9-4\Ex 9-4.apj or \Accessing Tutor\ Metric\Ch9\Ex 9-4\Ex 9-4.apj in the Look in field. Click Open to dismiss the Select Project dialog box. Click OK to dismiss the Add Project dialog box. Click Close to dismiss the Project Browser.

3. Select the Constructs tab of the Project Navigator, click the Constructs category, right-click, and choose New > Construct. Type Stair in the Name field of the Add Construct dialog box. Check Level 1 [G] and 2 [1] for Division 1 of the construct. Verify that the Open in drawing editor checkbox is checked. Click OK to dismiss the Add Construct dialog box.

4. Select the Construct tab of the Project Navigator, select the Floor 1 construct, right-click, and choose Xref Overlay from the shortcut menu.

5. Select **Zoom Extents** on the Zoom flyout menu of the Navigation bar to view the existing walls of Floor 1.

6. Right-click the Object Snap button on the status bar and choose Settings from the shortcut menu. Edit the Object Snap tab of the Drafting **Settings** dialog box as follows: clear all object snap modes and check the **Intersection** object snap. Check the **Object Snap On (F3)** checkbox and clear **Object Snap Tracking On (F11)**. Select the **Dynamic Input** tab and verify that the **Enable Pointer Input, Enable Dimension Input where possible**, **Show command prompting and command input near the crosshairs** checkboxes are checked. Click **OK** to dismiss the Drafting Settings dialog box.

7. Toggle ON Polar Tracking, Dynamic Input and Quick Properties. Toggle OFF Snap Mode and Grid Display on the status bar.

8. Right-click the command window, choose **Options**, and select the **AEC Object Settings** tab of the **Options** dialog box. Clear the **Presentation Format (No Cut Lines or Path)** checkbox and set Node Osnap = **Flight & Landing Corners**, Measure Stair Height = **Finished Floor to Floor**, and Calculator Limits = **Relaxed**. Click **OK** to dismiss the Options dialog box.

9. In this step, you will fade the display of the reference file. Choose Insert tab. Expand the Reference panel and choose the Xref fading slide bar and adjust to 25 as shown in Figure 9.63 to fade the display of the walls in the reference file.

10. If the Design tool palette set is not displayed, choose the Home tab. Choose the Tools command to display the Design palette set. To create a straight stair for a finish floor-to-finish floor rise equal to 8'-10 1/2" [2705], select the **Stair** command from the **Design** palette. Edit the **Properties** palette for the stair as follows: Basic-General: Style = **Standard**, Shape = **Straight**, Vertical Orientation = **Up**, Terminate with = **Landing**; Dimensions: A-Width = **3'-0" [914]**, B-Height = **8'-10 1/2" [2705]**, and Justify = **Left**. Select **Calculation Rules**, toggle ON **Riser Count**, set Riser Count = **16**,

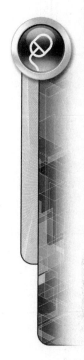

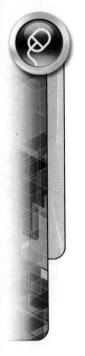

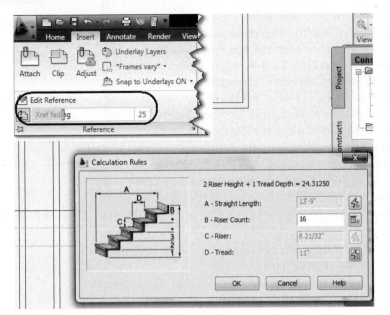

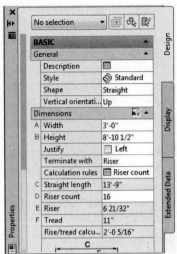

FIGURE 9.63 *Calculation Rules dialog box*

and verify that other fields are set to **Automatic** as shown in Figure 9.63. Click **OK** to dismiss the **Calculation Rules** dialog box. In the Advanced section verify Floor Settings are A-Top Offset = **0** and C-Bottom Offset = **0**. Verify that the Flight Height: Minimum limit = **None** and Flight Maximum limit = **None**. Respond to the following workspace prompts:

Flight start point or: *(Select the wall near p1 as shown in Figure 9.64 using the Intersection object snap.)*

Flight endpoint or: *(Select beyond the wall near p2 using the Polar tracking 90° angle beyond the stair bounding box as shown in Figure 9.64.)*

Flight start point or: ESC *(Press ESC to end the command.)*

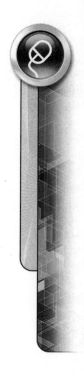

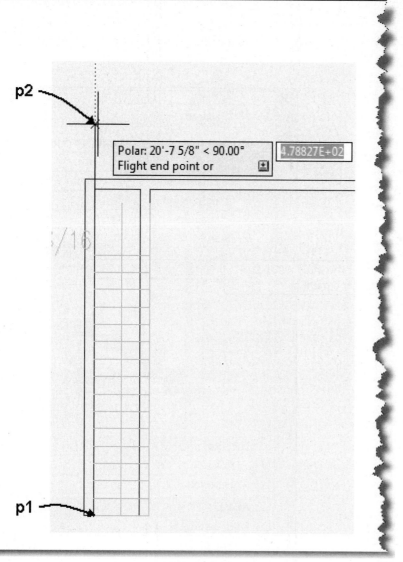

Polar: 20'-7 5/8" < 90.00°
Flight end point or

FIGURE 9.64 *Start flight and end flight locations for stair placement*

11. To create a stair that goes down with automatic landing placement, select the **Stair** command from the **Design** palette. Edit the **Properties** palette to make Terminate with = **Riser** in order to display the **Vertical Orientation** field. Edit the **Properties** palette for the stair as follows: Basic-General: Style = **Standard**, Shape = **Straight**, Vertical Orientation = **Down**; Dimensions: A-Width = **3'-0" [914]**, B-Height = **14'-0" [4267]**, and Justify = **Left**. Select **Calculation Rules** and toggle ON **Automatic** for all options. Click **OK** to dismiss the **Calculation Rules** dialog box. Verify that Calculation rules = **Height**. Verify that Advanced-Floor Settings are A-Top Offset = **0** and Bottom Offset = **0**. Set Flight Height: Minimum Limit type = **Height**, Minimum height = **3'-0" [914]**, Maximum Limit type = **Height**, and Maximum height = **12'-0" [3658]**. Respond to the following workspace prompts:

Flight start point or: *(Select the wall near p1 as shown in Figure 9.65 using the Intersection object snap.)*

Flight endpoint or: *(Select beyond the stair bounding box near p2 as shown in Figure 9.65.)*

(The landing is placed since height exceeds the maximum height of 12'.)

Flight start point or: ESC *(Press ESC to end the command.)*

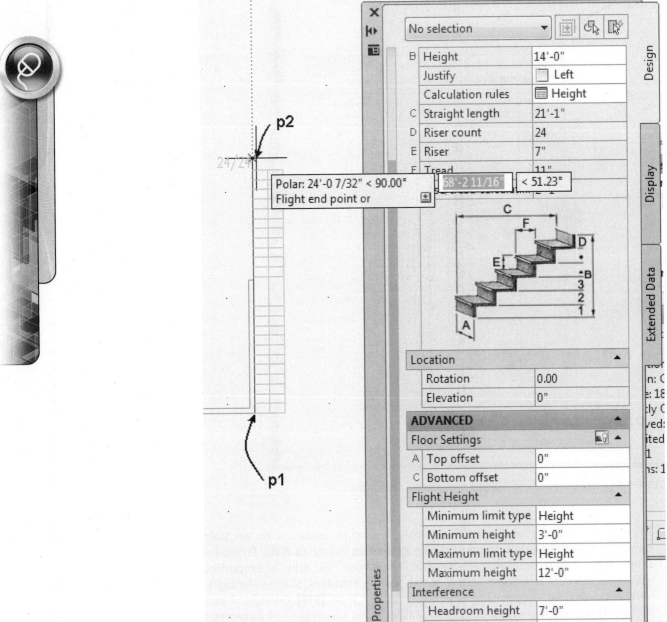

FIGURE 9.65 *Start and end flight locations of a down flight*

12. Select **SE Isometric** from the **View** Cube.

13. Select **Zoom Window** from the **Zoom** flyout menu of the **Navigation** bar. Respond to workspace prompts as shown below:

> Specify first corner: 24',6' [19420, 18000] ENTER *(First corner specified.)*
>
> Specify opposite corner: 26',18' [23930, 17000] ENTER *(Opposite corner specified.)*

14. Press F3 to turn OFF object snaps.

15. The wall shown at **p1** in Figure 9.66 will be trimmed to the stair using the **RoofLine** command; therefore, the wall endpoint must be located within the limits of the stair. Select a wall and choose **Edit Reference In Place** from the Edit panel of the External Reference contextual tab. Verify that the **Automatically select all nested objects** option is selected. Click **OK** to dismiss the Reference Edit dialog box.

16. In this step, you will move the start point of the wall to within the stair. Select the wall at **p1** as shown in Figure 9.66 to display the grips of the wall. Select the wall start grip as shown at the right in Figure 9.66. Move the cursor to display Polar > tracking tip, press TAB if necessary to highlight, and modify the dynamic dimension as shown at the right in Figure 9.66. Type **3"** **[75]** in the workspace and press ENTER to move the wall end under the stair.

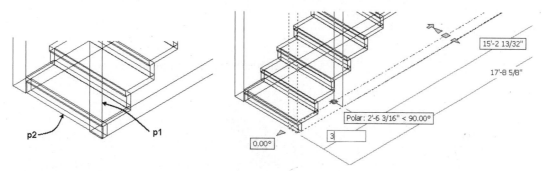

FIGURE 9.66 *Grip of wall selected*

17. Select the wall at **p1** as shown in Figure 9.66 and choose **Modify Roof Line** from the Roof/Floor Line flyout menu in the Modify panel of the Wall contextual tab. Respond to the workspace prompts as shown below.

> RoofLine: *(Choose* Auto project *from the workspace prompts.)*
>
> Select objects: 1 found *(Select the stair at* p2 *as shown in Figure 9.66.)*
>
> Select objects: ENTER
>
> RoofLine: ESC *(Press* ESC *to end the command; the wall is trimmed by the stair as shown in Figure 9.67.)*

18. Choose **Save Changes** from the **Edit Reference** panel of the **Wall** contextual tab. Click **OK** to save all changes and dismiss the AutoCAD dialog box. Click the **Manage Xrefs** button located in the lower-right corner of the **Drawing Window** status bar to open the **External References** palette. Select **FLOOR 1**, right-click, and choose **Detach**. Close the External References palette. Save and close the Stair drawing.

19. Select the Construct tab of the Project Navigator and double-click **Slab 1** construct.

20. Select the Stair construct in the Project Navigator, right-click, and choose Xref Overlay.

21. Select the slab and choose Add from the Interference flyout menu in the Modify panel of the Slab contextual tab. Respond to the workspace prompts as follows:

> Select objects to add: *(Select the stair at* p1 *as shown in Figure 9.67.)*

Select objects to add: ENTER *(Press ENTER to end selection.)*
1 found

Enter shrinkwrap effect [Additive/Subtractive]: S *(Choose Subtractive option.)*

22. Select Stair in the External References palette, right-click, and choose Unload. The Unload option will retain interference with the stair.

23. Save and Close Slab 1 drawing.

24. To view the stairs in the View drawing, select the **Views** tab of the Project Navigator. Select the **Views** category, right-click, and choose **Regenerate**. Double-click the **Floor 1 View** drawing. Choose SE Isometric from the View Cube. Save and close the drawing.

25. Select **Project** tab of the Project Navigator. Choose **Close Current Project** of the toolbar located at the bottom of the Project tab. Choose Close all project files of the Project Browser – Close Project Files dialog box. Click **Yes** in the AutoCAD dialog box to save all drawing files.

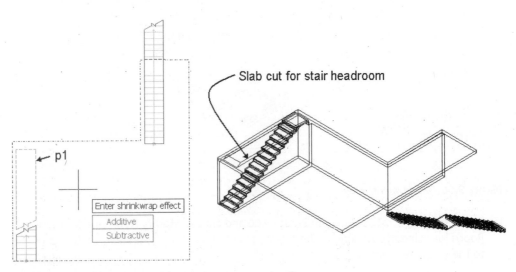

FIGURE 9.67 *RoofLine command applied to a stair and wall*

PROJECT TUTORIAL 9-5: CREATING STAIR TOWER WITH A U-SHAPED STAIR

1. Verify that the Accessing Tutor\Imperial or Accessing Tutor\Metric content for Chapter 9 of the CengageBrain http://www.cengagebrain.com has been downloaded to your Accessing Tutor folder on your computer as described in Organizing Tutorial Directories in the Preface.

2. Open Autodesk AutoCAD Architecture 2012 and select **QNew** from the Quick Access toolbar. Choose **Project Browser** from the Quick Access toolbar. Use the **Project Selector** drop-down list to navigate to your *Student\Ch9* directory. Select **New Project** to open the **Add Project** dialog box. Type **EX 9-5** in the Project **Number** field and type **Ex 9-5** in the Project **Name** field. Check the **Create from template project** check box, choose **Browse,** and edit the path to *\Accessing Tutor \Imperial\Ch9\Ex 9-5\Ex 9-5.apj* or *\Accessing Tutor\Metric\Ch9\Ex 9-5\Ex 9-5.apj* in the **Look in** field. Click **Open** to dismiss the Select Project dialog box. Click **OK** to dismiss the Add Project dialog box. Click **Close** to dismiss the Project Browser.

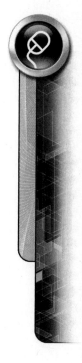

3. Select the **Constructs** tab of the **Project Navigator**, click the **Constructs** category, right-click, and choose **New > Construct**. Type **Stair** in the name field of the Add Construct dialog box. Check Level **1, 2, 3,** and **4 [G, 1, 2 and 3]** for **Division 1** of the construct. Verify that Open in drawing editor is checked. Click **OK** to dismiss the Add Construct dialog box.

4. Select the **Constructs** tab of the **Project Navigator**, select the **Floor 1** construct, right-click, and choose **Xref Overlay** from the shortcut menu.

5. Right-click the Object Snap button of the status bar and choose **Settings** from the shortcut menu. Edit the **Object Snap** tab of the **Drafting Settings** dialog box as follows: clear all object snap modes and check the **Endpoint** object snap mode. Check the **Object Snap On (F3)** checkbox and clear **Object Snap Tracking On (F11)**. Select the **Dynamic Input** tab and check the **Enable Pointer Input, Enable Dimension Input where possible,** and **Show command prompting and command input near crosshairs** checkboxes. Click **OK** to dismiss the Drafting Settings dialog box.

6. Toggle Polar Tracking, Dynamic Input and Quick Properties ON and toggle OFF Snap Mode and Grid Mode on the status bar.

7. Select **Zoom Window** from the **Zoom** flyout menu of the **Navigation** bar. Respond to the workspace prompts as follows to view the stairwell:

   Specify first corner: 142',40' [28000, 10000] ENTER
   (Coordinates of the zoom window are specified.)

   Specify opposite corner: 155',60' [34000, 1500] ENTER
   (Coordinates of the zoom window are specified.)

8. Choose the Home tab. To create a U-shaped stair for a finish floor-to-finished rise equal to **12' [3658]** and select the **Stair** command from the **Build** panel of the Home tab. Edit the **Properties** palette for the stair as follows: Basic-General: Style = **Standard**, Shape = **U-shaped**, Turn type = **1/2 landing**, Horizontal Orientation = **Clockwise**, Vertical Orientation = **Up**; Dimensions: A-Width = **44" [1118]**, B-Height = **12' [3658]**, Justify = **Outside**, Terminate with = **Landing**; select **Calculation Rules** and verify that all options are set to **Automatic**. Click **OK** to dismiss the Calculation Rules dialog box. Verify that Advanced-Alignment type = **Tread to tread**, Alignment offset = **0**, Extend alignment = **Lower flight**, and Uneven tread on = **Lower flight**. Floor Settings are: A-Top Offset = **0** and C-Bottom Offset = **0**. Flight Height: Minimum Limit Type = **None**, Maximum Limit Type = **Height**, and Maximum Height = **12' [3658]**; Interference: Headroom height = **7'-0" [2100]**, Inside clearance = **0**, and Outside clearance = **1" [25]**. Respond to the following workspace prompts:

   Flight start point or: *(Select the wall near p1 using the Endpoint object snap as shown in Figure 9.68.)*

   Flight endpoint or: *(Select the wall near p2 using the Endpoint object snap as shown in Figure 9.68.)*

   Flight start point or: ESC *(Press ESC to end the command.)*

9. Choose **Move** from the Modify panel in the Home tab. Respond to the workspace prompts as follows:

   Select objects: *(Select the stair at* p3 *as shown in Figure 9.68.)* 1 found

   Select objects: *(Press ENTER to end selection.)*

   Specify base point or [displacement] <Displacement>: *(Select a point near p4 as shown in Figure 9.68.)*

   Specify second point or <use first point as displacement>: *(Select a point near p5 as shown in Figure 9.68.)*

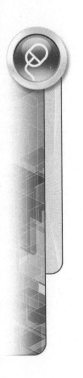

10. Select the Content Browser from the Tools flyout of the Build panel of the Home tab and open Design Tool Catalog – Imperial > Stairs and Railings > Railings [Design Tool Catalog - Metric] >Stairs and Railings > Railings. Select the i-drop for Guardrail-Rect Balusters from page 2 of the Railings category and drag this style into the workspace. Edit the Properties palette for the railing as follows: Attached to = Stair, Perpendicular posts = No, Maintain sloping posts at landings = No, Side offset = 2" [50], and Automatic placement = yes. Select the stair near p6 as shown in the command sequence below and in Figure 9.68.

> Select a stair or [Style/Attach/Match/Undo]: *(Select the stair near p6 as shown in Figure 9.68.)*
>
> Select a stair or [Style/Attach/Match/Undo]: ENTER *(Press ENTER to end selection.)*

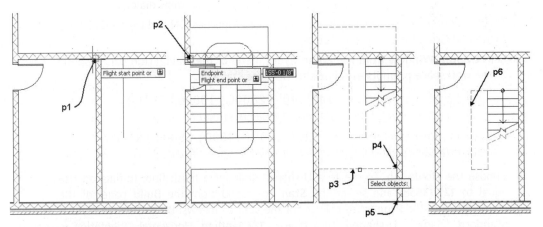

FIGURE 9.68 *Location of flight points of U-shaped stair*

11. Select **SW Isometric** from the **View Cube.** To turn off the surface hatch and choose **Surface Hatch Toggle** from the **Drawing Window** status bar (see Figure 9.69).

12. Select the Stair, right-click, and choose **Stair Tower Generate** from the shortcut menu. Respond to the workspace prompts as follows:

> Select railings and slabs: *(Select the railing at p1 as shown in Figure 9.69.) 1 found*
>
> Select railings and slabs: ENTER *(Press ENTER to end selection and open the* Select Levels *dialog box to verify that levels* 1[G], 2 [1], and 3[2] *are checked; click* OK *to dismiss the Select Levels dialog box.)*
>
> *(The stair tower is generated as shown at the right in Figure 9.69.)*

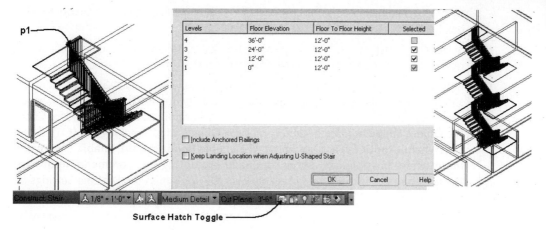

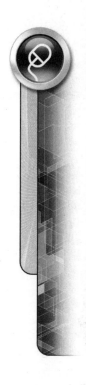

FIGURE 9.69 *Stair and railing generated for a stair tower*

13. Select **Manage Xrefs** from the Drawing Window status bar. Select **Floor 1**, right-click, and select **Detach** from the shortcut menu. Save and close the Stair drawing.

14. Select the Project Navigator of the Quick Access toolbar. Select the **Views** tab. Choose the **Views** category, right-click, and choose **Regenerate**. Double-click **Floor 2 View** to open the drawing. Choose **Medium Detail Intermediate Level** from the **Display Configuration** flyout menu. Double-click **Floor 3 View** to open the drawing. Choose **Medium Detail Intermediate Level** from the **Display Configuration** flyout menu.

15. Select the **Views** tab. Double-click the **Floor 4 View** drawing. Choose **Medium Detail Top Level** from the **Display Configuration** flyout menu. Double-click the **Floor 1 View** drawing on the **Views** tab. Choose **Medium Detail** from the **Display Configuration** flyout menu.

16. Select the **Project** tab of the Project Navigator. Choose the **Close Current Project** from the toolbar located at the bottom of the Project tab. Save and close all project drawings.

PROJECT TUTORIAL 9-6: INSERTING A RESIDENTIAL MULTI-LANDING STAIR

1. Verify that the Accessing Tutor\Imperial or Accessing Tutor\Metric content for Chapter 9 of the CengageBrain http://www.cengagebrain.com has been downloaded to your Accessing Tutor folder on your computer as described in Organizing Tutorial Directories in the Preface.

2. Open Autodesk AutoCAD Architecture 2012 and select QNew to start a new drawing. Select Project Browser from the Quick Access toolbar. Use the Project Selector drop-down list to navigate to your Student\Ch9 directory. Select New Project to open the Add Project dialog box. Type **16** in the Project Number field and type **Ex 9-6** in the Project Name field. Check the Create from template project checkbox, choose Browse, edit the path to \Accessing Tutor\Imperial\Ch9\Ex 9-6\Ex 9-6.apj or \Accessing Tutor\Metric\Ch9\Ex 9-6\Ex 9-6.apj in the Look in field. Click Open to dismiss the Select Project dialog box. Click OK to dismiss the Add Project dialog box. Click Close to dismiss the Project Browser.

3. Select the Constructs tab of the Project Navigator, click the Constructs category, right-click, and choose New > Construct. Type Stair in the name field of the Add Construct dialog box. Check Level 1 [G] and 2 for Division 1 of the construct. Verify

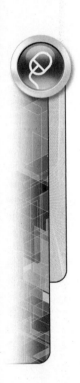

that Open in drawing editor is checked. Click OK to dismiss the Add Construct dialog box.

4. Select the Constructs tab of the Project Navigator, select the Basement construct, right-click, and choose Xref Overlay from the shortcut menu.

5. Right-click the Object Snap button of the status bar and choose Settings from the shortcut menu. Edit the **Object Snap** tab of the **Drafting Settings** dialog box as follows: clear all object snap modes and check the **Endpoint** object snap mode. Check the **Object Snap On (F3)** checkbox and clear **Object Snap Tracking On (F11)**. Select the **Dynamic Input** tab and verify that the **Enable Pointer Input**, **Enable Dimension Input where possible**, and **Show command prompting and command input near the crosshairs** checkboxes are checked. Click **OK** to dismiss the Drafting Settings dialog box.

6. Toggle Polar Tracking on the status bar. Toggle OFF Snap Mode and Grid Display on the status bar.

7. Select **Zoom Extents** from the **Zoom** flyout menu of the Navigation bar.

8. To import the Residential stair style from a resource drawing, choose Stair Styles from the right-click menu of the Stair tool of the Design palette. Choose **Open** from the Style Manager menu bar. Navigate to *Accessing Tutor\Imperial\Support* or *Accessing Tutor\Metric\Support* and choose the **Lab 9-2** drawing.

9. Expand the Architectural Objects category of the *Lab 9-2.dwg* in the left pane. Expand the Stair Styles category, select the **Residential** stair style, right-click, and choose **Copy**. Select the current drawing, right-click, and choose **Paste**. Click **OK** to dismiss the Style Manager.

10. Choose the Home tab. Choose the Stair tool from the Build panel. To create a Multi-landing stair, select **Stair** from the **Design** palette and edit the **Design** tab of the **Properties** palette as follows: Basic-General: Style = **Residential**, Shape = **Multi-landing**, Turn type = **1/2 Landing**, Terminate with = **Riser**, Vertical Orientation = **Up**; Dimensions: A-Width = **3'-0" [914]**, B-Height = **8'-11 1/2" [2731]**, Justify = **Left**, and Flight length = **Distance**. Select **Calculation Rules** and verify that all fields are set to **Automatic** except B-Riser Count and D-Tread. Set B-Riser Count = **14** and D-Tread = **10" [254]** and then click **OK** to dismiss the Calculation Rules dialog box. Verify that the Advanced-Floor Settings are: A-Top Offset = **0** and C-Bottom Offset = **0**; Flight Height: Minimum Limit Type = **None**, Maximum Limit Type = **None**. Respond to the following workspace prompts:

    ```
    Flight start point or: near to ([SHIFT + right-click and
    choose Nearest from the Object Snap shortcut menu.] _ Select
    a point near p1 as shown in Figure 9.70.)

    Flight endpoint or: (Select a point near p2 using the
    Endpoint object snap as shown in Figure 9.71.)
    ```

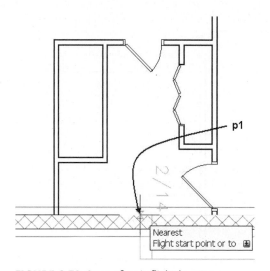

FIGURE 9.70 *Start of stair flight location*

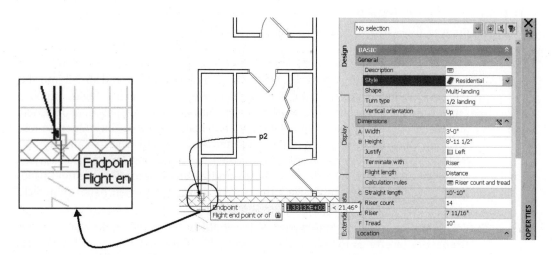

FIGURE 9.71 *Location of end of first flight*

Flight start point or: *(Select a point near* p3 *using the Endpoint object snap as shown at the left in Figure 9.72.)*

Flight endpoint or: *(Select a point near* p4 *beyond the stair bounding box as shown at the right in Figure 9.72.)*

Flight start point or: ENTER *(Press ENTER to end the command.)*

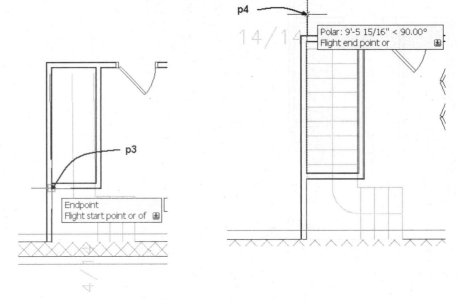

FIGURE 9.72 *Location of the start and end of the second flight*

11. In this step you will import the Handrail and Guardrail styles from Lab 9-3 into the current drawing. Right-click the Railing tool and choose Railing Styles. Choose **Open** from the menu bar of the Style Manager. Navigate to *Accessing Tutor\Imperial\ Support* or [*Accessing Tutor\Metric\Support*] and choose **Lab 9-3**. Click **Open** to dismiss the Open drawing dialog box. Expand the Architectural Objects category and Railings Styles category in the left pane for the Lab 9-3 drawing. Select Railing Styles in the left pane, press the CTRL key, select **Handrail and Guardrail** in the right pane, right-click, and choose **Copy**. Select the current drawing in the left pane, right-click, and choose **Paste**. Click **OK** to dismiss the Style Manager.

12. Select **Railing** from the Stair flyout of the **Build** panel of the Home tab. Edit the **Properties** palette as follows: Style = **Handrail**, Perpendicular Posts = **No**, Attach to = **Stair**, Maintain Sloping Post at Landings = **No**, Side offset = **2″ [50]**, and Automatic Placement = **Yes**. Respond to the command prompts as shown below to place the railing.

   Select a stair or: *(Select the stair at p1 as shown in Figure 9.73.)*

   Select a stair or: ESC *(Press ESC to end the command.)*

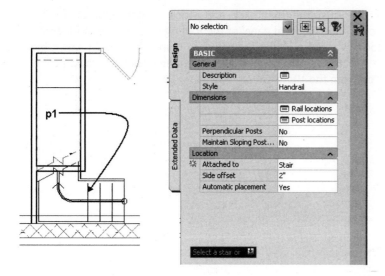
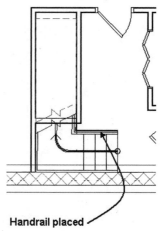

Handrail placed

FIGURE 9.73 *Selection of stair for automatic placement of a railing*

13. Select **Railing** from the Stair flyout of the **Build** panel of the Home tab and edit the **Properties** palette as follows: Style = **Guardrail**, Perpendicular Posts = **No**, Maintain sloping posts at landing = **No**; Location: Attach to = **Stair Flight** and Side Offset = **2" [50]**. Respond to the workspace prompts as follows:

Select a stair or: *(Select the stair at* **p1** *as shown in Figure 9.74.)*

Select a stair or: ESC *(Press ESC to end the command.)*

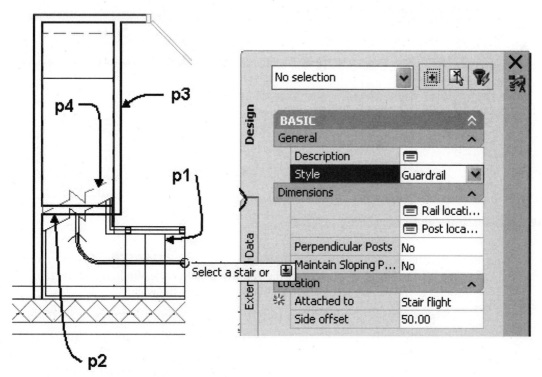

FIGURE 9.74 *Selection of stair for location of railing*

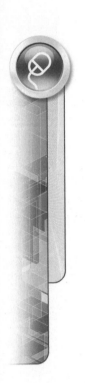

14. Select the handrail and guardrail, right-click, and choose **Isolate Objects** from the Isolate Objects flyout of the **General** panel of the **Railing** tab. Choose **SW Isometric** from the View Cube to view the railings as shown in Figure 9.75. Choose **End Object Isolation** from the **Isolate Objects** flyout in the General panel of the Railing contextual tab. Choose **Top** from the **View** Cube.

FIGURE 9.75 *Railings added to stair*

15. After placing the stair it is obvious the interior wall in the basement should be extended to the stair. To modify the interior wall, you must edit the reference file. Select the wall at **p1** as shown in Figure 9.76 and choose **Edit Reference In Place** from the Edit panel of the External Reference contextual tab. Select the **Prompt to select nested objects** radio button and click **OK** to dismiss the Reference Edit dialog box. Select the wall at **p1** and **p2** as shown in Figure 9.76. Press ENTER to end the command. The **External Reference** contextual tab is displayed during the edit.

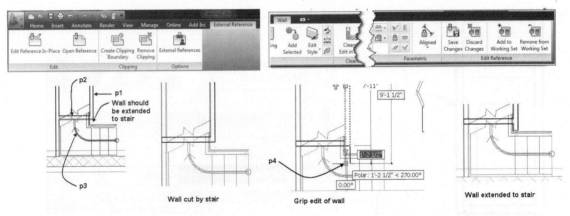

FIGURE 9.76 *Selection of walls for roof line edit*

16. Select the wall shown at **p2** in Figure 9.76 and choose **Modify Roof Line** from the **Roof/Floor Line** flyout menu in the **Modify** panel of the **Wall** contextual tab. Respond to the workspace prompt as follows to modify the wall:

RoofLine: *(Choose* **Auto project** *from the workspace options.)*

Select objects: 1 found *(Select the stair at* **p3** *as shown in Figure 9.76.)*

Select objects: ENTER *(Press ENTER to end the selection.)*

```
[1] Wall cut line(s) converted.
RoofLine: ESC (Press ESC to end the command.)
```

17. Prior to extending the wall at p1 (see Figure 9.76), the **Autosnap to Grip Edit walls** setting must be turned off. The Autosnap to Grip Edit walls setting is currently set to **6″ [75]**, which prohibits the grip edit of a wall for distances less than 6″ [75]. In this step you will temporarily turn off Autosnap to allow the wall edit. Right-click the command line and choose **Options**. Choose the **AEC Object Settings** tab and clear **Autosnap Grip Edited Wall Justification Lines**. Click **OK** to dismiss the Options dialog box.

18. Select the wall at **p1**, as shown in Figure 9.76, to display the grips. Choose the **Start** grip shown at **p4** and drag the grip down, then type **5 1/2″ [140]** (Enter) in the dynamic dimension as shown in Figure 9.76. Press ESC to end the edit and clear the selection. Right-click the command line and choose **Options**. Choose the **AEC Object Settings** tab and check **Autosnap Grip Edited Wall Baselines**.

19. Select **Save Changes** on the Edit Reference panel of the Wall contextual tab. Click **OK** to dismiss the AutoCAD dialog box.

20. Select **Manage Xref** in the **Drawing Window** status bar as shown in Figure 9.77 to open the External References palette. Select **Basement**, right-click, and choose **Detach**. Close the External References palette.

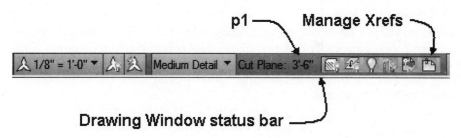

FIGURE 9.77 *Manage Xref and Global Cut Plane of the Drawing Window status bar*

21. Select **NE Isometric View** from the **View Cube**. Select **Hidden** from the **Visual Styles** flyout menu of the **Visual Styles** panel in the **View** tab. Save and close the drawing.

22. Select the **Views** tab of the **Project Navigator**. Select the **Views** category, then right-click and choose **Regenerate**. Double-click the **Basement View** drawing. Choose **Zoom Extents** from the Zoom flyout menu of the Navigation bar. Choose **High Detail** from the Display Configuration flyout menu. Choose the Global Cut Plane and refer to p1 as shown in Figure 9.77. Verify the Cut Height to **5′-0″ [1500]**. Click **OK** to dismiss the Global Cut Plane dialog box. Select **NE Isometric View** from the **View Cube**. Choose **Realistic** from the Visual Styles flyout of the Visual Styles pane in the **View tab** to view the stair as shown at right in Figure 9.78.

23. In the **Views** tab of the **Project Navigator,** double-click **Floor 2 View** to open the drawing. Verify **High Detail** display configuration is current. Choose the Global Cut Plane to open the Global Cut Plane dialog box. Edit the Distance Below Range to **-7′ [-2135]**. Click **OK** to dismiss the Global Cut Plane dialog box. Save and close the drawing.

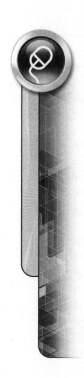

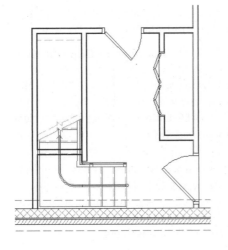

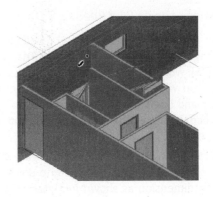

Plan view of stair

NE Isometric view of stair

FIGURE 9.78 *Stair created for a basement*

24. Select the Sheets tab. Expand the Plans subset of the Architectural category. Double-click A-4 Floor Plan Level 2 drawing.

25. In this step you will adjust the Display Configuration and Cut Plane values to match the Floor 2 View.dwg. Verify that High Detail display configuration is current. Click the **Cut Plane** value in the Drawing Window status bar to open the **Global Cut Plane** dialog box. Edit the Display Below Range to **-7' [-2135]**. Click OK to dismiss the Global Cut Plane dialog box.

26. Select Close Current Project from the toolbar of the Project tab. Choose Close all project files. Save all project drawings.

NOTE

Refer to **e Projects** of Chapter 9 for download from the Student Companion Site of CengageBrain http://www.cengagebrain.com described in the Preface. **The** Project Ex 9-7 Creating a Custom Stair is included in the e Project document.

SUMMARY

- Stairs are created in a drawing with the **StairAdd** command.
- Design of the number of risers, tread dimension, and run of the stair is set in the Properties palette.
- Stairs are generated in the drawing from start of the flight up or down from the $Z = 0$ coordinate.
- The shapes of a stair can be U-shaped, U-shaped winder, Multi-landing, Spiral, and Straight.
- The start point and center of a spiral stair can be edited with grips.
- The Properties palette is used to change the size, components, and style of a stair.
- The **Style Manager** is used to create, edit, import, export, and purge stair styles from a drawing.

- The **Calculation Rules** dialog box allows you to constrain the stair by the following properties: A-Straight Length, B-Riser Count, C-Riser, and D-Tread dimensions.

- The display of the lower or upper half of the stair and railing is controlled in the **Display Properties** dialog box for a stair style.

- The **RailingAdd** command is used to place a railing attached to a stair or free-standing railing on a balcony.

- A stair is placed on the A-Flor-Strs layer and railings are placed on the A-Flor-Hral layer.

- Railing heights and posts extensions can be edited to create deck post and guardrails.

- Railing styles allows you to edit the size, location, and profile used for the railing posts, handrail, guardrail and balusters.

Refer to e **Review Questions** folder which includes review questions for Chapter 9 of Student Companion Site of **CengageBrain** http://www.cengagebrain.com described in the Preface.

NOTE

Using and Creating Symbols

INTRODUCTION

AutoCAD Architecture includes hundreds of symbols created for two-dimensional and three-dimensional representations of building components. The two-dimensional symbols include the plan and elevation views. The symbols are selected from the Design Tool Catalog or the DesignCenter and include blocks, multi-view blocks, and mask blocks. Sample multi-view blocks are included in the FF + E palette of the Design palette group.

OBJECTIVES

After completing this chapter, you will be able to

- Set the scale of a drawing for inserting symbols.
- Use the DesignCenter and the Content Browser to insert symbols in a drawing.
- Insert and modify multi-view blocks.
- Define the display representation for a multi-view block and mask blocks.
- Create and insert mask blocks.
- Attach mask blocks to AutoCAD Architecture objects to control display.

SETTING THE SCALE AND LAYER FOR SYMBOLS AND ANNOTATION

Prior to symbols being inserted in the drawing, the scale of the plotted drawing should be established. Some of the symbols are created at a specific size, while others are sized according to the scale of the drawing. For example, an electrical receptacle is sized according to the scale of the drawing, and a cabinet will be inserted at its actual size. If a receptacle symbol is placed in a drawing with the scale set to 1/4" = 1'-0" [1:50], it is drawn with a 6" [152] diameter circle. This symbol, if placed in a drawing with the 1/8" = 1'-0" [1:100] scale, would be drawn with a 12" [305] diameter circle because the scale has changed. The scale of a drawing sets the scale factor, which is used as a multiplier for selected symbols and annotation. The scale factor is the ratio between the size of the AutoCAD entity and its display printed on paper. The typical architectural scales can be specified from the Annotation Scale flyout of the Drawing window status bar.

AutoCAD Architecture elements placed in a drawing are automatically placed on the appropriate layer according to the layer key style. Therefore, if the AIA 256 Color layer key style is used, a refrigerator is inserted on the A-Flor-Appl [A-734G] layer. The *Current* layer key style places the object on the layer that is current at the time of insertion. Table 10.1 shows the layer names for an appliance when the respective layer key style options are selected.

TABLE 10.1 *Layer names for an appliance using layer key styles*

Layer for an Appliance	Units	Layer Key Style
A-Flor-Appl	Imperial	AIA (256 Color)
A-734G	Metric	BS1192 Cisfb

Using a layer standard enhances the drawing because the display of similar objects can be controlled as a group in the drawing. AutoCAD Architecture places the objects on the assigned layer automatically when the object is inserted in the drawing. If the layer key style is changed in the middle of a drawing, all previous objects remain on the layer of their initial insertion. Therefore, you should determine the layer key style prior to beginning the drawing.

USING THE CONTENT BROWSER

Symbols can be selected from the Design Tool Catalogs of the **Content Browser**. Access the **Content Browser** from the **Tools** flyout menu of the Build panel in the Home tab or the Design palette shown in Figure 10.1. (Symbols are organized in catalogs as shown in Table 10.3.) Therefore, to place plumbing fixtures, select from the Mechanical [Bathroom] catalog. Cabinets are placed in the drawing from the Furnishings [Kitchen Fittings] folder. The Design Tool Catalog shown in Figure 10.2 includes a preview of the symbol. You can select the i-drop of the symbol and drag the symbol into the drawing, or you can drag the i-drop of the symbol onto a custom tool palette. A sample of symbols available from the Content Browser are included on the FF + E palette of the Design palette group shown in Figure 10.1. Storing and publishing tool palettes will be presented later in this chapter.

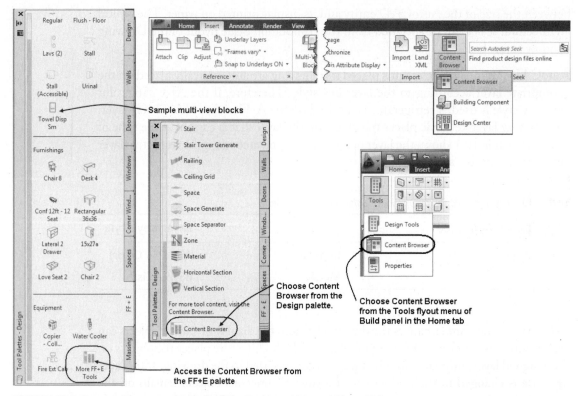

FIGURE 10.1 *Content Browser commands of the Design palette and Home Tab*

FIGURE 10.2 *Design Tool Catalog—Imperial of the Content Browser*

USING THE DESIGNCENTER

Architectural symbols can also be inserted in a drawing from the DesignCenter. Access the AutoCAD DesignCenter as shown in Table 10.2. The AutoCAD Design-Center is modified to include the display of the **AEC Content** view. Figure 10.3 shows the DesignCenter window and the commands of the DesignCenter.

TABLE 10.2 *DesignCenter command access*

Command prompt	ADCENTER
Ribbon	Select DesignCenter from the Content panel of Insert tab.
Keyboard shortcut	CTRL + 2

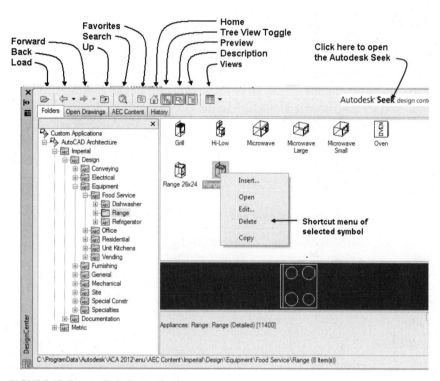

FIGURE 10.3 *AutoCAD DesignCenter*

The tabs located below the **DesignCenter** toolbar allow you to quickly edit the Tree View to access directories and the Internet for symbols. The toolbar for navigation within the DesignCenter is shown in Figure 10.3. The tabs are described below.

Folders—The Folders tab edits the Tree View to display the folder level of AutoCAD Architecture.

Open Drawings—The Open Drawings tab displays the Tree View of the current open drawing. Content for the current drawing includes Blocks, Dimstyles, Layers, Layouts, Linetypes, Textstyles, and Xrefs.

AEC Content—The AEC Content tab opens the Tree View to the Custom Applications folder.

History—The History tab lists the drawing previously accessed in the DesignCenter.

DC Online—The DC Online tab opens the DesignCenter Online, allowing access to additional symbols available via the Internet.

The **AEC Content** view includes all the symbols and documentation of AutoCAD Architecture. When the DesignCenter is opened from the Content panel of the

Insert tab, the AutoCAD DesignCenter opens to the AutoCAD Architecture node in the Tree View as shown in Figure 10.3. The symbols are located in the *{Vista: C:\ProgramData\Autodesk\ACA 2012\enu\AEC Content}* directory. The Imperial and Metric directories include Design and Documentation subdirectories. You can set the path of the default symbol menu for the DesignCenter on the AEC Content tab of the Options dialog box. The **AEC Content** tab includes a **Browse** button, which allows you to edit the path to the Content Browser and DesignCenter.

INSERTING SYMBOLS

When you select symbols from the **Content Browser,** click and drag the i-drop onto the workspace of the drawing. Release the left mouse button while over the workspace and then left-click in the workspace to specify the insert point. To select a symbol from the DesignCenter, double-click the symbol displayed in the right pane of the DesignCenter and specify the insertion point in the workspace. The dynamic preview of the symbol is displayed at the cursor as you are prompted to specify the insertion point of the symbol. The symbol rotation and scale are preset; however, you can choose options from the drop-down list of the dynamic input prompt to change rotation and scale prior to specifying the insert point.

TABLE 10.3 *Content folders of the DesignCenter and Design Tool Catalog*

Design Tool Catalog—Imperial	Imperial\Design Folder of the DesignCenter	Metric\Design Folder of the DesignCenter	Design Tool Catalog—Metric
Conveying	Conveying	Bathroom	Bathroom
Curtain Walls			Curtain Walls
Doors and Windows			Doors and Windows
Electrical	Electrical	Domestic Furniture	Domestic Furniture
Equipment	Equipment	Electrical Services	Electrical Services
Furnishing	Furnishing	Kitchen Fittings	Kitchen Fittings
General	General	Office Furniture	Office Furniture
Mechanical	Mechanical	Piped & Ducted Services	Piped and Ducted Services
Roof Slabs and Slabs			Roof Slabs and Slabs
Site	Site	Site	Site
Spaces			Spaces
Special Constr	Special Constr		
Specialties	Specialties		
Stairs and Railings			Stairs and Railings
Structural			Structural Members
Walls			Walls

Symbols inserted from the DesignCenter or Content Browser are multi-view or mask block styles that will be listed in the Multi-Purpose Objects\Mask Block Definitions and Multi-Purpose Objects\Multi-View Block Definitions folders of the Style Manager.

Steps to Insert Symbols into a Drawing from the Content Browser

1. Right-click the Object Snap button on the status bar and choose **Settings** from the shortcut menu.

2. Select desired Running Object Snap modes from the **Drafting Settings** dialog box.

3. Click **OK** to dismiss the **Drafting Settings** dialog box.

4. Select a scale from the Annotation Scale flyout of the Drawing window status bar.

5. Choose the **Content Browser** from the Design tool palette.

6. Open the **Design Tool Catalog—Imperial** or **Design Tool Catalog—Metric** and expand the tool categories displayed in the left pane.

7. Click the i-drop of the desired tool and drag the i-drop into the workspace. Prior to specifying the insert point, edit the Properties palette if necessary and then left-click in the drawing to specify the insertion point. Press ENTER to end the command.

PROPERTIES OF MULTI-VIEW BLOCKS

Many of the symbols inserted from the Design Tool Catalog or DesignCenter are multi-view blocks. Multi-view blocks have special display control properties, which allow the block to be displayed or not displayed when viewed from specified directions. A multi-view block of a duplex outlet can be defined to display only in the top viewing position and not in other viewing positions. Controlling display according to viewing direction is defined for each display representation.

When you insert symbols from the Content Browser or DesignCenter, the name of the multi-view block is displayed in the Properties palette, and you are prompted to specify the insertion point. The options included in the drop-down list of Dynamic Input are as follows: Name, X scale, Y scale, Z scale, Rotation, Match, and Base point.

Additional features of the multi-view block can be specified in the Properties palette prior to specifying the insertion point. The symbol is attached to the pointer scaled and rotated as specified in the Properties palette. Prior to specifying the insertion point, you can return to the Properties palette and edit the scale, rotation, and other features of the multi-view block. The options in the Design tab of the Properties palette of a multi-view block displayed during insertion are shown below.

Design Tab—Basic

General

Description—The Description button opens the **Description** dialog box, which allows you to add additional text.

Definition—The Definition field lists the name of the multi-view block. The drop-down list displays all blocks inserted in the drawing. A multi-view block can be selected from the drop-down list and inserted into the drawing.

Bound Spaces—The Bound Spaces field allows you to specify if the symbol can be used to limit the boundary of spaces.

Scale

Specify on Screen—The Specify on Screen field includes **Yes** and **No** options. If the **No** option is selected, you can type a scale factor in this field or in the command line. The **Yes** option prevents you from entering values in the

Properties palette. Scale factors can also be specified by selecting points in the drawing area.

X—The X field displays the current scale along the X axis.

Y—The Y field displays the current scale along the Y axis.

Z—The Z field displays the current scale along the Z axis.

Location

Specify Rotation on Screen—The Specify Rotation on Screen field includes **Yes** and **No** options. If the **No** option is selected, you type an angle in this field or in the command line. Rotation angles can also be specified by selecting points in the workspace with the cursor.

Elevation—The multi-view block is inserted at $Z = 0$ coordinate elevation. Editing the elevation value can change the elevation of existing multi-view blocks.

Design Tab—Advanced

Insertion Offsets—The Insertion offsets button opens the **Multi-view Block Offsets** dialog box as shown in Figure 10.4.

NOTE

The insertion point offsets for all views defined for the multi-view block should be edited. If all insertion point offsets are not edited, for each view the block will be positioned differently when viewed in the plan versus the model view of the block.

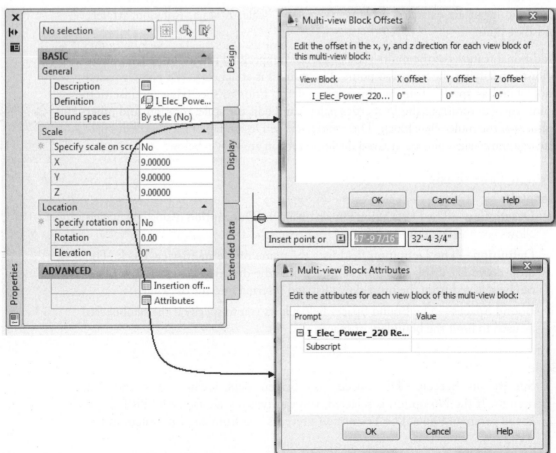

FIGURE 10.4 *Multi-view Block Offsets and Multi-view Block Attributes dialog boxes*

Attributes—The Attributes button opens the **Multi-view Block Attributes** dialog box shown in Figure 10.4. You can add the values for attributes for a block by entering text in the **Value** column.

Editing Multi-View Blocks Using Properties

Selecting a multi-view block, right-clicking, and choosing **Properties** displays the features of an existing multi-view block. The Properties palette can be used to edit the features of the multi-view block. The following additional fields are added to the Properties palette for existing blocks.

Design Tab—Basic

General

Layer—The Layer field displays the name of the layer of the multi-view block. The layer drop-down list displays the layers of the drawing.

Location

Additional information—The Additional information button opens the Location dialog box shown in Figure 10.5, which lists the x, y, and z coordinates of the multi-view block, and normal extrusion and rotation.

Design Tab—Advanced

Insertion Offsets—The **Insertion Offsets** button opens the **Multi-view Block Offsets** dialog box, which displays each of the existing view blocks included in the multi-view block. The **Multi-view Block Offsets** dialog box shown in Figure 10.5 includes view blocks designed for model, plan, front, left, and right views. The **X offset**, **Y offset**, and **Z offset** values allow you to modify the location of the insertion point for placing the multi-view block. The insertion point offset for each view of the multi-view block should be edited to maintain the objects insertion point in each view direction.

If the insertion point offsets for plan and model view blocks are not edited with the same values, the view block when viewed in the plan versus the model view will be positioned differently based upon the respective insertion points.

NOTE

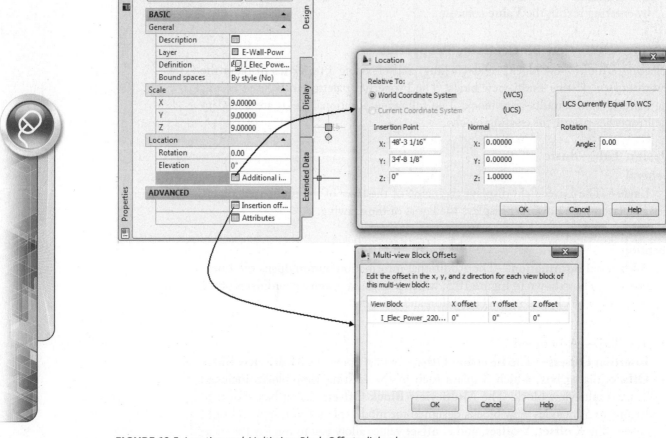

FIGURE 10.5 *Location and Multi-view Block Offsets dialog boxes*

NOTE

Access Instructional Video 10.1 Inserting Multi-view Blocks from the Instructional Video category of the Student Companion site of CengageBrain http://www.cengagebrain.com described in the Preface.

TUTORIAL 10-0: INSERTING MULTI-VIEW BLOCKS

1. Download Chapter 10 files from the Accessing Tutor\Imperial or Accessing Tutor\ Metric category of the Student Companion site of CengageBrain http://www. cengagebrain.com described in the Preface to your Accessing Tutor folder.

2. Choose **Open** from the Quick Access toolbar and open *Ex 10-0.dwg* from the *Accessing Tutor\Imperial\Ch 10* or *Accessing Tutor\Metric\Ch 10* folders.

3. Choose **Save As** > **AutoCAD Drawing** from the Application menu and save the drawing as **Lab 10-0** in your student directory.

4. Verify that the **Model** toggle is selected in the **Application** status bar. Verify that only Allow/Disallow Dynamic UCS and Dynamic Input are toggled ON on the status bar. Verify that Annotation Scale = 1/8" = 1'-0" [1:100]. Right-click the Dynamic Input toggle and choose **Settings**. Verify that the **Enable Pointer Input**, **Enable Dimensional Input where possible**, and **Show command prompting and command input near the crosshairs** checkboxes are checked. Click **OK** to dismiss all dialog boxes.

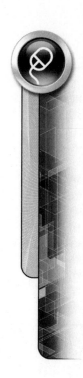

5. Right-click the tool palette titlebar and choose **New Palette**. Overtype **Furniture** as the name of the new palette.

6. Choose the **Content Browser** from the Tools flyout menu of the Build panel.

7. Open the **Design Tool Catalog—Imperial** to **Furnishing > Furniture > Sofa [Design Tool Catalog—Metric** to **Domestic Furniture > Sofas]**. Drag the i-drop of the **Long [3D Sofa – 3 Seat]** onto the new Furniture palette.

8. Choose the **Long [3D Sofa – 3 Seat]** sofa from the new palette. Move the cursor over the workspace and verify in the Properties palette that Specify scale on screen = **No**, X = **1 [900]**, Y = **1 [1800]**, Z = **1 [450]**, and Specify rotation on screen = **No**. Respond to the workspace prompts as follows:

   Insert point or: *(Choose* Rotation *from the down arrow of the dynamic input and type* -90 *to change the rotation. Select a point near* p1 *as shown at the left in Figure 10.6.)*

   Insert point or: *(Choose* Rotation *from the down arrow of the dynamic input and type* 90 *to change the rotation. Select a point near* p2 *as shown at the left in Figure 10.6.)*

   Insert point or: ESC *(Press ESC to end the command.)*

9. Open the Design Tool Catalog—Imperial to Furnishings > Furniture > Table [Design Tool Catalog—Metric to Office Furniture > Tables]. Drag the i-drop of Square 36 × 36 [3D Rectangular Table] onto the new palette.

10. Choose the **Square 36 × 36 [3D Rectangular]** from the new palette. Open the Properties palette and edit Specify scale on screen to **No**, X = **36" [600]**, Y = **36" [750]**, Z = **30" [750]**, and Specify rotation on screen to **No**. Move the cursor over the workspace and respond to the workspace prompts as follows:

    Insert point or: *(Select a point near* p3 *as shown in* Figure 10.6.*)* Insert point or: *(Select a point near* p4 *as shown in Figure 10.6.)*

    Insert point or: ESC *(Press* ESC *to end the command.)*

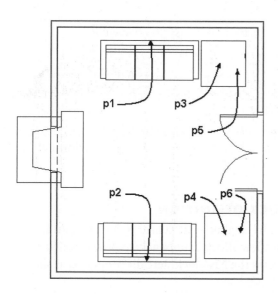

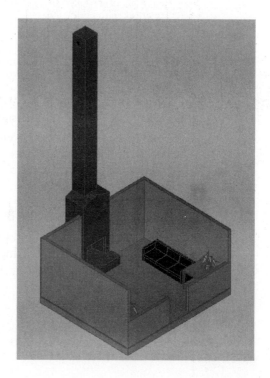

FIGURE 10.6 *Location of multi-view blocks*

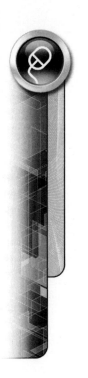

11. In this step you will search for a tool. Choose **Design Tool Catalog—Imperial [Design Tool Catalog—Metric]**. Type **Plant** in the Search field of the Design Tool Catalog—Imperial [Design Tool Catalog - Metric]. Click **Go** to view the results of the search as shown in Figure 10.7.

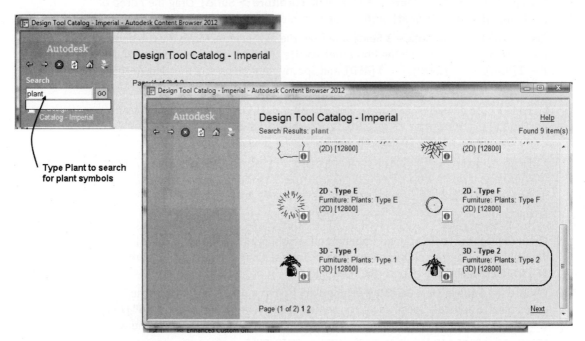

FIGURE 10.7 *Search tool applied to located specific multi-view blocks*

12. Drag the i-drop of the **3D Type 2 [3D Plant B]** in the Content Browser to the new palette. Select the **3D Type 2 [3D Plant B]**, open the Properties palette, and edit Specify scale on screen to **No, X = 1 [300], Y = 1 [300], Z = 1 [300]**, and **set** Specify rotation on screen to **No**. Respond to the workspace prompts as follows:

> Insert point or: (Select a point near p5 as shown in Figure 10.6.)
>
> Insert point or: (Select a point near p6 as shown in Figure 10.6.)
>
> Insert point or: ESC (Press ESC to end the command.)

13. Select the plants placed in step 11, right-click, and choose Properties. Edit the Elevation field to 30" [750] in the Design tab.

14. Choose **SE Isometric** from the View Cube. Choose **Realistic** from the Visual Styles flyout of the Visual Styles panel in the View tab to view the room as shown at the right in Figure 10.6. Save and close the drawing.

EDITING THE MULTI-VIEW BLOCK DEFINITION

The Properties palette can be used to edit attributes, dimensions, location, and rotation of a multi-view block. The Multi-View Block contextual tab provides edit tools for the style of a Multi-View block. Access the MvBlockDefEdit command as shown in Table 10.4.

TABLE 10.4 *Edit Multi-View Block Definition command access*

Command prompt	MVBLOCKDEFEDIT
Ribbon	Select a multi-view block and choose Edit Style from the General panel of the Multi-View Block contextual tab to open the Multi-View Block Definition Properties dialog box.
Shortcut menu	Select a multi-view block, right-click, and choose Edit Multi-View Block Definition.

When you select a multi-view block and choose **Edit Style** from the **General** panel of the Multi-View Block contextual tab, the **Multi-View Block Definition Properties** dialog box opens as shown in Figure 10.8. The View tab of the Multi-View Block Definition Properties allows you to edit the definition of view blocks used to create the multi-view block. The display representations and views assigned to the view blocks can be edited in the Multi-View Block Definition Properties dialog box. The Multi-View Block Definition Properties dialog box includes the **General, View Blocks**, and **Classifications** tabs shown in Figure 10.8. Since the contents of the General and Classifications tabs are similar to the same tabs object styles, only the View Blocks tab is described below.

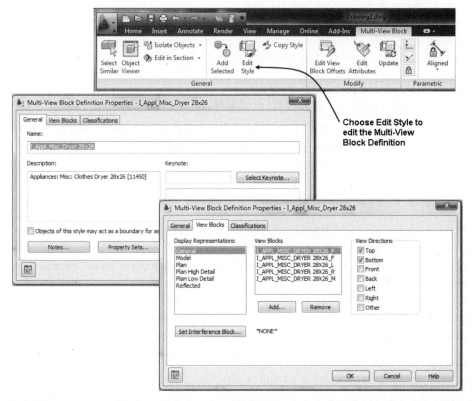

FIGURE 10.8 *General and View Blocks tabs of the Multi-View Block Definition Properties dialog box*

View Blocks Tab

The **View Blocks** tab shown in Figure 10.8 includes a list of the display representations of the view blocks assigned to the multi-view block.

Display Representations—The Display Representations list includes each of the display representations used by the multi-view block. Multi-view blocks can be defined for the General, Model, Plan, Plan High Detail, Plan Low Detail, and Reflected display representations.

View Blocks—The View Blocks list is used to define the multi-view block's appearance for each of the display representations. Each of the names in the **Display Representations** list box represents a display method for controlling the view of a single block. Some of the view blocks defined in the AutoCAD Architecture folders have a suffix to the block name to indicate the viewing direction that will apply the block. In Figure 10.8, suffixes used were: F (front), L (left), M (model), and P (plan). However, the same view block can be defined for each of the display representations and view. The **View Blocks** list varies according to the display representation selected.

Add—The Add button opens the **Select a Block** dialog box, which lists the blocks of the drawing. If you select a block from this list, it is assigned to the selected display representation and view directions for the multi-view block.

Remove—To remove a view block, select it from the view block list and then click the **Remove** button. View blocks are added and removed for each of the display representations.

View Directions—The View Directions checkbox allows you to turn on or off the visibility of the block when the drawing is viewed from the direction listed.

Set Interference Block—The Set Interference Block button opens the **Select a Block** dialog box.

CREATING A MULTI-VIEW BLOCK

Most multi-view blocks are developed from a three-dimensional object. The multi-view block is defined from one or more view blocks. A view block is an AutoCAD block created from an elevation or plan View drawing of a three-dimensional object. Therefore, view blocks can be created for the top, back, front, right side, left side, bottom, and model views of the three-dimensional object. The tools used to create elevations, discussed in Chapter 12, can be used to develop each of the view blocks. You can include unique insertion points on each of the view blocks. The insertion points are created on the Defpoints layer as AutoCAD POINT objects. The view blocks shown in Figure 10.9 are the components of the Dryer 28 × 26 multi-view block. View blocks are created as a projection drawing from a three-dimensional object; however, lines can be added or deleted based upon the representation conventions for that view. The left view of the dryer shown in Figure 10.9 does not include hidden lines to represent the top plane because the conventional representation of a dryer does not include the hidden lines.

Multi-view blocks are created in the Multi-Purpose Objects folder of the **Style Manager**. The name and definition of the multi-view block is specified in the Style Manager. The definition requires reference to other multi-view blocks or view blocks in a drawing. The multi-view block definition includes control of the visibility of the block according to the viewing direction. See Table 10.5 for **Style Manager** command access.

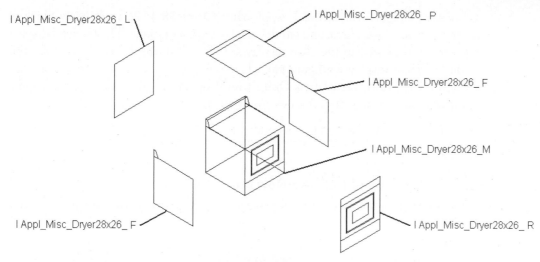

FIGURE 10.9 *View blocks used to create a multi-view block*

TABLE 10.5 *Style Manager access to create a multi-view block*

Command prompt	AECSTYLEMANAGER
Ribbon	Select a multi-view block and choose Multi-View Block Definitions from the Edit Style flyout of the General panel to open the Style Manager.

After you create a name for the multi-view block, you create view blocks and assign the view blocks to a display representation in the View Blocks tab. View blocks are created from blocks or other multi-view blocks and are defined for one or more view directions. The view blocks of the I_Appl_Misc_Dryer 28 × 26 multi-view block are shown in Figure 10.9. Each of these blocks is assigned to one or more display representations and view directions in the Multi-View Block Definition Properties dialog box.

> You can select a multi-view block and choose Copy Style from the General panel of the Multi-View Block tab; see Figure 10.8 to create a copy of the style from the selected multi-view block.
>
> **NOTE**

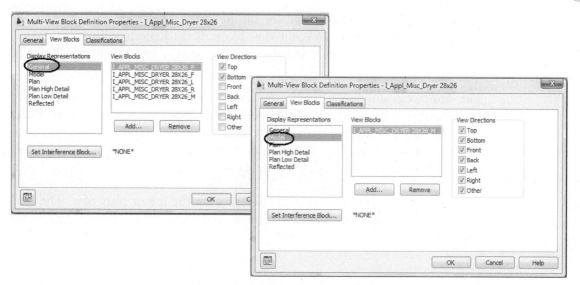

FIGURE 10.10 *View block settings for the General and Model display representations*

As shown in Figure 10.10, the I_Appl_Misc_Dryer 28 × 26 multi-view block is defined only for the General and Model display representations. The General display representation is used in the Plan display set. The Plan display set is used for the Medium Detail display configuration when the drawing is viewed from the top. Therefore, the **Multi-View Block Definition Properties** dialog box links the display representation to a view block for describing the dryer.

The following steps outline the procedure to create a new multi-view block in the Style Manager.

Steps to Create and Define a New Multi-View Block

1. Select Manage tab. Choose Style Manager from the Style & Display panel. Expand Multi-Purpose Objects. Select Multi-View Block Definitions from the left pane, right-click, and choose New Style from the shortcut menu.

2. Overtype a name for the new multi-view block.

3. Edit the tabs of the right pane to define the multi-view block. Select a display representation, view block, and view direction. Click the **Add** button to add blocks as view blocks.

Defining the Properties of a New Multi-View Block

After you create a name for the new multi-view block, three edit tabs of the right pane are displayed. If you select a multi-view block, right-click, and choose **Edit Multi-View Block Definition** from the shortcut menu, the **Multi-View Block Definition Properties** dialog box opens. The Multi-View Block Definition Properties dialog box for the I_Appl_Misc_Dryer 28 × 26 multi-view block is shown in Figure 10.10. The **View Blocks** tab shown at the left in the figure lists five view blocks defined for the General display representation. When this multi-view block was defined, the General display representation was selected and the **Add** button was used to specify each of the four view blocks. A view direction was specified for each of the view blocks selected for the General display representation. The results of this edit display the I_Appl_Misc_Dryer 28 × 26_P view block when the General display representation is used and the drawing is viewed from the Top or Bottom. In contrast, the Model display representation shown at the right in the figure includes only the I_Appl_Misc_Dryer 28 × 26_M view block, and it is used for all view directions when the Model display representation is used.

There are no view blocks defined for the Reflected display representation of the I_Appl_Misc_Dryer 28 × 26 multi-view block. If the Reflected display configuration is assigned to a viewport and the view direction is top, the dryer will not be displayed.

The view blocks and display representations of a multi-view block should be defined based upon the intended use of the multi-view block. The Configurations folder of the Display Manager allows you to identify the display representation set that is used for the associated view direction as shown in Figure 10.11. The Sets folder allows you to identify the display representation used per object type within a display set as shown in Figure 10.12. The Medium Detail display configuration uses the Plan display representation set when viewed from the top or plan view. The Plan display representation set includes General and Plan display representations for multi-view blocks.

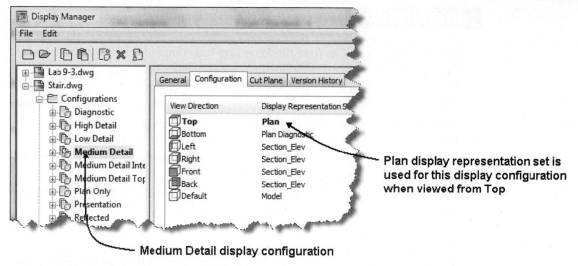

FIGURE 10.11 *Display configurations of the Display Manager*

Symbols created for a floor plan should be assigned the **General** display representation. If you are developing an electrical plan to include receptacles, switches, and incandescent lights, all of these symbols must share a common display representation or set. A General display representation is defined for the incandescent light symbol, switch, and receptacle. Therefore, the switch, receptacle, and incandescent light symbols are displayed when the **Medium Detail** display configuration is applied to a viewport. Symbols inserted in a ceiling grid may only include in their definition a reflected display representation and may not display when the Medium Detail display configuration is current.

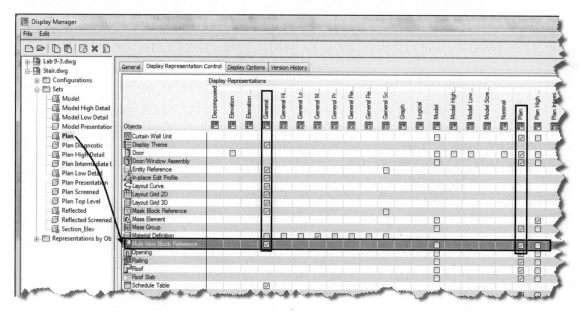

FIGURE 10.12 *Display representations of display sets*

TUTORIAL 10-1: MODIFYING THE MULTI-VIEW BLOCK DEFINITION

1. Download Chapter 8 files from the Accessing Tutor\Imperial category of the Student Companion site of CengageBrain http://www.cengagebrain.com described in the Preface to your Accessing Tutor folder.

2. Open the Ex 10-1.dwg from the Accessing Tutor\Imperial\Ch10. This tutorial is Imperial-only since the purpose of this tutorial is to modify the display of the incandescent light symbol included in the Design Tool Catalog—Imperial catalog.

3. Choose Save As > AutoCAD Drawing from the Application menu and save the drawing as Lab 10-1 in your student directory.

4. Select 1/4" = 1'-0" from the Annotation Scale flyout menu of the Drawing Window status bar.

5. Right-click the Object Snap button on the status bar, choose Settings, click the Clear All button, check Nearest object snap mode, and check Object Snap On (F3) from the Drafting Settings dialog box. Click OK to dismiss the Drafting Settings dialog box. Toggle Dynamic Input ON in the status bar.

6. Right-click the titlebar of the Design tool palette and choose New Palette. Overtype the name Electrical to name the new palette.

7. Select Content Browser from the Tools flyout of the Build panel in the Home tab. Open the Design Tool Catalog—Imperial to the Electrical > Power > Switch palette as shown in Figure 10.13.

8. Drag the i-drop for the Single Pole switch symbol onto the new custom palette named Electrical. Select the Single Pole switch tool from the palette and move the cursor over the drawing area to determine the default rotation angle.

9. Edit the Properties palette. Select No for the Specify rotation on screen option in the Properties palette. Type 90 in the rotation field in the Design tab of the Properties palette.

10. Move the cursor over the workspace and select a location along the bottom wall using the Nearest object snap near p1 as shown in Figure 10.13. Press ESC to end the command.

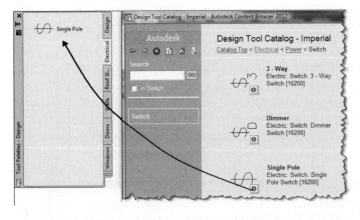

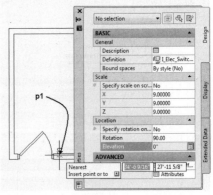

FIGURE 10.13 *Selecting Single Pole Switch from the Design Tool Catalog—Imperial*

11. Open the **Electrical > Lighting > Incandescent** category of the Design Tool Catalog—Imperial. Drag the i-drop for the **Ceiling** symbol onto the **Electrical** tool palette of the drawing. Select the **Ceiling** symbol from the tool palette and place the symbol near the middle of the room as shown in Figure 10.14. Press ESC to end the command.

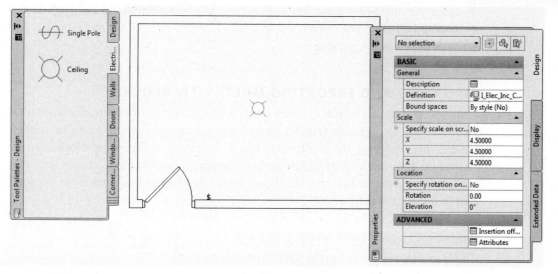

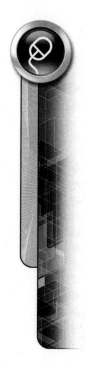

FIGURE 10.14 *Incandescent light symbol placed near center of room*

12. Select the **Reflected** display configuration from the Drawing window status bar. The incandescent light symbol is displayed; however, the switch disappears.

13. Select the **High Detail** display configuration to view the switch and incandescent light. In this step you will view the definition of Switch multi-view block. Select the Switch, right-click, and choose **Edit Style** from the General panel of the Multi-View Block contextual tab.

14. Select the View Blocks tab of the Multi-View Block Definition Properties I_Elec_ Switch_Single Pole dialog box. Select a display representation in the left window to display the view block applied for the View Direction that is checked. Therefore, the I_Elec_Switch_Single Pole_P view block will be displayed when the workspace is viewed from the Top or Bottom as shown in Figure 10.15.

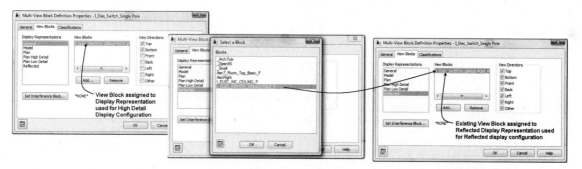

FIGURE 10.15 *View block added to the Reflected display representation*

15. To view the switch with the Reflected display configuration, choose Reflected display representation from the View tab. Select the **Reflected** display representation at the left and click the **Add** button to open the **Select a Block** dialog box. Select the **I_ELEC_SWITCH_SINGLE_POLE_P** block from the **Select a Block** dialog box as shown at the right in Figure 10.15. (Note: the blocks listed in the **Select a Block** dialog box of your drawing may include additional blocks not shown in Figure 10.15.) Click **OK** to dismiss the **Select a Block** dialog box. Verify that all **View Directions** are checked ON for the **Reflected** display configuration. Click **OK** to dismiss the Multi-view Block Definition Properties dialog box.

16. Since the multi-view block has been edited, select the **Reflected** and **High Detail** display configurations to view the switch.

17. Save and close the drawing.

IMPORTING AND EXPORTING MULTI-VIEW BLOCKS

Multi-view blocks allow you to insert symbols in a drawing with more display control than with blocks. You can transfer multi-view blocks from one drawing to another drawing by exporting the multi-view block definition. Multi-view blocks can be transferred across drawings in the Style Manager. The AecToolCatalogGenerator tool presented later in this chapter allows you to develop tool palettes from the tools used in one or more drawings.

Steps to Export Multi-View Blocks

1. Open a drawing that has the desired multi-view blocks.

2. Select a multi-view block. Choose Multi-view Block Definitions from the Edit Style flyout of the General panel.

3. The **Style Manager** opens to the **Multi-Purpose Objects** category of the current drawing in the left pane.

4. Select the **Multi-View Block Definitions** category, select multi-view blocks for export, right-click, and choose **Copy** as shown in Figure 10.16.

5. Select **Open** from the Style Manager toolbar to display the **Open Drawing** dialog box. Edit the directory and select the target drawing for the multi-view blocks. Click **Open** to dismiss the Open Drawing dialog box.

6. Select the target drawing in the left pane, right-click, and choose **Paste** to paste the multi-view blocks into the target drawing.

If a target drawing consists of multi-view blocks with names identical to the source drawing, the **Import/Export Duplicate Names Found** dialog box is displayed when you paste the multi-view blocks into the target drawing. The **Import/Export Duplicate Names Found** dialog box includes options to Leave Existing, Overwrite Existing, or Rename to Unique the duplicate multi-view blocks.

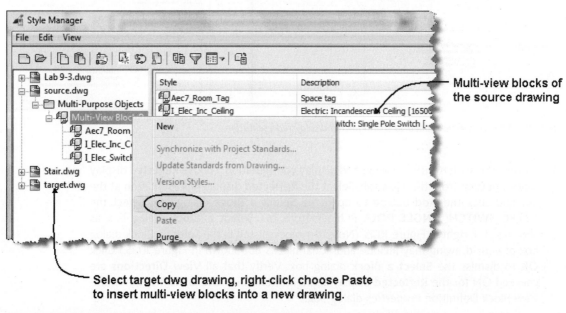

FIGURE 10.16 *Copying the multi-view block definitions of a drawing in the Style Manager*

INSERTING NEW MULTI-VIEW BLOCKS

Multi-view blocks are usually inserted from the Design Tool Catalog or Design-Center; however, new multi-view blocks or imported multi-view blocks will **not** be included in these resources. The **Multi-View Block Reference** tool can be accessed from the Ribbon or Content Browser as described in Table 10.6. The AutoCAD **INSERT** command cannot be used to insert multi-view blocks.

TABLE 10.6 *Multi-View Block Reference command access*

Command prompt	MVBLOCKADD
Ribbon	Select the Insert tab. Choose Multi-View Block from the Block panel (refer to Figure 10.17).
Content Browser	Open the Stock Tool Catalog > Helper Tools category and select Multi-View Block Reference.

When you select the **Multi-View Block Reference** tool, from the Block panel of the Insert tab you can select the name of new multi-view blocks from the Definition drop-down list in the Design tab of the Properties palette as shown in Figure 10.17.

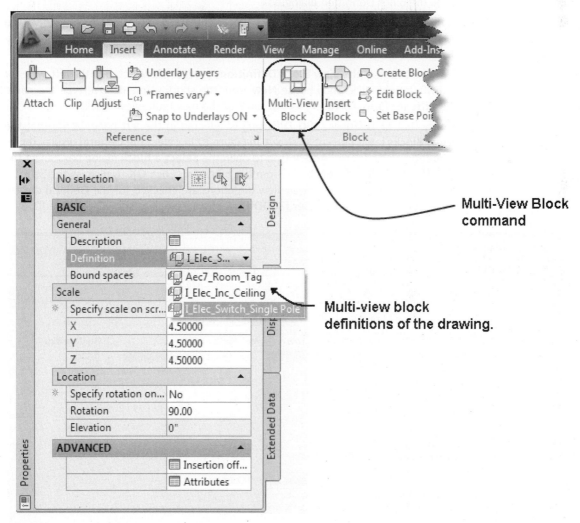

FIGURE 10.17 *Using the Multi-View Block Reference command*

TUTORIAL 10-2: CREATING A MULTI-VIEW BLOCK DEFINITION

1. Verify that the Accessing Tutor\Imperial or Accessing Tutor\Metric content for Chapter 10 of the CengageBrain http://www.cengagebrain.com has been downloaded to your Accessing Tutor folder on your computer as described in Organizing Tutorial Directories in the Preface.

2. Open the Ex 10-2.dwg from the Accessing Tutor\Imperial\Ch10 or Accessing Tutor\Metric\Ch10 folders. Save the drawing to your student folder as Lab 10-2. In this tutorial you will create a multi-view block.

3. Select the Home tab. To create a circle for the view block, select Circle from the Draw panel. Respond to the workspace prompts as shown below:

```
Specify center point for circle or: 35',33' [6100,10000] ENTER
Specify radius of circle or: 6 [150] ENTER
```

4. In this step you create an AutoCAD block that will be used in the Multi-View block definition. To create a block, select Insert tab. Choose Create Block from the Block panel. Type **12Diffuser [300Diffuser]** in the **Name** field, specify the Base point **X** insertion point = **35'[6100]**, and specify the **Y** insertion point = **33'[10000]**. Click the **Select Objects** button and then select the circle drawn in Step 2. Press ENTER to end the selection of objects and return to the **Block Definition** dialog box. Select the **Delete** radio button to delete the circle when the block is defined. Click **OK** to dismiss the Block Definition dialog box.

5. To apply the AutoCAD block as a view block for the new multi-view block, choose the Manage tab. Choose Style Manager from the Style & Display panel. Expand Multi-Purpose Objects **Multi-View Block Definitions**.

6. Verify that the **Multi-view Block Definitions** category is selected in the left pane for Lab 10-2, right-click, and choose **New** from the shortcut menu. Overtype the name **12_Diffuser [300_Diffuser]** as the name of the new multi-view block.

7. Verify that **12_Diffuser [300_Diffuser]** is selected in the left pane. Verify that the **View Blocks** tab is current as shown in Figure 10.18, select the **General** display representation, and click the **Add** button to open the **Select a Block** dialog box. Select the **12Diffuser [300Diffuser]** block from the block list. Click **OK** to dismiss the Select a Block dialog box. Verify that all view directions are checked for the **General** display representation and **12Diffuser [300Diffuser]** view block as shown in Figure 10.18.

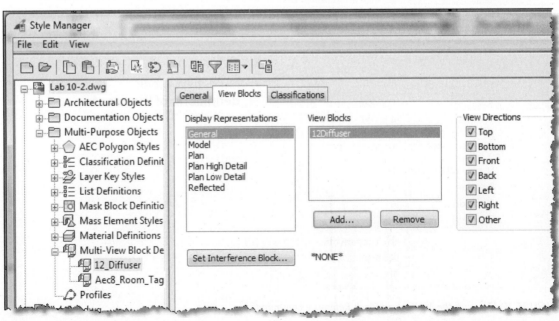

FIGURE 10.18 *Selecting a block for the view block*

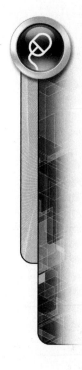

8. Click **OK** to dismiss the Style Manager.

9. Select the **Insert tab**. Choose **Multi-View Block** from the Block panel. Select the **Definition** field in the Design tab of the Properties palette and verify that **12_ Diffuser [300_Diffuser]** is current in the drop-down list of multi-view blocks. Select the Specify scale on screen = No, X = **1**, Y = **1**, and Z = **1**. Respond to the workspace prompts as follows:

> Insert point or: *(Select the location for the diffuser near* p1 *as shown in Figure 10.19.)*
>
> Insert point or: ENTER *(Press* ENTER *to end the command.)*

10. Since drawings must be saved prior to placing content from a drawing onto tool palettes, choose Save from the Quick Access toolbar. Select and drag the 12_Diffuser [300_Diffuser] multi-view block onto the Electrical palette.

11. To complete the tutorial, choose **Design Tool Catalog—Imperial [Design Tool Catalog—Metric]**, expand the **Electrical\Power\Receptacle [Electrical Services > Power Outlets]** folder, and drag the i-drop for the **Duplex Recpt [BS1363 Socket – Any Type]** symbol onto the Electrical tool palette. Select the **Duplex Recpt [BS1363 Socket – Any Type]** symbol from the Electrical tool palette. Edit Specify rotation on screen to **Yes** and add the symbol at points **p2** to **p9** as shown in Figure 10.19.

12. Save and close the drawing.

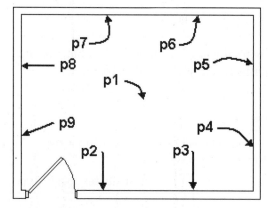

FIGURE 10.19 *Locations of receptacles*

IN-PLACE EDITING OF VIEW BLOCK OFFSETS

You can edit the location of insertion offsets and attribute locations of view blocks within the multi-view block in the workspace without accessing the Properties palette. The **Edit View Block Offset** command is located on the Modify panel in the Multi-View Block tab as shown in Figure 10.20 and Table 10.7. Enable Dimension Input where possible should be toggled ON during in-place editing. When this command is executed, the location grip is temporarily used to edit the offset.

TABLE 10.7 *Edit View Block Offsets command access*

Command prompt	**MVBLOCKEDITVIEWBLOCKOFFSETS**
Ribbon	Select a multi-view block; choose Edit View Block Offsets of the Modify panel in the Multi-View Block contextual tab (refer to Figure 10.20).
Shortcut menu	Select a multi-view block, right-click, and choose Edit View Block Offsets.

Steps to In-Place Edit View Block Insertion Offsets

1. Select a multi-view block and choose **Edit View Block Offsets** from the Modify panel in the Multi-View Block tab to display the grip of the insertion point.

2. Select the grip and move the cursor to display the dynamic dimensions as shown in Figure 10.20. Press TAB to specify a horizontal or vertical dimension for edit as shown in Figure 10.20 and overtype the desired distance for edit.

3. Press ESC to end edit and remove block selection. Choose Insertion offsets in the Properties palette to open the Multi-view Block Offsets dialog box shown at the right in Figure 10.20.

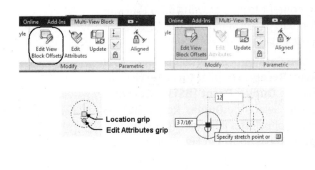
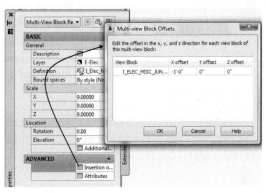

FIGURE 10.20 *Editing Insertion Offsets in the workspace*

You can change the location of the attributes if you select the Edit Attributes located on the Modify panel in the Multi-View Block tab; refer to Table 10.8.

TABLE 10.8 *Edit Attribute Orientations command access*

Ribbon	Select a multi-view block and choose Edit Attributes from the Modify panel in the Multi-View Block tab.
Toggle Grip	Select a multi-view block and select the Edit Attributes toggle grip.

Steps to an In-Place Edit of Attribute Orientations

1. Select a multi-view block to display the grips of the attribute.

2. Select the Edit Attributes command on the Modify panel in the Multi-View Block tab.

3. Select the Location or Rotation grips and move the cursor to display the dynamic dimensions as shown in Figure 10.21.

4. Press TAB to specify a horizontal or vertical dimension for edit as shown in Figure 10.21 and overtype the desired distance for edit.

5. Press ESC to end the edit.

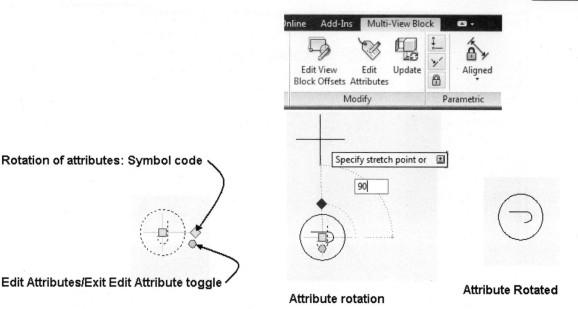

Rotation of attributes: Symbol code

Edit Attributes/Exit Edit Attribute toggle

Attribute rotation

Attribute Rotated

FIGURE 10.21 *Editing the orientations of attributes in the workspace*

INSERTING LAYOUTS WITH MULTIPLE FIXTURES

The Content Browser and the DesignCenter include layouts that consist of several fixtures and symbols appropriate for restroom layouts. The layouts are located in the Design Tool Catalog—Imperial > Mechanical > Plumbing Fixtures > Layouts folder and the Design Tool Catalog—Metric > Bathroom > Layouts of the Content Browser. Multiple lavatories are spaced evenly in a counter as shown in Figure 10.22. The symbols are three-dimensional AutoCAD blocks, and they can be used in the development of interior elevations. When you drag the i-drop from the Content Browser catalog to the drawing, the **Insert** dialog box opens, allowing you to set the insertion point, scale, and rotation, and toggle **Explode**. When the layouts are inserted in the drawing, you cannot adjust the length of the counter position of other fixtures unless you check **Explode** during insertion or select **Edit Block in-Place** from the shortcut menu to edit the inserted block. If you check **Explode**, the changes apply only to the selected block. If you apply **Edit Block in-Place** or the **Block Editor** to the layout, when you save your changes, you are editing the original block definition, and future insertions of the layout will be changed to comply to the changes made in the current drawing.

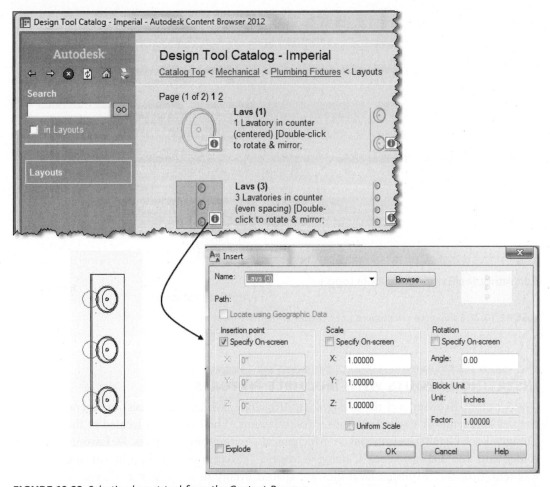

FIGURE 10.22 *Selecting layout tool from the Content Browser*

When you Explode the layout upon insertion or use Edit Block in-Place, you can edit the position of the components. The layout shown in Figure 10.23 has been exploded upon insertion. The layout consists of a custom wall style that serves as the counter. The counter has wall grips, which are displayed in Figure 10.23. The grips can be stretched to fit the design of a restroom according to the existing room shape as shown in Figure 10.23. The lavatories are anchored to a layout curve to evenly space the fixtures. Each component of the original layout can then be manipulated to create customized layouts of fixtures.

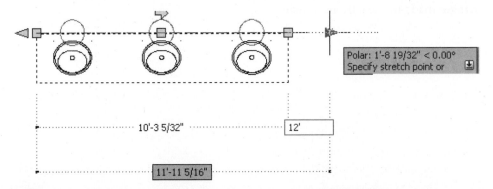

FIGURE 10.23 *Editing a layout multi-view block from the Design Tool Catalog—Imperial*

The grips of the anchor nodes of the layout curve are shown in Figure 10.24. The anchor nodes are located on the G-Grid-Nplt [A-SanitaryFixture-C] layer, a no plot layer. When you select a node, the properties of the layout curve are shown in the Properties palette shown at the right in Figure 10.24. The shortcut menu of the selected node also allows you to add or remove a node and change the layout to manual, repeat, or space evenly. You can select the (+) and (−) grips to edit the layout curve in the workspace as shown in Figure 10.24. (Layout curves are discussed in Chapter 13, "Drawing Commercial Structures.")

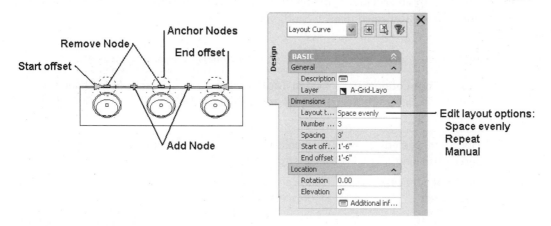

FIGURE 10.24 *Editing a layout multi-view block*

Viewing Multi-View Blocks with Reflected Display Representation

The majority of the multi-view blocks in Design Tool Catalog include a General display representation, which displays in plan view. Some multi-view blocks are designed as symbols for reflected ceiling plans. The multi-view blocks, listed in Table 10.9, are visible with the Reflected display configuration and are defined with view blocks that are displayed when the Reflected display configuration is current. Therefore, if a multi-view block with only reflected display representation is inserted when the Medium Detail display configuration is current, the symbol will not be displayed. The procedure to modify display representation settings was presented in Tutorial 10-2.

> **TIP** Prior to inserting multi-view blocks with Reflected only display representations, select **Reflected** from the Display Configuration flyout menu of the Drawing Window status bar.

TABLE 10.9 *Multi-view blocks with Reflected display configuration*

Design Tool Catalog—Imperial and DesignCenter\Imperial\Design Locations	Description	Display Representation
Electrical > Lighting	Incandescent, Fluorescent, and track lights	Reflected
Special Construction\ > Detection and Alarm	Exit Signs, Heat Detector and Smoke Detector, and alarms	Reflected only
Design Tool Catalog—Imperial and DesignCenter\Imperial\Design Locations		
Electrical Services > Fluorescent	Fluorescent luminaires	Reflected
Piped & Ducted Services > Ceiling	Diffusers and ceiling access	Reflected only

CREATING MASK BLOCKS

Some symbols in the DesignCenter and the Design Tool Catalog—Imperial are mask blocks. These symbols can be used to mask or block the display of other AutoCAD Architecture objects that are located within the mask block boundary. Mask blocks are often used in reflected ceiling plans or other plan views to block the display of ceiling grids or other objects. Mask blocks are in the folders as shown in Table 10.10 and include symbols for ceiling mounted devices. The mask blocks are visible in the Reflected display configuration.

TABLE 10.10 *Folders with mask block content using Reflected display representation*

Design Tool Catalog—Metric and DesignCenter\Metric\Design	Design Tool Catalog—Imperial and DesignCenter\Imperial\DesignDesign	Description	Display Configuration Required
Piped and Ducted Services\Ceiling	Mechanical\Air Distribution	Diffusers	Reflected
Electrical Services\Fluorescent	Electrical\Lighting\Fluorescent	Fluorescent lights	Reflected

The mask blocks of the Fluorescent and Air Distribution folders include fluorescent lights and Access Panels, Diffuser Returns, and Diffuser Supplies. If you select the symbol by double-clicking the symbol in the DesignCenter or dragging the i-drop from the tool catalog, you are prompted for the insertion point. These symbols are designed to fit in a suspended ceiling grid. However, if you drag the symbols to the workspace from the DesignCenter, you are prompted to select a layout node. The AutoCAD Architecture ceiling grid includes layout nodes; therefore, when you select the grid to insert the symbol, it will snap in the grid without the use of object snaps. (Ceiling grids and other AutoCAD Architecture grids will be discussed in Chapter 13.)

Mask blocks do not automatically block all entities that are within their boundaries. The mask block must be attached to the objects you want to block before it performs the mask function. The fluorescent fixture shown at **p1** in Figure 10.25 was attached to the ceiling grid at **p2**. The resulting mask of the ceiling grid is shown at **p3** on the right in Figure 10.25. The polyline representing the raceway that crosses the fluorescent fixture is not masked by the fixture **p4** because it is not an AEC object.

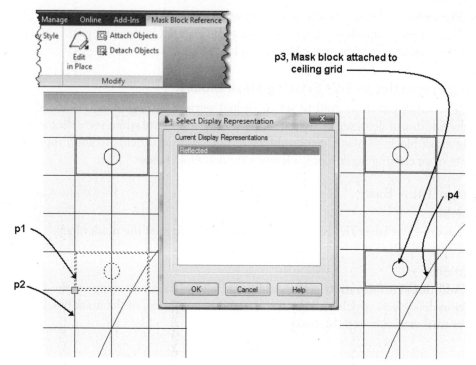

FIGURE 10.25 *Applying mask blocks*

PROPERTIES OF THE MASK BLOCK

When you insert a mask block from the DesignCenter, the Properties palette displays the name of the mask block and other properties. The Properties palette can be edited to change the description, scale, and rotation.

Design Tag—Basic
General

Description—The Description button opens the **Description** dialog box, which allows you to add additional text to describe the mask block.

Definition—The Definition field lists the name of the mask block. The drop-down list displays all mask blocks previously inserted in the drawing.

Bound spaces—The Bound spaces field consists of Yes, No, and By Style options. Multi-view blocks can function as a boundary for AEC Spaces.

Scale

Specify scale on screen—The Specify scale on screen field includes a Yes and No option. If the **No** option is selected, you can type a scale factor in this field or the command line. Scale factors for X, Y, and Z directions can be specified in the Properties palette.

Location

Specify rotation on screen—The Specify rotation on screen field includes **Yes** and **No** options. If the **No** option is selected, you type an angle in this field or the workspace.

Elevation—The multi-view block is inserted at $Z = 0$ coordinate elevation. The Node object snap allows the mask block to be inserted in a ceiling grid. The elevation value of existing mask blocks can be changed.

Using Properties to Edit Existing Mask Blocks

The Properties palette is used to set the initial settings during insertion of a mask block. Additional fields are displayed in the Properties palette when you select an existing mask block such as a ceiling diffuser. The additional fields to the Properties palette of an existing mask block reference are described below.

Design Tab—Basic

General

> **Layer**—The Layer field displays the name of the layer of the mask block.

Location

> **Additional information**—The Additional information button opens the **Location** dialog box, which includes the x, y, and z coordinates of the mask block, and normal extrusion and rotation.

ATTACHING OBJECTS TO MASK BLOCKS

If the mask blocks have been inserted in the drawing, they perform their mask function if attached to selected AutoCAD Architecture objects.

TIP

> Mask blocks will only block the display of selected AutoCAD Architecture objects when attached to the objects.

Attached objects can also be detached to return the visibility of the object. The **Attach Objects to Mask** command (**MaskAttach**) is used to associate the mask block with the selected AutoCAD Architecture object. See Table 10.11 for **Mask-Attach** command access.

TABLE 10.11 *MaskAttach command access*

Command prompt	MASKATTACH
Ribbon	Select a mask block and choose Attach Objects from the Modify panel in the Mask Block Reference contextual tab.
Shortcut menu	Select a mask block, right-click, and choose Attach Objects.

The workspace sequence for attaching the mask block to the ceiling grid using the **MaskAttach** command is shown below:

> *(Select a mask block at* p1 *as shown in Figure 10.25 and choose* Attach Objects *from the Modify panel in the Mask Block Reference contextual tab.)*
>
> Select AEC entity to be masked: *(Select the ceiling grid at* p2 *to attach the ceiling grid to the mask block.)*
>
> [1] Masks attached.
>
> *(The* Select Display Representation *dialog box opens, listing the current display representation as shown in Figure 10.25; click* OK *to dismiss the dialog box and mask the ceiling grid as shown at* p3*.)*

The result of attaching the ceiling grid to the mask block is shown at the right in Figure 10.25. If the fixture is moved to another location, it will continue to block the display of the ceiling grid. The AutoCAD Architecture objects that are attached to a mask block can be identified by applying the **LIST** command to the mask block.

Detaching Mask Blocks

Mask blocks can be detached from an AutoCAD Architecture object by the **Detach Objects** of the Modify panel in the Mask Block Reference contextual table. See Table 10.12 for the **Detach Objects** command access.

TABLE 10.12 *MaskDetach command access*

Command prompt	**MASKDETACH**
Ribbon	Select a mask block and choose Detach Objects from the Modify panel in the Mask Block Reference contextual tab.
Shortcut menu	Select a mask block, right-click, and choose Detach Objects.

The workspace sequence for detaching a mask block from a wall using the Mask-Detach command is shown below:

(Select a mask block at **p1** *in Figure 10.26 [with mask applied] and choose* **Detach Objects** *from the* **Modify** *panel in the Mask Block Reference contextual tab.)*

Select AEC object to be detached: *(Select the ceiling grid at* **p2**.*)*

1 mask(s) detached.

Detaching the block returns the visibility of the ceiling grid through the symbol as shown at the right in Figure 10.26.

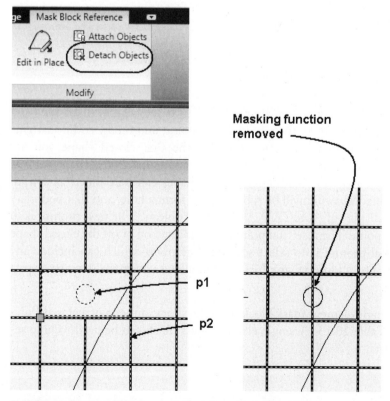

FIGURE 10.26 *Detaching the mask block from the ceiling grid*

CREATING MASK BLOCKS

Mask blocks are included in the Design Tool Catalog and DesignCenter; however, any closed polyline can be used to create a mask block. Custom mask blocks are created in the Style Manager.

Steps to Create a Mask Block

1. Draw a closed Polyline. Choose the Manage tab. Select **Style Manager** from the **Style & Display** panel in the **Manage** tab.

2. Double-click the **Multi-Purpose Objects** folder in the left pane.

3. Select the **Mask Block Definitions** folder in the left pane, right-click, and choose **New**.

4. Overtype a name, **MyMaskBlock**.

5. Select the name of the new mask block, right-click, and choose **Set From**. Respond to the workspace prompts as shown below:

> Select a closed polyline, spline, ellipse, or circle: *(Select the polyline at* **p1** *as shown in Figure 10.27.)*
>
> Add another ring? [Yes/No] <No>: ENTER *(Press* ENTER *to end the selection.)*
>
> Insertion base point: *(Select a point near* **p2** *as shown in Figure 10.27.)*
>
> Select additional graphics. *(Select additional entities or press* ENTER *to end the selection.)*

6. Double-click the new name of the mask block to display the edit tabs in the right pane of the Style Manager. The content of the General, Classification, and Display Properties tabs is similar to other object styles.

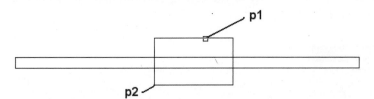

FIGURE 10.27 *Selecting a closed polyline for the mask block*

In the command sequence above, you are prompted to select a closed shape such as a polyline, spline, ellipse, or circle, then an insertion point, and finally additional graphics to define the mask block. After selecting the first closed shape you are prompted: **Add another ring?** If you respond **Yes**, additional closed shapes can be added to the mask block. The additional closed shapes can be defined as a void. Closed shapes defined as a void will be subtracted or form a hole, and that boundary will not block the display of AutoCAD Architecture objects. The next prompt is to select an insertion point. The insertion point can be located on or off the closed shape geometry. The final prompt is to select additional graphics, which can include any AutoCAD entity. The additional graphics prompt allows you to include text as part of the mask block.

You can create a single element mask block from the shortcut menu of a closed Polyline as shown in Table 10.13. A single element mass block can be used to hide selected AEC Objects.

TABLE 10.13 *Convert to Mask block command access*

Command prompt	**CONVERTLINEWORKTOMASKBLOCK**
Tool Palette	Select a Polyline, right-click, and choose Convert to > Mask Block.

Since the DIMBREAK command of AutoCAD is not available, a mask block can be used for this function. A mask block was created as shown in the following command sequence and Figure 10.28 to break the extension line of the dimension.

> (Choose a closed Polyline at p1 as shown in Figure 10.28, right-click, and choose Convert to > Mask Block. The Convert to Mask Block dialog box opens. Choose the Select AEC Objects to Mask option and Erase layout geometry. Click OK to dismiss the dialog box.)
>
> Select AEC object to be masked: 1 found (Select the extension line at p2.)
>
> Select AEC object to be masked: ENTER

If the mask block is placed on a layer with no plot property, the geometry of the mask block will not plot as shown in the lower-left corner of Figure 10.28. The content of the G-Anno-Nplt layer will not plot.

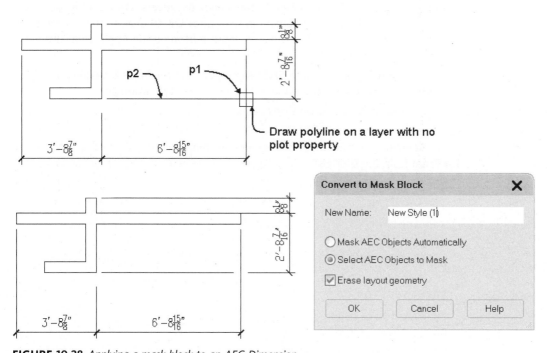

FIGURE 10.28 *Applying a mask block to an AEC Dimension*

INSERTING NEW MASK BLOCKS

Mask blocks of AutoCAD Architecture will be inserted from the Design Tool Catalog; however, new mask blocks that you create or imported mask blocks will not be included in the tool catalog or tool palettes. To create a tool for new mask blocks drag the geometry onto a tool palette. You can open the **Style Manager** and drag a new mask block from the Style Manager to a tool palette. The mask block can then be inserted into the drawing from the tool palette. The **Mask Block** tool can be accessed as described in Table 10.14. When you click the **Mask Block** tool, you can specify the mask block in the Definition field of the Properties palette.

TABLE 10.14 *MaskAdd command access*

Command prompt	**MASKADD**
Tool Catalog	Select the Stock Tool Catalog > Drafting Tools and choose Mask Block.

TUTORIAL 10-3: CREATING MASK BLOCKS AND USING THE DESIGNCENTER

1. Verify that the Accessing Tutor\Imperial or Accessing Tutor\Metric content for Chapter 10 of the CengageBrain http://www.cengagebrain.com has been downloaded to your Accessing Tutor folder on your computer as described in Organizing Tutorial Directories in the Preface.

2. Open Ex 10-3.dwg from the Accessing Tutor\Imperial\Ch10\ or Accessing Tutor\Metric\Ch10 folders.

3. Choose **Save As > AutoCAD Drawing** from the Application menu and save the drawing as **Lab 10-3** in your student directory.

4. Choose **1/4" = 1' -0" [1:50]** from the **Drawing Window** status bar **Annotation Scale** flyout menu.

5. Move the cursor over the Object Snap button on the status bar, right-click, and choose Settings. Click the Clear all button, check the **Intersection** and **Node** object snap modes, and check Object Snap On (F3). Click **OK** to dismiss the Drafting Settings dialog box. Verify that the Dynamic Input button on the Application status bar is pushed in.

6. Choose the **Manage** tab. Choose **Style Manager** from the **Style** & **Display** panel. Expand **Multi-Purpose Objects** in the left pane of the **Style Manager**.

7. Choose the **Mask Block Definitions** category, right-click, and choose **New** from the shortcut menu. Overtype **Speaker** as the name of the new mask block.

8. Select **Speaker**, the new mask block name, in the left pane, right-click, and choose **Set From** from the shortcut menu. Respond to the workspace prompts as follows.

   Select a closed polyline, spline, ellipse, or circle: *(Select the rectangle at* **p1** *as shown in the drawing file.)*

   Add another ring? [Yes/No] <No>: ENTER *(Press ENTER to respond* **No**.*)*

   Insertion basepoint: *(Select the lower-left corner of the rectangle at* **p2** *as shown in the drawing file using the* Intersection *object snap mode.)*

   Select additional graphics: 1 found *(Select the rectangle at* **p3** *as shown in the drawing file.)*

   Select additional graphics: 1 found, 2 total *(Select the hexagon at* **p4** *as shown in the drawing file.)*

   Select additional graphics: 1 found, 3 total *(Select the letter* S *at* **p5** *as shown in the drawing file.)*

   Select additional graphics: ENTER *(Press ENTER to end selection.)*

9. Click **OK** to dismiss the Style Manager.

10. Select Zoom **Extents** from the **Zoom** flyout menu of the **Navigation bar**.

11. Right-click the titlebar of the Tool Palette and choose New Palette. Overtype **Blocks** as the name of the tool palette. Choose the Home tab. Choose the Content Browser

from the **Tools** flyout of the **Build** panel. Open the Stock Tool Catalog to Drafting Tools. Choose page 2; drag the i-drop for Mask Block onto the Blocks tool palette.

12. Select the **Mask Block** tool. Edit the **Properties** palette: Definition = **Speaker**, Specify scale on screen = **No**, **X = 1**, **Y = 1**, **Z = 1**, Specify rotation on screen = **No**, and Rotation = **0**. Respond to the workspace prompts as follows:

> Insert point or: *(Select the ceiling grid at* **p1** *as shown in Figure 10.29 to place the mask block. Press ESC to end the command.)*

13. Select the Insert tab. Choose **DesignCenter** from the Content panel to open the Design Center.

14. Select the **AEC Content** tab and expand **Imperial\Design\Electrical\Lighting\ Fluorescent [Metric\Design\Electrical Services\Fluorescent\600×1200]** in the DesignCenter.

15. Double-click the **2 × 4 [600 × 1200]** fluorescent light symbol from the DesignCenter. Respond to the workspace prompts as shown below:

> Select Layout Node: *(Select the ceiling grid at* **p2** *as shown in Figure 10.29 to place the mask block.)*

> Insert point or: *(Select a ceiling grid node at* **p3** *as shown in Figure 10.29 to place the mask block.)*

> *(Insert additional 2 × 4 [600 × 1200] light symbols as shown in Figure 10.30.)*

16. Select the 2 × 4 [600 × 1200] light symbol and choose **Select Similar** from the General panel of the Multi-View Block tab to select all the 2 × 4 [600 × 1200] lights. Select the **Speaker** symbol and choose **Attach Objects** from the Modify panel of the Mask Block Reference. Respond to the workspace prompts as shown below:

> Select AEC object to be masked: *(Select the ceiling grid.)(Click* OK *to dismiss the* Select Display Representation *dialog box.)*

> *(Mask blocks hide the ceiling grid as shown in Figure 10.30.)*

17. Save and close the drawing.

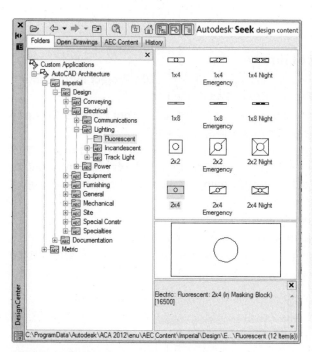

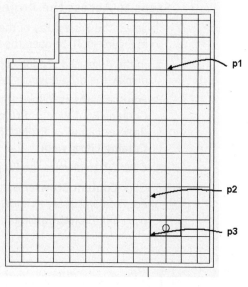

FIGURE 10.29 *Insertion point of mask blocks*

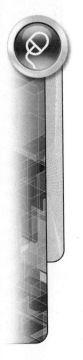

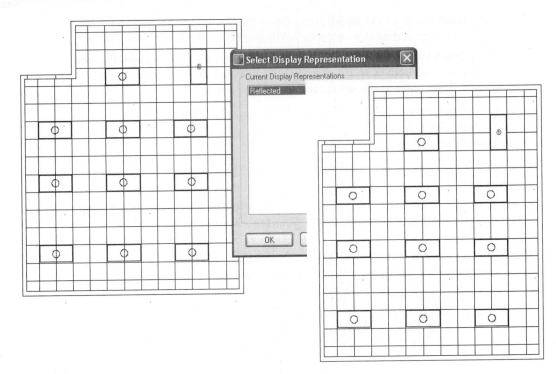

FIGURE 10.30 *Mask block hides ceiling grid*

CREATING A SELF-CONTAINED PROJECT

Throughout the tutorials of this book you have developed styles and display settings. Some of the tools developed were placed on tool palettes. Palette content can be placed in catalogs of a project library. There are three settings for tool palettes: None, Shared workspace catalog, and Per user workspace. The Per user workspace setting retains the custom palettes in a folder on the local computer in the Documents and Settings\per user login, whereas the Shared workspace catalog can retain the path to the tool catalog library within the user's project profile. Therefore, the Shared workspace catalog allows transfer of the tool catalog with the project when placed on other computers.

Steps to Access the Project Library

1. Select the Project tab of the Project Navigator. Click the **Edit Project** button, shown in Figure 10.31, to open the **Project Properties** dialog box.

2. In the Advanced section click in the right margin of **Project Tool Palettes Group** to display the drop-down list, select Shared workspace catalog.

3. Edit the paths as follows: Tool Palette Group = **Project Folder\ Project name**, Tool Content Root Path = **Project Folder\ Project name\Standards\Content**, Content Browser Library = **Project Folder\ Project name\Standards\ToolCatalogs\Project Catalog Library.cbl**.

4. Click OK to dismiss the Project Properties.

5. Click **Content Browser** on the Project tab of the Project Navigator to open the Project Catalog Library as shown in Figure 10.32. Tools from custom tool palettes can be placed in Project Catalog.

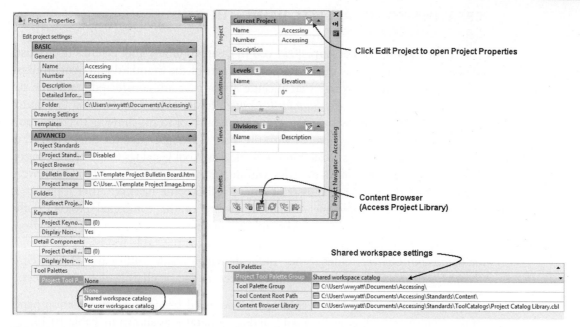

FIGURE 10.31 *Editing the Project Properties to access the Project Catalog Library*

Projects created from the project templates provided include a Standards folder. The Standards folder is placed in the project for CAD Management to store palette information and drawing content that can be used in the project. The Project Library and Project Catalog is created within the Standards\Tool Catalogs folder of a project. The Content Browser button located on the Project Navigator (see Figure 10.31) provides access to this tool catalog by default. Drawings that include styles for standardization can be placed in the Content folder and used during project synchronization. You can use the **AecToolCatalogGenerator** command to create tools from objects that have been used in a standard drawing or all drawings of a project. The steps to creating a self-contained project are described below.

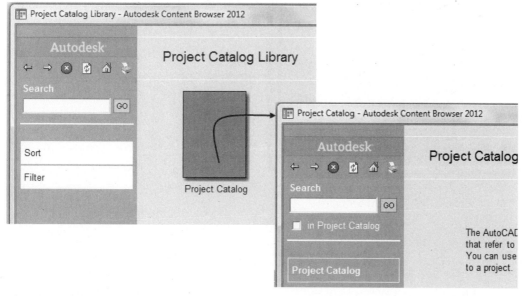

FIGURE 10.32 *Project Catalog Library*

CREATING TOOL PALETTES USING THE TOOL CATALOG GENERATOR

The **Tool Catalog Generator** allows you to create tools and tool palettes from existing drawings. This tool is most often used as part of the Standard and Synchronization. One or more drawings that include styles that are considered standards for projects can be placed in the *Standards\Content* folder. The Tool Catalog Generator can be applied to the drawings that serve as standards or to any drawing to create tools and tool palettes. Tools generated with the Tool Catalog Generator can be placed in the Project Catalog that is located in the Standards\Tool Catalogs folder of the project. The tools created are linked to styles within the project; therefore, if you remove a drawing from the project, the associated tool will become inactive. After a project is completed, you can copy one or more drawings of the completed project to the *Project title\Standards\Content* folder of the project and generate the tools from these completed drawings. The source definition for the tools resides in the *Project title\Standards\Content* folder. Since these drawings reside within the *Project Title\ Standards\Content* folder, they become resources for the synchronization of a project.

The **Tool Catalog Generator** command is accessed from the Customization panel of the Manage tab. Access the Tool Catalog Generator as shown in Table 10.15.

TABLE 10.15 *Tool Catalog Generator command access*

Command	AECTOOLCATALOGGENERATOR
Ribbon	Select the Manage tab. Choose Generate Tool Catalog from the Customization slide-out panel.

When you select the **AecToolCatalogGenerator** command, the **Populate Tool Catalog from Content Drawings** dialog box opens as shown in Figure 10.33. The Populate Tool Catalog from Content Drawings dialog box allows you to specify the source drawing to extract the tools and add to specify the name and location of the new catalog. You can specify whether the tools are created on palettes and which tools to include in the new catalog.

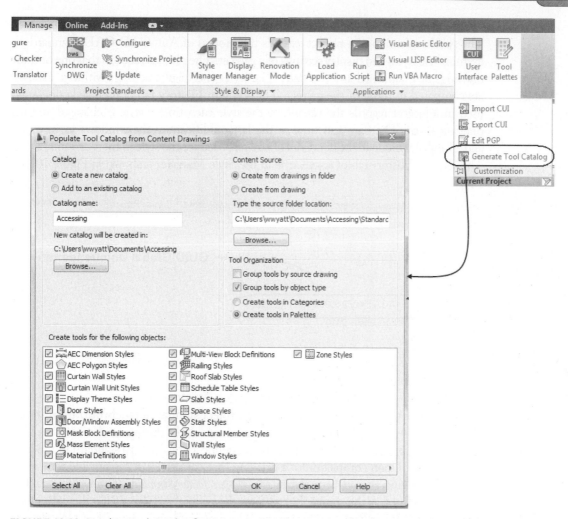

FIGURE 10.33 *Populate Tool Catalog from Content Drawings*

The tools can be organized into groups by object type or by source drawing. The tools can be placed on tool palettes. The tool catalog is created as a file with the *.atc* extension.

Steps to Create Tool Palettes for a Self-Contained Project

1. Choose Manage tab. Choose Tool Catalog Generator from the Customization panel.

2. Choose the **Add to an existing catalog** radio button. Edit the name and path to the current *Project Name\Standards\Tool Catalogs\Project Catalog.atc*.

3. Specify the source and path to drawings in the Content Source section. Edit the path to the Project Name\Standards\Content. Choose the options of **Tool Organization** as shown in Figure 10.33 to specify how to group the tools.

You can assign the palettes to the project palette group or drag palettes from the Content Browser to the tool palette. Right-click over the titlebar of the tool palette and choose Customize Palettes to add project palettes to the project group. When the project is closed, the palette group is not displayed.

TIP

SYNCHRONIZATION OF A PROJECT

The synchronization of a project extends the process of AutoCAD CAD Standards to examine the styles and display settings of a project. The process requires that one or more drawings be established as a standard. The styles and display settings within a standard are compared to those of the project. The **Version History** tab of each style or display used in a project records the version of the style each time a style is changed and the drawing saved. The date and time of the last modification, the windows login name, and a Global Unique Identifier (GUID) are recorded for the style in the Version History. The Version History tab of a wall style in the Style Manager is shown in Figure 10.34.

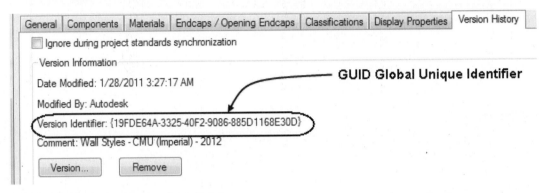

FIGURE 10.34 *Version History of a wall style*

You can check the checkbox **Ignore during project standards synchronization** shown in Figure 10.34 to remove a style from the styles considered during synchronization. The standard for a project is established when you define a drawing as a standard in the configure process. Upon completion of a drawing that consists of styles suitable for a standard, the drawing should be saved to the *Project folder\Standards\Content* folder. The drawing is defined as the source for styles for the palette tools by setting the path of the **Tool Content Root Path** to the drawing. The drawing standard specified for styles can also be applied as the standard for display settings. The drawings used as a standard can be located within a network or *Standards\Content* folder of a project template. The Version History tab of a display is shown in Figure 10.35.

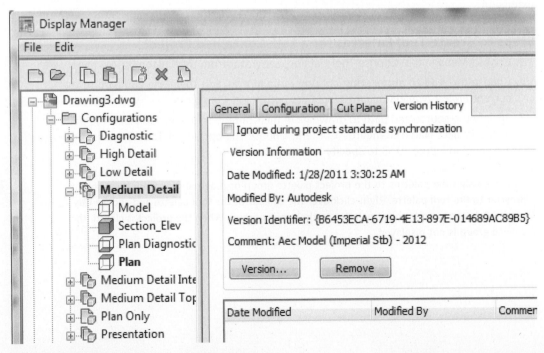

FIGURE 10.35 *Version History tab of display settings*

Synchronization can be specified to occur as Automatic, Semi-Automatic, or Manual. The Automatic option synchronizes the project when it is opened or when there are differences in the project from the standard, whereas the semi-automatic option checks the project when opened and you are prompted if differences between the standard and the project develop. The Manual option requires the user to specify when to synchronize the project and the user is prompted when differences occur in the project from the standard.

Steps to Create a Standard and Synchronization

1. Select **Project Browser** from the Quick Access toolbar. Open a project.
2. Identify the location and name of drawings considered as standard.
3. Select the Manage tab. Choose Configure from the Project Standards panel to open the **Configure Standards** dialog box as shown in Figure 10.36. Check Enable project standards for project.

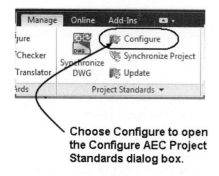

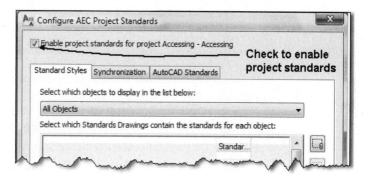

FIGURE 10.36 *Project Properties dialog box*

1. Select the Standard Styles tab of the Configure Standards dialog box.
2. Click the Add Drawing button as shown at the right in Figure 10.37. Select the standard drawing identified in step 1 from the Select Standards Drawing dialog box. The drawing may reside within the project folder *Standards\Content* or at a location within a network.
3. The drawing specified is listed in the Standards Drawings column shown in Figure 10.37. Select the header of the drawing column, right-click, and choose Select Column to specify all styles of the drawing to serve as a standard.
4. Select the Synchronization tab and choose the synchronization behavior as shown in Figure 10.37.

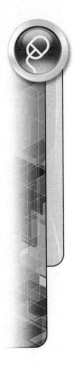

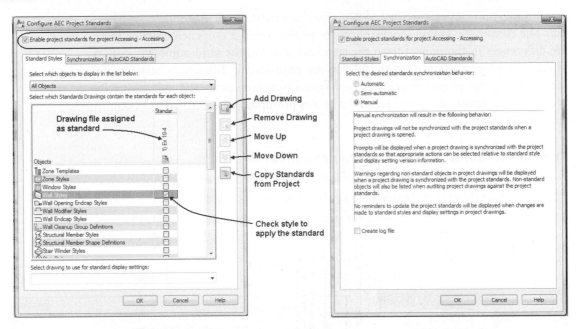

FIGURE 10.37 *Synchronization tab of Configure AEC Project Standards*

5. Select the **AutoCAD Standards** tab to specify drawings for AutoCAD CAD standard comparison. (Note: To create AutoCAD drawing standard files, open the drawing and choose **Save As > AutoCAD Drawing Standard** from the Application menu. Verify **Files of type** filter to **AutoCAD Drawing Standards *.dws.**)

6. Select OK to dismiss the Configure Standards dialog box.

7. Select **Manage tab.** Choose **Synchronize DWG from the Project Standards** panel.

TUTORIAL 10-4: INSERTING MULTI-VIEW BLOCKS

1. Verify that the Accessing Tutor\Imperial or Accessing Tutor\Metric content for Chapter 10 of the CengageBrain http://www.cengagebrain.com has been downloaded to your Accessing Tutor folder on your computer as described in Organizing Tutorial Directories in the Preface.

2. Open Autodesk AutoCAD Architecture 2012 and select Project Browser from the Quick Access toolbar. Use the Project Selector drop-down list to navigate to your student Ch 10 folder. Select New Project to open the Add Project dialog box. **Type 17** in the Project Number field and type **Ex 10-4** in the Project Name field. Check Create from template project, choose Browse, and edit the path to \Accessing Tutor\ Imperial\Ch10\Ex 10-4\Ex. 10-4.apj or \Accessing Tutor\Metric\Ch10\Ex 10-4\ Ex.10-4.apj in the Look in field. Click Open to dismiss the Select Project dialog box. Click OK to dismiss the Add Project dialog box. Click Close to dismiss the Project Browser.

3. Select the Constructs tab of the Project Navigator. Double-click Floor 2 to open the Floor 2 drawing.

4. Select Zoom **Window** from the Zoom flyout of the Navigation bar and respond to the following workspace prompts.

```
Specify first corner: 135',60' [41000, 18000] ENTER
Specify opposite corner: 145',50' [45000, 14000] ENTER
```

5. Right-click the Object Snap button on the status bar; then choose **Settings** from the shortcut menu. Click the **Clear All** button and then select the **Endpoint** and **Intersection** object snap modes. Click **OK** to dismiss the Drafting Settings dialog box.

6. Toggle Dynamic Input, Object Snap, and Polar Tracking ON in the status bar. Turn OFF Snap Mode, Grid Display, and Object Snap Tracking.

7. Right-click over the tool palette titlebar and choose New Palette. Overtype the name **Symbols** to name the new palette.

8. Select the Home tab. Choose Content Browser from the Tools flyout menu in the Build panel. Open **Design Tool Catalog—Imperial [Design Tool Catalog—Metric]** and expand the **Mechanical > Plumbing Fixtures\Toilet [Bathroom > Toilet]** folder as shown in Figure 10.38.

9. Drag the i-drop of **Tank 1 [3D Toilet – Modern A]** onto the Symbols palette. Select **Tank 1 [3D Toilet – Modern A]** from the tool palette.

10. Click the **Tank 1 [3D Toilet – Modern A]** tool and move the cursor to the drawing area to determine the default symbol rotation as shown in Figure 10.38.

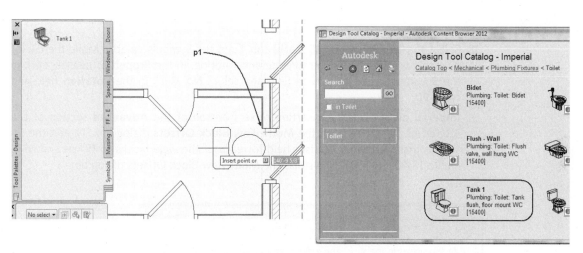
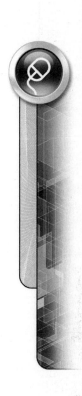

FIGURE 10.38 *Default rotation of the symbol when placing Tank 1*

11. Move the cursor over the Properties palette, click in the **Specify rotation on screen** field, set the field to **No**, click in the **Rotation** field, and type **180**. Verify that **Specify Scale on Screen** is set to **No**.

12. Move the cursor over the workspace and select point **p1** at the edge of the window shown in Figure 10.38 using the **Endpoint** object snap mode to locate the toilet. Press ESC to end insertion.

13. Select the **Content Browser** from the Windows taskbar and open the **Design Tool Catalog–Imperial > Mechanical > Plumbing Fixtures > Lavatory [Design Tool Catalog—Metric > Bathroom > Basin]** category. Navigate to page 2 [1] of the catalog and drag the i-drop for the **Oval-3D [Rectangular]** symbol onto the Symbols tool palette.

14. Select the **Oval-3D [Rectangular]** tool from the tool palette and move the cursor to the drawing area to identify the default rotation.

15. Move the cursor to the Properties palette, click in the **Specify rotation on screen** field in the Design tab of the Properties palette, and select **No** from the drop-down list. Click in the **Rotation** field and type **180**.

16. Move the cursor over the workspace, pause over **p2** as shown in Figure 10.39 until the Endpoint Osnap marker is displayed, and then click to select the location of the lavatory. Press ESC to end insertion.

17. Select the **Content Browser** from the Windows taskbar and open the **Design Tool Catalog—Imperial > Mechanical > Plumbing Fixtures > Bath [Design Tool Catalog–Metric > Bathroom > Bath]** category. Drag the i-drop for the **Tub 30 × 60 [3D Bath]** symbol onto the Symbols palette.

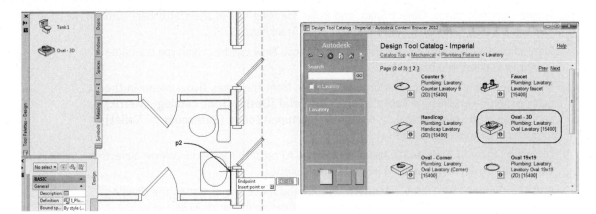

FIGURE 10.39 *Selecting Endpoint for placement of lavatory*

18. Select the **Tub 30 × 60 [3D Bath]** tool from the Symbols palette. Move the cursor over the workspace to determine default rotation. In the Properties palette, click in the Specify rotation on screen field and select **No**. Click in the **Rotation** field and type **-90**.

19. **Imperial only**: Click the Insertion offsets button of the **Advanced** section of the Properties palette to open the **Multi-view Block Offsets** dialog box. To edit the insertion point, edit the Y Offset field for each of the view blocks to **15"** as shown in Figure 10.40. Click **OK** to dismiss the Multi-view Block Offsets dialog box.

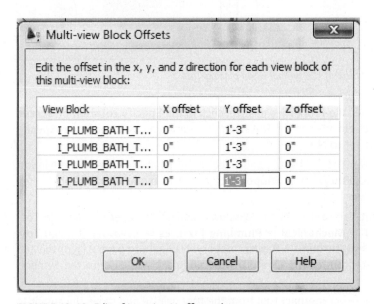

FIGURE 10.40 *Edit of insertion Y offset values*

20. **Metric only**: Edit the X scale factor to **1500** in the Properties palette.

21. Insert the tub using the **Endpoint** object snap at **p1** as shown in Figure 10.41. Press ESC to end the insertion.

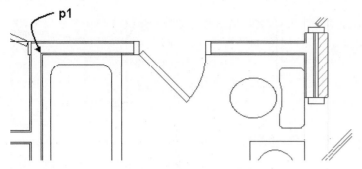

FIGURE 10.41 *Insertion point of tub*

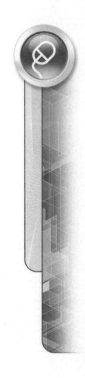

22. Select **Zoom Window** from the **Zoom** flyout menu of the Navigation bar. Respond to the workspace prompts as shown below:

   Specify first corner: 86',60' [26200,18200] ENTER (Type coordinates for the first corner of the zoom window.)

   Specify opposite corner: 108',36' [33000, 11000] ENTER (Type coordinates for the opposite corner of the zoom window.)

23. Select the Content Browser from the Windows taskbar. Select Design Tool Catalog— Imperial > Equipment > Food Service > Refrigerator [Design Tool Catalog - Metric > Kitchen Fittings > Appliances]. Drag the Top-Bot 31 × 28 [Fridge-freezer A] i-drop onto the Symbols palette. Select the Top-Bot 31 × 28 [Fridge-freezer A] tool from the tool palette and insert the refrigerator at A as shown in Figure 10.42.

24. In this step you will adjust the position of the refrigerator using parametric constraints. Select the refrigerator to display Multi-View Block tab. Choose Aligned from the Parametric panel. Respond to the workspace prompts as follows:

   Select first line: (Move the cursor to the wall at p1 in Figure 10.42; select the wall when the red assistant line is displayed.)

   Select second line to make parallel: (Move the cursor to the refrigerator at p2 as shown in Figure 10.42; select the refrigerator when the red assistant line is displayed at p2.)

   Specify dimension line location: (Select a point near p3 as shown in Figure 10.42.)

   Dimension text = 11 7/16": (Overtype 1[25] to specify the parametric dimension value.)

25. Select the refrigerator to display Multi-View Block tab. Choose Aligned from the Parametric panel. Respond to the workspace prompts as follows:

   Select first line: (Move the cursor to the wall at p4 in Figure 10.42; select the wall when the red assistant line is displayed.)

   Select second line to make parallel: (Move the cursor to the refrigerator at p5 as shown in Figure 10.42; select the refrigerator when the red assistant line is displayed at p2.)

   Specify dimension line location: (Select a point near p6 as shown in Figure 10.42.)

   Dimension text = 10 1/2": (Overtype 0[0] to specify the parametric dimension value.)

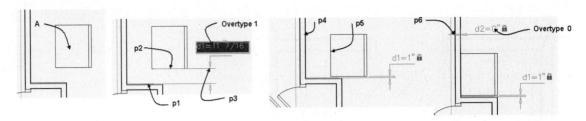

FIGURE 10.42 *Placing the refrigerator using parametric constraints*

26. Select Content Browser from the Windows taskbar and open Design Tool Catalog— Imperial > Furnishing > Casework > Base Cabinet [Design Tool Catalog – Metric > Kitchen Fittings > Units 600 mm Deep]. Drag the 15 in Wide [3D Kitchen Unit 600 mm Style A] i-drop onto the Symbols palette. Select the 15 in Wide [3D Kitchen Unit 600 mm Style A] tool from the tool palette and insert the cabinet at B as shown in Figure 10.43. Metric only: Click in the Properties palette to edit the Y scale to [350].

27. Select Content Browser from the Windows taskbar and open Design Tool Catalog— Imperial > Equipment > Food Service > Range [Design Tool Catalog–Metric > Kitchen Fittings > Appliances] category. Drag the Range 30 × 26 [Cooker] i-drop onto the Symbols palette. Select the Range 30 × 26 [Cooker] tool from the tool palette and insert the range at C as shown in Figure 10.43. Press ESC to end the command.

28. Imperial only: Select Content Browser from the Windows taskbar and open the Design Tool Catalog–Imperial > Furnishing > Casework > Base Cabinet category. Drag the 21 in Wide i-drop onto the Symbols palette.

29. Select the 21 in Wide [3D Kitchen Unit 600 mm Style A] tool from the tool palette, rotate -90 in the Properties palette, and insert the cabinet at the D location(s) as shown in Figure 10.43. Metric only: Click in the Properties palette to edit the Y scale to [550]. Press ESC to end the command.

30. Imperial only: Select Content Browser from the Windows taskbar and open the Design Tool Catalog–Imperial > Furnishing > Casework > Base Cabinet category. Drag the 36 in Wide i-drop onto the Symbols palette.

31. Select the 36 in Wide [3D Kitchen Unit 600 mm Style A] tool from the tool palette and insert the cabinet (rotate -90) at E as shown in Figure 10.43. Metric only: Click in the Properties palette to edit the Y scale to [900]. Press ESC to end the command.

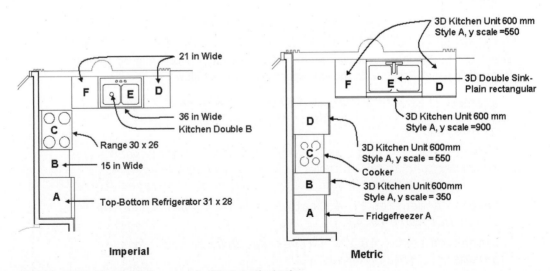

FIGURE 10.43 *Placing appliances and cabinets in the kitchen*

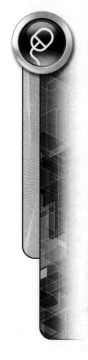

32. Select the **21 in Wide [3D Kitchen Unit 600 mm Style A]** tool from the tool palette and insert the cabinet (rotate -90) at **F** as shown in Figure 10.43. **Metric only**: Click in the Properties palette to edit the Y scale to [550]. Press ESC to end the command.

33. Select **Content Browser** from the Windows taskbar and open the **Design Tool Catalog—Imperial > Mechanical > Plumbing Fixtures > Sink [Design Tool Catalog—Metric > Kitchen Fittings > Sinks]** category. Drag the **Kitchen-Double B [3D Double Sink Plainrectangular]** i-drop onto the Symbols palette. Select the **Kitchen-Double B [3D Double Sink Plainrectangular]** tool from the tool palette, edit its elevation to **37 [950]** and its rotation to **-90**, SHIFT + right-click, choose the **Midpoint** object snap mode, and insert the sink at the midpoint of the 36" [900] wide base at **E** as shown in Figure 10.43. Select the sink and click the **Insertion offsets** button in the **Advanced** section of Properties. Edit the X offset to **1 [25]** for each of the 6 [5] view blocks. Click **OK** to dismiss the dialog box.

34. Open the **Design Tool Catalog—Imperial > Walls > Casework [Design Tool Catalog—Metric > Walls > Casework]** category. Drag the i-drop for **Casework – 36 (Counter) [Casework 950 (Counter)]** onto the Symbols palette. To import the style into the drawing, select the **Casework – 36 (Counter) [Casework 950 (Counter)]** tool, right-click, and choose Import **Casework – 36 (Counter) [Casework 950 (Counter)]**.

35. Select the **Casework – 36 (Counter) [Casework 950 (Counter)]** tool, right-click, and choose **Wall Styles** to open the **Style Manager**. Double-click the **Casework 36 (Counter) [Casework 950 (Counter)]** wall style.

36. Select the **Endcaps/Opening Endcaps** tab and change the **Wall Endcap Style** to **Standard**. To adjust the height of the counter, select the **Components** tab and edit the **Case-Counter**: Bottom Elevation Offset = **2' -10 1/2" [875]** from **Baseline**.

37. Click **OK** to close the Style Manager.

38. Verify that **High Detail** display configuration and **1/4" = 1' -0" [1:50]** Annotation Scale are current as shown in the **Drawing Window** status bar flyout menu.

39. Right-click the Object Snap toggle of the Application status bar; choose Settings. Clear the checkbox for Allow general object snap settings to act upon wall justification line. Choose OK to dismiss the Drafting Settings dialog box.

40. Select the **Casework – 36 (Counter) [Casework 950 (Counter)]** tool from the **Symbols** palette. Edit the **Properties** palette to Style = **Casework 36 (Counter) [Casework 950 (Counter)]**, Base Height = **8' -0 [2500]**, and Justify = **Left**. Set the Cleanup Group definition = **Standard**. Draw the counter top wall segments as shown below:

```
Start point or: (Select p1 as shown in Figure 10.44.)

Endpoint or: (Select p2 as shown in Figure 10.44.)

Endpoint or: Enter (Pressing Enter will end the command; the
next Enter will repeat the last command.)

Start point or: (Select p3 as shown in Figure 10.44.)

Endpoint or: (Select p4 as shown in Figure 10.44.)

Endpoint or: (Select p5 as shown in Figure 10.44.)

Endpoint or: Enter (Pressing Enter will end the command.)
```

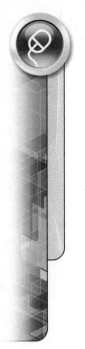

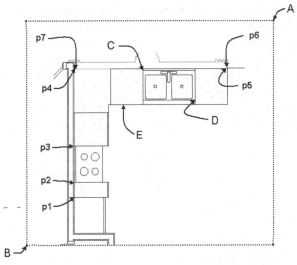

FIGURE 10.44 *Selection points for counter top wall style*

41. Open the Design Tool Catalog—Imperial > Walls > Casework [Design Tool Catalog—Metric > Walls > Casework] category. Drag Casework – 42 (Bar) [Casework 1050 Bar] into the drawing to start the WallAdd command. Edit the Properties palette: set Style = Casework 42 (Bar) [Casework 1050 Bar], Width = 0 [0], Base height = 8' -0" [2500], and Justify = Left. Respond to the command prompts to create the bar counter top.

> Start point or: *(Select p6 as shown in Figure 10.44.)*
>
> Endpoint or: *(Select p7 as shown in Figure 10.44. Press ESC to end the command.)*

42. Close the Content Browser. Select the bar wall, right-click, and choose **Edit Wall Style**. Choose the **Endcap/Opening Endcap** tab and change the Endcap style to **Standard** for **Wall Endcap Style**. Click **OK** to dismiss the Wall Style Properties dialog box.

43. The Casework 36 (Counter) [Casework 950 (Counter)] and Casework 42 (Bar) [Casework 1050 Bar] wall styles are hatched. To turn off the hatch of the Casework 36 (Counter) [Casework 950 (Counter)] wall style, select the **Casework 36 (Counter) [Casework 950 (Counter)]**, right-click, and choose **Properties**. Select the **Display** tab, verify that **Display controlled by:** Wall Style Casework 36 (Counter) [Casework 950 (Counter)], click in the **Display Component** field, and click off the light bulb for **Hatch 1 (Case-Counter)**. Click **OK** to dismiss the **Modify Display Component at the Style Override Level** message box. Click in the **Display Component** field and click off the light bulb for **Hatch 2 (Case-Backsplash)**. Press ESC to end the selection.

44. To turn off the hatch of the Casework 42 (Bar) [Casework 1050 Bar] wall style, select the **Casework 42 (Bar) [Casework 1050 Bar]**, right-click, and choose **Properties**. Select the **Display** tab, verify that **Display controlled by:** Wall Style: Casework 42 (Bar) [Casework 1050 Bar], click in the **Display Component** field, and click off the light bulb for **Hatch 1 (Stud)**. Click **OK** to dismiss the **Modify Display Component at the Style Override Level** message box. Click in the **Display Component** field and click off the light bulb for **Hatch 2 (Case-Counter)**.

45. Press Escape to clear selection. Create a crossing selection to include the objects from point A to point B as shown in Figure 10.44, right-click and choose Isolate Objects > Isolate Objects. Select **SE Isometric** from the **View** Cube. Select **Realistic** from the visual style flyout of the Visual Styles panel in the View tab. The Sink is not subtracted from the counter top.

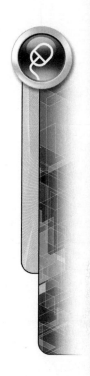

46. To subtract the sink from the counter, choose Top from the View Cube. Choose 2D Wireframe from the Visual style flyout of the Visual Styles panel in the View tab.

47. Select the **Rectangle** command from the Draw panel of the Home tab. Respond to the command line as follows:

        Specify first corner point or [Chamfer/Elevation/Fillet/
        Thickness/Width]: (Select the endpoint of the sink at C as
        shown in Figure 10.44.)

        Specify other corner point or [Area/Dimensions/Rotation]:
        (Select the endpoint of the sink at D as shown in Figure 10.44.)

48. Select the rectangle drawn in the previous step, right-click, and choose Convert to > Mass Element. Respond to the workspace prompts as follows.

        Erase selected linework? [Yes/No] <No>: Y

        Specify extrusion height <3'-0">: -37 [-950]

        (1) new mass element(s) created.

49. Select the counter at E shown in Figure 10.44, right-click, and choose Body Modifiers > Add from the shortcut menu. Respond to the workspace prompts as follows:

        Select objects to apply as body modifiers: (Select the mass
        element created in the previous step.)1 found

        Select objects to apply as body modifiers: ENTER (Press ENTER
        to end selection and open the Add Body Modifier dialog box.
        Edit Wall Component = Case-Counter, Operation =
        Subtractive, and check Erase Select Objects. Click OK to
        dismiss the Add Body Modifier dialog box.)

        (1) Body Modifiers added to the selected object.

50. Choose SE Isometric from the View Cube. Select **Realistic** from the **Visual Styles** flyout menu of the Visual Styles panel in the **View** tab to view the complete kitchen as shown in Figure 10.45.

51. Press Escape to clear all selection, right-click, and choose Isolate Objects > End Object Isolation.

52. Click Close Current Project located on the Projects tab. Click Close all project files and save all project drawings.

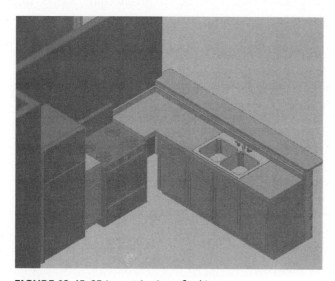

FIGURE 10.45 *SE Isometric view of cabinets*

NOTE

Refer to **e Tutorials** of Chapter 10 for download from the Student Companion Site of CengageBrain http://www.cengagebrain.com described in the Preface. **Tutorial 10-5 Creating Tool Palettes for a Project is provided.**

NOTE

Refer to **e Projects** of Chapter 10 for download from the Student Companion Site of CengageBrain http://www.cengagebrain.com described in the Preface. **Project 10-6 Inserting Symbols** is provided.

SUMMARY

- The scale for symbols and annotation is set with the AecDwgSetup command and Annotation Scale.
- Define a layer standard for the drawing by selecting the Layering tab of the Drawing Setup dialog box.
- Symbols are inserted in the drawing from the AEC Content tab of the DesignCenter.
- The content path is set in the AEC Content tab of the Options dialog box.
- Multi-view blocks include display control based upon the viewing direction and the display representation.
- View blocks can be created from other multi-view blocks or blocks.
- The Multi-View Block Reference Properties dialog box allows you to edit the location of the insertion point, insert attribute data, scale, rotate, and edit the style of the multi-view block.
- Display representations for multi-view blocks can be set to General, Model, or Reflected.
- Multi-view blocks can be imported from other drawings or exported to other drawings.
- Mask blocks are created to block the display of other AutoCAD Architecture objects.
- Mask blocks are created from closed polylines.
- Mask blocks must be attached to other AutoCAD Architecture objects to perform their blocking function.

NOTE

Refer to e **Review Questions** folder which includes review questions for chapter 10 of Student Companion Site of **CengageBrain** http://www.cengagebrain.com described in the Preface.

Annotating and Documenting the Drawing

INTRODUCTION

The annotation and documentation of a drawing includes the dimensions, notes, schedules and associated symbols for specifying the size and location of architectural features. The symbols, dimensions, and leaders for annotating the drawing are selected from the tool palettes of the Document palette set. The Document palette set includes Annotation, Dimensions, Callouts, Tags, Scheduling, and Themes tool palettes. The commands of the Callouts palette and keynoting commands will be presented in Chapter 12.

The Tags and Scheduling palettes include commands for creating schedules. The content of tags and schedules is defined in the schedule table style. Each schedule table style utilizes one or more property sets, which are defined for objects listed in the schedule. Schedules can be created, which automatically update when changes are made in the drawing. The content of a schedule can be exported to external database files to enhance the use of the schedule data.

OBJECTIVES

After completing this chapter, you will be able to

- Place AEC Dimensions, text, and leaders.

- Create revision clouds using various styles of revision clouds.

- Create tags for doors, windows, rooms, and walls.

- Create AEC Space objects for room and place finish tags.

- Create door and window schedules using tags and schedule styles.

- Apply Display Themes to a drawing.

- Import property sets for export to Green Building Studio web service.

PLACING ANNOTATIONS ON A DRAWING

The content of annotation is unique to a drawing; therefore, it is usually placed on the View drawings in a project. The scale of View drawings is specified in the properties of the view when placed on a sheet. Prior to the placing of annotations in any project or non-project drawing, the annotation scale and display configuration must be set in the Drawing Window status bar.

The commands of the Annotate tab and the **Annotation** palette shown in Figure 11.1 represent a sample of the commands available in the Documentation Tool Catalog – Imperial and the Documentation Tool Catalog – Metric of the **Content Browser**. The annotation is inserted at the scale specified in the annotation scale flyout located in the Drawing Window status bar. The majority of blocks and text located on the Annotation tool palette insert with the annotative property. The tools of the Dimensions tool palette presented later in this chapter include the annotative property.

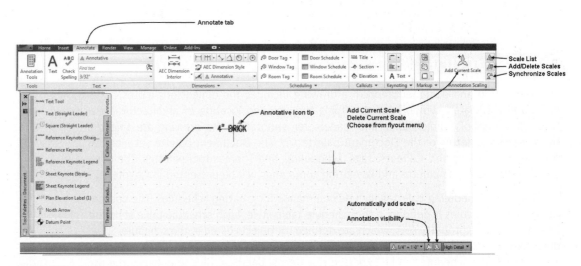

FIGURE 11.1 *Tools of the Annotate tab and the Annotation tool palette*

The scale representation of the annotative block, text, or AEC Dimension is specified and saved according to the annotation scale current for the model space, viewport, or model view. When the **Automatically add scales** toggle located in the Drawing Window status bar shown in Figure 11.1 is ON, the current scale representation of the annotative content added will be assigned. If you place annotation when **Automatically add scales** is OFF the current annotative scale will be assigned and will not resize to future scale representations as the annotation scale changes. A scale representation is saved with the annotative content each time the annotation scale changes. The commands of the Annotation Scaling panel shown in Figure 11.1 and described in Table 11.1 allow you to add or delete scale representations.

TABLE 11.1 *Annotation Scaling panel of the Annotate tab*

Ribbon	Purpose
Add Current Scale	If Automatically Add Scales is toggled OFF in the Drawing Window status bar this command opens the Annotation Object Scale dialog box, which allows you to add scale representation for the selected annotation.
Delete Current Scale	Choose Delete Current Scale from the flyout menu of the Annotation Scaling panel. The current scale representation for the selected annotation will be removed.
Scale List	Select Scale List to open the Edit Scale List dialog box. This command allows you to create new scales for annotation.
Add/Delete Scales	The Add/Delete Scales option opens the Annotation Object Scale dialog box as shown in Figure 11.2. You can choose the Add button for additional scale representations from the scale list of the drawing. Select a scale from the Object Scale List and choose Delete to remove the scale representation from the annotation.
Synchronize Scale Positions	The Synchronize Multiple Scale Positions modifies the position of scale representation consistent with the scale settings.

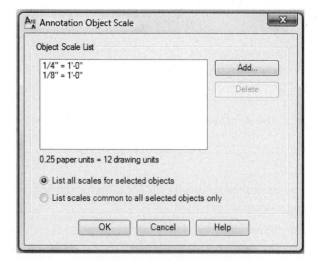

FIGURE 11.2 *Annotation Object Scale dialog box*

When **Automatically Add Scales** is ON, scale representations for each Annotative Scale are defined. Scale representations are not created when you apply the Zoom command or change the zoom using the wheel of the mouse.

When you move your cursor over a block or text that has the annotative property, a scale icon will display as shown in Figure 11.1. When the **Annotative Visibility** toggle located in the Drawing Window status bar is toggled ON only the annotation placed and assigned the scale representation that matches the current annotation scale will be displayed. The **Annotative Visibility** toggle allows you to annotate a building component within two viewports with unique annotation displayed as defined by a link between the scale representation and annotation scale.

Steps to Setting Annotation Scale

1. Choose the Annotate tab. Choose the Annotation Tools command of the Tools panel to display the **Annotation** palette.
2. Click Model toggle on the Drawing Window status bar.
3. Toggle ON **Annotation Visibility** and **Automatically Add Scales**.
4. Select a scale from the Annotation Scale flyout of the Drawing Window status bar.
5. Add annotation content from the Annotation palette.
6. Toggle OFF Automatically Add Scales.

Placing Text

The **Text Tool** tool, accessed as shown in Table 11.2, inserts Mtext scaled similarly to dimensions and leaders. The text style, font, height, and layer are preset by the command. Text is placed on the A-Anno-Note [A000T] layer. The text height is determined from the scale factor specified in **Drawing Setup**, which is multiplied by the annotation plot size of 3/32" [3.5]. Therefore, if you are placing text with the drawing scale set to 1/4" = 1'-0" [1:50], the scale factor (48) [50] is multiplied by 3/32" [3.5] to obtain a text height for the scale. The Model Text Height = 4½" for text placed using the 1/4" = 1'-0" [1:50] Annotation Scale as shown in the Properties palette of Figure 11.3.

NOTE	Text placed in the drawing from the commands located on the Text panel does not include controls for layer.

TABLE 11.2 *Text Tool command access*

Tool palette	Select Text Tool from the Annotation tool palette of the Document palette set.

When you select **Text Tool**, you are prompted to specify the location for the anticipated text as shown in the following workspace prompts:

```
Select MText insertion point: (Select a point to specify the
location of the text.)
Select text width <0>: (Press ENTER to accept text width.)
Enter first line of text <Mtext>: (Type text in the dynamic
prompt or press ENTER to open the Mtext window.)
```

The Mtext formatting window shown in Figure 11.3 allows you to create paragraph-style text and modify the style, font, and text height. The Display menu within the text window allows you to change the justification and case. The Symbol option of the Display menu provides access to special characters. After text is placed, you can double-click the text to open the **Text Formatting** toolbar and editor.

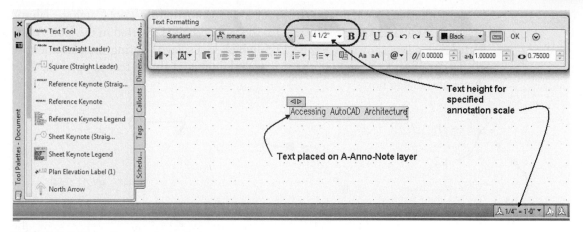

FIGURE 11.3 *Text editor for placing text*

Placing Leaders in the Drawing

Leaders are placed in a drawing to identify building components and sizes. Leaders are part of the tools used to dimension a drawing. The **Text (Straight Leader)** tool can be selected from the Annotation tool palette as shown in Table 11.3.

TABLE 11.3 *Text (Straight Leader) tool access*

Tool palette	Select Text (Straight Leader) from the Annotation palette, as shown in Figure 11.3.

When you select the **Text (Straight Leader)** tool, you are prompted in the command sequence to select the start point of the leader and additional vertices of the leader. You are prompted to specify the text width and then insert the text after a null response to the vertices prompt as shown in Figure 11.4 and the following workspace sequence:

> Specify first point of leader line: *(Specify the start point of the leader near* **p1** *as shown in Figure 11.4.)*
>
> Specify next point of leader line: *(Specify leader vertex near* **p2** *as shown in Figure 11.4.)*
>
> Specify next point of leader line: ENTER
>
> Select text width <0>: ENTER
>
> Enter first line of text <Mtext>: *(Type text in the pointer input or press* ENTER *to enter text in the Multiline Text Editor. Click* OK *to dismiss the Multiline Text Editor.)*

The geometry and text of the leader are annotative. The scale is specified in the Annotation Scale of the Drawing Window status bar. The Text (Straight Leader) consists of a multi-leader style; therefore, you can select the leader, right-click, and choose **Add Leader** to add additional leaders as shown at p3 and p4 in Figure 11.4. The position of the leader can be edited by the grips shown in Figure 11.4. Additional leaders that utilize multi-leader styles can be placed in the drawing from the Documentation Tool Catalog – Imperial or Documentation Tool Catalog – Metric of the **Content Browser** shown in Figure 11.5. The leader tools included in the

Documentation Tool Catalog – Imperial and Documentation Tool Catalog – Metric create an AutoCAD block. The text for the tag symbols is inserted in the **Edit Attributes** dialog box shown in Figure 11.5, which is displayed during the placement of the leader. To edit the text of the tag symbol double-click the text to open the Edit Attributes dialog box.

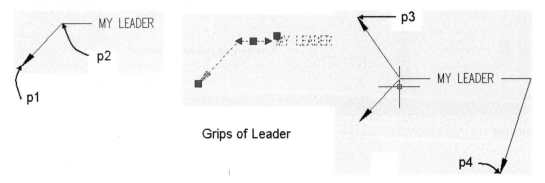

FIGURE 11.4 *Placing a leader on a drawing*

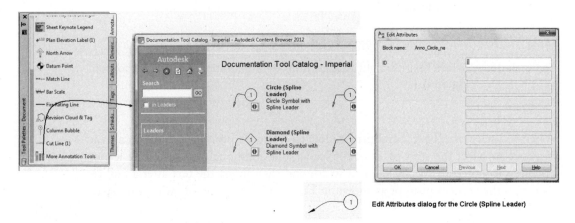

FIGURE 11.5 *Leaders category of the Documentation Tool Catalog – Imperial*

CREATING AEC LINEAR DIMENSIONS

AEC Dimensions are designed for dimensioning plan View drawings such as floor and foundation plans. When you work in a project, dimensions are usually placed in View drawings because this drawing may consist of one or more Construct drawings to create the floor plan. If you place the dimensions in the View drawing, the Construct drawings can be used in the model without the obstructions of the dimensions. The View drawing can be developed into a floor plan by attaching deck or porch plans to the main floor. Working within the View drawing, you can use Object Snap modes to locate features in the Construct drawings for the dimensions. The primary source of dimensioning tools resides on the Dimensions palette shown in Figure 11.7. Access the **DimAdd** command as shown in Table 11.4.

TABLE 11.4 *Accessing AEC Dimensions*

Command prompt	DIMADD
Ribbon	Select an AEC object and choose Dimension from the General panel of the contextual tab for the object.
Shortcut menu	Select an AEC object, right-click, and choose AEC Dimension.
Tool palette	Select AEC Dimension Exterior R.O., AEC Dimension – Exterior, or AEC Dimension – Interior from the Dimensions palette as shown in Figure 11.7. [The AEC Dimension Exterior R.O. is not included on the Dimensions palette of the metric installation.]

There are sample AEC Dimension styles (see Table 11.5) located on the Dimensions tool palette. The AEC Dimensions are annotative dimensions; therefore, additional scale representations can be assigned to the dimension. The Annotation Scale and the display configuration should be set prior to adding the dimensions. The settings such as extension line locations and the chains of the dimension are defined per display configuration. As shown in Table 11.5, the style controls how openings are dimensioned.

TABLE 11.5 *AEC Dimension Styles*

AEC Dimension Style	Chain	Content of Edit	Scale	Purpose
AEC Dimension Exterior R.O. (Imperial only)	3	Display Representation	Annotative	Exterior wall and rough openings dimensioned.
AEC Dimension – Exterior	3	Display Representation	Annotative	Exterior wall and centers of openings dimensioned.
AEC Dimension – Interior	1	Display Representation	Annotative	Interior walls dimensioned to center, left, or right edge.

The size of dimension geometry, such as text and arrowheads, in the AEC Dimension is controlled by Annotative Scale representation, whereas the content regarding points included in the dimension as specified during the edit of the AEC Dimension is defined per display representation as shown in Table 11.6.

TABLE 11.6 *Recommended Annotation Scales for display configurations*

AEC Dimensioning Style	Annotation Scale	Display Configuration
Annotative	1/4" = 1'-0" [1:50]	High Detail
Annotative	1/8" = 1'-0" [1:100]	Medium Detail, Presentation, Reflected, Reflected Screened, Screened, Standard
Annotative	1/16" = 1'-0" [1:200]	Low Detail

Linear dimensions are placed on the A-Anno-Dims [A000D] layer.

The advantage of the AEC Dimension is that, if the location or size of wall components is changed, the dimension is modified to reflect the change. Therefore, if you add, delete, or move a window in the wall, the dimensions for the object will update to reflect the change. AEC Dimensions are tied to the logical points of the AEC object. When you apply an **AEC Dimension** within a View drawing, you select the walls of a reference Construct drawing. If the walls of the Construct drawing are edited, the dimension in the View drawing will reflect the changes when the reference file is reloaded. AEC Dimensions can be applied to AEC objects or AutoCAD linework.

TIP	Choose the Dialog Box Launcher arrow on the Reference panel in the Insert tab to open the External Reference palette and reload reference files.

NOTE	Access Instructional Video **11.1 Creating an AEC Dimension** from the Instructional Video category of the Student Companion site of CengageBrain http://www.cengagebrain.com described in the Preface.

Inserting AEC Dimensions

When you select an **AEC Dimension** style tool from the Dimension tool palette, you are prompted to select an AEC object or choose the Pick point option. The AEC Dimension was applied to the wall at the left in Figure 11.6. The following workspace prompts allow you to specify the walls for dimensioning and the location of the dimension line:

> Select Objects or: *(Select the wall at* **p1** *as shown at the left in Figure 11.6.)* 1 found
>
> Select Objects or: ENTER *(Press* ENTER *to end selection.)*
>
> Specify insert point or *(Select a point near* **p2** *to specify the dimension line location.)*

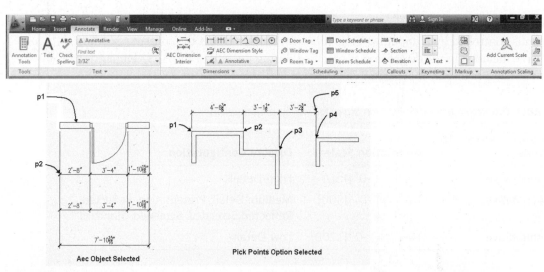

FIGURE 11.6 *Placing an AEC Dimension*

If the Pick points option is selected, you can choose the dimension points as shown in the following workspace prompts and at the right in Figure 11.6:

```
Select Objects or: p (Pick points option selected.)
pick points: (Select the intersection at p1 as shown at right
in Figure 11.6.)
pick points: (Select the intersection at p2 as shown at right
in Figure 11.6.)
pick points: (Select the intersection at p3 as shown at right
in Figure 11.6.)
pick points: (Select the intersection at p4 as shown at right
in Figure 11.6.)
pick points: ENTER (Press ENTER to end the selection.)
Specify insert point or [Style/Rotation/Align]: (Select a
point near p5 as shown at right in Figure 11.6 to locate the
dimension line.)
4 nonassociative points added
```

The dimension line aligns with the movement of the cursor relative to the objects. However, you may select Rotation or Align from the command options to specify the orientation of the dimension line. Additional objects can be selected with a crossing or window selection. If you select a wall, the windows and doors will be included in the dimension. The default dimension style applied during the dimension operation is defined by the AEC Dimension style. This style also provides control of the units and number of dimension strings.

Angular and Radial tools can be accessed from the Dimensions palette shown in Figure 11.7. Choose More Dimension Tools from the Dimensions palette to access additional dimension commands from the Miscellaneous folder of the Documentation Tool Catalog – Imperial or Documentation Tool Catalog – Metric of the Content Browser. Linear and angular dimension tools can be used to place dimensions on the A-Anno-Dims [A000D] layer that are scaled according to the viewport scale specified in **Drawing Setup**. The commands of the AutoCAD Dimensioning toolbar allow you to specify an annotative dimension style available from the Dimension Style drop-down window.

CAUTION

AutoCAD dimensioning commands are provided on the Dimensions panel as shown in Figure 11.7. AutoCAD commands place the dimensions on the current layer using the current dimensioning style settings.

Editing an AEC Dimension Style

When you insert AEC Dimensions, the style is preset in the properties of the tool. The Exterior-Center of Opening, Exterior-Rough Opening (Imperial only), and Interior Partitions styles are accessed from the tools on the Dimensions tool palette. The Standard AEC dimensioning style is included in drawings developed from the Imperial and Metric AEC Model templates. If you select a wall, right-click, and choose AEC Dimension, the dimension will be placed with the Standard style. You can select one of the AEC Dimension tools on the Dimensions tool palette to insert an AEC Dimension with a preset AEC Dimension style.

TIP	Right-click an AEC Dimension tool of the Dimensions palette, choose Re-import to import the original settings of Dimension style from the tool catalog.

The properties of the AEC Dimension tools of the tool palette can be viewed and edited if you right-click the tool and choose **Properties** as shown in Figure 11.7.

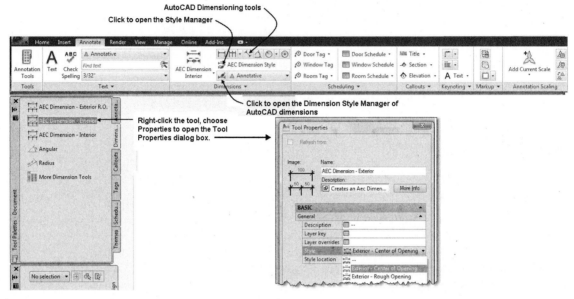

FIGURE 11.7 *Properties of AEC Dimension tools*

Additional styles can be created, or you can edit any AEC Dimension style in the Style Manager. Dimension styles are located in the *Documentation Objects > AEC Dimension Styles* folder. The dimension style determines what is included in the dimension and the number of dimension strings. The dimension string can be defined to include only selected components such as outer boundaries of the wall, wall lengths, width of wall components, wall intersections, and openings in the wall. The style of an AEC Dimension can include up to 11 dimension strings; therefore, before placing AEC Dimensions, you should verify the style in the Properties palette and refine that style to include the necessary components.

Refer to Table 11.7 for **AecDimStyle** command access.

TABLE 11.7 *Edit AEC Dimension Style command access*

Command prompt	AECDIMSTYLE
Ribbon	Select AEC Dimension Style from the Dimensions panel of the Annotate tab.
Tool palette	Right-click an AEC Dimension tool on the Dimensions palette, choose AEC Dimensions to open the Style Manager.

When you access the **AecDimStyle** command, the Style Manager opens to the *Documentation Objects > AEC Dimension Styles* folder. Additional styles are located in the AEC Dimension Styles – Imperial and AEC Dimension Styles – Metric folders of C:\ProgramData\Autodesk\ACA 2012\enu\Styles\Imperial[Metric] folders.

The **AEC Dimension Style Properties** dialog box can also be accessed directly for a selected dimension when you select the **AecDimStyleEdit** command. Access the **AecDimStyleEdit** command as shown in Table 11.8.

TABLE 11.8 *Edit AEC Dimension Style command access*

Command prompt	AECDIMSTYLEEDIT
Ribbon	Select an AEC Dimension and choose Edit Style from the General panel of the AEC Dimension tab.
Shortcut menu	Select an AEC Dimension, right-click, and choose Edit AEC Dimension Style.

In the following tutorial you will explore the settings of AEC Dimension Styles.

TUTORIAL 11-0: EXPLORING AEC DIMENSION STYLES

1. Download Chapter 11 files from the Accessing Tutor\Imperial or Accessing Tutor\ Metric category of the Student Companion site of CengageBrain **http://www. cengagebrain.com** described in the Preface to your Accessing Tutor folder.

2. Open Accessing Tutor\Imperial\Ch11\Ex 11-0.dwg or Accessing Tutor\Metric\Ch11\ Ex 11-0.dwg. Save the drawing to your student folder as Lab 11-0.dwg.

3. Choose the **Annotate** tab. Double-click the Annotation Tools command of the Tools panel to display the Document palette set.

4. Select the interior wall at **p1** as shown in Figure 11.8.

5. Choose **Dimension** from the General panel of the Wall contextual tab. Move the cursor to **p2** as shown in Figure 11.8 to specify the location of the dimension.

6. Press Escape to clear selection. Select the dimension placed at **p2** as shown in Figure 11.8. The style of this AEC dimension is listed as Standard in the Properties palette.

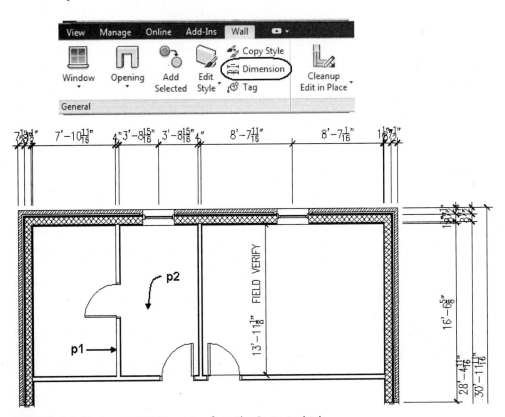

FIGURE 11.8 *Placing an AEC Dimension from the Contextual tab*

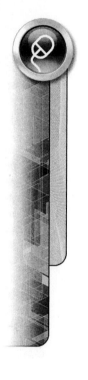

7. Choose the Dimensions palette of the Document palette group. The following three AEC dimensions are included on the palette: AEC Dimension - Exterior, AEC Dimension - Exterior R.O. (Imperial only), and AEC Dimension - Interior Dimensions.

8. Right-click the **AEC Dimension – Exterior** dimension of the Dimensions palette shown at **p3** in Figure 11.9, choose Properties from the shortcut menu to open the Tool Properties palette.

9. The AEC Dimension – Exterior tool applies the Exterior – Center of Opening AEC dimension style as shown at **p4** as shown in Figure 11.9.

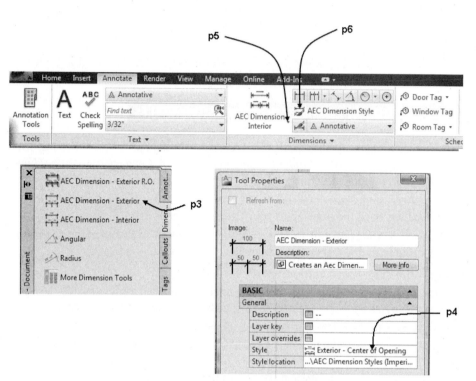

FIGURE 11.9 *AEC Dimension style of palette dimension tools*

10. Choose OK to dismiss the Tool Properties dialog box.

11. Verify the Annotate tab is current. Choose the flyout at **p5** as shown in Figure 11.9 to view the three available AEC Dimension tools.

12. Click at **p6** as shown in Figure 11.9 to open the Style Manager to view the AEC Dimension Styles of the drawing.

13. Double-click the Exterior – Center of Opening of the left pane in the Style Manager to display the tabs of the right pane.

14. The tabs of the right pane include General, Chains, Classifications, Display Properties, and Version History. The contents of the General, Classifications and Version History tabs are similar to other object styles; however, the contents of the Chains and Display Properties tabs are unique, so they are described in the following steps.

15. Choose the Chains tab of the right pane. This AEC Dimension style includes three chains as shown in Figure 11.10.

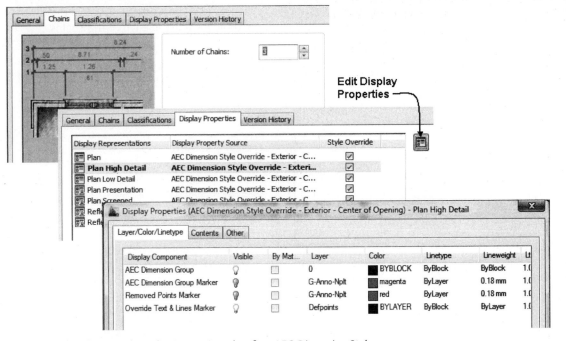

FIGURE 11.10 *Chains and Display Properties tabs of an AEC Dimension Style*

16. Choose the Display Properties tab.

17. The Display Properties tab, shown in Figure 11.10, includes a list of the Display Representations, Display Property Source, and Style Override fields. Each display representation is assigned a dimension style, as shown in Figure 11.10; therefore, the properties of the dimension style are linked to the display configuration.

18. Choose **Edit Display Properties** button as shown in Figure 11.10.

19. Choose the Layer/Color/Linetype tab as shown in Figure 11.10. This tab allows you to turn off or on the display of the components of the dimension. The AEC Dimension Group consists of the dimension chains of the dimension. The AEC Dimension Group component is displayed for all display representations. The AEC Dimension Group Marker and Remove Points Marker, if turned on, can be selected to edit the grips of the dimension. The Removed Points Marker component is usually not visible since it is used to represent points that have been removed from the chain.

20. The Override Text & Lines Marker component consists of an overline placed over text when it has been overridden as shown in Figure 11.11.

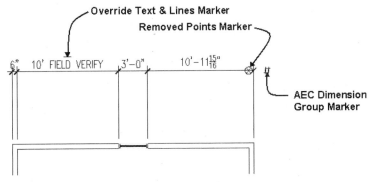

FIGURE 11.11 *AEC Dimension Group Marker and Removed Points Marker display toggled ON*

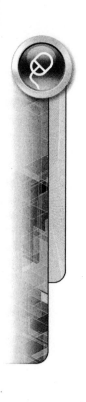

21. Choose the Contents tab as shown in Figure 11.12. The Contents tab is the most significant tab because it lists the contents to be included in each of the chains.

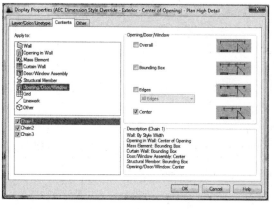 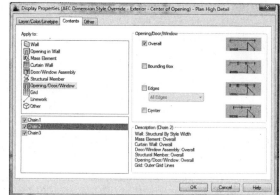

FIGURE 11.12 *Contents tab of an AEC Dimension Style*

22. The **Apply to** window lists the objects included in each of the selected chains. This tab is edited for a selected chain by selecting an object in the Apply to window and then editing the aspects of the object to be included in the dimension at the right.

23. Select Chain 1 and choose the Opening/Door/Window object of the Apply to list. The checkboxes located in the right panel refer to the building object dimension points for the object selected in the Apply to window at left. Notice the Center is checked. Therefore the chain 1 will be used to locate the Center of openings, doors, and windows.

24. Choose Chain 2 listed in the lower left pane, notice only the Overall component is checked for dimensioning the Opening/Door/Window objects with this chain.

25. Choose the Wall object of the Apply to window. Choose Chain 1, notice the Wall Width is checked. Choose the flyout of the Wall Width.

26. The **Wall width** options include: Overall, All Components, By Style, Structural By Style, and Center. The wall width is defined By Style. The By Style option references the settings of the wall style. In the next step you will edit the wall style to change the components included in the dimension. Choose OK to dismiss all dialog boxes.

27. Select the wall at **p1** as shown in Figure 11.13. Choose Edit Style from the Edit Style flyout of General panel. Choose the Components tab. Clear the check box for GWB as shown in Figure 11.13.

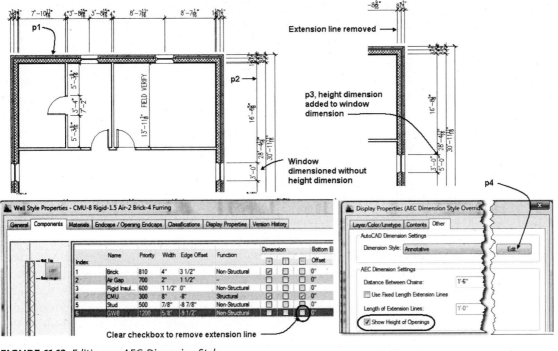

FIGURE 11.13 *Editing an AEC Dimension Style*

28. Choose OK to dismiss all dialog boxes and view the dimension without the extension line locating the GWB.

29. An AEC Dimension specified as **Structural By Style** will only include components defined as structural in the wall style definition. The **Center** option will dimension to the wall center without respect to dimension or structural settings within the wall style.

30. Select the AEC Dimension shown at **p2** in Figure 11.13. Choose Edit Style from the Edit Style flyout of the General panel of the AEC Dimension tab. Choose the **Display Properties** tab. Choose the **Edit Display Properties** button. Choose the **Other** tab. Check the **Show Height of Opening** to specify the height of openings with the dimension string as shown at **p3** in Figure 11.13.

31. The Other tab lists the AutoCAD Dimension style used in the AEC Dimension. Choose the Edit button shown at **p4** in Figure 11.13. Open the Edit Dimension Style: Annotative dialog box. This option allows you to modify the dimension style.

32. Choose OK to dismiss all dialog boxes.

33. All AEC Dimensions utilize the **Annotative** dimension style; therefore, when the annotation scale changes, the size of text and other geometry associated with the dimension change proportionately as required by the annotation scale. In previous releases of Architectural Desktop the size of the geometry of AEC Dimensions was controlled by the dimension style defined for a display representation. When you add or remove extension lines for an AEC Dimension, the edits are saved to the AEC Dimension that is current for the Display Representation and Display Configuration. Therefore in the next step you will set the Display Configuration and Annotation Scale in the Drawing Window status bar.

34. Choose the **1/4" = 1'-0" [1:50]** scale from the Annotation Scale flyout menu and **High Detail** from the Display Configuration flyout menu of the Open Drawing Window status bar.

35. Verify that Automatically Add Scales and Annotation Visibility are toggled ON in the Drawing Window.

36. Save and Close the drawing.

Editing AEC Dimensions

The features included in an AEC Dimension can be defined in the AEC Dimension Style dialog box; however, grips of the dimension provide additional editing. The grip locations and functions of AEC Dimensions differ from the grips of AutoCAD dimensions. In addition to the grips, the shortcut menu of a selected AEC Dimension allows you to add or remove objects, add or remove extension lines, edit in place, and override the text.

Grip Editing AEC Dimensions

When you select an AEC Dimension, the **Move All Chains** and **Edit in Place** trigger grips are displayed as shown in Figure 11.14. When you select the **Move All Chains** grip, you can stretch the dimension group to a different location. If you select the **Edit In-Place** trigger grip, the grips of the dimension are displayed as shown at the right in Figure 11.14. You can select the grips of the text and move the text. The grips allow you to edit the location of the text, the location of each dimension line, and the location of each extension line. When you edit the grips, an override to the style is created. When you finish editing the grips, reselect the **Edit In-Place/Exit Edit In-Place** trigger grip to end the edit and save changes. You can choose Edit in Place from the ribbon or the shortcut menu to open an Edit in Place session.

Edit in Place changes to an AEC Dimension can be removed by selecting from the Remove Override options of its shortcut menu or the ribbon (see Table 11.9). If text locations have been edited choose the **All Text Positions** option of **Remove Override** shown in the Modify panel to return the text to its location just prior to the edit in place operation. The **All Components** option of the **Remove Overrides** will return text and extension lines to their locations prior to the edit in place operation. The **All Extension Lines** option of **Remove Overrides** will remove modifications to the extension lines prior to the edit in place operation. The Reverse command located on the Modify panel allows you to reverse or flip the wall component dimensioned. This command requires dimensioned points located on the interior and exterior sides. Wall components are specified for the positive or negative side of the component in the wall style.

TABLE 11.9 *Editing AEC Dimensions command access*

Remove Override	Select a dimension that includes text position edits, right-click, and choose Remove Override > All Text Positions.
	Select a dimension that includes extension line edits, right-click, and choose Remove Override > All Extension Lines.
	Select a dimension that includes extension line and text edits, right-click, and choose Remove Override > All Components.
Edit in Place	Select an AEC Dimension and choose Edit in Place from the Modify panel.
Reverse	Select an AEC Dimension that includes interior dimension points specified for positive or negative component side and choose Reverse from the Modify panel.

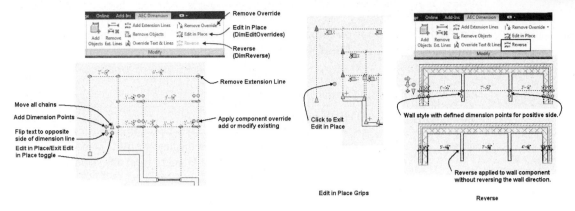

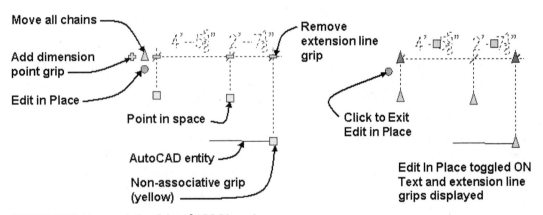

FIGURE 11.14 *Grip editing of AEC Dimensions*

Nonassociative AEC Dimensions are placed when you choose the Pick points option without choosing an object or entity. The grips of a nonassociative extension line include a yellow square grip at the origin of the points selected. You can move, mirror, rotate, scale, and stretch this grip, and the dimension will update to the new location. The Pick points option of an AEC Dimension was used to select points not associated with entities or objects in Figure 11.15.

FIGURE 11.15 *Nonassociative Grips of AEC Dimensions*

Editing AEC Dimensions with the Ribbon and Shortcut Menu

The AEC Dimension contextual tab includes General, Modify, and Annotation Scaling panels. The Modify panel includes the following options: **Add Objects**, **Remove Objects**, **Add Extension Lines**, **Remove Extension Lines**, **Override Text & Lines**, **Edit in Place**, and **Remove Override**. The General panel includes the Edit Style tool which can be selected to edit the style in the AEC Dimension Style Properties dialog box. The Save As tool allows you to copy the style of the AEC Dimension and assign the style to the selected AEC Dimension. The shortcut menu of a selected AEC Dimension includes the Edit options of the Modify panel.

Prior to editing AEC Dimensions, verify that the display configuration and the Annotation Scale of the drawing are set as per Table 11.6.

CAUTION

Attaching Building Objects to AEC Dimensions

The **Add Objects** option allows you to add building objects to an existing AEC Dimension. Access the **DimObjectsAdd** command as shown in Table 11.10.

TABLE 11.10 *DimObjectsAdd command access*

Command prompt	DIMOBJECTSADD
Ribbon	Select an AEC Dimension to display the AEC Dimension contextual tab. Choose Add Objects from the Modify panel as shown in Figure 11.16.
Shortcut menu	Select an AEC Dimension, right-click, and choose Add Objects.

The wall at **p1** was attached to the AEC Dimension as shown in the following workspace sequence:

(Select the dimension and choose **Add Objects** *from the Modify panel of the AEC Dimension tab.)*

    Select Building Elements: 1 found *(Select the wall at* **p1** *as shown at the left in Figure 11.16.)*

    Select Building Elements: ENTER *(Press* ENTER *to end selection.)*

    (Dimensions are added as shown at the right in Figure 11.16.)

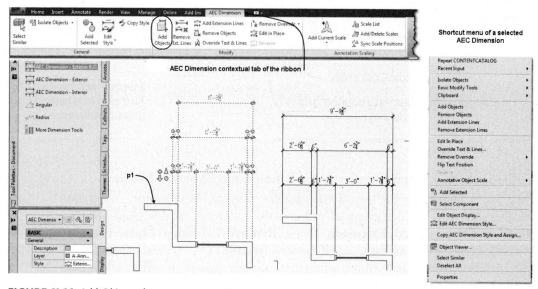

FIGURE 11.16 *Add Objects shortcut menu option*

Removing Building Objects

Building objects attached to an AEC Dimension can be removed. See Table 11.11 for **DimObjectsRemove** command access.

TABLE 11.11 *DimObjectsRemove command access*

Command prompt	**DIMOBJECTSREMOVE**
Ribbon	Select an AEC Dimension and choose Remove Objects from the Modify panel.
Shortcut menu	Select an AEC Dimension, right-click, and choose Remove Objects.

The command prompts shown below removed the building objects from the dimension as shown at the right in Figure 11.17:

(Select the dimension and choose **Remove Objects** *from the Modify panel of the AEC Dimension tab.)*

Select the AEC Dimension shown in Figure 11.17. Choose Remove Objects from the Modify panel of the AEC Dimension contextual tab.

Select Building Elements: *(Select the wall at* **p1** *as shown at the left in Figure 11.17.)*

Select Building Elements: ENTER 1 Removed *(Press ENTER to end the selection.) (The dimension is removed as shown at the right in Figure 11.17.)*

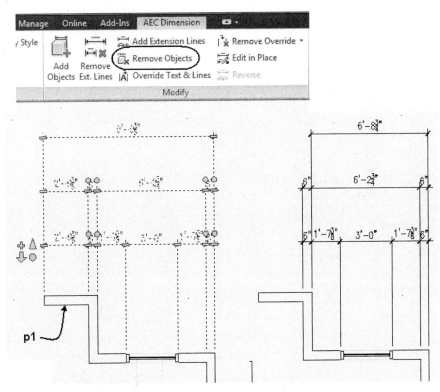

FIGURE 11.17 *Points removed from an AEC Dimension*

If you type **DIMOBJECTSREMOVE** in the command line, you must select the dimension group prior to selecting the building elements. However, if you choose **Remove Objects** from the shortcut menu of the dimension, the selection of the dimension group is not required.

Adding Extension Lines

Additional extension lines can be added to building objects or at locations not associated with objects or entities. If the point selected is placed on an object or entity, it is associative to the dimension, but if a point in free space is chosen, it is nonassociative. The grip of nonassociative points is yellow. Access the **DimExtLinesAdd** command as shown in Table 11.12.

TABLE 11.12 *DimExtLinesAdd command access*

Command prompt	DIMEXTLINESADD
Ribbon	Select an AEC Dimension and choose Add Extension Lines from the Modify panel.
Shortcut menu	Select an AEC Dimension, right-click, and choose Add Extension Lines.

The point at the left end of the wall shown at the left in Figure 11.18 was added to the dimension group using the following command sequence:

(Select the dimension, choose **Add Extension Lines** *from the Modify panel of the AEC Dimension tab.)*

        Pick points: *(Select a point at* p1 *as shown at the right in Figure 11.18.)*

        Pick points: ENTER *(Press ENTER to end the selection.)*

        Select Dimension chain: *(Select the dimension chain at* p2 *as shown at the right in Figure 11.18.)*

        1 added

        (The dimension point is added to the chain as shown at the right.)

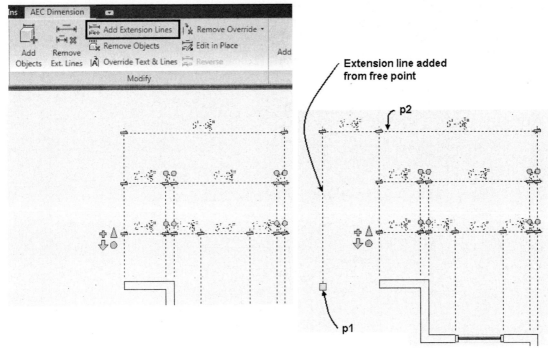

FIGURE 11.18 *Extension line added to an AEC Dimension*

Removing Extension Lines

Extension lines can be removed to revise the dimension to include only selected points on the building objects. An extension line can be removed by selecting the Remove extension line grip or by Remove Extension Lines (**DimExtLinesRemove**) command. Access the **DimExtLinesRemove** command as shown in Table 11.13.

TABLE 11.13 *DimExtLinesRemove command access*

Command prompt	**DIMEXTLINESREMOVE**
Ribbon	Select an AEC Dimension choose Remove Ext. Lines from the Modify panel of the AEC Dimension tab.
Shortcut menu	Select an AEC Dimension, right-click, and choose Remove Extension Lines.

The points representing the width of the wall shown at the left in Figure 11.19 were deleted from the dimension group as shown in the following workspace sequence:

> *(Select the dimension and choose* Remove Ext.
>
> Lines *from the Modify panel of the AEC Dimension tab.)*
>
> Select extension lines to remove dimension points: 1 found
> *(Select the extension line at* p1 *as shown in Figure 11.19.)*
>
> Select extension lines to remove dimension points: 1 found
> *(Select the extension line at* p2 *as shown in Figure 11.19.)*

Select extension lines to remove dimension points: ENTER
(Press ENTER to end the selection.)

(The dimension is revised as shown at the right in Figure 11.19.)

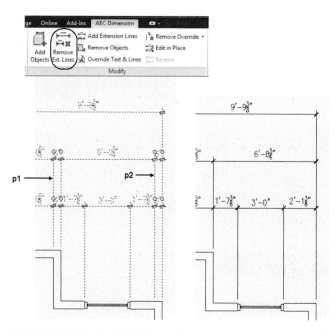

FIGURE 11.19 *Revised AEC Dimensions*

Overriding the Text of AEC Dimensions

The display and contents of the text used in an AEC Dimension can be overridden. Access the **DimTextOverride** command as shown in Table 11.14.

TABLE 11.14 *Override Text & Lines command access*

Command prompt	DIMTEXTOVERRIDE
Ribbon	Select an AEC Dimension choose Override Text & Lines from the Modify panel of the AEC Dimension tab.
Shortcut menu	Select an AEC Dimension, right-click, and choose Override Text & Lines.

The **DimTextOverride** command opens the **Override Text & Lines** dialog box, which allows you to add notes or change the dimension text. The "FIELD VERIFY" note was added in the following workspace sequence to change the dimension shown in Figure 11.20 at the right.

Select the AEC Dimension and choose Override Text & Lines from the Modify panel.

Select dimension text to change: *(Select a dimension to edit.)*

(The Override Text & Lines *dialog box opens as shown in Figure 11.20.)*

[hide Text/hide text and Lines/Override text/Underline/ Prefix/ Suffix/Remove override]: DBOX

(Type Field Verify *in the* Text Override *field of the Override Text & Lines dialog box. Click* OK *to dismiss the dialog box. To indicate that the text is overridden, a line is displayed above the edited text.)*

The overline can be turned off by turning off the **Override Text & Lines Marker** display component in the Display Properties dialog box of the dimension style.

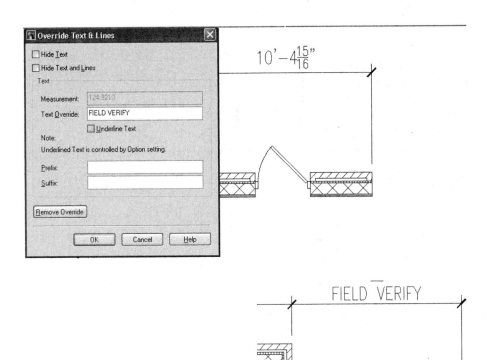

FIGURE 11.20 *Overriding the text of the AEC Dimension*

Using the AEC Dimension Style Wizard

The **AEC Dimension Style Wizard** allows you to change display features for each AEC Dimension Style of the drawing. Refer to Table 11.15 to access the **AEC Dimension Style Wizard** command.

TABLE 11.15 *AEC Dimension Wizard command access*

Command prompt	**AECDIMWIZARD**
Ribbon	Select AEC Dimension Display Wizard from the Edit Style flyout of the General panel.

The **AEC Dimension Display Wizard** dialog box consists of four pages as shown in Figures 11.21 and 11.22. The **AEC Dimension Display Wizard** dialog box allows you to change the display representation, layer, size, and color of the geometry used in the AEC Dimension. The pages of the **AEC Dimension Display Wizard** dialog box are described in the following list:

Select Style—The Select Style page includes a drop-down list of AEC Dimension styles of the drawing.

Lines and Arrows—The Lines and Arrows page allows you to specify the display representation of arrowhead block dimension and extension lines. The distance between the dimension chains and the distance the extension lines extend beyond the dimension line can be specified as shown in Figure 11.21.

Text—The Text page allows you to specify for the display representation the text style, text height, and text round off values as shown in Figure 11.22.

Color and Layer—The Color and Layer page allows you to specify for the display representation color of the text dimension lines, extension lines, and the layer assignment. If the layer is set to Layer 0 the layer key assigns the layer.

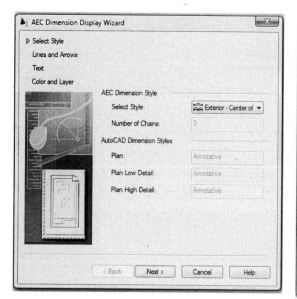
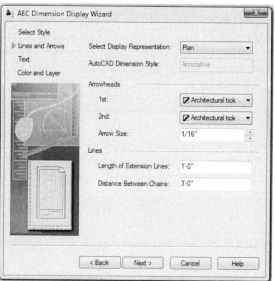

FIGURE 11.21 *Selecting Style and Lines and Arrows pages of the AEC Dimension Display Wizard dialog box*

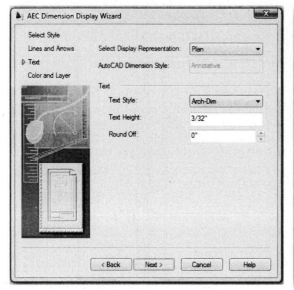
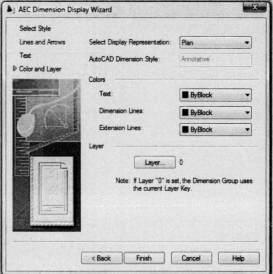

FIGURE 11.22 *Text and Color and Layer pages of AEC Dimension Display Wizard dialog box*

TUTORIAL 11-1: INSERTING AEC DIMENSIONS

1. Verify that the Accessing Tutor\Imperial or Accessing Tutor\Metric content for Chapter 11 of the CengageBrain has been downloaded to your Accessing Tutor folder on your computer as described in Organizing Tutorial Directories in the Preface.

2. Open Accessing Tutor\Imperial\Ch11\Ex 11-1.dwg or Accessing Tutor\Metric\Ch11\Ex 11-1.dwg. Save the drawing to your student folder as Lab 11-1.dwg.

3. Right-click the Object Snap toggle of the **Application** status bar to open the **Drafting Settings** dialog box. Check only the **Endpoint** object snap mode. Click **OK** to dismiss the dialog box. Toggle ON Dynamic Input in the Application status bar.

4. Choose the **Annotate** tab. Double-click the Annotation Tools command of the Tools panel to display the Document palette set.

5. Choose the **1/4″ = 1′-0″ [1:50]** scale from the Annotation Scale flyout menu and **High Detail** from the Display Configuration flyout menu of the Open Drawing Window status bar.

6. Verify that Automatically Add Scales and Annotation Visibility are toggled ON in the Drawing Window status bar.

7. Choose **AEC Dimension – Exterior** [AEC Dimension – Exterior] from the Dimensions tool palette. Respond to the workspace prompts as follows:

   Select Objects or: *(Select the wall at* **p1** *as shown in Figure 11.23 to specify the dimension line alignment.)*

   Select Objects or: *(Select the wall at* **p2** *as shown in Figure 11.23.)*

   Select Objects or: *(Select the wall at* **p3** *as shown in Figure 11.23.)*

   Select Objects or: *(Right-click and click* ENTER.*)*

   Specify insert point or: *(Choose a point near* **p4** *to specify the dimension line location as shown in Figure 11.23.)*

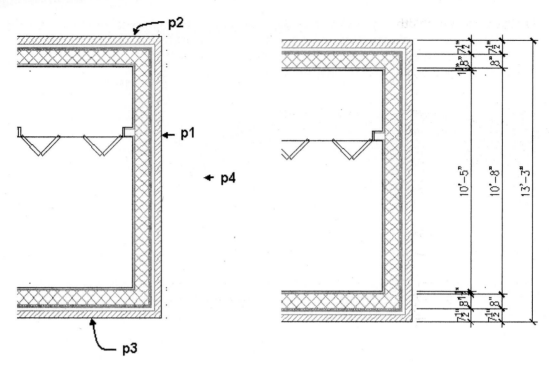

FIGURE 11.23 *AEC Dimension – Exterior placed*

8. Select the AEC Dimension placed in step 6 and choose **Edit Style** from the General panel of AEC Dimension tab. Choose the **Display Properties** tab of the AEC Dimension Style Properties – Exterior – Center of Opening dialog box. Verify that **Plan High Detail** is selected and choose the **Edit Display Properties** button at the right. Choose the **Contents** tab. Choose **Chain 1** and **Wall** in the Apply to window as shown in Figure 11.24. Verify that the **Wall Width** is specified as **By Style**. Click **OK** to dismiss all dialog boxes.

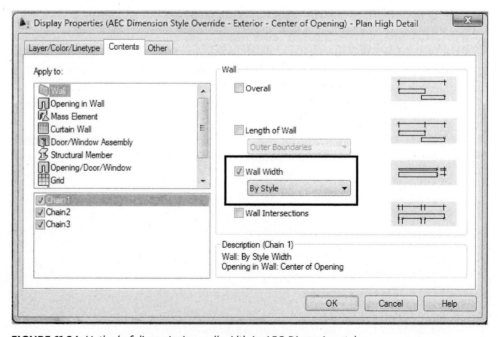

FIGURE 11.24 *Method of dimensioning wall width in AEC Dimension style*

9. Since the **Wall Width** is specified as **By Style** you can control the content of the dimension in the wall style. Select the exterior wall, right-click, and choose **Edit Wall Style**. Choose the **Components** tab and clear the **GWB** checkbox as shown in Figure 11.25. Click **OK** to dismiss all dialog boxes.

Index	Name	Priority	Width	Edge Of...	Function	Dimension			Bottom Ele
						▣	▣	▣	Offset
1	Brick	810	4"	3 1/2"	Non-Struct...	☑	☐	☐	0"
2	Air Gap	700	2"	1 1/2"	--	☐	☐	☐	0"
3	Rigid Insu...	600	1 1/2"	0"	Non-Struct...	☐	☐	☐	0"
4	CMU	300	8"	-8"	Structural	☑	☐	☑	0"
5	Stud	500	7/8"	-8 7/8"	Non-Struct...	☐	☐	☐	0"
6	GWB	1200	5/8"	-9 1/2"	Non-Struct...	☐	☐	☑	0"

Clear the GWB

Gypsum wall board extension line

Gypsum Wall Board extension line removed

FIGURE 11.25 *Removing GWB from the dimension definition in the wall style*

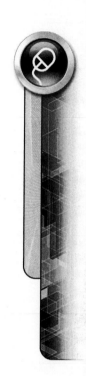

10. Select the **AEC Dimension − Interior** tool from the **Dimensions** tool palette. Respond to the workspace prompts as follows:

> Select Objects or: *(Select the wall at* **p1** *as shown in Figure 11.26 to specify the dimension line alignment.)*
>
> Select Objects or: *(Create a crossing selection by selecting a point at* **p2** *as shown in Figure 11.26.)*
>
> Specify opposite corner: *(Select a point at* **p3** *as shown in Figure 11.26.)*
>
> Select Objects or: *(Right-click and press* ENTER.*)*
>
> Specify insert point or: *(Choose a point near* **p4** *to specify the dimension line location as shown in Figure 11.26.)*

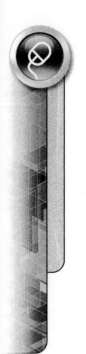

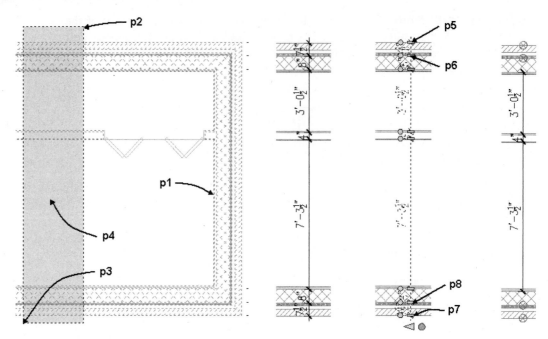

FIGURE 11.26 *Applying the Interior AEC Dimension*

11. Select the Interior dimension placed in step 9 to display the grips. Select the **Remove Extension Line** grip at points **p5**, **p6**, **p7**, and **p8** as shown in Figure 11.26. Remove Points Markers will be displayed since extension lines have been removed; however, the points will not plot since they are placed on the G-Anno-Nplt layer. Press ESC to clear the grips.

12. Select the **AEC Dimension – Interior** from the **Dimensions** tool palette. Respond to the workspace prompts as follows:

Select Objects or: *(Choose the down arrow and choose the* Pick points *option from the workspace.)*

Pick points: *(Choose the Endpoint of the column at* p1 *as shown in Figure 11.27.)*

Pick points: *(Choose the Endpoint of the cabinet at* p2 *as shown in Figure 11.27.)*

Pick points: *(Right-click)*

Specify insert point or: *(Select a point near* p3 *as shown in Figure 11.27.)*

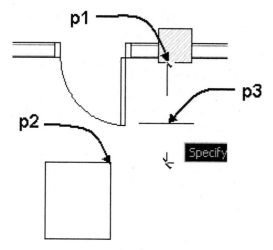

FIGURE 11.27 *Locating a multi-view block using AEC Dimensions*

13. Choose the 1/8″ = 1′-0″ [1:100] Annotation Scale from the Drawing Window status bar. The height of text changes to reflect the scale representation.

14. Choose the Medium Detail display configuration from the Drawing Window status bar. The new display configuration removes the previous extension line edits.

15. Choose High Detail display configuration and 1/4″ = 1′-0″ [1:50] Annotation Scale. Select the exterior dimension and each of the interior dimensions, right-click, and choose Annotative Object Scale > Add/Delete Scales to open the Annotation Object Scale dialog box. Choose the 1/8″ = 1′-0″ [1:100] from the Object Scale List and click the Delete button at the right to remove this scale representation from the dimensions. Click OK to dismiss the Annotation Object Scale dialog box.

16. Save and close the drawing.

PLACING STRAIGHT CUT LINES

Straight cut lines are placed in the drawing by selecting **Cut Line (1)** from the Annotation tool palette (refer to Table 11.16). When you select **Cut Line (1)**, you are prompted to select the start and end of the straight cut line. After specifying the endpoints of the cut line, you are prompted to select the break line extents, which specify the content to mask as shown in Figure 11.28. The symbol is not an annotative block and the size of the Z portion is scaled by the value of the Annotation Scale displayed in the Drawing Window status bar.

TABLE 11.16 *Cut Line (1) tool command access*

Tool palette	Select Cut Line (1) from the Annotation tool palette.

The following workspace sequence trimmed the walls shown in Figure 11.28 when the Cut Line (1) tool was applied:

```
Specify first point of break line: (Select a point near p1 as
shown at the left in Figure 11.28.)

Specify second point of break line: (Select a point near p2 as
shown at the left in Figure 11.28.)
```

Specify break line extents: *(Select a point near* **p3** *as shown at the left in Figure 11.28.)*

(Walls are trimmed as shown at the center in Figure 11.28.)

After placing the break you can select the cut line using a crossing selection as shown at the right in Figure 11.28 to display its grips and modify the cut line and contents of the break. Choose More Annotation Tools to access additional break and cut lines from the Break Marks category of the Documentation Tool Catalog – Imperial and the Documentation Tool Catalog – Metric catalogs of the **Content Browser**.

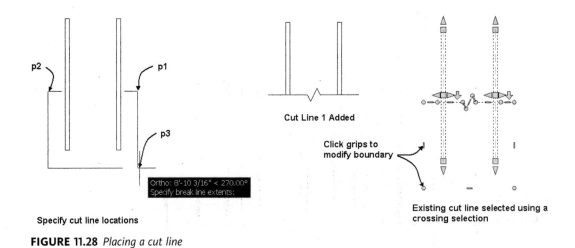

Specify cut line locations

FIGURE 11.28 *Placing a cut line*

REVISION CLOUDS

The revision cloud allows you to identify an area of the drawing that has been revised or add a note that applies to this portion of the building. The revision cloud is created freestyle with a series of arcs connected to enclose an area of the drawing. Access the **Revision Cloud & Tag** tool from the Annotation palette of the Document palette group. Refer to Table 11.17 for command access.

TABLE 11.17 *Revision Cloud & Tag tool access*

Tool palette	Select Revision Cloud & Tag from the Annotation tool palette of the Document palette group.

When you select **Revision Cloud & Tag** from the Annotation tool palette, the width of the arc is adjusted according to the Annotation Scale factor defined by the Drawing Setup. The cloud is drawn with arcs scaled to the Annotation Scale. To place a revision cloud, select **Revision Cloud & Tag** from the Annotation tool palette and then specify the start point. Move the cursor from the start point in a circular direction. When the cursor is returned near the start point, the final arc will close and you will be prompted to specify the callout location. Additional styles of revision clouds can be inserted from the Revision Clouds folder of Documentation Tool Catalog – Imperial or Documentation Tool Catalog – Metric catalogs of the Content Browser.

The following workspace sequence was used to create the revision cloud shown in Figure 11.29:

> Specify cloud starting point or: *(Select point* p1 *in Figure 11.29 to start the revision cloud.)*
>
> *(Move the cursor in a counterclockwise direction as shown in Figure 11.29 to form the enclosure and close the revision cloud.)*
>
> Specify centerpoint of revision tag <None>: *(Select a point near* p2 *as shown in Figure 11.29.)*
>
> *(Type the number of the detail in the* **Edit Attributes** *dialog box; click* **OK** *to dismiss the dialog box.)*

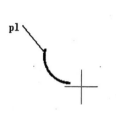 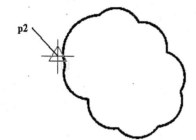 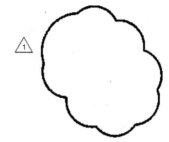

FIGURE 11.29 *Creating a revision cloud*

> Revision clouds should be drawn counterclockwise. Revision clouds drawn in a clockwise direction will create arc endpoints pointing out from the enclosure.
>
> **NOTE**

INSERTING CHASES

The **Chase** symbol can be used to represent chimneys, ducts, or wall niches. Representations of chases can be placed in the drawing from the **Chase** category of the Documentation Tool Catalog – Imperial [Documentation Tool Catalog – Metric]. Refer to Table 11.18 and Figure 11.30.

TABLE 11.18 *Chases command access*

DesignCenter	Select the AEC Content tab of the DesignCenter and select *AutoCAD Architecture\Imperial\Documentation\Chases* or *AutoCAD Architecture\Metric\Documentation\Chases*.
Content Browser	Open the Documentation Tool Catalog – Imperial or Documentation Tool Catalog – Metric and choose the Chases category.

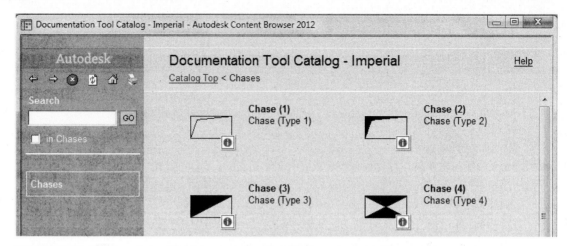

FIGURE 11.30 *Chase symbols of the Documentation Tool Catalog – Imperial Chase category*

When you drag the i-drop for a chase symbol from the **Content Browser**, you are prompted to specify the insertion point and the **Add MV-Block with Interference** dialog box opens. The interference of the chase with walls, spaces, slabs, and roof slabs can be specified as additive, subtractive, and ignore. The ignore interference condition is active only for walls. The size and description of the chase is specified in the **Add MV-Block with Interference** dialog box as shown in Figure 11.31. The size is specified by editing the scale values of the unit block in the **Add MV-Block with Interference** dialog box.

The options of the Add MV-Block with Interference dialog box are described in the following list:

Description—A Description can be typed in this field to describe the chase.

INSERTION POINT:

Pick Point—Select the Pick Point button to specify on screen the insertion point.

X—The X absolute coordinate value can be typed to specify the X location of the insertion point of the chase.

Y—The Y absolute coordinate value can be typed to specify the Y location of the insertion point of the chase.

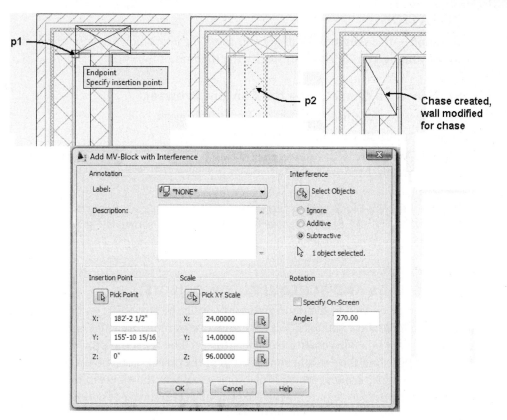

FIGURE 11.31 *Add MV-Block with Interference dialog box*

Z—The Z absolute coordinate value can be typed to specify the Z location of the insertion point of the chase.

SCALE:

Pick XY Scale—Clicking the **Pick XY Scale** button returns you to the workspace. When you return to the workspace the cursor rubber-bands from the insertion point and you can select a point diagonally located from the insertion point to size the chase.

X—Specify the X scale dimension to size the unit block. Therefore, if 24 is typed in this field the chase will be 24 units wide in the X direction.

Y—Specify the Y scale factor to size the unit block in the Y direction.

Z—Specify the Z scale factor to size the unit block in the Z direction.

INTERFERENCE:

Interference Select Objects—The Select Objects button of the Interference section returns you to the drawing area, allowing you to select AEC objects that may interfere with the chase. **Ignore Additive** and **Subtract** Boolean operations can be applied to the interference between the chase and the selected AEC objects. The chase inserted as shown in Figure 11.31 is applied to subtract from the wall and the shrinkwrap is excluded. The **Ignore** option breaks the wall and the shrinkwrap is applied without regard to the chase, whereas the **Additive** option breaks the wall and includes the shrinkwrap with the chase.

ROTATION:

Rotation—The **Angle** field allows you to type the rotation angle or check the **Specify on Screen** checkbox to specify the rotation of the chase.

Steps to Insert a Chase Using the Content Browser

1. Drag the i-drop for **Chase (6)** from the **Content Browser**.
2. Select an insertion point at **p1** as shown in Figure 11.31.
3. The **Add MV-Block with Interference** dialog box opens; click the **Select Objects** button and select the wall at **p2** as shown in Figure 11.31. Press ENTER to end the selection.
4. Edit the **Add MV-Block with Interference** dialog box: specify Rotation = 270 and Y Scale value = 1'-2". Specify Subtractive interference as shown in Figure 11.31.
5. Click **OK** to close the dialog box; the chase is created as shown in Figure 11.31.

CREATING TAGS AND SCHEDULES FOR OBJECTS

Schedules can be created to list doors, windows, walls, or any object in the drawing. The schedule tools allow you to collect data from one or more drawings and to export the schedule data to a Microsoft Excel spreadsheet. The schedule provides detailed information regarding the size and construction of the object. Each object of a schedule is usually assigned a mark, which is placed as a tag near the object in the drawing. Schedules can be used to determine the quantity of a certain type of building component repeated throughout the building.

Tags and schedules can be developed within Construct drawings, or if project-based tags are used, the schedule can be developed based upon all drawings of the project. When you insert a tag for an object, the **properties** from the object associated with the **tag** are inserted in the drawing. The tag is a multi-view block with property sets and properties defined in its attributes as shown in Table 11.19.

TABLE 11.19 *Property Sets and Properties inserted as tags of attributes*

Tag	Property Set	Property
Door Tag	DOOROBJECTS	NUMBER
	DOORSTYLE	TYPE
	FRAMESTYLES	TYPE
	MANUFACTURERSTYLES	COST
Door Tag – Project Based	DOOROBJECTS	ROOMNUMBER
	DOOROBJECTS	NUMBERSUFFIX
	DOORSTYLE	TYPE
	FRAMESTYLES	TYPE
	MANUFACTURERSTYLES	COST
Window Tag	WINDOWOBJECTS	NUMBER
	WINDOWSTYLES	TYPE
Room Tag	SPACEOBJECTS	NAME
	SPACEOBECTS	NUMBER

Continued

TABLE 11.19 *Continued*

Tag	Property Set	Property
Room Tag – Project Based	SPACEOBJECTS	NAME
	SPACE OBJECTS	NUMBERPROJECTBASED
	ROOMFINISHOBJECTS	BASECOLOR
Room Tag BOMA	SPACEOBJECTS	NUMBER
	SPACEOBJECTS	NAME
	SPACESTYLES	NETAREA
Wall Tag (Leader)	WALLSTYLES	TYPE
	WALLOBJECTS	STYLE

The data defined for the **properties** are extracted and placed in the schedule when the schedule is inserted. The content of each schedule table is defined in the schedule table style. The content of a door schedule table is shown in the **Columns tab** of the **Schedule Table Style Properties** dialog box for a Door Schedule Style as shown in Figure 11.32. The content includes properties and property sets. A property set is a group of properties. Property set data of an object is displayed on the Extended Data tab of the Properties palette. Schedule table styles are located in the **Schedule Table Styles** folder of **Documentation Objects** in the Style Manager. Select a schedule table style in the left pane of the Style Manager and choose the **Columns** tab to view the properties and property sets specified for the schedule table. The Property Sets of the sample schedule tables are shown in Table 11.20.

TABLE 11.20 *Property Sets included in sample schedules*

Schedule	Property Set
Door Schedule	DoorObjects
	DoorStyles
	FrameStyles
Door Schedule Project Based	DoorObjects
	DoorStyles
	FrameStyles
Window Schedule	WindowObjects
	WindowStyles
Wall Schedule	WallStyles
	WallObjects
Space List BOMA	SpaceObjects
Space Inventory	GeoObjects
	SpaceStyles
Room Finish Schedule	SpaceObjects
	RoomFinishObjects
	SpaceStyles

Properties consist of data such as height, width, hardware, or materials. Some data for the properties are **Automatic**, extracted from the object in the drawing. However, other data are **Manual**, inserted by editing the **Edit Property Set Data** dialog box. Manual and Automatic data are shown in Figure 11.32. When a schedule table is inserted, the property data from the property sets of the drawing is extracted and included in the schedule.

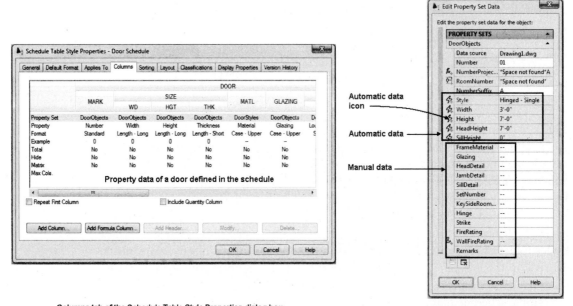

Columns tab of the Schedule Table Style Properties dialog box

Property data of door displayed in the Edit
Property Set Data dialog box when door is tagged

FIGURE 11.32 *Property sets of the Schedule Table Style Properties dialog box*

The AEC Content tab of the Options dialog box allows you to toggle ON Display Edit Property Data Dialog During Tag Insertion. Therefore, upon insertion of the tag you can enter the data for the schedule in the Edit Property Set Data dialog box.

Door and window schedules are located on the Scheduling tool palette and sample tags are located on the Tags palette and Scheduling panel of the Annotate tab as shown in Figure 11.33. Additional Schedule tables and tags can be accessed from tool catalogs when you choose More Scheduling Tools and More Tag Tools. Schedule tables included in the tool catalogs can be imported within the Style Manager from *Schedule Table (Imperial).dwg* or *Schedule Table (Metric).dwg* of the Imperial or Metric folders in the *{Vista – C:\ProgramData\Autodesk\ACA 2012\enu\Styles}* directory. Choose **Content** in the Places panel to navigate to the enu folder. Schedules are located in the Documentation Objects\Schedule Table Styles folder.

The following schedule tables can be imported from the tool catalogs: Beam Schedule, Door Schedule, Door Schedule Project Based, Equipment Schedule, Footing Schedule, Furniture Schedule, Pier Schedule, Room Finish Matrix, Room Finish Matrix Project Based, Room Finish Schedule, Room Finish Schedule Project Based, Room Schedule, Room Schedule Project Based, Space Inventory, Space

List-BOMA, Wall Schedule, and Window Schedule. You can also modify these schedule table styles or create new ones to meet your schedule needs.

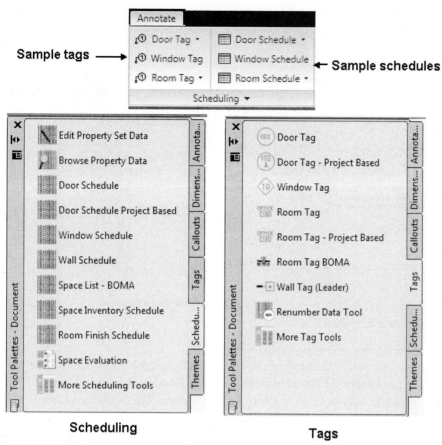

FIGURE 11.33 *Scheduling and Tags palettes*

In summary, the schedule is created based upon the format defined in schedule table style, which extracts property information from the schedule data associated with the objects. When tags are inserted in the drawing, the properties for the tagged object are inserted. After a schedule is inserted, it can be dynamically updated as the drawing changes.

Properties of Objects and Object Styles

When an object is tagged, the properties associated with the tag are inserted into the drawing. When you place a door tag, the **DoorObject** property and the **DoorStyle** property are inserted as shown in Table 11.19. The **DoorObject** property set is displayed in the upper half of the **Extended Data** tab as shown in Figure 11.34, while the style-based properties associated with a door style are displayed in the lower half. When the door is tagged, the **Edit Property Set Data** palette opens and allows you to edit the property data. After the door is tagged you can edit the properties in the **Extended Data Tab** of the Properties palette. The style-based properties can be edited by selecting the **Edit Style Property Set Data** button, shown at the left in Figure 11.34, to open the **Edit Property Set Data** dialog box. Style property set data can then be entered in the **Edit Property Set Data** dialog box. The Extended Data tab of the Properties palette for a door that has not been tagged is shown at the left in Figure 11.34. After the door is tagged, the DoorObject properties are added to the Extended Data tab as shown at the right.

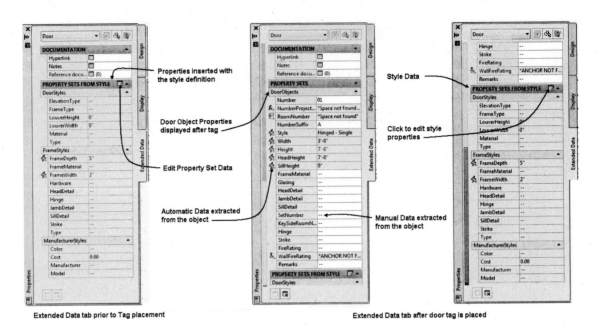

Extended Data tab prior to Tag placement   Extended Data tab after door tag is placed

FIGURE 11.34 *Extended Data tab for a door object*

If a drawing consists of Standard style doors without tags, the Properties palette will display an empty **Extended Data Property** field as shown on the left in Figure 11.35. The door shown in Figure 11.35 is a Standard style door. The Extended Data Property fields for doors inserted from the door styles of the Doors palette will include style properties. If you paste the **DoorObjects** property set into the drawing using the Style Manager, the **Add property sets** button becomes active in the **Extended Data** tab of the Properties palette as shown at the right in Figure 11.35. If you click the **Add property sets** button, the **Add Property Sets** dialog box opens, listing the property sets appropriate for the door object.

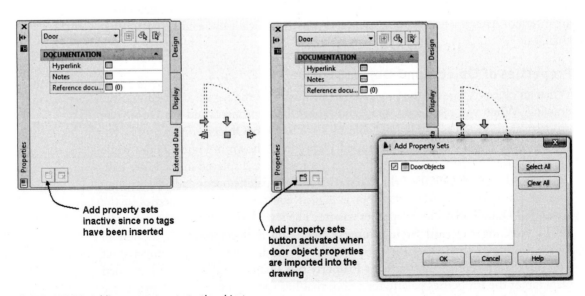

FIGURE 11.35 *Adding property sets to the object*

The RoomFinishObjects and SpaceObjects sets are the only property sets included in the drawing prior to placing tags. Property sets are included in the *Schedule Table (Imperial).dwg* or *Schedule Table (Metric).dwg* files of the Documentation Objects\ Property Set Definitions category. These files are located in the Imperial and Metric folders of the *{Vista – C:\ProgramData\Autodesk\ACA 2012\enu\Styles}* directory. Choose Content in the Places panel to navigate to the enu folder. The property sets shown in Table 11.21 are inserted in the drawing when a door tag is placed.

TABLE 11.21 *Property Sets*

Property Sets	Objects/Styles and Definition	Applies to
DoorObject	Object	Door, Door/Window Assembly
DoorStyles	Styles and Definitions	Door Style, Door/Window Assembly
FrameStyles	Styles and Definitions	Door Style, Door/Window Assembly Style, Window Styles
ManufacturerStyles	Styles and Definitions	All Objects

Placing Door Tags

Door tags can be applied to door objects by selecting the **Door Tag** or **Door Tag Project Based** command on the Scheduling palette or the Door Tag flyout menu in the Scheduling panel. The tag is a multi-view block with schedule properties defined as its attribute. The door tag can be applied in the View drawing, although the door exists within the Construct drawings attached as a reference to create the view. Tags can be applied in Construct drawings; however, if annotation is placed only in View drawings the construct drawing can be inserted in a model view drawing without the annotation clutter. When you place a door or window tag, the **AecScheduleTag** command is used to place the tag with preset symbol, leader, and dimstyle values. Door tags are restricted multi-view blocks; therefore, you must select a door object to place a door tag. Tags are annotative multi-view blocks; therefore, the size of the mark is determined by the current annotative scale when the tag is placed. The steps to placing a door tag are shown below.

Steps to Place a Door Tag

1. Set the **Annotation Scale** in the Drawing Window status bar.
2. Select the Annotate tab. Select the **Door Tag** tool from the Tags tool palette or the Scheduling panel of the Annotate tab in the ribbon.
3. You are prompted to select an object to tag; select a door at **p1** as shown in Figure 11.36.
4. Specify the location of the tag near the door at **p2** as shown in Figure 11.36.
5. The **Edit Property Set Data** dialog box opens as shown in Figure 11.36; make the necessary changes.
6. Click **OK** to dismiss the **Edit Property Set Data** dialog box.
7. The door tag is placed in the drawing. Press ESC to end the command.

The location of the tag can be selected with the cursor, or you can press ENTER to center the tag about the door. The **Edit Property Set Data** dialog box opens each time you select an additional door. Tags for multiple doors can be placed by selecting the **Multiple** option of the command. The **Multiple** option allows you to select additional doors without reopening the **Edit Property Set Data** dialog box for each insertion. If you select Multiple, you can select the entire drawing and all doors will be tagged.

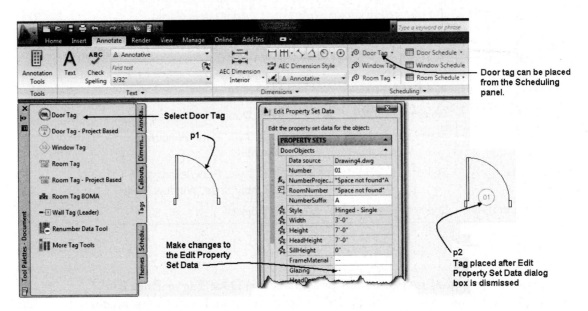

FIGURE 11.36 *Door tag placed in the drawing*

The Multiple option can be selected by pressing the down arrow, choosing **Multiple** if dynamic prompting is active, or typing **M** in the command after you close the **Edit Property Set Data** dialog box for the first tag as shown in the following workspace sequence. The position of the additional tags is defined based upon the position of the first tag; therefore, if you place the first door tag in the center of the opening, the multiple insertions of the remaining tags will be placed in a similar position. The door tag is placed on the **A-Door-Iden** [A315T] layer, which has the color 132 (green).

(Select the **Door Tag** *tool from the Tags tool palette.)*

Select object to tag: *(Select a door.)*

Specify location of tag <Centered>: *(Select a location for the tag or press* ENTER *to center the tag about the object.)*

(The **Edit Property Set Data** *dialog box opens; edit the properties and click* OK *to dismiss the dialog box.)*

Select object to tag [Multiple]: M ENTER *(Press the down arrow to display options; type* M *at the cursor and input* ENTER.*)*

Select objects to tag: *(Select additional doors, press* ENTER *to end the selection, and edit the* **Edit Property Set Data** *dialog box.)*

Placing Window Tags

A window tag is also located on the Tags tool palette or the Scheduling panel of the ribbon as shown in Figure 11.36. Choose the More Tag Tools of the Tags palette to open the Documentation Tool Catalog – Imperial [Documentation Tool Catalog – Metric] as shown in Figure 11.37. Tags are annotative multi-view blocks; therefore, if the Annotation Scale is changed the size of the tag will change.

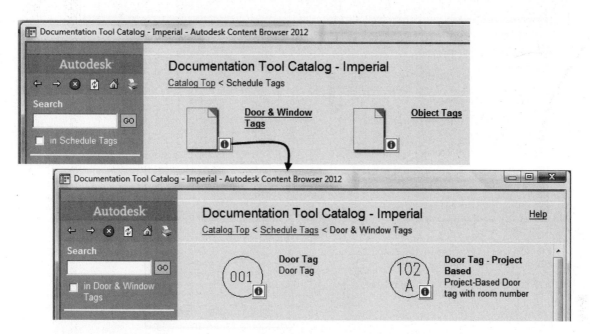

FIGURE 11.37 *Door and Window tags of the Documentation Tool Catalog – Imperial*

Window tags are multi-view blocks that are restricted to tagging only windows. Therefore, you cannot insert a window tag unless a window exists in the drawing. The window tag is placed on the **A-Glaz-Iden [A314T]** layer, which has the color 152 (a hue of blue).

Steps to Place a Window Tag

1. Select the **Window Tag** from the Tags palette or the Scheduling panel in the Annotate tab.

2. Select a window at **p1** as shown at the left in Figure 11.38 for the placement of a tag.

3. Select a location for the window tag (**p2** as shown at the right in Figure 11.38). The **Edit Property Set Data** dialog box opens as shown in Figure 11.38.

4. Edit the **Edit Property Set Data** dialog box and click **OK** to close the dialog box and display the window tag as shown at the right in Figure 11.38.

5. Continue to select additional windows to place additional tags or press ENTER to end the command.

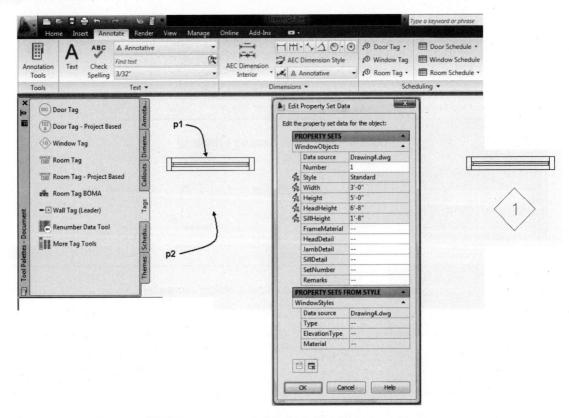

FIGURE 11.38 *Placing a window tag*

Room and Finish Tags

Room tags can be placed by selecting **Room Tag** from the Tags palette or the Scheduling panel. Room tags are located in the Schedule Tags > Room and Finish Tags category of Documentation Tool Catalog – Imperial or the Documentation Tool Catalog – Metric catalogs. When you select the **Room Tag** tool from the Tags palette, the **AecScheduleTag** command is used to place a style of room tag. To place a room tag, you must select a space object to tag the room. The space object includes data regarding the length, width, and height of the room. The room and finish tags are specifically designed for use with AEC Space objects created with the **SpaceAdd** command. When tags are applied to an AEC Space object, the text typed in the **Name** field of the Design tab in the Properties palette will be displayed as the room name.

Object and Structural Tags

Tags for structural components, equipment, and furniture can be placed in the drawing to identify a mark for the development of a schedule. The object tags listed in Table 11.22 can be accessed when you click the More Tag Tools button of the Tags palette. The More Tag Tools button opens the Schedule Tags category of the Documentation Tool Catalog – Imperial [Documentation Tool Catalog – Metric], which includes the Object Tags and Structural Tags folders.

The Object Tags folder includes tags for furniture and equipment. The Structural Tags extract includes the StructuralMemberStyles, Style, and StucturalMemberObjects Number properties. Therefore, when you create a structural member style from the Structural Catalog, the name assigned to the member can be extracted and included in the tag. Structural tags are restricted to structural members created from the **Structural Catalog**, discussed in Chapter 13, "Drawing Commercial Structures." The Equipment tag is designed to tag block references or multi-view block references such as equipment, furniture, or appliances. Table 11.22 lists each tag and its layer and color properties. Object tags are placed using the same procedure as other tags; therefore, only objects of the type specified for the tag can be selected.

TABLE 11.22 *Object and Structural tags*

Tag	Tag Purpose	Layer	Color
Beam Tags	Structural Beam	S-Beam-Iden [S282T]	131 hue of cyan
Brace Tags	Structural Brace	S-Cols-Brce-Iden [S289T]	151 hue of blue
Column Tags	Structural Columns	S-Cols-Iden [S280T]	171 hue of blue
Equipment Leader	Multi-view block	A-Eqpm-Iden [A850T]	132 hue of cyan
Equipment Tag	Multi-view block	A-Eqpm-Iden [A850T]	132 hue of cyan
Equipment Tag Scale Dependent	Multi-view block	A-Eqpm-Iden [A850T]	132 hue of cyan
Furniture	Furniture	I-Furn-Iden [A780T]	232 hue of magenta
Furniture Leader	Furniture	I-Furn-Iden [A780T]	232 hue of magenta
Furniture Tag Scale Dependent	Furniture	I-Furn-Iden [A780T]	232 hue of magenta

Wall Tags

Wall tags are used to identify different types of wall construction. A wall tag identifies a wall that is detailed in a different location or in a wall schedule. A wall tag is located on the Tags palette and the Scheduling panel in the Annotate tab (see Figure 11.39). Additional wall tags are located in the Wall Tags category of the Documentation Tool Catalog – Imperial or Documentation Tool Catalog – Metric. Tagging a wall allows you to create a schedule of the walls that lists the height, width, surface area, and volume.

Wall tags can be attached to a wall with a leader or placed near the wall. The wall tag is placed on the **A-Wall-Iden [A210T]** layer, which has the color 211 (magenta). The text or value placed in the tag is extracted from the wall style. After you place the tag, you can assign a value for the tag by editing the property set data for the wall. Click the **Edit Style Property Set Data** button on the Extended Data tab of the **Properties** palette to open the **Edit Property Set Data** dialog box as shown in Figure 11.40. The **Type** property for the style can be added in the **Edit Property Set Data** dialog box. The procedure to place a Wall Tag (Leader) is shown in the following steps.

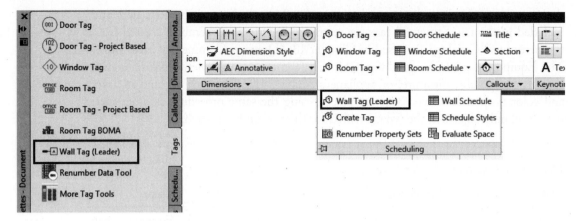

FIGURE 11.39 *Wall Tag (Leader) of the Tags palette and Scheduling panel*

Steps to Place Wall Tags

Select **Wall Tag (Leader)** from the **Tags** palette. Respond to the following workspace prompts to place the tag:

> Select object to tag: *(Select the wall at point* **p1** *as shown in Figure 11.40.)*
>
> Specify next point of leader line or: *(Select a point near* **p2** *as shown in Figure 11.40 to end the leader.)*
>
> Specify next point of leader line or <End leader>: ENTER *(Press* ENTER *to end the leader location.)*
>
> *(Click* OK *to dismiss the* Edit Property Set Data *dialog box.)*
>
> Select object to tag: ENTER *(Press* ENTER *to end the command.)*

1. To add text to the tag, select the wall, right-click, and choose **Properties**. Choose the **Extended Data** tab. Click the **Edit Style Property Set Data** button (see Figure 11.40) to open the **Edit Property Set Data** dialog box.

2. Type the text for the wall tag in the **Type** field.

3. Click **OK** to dismiss the **Edit Property Set Data** dialog box and press ESC to clear the selection.

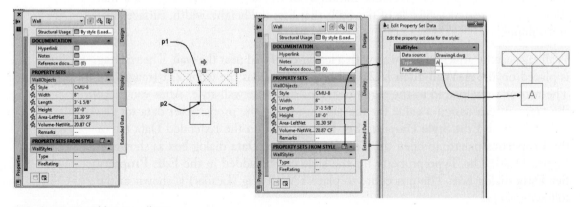

FIGURE 11.40 *Adding a wall tag*

EDITING TAGS AND SCHEDULE DATA

When tags are inserted in the drawing, the object-based property information for a schedule can be typed in the **Edit Property Set Data** dialog box that opens when the tag is placed. The object-based property data are also displayed on the **Extended Data** tab of the Properties palette for the object. To add style-based data, click the **Edit Style Property Set Data** button to open the **Edit Property Set Data** dialog box. Then you can edit property data for the object style.

ADDING A SCHEDULE TABLE

The Scheduling tool palette includes the Door Schedule, Door Schedule Project Based, Window Schedule, Wall Schedule, Space List-BOMA, Space Inventory Schedule, and Room Finish Schedule. Sample schedules can be added from the Scheduling panel of the Annotate tab. Prior to inserting a schedule table, tag the objects related to the table. The process of adding the tag and editing the **Edit Property Set Data** dialog box creates most of the data required for the schedule. Access the **ScheduleAdd** command as shown in Table 11.23 to insert a schedule style. The content of the schedule includes properties and property sets assigned to that object. To add a schedule table, click one of the schedule tables included on the Scheduling palette as shown in Figure 11.41.

TABLE 11.23 *ScheduleAdd command access*

Command prompt	SCHEDULEADD
Tool palette	Select a schedule table from the Scheduling palette.
Ribbon	Select a schedule table from the Scheduling panel of the Annotate tab.

When you select a schedule table from the Scheduling palette, you are prompted in the command line to select objects for the schedule or press ENTER to apply the schedule to an external drawing. Only objects of the type designed for the schedule can be selected and applied to the schedule. Therefore, you can select all the objects of the drawing and the selection will be filtered to include only the objects of the type designed for the schedule.

The **ScheduleAdd** command is used to add schedules to the drawing. When you select the **Window Schedule** command from the Scheduling palette, the ScheduleAdd command is used to insert the window schedule style. The ScheduleAdd command opens the Properties palette, which allows you to control the objects included in the schedule, size of the schedule, and options regarding the update of the schedule.

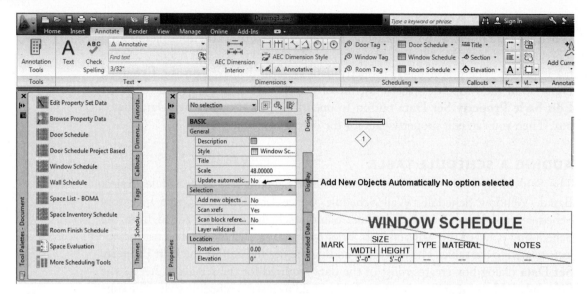

FIGURE 11.41 *Window Schedule tool and Window Schedule*

The Properties palette of the **ScheduleAdd** command consists of the following options.

Design Tab—Basic

General

> **Description**—The Description button opens the **Description** dialog box, which allows you to add text describing the schedule table.

> **Style**—The Style list includes all schedule table styles that have been inserted in the drawing.

> **Title**—The Title option allows you to type a unique schedule title.

> **Scale**—The default Scale value is equal to the specified value in the **Scale** tab of the **Drawing Setup** dialog box.

> **Update automatically**—The **Yes** option of this property allows changes to the drawing to be automatically reflected in the schedule. If the object is changed after the schedule is inserted when the **No** option of this property is selected, a diagonal line is drawn through the schedule as shown in Figure 11.41. The diagonal line can be removed by updating the schedule.

Selection

> **Add new objects automatically**—If the **Yes** option is specified for Add new objects automatically, objects added after the schedule is placed will automatically be included in the schedule.

> **Scan xrefs**—The **Yes** option of Scan xrefs allows the search of objects to include the content of external reference files attached to the drawing.

> **Scan block reference**—If the Scan block reference drawings option is set to **Yes**, objects included in blocks that are inserted in the current drawing will be included in the schedule selection set.

> **Layer wildcard**—The Layer wildcard field allows you to create a layer filter to limit the search for objects that apply to the schedule. The asterisk wildcard includes all layers of the drawing.

Location

Rotation—The Rotation field is active if you are editing an existing schedule. The angle of rotation of the schedule table is displayed in this field.

Elevation—The Elevation field is active if you are editing an existing schedule. The elevation of the schedule table is displayed in this field.

After editing the Properties palette for the schedule table and selecting objects, you are prompted to specify the location for the schedule in the drawing area. When you specify the location of the schedule, you can specify its size by selecting the location of the upper-left and lower-right corners, or you can accept the default size by pressing ENTER. If you select the default size, the schedule size is scaled to the scale value specified by Annotation Scale as shown in the Drawing Window status bar.

USING SCHEDULE TABLE STYLES

Schedule tables are used to organize and present the data inserted in the **Edit Property Set Data** dialog box in the form of a schedule. Most schedules include a mark column, which is a letter or number identifying such objects as a door or window. The mark is defined when a tag is placed in the drawing. The tables included allow you to create door, window, space, and room finish schedules. Additional tables can be accessed if you choose the More Scheduling Tools button of the Scheduling palette. Schedule tables are located in the Schedule Tables folder of the Documentation Tool Catalog – Imperial [Documentation Tool Catalog – Metric] catalogs of the Content Browser. In the Style Manager, schedule tables are located in the *Schedule Tables (Imperial).dwg* or *Schedule Tables (Metric).dwg* of the Imperial or Metric folders of the *{Vista – C:\ProgramData\Autodesk\ACA 2012\enu\Styles}* directory. Schedule tables can be developed to display the schedule data in the desired order. Access the schedule table style as shown in Table 11.24.

Schedule tables can be placed on the Scheduling palette or on a new palette by opening the Content Browser and dragging the i-drop of the schedule table from the Documentation Tool Catalog – Imperial or the Documentation Tool Catalog – Metric of the Content Browser.	**TIP**

TABLE 11.24 *Schedule Table Styles command access*

Command prompt	SCHEDULESTYLE
Ribbon	Select Schedule Styles from the Scheduling slide out panel to open the Style Manager to the Schedule Table Styles category.
Ribbon	Select a Schedule table and choose Edit Style from the General panel.
Palette	Select a schedule table tool on the Scheduling tool palette, right-click, and choose Edit Schedule Table Styles.

Sample schedule tables of the Scheduling palette are shown in Figure 11.42. The **ScheduleStyleEdit** command allows you to edit the format of the schedule according to user preferences.

DOOR AND FRAME SCHEDULE

MARK	DOOR							FRAME						FIRE RATING LABEL	HARDWARE		NOTES
	SIZE			MATL	GLAZING	LOUVER		MATL	EL	DETAIL				SET NO	KEYSIDE RM NO	
	WD	HGT	THK			WD	HGT			HEAD	JAMB	SILL				

WINDOW SCHEDULE

| MARK | SIZE | | TYPE | MATERIAL | NOTES |
| | WIDTH | HEIGHT | | | |

SPACE INVENTORY

| SPACE | LOCATION | | | | | | AREA | QTY |
| | SITE | BUILDING | FLOOR | ZONE | DEPARTMENT | OWNER | | |

ROOM FINISH SCHEDULE

| ROOM NO | ROOM NAME | FLOOR | WALLS | | | | CEILING | | NOTES |
| | | | N | S | E | W | MATL | HEIGHT | |

FIGURE 11.42 *Schedules of the Scheduling tool palette*

When you select a schedule table, right-click, and select **Edit Schedule Table Style**, the **Schedule Table Style Properties** dialog box opens as shown in Figure 11.43. The tabs of the Schedule Table Style Properties dialog box allow you to define the content of the schedule. A description of these tabs follows.

General Tab

Name—The Name of the schedule table is displayed in this field.

Description—A Description of the schedule can be typed in this field.

Notes—The Notes button allows you to add notes and files in the **Notes** and Reference Docs tabs.

Default Format Tab

The Default Format tab is shown in Figure 11.43.

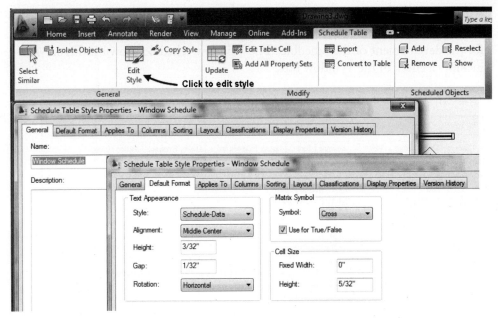

FIGURE 11.43 *General and Default Format tabs of the Schedule Table Style Properties – Window Schedule dialog box*

Text Appearance

Style—The Style drop-down list displays the text styles of the drawing.

Alignment—The Alignment options define the justification of the text in the schedule. The Middle Center alignment centers the text horizontally and vertically within the cell of the schedule.

Height—The Height field allows you to define the text height which will be scaled according to the scale factor of the **Scale** tab of the **Drawing Setup** dialog box.

Gap—The Gap field defines the distance between rows of text and the schedule table lines.

Rotation—The Rotation list includes the Horizontal and Vertical options that control the orientation of the text.

Matrix Symbol

Symbol—The Symbol list allows you to specify one of the following symbols used in a matrix schedule: Check, Dot, Cross, and Slash.

Use for True/False—The Use for True/False checkbox if selected applies the matrix symbol when the option of the schedule applies.

Cell Size

Fixed Width—The Fixed Width of a cell can be set to a specific distance. If the Fixed Width is set to zero, the width varies with the width of text necessary for the cell.

Height—The Height field is the vertical dimension of the cell.

Applies To Tab

The **Applies To** tab includes a list of all objects to which the schedule can be applied for the development of the schedule. The Window schedule shown in Figure 11.44 includes a check in the Window box.

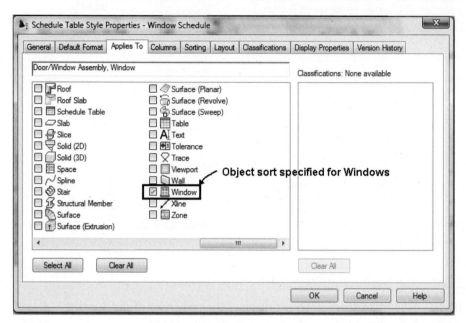

FIGURE 11.44 *Applies To tab of the Schedule Table Style Properties – Window Schedule dialog box*

Columns Tab

The Columns tab allows you to preview the list of properties to be included in the table. It includes buttons for creating and modifying each component. The Columns tab shown in Figure 11.45 lists the properties of the Window Schedule. This tab allows you to identify the property set and property used in each column of the schedule to extract the data from the object. Therefore, the **Mark** column will display the value of the **Number** property for each window object. The window schedule is developed from the WindowObjects and WindowStyles property sets as shown in Figure 11.45.

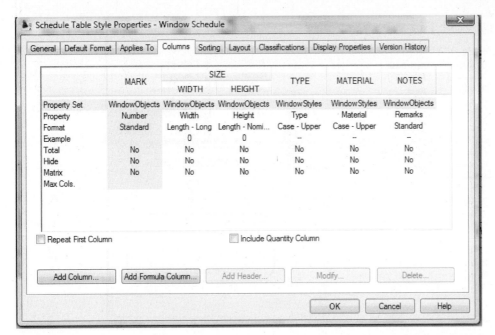

| | MARK | SIZE | | TYPE | MATERIAL | NOTES |
		WIDTH	HEIGHT			
Property Set	WindowObjects	WindowObjects	WindowObjects	WindowStyles	WindowStyles	WindowObjects
Property	Number	Width	Height	Type	Material	Remarks
Format	Standard	Length - Long	Length - Nomi...	Case - Upper	Case - Upper	Standard
Example		0	0	–	--	--
Total	No	No	No	No	No	No
Hide	No	No	No	No	No	No
Matrix	No	No	No	No	No	No
Max Cols.						

Schedule Table Style Properties - Window Schedule

General | Default Format | Applies To | Columns | Sorting | Layout | Classifications | Display Properties | Version History

☐ Repeat First Column   ☐ Include Quantity Column

[Add Column...] [Add Formula Column...] [Add Header...] [Modify...] [Delete...]

[OK] [Cancel] [Help]

FIGURE 11.45 *The Columns tab of the Schedule Table Style Properties – Window Schedule dialog box*

The elements of the Columns tab include the following:

Repeat First Column—The Repeat First Column checkbox allows you to insert a copy of the first column on the right side of the schedule.

Repeating the first column assists the reader when the chart is extremely wide. Using Repeat First Column can allow the Mark column to be placed on the left and right margins of the schedule.

Include Quantity Column—The Include Quantity Column checkbox allows you to insert a column that lists the number of repetitions of a building component in the schedule.

Add Column—The Add Column button allows you to add a column to the schedule. The Add Column button opens the **Add Column** dialog box shown in Figure 11.46. The Add Column dialog box allows you to define the contents of the new column.

The Add Column dialog box allows you to select the property to be included in a new column. The Add Column dialog box shown in Figure 11.46 includes the **Property Set/Properties** list, **Column Properties**, and **Column Position** sections described below.

Property Set/Properties list—The Property Set/Properties list includes a comprehensive list of the properties of the drawing available for the schedule. When you select a property from the list, you are adding that property to the schedule. The position and properties for the new column are defined in the sections located at the right in the dialog box.

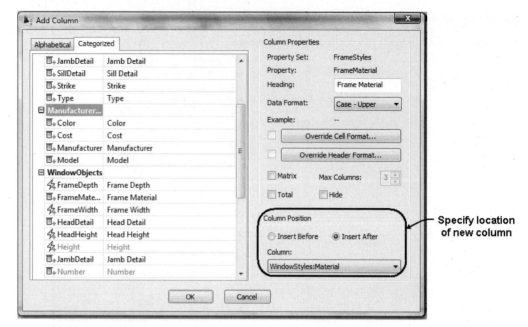

FIGURE 11.46 *Add Column dialog box*

Column Properties

Heading—The Heading field specifies the title of the heading for the new column in the schedule.

Data Format—The Data Format allows you to specify how the data of the column will be displayed. Data Formats can be selected from the drop-down list.

Override Cell Format—The Override Cell Format button opens the **Cell Format Override** dialog box, which allows you to specify the text size and properties as shown in Figure 11.47. The Cell Format Override dialog box allows you to specify the text appearance including style alignment height gap and rotation for the heading of the column. If a matrix schedule is specified you can select the symbol used in the matrix. The width of the cell can be specified in the **Cell Size Fixed Width** field. If the width is set to zero, the width will vary according to the needs of the text width.

Matrix—The Matrix checkbox allows you to display the data for the property in the matrix format.

Total—The Total option allows you to generate a total for the property.

Hide—The Hide option hides the property from the table.

Column Position—This section allows you to define the position of the column relative to other columns. In Figure 11.46 the new column will be inserted after the **WindowStyles:Material** column. The radio buttons allow you to position the new column before or after an existing column. The **Column** list shows the properties of the schedule that can be selected to set the relative position of the new column.

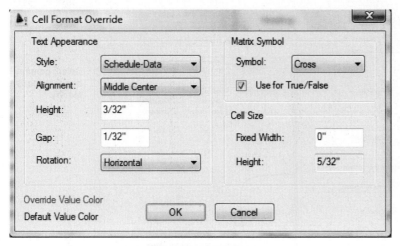

FIGURE 11.47 *Cell Format Override dialog box*

Add Formula Column—The Add Formula Column button opens the Add Formula Column dialog box. The Add Formula Column dialog box allows you to create a formula that includes quantities, VBScripts that are applied to Property Definitions of the objects included in the schedule.

Add Header—The Add Header button of the Columns tab allows you to create a header for a column. The header titled SIZE was created as shown in Figure 11.49 by selecting the **Width** and **Height** column titles and then selecting the Add Header button. Holding down CTRL when you select the column titles allows you to select more than one title. The title of the header can then be typed in the header space. You can delete headers by clicking the header and then clicking the **Delete** button.

Modify—If you select the title of a column and then select the Modify button the **Modify Column** dialog box opens as shown in Figure 11.48. The Modify Column dialog box allows you to edit the contents of the column. As shown in Figure 11.48, the **Type** column can be edited to change the heading data format or matrix style or to include a total.

Delete—The Delete button allows you to delete a selected column from the schedule.

You can reposition a column within the schedule by selecting the column title in the Columns tab and dragging the column title to the left or right. Releasing the mouse button over another column repositions the selected column within the schedule as shown in Figure 11.49.

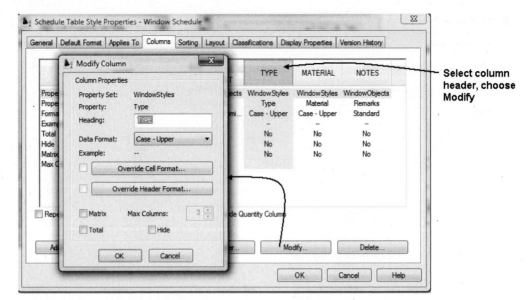

Select column
header, choose
Modify

FIGURE 11.48 *Modify Column dialog box*

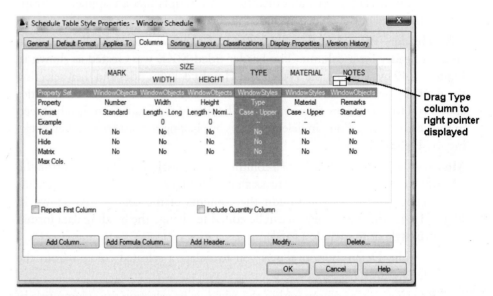

Drag Type
column to
right pointer
displayed

FIGURE 11.49 *Moving the position of a column in a Schedule Table Style*

Sorting Tab

The Sorting tab allows you specify how the schedule data are displayed in the schedule. In Figure 11.50, the schedule data are sorted according to the **WindowObjects: Number** property in ascending order. You can define the sorting of the properties by selecting one of the following four buttons: **Add, Remove, Move Up**, and **Move Down**.

> **Add**—The Add button opens the **Select Property** dialog box, which lists other properties that can be defined for sorting.

> **Remove**—To remove a property for sorting select a property from the property list and then select the Remove button.

Move Up—If more than one property is used for sorting you can select a property and then select the Move Up button and that property will move to the top of the list.

Move Down—When a property has been moved up in priority in the property list you can select the property and then select the Move Down button and that property will be moved down to lower priority when the sort is executed.

Sort Order—The Sort Order can be specified as **Ascending** or **Descending**.

Layout Tab

The Layout tab allows you to define the text size for the schedule title, column headers, and matrix headers. The Layout tab shown in Figure 11.50 includes a **Table Title** field and buttons to define the format of the title, column headers, and matrix headers.

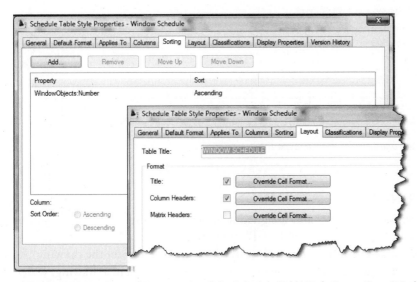

FIGURE 11.50 *Sorting and Layout tabs of the Schedule Table Style Properties – Window Schedule dialog box*

The **Table Title** field allows you to enter the text for the title of the schedule, and the remaining portion of the dialog box allows you to define the text height and other properties of the title and headers. You can edit the appearance of the title, column headers, and matrix headers by selecting the **Override Cell Format** buttons. When you select one of the **Override Cell Format** buttons, the **Cell Format Override** dialog box opens. This dialog box allows you to increase the text height and properties of the title and headers.

Classifications Tab

The Classifications tab allows you to add classification definitions to property sets for the schedule.

Display Properties Tab

The Display Properties and Version History tabs are similar to most objects as shown in Figure 11.51. The Version History tab lists the date modified of the style used when a project is synchronized.

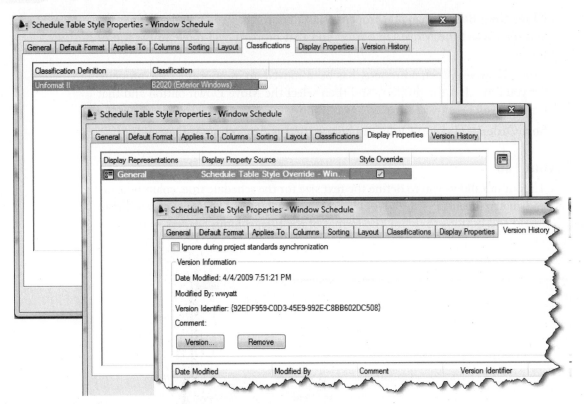

FIGURE 11.51 *Display Properties tab and Layer/Color/Linetype tab*

RENUMBERING TAGS

Tags are placed on a drawing and assigned a mark in a manner that allows you to predict the location of the next number. Placing tag numbers or letters in a drawing in a systematic manner makes it easier for the craftsman to interpret the drawing. As the design develops, additional windows or doors can be inserted in a drawing, which creates disorder in the tag numbering sequence. Therefore, the **Renumber Data Tool** located on the Tags tool palette can be used to renumber existing tags. Access the **Renumber Data Tool** command as shown in Table 11.25 and Figure 11.52.

TABLE 11.25 *Renumber Data Tool command access*

Command prompt	PROPERTYRENUMBERDATA
Ribbon	Select Renumber Property Sets from the slide out of the Scheduling panel.
Palette	Choose Renumber Data Tool from the Tags palette.

When you select the **Renumber Data Tool** command, the **Data Renumber** dialog box opens, as shown in Figure 11.52. This dialog box allows you to edit the property, start number, and increment for changing the tag number. The settings shown in Figure 11.52 would select the first tag and change its value to 1, and each successive tag selected would be increased by one.

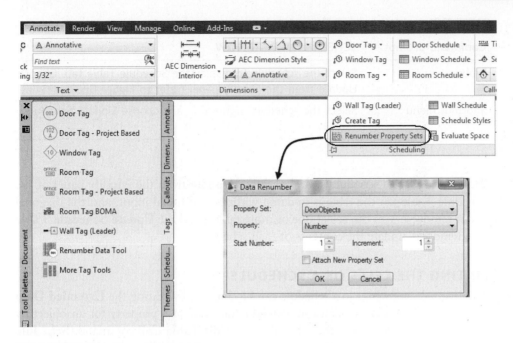

FIGURE 11.52 *Data Renumber dialog box*

The **Data Renumber** dialog box consists of the following options:

Property Set—The Property Set drop-down list displays property sets defined for the drawing.

Property—The Property drop-down list displays the property specified for the renumbering operation.

Start Number—The Start Number field allows you to specify the beginning number of the renumbering operation.

Increment—The Increment value increases the tag number by adding the increment number to the previous tag number.

Attach New Property Set—When you check the Attach New Property Set checkbox, the property of the objects in the drawing will change as you renumber them.

UPDATING A SCHEDULE

The content of a drawing often changes after a schedule is developed and objects are tagged. If the **Add New Objects Automatically** option is set to **Yes** in the Properties palette when a schedule is placed, the additional objects added to the drawing will automatically be inserted in a schedule. When the schedule is created, the **Add New Objects Automatically** can be set to **No** as shown in Figure 11.41 in the Properties palette; the objects are not included in the schedule. When additional objects are added to this schedule, a diagonal line is drawn across the schedule to indicate that it is not current, as shown in Figure 11.41. This diagonal line indicates that the table needs to be updated; it will be removed when the table is updated with the **Update Schedule Table** command (**ScheduleUpdateNow**). Access the **Update Schedule Table (ScheduleUpdateNow)** command as follows (Table 11.26).

TABLE 11.26 *Update Schedule Table command access*

Command prompt	SCHEDULEUPDATENOW
Ribbon	Select the schedule to display the Schedule Table tab. Choose Update from the Modify panel (refer to Figure 11.53).
Shortcut menu	Select the schedule, right-click, and choose Update Schedule Table.

When you select a schedule and choose the **Update** from the Modify panel of the Schedule Table tab the table is updated without additional input, and the diagonal line is removed. You can select one or more tables, and the **Update Schedule Table** command will update the contents of each schedule.

EDITING THE CELLS OF A SCHEDULE

The manual data values in a schedule can be changed by editing the **Extended Data** of the Properties palette. You can also edit the value of a property for an object by selecting the cell directly in the schedule. The **Edit Table Cell** command (**Schedule CellEdit**) is used to edit the data directly in the schedule. This command allows you to edit cells with manual data. Cells with Automatic data extracted from the object cannot be edited with this command because the object must be changed to change the automatic data. Access the **Edit Table Cell** command as shown in Table 11.27.

TABLE 11.27 *Edit Table Cell command access*

Command prompt	SCHEDULECELLEDIT
Ribbon	Select the schedule and choose Edit Table Cell from the Modify panel of the Schedule Table tab.
Shortcut menu	Select the schedule, right-click, and choose Edit Table Cell.

When you select the **ScheduleCellEdit** command, you are prompted to select a cell within a schedule. However, prior to selecting a cell, you can move your cursor over a cell, and the object represented by the cell will be highlighted in the drawing. If you press CTRL when you select a cell, the graphic screen will zoom to the object represented by the cell.

Selecting the cell in the schedule opens the **Edit Referenced Property Set Data** dialog box or message box depending upon the type of property data. The edit for manual property data, automatic property data, and style-based property data is shown in Figure 11.53.

The **Edit Referenced Property Set Data** dialog box allows you to enter a new value in the schedule. Included in the dialog box are the names of the property and property set that are defined for the cell. To change the value of the cell, overtype a new value and click the **OK** button; the schedule is then changed.

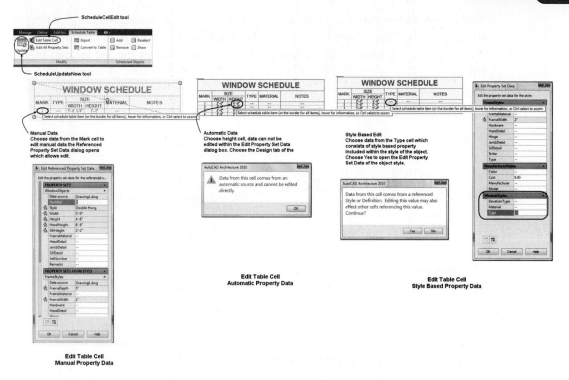

FIGURE 11.53 *Editing a cell of the schedule in the Edit Schedule Property dialog box*

USING BROWSE PROPERTY DATA

The Browse Property Data tool located on the Scheduling tool palette allows you to view and edit the values for the properties of all objects of a specific type in a single dialog box. The values for the properties can be changed in the **Browse Property Data** dialog box. Access the **Browse Property Data** command as shown in Table 11.28.

TABLE 11.28 *PropertyDataBrowse command access*

Command prompt	PROPERTYDATABROWSE
Tool palette	Select the Browse Property Data command from the Scheduling tool palette.

When you select the **Browse Property Data** command from the Scheduling palette, the **Browse Property Data** dialog box opens as shown in Figure 11.54. This dialog box has been edited to specify the **WindowObjects** property set definition and filter out objects other than windows. Therefore, only windows are listed in the left pane. If you select a window in the left pane, its properties will be displayed in the right pane, as shown at the right in Figure 11.54. You can select additional windows if you hold down CTRL during selection. If you have selected more than one window in the left pane, you can edit properties in the right pane that will be applied to all windows selected. The remark "Insulated Glass" was added to all windows of the schedule in Figure 11.55.

The **Zoom to** button in the bottom center of the dialog box zooms the drawing to the location of the selected window. The **Highlight** checkbox in the dialog box highlights the window in the drawing when it is selected in the list located in the left pane.

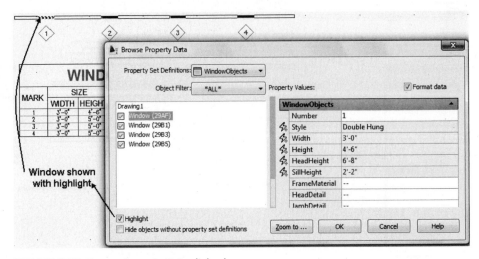

FIGURE 11.54 *Browse Property Data dialog box*

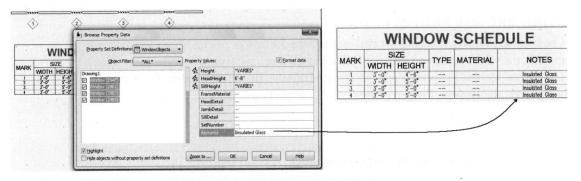

FIGURE 11.55 *Editing window properties in the Browse Property Data dialog box*

TUTORIAL 11-2: PLACING TAGS FOR WINDOWS AND CREATING WINDOW SCHEDULES

1. Verify that the Accessing Tutor\Imperial or Accessing Tutor\Metric content for Chapter 11 of the CengageBrain has been downloaded to your Accessing Tutor folder on your computer as described in Organizing Tutorial Directories in the Preface.

2. Open Ex 11-2.dwg from your Accessing Tutor\Imperial\Ch11 or Accessing Tutor\Metric\Ch11 directory.

3. Select **SaveAs > AutoCAD Drawing** from the Application menu and save the drawing as **Lab 11-2** in your student directory. Toggle OFF Object Snap in the status bar.

4. Verify that Annotation Scale = 1/8" = 1'- 0" [1:100] and that Annotation Visibility and Automatically Add Scales are toggled ON.

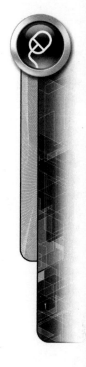

5. Choose the Annotate tab. Double-click the Annotation Tools command of the Tools panel to display the Document palette group.

6. Select **Window Tag** from the **Scheduling** panel of the Annotate tab. Respond to the following workspace prompts to tag the windows:

> Select object to tag: *(Select window at* p1, *the sash of the window, as shown in Figure 11.56 Note: If you select the lines representing the window that are coincident with the wall lines, the tag will not recognize the window selection and the command will be terminated.)*
>
> Specify location of tag <Centered>: *(Select a point near* p2 *as shown in Figure 11.56.)*
>
> *(Verify Number = 1 in the* **Edit Property Set Data** *dialog box. Click* OK *to dismiss the* **Edit Property Set Data** *dialog box.)*
>
> Select object to tag [Multiple]: *(Select window at* p3 *as shown in Figure 11.56.)*
>
> Specify location of tag <Centered>: *(Select a point near* p4 *as shown in Figure 11.56.)*
>
> *(Verify Number = 2 in the* **Edit Property Set Data** *dialog box. Click* OK *to dismiss the* **Edit Property Set Data** *dialog box.)*
>
> *(Continue to tag the remaining windows as shown in Figure 11.56.)*

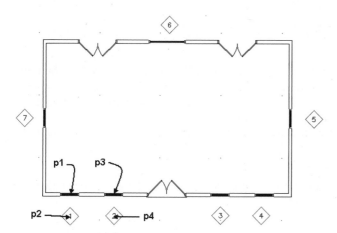

FIGURE 11.56 *Placing window tags*

7. Select **Window Schedule** from the **Scheduling** panel of the Annotate tab as shown in Figure 11.57. Verify the following settings in the **Properties** palette: Style = **Window Schedule**, Scale = **96 [100]**, Update automatically = **Yes**, Add new objects automatically = **Yes**, Scan Xrefs = **Yes**, Scan block references = **Yes**, and Layer wildcard = *****. Move the cursor from the Properties palette to the workspace. Respond to the workspace prompts as follows:

> Select objects or ENTER to schedule external drawing: all ENTER *(Type* all *to select all window objects in the drawing.)* 11 found 3 were not in current space.
>
> Select objects or ENTER to schedule external drawing: ENTER *(Press* ENTER *to end the selection.)*

Upper-left corner of table: *(Select a point near* **p1** *as shown in Figure 11.57.)*

Lower-right corner (or RETURN): ENTER *(Press* ENTER *to insert the schedule scaled to the drawing scale. The schedule is inserted as shown in Figure 11.57.)*

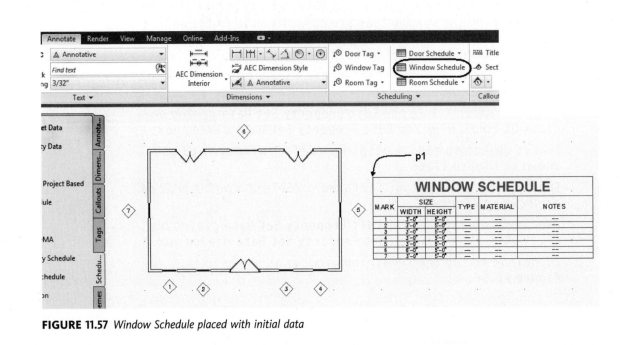

FIGURE 11.57 *Window Schedule placed with initial data*

8. Choose the Home tab. Click the Tools command of the Build panel. Choose the Windows palette. Select the **Casement** tool from the **Windows** palette. Edit the Width = **3' [900]**, Height = **4'-0" [1200]**, Measure to = **Outside of Frame**, Swing angle = **45**, Position along wall = **Offset**, Automatic offset = **3'-0" [900]**, Vertical alignment = **Head**, and Head height = **6'-8" [2050]**. Add two casement windows on the left wall of the building as shown at **p1** and **p2** in Figure 11.58.

> **NOTE** When casement windows are added to the drawing, question marks are added to the schedule.

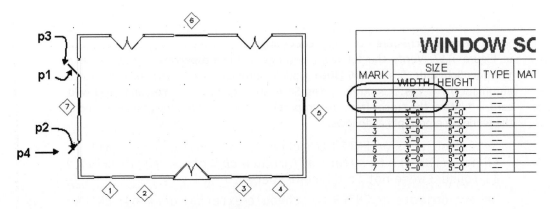

FIGURE 11.58 *Casement windows added to floor plan*

9. To remove the question marks of the first two schedule entries, choose the Annotate tab and select the **Window Tag** from the Scheduling panel. Respond to the following workspace prompts to tag the windows:

> Select object to tag: *(Select window at* p1 *as shown in Figure 11.58.)*
>
> Specify location of tag <Centered>: *(Select a point near* p3 *as shown in Figure 11.58* to locate the tag.*)*
>
> *(Verify that* 8 *is the value of the* **Number** *field of the* **Edit Property Set Data** *dialog box. Click* OK *to dismiss the Edit Property Set Data dialog box.)*
>
> Select object to tag [Multiple]: *(Select window at* p2 *as shown in Figure 11.58.)*
>
> Specify location of tag <Centered>: *(Select a point near* p4 *as shown in Figure 11.58.)*
>
> *(Verify that* 9 *is the value of the* **Number** *field of the* **Edit Property Set Data** *dialog box. Click* OK *to dismiss the Edit Property Set Data dialog box.)*
>
> *(Press* ESC *to end the command; question marks are removed from schedule as shown in Figure 11.59.)*

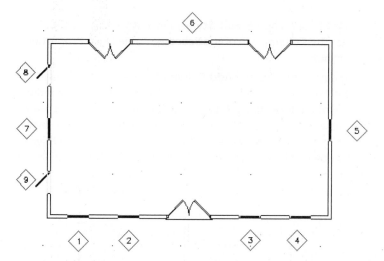

FIGURE 11.59 *Mark data from the window tag automatically updated in window schedule*

10. Select the **Renumber Property Sets** from the slide out of the Scheduling panel. Edit the **Data Renumber** dialog box as follows: Property Set = **Window Objects**, Property = **Number**, Start Number = **7**, Increment = **1**, and clear the **Attach New Property Set** checkbox. Click **OK** to dismiss the Data Renumber dialog box. Select the windows of the left wall using the following workspace prompts (see Figure 11.60).

> Select object to renumber to 7: *(Select window at* p1.*)*
>
> Select object to renumber to 8: *(Select window at* p2.*)*
>
> Select object to renumber to 9: *(Select window at* p3.*)*
>
> Select object to renumber to 10: ENTER *(Press* ENTER *to end the selection.)*

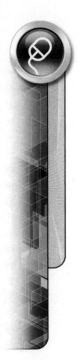

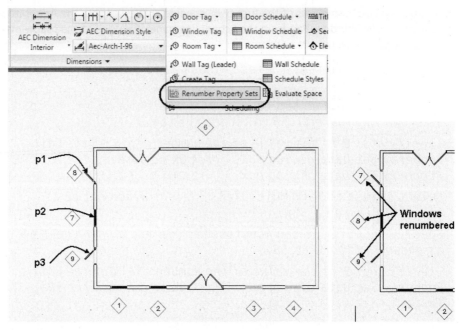

FIGURE 11.60 *Tags numbered using the Renumbered Data tool*

11. Select the Window Schedule and choose **Edit Style** from the General panel of the Schedule Table tab. Select the **Columns** tab of the **Schedule Table Style Properties – Window Schedule** dialog box.

12. Select the **Add Column** button to open the **Add Column** dialog box. Scroll down the list of properties in the left column and select **FrameDepth** from the **WindowObjects** property set. Overtype **FRAME DEPTH** in the **Heading** field of the **Column Properties** section at the right. Set Data Format = **Length-Long**. Select the **Insert After** radio button of the **Column Position** section and select **WindowObjects: Remarks** from the drop-down list as shown in Figure 11.61. Click **OK** to dismiss the **Add Column** dialog box.

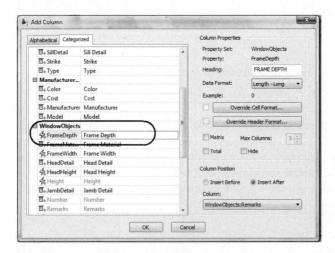

FIGURE 11.61 *Adding a Frame Depth column to the window schedule*

13. Click the **Add Column** button to open the **Add Column** dialog box. Scroll down the left column of properties and select **FrameMaterial** from the **WindowObjects** property set. Overtype **FRAME MATERIAL** in the **Column Properties** section. Verify that Data Format = **Case-Upper**. Select the **Insert After** radio button of the **Column Position** section and select **WindowObjects:FrameDepth** from the **Column** drop-down list as shown in Figure 11.62. Click **OK** to dismiss the **Add Column** dialog box.

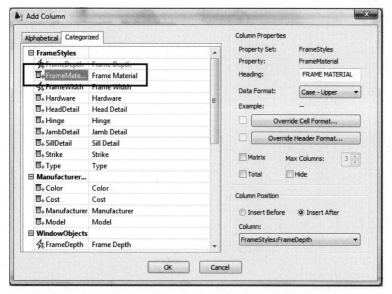

FIGURE 11.62 *Adding a Frame Material column to the window schedule*

14. In this step a header is added for the new columns added in the previous steps. Hold down CTRL and select the **FRAME DEPTH** and **FRAME MATERIAL** column headings. Click the **Add Header** button and type **FRAME** in the **Add Header** dialog box as shown in Figure 11.63. Click **OK** to dismiss all dialog boxes.

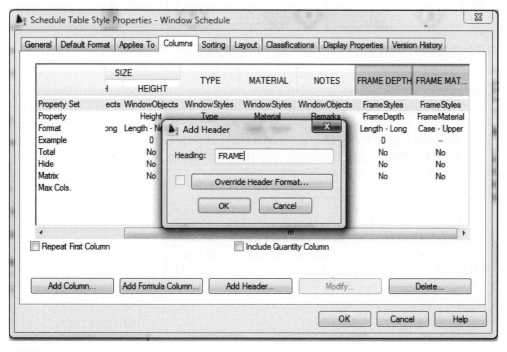

FIGURE 11.63 *Adding a header*

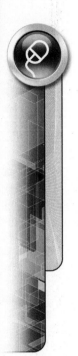

15. Manual data for a window style are added in this step. Select the window tagged number **1** in the floor plan, right-click, and choose **Properties**. Choose the **Extended Data** tab. Click the **Edit Style Property Set Data** button shown at **p1** in Figure 11.64 to open the **Edit Property Set Data** dialog box. Scroll down to the Window Styles property set and type **DOUBLE HUNG** in the **Type** field and **VINYL CLAD** in the **Material** field as shown in Figure 11.64. Click **OK** to dismiss the **Edit Property Set Data** dialog box. Press ESC to end the selection. The data are added to the schedule for all windows of the style.

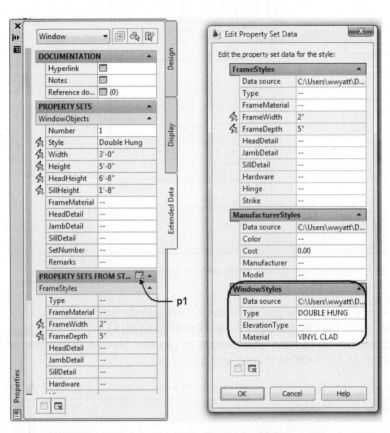

FIGURE 11.64 *Inserting style-based data for the double hung window*

16. Double-click the window tagged number **9** to open the **Properties palette**. Choose the **Extended Data** tab. Choose the **Edit Style Property Set Data** button. Scroll down to the Window Styles property set and type **SINGLE CASEMENT** in the **Type** field and type **VINYL CLAD** in the **Material** field. Click **OK** to dismiss the **Edit Property Set Data** dialog box. Press ESC to end selection. Type and material data are added to the schedule as shown in Figure 11.65.

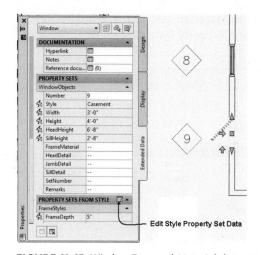

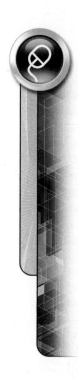

FIGURE 11.65 *Window Type and Material data entered in the schedule from style-based properties*

17. Select the Window Schedule and select **Edit Table Cell** from the **Modify** tab of the Schedule Table tab. Click the **(- -)** in the **Type** column for the number **6** window at **p1** in Figure 11.66. Click **Yes** to dismiss the AutoCAD **Architecture** warning box and open the **Edit Property Set Data** dialog box. Type **PICTURE ARCHED** in the **Type** field of the **Window Styles** property set. Type **VINYL CLAD** in the **Material** field of the Window Styles property set. Click **OK** to dismiss the Edit Property Set Data dialog box. Press ESC to end the command.

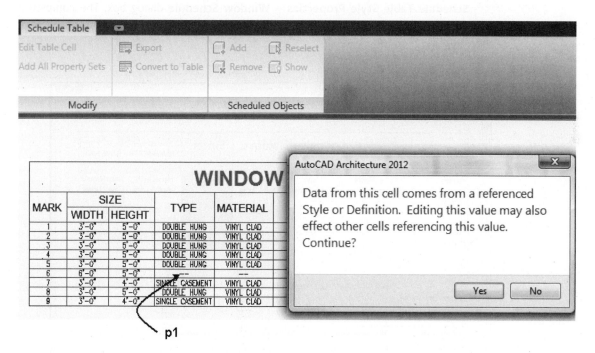

FIGURE 11.66 *AutoCAD Architecture message box for editing style-based data*

18. Select **Browse Property Data** from the Scheduling tool palette. Edit the Property Set Definitions = **Window Objects** and Object Filter = **Window** in the **Browse Property Data** dialog box. Hold down CTRL and select all the windows as shown in

Figure 11.67. Type **Vinyl Clad** in the **FrameMaterial** field of the **WindowObjects** property set. Click **OK** to dismiss the Browse Property Data dialog box.

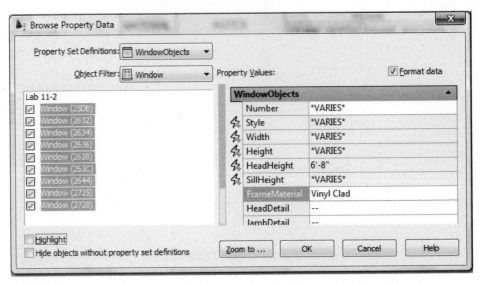

FIGURE 11.67 *Editing the Frame Material in the Browse Property Data dialog box*

19. Select the window schedule and choose **Edit Style** from the General panel of the Schedule Table tab. Select the **Columns** tab. Select the **Notes [Remarks]** heading and drag the heading to the right as shown in Figure 11.68. Click **OK** to dismiss the **Schedule Table Style Properties – Window Schedule** dialog box. The completed window schedule is shown in Figure 11.69.

20. Save and close the drawing.

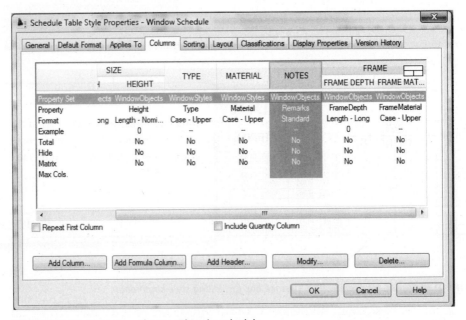

FIGURE 11.68 *Moving a column within the schedule*

WINDOW SCHEDULE

MARK	SIZE		TYPE	MATERIAL	FRAME		NOTES
	WIDTH	HEIGHT			FRAME DEPTH	FRAME MATERIAL	
1	3'-0"	5'-0"	DOUBLE HUNG	VINYL CLAD	5"	VINYL CLAD	--
2	3'-0"	5'-0"	DOUBLE HUNG	VINYL CLAD	5"	VINYL CLAD	--
3	3'-0"	5'-0"	DOUBLE HUNG	VINYL CLAD	5"	VINYL CLAD	--
4	3'-0"	5'-0"	DOUBLE HUNG	VINYL CLAD	5"	VINYL CLAD	--
5	3'-0"	5'-0"	DOUBLE HUNG	VINYL CLAD	5"	VINYL CLAD	--
6	6'-0"	5'-0"	PICTURE ARCHED	VINYL CLAD	5"	VINYL CLAD	--
7	3'-0"	4'-0"	SINGLE CASEMENT	VINYL CLAD	5"	VINYL CLAD	--
8	3'-0"	5'-0"	DOUBLE HUNG	VINYL CLAD	5"	VINYL CLAD	--
9	3'-0"	4'-0"	SINGLE CASEMENT	VINYL CLAD	5"	VINYL CLAD	--

Imperial Window Schedule

WINDOW SCHEDULE

MARK	SIZE		TYPE	MATERIAL	FRAME		REMARKS
	WIDTH	HEIGHT			FRAME DEPTH	Frame Material	
1	1010	1510	DOUBLE HUNG	VINYL CLAD	100	VINYL CLAD	--
2	1010	1510	DOUBLE HUNG	VINYL CLAD	100	VINYL CLAD	--
3	1010	1510	DOUBLE HUNG	VINYL CLAD	100	VINYL CLAD	--
4	1010	1510	DOUBLE HUNG	VINYL CLAD	100	VINYL CLAD	--
5	1010	1510	DOUBLE HUNG	VINYL CLAD	100	VINYL CLAD	--
6	1800	1510	PICTURE ARCHED	VINYL CLAD	100	VINYL CLAD	--
7	900	1200	SINGLE CASEMENT	VINYL CLAD	100	VINYL CLAD	--
8	1010	1510	DOUBLE HUNG	VINYL CLAD	100	VINYL CLAD	--
9	900	1200	SINGLE CASEMENT	VINYL CLAD	100	VINYL CLAD	--

Metric Window Schedule

FIGURE 11.69 *Completed window schedules*

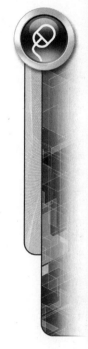

CHANGING THE SELECTION SET OF A SCHEDULE

You can revise a schedule by adding or removing objects from the schedule. The **Scheduled Objects** panel includes the following commands to modify the objects included in the schedule: **Add, Remove, Reselect,** and **Show** objects. The commands are described in the following list:

> **Add**—The Add option (**ScheduleSelectionAdd**) allows you to select additional objects in the drawing to add to the selected schedule.

> **Remove**—The Remove option (**ScheduleSelectionRemove**) allows you to remove from a schedule an object that is listed in the schedule.

> **Reselect**—The Select option (**ScheduleSelectionReselect**) allows you to reselect the objects to be included in the schedule.

> **Show**—The Show option (**ScheduleSelectionShow**) allows you to select the text within one or more cells in the schedule and the associated object in the drawing will be highlighted.

Each of these commands is included on the shortcut menu of a selected schedule. If Automatic Update is not being used for a schedule, these commands allow you to selectively revise the objects included in the schedule.

Adding Objects to an Existing Schedule

When you select schedule and choose **Add** from the Scheduled Objects panel of the Schedule Table the **ScheduleSelectionAdd** command is selected as shown in Table 11.29.

TABLE 11.29 *ScheduleSelectionAdd command access*

Command prompt	SCHEDULESELECTIONADD
Ribbon	Select a schedule and choose Add from the Scheduled Objects panel of the Schedule Table tab.
Shortcut menu	Select a schedule, right-click, and choose Selection > Add from the shortcut menu.

The window schedule shown in Figure 11.70 includes three windows. The following workspace sequence will add the fourth window to the existing schedule.

(Select the window schedule at **p1** *as shown at the left in Figure 11.70 and choose* **Add** *from the Scheduled Objects panel.)*

Select objects or ENTER to schedule external drawing: 1 found *(Select the window at* **p2** *as shown at the left in Figure 11.70.)*

Select objects or ENTER to schedule external drawing: ENTER *(Press* ENTER *to end the selection.)* 1 object(s) added.

(An additional window is added to the schedule as shown at the right in Figure 11.70.)

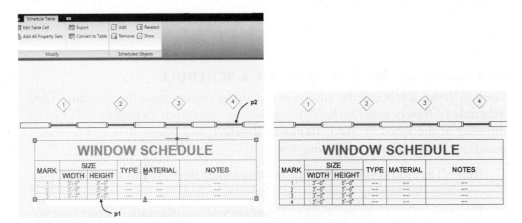

FIGURE 11.70 *Window Schedule developed after adding to the selection*

Removing Objects from an Existing Schedule

When you choose a schedule and choose **Remove** from the Selected Objects panel the **ScheduleSelectionRemove** command is selected as shown in Table 11.30.

TABLE 11.30 *ScheduleSelectionRemove command access*

Command prompt	SCHEDULESELECTIONREMOVE
Ribbon	Select a schedule and choose Remove from the Scheduled Objects panel of the Schedule Table tab.
Shortcut menu	Select the schedule, right-click, and choose Selection > Remove.

The window schedule shown in Figure 11.71 includes four windows. The following workspace sequence was used to remove windows from the schedule.

> *(Select the window schedule at* **p1** *as shown at the left in Figure 11.71; choose* **Remove** *from the Scheduled Objects panel of the Schedule Table tab.)*
>
> Select objects: *(Select the window at* **p2** *as shown in Figure 11.71.)*
>
> Select objects: *(Select the window at* **p3** *as shown in Figure 11.71.)*
>
> Select objects: ENTER *(Press ENTER to end the selection. Windows are removed from the schedule as shown at the right in Figure 11.71.)*

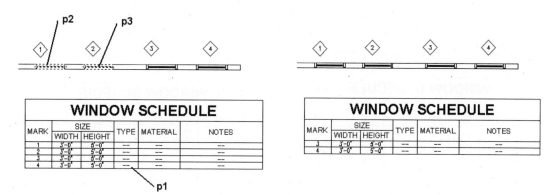

FIGURE 11.71 *Revised schedule based upon remaining objects*

Reselecting Objects for an Existing Schedule

The **Reselect** option allows you to reselect the objects you want included in the schedule. Only those objects selected in the new selection set will be included in the schedule. Objects in the schedule prior to your selecting this command and not included in the reselection will be removed from the schedule. Select a schedule and choose **Reselect** (**ScheduleSelectionReselect**) command from the Scheduled Objects panel as shown in Table 11.31.

TABLE 11.31 *ScheduleSelectionReselect command access*

Command prompt	SCHEDULESELECTIONRESELECT
Ribbon	Select a schedule and choose Reselect from the Scheduled Objects panel in the Schedule Table tab.
Shortcut menu	Select the schedule, right-click, and choose Selection > Reselect.

The window schedule shown in Figure 11.72 includes four windows. The following workspace sequence was used to reselect the windows for the schedule.

(Select the Window Schedule at **p1** *as shown at the left in Figure 11.72, choose* **Reselect** *from the Scheduled Objects panel of the Schedule Table tab.)*

Select objects or ENTER to schedule external drawing: 1 found
(Select window at **p2** *as shown at the left in Figure 11.72.)*

Select objects or ENTER to schedule external drawing: 1 found
(Select window at **p3** *as shown at the left in Figure 11.72.)*

Select objects or ENTER to schedule external drawing: 1 found
(Select window at **p4** *as shown at the left in Figure 11.72.)*

Select objects or ENTER to schedule external drawing: ENTER
(Press ENTER to end the selection. The revised schedule is shown at the right in Figure 11.72.)

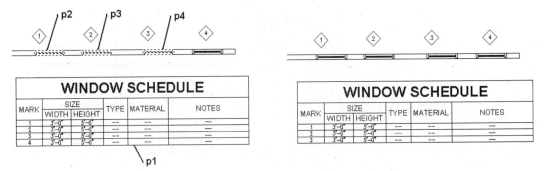

FIGURE 11.72 *Reselect option to change the selection set of a schedule*

Finding Objects in the Drawing Listed in the Schedule

The **Show** command of the Scheduled Objects panel of the Schedule Table allows you to select an object in the schedule, and it will be highlighted in the drawing. This option allows you to quickly locate an object in the drawing that has been included in the schedule. Access the **ScheduleSelectionShow** command as shown in Table 11.32.

TABLE 11.32 *Schedule Selection Show command access*

Command prompt	SCHEDULESELECTIONSHOW
Ribbon	Select a schedule table and choose Show from the Scheduled Objects panel of the Schedule Table tab.
Shortcut menu	Select a schedule table, right-click, and choose Selection > Show.

When you select a schedule and choose **Show**, you are prompted to select the text in the schedule. When you select the text, the object will be highlighted in the drawing. The workspace prompt includes a **CTRL-select** option that zooms the display to the object associated with the text selected in the schedule.

The command line prompt also allows you to select the border of the schedule to highlight all items in the drawing associated with the schedule. The following workspace sequence was used to zoom to a selected item of the schedule.

(Select the window schedule at p1 *as shown in Figure 11.73, and choose* Show *from the Scheduled Objects panel.)*

Select schedule table item (or the border for all items) or CTRL-select to zoom: *(Hold* CTRL *down and select the window mark at* p2 *as shown in Figure 11.73; Release the CTRL key.)*

(The Zoom display of the window is shown at the right in Figure 11.73.)

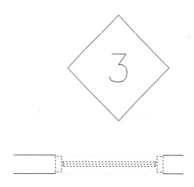

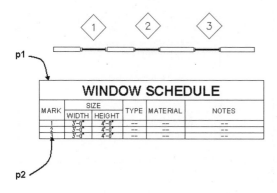

Zoom display of selected item

FIGURE 11.73 *Showing a selection of the schedule table*

EXPORTING SCHEDULE DATA

The data of the schedules can be exported to a text file or spreadsheet. If the data are exported to a spreadsheet, the data can then be used in cost estimates. The command to export schedule data is **ScheduleExport**. Select a schedule and choose **Export** from the Modify panel to access the **ScheduleExport** command (see Table 11.33).

TABLE 11.33 *Schedule Export command access*

Command prompt	SCHEDULEEXPORT
Ribbon	Select a schedule and choose Export from the Modify panel in the Schedule Table tab.
Shortcut menu	Select a schedule, right-click, and choose Export.

When you select **Export** from the Modify panel of the Schedule Table tab of a selected schedule, the **Export Schedule Table** dialog box opens, which allows you to define the output and input for exporting the schedule data. The **Export Schedule Table** dialog box shown in Figure 11.74 includes an **Output** and **Input** section. The options of the **Input** section are shown inactive since input is preset to the schedule selected. The options of the **Output** section are described below.

Output

SaveAs Type—The SaveAs Type list allows you to select the file format for the exported data. The type of file can be selected from one of the following: Microsoft

Excel 97 or 2003 (*.xls*), Text (Tab delimited) (*.txt), and CSV (Comma delimited) (*.csv*).

File Name—The File Name field allows you to type the name of the file for the data. The **Browse** button opens the file dialog box, which allows you to specify the directory for the data file.

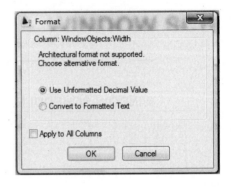

FIGURE 11.74 *Export Schedule Table dialog box*

Editing the Export Schedule Table dialog box specifies the format for the data file. When you click **OK** the format of the data in the schedule may cause the **Format** dialog box to open as shown at the right in Figure 11.74. The **Format** dialog box allows you to select the decimal or text format for the data of cells in an Excel spreadsheet. The options of the **Format** dialog box are described in the following list:

Use Unformatted Decimal Value—The Use Unformatted Decimal Value option converts feet and inches to decimal inches.

Convert to Formatted Text—The Convert to Formatted Text option retains the feet and inches format in data.

After you click **OK** the file will be created. You can check the file by opening the file in the Excel software program.

Steps to Export Schedules to Data Files

1. Select the schedule and choose **Export** from the Modify panel of the Schedule Table tab.
2. Select **Microsoft Excel 97 or 2003** as the **SaveAs Type** and type the name of the data file in the **File Name** field of the **Export Schedule Table** dialog box as shown in Figure 11.75. Click the **Browse** button and specify your student directory.
3. Click **OK** to dismiss the Export Schedule Table dialog box.
4. You can check the file by launching the Excel program and opening the file.

FIGURE 11.75 *Export of Myschedule Table dialog box*

The Context menu of a schedule includes Convert to Table, which allows you to convert a schedule to an AutoCAD table. The converted table cannot be updated; however, you can customize the converted table with additional columns and rows.

Creating Schedules for Project Data

The benefits of using project-based schedules can be realized if project-based tags for spaces and doors are used in drawings included in a project. This technique provides flexibility in maintaining an accurate and up-to-date comprehensive schedule. The project-based door tags increment the door number using the room number as a prefix for the door identifier. The Room Tag – Project Based or the Room Tag – Project Based – Scale Dependent creates a room identifier prefixed by the level. The tag is annotative; therefore, its size varies with the Annotation Scale settings. (In previous releases, scale-dependent tags were adjusted by the display configuration setting.) Therefore, room tags for the first level would increment from 111 to 112, and the second level would increment from 201 to 202. When the Door Tag – Project Based or Door Tag – Project Based Scale Dependent is inserted, the door identifier includes the room number suffixed by a letter. Since no two rooms share the same number across levels, each door is assigned a unique identifier. Therefore, if the door leaf is placed over a space, the door identifier will be numbered based upon the associated space. The Property Data Location grip shown in Figure 11.76 determines which space to link the door. Therefore, the leaf of exterior doors that swing out to comply with egress are not linked to a space. You can select the Property Data Location grip and drag the grip to a point over the interior space to link the door identifier to the space as shown at the right in Figure 11.76.

Property Data Location

Property Data Location

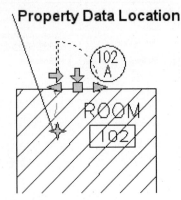

FIGURE 11.76 *Door Tag – Project Based*

The comprehensive schedule is developed in a Sheet drawing from the data located in View drawings. After placing a schedule in the Sheet drawing, toggle ON **Schedule external drawing** and specify the View drawings to include in the **External drawing** field of the **Schedule** properties palette as shown in Figure 11.77. When you update the schedule, the data from the View drawings specified will be displayed in the schedule.

Steps to Create a Comprehensive Project-Based Schedule

1. Select the Project tab of the Project Navigator. Click the Edit Levels button to open the Levels dialog box. Verify that an ID number for each level is specified. Open the Construct drawing and create spaces for each room. Save and close the drawing.

2. Create a View drawing that consists of the Construct drawing. Tag each space using the Room Tag – Project Based or the Room Tag – Project Based – Scale.

3. Tag each door in the View drawing using the Door Tag – Project Based or Door Tag – Project Based Scale Dependent option.

4. Create a new Sheet drawing in the Schedules and Diagrams category of the **Sheets** tab in the **Project Navigator**. Open the new Sheet drawing.

5. Select the Door Schedule – Project Based tool from the Scheduling tool palette. Press ENTER to schedule an external drawing and insert the schedule table onto the sheet.

6. Select the schedule, right-click, and edit the **External Source** section of the **Properties** palette as follows: toggle **Yes** for **Schedule external drawings** and edit **External drawing path** to include the View drawings of the project, as shown in Figure 11.77.

7. Select the schedule, right-click, and select **Update Schedule Table** from the shortcut menu.

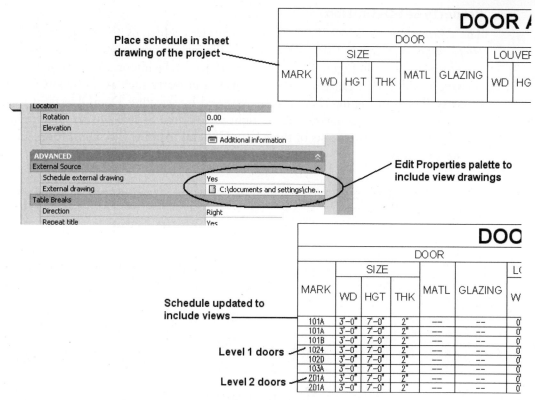

FIGURE 11.77 *Creating a comprehensive project door schedule*

INSERTING PROPERTIES WITHOUT TAGS

When tags are placed in a drawing, the multi-view block consists of attributes. The tags of the attributes specify the property set and property. Therefore, when an object is tagged, the property is inserted in the drawing with the multi-view block. The properties can then be used to develop schedules. The attribute tags of the Window tag multi-view block are shown in Figure 11.78. **Property sets** can be imported into a drawing in the Style Manager. The property set and properties are imported in the Style Manager to support the properties defined in the schedules. When you import property sets into a drawing, schedules can be developed for any object of the drawing without tagging each object. New tags can be developed that specify the property set and properties. The master set of property sets is located within the Documentation folder of the *Schedule Tables (Imperial).dwg* file located in the *{Vista – C:\ProgramData\ Autodesk\ACA 2012\enu\Styles\Imperial}* directory.

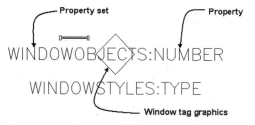

FIGURE 11.78 *Attribute tags of a multi-view block*

Property Set Definitions

The *Documentation Objects\Property Set Definitions* folder of the *Schedule Tables (Imperial).dwg [Schedule Tables (Metric).dwg]* file includes the comprehensive list of properties as shown in Figure 11.79. The properties can be imported into the current drawing in the Style Manager. The content of a property such as the DoorObject property shown in Figure 11.79 can be reviewed by selecting DoorObjects in the left pane and edit the properties in the right pane of the Style Manager. The right pane for a property set consists of three tabs: **General, Applies To, Definition**, and **Version History**.

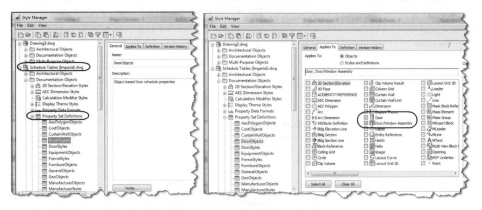

FIGURE 11.79 *Property Set Definitions of the Schedule Tables (Imperial).dwg*

General Tab

The General tab shown in Figure 11.79 allows you to type a name and description of the property set.

Applies to Tab

The Applies To tab includes options to specify to which type of object the property set will be applied when the schedule is developed. The options of the Applies To tab are described in the following list:

> **Applies To Objects**—The **Objects** radio button applies the data to the object without regard to style.
>
> **Applies To Styles and Definitions**—The **Styles and Definitions** radio button applies the properties to a specific style.
>
> **Object list**—The object list includes all objects. The DoorObject property can be applied only to doors and door/window assemblies as shown in Figure 11.79.

Definition Tab

The Definition tab lists all the properties of a property set. The Definition tab shown in Figure 11.80 lists the properties for a DoorObject. The buttons shown at the right in Figure 11.80 allow you to add additional properties to the DoorObject property set. You can choose the **Add Automatic Property Definition** button to choose an additional property from a comprehensive list of automatic properties available for door objects.

Name—Click the Name button to edit the name of the property.

Description—Type a Description in the field to specify a description for the property.

Edit Source—Click the Edit Source button to open the **Automatic Property Source** list of data sources for the property. The Edit Source button is active for automatic properties.

Type—The Type of property can be auto increment-character, auto increment-integer, integer real text, and true/false. Auto increment-character allows the letter of the tag to increase in the alphabet as you tag additional objects.

Source—The Source column lists the source for the object property of the automatic data.

Default—A default value will be displayed if no value is typed in the dialog box. A data field can be defined as the default value for the schedule.

Format—The Format list includes the following *Property Data Format*, which controls whether the values are displayed in feet and inches, decimal, or whole number formats. You can view the properties of a data format by editing the data format in the Style Manager. Data formats are located in the Documentation\Property Data Formats folder. The data format controls whether the data are displayed in all caps and the precision of the data. The Length-long format sets the precision to 1/4" while the Length-short format displays the data with 1/16" precision. Refer to the Property Data Format folder of the Style Manager.

Example—The Example column lists an example for the object property.

Visible—If Visible is checked, the property will be listed in the Edit Property Set Data and the Extended Data tab of Properties.

Order—The Order field allows you to specify the position of the property within the Edit Property Set Data dialog box.

The buttons located in the upper-right corner include the following:

Add Manual Property Definition—Select Add Manual Property Definition to create a new property in the **New Property** dialog box. The **Start With** field allows you to copy the attributes of existing properties.

Add Automatic Property Definition—Select Add Automatic Property Definition to specify an automatic property for the new property.

Add Formula Property Definition—Select Add Formula Property Definition to define a property based upon a formula that applies arithmetic operations on other property definitions.

Add Location Property Definition—Select Add Location Property Definition to open the **Location Property Definition** dialog box to define the location property definition.

Add Classification Property Definition—If Classifications are attached to the drawing, select Add Classification Property Definition to add reference to a classification.

Add Material Property Definition—Select Add Material Property Definition to include materials in the property definition.

Add Project Property Definition—Select Add Project Property Definition to reference a project to the property.

Add Anchor Property Definition—The Add Anchor Property Definition allows you to specify a property definition for the anchor of an object.

Add Graphic Property Definition—The Add Graphic Property Definition allows you to specify a property definition to a block or image.

Remove—Select Remove to remove a property from the property set.

Property sets allow you to control the text of a tag. When you place a door tag, the Number property of the DoorObjects property set is used as the tag. The default property Type for the Number object is Auto Increment-Integer and the Format is Number-Object. Therefore, you can edit the property Type to **Text** and tag doors with alphanumeric characters. If you set the Type to **Auto Increment-Character**, you can place tags that increment like their numeric counterparts.

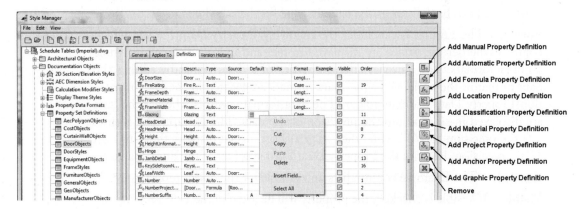

FIGURE 11.80 *Definition tab of the Property Set Definition Properties dialog box*

Version History

The Version History tab lists the date on which the style was modified, which is referenced during project synchronization.

INSERTING DATA FIELDS IN A SCHEDULE

You can define a data field as a default value for a property in the property set definition. The shortcut menu of the Default value of a property includes **Insert Field** as shown in Figure 11.80. When the **Insert Field** option is selected, the **Field** dialog box opens as shown in Figure 11.81. The **Field** dialog box lists field categories and field names. You can select a field name, such as File name or date, and this data will be displayed in the schedule as a default value for the property.

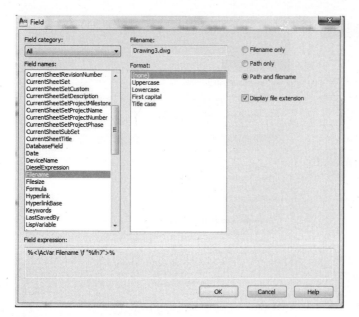

FIGURE 11.81 *Field dialog box*

IMPORTING PROPERTIES FOR BUILDING INFORMATION MODELING AND GREEN BUILDING STUDIO

Properties can be imported from external drawings to develop a building information modeling (BIM) data or to create a gbxml data file from the AutoCAD Architecture model. The United States General Services Administration (GSA) has validated AutoCAD Architecture as a building information modeling (BIM) authoring application. The property sets included in the GSA Model (Imperial Stb) or GSA Mode (Metric Stb) templates are imported into the drawing files to create the BIM data. In addition, you may export the content of a drawing file to create a green building extensible markup language (gbxml) file which may be used by Green Building Studio for energy analysis. The gbxml file contains data regarding the walls, doors, and slabs without the associated graphics.

Green Building Studio is a web-based energy analysis service and the Autodesk Ecotect Analysis software analyzes the whole building performance based upon the building model information extracted from a gbxml file. The results of the Green Building Studio service is shown at right in Figure 11.82. You may revise your model and export the model from AutoCAD Architecture or AutoCAD MEP to a gbxml file which can be imported into Green Building Studio or Ecotect. Prior to exporting the model from AutoCAD Architecture, the property sets located in the gbxml Property Set Definitions (Imperial).dwg or gbxml Property Set Definitions (Metric).dwg must be imported from the C:\ProgramData\Autodesk\ACA 2012\enu\Styles \Imperial\ or C:\ProgramData\Autodesk\ACA 2012\enu\Styles\Metric to the current drawing.

The properties must be added to walls, spaces, doors, and windows prior to export. The following property sets are available from the gbxml Property Set Definitions (Imperial).dwg or gbxml Property Set Definitions (Metric).dwg file: ThermalProperties, SpaceEngineeringObjects, and ZoneEngineeringObjects. The procedure for developing the gbxml file from an AutoCAD Architecture building is described in eTutorial Ex 11-8.

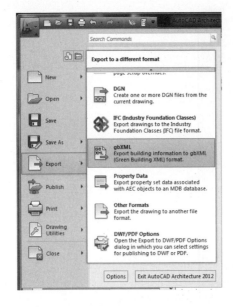

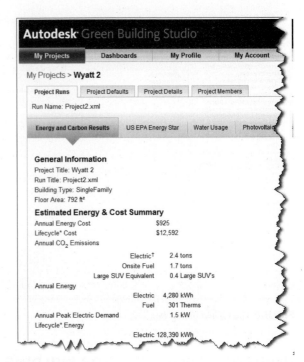

FIGURE 11.82 *Exporting to GBXML for Green Building Studio*

The GBXMLExport command is used to create the gbxml file from the model. Access the GBXMLEXPORT command as shown at left in Figure 11.82 and Table 11.34.

TABLE 11.34 *GBXLExport command access*

Command prompt	**GBXMLEXPORT**
Ribbon	Choose Export > **gbXML** from the Application menu.

NOTE Refer to e**Tutorials of Chapter** 11 for download from the Student Companion Site of **CengageBrain.com** described in the Preface includes: eTutorial 11-8 Importing Property Sets for Green Building Studio.

TUTORIAL 11-3: PLACING ALPHABETIC TAGS AND QUANTITIES FOR DOOR SCHEDULES

1. Verify that the Accessing Tutor\Imperial or Accessing Tutor\Metric content for Chapter 11 of the CengageBrain has been downloaded to your Accessing Tutor folder on your computer as described in Organizing Tutorial Directories in the Preface.

2. Open Ex 11-3.dwg from the Accessing Tutor\Imperial\Ch11 or Accessing Tutor\Metric\ Ch 11 directory.

3. Select **Save As > AutoCAD Drawing** from the Application menu and save the drawing as **Lab 11-3** in your student directory.

4. Verify that Annotation Scale = 1/8" = 1'-0" [1:100] and that Annotation Visibility and Automatically Add Scales are toggled ON.

This tutorial will modify the property used in the door tag to place text tags. In the following steps you will insert the door tag from the Tags tool palette, insert the schedule, alter the property set definition, and then place tags and modify the schedule.

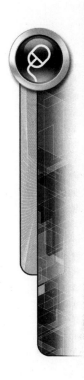

5. Select the Annotate tab and double-click the Annotation Tools of the Tool panel to display the Document palette set.

6. Insert the **DoorObject** property into the drawing by placing a door tag. Select **Door Tag** from the Scheduling panel **and** respond to the following workspace prompts to tag the front door:

> Select object to tag: *(Select the front door at* **p1** *as shown in Figure 11.83.)*
>
> Specify location of tag <Centered>: *(Select a location at* **p2** *as shown in Figure 11.83.)*
>
> *(Edit the Number property =* 01 *in the* **Edit Property Set Data** *dialog box. Click* OK *to dismiss the Edit Property Set Data dialog box.)*
>
> Select object to tag [Multiple]: ESC *(Press ESC to end the command.)*

7. Select **Door Schedule** from the Door Schedule flyout of the Scheduling panel. Edit the Design tab of the Properties palette as follows: Style = **Door Schedule**, Scale = **96.00 [100]**, Update automatically = **yes**, Add new objects automatically = **yes**, Scan xrefs = **yes**, Scan block references = **yes**, and Layer wildcard = *****. Respond to the following workspace prompts to place the schedule as shown in Figure 11.83.

> Select objects or ENTER to schedule external drawing: all ENTER
>
> Select objects or ENTER to schedule external drawing: ENTER *(Press ENTER to end the selection.)*
>
> Upper-left corner of table: *(Select a point near* **p3** *as shown in Figure 11.83.)*
>
> Lower-right corner (or RETURN): ENTER *(Press ENTER to insert the schedule scaled to the drawing scale.)*

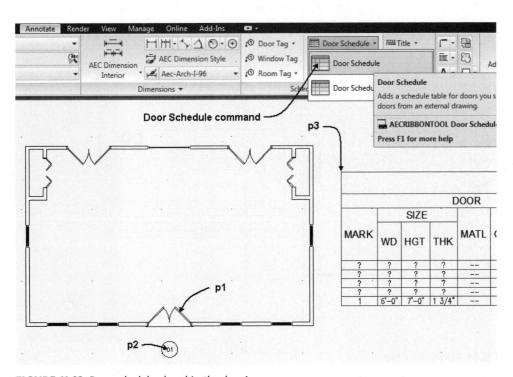

FIGURE 11.83 *Door schedule placed in the drawing*

8. Select the Manage tab. Select **Style Manager** from the Style & Display panel. Expand **Documentation Objects > Property Set Definitions** in the left pane. (The Number property is inserted in the drawing when the Door Tag is placed; this step edits the data type of the property.) Select the **DoorObjects** property in the left pane to open the edit tabs of the right pane. Select the **Definition** tab, scroll down the list of properties, and select **Number**. Set Edit Type = **Text**, Default = **A**, and Format = **Case-Upper**. Click **OK** to dismiss the Style Manager.

9. Select the Annotate tab. Select **Door Tag** from the Door Tag flyout in the Scheduling panel. Respond to the command line prompts as shown below to tag the doors.

> Select object to tag: *(Select the door at* **p1** *as shown in Figure 11.84.)*
>
> Specify location of tag <Centered>: *(Select a point near* **p2** *as shown in Figure 11.84.)*
>
> *(Verify that* **A** *is the* **Number** *value of the* **Edit Property Set Data** *dialog box. Click* **OK** *to dismiss the Edit Property Set Data dialog box.)*
>
> Select object to tag [Multiple]: **M** ENTER *(Use a crossing window to select all the doors of the floor plan as shown in Figure 11.84.)*
>
> Select objects to tag: *(Select a point near* **p3** *shown in Figure 11.84.)*
>
> Specify opposite corner: *(Select a point near* **p4** *shown in Figure 11.84.)* 9 found
>
> 6 were filtered out.
>
> Select objects to tag: ENTER *(Press ENTER to end the selection.)*
>
> *(Click* **OK** *to dismiss the Edit Property Set Data dialog box.)*
>
> Select object to tag: ENTER *(Press ENTER to end the command.)*

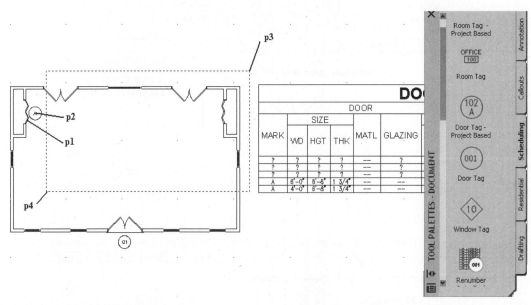

FIGURE 11.84 *Placing door tags*

10. Select the tags to display their grips and move the tags as necessary to improve clarity.

11. The outcome of this tutorial is to create a schedule that lists all doors of the same type with the same mark. Select the doors at **p1** and **p2**, as shown in Figure 11.85, right-click, and choose **Properties** to open the Properties palette. Select the **Extended Data** tab and edit the **Number** property to **B, press Enter**. Move the cursor over the workspace, left-click to move the focus from the Properties palette, and press ESC to clear the selection.

12. Double-click the door at **p3**, as shown in Figure 11.85, to open the Properties palette. Select the **Extended Data** tab and edit the **Number** property to **C, press Enter**. Move the cursor from the Properties palette and left-click in the workspace to change the focus to the workspace. Press ESC to clear the selection.

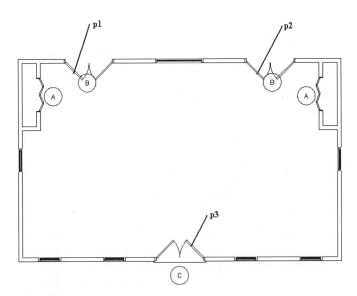

MARK		SIZE		DOOR	
	WD	HGT	THK		MATL
A	4'-0"	6'-8"	1 3/4"		--
A	4'-0"	6'-8"	1 3/4"		--
B	6'-0"	7'-0"	1 3/4"		--
B	6'-0"	7'-0"	1 3/4"		--
C	6'-0"	6'-8"	1 3/4"		--

FIGURE 11.85 *Door tags edited*

13. Select **Zoom Extents** from the **Zoom** flyout menu of the Navigation bar.

14. Select the schedule, and choose **Edit Style** from the General panel of the Schedule Table tab. Select the **Columns** tab and check **Include Quantity Column**. Referring to Figure 11.86, hold down CTRL and select the following headings:

- DOOR: MATL
- DOOR: GLAZING
- DOOR: LOUVER:WD HGT
- FRAME: MATL EL
- FRAME: DETAIL HEAD JAMB SILL
- FIRE RATING
- HARDWARE: SET NO KEYSIDE RM NO

15. Select **Delete** to remove the columns. Click **OK** to dismiss the **Remove Columns/ Headers** dialog box. Click **OK** to dismiss the Schedule Table Style Properties – Door Schedule dialog box.

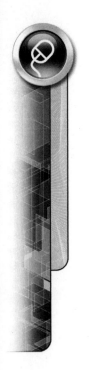

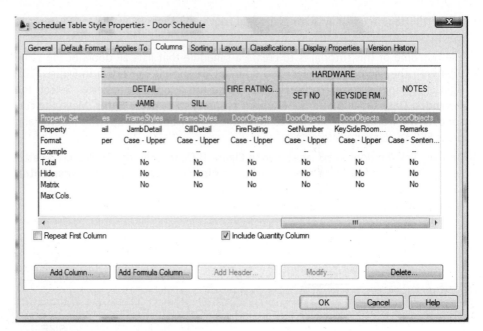

					HARDWARE		
		DETAIL		FIRE RATING...	SET NO	KEYSIDE RM...	NOTES
		JAMB	SILL				
Property Set	es	FrameStyles	FrameStyles	DoorObjects	DoorObjects	DoorObjects	DoorObjects
Property	ail	JambDetail	SillDetail	FireRating	SetNumber	KeySideRoom...	Remarks
Format	per	Case - Upper	Case - Upper	Case - Upper	Case - Upper	Case - Upper	Case - Senten...
Example		--	--	--	--	--	--
Total		No	No	No	No	No	No
Hide		No	No	No	No	No	No
Matrix		No	No	No	No	No	No
Max Cols.							

☐ Repeat First Column   ☑ Include Quantity Column

[Add Column...] [Add Formula Column...] [Add Header...] [Modify...] [Delete...]

[OK] [Cancel] [Help]

FIGURE 11.86 *Deleting columns from a schedule*

16. The schedule shown in Figure 11.85 lists door A with incorrect thickness. Select the schedule and choose **Show** from the Scheduled Objects panel. Select **Mark A** in the schedule to identify where the doors with incorrect thickness are located in the drawing. Press ESC to end the selection.

17. Door thickness is a style-based property; select door **A** at **p1**, as shown in Figure 11.87, right-click, and choose **Edit Door Style** from the shortcut menu. Select the **Dimensions** tab of the **Door Style Properties – Bifold – Double** dialog box; edit Door Thickness = **1-1/4" [30]**. Click **OK** to dismiss the Door Style Properties – Bifold – Double dialog box.

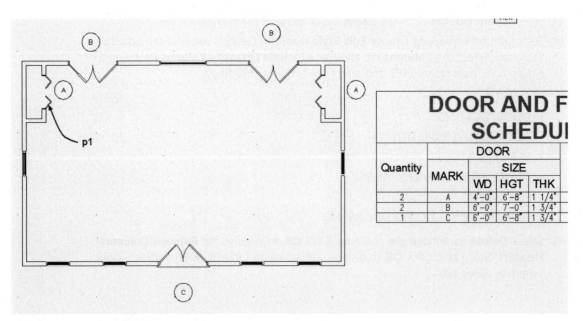

FIGURE 11.87 *Door schedule revised by style*

18. Select the schedule and choose **Export** from the Modify panel of the Schedule Table tab to open the **Export Schedule Table** dialog box shown in Figure 11.88. Select the **Browse** button, navigate to your student folder, and edit File name = **Lab 11-3**, Save As Type = **Microsoft Excel 2003 (*xls)**, and save in your student directory. Click **Save** to complete the selection and dismiss the **Create File** dialog box.

19. Click **OK** to dismiss the **Export Schedule Table** dialog box. **Imperial only:** Verify that the **Convert to Formatted Text** radio button is selected and **Apply to All Columns** is checked in the **Format** dialog box as shown in Figure 11.88. Click **OK** to dismiss the Format dialog box.

20. Save and close the drawing.

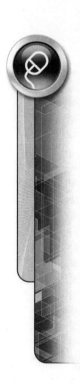

FIGURE 11.88 *Door schedule exported to the Excel program*

USING DISPLAY THEMES

Display themes allow you to assign a display to all objects with specific property definitions. Sample display theme tools are on the **Themes** palette of the **Document** palette set. Each of the tools uses a **Display Theme Style** that specifies a theme rule that compares the value of a property defined in the property set to that of the one specified in the rule of the display theme style. The rules of display are defined in the **Design Rules** tab of the **Display Theme Style** section of the Style Manager as shown in Figure 11.89. The Design Rules tab consists of a **Theme Settings** window that lists the Index number for each theme. The theme setting allows you to specify visibility, color, hatch, and scale properties for the display of the theme. A theme setting may consist of a color such as red, so that when a theme rule is satisfied, the related object will be displayed in red.

The rules for the theme are specified in the **Theme Rules for selected Theme Setting** section as shown in Figure 11.89. The theme rules shown will apply the theme settings defined by Index 1 to all wall styles with a value of **1 HOUR**.

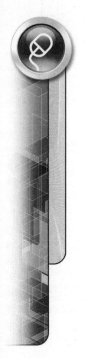

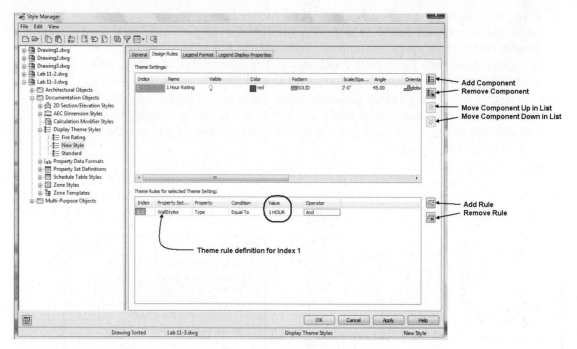

FIGURE 11.89 *Theme rules for theme setting*

After a display theme style is defined in the Style Manager, you can drag the style name from the Style Manager onto a tool palette. When you select the display theme tool from the tool palette, you are prompted to specify the insertion point. Upon insertion of the display theme table, the objects of the drawing will be displayed based upon the rule specified in the style.

To turn off the display of the display theme, select the display theme table, right-click, and choose **Disable Display Theme**. Only one display theme can be applied to a drawing; therefore, if other display theme tables have previously been inserted, a diagonal line will be drawn through the table to indicate that it is inactive as shown in Figure 11.90.

FIGURE 11.90 *Inactive display theme table*

PROJECT TUTORIAL 11-4: INSERTING AEC DIMENSIONS

1. Verify that the Accessing Tutor\Imperial or Accessing Tutor\Metric content for Chapter 11 of the CengageBrain has been downloaded to your Accessing Tutor folder on your computer as described in Organizing Tutorial Directories in the Preface.

2. Open AutoCAD Architecture and select **Project Browser** from the Quick Access toolbar. Use the **Project Selector** drop-down list to navigate to your *Student\Ch 11* directory. Select **New Project** to open the **Add Project** dialog box. Type **20** in the

Project Number field and type **Ex 11-4** in the **Project Name** field. Check **Create from template project**, choose **Browse**, and edit the path to *Accessing Tutor\ Imperial\Ch11\Ex 11-4\Ex 11-4.apj* or *Accessing Tutor\Metric\Ch11\Ex 11-4\Ex 11-4.apj* in the **Look in** field. Click **Open** to dismiss the Select Project dialog box. Click **OK** to dismiss the Add Project dialog box. Click **Close** to dismiss the Project Browser.

3. Select the **Views** tab of the **Project Navigator**, choose the **Views** category, right-click, and choose **Regenerate** to update external reference drawings to the view drawings. Double-click **Floor 2 View** to open the drawing.

4. Toggle OFF Snap Mode, Grid Display, Polar Tracking, Object Snap, and Object Snap Tracking and toggle ON Ortho Mode and Dynamic Input on the status bar.

5. Choose the Annotate tab and double-click Annotation Tools to display the Document palette group.

6. The Floor 2 View drawing must be setup for annotation. Verify the Drawing Window status bar settings as follows: Annotation Scale is 1/4″ = 1′-0″ [1:50], and Annotation Visibility and Automatically Add Scales are toggled ON. Verify that **High Detail** display configuration is current.

7. Select the Dimensions tool palette, choose the **AEC Dimension Exterior** tool, and respond to the workspace prompts as follows:

```
Select Objects or: (Select the wall at p1 as shown in
Figure 11.91.)
Select Objects or: (Select the wall at p2 as shown in
Figure 11.91.)
Select Objects or: (Select the wall at p3 as shown in
Figure 11.91.)
Select Objects or: (Select the wall at p4 as shown in
Figure 11.91.)
Select Objects or: (Select the wall at p5 as shown in
Figure 11.91.)
Select Objects or: (Select the wall at p6 as shown in
Figure 11.91.)
Select Objects or: ENTER (Press ENTER to end the selection.)
Specify insert point or: (Move the cursor to display a
horizontal dimension and select a point near p7 as shown in
Figure 11.91.)
(An AEC Dimension is placed as shown in Figure 11.91.)
```

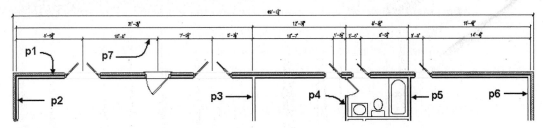

FIGURE 11.91 *Walls selection for an AEC Dimension*

8. Select **AEC Dimension – Exterior** from the **Dimensions** tool palette (Document palette set) and respond to the workspace prompts as follows:

> Select Objects or: *(To specify dimension alignment, select the wall at* **p1** *to specify as shown in Figure 11.92.)*
>
> Select Objects or: *(Select the wall at* **p2** *as shown in Figure 11.92.)*
>
> Select Objects or: *(Select the wall at* **p3** *as shown in Figure 11.92.)*
>
> Select Objects or: *(Select the wall at* **p4** *as shown in Figure 11.92.)*
>
> Select Objects or: *(Select the wall at* **p5** *as shown in Figure 11.92.)*
>
> Select Objects or: ENTER *(Press ENTER to end the selection.)*
>
> Specify insert point or *(Move the cursor to display a vertical dimension and select a point near* **p6** *as shown in Figure 11.92.)*

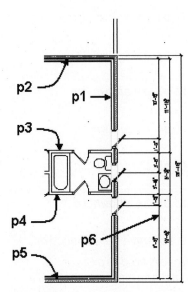

FIGURE 11.92 *Walls selection for an AEC Dimension*

9. Select the **AEC Dimension – Interior** tool from the **Dimensions** tool palette of the **Document** palette set and respond to the workspace prompts as follows:

> Select Objects or: *(To specify dimension alignment, select the wall at* **p1** *to specify as shown in Figure 11.93.)*
>
> Select Objects or: *(Select the wall at* **p2** *as shown in Figure 11.93.)*
>
> Select Objects or: *(Select the wall at* **p3** *as shown in Figure 11.93.)*
>
> Select Objects or: *(Create a crossing selection and select a point near* **p4** *as shown in Figure 11.93.)*

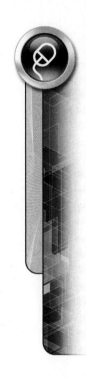

Specify opposite corner: *(Choose a point near* **p5** *as shown in Figure 11.93.)*

Select Objects or: ENTER *(Press ENTER to end selection.)*

Specify insert point or *(Move the cursor to display a horizontal dimension and select a point near* **p6** *as shown in Figure 11.93.)*

10. Select the AEC Dimension placed in step 9 and choose **Edit Style** from the General panel. Choose the **Display Properties** tab of the **AEC Dimension Style Properties – Interior Partitions** dialog box. Verify that **Plan High Detail** is the current display representation. Click the **Edit Display Properties** button. Choose the **Contents** tab. Choose the **Opening/Door/Window** selection in the **Apply to** window and clear the **Center** checkbox for the dimension chain. Choose the **Opening in wall** selection of the **Apply to** window and clear the **Center of opening** checkbox for the dimension chain. Click **OK** to dismiss all the dialog boxes.

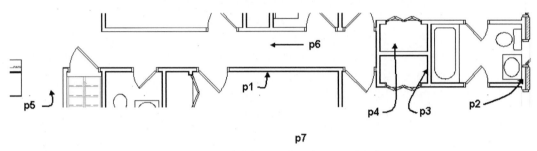

FIGURE 11.93 *Walls selection for an AEC Dimension*

11. Select the Interior dimension placed in step 9 and choose **Remove Objects** from the **Modify** panel of the AEC Dimension tab. Respond to the workspace prompts as follows:

Select Building Elements: *(Choose the stair at* **p1** *shown in Figure 11.94.)*

Select Building Elements: *(Press ENTER to end the command.)*

12. An additional extension line can be added to the Interior dimension to specify the end of the wall at point **p2** as shown in Figure 11.94. Select the Interior dimension and choose **Add Extension Lines** from the **Modify** panel of the AEC Dimension tab. Respond to the workspace prompts as follows:

Pick dimension points on screen to add: *(Hold the SHIFT key down, right-click, and choose* Endpoint *to choose point* **p2** *as shown in Figure 11.94.)*

Pick dimension points on screen to add: ENTER *(Press ENTER to end selection.)*

Select Dimension chain: *(Select the dimension line at* **p3** *as shown in Figure 11.94.)*

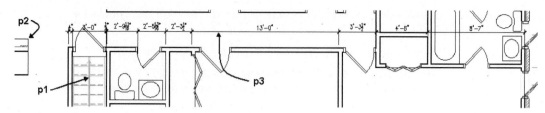

FIGURE 11.94 *Walls selection for an AEC Dimension*

13. In this step you will add an extension line to Chain 3 from the outside surface of the stud. Select the AEC Dimension of the exterior wall at **p3** as shown in Figure 11.95 and choose Add Extension Lines from the Modify panel. Respond to the workspace prompts as follows:

    Pick dimension points on screen to add: *(SHIFT + right-click, choose the Endpoint object snap, and select the outside stud surface at* **p2** *as shown in Figure 11.95.)*

    Pick dimension points on screen to add : ENTER: *(Press ENTER to end the command.)*

    Select Dimension chain: *(Choose the dimension line at* p3 *as shown in Figure 11.95.)*

    (Press ESC to end the command and display the extension line as shown at the right in Figure 11.95.)

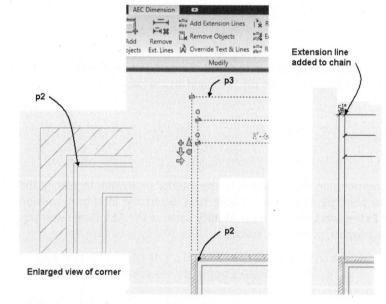

FIGURE 11.95 *Extension line added to AEC Dimension*

14. Dimension the remaining exterior walls. Add extension lines to represent the outside of the stud as developed in step 12 for each corner as shown in *Figure 11.96.*

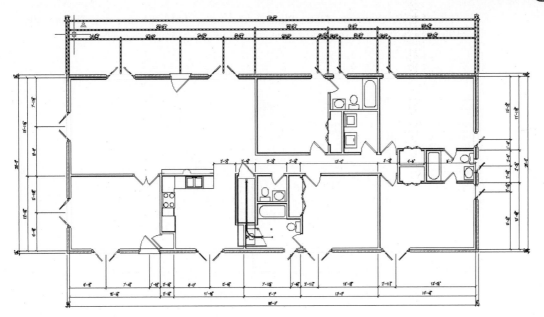

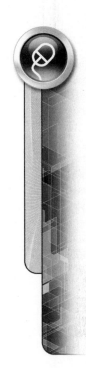

FIGURE 11.96 *AEC Dimensions added to plan*

15. Save and close the drawing.

16. Select the **Sheets** tab and expand Architectural and Plans categories. Double-click the **A-4 Floor Plan Level 2** drawing of the **Project Navigator**.

17. This sheet was created in a previous tutorial; therefore, the viewport is not sized to display the dimensions. In this step you will delete the viewport to prepare the sheet for a model space view and viewport. Double-click outside the viewport near the border of the sheet to set Paper Space current. Select the viewport, right-click, and choose Basic Modify Tools > Delete. Save the drawing.

18. The Viewport has been deleted but **Floor 2 View** remains attached. Choose **Manage Xrefs** icon in the lower-right corner of the Drawing Window status bar to open the **External References** palette. Choose **Floor 2 View** in the file list, right-click, and choose **Detach**. Close the External References palette.

19. Select the Views tab of the Project Navigator. Double-click **Floor 2 View** to open the drawing. Choose Zoom > All from the Zoom flyout of the Navigation toolbar. Select the **Floor 2 View** drawing title in the Project Navigator, right-click, and choose **New Model Space View** to open the **Add Model Space View** dialog box shown in

Figure 11.97. Overtype **Floor Plan** in the name field. Choose the **Define View Window** button and respond to the command line prompts as follows:

FIGURE 11.97 *AEC Dimensions added to plan*

> *Specify first corner: (Select a point near **p1** as shown in Figure 11.97.)*
>
> *Specify opposite corner: (Select a point near **p2** as shown in Figure 11.97.)*
>
> *(Click OK to dismiss the* Add Model Space View *dialog box.)*

20. Save and close the drawing.

21. Verify that **A-4 Floor Plan Level 2** sheet is open.

22. Select the Views tab. Select the **Floor Plan** model space view, right-click, and choose **Place on Sheet**. Move the cursor to the sheet and click near point **p1** as shown in Figure 11.98.

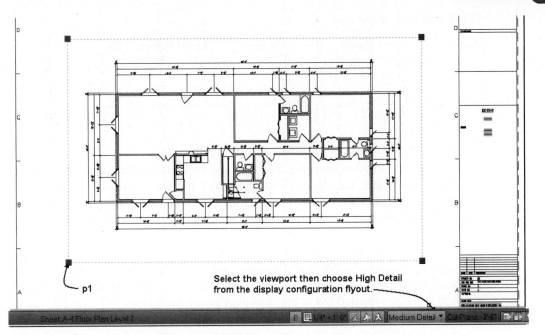

Select the viewport then choose High Detail
from the display configuration flyout.

Sheet A-4 Floor Plan Level 2

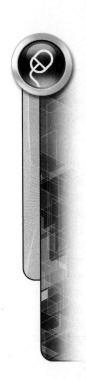

FIGURE 11.98 *Edit viewport properties of a sheet*

23. Select the viewport to display its grips. Select High Detail display configuration from the flyout. Toggle ON Annotation Visibility in the Drawing Window status bar.

24. Select the Cut Plane value in the Drawing Window status bar. Verify that the Cut Height = 3'-6" [1400] and Display Range Below = – 7' [–2135]. Click **OK** to dismiss the Global Cut Plane dialog box.

25. Save and close the drawing. Choose the Project tab of the Project Navigator. Choose **Close Current Project**. Choose Close all project files. Choose Yes to save all project drawings.

TUTORIAL 11-5: CREATING TAGS AND SCHEDULES IN A PROJECT

1. Verify that the Accessing Tutor\Imperial or Accessing Tutor\Metric content for Chapter 11 of the CengageBrain has been downloaded to your Accessing Tutor folder on your computer as described in Organizing Tutorial Directories in the Preface.

2. Open AutoCAD Architecture and select **Project Browser** from the Quick Access toolbar. Use the **Project Selector** drop-down list to navigate to the *Accessing Tutor\ Ch 11* directory. Select **New Project** to open the **Add Project** dialog box. Type **Ex 11-5** in the **Project Number** field and type **Ex 11-5** in the **Project Name** field. Check the **Create from template project** checkbox. Choose **Browse** and edit the path to *Accessing Tutor\Imperial\Ch11\Ex11-5\Ex 11-5.apj* or *Accessing Tutor\Metric\Ch11\Ex11-5\ Ex 11-5.apj* in the **Look in** field. Click **Open** to close the Select Project dialog box. Click **OK** to dismiss the Add Project dialog box. Click **Close** to dismiss the Project Browser.

3. Select the **Project** tab and click the **Edit Levels** button to open the **Levels** dialog box. The room numbers and door numbers assigned in this tutorial will reference the ID number listed in the Levels dialog box. Edit and verify the data as shown in Table 11.35. Click **OK** to dismiss the Levels dialog box.

TABLE 11.35 *ID of project levels*

Imperial	Level Name	ID	Metric	Level Name	ID
	1	1		G	1
	2	2		1	2

4. Select the Constructs tab of the Project Navigator.

5. Double-click Floor 1 to open the drawing. Right-click the Dynamic Input toggle of the Application status bar, choose Settings. Select the checkboxes of Enable Pointer Input, Enable Dimension Input where possible, and Show command prompting and command input near the crosshairs. Select OK to dismiss the Drafting Settings dialog box.

6. Select the High Detail display configuration from the Display Configuration flyout menu of the Drawing Window status bar. Select the 1/4" = 1'-0" [1:50] Annotation Scale from the flyout scale list. Verify that Annotation Visibility and Automatically Add Scales are toggled ON in the Drawing Window status bar.

7. Toggle OFF Snap Mode, Grid Display, Ortho Mode, Object Snap, Object Snap Tracking, Allow Disallow Dynamic UCS, and Show Hide Lineweight on the Application status bar. Toggle ON Dynamic Input, Polar Tracking, Quick Properties on the status bar.

8. Choose the Home tab. Choose Tools of the Build panel if the Design palette set is not displayed.

9. Choose the Spaces palette of the Design palette group.

10. Choose the Lobby space from the Spaces tool palette. Set Tag = None, Create type = Generate, and Geometry Type = Extrusion. Select the space at p1 as shown in Figure 11.99. Press ESC to end the command.

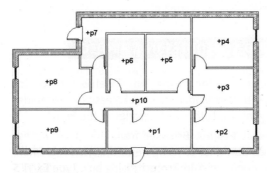

FIGURE 11.99 *Spaces for level 1*

11. Choose the **Office (Small)** space from the Space tool palette. Verify that Create type = Generate and Tag = None in the Properties palette. Select the spaces in the rooms at **p2**, **p3**, **p4**, **p8**, and **p9** as shown in Figure 11.99. Press ESC to end the command.

12. Choose the **Restroom—Women (Medium).** Verify that Create type = Generate and Tag = None in the Properties palette. Select the space in room **p5** as shown in Figure 11.99. Press ESC to end the command.

13. Choose the **Restroom—Men (Medium)** space tool from the Spaces tool palette. Verify that Create type = Generate and Tag = None in the Properties palette. Select the space in room **p6** as shown in Figure 11.99. Press ESC to end the command.

14. Choose the **Stairway** space tool from the Spaces tool palette. Verify that Create type = Generate and Tag = None in the Properties palette. Select the space in room **p7** as shown in Figure 11.99. Press ESC to end the command.

15. Choose the **Corridor** space tool from the Spaces tool palette. Verify Create type = Generate and Tag = None in the Properties palette. Select the space in room **p10** as shown in Figure 11.99. Press ESC to end the command.

16. Save and close the drawing.

17. Select the **Views** tab. Open the **Floor Plan 1** View drawing.

18. Choose the **View unreconciled new layers in the Layer Properties Manager** hyperlink of the **Unreconciled New Layers** message box located in the lower-right corner of the Drawing Window status bar. In the Layer Properties Manager, hold the CTRL key down and select all layers, right-click, and choose **Reconcile Layer** from the shortcut menu. Close Layer Properties Manager.

19. Choose **1/4" = 1"-0" [1:50]** Annotation Scale and the **Low Detail** display configuration from the **Drawing Window** status bar. Low Detail display configuration is chosen to facilitate selection of spaces. Verify that Annotation Visibility and Automatically Add Scales are toggled ON. Toggle OFF Snap Mode and Object Snap in the Application status bar.

20. Choose the **Annotate** tab. Choose **Room Tag – Project Based** from the Room Tag flyout menu of the Scheduling panel and respond to the workspace prompts as follows:

    Select object to tag: *(Select the User Single hatch pattern of the space at* p1 *as shown in Figure 11.99.)*

    Specify location of tag <Centered>: *(Select a location in the middle of the room to open the* **Edit Property Set Data** *dialog box. Scroll down to* **Space Objects** *and verify as follows: NumberProjectBased = 101 and Increment = 01. Click OK to dismiss the dialog box.)*

    Select object to tag: *(Press the down arrow, choose* Multiple, *and choose the remaining spaces in the following order:* p2, p3, p4, p5, p6, p7, p8, p9, *and* p10 *as shown in Figure 11.99. Press ENTER to end selection. Click OK to dismiss the Edit Property Set Data dialog box. Press ESC to end the command.)*

21. Select the tag for the spaces placed at **p7** and **p10** as shown in Figure 11.99. Drag each tag to the middle of the space.

22. When the spaces are tagged, property data are added to the external reference file, Floor 1. To reload Floor 1, choose Manage Xrefs from the lower-right corner of the Drawing Window status bar. Hold CTRL + select Floor 1 and Stair, right-click, and choose Reload. Close the External References palette.

23. Choose the **Door Tag – Project Based** from the **Scheduling** panel of the Annotate tab and respond to the workspace prompts as follows:

    Select object to tag: *(Select the door at* p1 *as shown in Figure 11.100.)*

    Specify location of tag <Centered>: ENTER *(Press ENTER to center the tag.)*

    (The **Edit Property Set Data** *dialog box opens; click OK to dismiss it.)*

    Select object to tag: *(Choose the down arrow and select* Multiple *from the workspace option list.)*

Select objects to tag: *(Create a crossing selection from points* **p2** *to* **p3** *as shown in Figure 11.100.)*

Select objects to tag: ENTER *(Press ENTER to end the selection. The AutoCAD warning dialog opens. Click* **No** *to refrain from creating a duplicate tag. Click* **OK** *to dismiss the Edit Property Set Data dialog box. Press ESC to end the command.)*

(The tags of the doors located at p1 and p4 of Figure 11.100 indicate Space not found. The **Door Tag – Project Based** *tag extracts space information to number the door. Since the door swing is not over a space, the tag cannot find a space to associate the door number. The Property Data Location grips of the doors will be adjusted later in this tutorial to tag the doors.)*

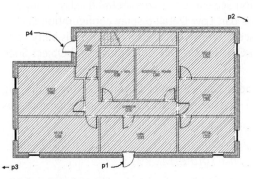

FIGURE 11.100 *Creating spaces for level 1*

24. Choose High Detail display configuration from the Drawing Window status bar.

25. Save and close the **Floor Plan 1** drawing.

26. Choose the Constructs tab. Double-click **Floor 2** to open the drawing.

27. Select the **High Detail** display configuration from the **Display Configuration** flyout menu of the Drawing Window status bar. Select the **1/4" = 1'-0" [1:50]** scale from the flyout scale list.

28. Select the Office (Small) space tool from the Spaces tool palette. Set Tag = None, Create type = Generate, and Geometry Type = Extrusion. Select the spaces in the rooms at **p1, p2, p3, p4, p6, p7**, and **p8** as shown in Figure 11.101. Press ENTER or ESC to end the command.

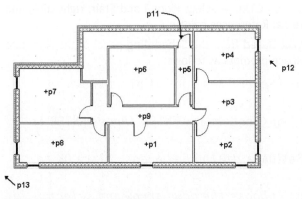

FIGURE 11.101 *Placing tags for level 2*

29. Select the Corridor space tool from the Spaces tool palette. Edit the Create type to Generate and Tag to None. Select a point near **p5** and **p9** as shown in Figure 11.101. Press ENTER or ESC to end the command. A space is not defined for stairway since this space spans two levels and was tagged in Floor 1.

30. Save and close the **Floor 2** Construct drawing. The Floor 2 drawing must be saved and closed to tag the View drawing in the subsequent steps.

31. Double-click **Floor Plan 2** View to open the drawing.

32. Since spaces were added to Floor 2, construct additional layers were added in the attached drawing. Choose the **View unreconciled new layers in the Layer Properties Manager** hyperlink of the **Unreconciled New Layers** message box located in the lower-right corner of the Drawing Window status bar. In the Layer Properties Manager, hold the CTRL key down and select all layers, right-click, and choose **Reconcile Layer** from the shortcut menu. Close the Layer Properties Manager.

33. Set Annotation Scale to **1/4″ = 1′-0″[1:50]** and Display Configuration = **Low Detail** in the Drawing Window status bar. Turn off object Snap Mode.

34. Select the Annotate tab. Choose **Room Tag – Project Based** from the Room Tag flyout of the Scheduling panel and respond to the workspace prompts as follows:

    > Select object to tag: *(Select the space at* p1 *as shown in Figure 11.101.)*
    >
    > Specify location of tag <Centered>: ENTER *(Press ENTER to center the tag in the space.)*
    >
    > *(The* **Edit Property Set Data** *dialog box opens; scroll down to SpaceObjects and verify that NumberProjectBased =* 201 *and Increment =* 01*. Click* OK *to dismiss the dialog box.)*
    >
    > Select object to tag: *(Press the down arrow and choose* Multiple*.)*
    >
    > Select objects to tag: *(Choose the remaining spaces in the following order:* p2, p3, p4, p5, p6, p7, p8, *and* p9 *as shown in Figure 11.101. Press ENTER to end the selection. Click* OK *to dismiss the Edit Property Set Data dialog box. Press ESC to end the command.)*

35. Select the tag for the space of **p9** as shown in Figure 11.101. Select the grip of the tag and drag the tag down into the corridor space.

36. Choose **Door Tag – Project Based** from the **Door Tag** flyout of the Scheduling panel and respond to the workspace prompts as follows:

    > Select object to tag: *(Select the door at* p11 *as shown in Figure 11.101.)*
    >
    > Specify location of tag <Centered>: ENTER *(Press ENTER to center the tag.)*
    >
    > *(The* **Edit Property Set Data** *dialog box opens; choose* OK *to dismiss the dialog box.)*
    >
    > Select object to tag: *(Press the down arrow and select* Multiple *from the workspace option list.)*
    >
    > Select objects to tag: *(Select the doors and create a crossing selection from point* p12 *to* p13 *as shown in Figure 11.101.)*

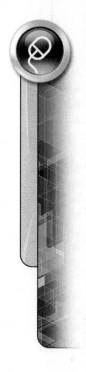

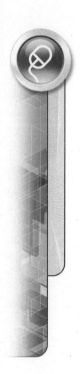

Select objects to tag: ENTER *(Press* Enter *to end the selection. The* Edit Property Set Data *dialog box opens; click* OK *to dismiss the Edit Property Set Data dialog box. Press* ESC *to end the command.)*

37. The insertion of tags in Floor Plan 2 view drawing added property data to the external reference drawing file, Floor 2. To reload the latest changes in the Floor 2 construct choose the **Xref Manager** in the lower-right corner of Drawing Window status bar. Hold the CTRL key and select the Floor 2 and Stair in the External References palette, right-click, and choose Reload. Close the External References palette.

38. To link the exterior doors to a space, select the door at **p11** as shown in Figure 11.101 and choose **Open Reference** from the External Reference tab. Select the door at **p1**, as shown in Figure 11.102, to display its grips. Drag the **Property Data Location** grip from point **p1** to a point near **p2** as shown in Figure 11.102. Press ESC to clear the grips.

39. Save and close the **Floor 2** Construct drawing.

40. Click **Reload FLOOR 2** of the **External Reference File Has Changed** tip located in the lower-right corner. The door mark updates to reflect the space as shown at the right in Figure 11.102.

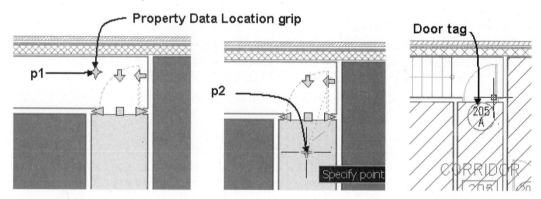

FIGURE 11.102 *Editing the Property Data Location*

41. Choose High Detail display configuration. Save and close the **Floor Plan 2** View drawing.

42. Double-click to open the **Floor 1** drawing in the Constructs tab of the Project Navigator. Refer to step 37 and drag the Property Data Location grip of each exterior door onto the interior space at **p1** and **p4** as shown in Figure 11.100. Save and close the drawing.

43. Open Floor Plan 1 view drawing in the Project Navigator. Verify that the door tags for the exterior doors are numbered. Save and close the Floor Plan 1 view drawing.

44. In this step you will create a model drawing with each floor plan attached as a reference; this drawing is used to extract data for a schedule. In the **Views** tab of the **Project Navigator**, select the **Views** category, right-click, and choose **New View Dwg > General**. Type **Model** as the name of the new drawing, select **Next**, and check Level **1 [G]** and Level **2** [1] for Division **1**. Click **Next** to verify that the **Floor 1**, **Floor 2, and Stair** constructs are checked. Click **Finish** to dismiss the Add General View dialog box.

45. Select the **Sheets** tab of the Project Navigator, expand the **Architectural** folder, and select the **Schedules and Diagrams** category. Right-click and choose **New > Sheet**.

Type **A-11** in the **Number** field and type **Schedules** in the **Sheet title** field. Choose the Open in drawing editor check box. Click **OK** to close the **New Sheet** dialog box.

46. Choose **Door Schedule Project Based** from the **Door Schedule** flyout in the **Scheduling** panel of the Annotate tab. Respond to the following workspace prompts to insert the schedule.

    Select objects or ENTER to schedule external drawing: ENTER *(Press ENTER to schedule external drawings.)*

    Upper-left corner of table: *(Select a point near p1 in the drawing as shown in Figure 11.103.)*

    Lower-right corner (or RETURN): ENTER *(Press ENTER to accept default size.)*

47. Select the schedule, right-click, and select **Properties** from the shortcut menu. Select the **Design** tab of the Properties palette, scroll down to the Advanced section, toggle **Schedule external drawing** to **Yes**, click the **External drawing** field, and select the *\Student\Ch 11\Ex 11-5\Views\Model.dwg* drawing.

48. Verify that the schedule is selected, then right-click and select **Update Schedule Table** to create the comprehensive schedule shown in Figure 11.103.

49. Choose the Project tab of the Project Navigator. Choose the Close Current Project command of the toolbar located on the Project tab. Choose Close all project files of the Project Browser – Close Project Files dialog box. Click Yes to save changes to all drawings of the AutoCAD message box.

FIGURE 11.103 *Schedule created for all floors of the building*

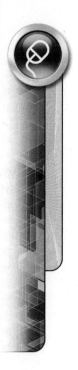

PROJECT TUTORIAL 11-6: USING DISPLAY THEMES

1. Verify that the Accessing Tutor\Imperial or Accessing Tutor\Metric content for Chapter 11 of the CengageBrain has been downloaded to your Accessing Tutor folder on your computer as described in Organizing Tutorial Directories in the Preface.

2. Open AutoCAD Architecture and verify that a drawing is open. Select **Project Browser** from the Quick Access toolbar. Use the **Project Selector** drop-down list to navigate to your student directory. Select **New Project** to open the **Add Project** dialog box. Type **Ex 11-6** in the **Project Number** field and **Ex 11-6** in the **Project Name** field. Check the **Create from template project** checkbox, choose **Browse,** and edit the path to *Accessing Tutor\Imperial\Ch11\Ex11-6\Ex 11-6.apj* or *Accessing Tutor\Metric\ Ch11\Ex11-6\Ex 11-6.apj* in the **Look in** field. Click **Open** to dismiss the **Select Project** dialog box. Click **OK** to dismiss the Add Project dialog box. Click Close to dismiss the Project Browser.

3. Select the **Constructs** tab of the **Project Navigator**. Double-click **Floor 1** to open the drawing.

4. Choose the **Home** tab. Choose the layer drop-down of the Layer Properties toolbar, toggle OFF A-Area-Spce [Aec-Space] and A-Area-Spce-Patt layers.

5. Choose the Tools button of the Build panel if necessary to display the Design palette set.

6. Select the **Stud 4 GWB-0.625 2 Layers Each Side [Stud-102 GWB-018 2 Layers Each Side]** tool of the **Walls** palette, right-click, and choose **Apply tool Properties to > Wall**. Select the walls at **p1, p2, p3, p4,** and **p5** as shown in Figure 11.104. Press ENTER to end the selection, but retain the selection for the next step.

7. Retain the selection of the walls and select the **Extended Data** tab of the Properties palette. Click the **Edit style property set data** button to open the **Edit Property Set Data** dialog box. Set Edit FireRating = **2 HOUR** and Type = **A** as shown in Figure 11.104. Click **OK** to dismiss the **Edit Property Set Data** dialog box. Press ESC to end the selection.

8. Save and close the Floor 1 drawing.

9. Choose the **Views** tab of the **Project Navigator**. Double-click **Floor Plan 1** View drawing to open the file.

10. Choose the Annotate tab and double-click the Annotation Tools button of the Tools panel. Choose the **Theme by Fire Rating** tool from the **Themes** palette of the **Document** palette set and respond to the command prompts as follows:

       Upper-left corner of display theme: *(Select a point at the left in the drawing.)*

       Lower-right corner of display theme (or RETURN): *(Press ENTER; 2 hour rated walls are shown in red.)*

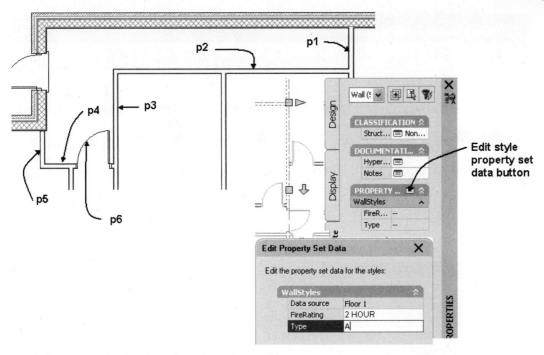

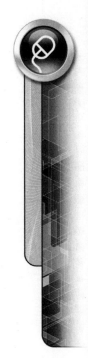

FIGURE 11.104 *Wall selection and edit for a theme table*

11. Choose the Home tab. Choose Layer Properties of the Layers panel to open the Layer Properties Manager. Expand the Xref category in the left pane. Choose Floor 1, toggle OFF Floor 1| A-Area-Spce [Floor1|Aec-Space] and Floor 1|A-Area-Spce-Patt layers. Close the Layer Properties Manager.

12. Select the Theme table, right-click, and choose Edit Display Theme Style to open the Display Theme Style Properties – Fire Rating dialog box.

13. Choose the Design Rules tab as shown in Figure 11.105. Select Index 2 in the top window. The theme rule for the Theme Setting is displayed in the lower window. The theme setting for 2 HOUR will display if the wall style FireRating property is equal to 2 HOUR or the DoorObjects FireRating property is equal to 2 HOUR as shown in the lower window of Figure 11.104. Click OK to dismiss the Display Theme Style Properties dialog box.

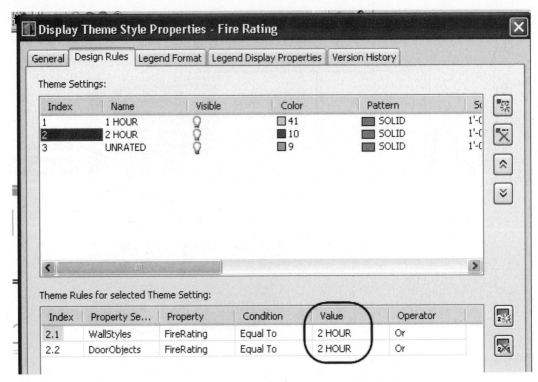

FIGURE 11.105 *Design Rules tab of the Display Theme Style Properties dialog box*

14. Save and close Floor Plan 1 view drawing.

15. Choose the Constructs tab and double-click Floor 1 construct.

16. Select the door at **p6** as shown in Figure 11.104. Choose the **Extended Data** tab of the Properties palette. Edit FireRating = **2 HOUR**. Save and close the drawing.

17. Choose the Views tab and double-click to reopen the Floor Plan 1 view drawing. The display theme rule has changed the door to a red color.

18. Choose the Project tab of the Project Navigator. Choose the **Close Current Project** command of the toolbar located on the Project tab. Choose Close all project files of the Project Browser – Close Project Files dialog box. Click Yes to save changes to all drawings of the AutoCAD message box.

NOTE

Refer to **e Projects** of Chapter 11 for download from the Student Companion Site of Cengage-Brain http://www.cengagebrain.com described in the Preface. The Project Ex 11-7 Creating Window Schedules is included in the e Project document.

SUMMARY

- The scale factor for annotation placed from tool palettes is set in Annotation Scale of the Drawing Window status bar. Tags are annotative multi-view blocks; therefore, the size of the tag is specified in the Annotation Scale of the Drawing Window status bar.

- Commands to place leaders, tags, and revision clouds are located on the Annotation tool palette.

- AEC Dimensions can be placed from the Dimensions palette, while non-AEC Dimensions are selected from the *Documentation\Miscellaneous* category of the Design Tool Catalog – Imperial or Design Tool Catalog – Metric catalog.

- Tags for objects can be placed in the drawing from the Tags tool palette or the Scheduling panel. Additional tags for doors, windows, walls, furniture, equipment, and structural members are located in the Schedule Tags category of the Documentation Tool Catalog – Imperial or Documentation Tool Catalog – Metric.

- The property set and properties for the schedule are inserted into the drawing when objects are tagged.

- Property Set Definitions for AEC objects can be imported into the Style Manager from tags or when you create a new property set definition in the Style Manager.

- The units and format of data in a schedule are displayed according to the Property Data Formats.

- The ScheduleStyleEdit command is used to define the components of a schedule.

- Additional schedules can be i-dropped from the Schedules category of the Documentation Tool Catalog – Imperial or Documentation Tool Catalog – Metric of the Content Browser.

Refer to e **Review Questions** folder which includes review questions for Chapter 11 of Student Companion Site of **CengageBrain** http://www.cengagebrain.com described in the Preface.

NOTE

CHAPTER
12

Creating Elevations, Sections, and Details

INTRODUCTION

This chapter includes the commands to extract elevations and sections from a building model and develop details. The use of the **Project Navigator** to create the building model and the callout tools used to create elevations and sections will be presented. The model is created as a View drawing in the Project Navigator by attaching each floor or building component as a construct. The Project Navigator provides for flexibility in design because it tracks changes that occur in Construct drawings and displays the latest version of a drawing in the model. Detail callouts, section marks, and elevation marks are used to link callouts to sheet numbers.

OBJECTIVES

Upon completion of this chapter, you will be able to

- Use the Project Navigator to create a 3D model of a building.

- Create elevations, sections, and details using the elevation and section marks tools located on the Callouts tool palette.

- Revise the display of the elevation based upon changes in the model using Refresh (2DSectionResultRefresh) and Regenerate (2DSectionResultUpdate) commands.

- Edit the dimensions and subdivision of the projection box for elevation or section objects.

- Create live sections and model views from pictorial views of a model.

- Modify lines of a section or an elevation using the 2DSectionResultEdit, 2DSectionEditComponent, and 2DSectionResultAddHatchBoundary commands.

- Create and modify 2D Section/Elevation styles.

- Create keynote and dimension details using the Detail Component Manager.

CREATING THE MODEL FOR ELEVATIONS AND SECTIONS

The elevation or section drawings of a building are developed from a model drawing. The model is constructed in the Project Navigator by attaching such Construct drawings as floor plans, decks, porches, or other detached components associated with each level. The levels specified in the Project Navigator as shown in Figure 12.1 determine the Z coordinate insertion point for the Construct drawings assigned to each level. The insertion point of the construct is located at the baseline of a wall. The siding wall component defined in the wall style of floor 1 extends below the baseline of the wall to cover the floor framing as shown in Figure 12.1. In this example, the FloorLine command was used in the Construct drawing to extend a wall component of the upper level down below its baseline to cover the floor framing. You can adjust the offset values for the FloorLine command based upon distances between the levels determined from elevation views of the model. The floor framing can be represented by a slab object with its thickness set equal to the joist and sill wood framing.

> Construct drawings are created as **Attached** reference files to the view drawing when you choose New View Dwg > General or New View Dwg > Detail, whereas construct drawings are created as **Overlay** reference files to the view drawing when New View Dwg > Section/ Elevation is selected.

TIP

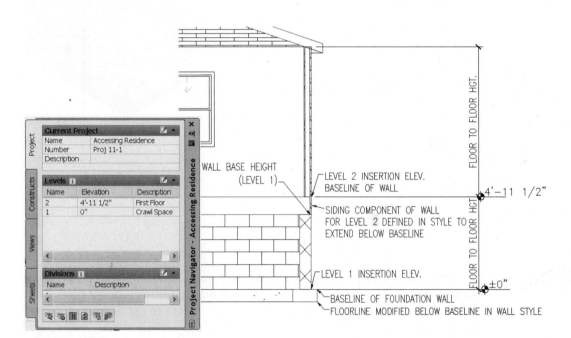

FIGURE 12.1 *Creating levels for a model*

The steps to creating a model using a project are listed below.

Steps to Create a Model

1. Open the **Project Browser** (select **Project Browser** from the Quick Access Toolbar) and set your project as current.
2. Click **Close** to dismiss the Project Browser and open the **Project Navigator**.

3. Click the **Levels** button of the **Project** tab as shown in Figure 12.1 to open the Levels dialog box. Specify values for floor elevation and finish floor to finish floor vertical distances for each level (foundation, floor 1, floor 2).

4. Select the **Views** tab.

5. Select the **Views** category, right-click, and choose **New View Dwg > General**.

6. Edit the **Add General View** dialog box: type **Model** in the drawing field of the **General** page, as shown in Figure 12.2. Click **Next** to open the **Context** page and check Level **1 [G]**, Level **2 [1]**, and Level **3 [2]** for Division **1**. Click **Next** to open the **Content** page and verify that all constructs are selected for the levels. Click **Finish** to dismiss the **Add General View** dialog box.

7. Double-click the **Model** View drawing in the **View** tab to open the drawing as shown at the right in Figure 12.2.

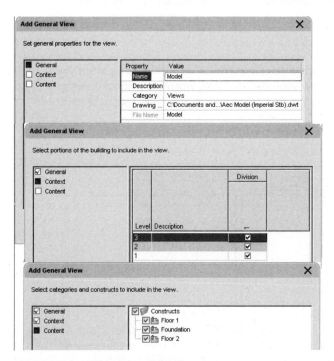

FIGURE 12.2 *Defining the model View drawing*

CREATING BUILDING ELEVATIONS AND SECTIONS

The model View drawing can include all levels and divisions of the building. After a model View drawing is created, you can open the drawing and place a callout mark. The callout tools, located on the **Callouts** palette of the **Document** palette set and shown in Figure 12.3, include tools for creating the elevation or section within the model drawing or creating additional View drawings for the elevations. The elevation tools of the Callouts palette are also included on the Callout panel of the Annotate tab in the ribbon and listed below:

- Elevation Mark A1: Elevation Mark
- Elevation Mark A2: Elevation Mark (w/ Sheet No.)

- Exterior Elevation Mark A3: Exterior Elevation Mark (Entire Building)
- Interior Elevation Mark B1: Interior Elevation Mark (4-Way) "1/2/3/4"
- Interior Elevation Mark B2: Interior Elevation Mark (4-Way) "N/E/S/W"

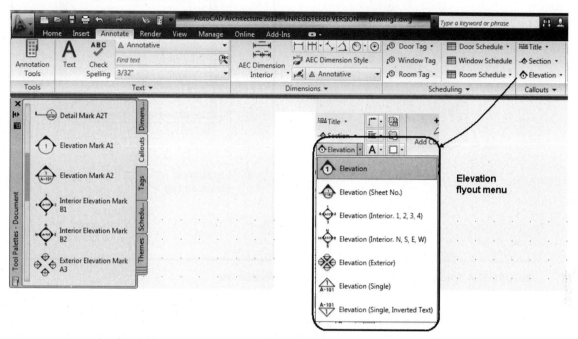

FIGURE 12.3 *Callouts tab of the Document tool palette set*

Each of the elevation mark tools of the Callouts palette opens the **Place Callout** dialog box, which includes options to place the mark as shown in Figure 12.5. When you select the elevation mark tool, you are prompted to specify the location and direction of the tag (see Figure 12.4), and then the **Place Callout** dialog box opens. The tag is an annotative tag that will include scale representations for the annotation scale. The building elevation line is created during the process of placing the callout if you check **Generate Section/Elevation**. The **Place Callout** dialog box allows you to create a new View drawing, replace an existing view elevation drawing, or add an elevation within the current drawing as shown in Figure 12.5. After you place the elevation mark, you are prompted to select the corner of the elevation region, which places the elevation building line and defines the view direction of the projection box for the elevation. The elevation or section line and projection box shown in Figure 12.13 specify the objects included in the elevation or section view.

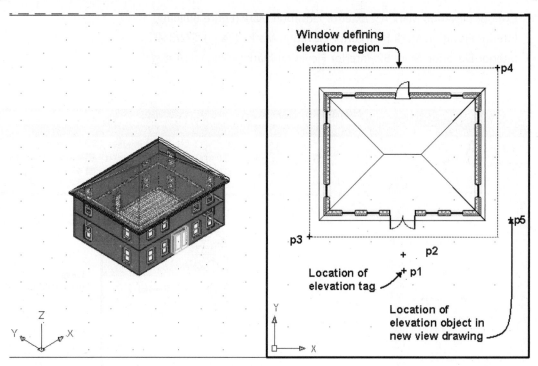

FIGURE 12.4 *Specify location of callout and building elevation line*

The following steps outline the process of placing Elevation Mark A1, developing the elevation as a separate View drawing, and finally placing the view on a sheet. When you place a callout, the tag block is inserted from the Callouts folder of the Documentation Tool Catalog. The tag and title blocks with associated attributes can be identified in the properties of each callout tool, as shown in Figure 12.6. The attributes of the callout include fields that extract view numbers or sheet names when placed. Therefore, when the callout is placed, a placeholder field is created that is displayed as a question mark until the view is placed on a sheet. The background of the field does not plot; only the view number is plotted. The **Place Callout** dialog box allows you to specify the name, scale, and titlemark of the view. When the view is placed on a sheet, the view number will be displayed on the sheet and on the elevation View drawing.

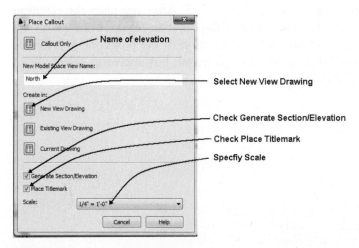

FIGURE 12.5 *Place Callout dialog box*

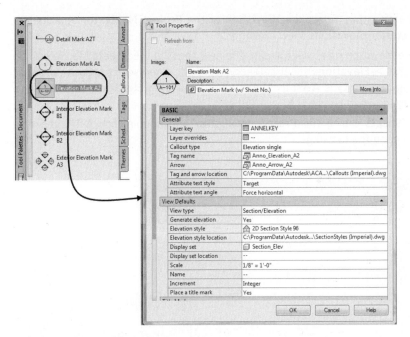

FIGURE 12.6 *Tool Properties of a callout tool*

Steps to Create an Elevation

1. Open the **Project Navigator** and select the **Views** tab.

2. Open a model View drawing, which consists of all levels.

3. Select **Elevation Mark A1** from the **Callouts** tool palette or the Elevation flyout menu of the Callouts panel in the Annotate tab shown in Figure 12.3 and respond to the workspace prompts as shown below:

   > Specify location of elevation tag: *(Select a point near* p1 *as shown in* Figure 12.4 *to place the callout.)*

   > Specify direction for elevation: *(Select a point near* p2 *as shown in* Figure 12.4 *to specify the direction of view.)*

4. The **Place Callout** dialog box opens as shown in Figure 12.5. Type **North** in the **New Model Space View Name** field, verify that **Generate Section/Elevation** and **Place Titlemark** are checked, set the scale to **1/4" = 1'-0" [1:50]**, and check **New View Drawing** to open the **Add Section/Elevation View** dialog box.

5. Edit the **Add Section/Elevation View** dialog box as follows: Type the Name = **Exterior Elevations** of the View drawing in the **General** page, click **Next**, and verify levels and constructs that are selected in the **Context** and **Content** pages as shown in Figure 12.7. Click **Finish** to dismiss the dialog box.

6. Respond to the remaining workspace prompts:

   > Specify first corner of elevation region: *(Select a point near* p3 *as shown in* Figure 12.4 *to define the start location of the projection box.)*

   > Specify opposite corner of elevation region: *(Select a point near* p4 *as shown in* Figure 12.4 *to define the size and content of the projection box.)*

   > Specify insertion point for the 2D elevation result: *(Select a point near* p5 *as shown in* Figure 12.4 *to locate the coordinates of the new elevation view.)*

   > *(The elevation is created in the new Exterior Elevations View drawing in a model view named North.)*

7. To view the elevation, select the **Views** tab of the **Project Navigator** and double-click the **Exterior Elevations** View drawing.

8. To place the elevation on a sheet, create the new sheet in the **Sheets** tab. Select the **Architectural > Elevations** sheet category, right-click, and select **New > Sheet** as shown in Figure 12.8. Type **A-111** for the number and **North Elevation** for the title of the sheet as shown in Figure 12.8.

9. Select the **Sheets** tab of the **Project Navigator** and double-click **A-111**, the new drawing, to open the drawing.

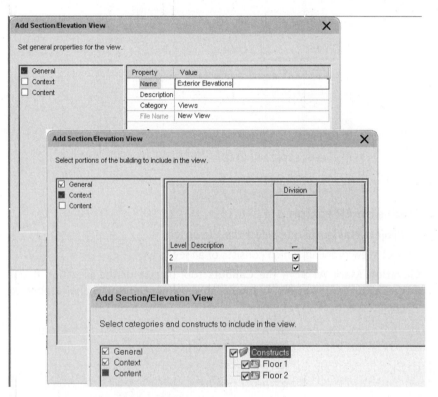

FIGURE 12.7 *Add Section/Elevation View dialog box*

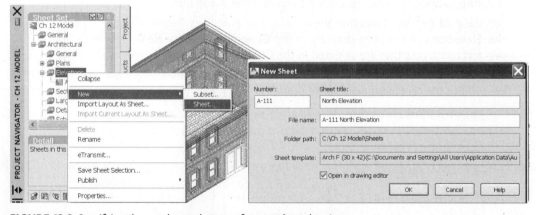

FIGURE 12.8 *Specifying the number and name of a new sheet drawing*

10. Select the **Views** tab, select the **North** view, right-click, and choose **Place on Sheet** as shown in Figure 12.9. Drag the view to the sheet and left-click to position the view on the sheet. The view number is placed in the title bubble of the elevation view as shown in Figure 12.10.

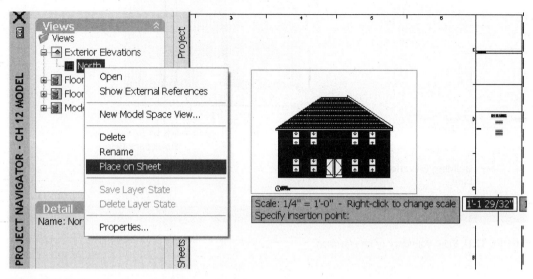

FIGURE 12.9 *Place the View drawing on a sheet*

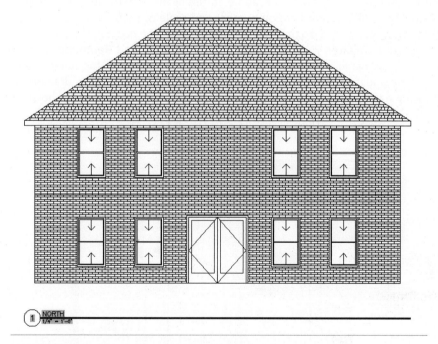

FIGURE 12.10 *Callout number placed in the elevation drawing*

When you create an elevation within a View drawing, you can select the elevation to display its view boundary. The grips of the view boundary are shown in Figure 12.11. You can select the grips of the view boundary and drag them to a different location to resize the view boundary. The scale of the elevation view is specified in the Place Callout dialog box when the view is created.

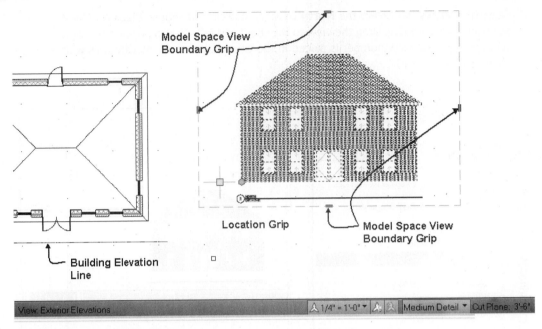

FIGURE 12.11 *View Boundary of an elevation*

Editing the Callout of a View

1. The callout number and view name can be edited in the Sheets tab of the Project Navigator. The callout identification number and scale of the view were assigned when the view was placed on the sheet. You can select the name in the Sheets tab of the Project Navigator, right-click, and choose **Rename and Renumber** to edit the number and name. When the view and sheet are reopened, the number and name will be updated. You can click and drag the callout of the view drawing to the drawing name in the Sheets tab to update the data of the callout as shown in Figure 12.12.

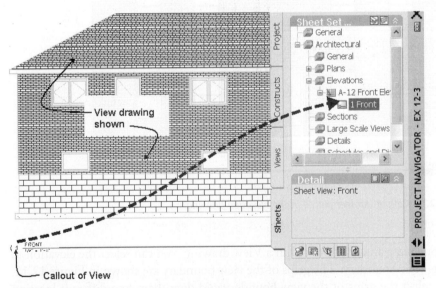

FIGURE 12.12 *Callout data extracted from the sheet field*

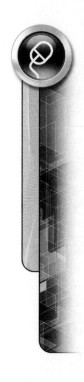

Editing the Building Section/Elevation Line Using Grips

After you create the elevation view, you can open the Construct drawings and make revisions. The elevation can be updated to reflect the changes. However, the building elevation line defines the viewing direction and size of the three-dimensional projection box for content of the elevation; therefore, only those objects that fall within the box will be included in the elevation. For the elevation to be captured correctly, the changes in the construct may require the size and location of the building elevation line to be changed. You can select the building elevation line in the View drawing to modify its location and size using grips. The size of the elevation projection box can be changed by dragging the grips located at the base of the projection box shown in Figure 12.13. The default height of the building elevation projection box is set equal to the height of the model space entities within the building line. If you select the trigger grip shown on the next page in Figure 12.13, the height of the projection box is displayed, and you can select the height and lower extension grips to edit the projection box.

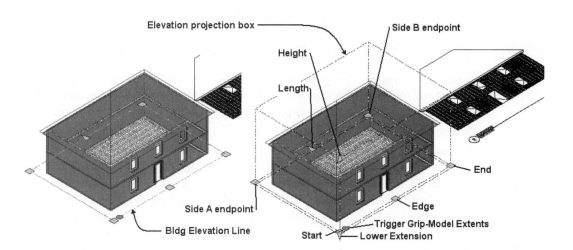

Use model extents for height toggled ON   **Use model extents for height toggled OFF**

FIGURE 12.13 *Elevation drawing in the View drawing*

TUTORIAL 12-0: CREATING ELEVATIONS AND ELEVATION STYLES

1. Download Chapter 12 files from the Accessing Tutor\Imperial or Accessing Tutor\ Metric category of the Student Companion site of CengageBrain http://www .cengagebrain.com described in the Preface to your Accessing Tutor folder.

2. Open Autodesk AutoCAD Architecture 2012 and open Accessing Tutor\Imperial\Ch12\ Ex 12-0.dwg or Accessing Tutor\Metric\Ch 12\Ex 12-0.dwg and save the drawing as **Lab 12-0** in your Student directory.

3. In this tutorial you will explore the settings of the Building Elevation/Section Line.

4. Select the building elevation/section line shown at **p1** in Figure 12.14. Verify the Properties palette is open. The General section of the Design tab lists the layer property as A-Elev-Line.

5. The Dimensions section indicates that the Use model extents for height are listed as Yes (refer to **p2** in Figure 12.14); therefore, the height property creates a projection box inclusive of the height of the objects selected for the elevation.

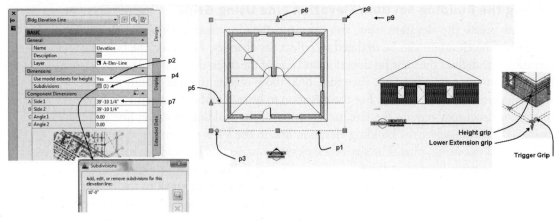

FIGURE 12.14 *Properties of the Building Elevation/Section line*

6. Choose the NO option to view the Height and Lower Extension fields of the Properties palette. When the Use model extents for height is set to No, you may enter specific distances to define the projection box.

7. Choose SW Isometric from the View Cube to view the Building Elevation/Section line in pictorial. An additional Height and Lower Extension grip are now displayed. Choose the Trigger Grip shown at **p3** to toggle Model extents for height to Yes.

8. Choose Top from the View Cube.

9. Choose the Subdivisions button shown at **p4** to open the Subdivisions dialog box. The existing Subdivision is located 10' from the Building Elevation/Section line shown at **p5**. User defined subdivisions allow you to adjust object display per subdivision. Therefore the lineweight of objects closer to the building section elevation line can be displayed by increasing lineweight or color of the entities within the subdivision.

10. The Component Dimensions section of the Properties palette displays the current settings of the building elevation/section line. In the following steps, you will modify the dimensions using the grips.

11. Choose the grip at **p6**, move the cursor up, and click to change the Side 1 and Side 2 dimensions as shown in the field at **p7**.

12. Choose the grip at **p8**, move the cursor right near **p9**. The Angle 2 and Side 2 dimensions are now different from the Angle 1 and Side 1 values; however, the Building Elevation/Section line remains parallel to the building. The grips allow you to create a custom projection box for the elevation.

13. A summary of the dimensions of the Building Elevation/Section line are shown in Figure 12.15.

14. Save and Close the drawing.

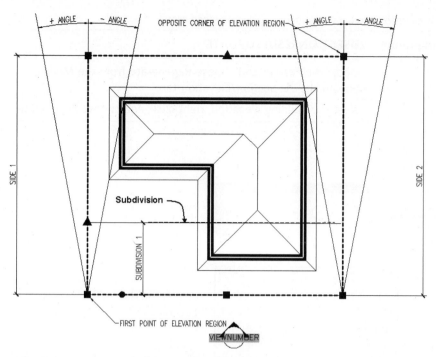

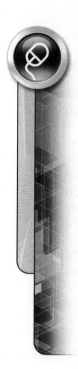

FIGURE 12.15 *Dimensions of the building elevation line*

REFRESHING AND REGENERATING ELEVATION/SECTIONS

Because elevations and sections are generated from a model drawing that can change as the design develops, the **Refresh (2DSectionResultRefresh)** and **Regenerate (2DSectionResultUpdate)** commands are provided to revise the elevation or section to the current state of the model. If windows and doors are added to the walls that were specified for an elevation, the Refresh command will revise the elevation to include the new windows and doors. The Refresh (2DSectionResultRefresh) command (refer to Table 12.1) updates the elevation based on the original settings of the elevation.

TABLE 12.1 *Refresh the section or elevation object command access*

Command prompt	2DSECTIONRESULTREFRESH
Ribbon	Select an elevation and choose Refresh from the Modify panel in the 2D Section/Elevation tab.
Shortcut menu	Select an elevation, right-click, and choose Refresh.

If additional objects have been added to the model drawing, the Refresh command updates the elevation without including new objects. Therefore, to include new objects, the **Regenerate (2DSectionResultUpdate)** command should be selected. Refer to Table 12.2 for Regenerate command access. The Regenerate command opens the Generate Section/Elevation dialog box shown at the left in Figure 12.16.

TABLE 12.2 *Regenerate section or elevation objects command access*

Command prompt	2DSECTIONRESULTUPDATE
Ribbon	Select an elevation and choose Regenerate from the Modify panel in the 2D Section/Elevation tab.
Shortcut	Select the building elevation, right-click, and choose Regenerate.

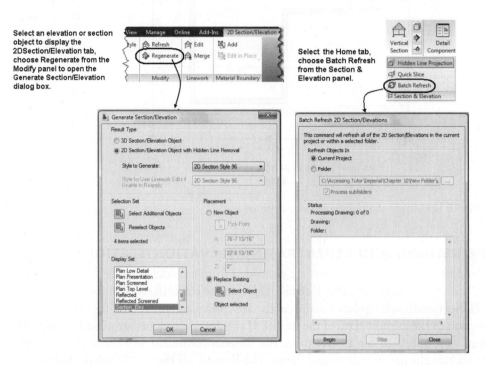

FIGURE 12.16 *Generate Section/Elevation and Batch Refresh 2D Section/Elevations dialog boxes*

The **RefreshSectionsElevations** command (see Table 12.3) allows you to refresh the content of all 2D sections and elevations within a folder or the current project. When you choose the command, the **Batch Refresh 2D Section/Elevations** dialog box opens as shown at the right in Figure 12.16. The dialog box allows you to specify the location of the folder and displays the content in the Status section.

TABLE 12.3 *Global refresh of elevations and sections*

Command prompt	RefreshSectionsElevations
Ribbon	Choose the Home tab. Extend the Section panel and choose Batch Refresh.

The **Generate Section/Elevation** dialog box allows you to change the elevation style, selection set of the object, and placement of the elevation.

The options of the Generate Section/Elevation dialog box are described below:

Result Type

3D Section/Elevation Object—The 3D Section/Elevation Object radio button creates a three-dimensional elevation object. The **Hidden Line Removal (CreateHLR)** command can be applied to this object to create a 2D representation of a 3D section/elevation object.

2D Section/Elevation Object with Hidden Line Removal—The 2D Section/Elevation Object with Hidden Line Removal radio button creates a two-dimensional elevation object.

Style to Generate—The Style to Generate drop-down list includes elevation styles defined in the drawing. Elevation styles can be created to control the display of components of the elevation object.

Style for User Linework Edits if Unable to Reapply—The Style for User Linework Edits if Unable to Reapply drop-down list allows you to assign a section/elevation style for linework edits. This option is active if lines of the elevation have been selected for edit.

Selection Set

Select Additional Objects—The Select Additional Objects button becomes active if objects have been selected previously using the **Select Objects** button. The **Select Objects** button changes to **Reselect Objects** after an object has been selected.

Select Objects—The Select Objects button returns the focus to the workspace, allowing you to select objects to be included in the elevation.

Display Set

Display Set—The Display Set lists the display representation sets. Select the **Section_Elev** display representation set to display objects appropriate for elevations or sections.

Placement

New Object—The New Object radio button allows you to create a new elevation object when objects are selected. Existing elevations can be updated if you select the **Replace Existing** radio button.

Pick Point—The Pick Point button returns you to the workspace and prompts you to specify the location for the elevation object.

X, Y, Z—The X, Y, and Z fields allow you to specify the location of the elevation using absolute coordinates.

Replace Existing—The Replace Existing radio button allows you to select and update an existing elevation object.

Select Object—The Select Object button returns you to the workspace to select an existing elevation.

Object not selected—The Object not selected message field indicates if an elevation object has not been selected for update.

Generating Elevations from a Building Section/Elevation Line

You can also update an elevation or create additional elevation views if you select a building elevation line, right-click, and choose **Generate 2D Elevation**. The **Generate Elevation** command (see Figure 12.17 and Table 12.4 for **BldgElevationLineGenerate** command access) opens the **Generate Section/Elevation** dialog box shown in Figure 12.16, which allows you to select new objects for the elevation, specify a new location, and select the elevation style for an elevation or a section. The BldgElevationLineGenerate command will develop an elevation from all entities selected. The objects selected should be represented by the SECTION_ELEV display representation set because it includes the typical representation of architectural objects for elevation views. The Elevation object is created on the **A-Elev** layer.

TABLE 12.4 *BldgElevationLineGenerate command access*

Command prompt	BLDGELEVATIONLINEGENERATE
Ribbon	Select the building elevation line and choose Generate Elevation from the Modify panel of the Building Elevation Line tab.
Shortcut	Select the building elevation line, right-click, and choose Generate 2D Elevation.

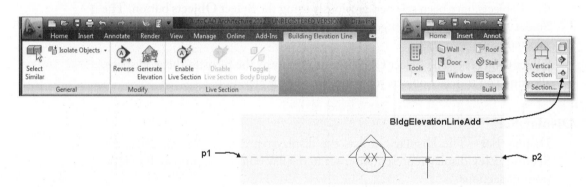

FIGURE 12.17 *Building Elevation Line tab*

Reversing the Building Section/Elevation Line

The **BldgElevationLineReverse** command can be used to change the viewing direction of an existing building elevation line. Access the BldgElevationLineReverse command as shown in Figure 12.17 and Table 12.5. When you select the building elevation line and choose **Reverse** from the Modify panel of the Building Elevation Line tab, the viewing direction changes 180 degrees.

TABLE 12.5 *BldgElevationLineReverse command access*

Command prompt	BLDGELEVATIONLINEREVERSE
Ribbon	Select the building elevation line and choose Reverse from the Modify panel in the Building Elevation Line tab.
Shortcut menu	Select the building elevation line, right-click, and choose Reverse.

Adding a Building Elevation Line

The Building Elevation Line located on the Section & Elevation panel of the Home tab allows you to develop an elevation independent of the tools of the Callouts palette. Access the **BldgElevationLineAdd** command as shown in Table 12.6.

TABLE 12.6 *Accessing the BldgElevationLineAdd command*

Command prompt	BldgElevationLineAdd
Ribbon	Select Building Elevation line from the Section & Elevation panel of the Home tab shown at the right in Figure 12.17.

When you choose the BldgElevationLineAdd command, you are prompted in the workspace as follows:

> Elevation line start point: (*Select a point* at p1 *as shown in Figure 12.17.*)
>
> Elevation line end point: (*Select a point* at **p2** *as shown in Figure 12.17.*)

After the BldgElevationLine is created, select the BldgElevationLine and choose Generate Elevation from the Modify panel of the Building Elevation Line tab to create an elevation object.

CONTROLLING DISPLAY OF AN ELEVATION OR SECTION

The display of the elevation can be controlled by its **Object Display** or by creating elevation styles. The Object Display and elevation styles control display by assigning display components that govern the appearance of the elevation. The material hatching displayed in the final elevation is extracted from the material assignments of the objects selected from the model. Therefore, if a wall style includes brick veneer with brick assigned as a material, the resulting elevation will be hatched in brick. Since the materials are usually assigned in a Construct drawing, to change the elevation you must edit the Construct drawing and update the elevation to implement a material hatch change. The display of material hatching in the elevation is controlled by the **Surface Hatch Linework** component of the elevation as shown in the **Display Properties** dialog box of Figure 12.18. If the Surface Hatch Linework component is turned off, the visibility of the Surface Hatch Linework in the elevation will be removed and only the defining lines of the elevation will be displayed.

Therefore, you can control the display of the components within the elevation by editing object display in the Style Manager or the Display tab of the Properties palette. The Display tab of the Properties palette provides the following options in the Display Controlled by flyout shown at **p1** in Figure 12.18:

This Object—This option is recommended for minor edits to override the display of the selected section/elevation.

2D Section/Elevation Style: 2D Section Style 96 [2D Section/Elevation Style: 2D Section Style 100]—This option allows you to edit the properties of the style that is applied to all sections and elevations sharing this style.

Drawing default settings—The Elevation object is displayed without regard to an elevation/section style.

When you select this Object or 2D Section/Elevation style from the Display controlled by list, you can click the Display component flyout (**p2** of Figure 12.18) to select a component for edit. The Subdivision 1 display component has been selected; therefore you can modify display in the Component Display Properties section. The color was modified to White as shown in the Properties palette.

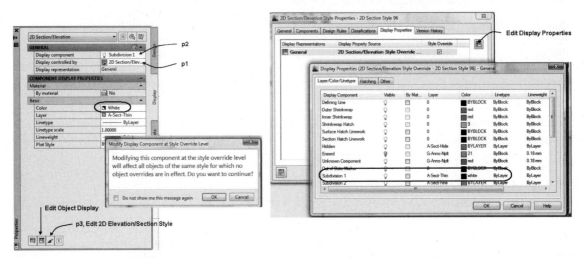

FIGURE 12.18 *Display Controlled by the Properties palette and the Display Properties dialog box*

You can choose Edit 2D Elevation/Section Style at **p3** in Figure 12.18 to open the 2D Section/Elevation Style Properties – 2D Section Style 96 dialog box. Choose the Edit Display Properties button to open the Display Properties dialog box. The display components are listed in the far left column of the **Layer/Color/Linetype** tab. When the elevation is created, the edges of objects in the model drawing are projected to create lines placed on the **Defining Line** component. The subdivisions listed in Figure 12.18 are assigned by creating subdivisions in the building elevation line as shown in Figure 12.15. Therefore, you could create a subdivision within 10 feet of the building elevation line. All objects projected from this zone are placed in the elevation object with the color, layer, linetype, and lineweight as defined by the Subdivision 1 in the **Layer/Color/Linetype** tab shown in Figure 12.18. The color of the objects within Subdivision 1 will be displayed white since it is defined in Display Properties dialog box of Figure 12.18. The lineweight of Subdivision 1 could be set wider to display the objects within Subdivision 1 with greater emphasis. The color, layer, linetype, and lineweight of each component listed can be changed in the **Layer/Color/Linetype** tab. The **Other** tab allows you to create additional custom display components.

STOP

Do Tutorial 12-1, "Creating a Model View Using the Project Navigator," at the end of the chapter.

Creating Elevation Styles to Control Elevation Display

Elevation styles can be created to apply the display settings for all elevations of the elevation style. When you create a 2D elevation using the elevation marks on the **Callouts** palette, the **2D Section Style 96 [2D Section Style 100]** elevation style is used to develop the elevation. Additional styles can be created to alter the display of the elevation by creating custom display components that have special layer, color, linetype, and lineweight properties. Elevation styles can be created to identify objects with specific colors within the model drawing and to assign the representation of the entities using the new elevation display components. When the elevation is developed with the new elevation style, all objects from the model with a specified color can be assigned to a new elevation display component. The display component can specify color, linetype, and line weight. Therefore, an elevation style could be created that represents all objects with a blue color in the model drawing with hidden lines in the elevation. The Standard elevation style will create an elevation without any recognition of colors within the model drawing.

Existing elevation styles can be modified in the **2D Section/Elevation Style** Properties dialog box (see Table 12.7 for command access).

TABLE 12.7 *2DSectionResultStyleEdit command access*

Command prompt	2DSECTIONRESULTSTYLEEDIT
Ribbon	Select an elevation and choose Edit Style from the General panel of the 2D Section/Elevation.
Shortcut menu	Select an elevation or a section, right-click, and choose Edit 2D Elevation/Section Style.

The General, Classifications, and Version History tabs are similar to other object styles. A description of the Components, Design Rules, and Display Properties tabs of the **2D Section/Elevation Style Properties** dialog box for a new 2D Section Style is outlined in the sections that follow.

Components Tab

The Components tab is shown in Figure 12.19.

Index—The Index specifies a number for the new component used in the display of the elevation.

Name—The Name specifies the name of the new component used to display the elevation.

Description—The Description field allows you to add a description to the component name.

Add/Remove—Click the Add or Remove button to create components for the list or delete them from the list.

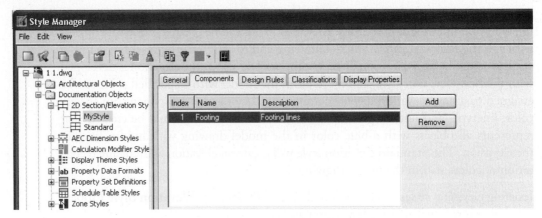

FIGURE 12.19 *Components tab of a 2D Section/Elevation Style in the Style Manager*

Design Rules Tab

The Design Rules tab is shown in Figure 12.20.

> **Rule**—The Rule column specifies the number of the rule used to define color and display method.

> **Color**—The Color column specifies the color number/name of objects in the model, which will be represented in the display method of the elevation.

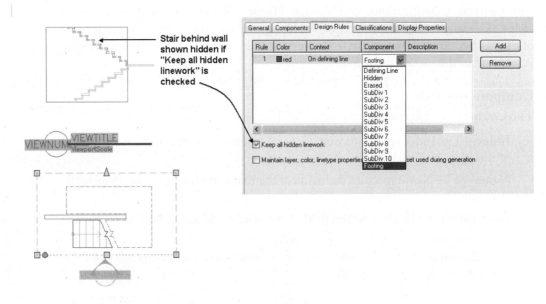

FIGURE 12.20 *Design Rules tab of the Elevation Style Properties dialog box*

> **Context**—The Context specifies to which component within the elevation display properties to apply the rule. The drop-down list of the **Context** column includes such components as: defining lines, within any subdivision, any visible, hidden, and within any of the subdivisions defined in the elevation.

> **Component**—The Component drop-down list of the column specifies the display components to assign the lines based upon the color specified rule.

> **Description**—Click in the Description field to add a description for each rule.

Keep all hidden linework—All hidden components are displayed in the elevation. The stair behind the wall shown in Figure 12.20 is shown with hidden lines since this toggle is checked.

Maintain layer, color, linetype properties from the display set used during generation—If this option is checked, the elevation will include the color and linetype of the material definition of the components. Check this option to generate an elevation with the colors assigned to the materials. When this checkbox is clear, all linework will assume the layer color of the elevation object.

Display Properties Tab

The Display Properties tab is shown in Figure 12.21. It lists the display representations available for the section or elevation object.

Display Representation—The name of the display representation used in the elevation object is listed in the Display Representation column.

Display Property Source—The current display property source is listed for the display representation. The property source can either be Drawing Default or 2D Section/Elevation Style Override.

Edit Display Properties—The Edit Display Properties button opens the **Display Properties (2D Section/Elevation Style Override—style name)** dialog box. The Display Properties (2D Section/Elevation Style Override—style name) dialog box consists of the **Layer/Color/Linetype**, **Hatching**, and **Other** tabs, which allow you to define the properties of the components. The **Other** tab allows you to create custom display components for the style.

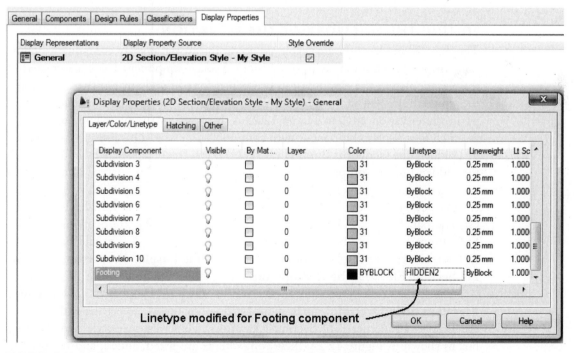

FIGURE 12.21 *Display Properties tab of a 2D Section/Elevation Style Properties dialog box*

Steps to Create and Edit an Elevation Style

The following steps outline the procedure to edit the **2D Section/Elevation Style Properties** dialog box to create an elevation style that controls object display based upon the color of the object. The steps create a style that will identify any red line in the model and assign it to the Footing custom display component. The Footing display component is shown in the elevation with hidden lines.

1. Select Manage tab. Choose Style Manager from the Style & Display panel. Expand the *Documentation Objects > 2D Section/Elevation* folder. Select the *2D Section/Elevation* folder, right-click, and choose **New**. Overtype **MyStyle** as the name of the new style.

2. Select the **Components** tab, click the **Add** button, and overtype **Footing** as the name of the custom component.

3. Select the **Display Properties** tab and check the **Style Override** box to open the **Display Properties (2D Section/Elevation Style Override—MyStyle)** dialog box. Click **OK** to dismiss the **Display Properties** dialog box.

4. Select the **Design Rules** tab (shown in Figure 12.20), click the **Add** button, and specify the color name/number. Click in the **Context** column and select a context, such as **Any Visible or Hidden**, of the object from the model that will be assigned unique display. Click in the **Component** column and specify the name of the display component, such as **Footing**, created in step 2.

5. Select the **Display Properties** tab as shown in Figure 12.21, verify that **Style Override** is checked, and click the **Edit Display Properties** button to open the **Display Properties (2D Section/Elevation Style Override—MyStyle)** dialog box. Select the **Layer/Color/Linetype** tab and edit the properties such as color or linetype of the Footing display component created in step 2. Click **OK** to dismiss the Display Properties (2D Section/Elevation Style Override—MyStyle) dialog box. Click **OK** to dismiss the 2D Section/Elevation Style Properties—MyStyle dialog box.

EDITING THE LINEWORK OF THE ELEVATION

The lines of an elevation are displayed according to the display component definition within the elevation style. If a custom display component has been created in the elevation style, the elevation can be opened for edit and any entity can be edited to assume the display defined by that custom display component. The line shown at **p1** in the elevation at A in Figure 12.23 is displayed as an object line because it is assigned to a display component with continuous linetype. This line can be changed to a hidden line (shown at B in Figure 12.23) if you create a display component with the hidden linetype property. The first step in changing the linetype of a line is to create a display component with the linetype property in the elevation style. You can create a new component by selecting the elevation and choosing **Edit Style** from the General panel of the 2D Section/Elevation tab to open the 2D Section/Elevation Style Properties dialog box. Select the components tab, click the **Add** button, and type a name for the display component. Choose the **Display Properties** tab of the **2D Section/Elevation Style Properties** dialog box shown in Figure 12.22. When you check the **Style Override** checkbox, the **Display Properties (2D Section/ Elevation Style Override—style name)** dialog box opens. Select the **Other** tab of the Display Properties (2D Section/Elevation Style Override—style name) dialog box as shown in Figure 12.22. New components are listed, added, and removed in the **Other** tab of the Display Properties (2D Section/Elevation Style Override—style name) dialog box. Display components created for display by object color are listed in the **Other** tab.

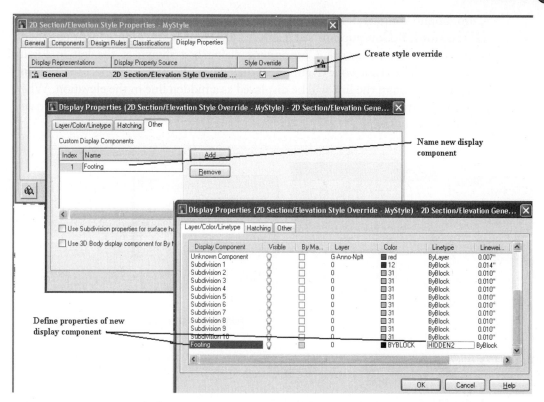

FIGURE 12.22 *Creating a display component*

After creating a name for the display component, select the **Layer/Color/Linetype** tab shown in Figure 12.22, and edit the properties of the new display component. **The Layer/Color/Linetype** tab of the Display Properties (2D Section/Elevation Style Override—style name) dialog box shown in Figure 12.22 includes the **Footing** component with hidden linetype.

You can open the elevation for edit by selecting the elevation and choosing **Edit** from the Linework panel of the 2D Section/Elevation tab (**2DSectionResultEdit** command; see Table 12.8). When the elevation is opened for edit, entities can be erased or modified. In addition, the hatch pattern is not displayed when the elevation is opened for edit. Upon completion of the changes, you save the changes and the display state is saved back to the elevation. This command allows you to change the linetype, color, or lineweight of selected entities. After opening the elevation for edit, you can erase any entity in the elevation.

TABLE 12.8 *Edit Linework command access*

Command prompt	2DSECTIONRESULTEDIT
Ribbon	Select an elevation and choose Edit from the Linework panel of the 2D Section/Elevation tab shown in Figure 12.23.
Shortcut menu	Select an elevation, right-click, and choose Linework > Edit.

The entities can be assigned to any of the display components or custom display components defined in the elevation style. After opening the elevation for edit, you can select an entity (shown at **p1** in Figure 12.23) of the elevation, and choose the **Modify**

Component command from the Profile panel of Edit in Place: Linework tab (**2DSectionEditComponent**; see Table 12.9). The Modify Component command opens the **Select Linework Component** dialog box shown at the left in Figure 12.23. The Select Linework Component dialog box allows you to select the **Footing** display component; the line will be displayed as a hidden line in the elevation. When you have finished editing the lines, select **Finish** to save the changes or **Cancel** to end the edit without saving.

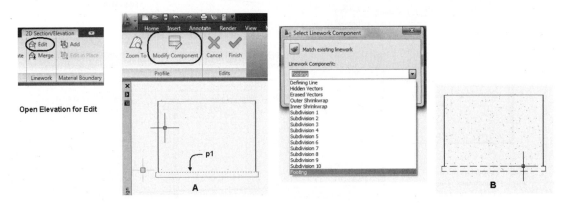

FIGURE 12.23 *Editing linework of an elevation*

TABLE 12.9 *Modify Component command access*

Command prompt	2DSECTIONEDITCOMPONENT
Ribbon	Select an elevation and choose Edit from the Linework panel. Choose an entity of the elevation. Choose Modify Component from the Profile panel of Edit-in Place Linework tab.
Shortcut menu	Select an elevation, right-click, and choose Linework > Edit. Choose an entity of the elevation, right-click, and choose Modify Component.

The entities of an elevation object can be changed to different display components of an elevation style. If you assign an entity to the Erased Vectors linework component, it will not be displayed in the elevation.

MERGING LINES TO THE ELEVATION

Lines and hatch patterns can be merged to the elevation object using the **Linework > Merge** command (see Table 12.10). This command allows you to add reference lines for the finish floor and finish ceiling of an elevation. The lines are merged to the elevation and assigned to a display component. Access the **Linework > Merge** (**2DSectionResultMerge**) command as follows.

TABLE 12.10 *Merge Linework command access*

Command prompt	2DSECTIONRESULTMERGE
Ribbon	Select an elevation and choose Merge from the Linework panel of the 2D Section/Elevation tab.
Shortcut menu	Select an elevation, right-click, and choose Linework > Merge.

When you select the elevation and choose **Merge** from the Linework panel of the 2D Section/Elevation tab, you are prompted to select the entities to merge with the elevation object as shown in the workspace prompt sequence below. In the sequence shown, a centerline was added to the elevation. Note that prior to beginning the Merge command shown below, a display component named Floor Line was created in the **Components** tab. The component was assigned center linetype in the Layer/Color/Linetype tab of the Display Properties (2D Section/Elevation Style Override—style name) dialog box.

> *(Select the elevation at* **p1** *and choose* **Merge** *from the Linework panel as shown at the left in Figure 12.24.)*
>
> Select objects to merge: 1 found *(Select the line at* **p2** *to merge it with the elevation object.)*
>
> Select objects to merge: ENTER *(Press* ENTER *to end the selection.)*
>
> *(Select the* **Floor** *Line display component from the drop-down list of the* **Select Linework Component** *dialog box as shown in Figure 12.24.)*
>
> *(The line is merged to elevation as shown at the right in Figure 12.24.)*

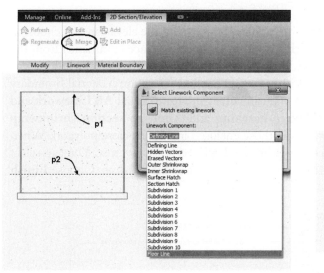

FIGURE 12.24 *Merging a line into an elevation*

MODIFYING THE MATERIAL HATCH PATTERN

The **Material Boundary > Add** command (**2DSectionResultAddHatchBoundary**; refer to Table 12.11) can be used to limit the area of material hatch displayed in an elevation. The material represented in the elevation is defined in the material definition of the objects selected for the elevation. Therefore, if you change a wall from brick to CMU, when the elevation is updated, the material displayed in the elevation will change. To limit the material hatch display, draw a polyline boundary where you want to highlight or mask the hatch in the elevation. The material display can be limited to within a closed polyline or erased within the closed polyline. The **2DSectionResultAddHatchBoundary** command allows you to specify which materials to limit or erase in the elevation.

TABLE 12.11 *Material Boundary > Add command access*

Command prompt	2DSECTIONRESULTADDHATCHBOUNDARY
Ribbon	Select an elevation and choose Add from the Material Boundary panel of the 2D Section/Elevation tab.
Shortcut menu	Select the elevation, right-click, and choose MaterialBoundary > Add.

A boundary was applied to the material representation of the brick masonry in the following workspace sequence. The **Material Boundary > Add** was set to **Limit**, and therefore the hatch was limited to within the polyline as shown at the right in Figure 12.25.

(Select the elevation and choose Add *from the Material Boundary panel.)*

Select a closed polyline for boundary: *(Select the polyline at* p1 *as shown in* Figure 12.25.*)*

Erase selected linework? [Yes/No] <No>: y *(Type* y *to erase the polyline; the* 2DSection/Elevation Material Boundary *dialog box opens. Select the* Erase *or* Limit *purpose and specify the material to apply to the boundary.)*

(The material pattern is limited to within the polyline as shown at the right in Figure 12.25.*)*

The boundary can be edited after it is applied by selecting the elevation and choosing **Edit in Place (2DSectionResultEditHatchBoundaryInPlace** command) from the Material Boundary panel. The Edit in Place: Material Boundary tab includes commands for editing the polyline. Within the 2D Section/Elevation Material Boundary dialog box shown in Figure 12.25 you can edit **Apply to** to equal **All Linework** and check **Apply to section shrinkwrap hatching** and **Apply to section shrinkwrap linework** to limit all graphics to only those enclosed inside the close polygon. When you edit the boundary in place, the grips are displayed at each vertex. You can select the grips or select from the ribbon to edit the vertex of the polyline. Choose **Finish** from the Edits panel to save and end the edit.

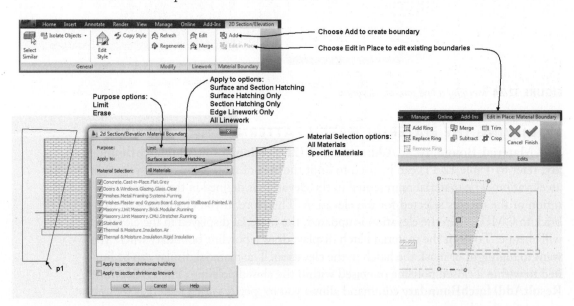

FIGURE 12.25 *Adding a material boundary to a hatch*

Do Tutorial 12-2, "Creating Elevations and Elevation Styles," at the end of the chapter.

STOP

CREATING SECTIONS OF THE MODEL USING CALLOUTS

The procedure to create a section is similar to creating an elevation. The section marks are located on the Callouts tool palette and the Section flyout of the Callouts panel as shown in Figure 12.26. The procedure to place the **Section Mark A1** is shown below.

Steps to Create a Section

1. Open the **Project Navigator** and select the **Views** tab.

2. Open the model drawing.

3. Select **Section** from the Section flyout of the **Callouts** panel of the Annotate tab in Figure 12.26, and respond to the workspace sequence as shown below:

> Specify first point of section line: *(Select a point near* p1 *as shown in* Figure 12.27 *to specify the start point of the cutting plane.)*
>
> Specify next point of line: *(Select a point near* p2 *to specify the endpoint of the cutting plane as shown in* Figure 12.27.*)*
>
> Specify next point of line or [Break]: ENTER *(Press* ENTER *to end the cutting plane line.)*
>
> Specify section extents: *(Select a point near* p3 *to open the* Place Callout *dialog box as shown in* Figure 12.27.*)*
>
> *(The* Place Callout *dialog box opens as shown in* Figure 12.27.*)*

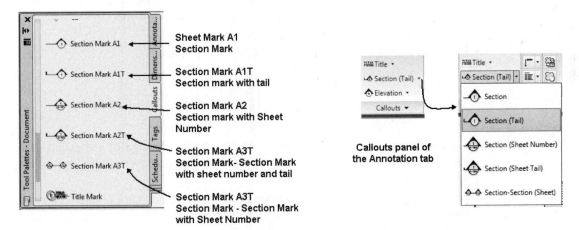

FIGURE 12.26 *Section marks located on the Callouts palette*

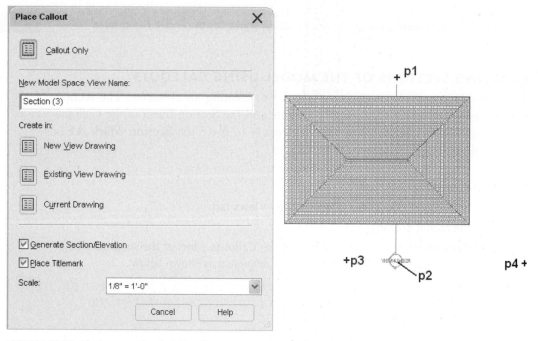

FIGURE 12.27 *Placing a section building line*

4. Type **Section A-A** in the **New Model Space View Name** field, verify that the **Generate Section/Elevation** and **Place Titlemark** checkboxes are checked, set the scale to **1/4″ = 1′-0″ [1:50]**, and check **New View Drawing** to open the **Add Section/Elevation View** dialog box.

5. Edit the **Add Section/Elevation View** dialog box, **General** page as follows: Type **Transverse** in the **Name** field; click **Next** to open the **Context** page and verify that Level **1 [G]** and level **2 [1]** are checked for division **1**; click **Next** to open the **Content** page and verify that all constructs are checked. Click **Finish** to dismiss the **Add Section/Elevation View** dialog box.

6. Continue responding to the workspace prompt as follows:

   Specify insertion point for the 2D section result: *(Select a point near* p4 *as shown in* Figure 12.27. *The section is created in the Transverse View drawing as a model space view named Section A-A.)*

7. To view the section, select the **Views** tab of the **Project Navigator** and double-click the **Transverse** View drawing. A question mark is placed in the mark to identify the section until it is placed on the sheet.

8. Select the **Sheets** tab, select the **Sections sheet** category, right-click, and select **New > Sheet**. Type **A-112** in the number field and **Sections** in the title field of the sheet. Click **OK** to dismiss the **New Sheet** dialog box.

9. Select the **Sheets** tab of the **Project Navigator** and double-click **A-112**, the new drawing.

10. Select the **Views** tab, expand the **Transverse** drawing, select the Section A-A view, right-click, and choose **Place on Sheet**. Drag the view to the sheet and left-click to position the view on the sheet. The view number is placed in the title bubble of the elevation view on the Sheet and View drawings. The Section View drawing is shown at the left in Figure 12.28.

When you place the section mark, you are prompted to specify the cutting plane line, view direction, section extents, and location of the section. The section can be created in the current drawing, an existing View drawing, or a new View drawing.

The display components of a 2D Section/Elevation object allow you to display material hatch of a wall by turning on Section Hatch Linework. The material definition of the objects will govern the hatch pattern. The Shrinkwrap and Surface Hatch are turned off in the elevation style of the section shown at the right in Figure 12.28. The display of each object cut in the section can be controlled if you edit the display properties of each material definition used in the object style. In the section shown at the right in Figure 12.28, the section hatch is turned on in the elevation style. The section hatch can be turned off in the wall style of the Construct drawing, which will display a section of the wall without material hatch representation. In Figure 12.29 the section hatch has been turned off for the standard wall on the left, whereas the section hatch for the brick masonry wall is turned on at the right.

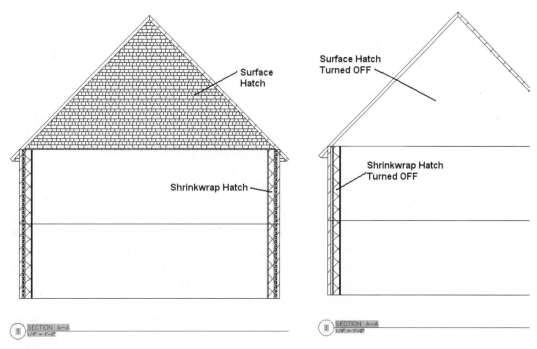

FIGURE 12.28 *Section View drawing*

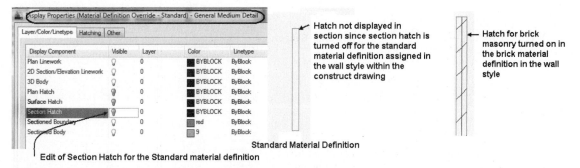

FIGURE 12.29 *Turning off section hatch in the material definition*

Creating Horizontal and Vertical Sections

During the design of a building, you can develop horizontal or vertical sections in View drawings. Access the **AecHorizontalSection** command as shown in Table 12.12. The command prompts shown below prompt you to specify the region for the section, cutting plane height, and depth of the section. The result of the command creates a building section line. The section is created when you select the building section line, right-click, and choose **Generate Section** or **Enable Live Section** as shown in Figure 12.30.

TABLE 12.12 *Horizontal Section tool from Design tool palette*

Command prompt	AECHORIZONTALSECTION
Ribbon	Select the Home tab. Choose Horizontal Section from the Section & Elevation panel.
Tool palette	Select the Horizontal Section tool from Design tool palette.

Steps to Create a Horizontal Section

1. Select the **Horizontal Section** tool from the Section & Elevation panel of the Home tab.

   Select corner for horizontal section: *(Select the drawing area at* p1 *as shown in* Figure 12.30.*)*

   Select corner for horizontal section: *(Select the drawing area at* p2 *as shown in* Figure 12.30.*)*

   Enter elevation of section cutting plane <3'-6"> [1000]: *(Press* ENTER *to accept default height.)*

   Enter depth of section <8'-0"> [2000]: *(Press* ENTER *to accept default depth and create the Building Section Line.)*

2. Select the building section line at **p3** as shown in Figure 12.30 and choose **Generate Section** from the Modify panel of the Building Section Line tab.

3. Click the **Select Objects** button in the **Selection Set** section and then create a crossing selection set by selecting a point near **p2** and **p1**. Press ENTER to open the **Generate Section/Elevation** dialog box.

4. Choose the **New Object** radio button, click the **Pick Point** button, and select a point near **p4** as shown in Figure 12.30. Click **OK** to dismiss the Generate Section Elevation dialog box.

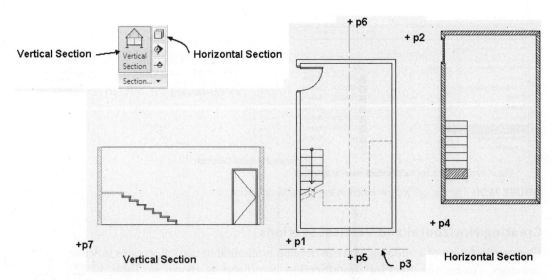

FIGURE 12.30 *Horizontal and Vertical Sections*

The **Vertical Section** tool allows you to create a section from a vertical cutting plane. The command prompts for the tool allow you to specify the location of the cutting plane and the depth of the projection box. Access the **Vertical Section** tool as shown in Table 12.13. A 2D section or live section can be developed from the vertical building section line.

TABLE 12.13 *Vertical Section tool access*

Command prompt	AECVERTICALSECTION
Ribbon	Select the Home tab. Choose the Vertical Section tool from the Section & Elevation panel.
Tool palette	Select the Vertical Section tool from Design tool palette.

Steps to Create a Vertical Section

1. Select the **Vertical Section** tool from the Section & Elevation panel of the Home tab. Respond to the workspace prompts as follows:

   Specify Start point: *(Select a point near* **p5** *as shown in* Figure 12.30.*)*

   Specify next point: *(Select a point near* **p6** *as shown in* Figure 12.30.*)*

   Specify next point: *(Press ENTER to end selection.)*

   Enter length <20'-0"> [5000]: *(Press ENTER to accept default depth.)*

2. Select the vertical building section line at **p5** and choose **Generate Section** from the **Modify** panel to open the Generate Section/Elevation dialog box.

3. Click the **Select Objects** button in the **Selection Set** section and then create a crossing selection set by selecting a point near **p2** and **p1** as shown in Figure 12.30. Press ENTER to return to the Generate Section/Elevation dialog box.

4. Choose the **New Object** radio button, click the **Pick Point** button, and select a point near **p7**. Click **OK** and a vertical section is created as shown at the left in Figure 12.30.

Creating a Live Section

A live section creates a view of the model that removes the display of the objects in front of the cutting plane line. A live section can be created from any building section line. Additional objects can be added to the drawing while the live section is active. The live section does not create a 2D section object. The live section is a tool to enable you to work on the model by removing the display of some objects. Prior to creating the live section, select a pictorial view of the building as shown in Figure 12.31, select the building section line, and choose **Enable Live Section** (see Table 12.14). The building components in front of the section line are removed from display, as shown in Figure 12.31.

TABLE 12.14 *LiveSectionEnable command access*

Command prompt	LIVESECTIONENABLE
Ribbon	Select the building section line and choose Enable Live Section of the Live Section panel in the Building Section Line tab.
Shortcut menu	Select the building section line, right-click, and choose Enable Live Section.

If a live section has been enabled, you can toggle ON **Sectioned Body Display** to display the outline of the components, turned off for the live section as shown at the right in Figure 12.31. Access **Sectioned Body Display** as shown in Table 12.15.

TABLE 12.15 *Toggle Sectioned Body Display command access*

Command prompt	AECTOGGLESECTIONEDBODY
Ribbon	Select the building section line and choose Toggle Body Display of the Live Section panel in the Building Section Line tab.
Shortcut menu	Select the building section line of a live section, right-click, and choose Toggle Sectioned Body Display.

The live section can be turned off by selecting the section line and choosing **Disable Live Section** (see Table 12.16).

TABLE 12.16 *LiveSectionDisable command access*

Command prompt	LIVESECTIONDISABLE
Ribbon	Select the building section line of a live section and choose Disable Live Section from the Live Section panel in the Building Section Line tab.
Shortcut menu	Select the building section line of a live section, right-click, and choose Disable Live Section.

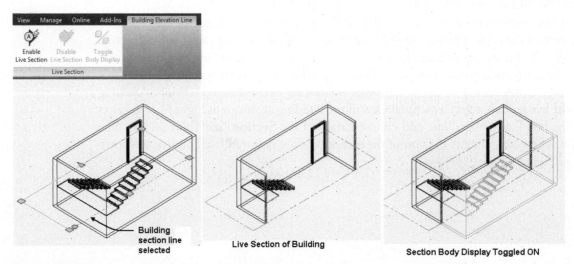

FIGURE 12.31 *Enabling a Live Section*

The **Hidden Line Projection (CreateHLR)** command will create a two-dimensional drawing of a view. The Hidden Line Projection command is located on the Section & Elevation panel of the Home tab. The **AECNapkin** command located in the Stock Tool Catalog > Helper Tools creates a block from a view of the model with a sketch display.

CREATING MODEL VIEWS WITH VISUAL STYLES

Model space views that capture a view of the model can be created. The model space view can be presented using Visual Styles with such features as shading or edge overhang. The model View drawing allows you to capture the visual style as applied to the model in orthographic or pictorial views. The **Named View** can be inserted on sheets within the Project Navigator. Access the **Named View** as shown in Table 12.17.

TABLE 12.17 *View command access*

Command prompt	VIEW
Ribbon	Choose the View tab. Choose View Manager from the View fly-out menu of the Appearance panel shown in Figure 12.32.

Prior to selecting the **View** command, set a pictorial view of the object and apply a visual style. When you choose the View command, the image will be saved as a named view that can be placed on a sheet.

Steps to Create a View with a Visual Style

1. Select the **Views** tab of the **Project Navigator**. Select a Views category, right-click, and choose **New View Dwg > General**. Specify the name, context, and content to create a View drawing of the model.
2. Open the View drawing.
3. Choose the View tab. Choose **View Manager** from the View flyout of the Appearance panel.
4. Click the **New** button of the View Manager to open the **New View** dialog box shown in Figure 12.32. Type a name in the **View** field. Click **OK** to dismiss all dialog boxes. Save and close the View drawing.
5. Select the **Sheets** tab. Choose a sheet category, right-click, and choose **New > Sheet**. Type a number and name for the new sheet. Open the sheet in the Project Navigator.
6. Select the **Views** tab of the Project Navigator. Select the new view name, right-click, and choose **Place on Sheet**. Select a location on the sheet. Save the Sheet drawing.

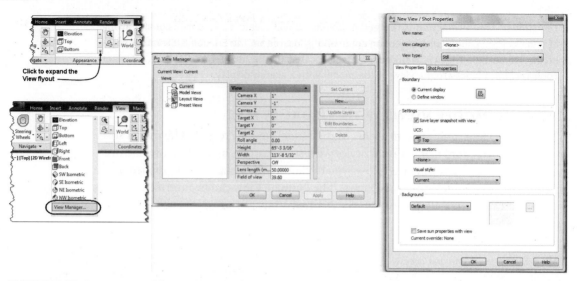

FIGURE 12.32 *Creating a Model View drawing*

STOP   Do Tutorial 12-3, "Creating 2D Sections and a Live Section," at the end of the chapter.

CREATING DETAILS

The tools to create details are located on the Callouts palette and the Callouts panel of the Annotation tab as shown in Figure 12.33. The Detail Boundary tools allow you to create a horizontal section appropriate for an enlarged plan view, whereas the Detail Mark tools create vertical sections. Each tool creates a section object that uses the 2D Section Background elevation style. This style places the geometry of the detail on the G-Anno-Nplt layer; therefore, the content will not plot. The geometry will display on the sheet and the view drawings. The geometry serves as a background for the development of a detail. The content of the detail can be added from the Detail Component Manager.

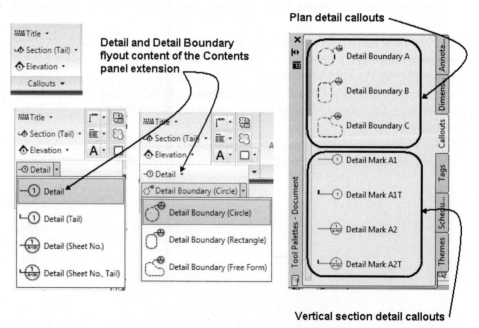

FIGURE 12.33 *Detail callouts of the Callouts palette and Callouts panel*

The following workspace steps were used to develop a detail of a bath:

1. Open a view drawing in the Project Navigator. Select the **Detail Boundary (Rectangle)** from the Detail Boundary flyout in the Callouts panel. Respond to the workspace prompts as follows:

   Specify one corner of detail box: *(Select a location,* **p1**, *as shown in* Figure 12.34.*)*

   Specify opposite corner of detail box: *(Select a location,* **p2**, *as shown in* Figure 12.34.*)*

   Specify first point of leader line on boundary: *(Select a location,* **p3**, *as shown in* Figure 12.34.*)*

   Specify next point of leader line <end line>: ENTER

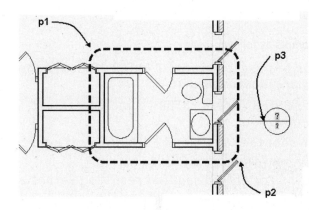
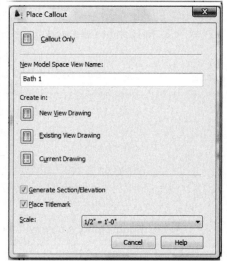

FIGURE 12.34 *Defining the boundary of the detail*

2. The Place Callout dialog box opens as shown in Figure 12.34. Type a name (Bath 1) in the New Model Space View field.
3. Click **New View Drawing** to open the Add Detail View dialog box shown in Figure 12.35.
4. Type a name for the view drawing (the Bath Details name) as shown in Figure 12.35.
5. Click **Next** to open the Context page; check divisions and levels for the view drawing.
6. Click **Next** to open the Content page; verify the constructs of the view drawing.
7. Click **Finish** to dismiss the Add Detail View dialog box.

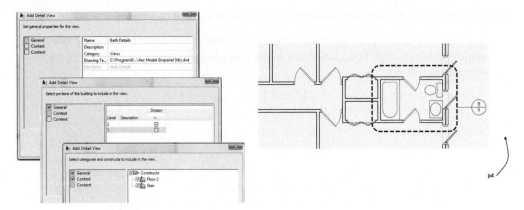

FIGURE 12.35 *Creating the View drawing and defining the location of detail for the new Model view drawing*

8. Respond to the workspace prompts as follows:

Specify elevation for plan section <6>: 6 ENTER *(Specify the vertical position of projection box.)*

Specify second point: ENTER *(Press ENTER to accept default position of second point.)*

Specify depth of plan section <8'-0">: 8' ENTER *(Specify depth of the plan section below the plan section.)*

Specify insertion point for the 2D section result: *(Select a point at* **p4** *as shown in* Figure 12.35 *to specify the location of the model space view.)*

The model space view will be located in the new view drawing at the position specified.

9. Double-click **Bath Details** from the **Views** tab to open the detail as shown in Figure 12.36.

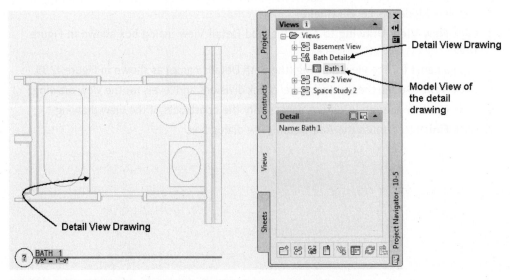

FIGURE 12.36 *Opening the Detail View drawing from the Views tab*

10. The detail of the model may allow the detail object to include lines useful to the detail. You can trace such lines using the AutoCAD line command. The lines should be drawn on the A-Detl-Wide [AEC-Detail-Wide] or A-Detl-Thin [AEC-Detail-Medium] layer. The A-Detl-Wide [AEC-Detail-Wide] or A-Detl-Thin [AEC-Detail-Medium] layers are inserted in the drawing when details from the Detail Component Manager are inserted. After inserting a sample detail, set A-Detl-Wide [AEC-Detail-Wide] or A-Detl-Thin [AEC-Detail-Medium] current and begin drawing lines. You can drag examples of such linework to a tool palette to create a tool with either the A-Detl-Wide [AEC-Detail-Wide] or A-Detl-Thin [AEC-Detail-Medium] layer assignment for future use.

The content of the detail is created on the A-Elev layer with the 2D Section/Elevation Style-2D Section Style Background style. The 2D Section/Elevation Style-2D Section Style Background style can be copied and the layer assignment of each display component can be modified to permit selected entities to print. This technique can be useful if an extensive quantity of lines from the detail object is useful to the detail. After modifying the section/elevation style, you could select the linework of the detail as shown in Figure 12.36 and change the display component assignment of this linework using the 2DSectionResultEdit and 2DSectionEditComponent commands.

In this chapter, detail and section view drawings have been presented. The detail object allows you to capture the physical arrangement of building components in the model for the development of the detail from a background. The Detail Component Manager includes 2D drawings that can be applied to detail or section objects.

Steps for modifying the 2D Elevation/Section Style Background of a Detail

1. Select the detail and choose **Copy Style** of the General panel in the 2D Section/Elevation tab to open the 2D Section/Elevation Style Override—2D Section Style Background(2).

2. Choose the General tab. Type a name for the new style.

3. Choose the **Display Properties** tab. Choose the **Edit Display Properties** button at the right. Choose the **Layer/Color/Linetype** tab.

4. Choose a display component, **Defining Line**, as shown in Figure 12.37. Choose the **Layer** field for the display component to open the Select Layer dialog box.

5. Choose a layer. A-Detl-Thin layer has been chosen in Figure 12.37.

6. Click **OK** to dismiss all dialog boxes.

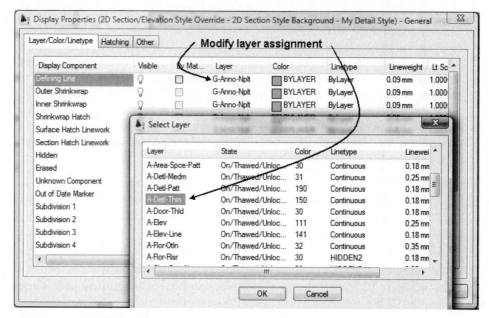

FIGURE 12.37 *Modifying the layer assignment for a display component of a detail*

7. If necessary, you can move lines to the Defining Line display component. Select the detail, right-click, and choose **Linework > Edit**.

8. Select an entity of the detail, right-click, and choose **Modify Component** to open the **Select Linework Component** dialog box as shown in Figure 12.38. Click the down arrow and choose **Defining Line**, the display component that was edited in step 4.

9. Lines changed to this display component will plot or print.

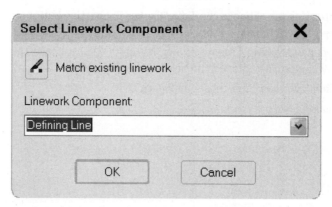

FIGURE 12.38 *Defining display component for a selected line*

USING THE DETAIL COMPONENT MANAGER

The **Detail Component Manager** provides access to 2D drawings that represent building components. The building components can be added to detail and section drawings to complete the drawings. The drawings can consist of fasteners, building units, and assemblies. Depending on the detail component, it can be inserted as an AutoCAD block, an entity, or a parametric block. The detail components may include top, elevation, or section views of building components with appropriate material hatch representation. The detail components are not AEC objects and do not interact with other detail components or AEC objects. All detail components can be accessed from the Detail Component Manager or the **Detailing** palettes. Access the **Detail Component Manager** as shown in Table 12.18.

TABLE 12.18 *Accessing the Detail Component Manager*

Command prompt	AECDTLCOMPMANAGER
Ribbon	Select the Home tab. Choose Detail Components from the Details panel.
Tool palette	Right-click the title bar of the Tool Palettes, choose Detailing to display the Detailing palette set. Select a detail tool from one of the palettes of the Detailing palette set, right-click, and choose Detail Component Manager.

The **Detail Component Manager** shown in Figure 12.39 includes components categorized in the MasterFormat™ 2004 [NBS Building] (National Building Services) divisions. The database for the details can be selected from the **Database** drop-down list shown in Figure 12.39. The database is specified in the AEC Content tab of the **Options** dialog box. The resources are located at *{Vista—C:\ProgramData\Autodesk\ACA 2012\enu\Details\Details (US)\ AecDtlComponents (US).mdb}* or *{Vista—C:\ProgramData \Autodesk\ACA 2012\enu\Details\Details (UK)\ AecDtlComponents (UK).mdb.}*

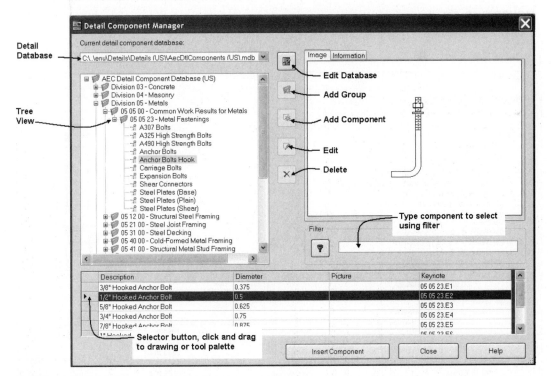

FIGURE 12.39 *Detail Component Manager*

The divisions of the Masterformat 2004 [NBS Building] are represented on the **Basic** palette of the Detailing palette set as shown in Figure 12.40. You can open the Detail Component Manager to a division by selecting a division tool on the Basic palette, right-clicking, and choosing **Detail Component Manager**. Division 22 of the Detail Component Manager is shown expanded in Figure 12.41. The properties of the selected component are displayed in the lower pane. The keynote number based upon Construction Specification Institute format is listed for each component. The keynote number can be identified with the keynote tools located on the Annotation tool palette. Details can be inserted from Detail Component Manager directly if you double-click the **Selector** at **p1** or select the component and then click the **Insert Component** button as shown in Figure 12.41. You can customize a selected detail component by clicking the **Edit Database** button shown in Figure 12.39 and editing the size and description of a component in the database.

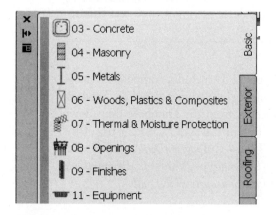
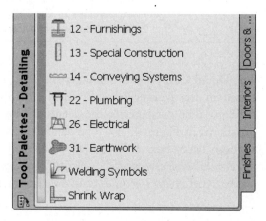

FIGURE 12.40 *Detailing tool palettes*

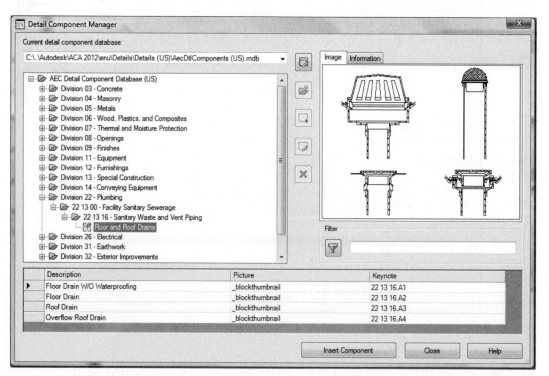

FIGURE 12.41 *Division 22-Plumbing*

When you select a tool from the Detail Component Manager, you can edit the properties of the tool in the Properties palette. You can edit such properties as view direction, content type, and other features of the detail. The properties of a tool can be preset if you right-click a detailing tool located on a palette and select **Properties** to open the **Tool Properties** dialog box. If you drag a detail component from the Detail Component Manager to a tool palette, as shown in Figure 12.42, you can set the properties in its Tool Properties dialog box for future applications. The Tool Properties dialog box of the Anchor Bolt tool shown in Figure 12.43 allows you to specify bolt type, view, and length.

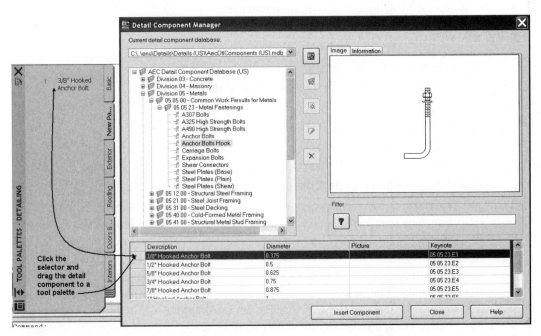

FIGURE 12.42 *Dragging a tool from the Detail Component Manager to a tool palette*

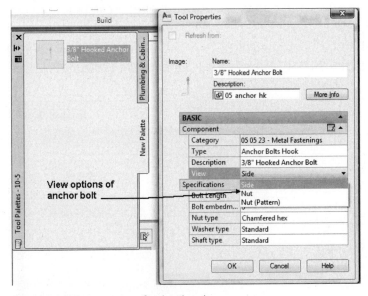

FIGURE 12.43 *Properties of a detail tool*

After specifying the settings, close the dialog box and insert the anchor bolt as specified. The settings in the Properties palette should be reviewed prior to inserting a detail since it controls the detail component. The procedure to insert components is dependent upon the setting in the Properties palette. The prompts to insert a detail differ according to the view and detail type. Detail components can consist of blocks, linework entities, or parametric drawing routines. Linework entities are used to represent components of variable length, such as structural members or reinforcement, while parametric drawing routines are often applied to specify the shape and size of components such as bolts. The types of detail components are classified based upon the insertion methods as follows:

- Stamp-Type (Multiple Insertion) Components
- Linear Repeating Pattern Components
- Boundary Filling Components
- Bookends-Type Components
- Countable Linear Repeating Pattern Components
- Rectangular Predefined-Depth Surface Components
- Dynamically-Sized Rectangular Components
- Bolt Components

Inserting a Stamp-Type Component

The procedure to insert a stamp type component such as a $2 \times 4 \, [50 \times 100]$ as shown in section view is accessed by specifying a **Section view** in the **Properties** of the tool as shown in Figure 12.44. The Section view option of this tool inserts the detail as a stamp-type component. The **Description** field shown in the Properties palette includes a drop-down list of lumber sizes. The tool can also insert lumber in plan or elevation views. The workspace prompts shown below allow you to rotate or flip the $2 \times 4 \, [50 \times 100]$ about the x or y axis to set its orientation. The block is inserted on the A-Detl-Wide [AEC-Detail-Wide] layer.

(Select the 06-Woods, Plastics and Composites *tool and review the settings in the Properties palette.* Metric only: *Choose the* G-Structural/Carcassing Metal/Timber edit Properties *palette and set Category =* Timber Framing, *Type =* Nominal Cut Timber, *and Description =* 50 × 100 mm Nominal.*)*

Insert point or: *(Select a point near* p1 *to insert the block.)*

Insert point or: ENTER *(Press* ENTER *to end the command.)*

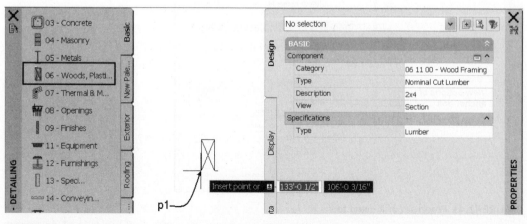

FIGURE 12.44 *Inserting a 2 × 4 from the Detail Component Manager*

Linear Repeating Pattern Components

A detail can also consist of a collection of repeating entities placed based on your response to the workspace prompts. A detail for welded wire fabric consists of a series of "X"s placed along a line as shown in Figure 12.45. To place this detail, open the Detail Component Manager to 03 22 00 Welded Wire Fabric Reinforcing > Wiremesh Reinforcing [E-In Situ Concrete/Large Precast Concrete > E30 Reinforcement For In Situ Concrete > Wiremesh Reinforcing] and drag the tool onto a tool palette or into the workspace. Edit the Properties palette to place the detail as described in the following steps.

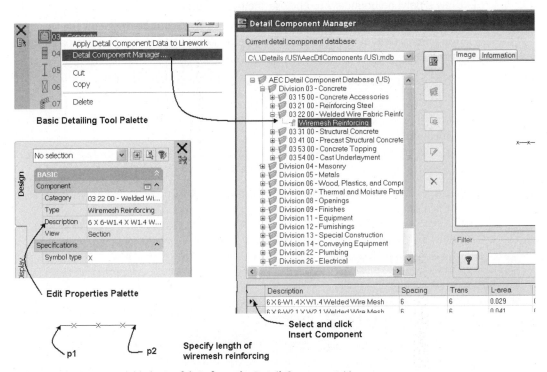

FIGURE 12.45 *Insert welded wire fabric from the Detail Component Manager*

Steps to Create Welded Wire Fabric

1. From the **Basic** tool palette of the **Detailing** palette set, select **03-Concrete**, right-click, and select **Detail Component Manager**. Expand the Division 3 in the right pane to 03 22 00 Welded Wire Fabric Reinforcing > Wiremesh Reinforcing as shown in Figure 12.45. **Metric only:** The location is [E-In Situ Concrete/Large Precast Concrete > E30 Reinforcement For In Situ Concrete > Wiremesh Reinforcing].

2. Double-click the **Selector** button for 6 × 6-W1.4 × W1.4 [150 × 150] Welded Wire Mesh, review the Properties palette, and respond to the workspace prompts as shown below to place the detail component:

   Start point: *(Select the start point of the wire fabric at* p1 *as shown in* Figure 12.45.*)*

   Endpoint: *(Select the endpoint of the wire fabric at* p2 *as shown in* Figure 12.45.*)*

   Start point: *(Press* ENTER *to end the command.)*

Boundary Filling Components

The Boundary Filling Components detail component applies a hatch pattern that represents material to a selected boundary. This type of detail component will prompt you to select a closed polyline that represents the boundary for the hatch. The 31-Earthwork tool of the Basic tool palette includes tools for a boundary filling component. The Properties palette of the 31-Earthwork tool, shown in Figure 12.46, allows you to select the fill options, such as gravel or sand. The steps to insert a boundary filling component are shown below.

Steps to Create a Gravel Boundary Filling Component

1. Draw a polyline boundary to enclose the gravel hatch.
2. From the **Basic** tool palette, select **31-Earthwork [D-Groundwork]**. Edit the **Description** field to **Gravel** in the **Properties** palette. Respond to the workspace prompts as shown below:

   Select objects to form the backfill boundary: 1 found *(Select the polyline at* **p1** *as shown in* Figure 12.46.)

   Select objects to form the backfill boundary: ENTER *(Press ENTER to end selection.)*

   Select a point within a boundary to backfill: *(Select inside the boundary at* **p2** *as shown in* Figure 12.46.)

   Select objects to form the backfill boundary: ENTER *(Press ENTER to end the command.)*

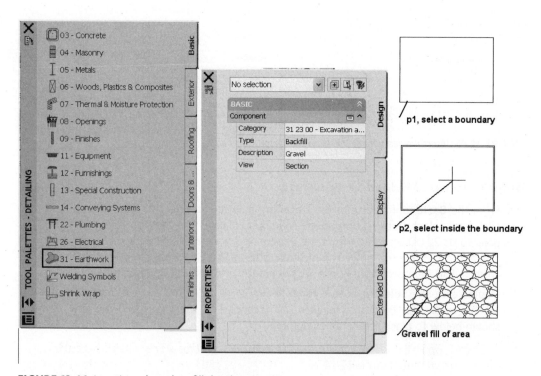

FIGURE 12.46 *Inserting a boundary fill detail component*

Bookends-Type Components

The **Bookends-Type** component starts and ends with unique geometry at each end. A repeated pattern fills the span between the two ends as shown in Figure 12.47. Therefore, objects such as vents and louvers are inserted by specifying the start and endpoints of the vent. The command options shown below include the options to flip the component about the x or y axis of the component.

(*Select* **Insert > Detail Component Manager** *and expand the Division 08 Openings >*
08 91 00 Louvers > Fixed Aluminum Louvers [L-Windows/Doors/Stairs>
L10-Windows/Rooflights/Screens/Louvres > Fixed Aluminum Louvres]. Select the **4"**
[100 mm] Fixed Aluminum Louver *detail and then click* **Insert Component**.)

> Start point or: *(Select the start of the louver at* **p1** *as shown*
> *in* Figure 12.47.*)*
>
> Endpoint or: *(Select the end of the louver at* **p2** *as shown in*
> Figure 12.47.*)*
>
> Start point or: ENTER *(Press* ENTER *to end the command.)*

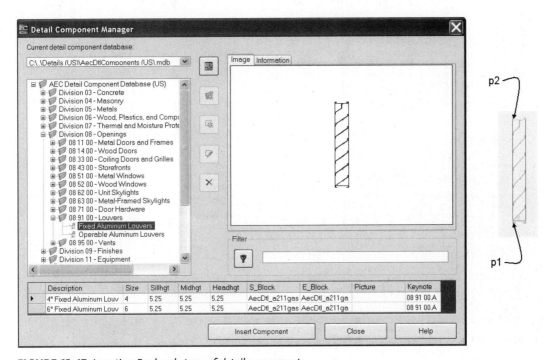

FIGURE 12.47 *Inserting Bookends type of detail component*

Countable Linear Repeating Pattern Components

The section view of masonry units, located on the 04-Masonry [F-Masonry] palette, is an example of the **Countable Linear Repeating Pattern**. This section view of the masonry unit can be specified from the drop-down list in the Properties palette or the **Tool Properties** dialog box of the tool as shown in Figure 12.48. This type of detail allows you to specify the number of courses of brick or block to insert in the detail. The following steps outline the procedure to place multiple courses of CMUs.

Steps to Create Masonry Units

1. Select **04-Masonry [F-Masonry]** from the **Basic** tool palette.

2. Edit the **Properties** palette as follows: Category = **Concrete Unit Masonry [F10 Brick/Block Walling]**, Type = **3 Core CMU [3 Core blocks]**, Description = **8" × 8" × 16" CMU [190 × 200 × 400mm-3 Core]**, and View = **Section**.

3. Respond to the workspace prompts as follows to place the concrete masonry units:

   Start point or: C *(Type C to specify the course count.)*

   Number of courses <1>: 6 ENTER *(Type the quantity of courses.)*

   Start point or: *(Select a point near* p1 *to specify the insertion point as shown in* Figure 12.48.*)*

   Pick point for direction or: *(Select a point above* p1 *as shown in* Figure 12.48 *to specify the direction.)*

   Start point or: *(Press ENTER to end the insertion.)*

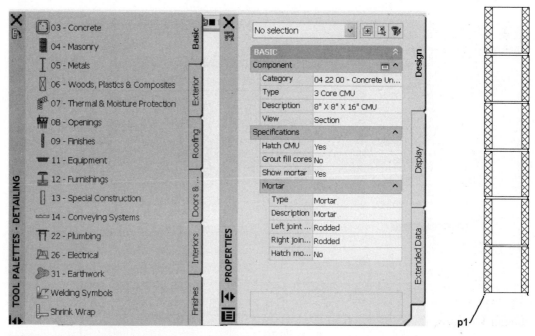

FIGURE 12.48 *Insert a countable repeating pattern component*

Rectangular Predefined-Depth Surface Components

Rectangular Predefined-Depth Surface components allow you to dynamically size a rectangular surface pattern about a baseline. The rectangular shape can be centered, left, or right about the baseline. The detail created in Figure 12.49 is developed about the baseline from **p1** to **p2**. The rectangle pattern is developed right-justified about the baseline. The component is accessed from Division 32-Exterior Improvements > 32 11 00 Base Courses > Base Courses [Q-Paving/Planting/Fencing/Site Furniture > Q20 Granular Sub-Bases to Roads/Pavings > Base Courses]. This detail component could be added to a palette by selecting the button at **p3** and dragging the cursor to the detailing tool palette as shown in Figure 12.49.

(Select **Insert > Detail Component Manager**, *expand Division 32-Exterior Improvements > 32 11 00 Base Courses > Base Courses [Q-Paving/Planting/Fencing/Site Furniture > Q20 Granular Sub-Bases to Roads/Pavings > Base Courses], and then click* **Insert Component**.*)*

Start point or: *(Select the start of baseline at* **p1** *as shown in* Figure 12.49.*)*

Endpoint or: *(Select the end of the baseline at* **p2** *as shown in* Figure 12.49.*)*

Start point or: *Cancel* *(Press* ENTER *to end the command.)*

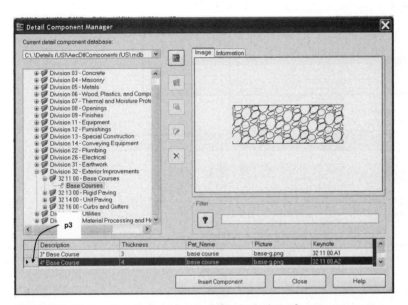
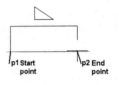

FIGURE 12.49 *Inserting Rectangular Predefined-Depth Surface Components*

Bolt Components

Bolt components are located in Division 05-Metals > 05 05 00 Common Work Results for Metals > 05 05 23—Metal Fastenings [G-Structural/Carcassing Metal/Timber > G 20-Carpentry/Timber Framing/First Fixing > First Fixing] of the Detail Component Manager. In Figure 12.50 the Anchor Bolts Hook category is expanded; click the selector button of a specific size listed and drag it to a tool palette. You can edit the Properties palette or the Tool Properties dialog box of the tool to specify view, length, bolt embedment, head type, nut type, and other features as shown in Figure 12.51. After editing the properties, insert the tool as shown in the following workspace sequence:

Projection point: *(Select a point at* **p1** *as shown in* Figure 12.50.*)*

Nut location: *(Select a point at* **p2** *as shown in* Figure 12.50.*)*

Hook location: *(Select a point at* **p3** *as shown in* Figure 12.50.*)*

Insert point or: *(Specify insert point and press* ESC *to end the command.)*

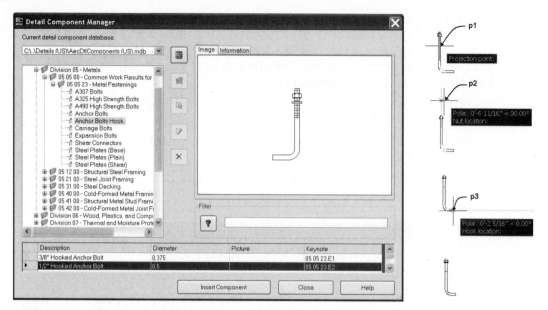

FIGURE 12.50 *Inserting an anchor bolt*

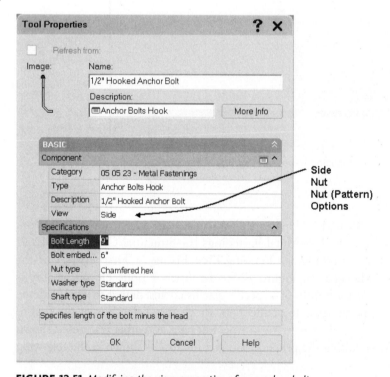

FIGURE 12.51 *Modifying the view properties of an anchor bolt*

Editing Details Using AEC Modify

Details can be edited as AutoCAD entities with the commands of the **Basic Modify Tools** shortcut menu. In addition, the **Extended Data tab** of the Properties palette of each detail component lists the properties of the detail component as shown in Figure 12.52. Additional editing tools are accessed from the shortcut menu of a detail component within the following options of **AEC Modify Tools**: Trim, Divide, Subtract,

Merge, and Crop. AEC Modify tools were presented in the "Modifying Nonassociative Spaces" section of Chapter 4. In addition, the **AEC Modify > Obscure** command is also available and is presented below.

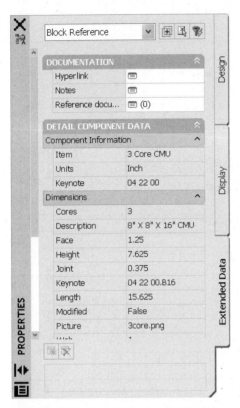

FIGURE 12.52 *Extended Data tab of the Properties palette*

Obscure

The **LineworkObscure** command will mask or hide an object or a detail component. This command can be used to hide a hatch pattern that is coincident with hatch patterns of other materials. Content that is hidden is placed on the A-Detl-Hide [Aec-Detail-Hidden] layer. Access the **LineworkObscure** command as shown in Table 12.19.

TABLE 12.19 *Accessing LineworkObscure command*

Command prompt	**LINEWORKOBSCURE**
Shortcut menu	Select an entity, right-click, and choose AEC Modify Tools > Obscure.

(Select the concrete hatch at p1 *as shown in* Figure 12.53, *right-click, and choose* AEC Modify Tools > Obscure.*)*

Command: LineworkObscure

Select obscuring object(s) or NONE to pick rectangle: 1 found *(Select rigid insulation hatch pattern at* p2 *as shown in* Figure 12.53.*)*

Select obscuring object(s) or NONE to pick rectangle: 1
found, 2 total *(Select rigid insulation hatch at* p3 *as shown
in* Figure 12.53.*)*

Select obscuring object(s) or NONE to pick rectangle: ENTER
(Press ENTER *to end the selection.)*

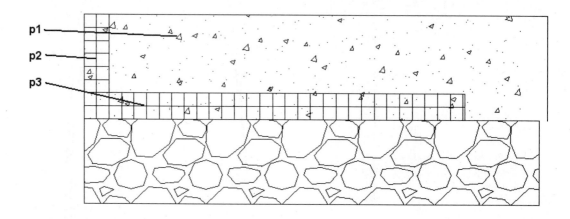

p1

p2

p3

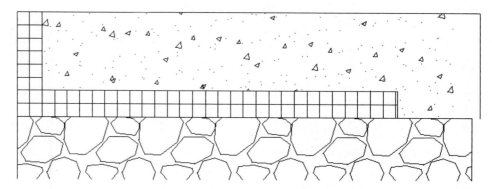

FIGURE 12.53 *Editing an area of the display for a hatch using the LineworkObscure command*

TIP

Details that have been modified can be pasted onto a detailing palette. You can create a
block that consists of multiple detail components and drag it onto a detailing palette. In addition, you can convert any entity to assume the keynote properties of division by selecting
the division tool, right-clicking, and selecting **Apply Detail Component Data to Linework**.

PLACING ANNOTATION WITH KEYNOTES

Keynotes annotate each component of a drawing with an identifying number that can
refer to a specification document. Keynotes that refer to a specification document are
reference keynotes. The specification document defines the attributes of the product
or required assembly techniques. The identifying number for detail components is
based upon the MasterFormat or National Building Specification. However, the
identifying number for assemblies is based upon the CSI Uniformat standard. The
master list of identifying numbers is included in an Access database file specified in
the **AEC Content** tab of the **Options** dialog box. The keynote number embedded in
the detail component, object, 2D linework, or material definition is accessed from the
AecKeynote field in the drawing.

The keynoting tools extract the field data from the object and display it as Mtext or
the attribute text within a block of a leader. The text-based tools used for reference

keynoting extract the keynote data using Mtext. The **Field** picks up the key from the **AecKeynote** field. In addition to reference keynoting, tools for sheet keynoting are located on the Annotation palette. Sheet keynoting inserts a key number based upon a sheet by sheet number system. The sheet keynote number is assigned after the keynotes are placed on a sheet. The keynoting tools shown on the Annotation tool palette of the Document palette set are shown in Figure 12.54. The keynote tools are also located in the Keynoting panel shown at the right in Figure 12.54. The keynoting tools extract the keynote information using text-based or block-based tools.

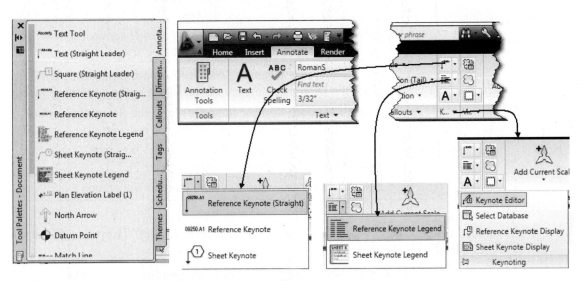

FIGURE 12.54 *Keynoting tools of the Annotation tool palette*

You may insert fields from the master list of identifying numbers included in an Access database file. The master list is accessed when you choose the Select Keynote button as shown in Figure 12.55.

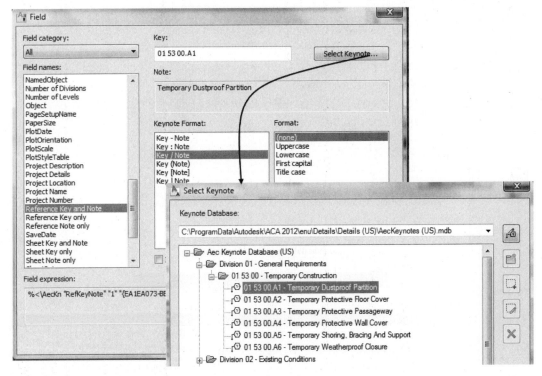

FIGURE 12.55 *AEC Keynote fields*

Reference Keynoting

The **Reference Keynote (Straight Leader) [Keynote (Straight Leader)]** tool and **Reference Keynote [Keynote]** (see Table 12.20) extract the keynote from the selected object as shown in the following workspace sequence. The keynote data for the CMU selected in the command sequence is extracted from the detail component. If Automatically add scales is on and Annotation scale is specified in the Drawing window status bar, the text and leader geometry is inserted in the drawing based upon the scale specified.

TABLE 12.20 *Accessing Reference Keynote (Straight Leader) and Reference Keynote [Keynote] command*

Ribbon	Select Reference Keynote (Straight) [Keynote (Straight Leader)] from the Reference Keynote Leader flyout in the Keynoting panel.
	Select Reference Keynote [Keynote] from the Reference Keynote Leader flyout in the Keynoting panel.
Tool palette	Select Reference Keynote (Straight Leader) [Keynote (Straight Leader)] from the Annotation tool palette.
	Select Reference Keynote [Keynote] from the Annotation tool palette.

The placement of a Reference Keynote (Straight Leader) **[Keynote (Straight Leader)]** is shown in the following workspace sequence:

(Select Reference Keynote (Straight Leader) [Keynote (Straight Leader)] *from the* Reference Keynote Leader flyout in the **Keynoting** panel.*)* Select object to keynote or press ENTER to select keynote manually: *(Select the masonry CMU at* p1 *as shown in* Figure 12.56.*)*

Select first point of leader: _mid of *(Select the location of the leader at* p2 *as shown in* Figure 12.56.*)*

Specify next point of leader line: *(Select the end of the leader line at* p3 *as shown in* Figure 12.56.*)*

Specify next point of leader line: ENTER *(Press* ENTER *to end the command.)*

Select text width <0> ENTER *(Press* ENTER *to accept the default.)*

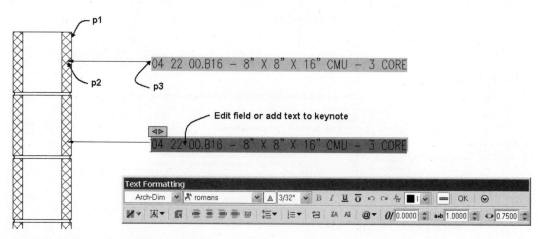

FIGURE 12.56 *Adding a reference keynote*

The Reference Keynote [Keynote] tool places the keynote in a similar position; however, no leader is inserted. If you press ENTER rather than select an object, the **Reference Keynote (Straight Leader) [Keynote (Straight Leader)]** or the **Reference Keynote [Keynote]** tool opens the **Select Keynotes** dialog box as shown in Figure 12.57. You can expand the MasterFormat [NBS Building] divisions and select a specific keynote in the Select Keynotes dialog box without selecting an entity or object.

The text of the leader is created in Mtext, which includes the **AecKeynoting** data field. You can double-click the text of the leader to edit the text. The text consists of a field as shown in Figure 12.57. If you select the field in the Mtext editor, right-click, and select **Edit Field**, the **Field** dialog box opens, which allows you to edit the format of the field. If the keynote is inserted as a placeholder, the keynote information is extracted from the selected object.

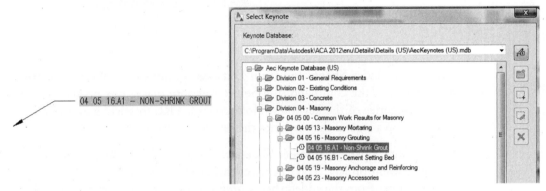

FIGURE 12.57 *Adding a keynote without specifying an object*

Creating a Keynote Legend

The **Reference Keynote Legend** tool (**KeynoteLegendAdd** command) can be selected from the Annotation tool palette to develop a legend of reference keynotes from the drawing. Refer to Table 12.21 to access the **KeynoteLegendAdd** command.

TABLE 12.21 *KeynoteLegendAdd command access table*

Ribbon	Select Reference Keynote Legend from the Keynote Legend flyout in the Keynoting panel.
Tool palette	Select the Reference Keynote Legend tool from the Annotation palette of the Document tool palette set.

When you select the **Reference Keynote Legend** tool from the Keynote Legend flyout in the Keynoting panel, you are prompted to select keynotes to include in the legend. You can create a legend of the keynotes selected, as shown in the following workspace sequence:

Select keynotes to include in the keynote legend or: *(Create a crossing selection window, select a point near* **p1***, then select a point near* **p2** *as shown in* Figure 12.58.*)*

Specify opposite corner: 3 found

Select keynotes to include in the keynote legend or: ENTER *(Press* ENTER *to end the selection.)*

Insertion point of table: *(Select a point near* **p3** *to specify the location of the table.)*

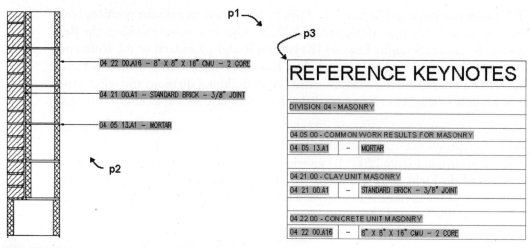

FIGURE 12.58 *Creating a reference keynote legend*

The command includes **Sheets** and **from Database** options. If you select the **from Database** option of the command, the **Select Keynote** dialog box opens. You can expand the categories of the keynotes. Hold down CTRL to select a group of keynotes and develop a legend that includes the selected keynotes.

The Sheets option opens the **Select Sheets to Keynote** dialog box, which list the sheets of the current project. You can select the sheets in the right pane, select **Add** to add the keynotes from the added sheets, and develop a legend that includes all the keynotes from the sheets as shown in Figure 12.59. This feature allows you to list all keynotes that were placed in the project on a single sheet.

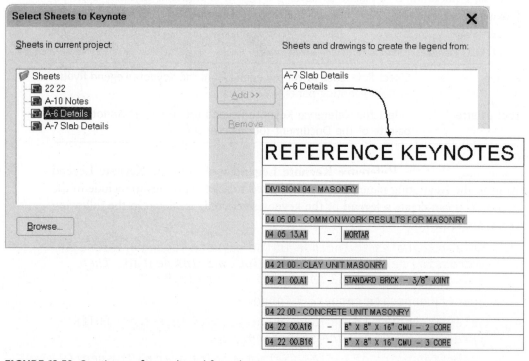

FIGURE 12.59 *Creating a reference legend from sheets*

Sheet Keynotes

The **Sheet Keynote (Straight Leader)** tool, accessed as shown in Table 12.22, inserts a leader and a specification number, which is linked to a specification listed on the current sheet. The identifying number is not inserted when the keynote is placed. A question mark is displayed to hold the place of the keynote number as shown at the left in Figure 12.60 until the sheet keynote legend is generated.

TABLE 12.22 *Accessing Sheet Keynote command*

Ribbon	Select Sheet Keynote from the Keynote Leader flyout in the Keynoting panel of the Annotate tab.
Tool palette	Select the Sheet Keynote (Straight Leader) tool from the Annotation tool palette.

The workspace prompt procedure to add a sheet keynote follows:

Select object to keynote or ENTER to select keynote manually: *(Select the detail component at* p1 *as shown in* Figure 12.60.*)*

Select first point of leader: *(Select the start of the leader at* p2 *as shown in* Figure 12.60.*)*

Specify next point of leader line: *(Specify the endpoint of the leader at* p3 *as shown in* Figure 12.60.*)*

Specify next point of leader line: ENTER *(Press ENTER to end the command.)*

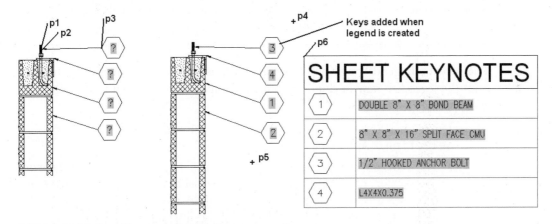

FIGURE 12.60 *Sheet Keynote and Sheet Keynote Legend*

Sheet Keynote Legend

The identifying number of the key is placed when you select the sheet keys to generate the sheet keynote legend. Access the **Sheet Keynote Legend** tool (**AecKeynoteLegendAdd**) from the Annotation tool palette as shown in Table 12.23.

TABLE 12.23 *Accessing the Sheet Keynote Legend command*

Ribbon	Select Sheet Keynote Legend from the Keynote Legend in the Keynoting panel of the Annotate tab.
Tool palette	Select the Sheet Keynote Legend from the Annotation tool palette.

If two keys are selected for the same type of detail component, the command recognizes the duplication and will repeat the key number. A keynote legend is placed using the following workspace prompts:

> (Select Sheet Keynote Legend *from the Annotation tool palette.)*
>
> Select keynotes to include in the keynote legend or: *(Create a crossing selection window and select point* **p4** *as shown in* Figure 12.60.*)*
>
> Specify opposite corner: *(Select a point near* **p5** *as shown in* Figure 12.60.*)* 15 found
>
> 10 were filtered out
>
> Select keynotes to include in the keynote legend or: ENTER *(Press ENTER to end the command.)*
>
> Insertion point of table: *(Select a point near* **p6** *as shown in* Figure 12.60.*)*

After a legend has been added, you can select the legend, right-click, and choose from the **Selection** shortcut flyout menu, which includes Add, Reselect, Apply Keys, and Show options. The Add option allows you to add additional sheet keys to the legend. After adding a key, the **KeynoteLegendApplyKeys** command can be used to assign an identifying key number to the new keys. Access the **KeyNoteLegendApplyKeys** command as shown in Table 12.24.

TABLE 12.24 *Keynote Legend Selection > Apply Keys command access*

Shortcut menu	Select the Sheet Keynote Legend, right-click, and choose Selection > Apply Keys.

The KeynoteLegendApplyKeys command will assign the key identifying number from the legend to new or duplicated detail components as shown in the following workspace sequence:

> (Select a Sheet Keynotes legend, right-click, and choose Selection > Apply Keys.)
>
> Select sheet keynotes to key: 1 found *(Select the keys in the drawing for the legend; the key numbers are updated from the legend as shown at the right in* Figure 12.61.*)*
>
> Select sheet keynotes to key: *(Press ENTER to end the selection.)*

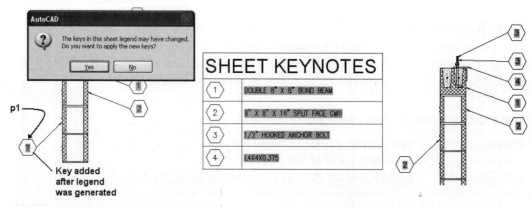

FIGURE 12.61 *Updating a sheet keynote*

Editing Legends

The selection set of a Reference or Sheet legend can be revised by selecting commands from the shortcut menu. The content of the legend is placed in a table. Therefore, you can insert additional rows and columns to the table. The size of the legend is defined by the current Annotation Scale when the table is placed. The shortcut menu of a sheet keynote legend includes **Add**, **Reselect**, and **Show** options. The **Keynote-LegendSelectionAdd** command (see Table 12.25) allows you to add additional keynotes to a table.

TABLE 12.25 *Sheet Keynote Legend Selection > Add command access*

Shortcut menu	Select the Sheet Keynote Legend, right-click, and choose Selection > Add.

The sheet key shown in Figure 12.61 was added after the legend was created in the following workspace sequence. After adding a key to the legend, choose the KeynoteLegendApplyKeys command (Table 12.24) to update the keys.

> *(Select a keynote legend, right-click, and choose* Selection > Add.*)*
>
> Select keynotes to add to the keynote legend or: 1 found *(Select the new sheet keynote at* p1 *as shown in* Figure 12.61.*)*
>
> Select keynotes to add to the keynote legend or: ENTER *(Press ENTER and the AutoCAD message box opens. Click* Yes *to add the key to the legend.)*

The **Selection > Reselect** command (see Table 12.26), located on the shortcut menu of a keynote legend, allows you to reselect the keynotes included in the schedule.

TABLE 12.26 *Reselecting the keys of a reference keynote legend*

Shortcut menu	Select a Reference Keynote Legend or a Sheet Keynote Legend, right-click, and choose Selection > Reselect.

The following workspace sequence revises the selection set of a legend:

> *(Select a legend, right-click, and choose* Selection > Reselect.*)*
>
> Select keynotes to include in the keynote legend or [Sheets/ from Database]: (Select the keys using a crossing window.)
>
> Specify opposite corner: 15 found

The **Locate Keynotes** command allows you to find the location of all keys referenced in the table. Access **Locate Keynotes** (**KeynoteLegendSelectionShow**) as shown in Table 12.27.

TABLE 12.27 *Show Selection of a Legend command access*

Command prompt	**KEYNOTELEGENDSELECTIONSHOW**
Shortcut menu	Select a Keynote Legend, right-click, and choose Selection > Show.

The following workspace sequence applies the KeyNoteLegendSelectionShow command to the legend:

(Select a keynote legend, right-click, and choose **Selection > Show.** *)*

Hover over keynote legend row to highlight corresponding keynotes or

CTRL-select to zoom: *(Move the cursor over keynote 2 as shown in* Figure 12.62. *)*

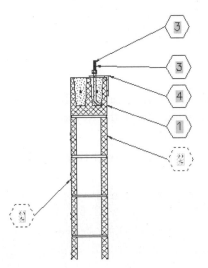

FIGURE 12.62 *Editing a Sheet Keynote Legend*

STOP   Do Tutorial 12-4, "Creating Details and Keynoting," at the end of the chapter.

TUTORIAL 12-1: CREATING A MODEL VIEW USING THE PROJECT NAVIGATOR

1. Verify that the Accessing Tutor\Imperial or Accessing Tutor\Metric content for Chapter 12 of the CengageBrain http://www.cengagebrain.com has been downloaded to your Accessing Tutor folder on your computer as described in Organizing Tutorial Directories in the Preface.

2. Open AutoCAD Architecture 2012 and select Project Browser from the Quick Access toolbar. Use the Project Selector drop-down list to navigate to your Student\Ch12 directory. Click New Project to open the Add Project dialog box. Type 22 in the Project Number field and type Ex 12-1 in the Project Name field. Check Create from template project, choose Browse, and edit the path to \Accessing Tutor\Imperial\ Ch12\Ex 12-1\ Ex 12-1.apj or \Accessing Tutor\Metric\Ch12\Ex 12-1\Ex 12-1.apj in the Look in field. Click Open to dismiss the Select Project dialog box. Click OK to dismiss the Add Project dialog box. Click Close to dismiss the Project Browser.

3. Select the **Project** tab. Click the **Edit Levels** button and edit the levels as shown in Table 12.28. Clear the **Auto-Adjust Elevation** checkbox. Click **OK** to dismiss the **Levels** dialog box. Click **Yes** to respond to the AutoCAD Architecture dialog box and regenerate all views from the level changes.

TABLE 12.28 *Level definition*

Name	Floor Elevation	Floor to Floor Height	ID	Description
2	8'-11-1/2" [2792]	8' [2438]	2	First Floor
1 [G]	0	8'-11-1/2" [2792]	1	Basement

4. Select the **Views** tab and select the **Add Category** command from the toolbar located at the bottom of the **Views** tab. Overtype **Exterior** as the name of the category. Select the **Exterior** category and then select the **Add View** command from the **Views** tab toolbar. Choose **Add a General View** option from the Add View dialog box. Verify **General** page of the Add General View dialog is open edit the following: Name = **Model**, Description = **Model View**, and then click **OK** to dismiss the **Description** dialog box. Choose *Open in the drawing editor* checkbox. Click **Next** and edit the **Context** page as follows: Check Division **1** for **Level 1 [G]** and **Level 2 [1]**. Click **Next** to display the **Content** page and verify that Constructs, Floor 2, Basement, and Stair are checked as shown in Figure 12.63. Click **Finish** to close the dialog box.

5. Verify the **Model** drawing is open. Select **Annotation Scale** from the Drawing Window status bar and choose the **1/4" = 1'-0" [1:50] scale**. Select **High Detail** from the Display Configuration flyout menu.

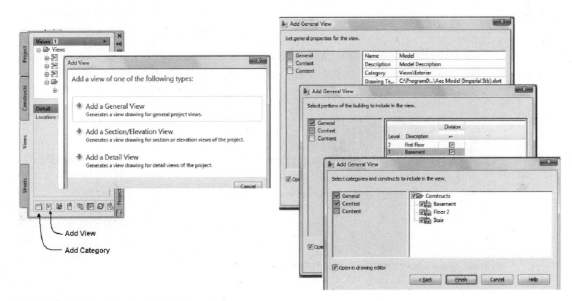
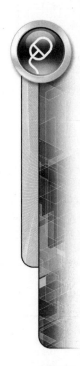

FIGURE 12.63 *Creating the Model View drawing*

6. Select **SW Isometric** from the **View Cube** to verify that Level **1[G]** and Level **2 [1]** are displayed. Select **Top** from the **View Cube**.

7. Choose the Annotate tab and double-click Annotation Tools of the Tools panel. Select **Elevation (Sheet No.)** from the Elevation flyout on the **Callouts** panel as shown in Figure 12.65. Respond to the workspace prompts as follows:

    Specify location of elevation tag: *(Select a point near* p1 *as shown in* Figure 12.64.*)*

    Specify direction for elevation: *(Select a point near* p2 *as shown in* Figure 12.64.*)*

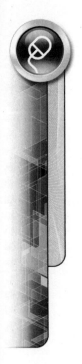

(Edit the **Place Callout** *dialog box as follows: Check* **Generate Section/Elevation** *and* **Place Titlemark**, *set scale to* 1/4" = 1'-0" [1:50], *type* **Front** *in the* **New Model Space View Name**, *and then click the* **New View Drawing** *button.)*

(Edit the **Add Section/Elevation View** *dialog box as shown in* Figure 12.65 *as follows: type* **Front** *in the* **Name** *field, edit Category to* Views\Exterior, *click* **Next**, *check level* 1 [G] *and level* 2 [1] *for the division* 1 *of the* **Context** *page as shown in* Figure 12.65, *click* **Next** *to open the* **Content** *page, and check* **Constructs**, **Floor 2**, **Basement**, *and* **Stair**. *Click* **Finish** *to close the dialog box.)*

Specify first corner of elevation region: *(Select a point near* **p3** *as shown in* Figure 12.64.*)*

Specify opposite corner of elevation region: *(Select a point near* **p4** *as shown in* Figure 12.64.*)*

Specify insertion point for the 2D elevation result: *(Select a point near* **p5** *as shown in* Figure 12.64.*)*

Save and close the Model drawing.

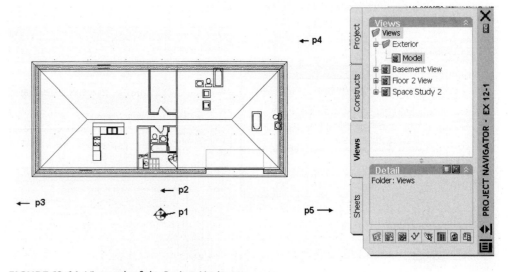

FIGURE 12.64 *Views tab of the Project Navigator*

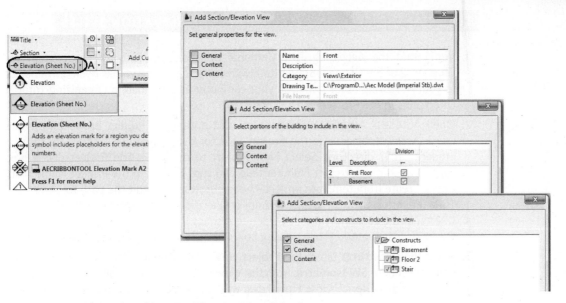

FIGURE 12.65 *Editing the Add Section/Elevation View dialog box*

8. Select the **Sheets** tab of the Project Navigator. Expand the **Architectural**, right-click the **Elevations** category, and choose **New > Sheet**. Edit the **New Sheet** dialog box as follows: Type Number = **A-12** and Sheet Title = **Front Elevation**. Choose the *Open in drawing editor* checkbox. Click **OK** to dismiss the New Sheet dialog box.

9. The A-12 Front Elevation drawing opens, including the display of the title block and border.

10. Select the **Views** tab, expand the Exterior category, select and drag the **Front** view into the **A-12 Front Elevation** sheet. Select a point near **p1** as shown in Figure 12.66.

11. Choose the Project tab. Choose Close Current Project from the toolbar located on the toolbar of the Projects tab. Click Close all project files and click **Yes** to save all drawings.

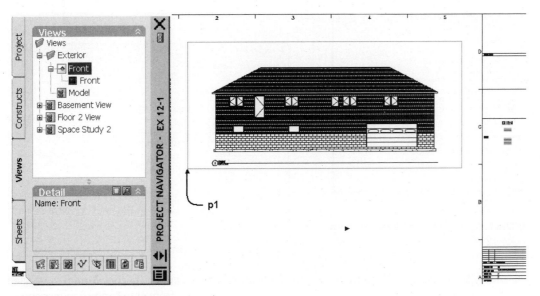

FIGURE 12.66 *Placing the Elevation on a sheet*

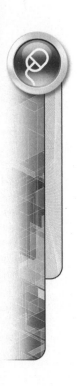

TUTORIAL 12-2: CREATING ELEVATIONS AND ELEVATION STYLES

1. Verify that the Accessing Tutor\Imperial or Accessing Tutor\Metric content for Chapter 12 of the CengageBrain http://www.cengagebrain.com has been downloaded to your Accessing Tutor folder on your computer as described in Organizing Tutorial Directories in the Preface.

2. Open Autodesk AutoCAD Architecture 2012 and select Project Browser from the Quick Access toolbar. Use the Project Selector drop-down list to navigate to your Student\Ch 12 directory. Select New Project to open the Add Project dialog box. Type Ex 12-2 in the Project Number field and type Ex 12-2 in the Project Name field. Check Create from template project, choose Browse, and edit the path to \Accessing Tutor\Imperial\Ch12\Ex 12-2\Ex 12-2.apj or \Accessing Tutor\Metric\Ch12\Ex 12-2\Ex 12-2.apj in the Look in field. Click Open to dismiss the Select Project dialog box. Click OK to dismiss the Add Project dialog box. Click Close to dismiss the Project Browser.

3. Select the **Constructs** tab of the **Project Navigator**. Double-click **Floor 1** to open the drawing. Choose **SW Isometric** from the **View Cube**. Elevations will be created for this building in this tutorial. Close Floor 1.dwg without saving changes to the drawing.

4. Select the **Views** tab and select the **Add Category** command from the toolbar located at the bottom of the **Views** tab. Overtype **Exterior** as the name of the category. Select the **Exterior** category and then select the **Add View** command from the **Views** tab toolbar. Click **Add a General View** in the Add View message box as shown in Figure 12.67, to open the **Add General View** dialog box, edit the following in the **General** page: Name = **Model** and Description = **Model View**. Click **OK** to dismiss the Description dialog box. Click **Next** and edit the **Context** page as follows: Check Division 1 for **Level 1 [G]**. Click **Next** to display the **Content** page and verify that **Constructs** and **Floor 1** are checked. Verify that **Open in drawing editor** is checked. Click **Finish** to dismiss the dialog box.

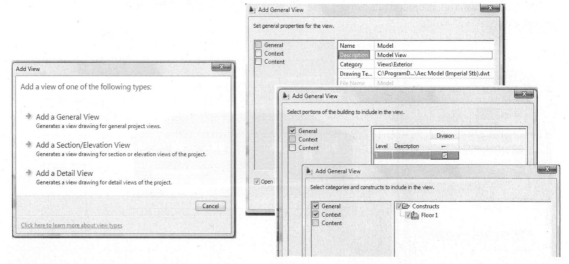

FIGURE 12.67 *Creating the Model View drawing*

5. Toggle OFF Snap Mode in the Application status bar. Verify that **Model** is toggled **ON** in the **Application** status bar. Select **Zoom All** from the **Zoom** flyout menu of the Navigation bar.

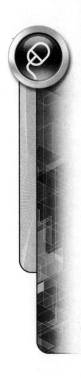

6. Select the Annotate tab. Select **Elevation (Exterior)** from the Elevation flyout in the **Callouts** panel. Respond to the workspace prompts as follows:

> `Specify first corner of elevation region:` *(Select a point near* `p1` *as shown in* Figure 12.68.*)*
>
> `Specify opposite corner of elevation region:` *(Select a point near* `p2` *as shown in* Figure 12.68.*)*
>
> *(Edit the* `Place Callout` *dialog box as follows: Check* `Generate Section/Elevation` *and* `Place Titlemark`*, set the scale to* `1/4" = 1'-0"` `[1:50]`*, and click* `Current Drawing`*.)*
>
> `Specify insertion point for the 2D elevation result:` *(Select a point near* `p3` *as shown in* Figure 12.68.*)*
>
> `Pick a point to specify the spacing and direction of elevations:` *(Move the cursor right and select a point near* `p4` *as shown in* Figure 12.68.*)*

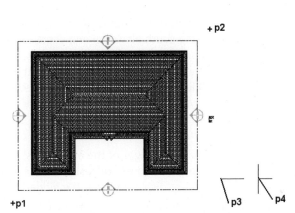

FIGURE 12.68 *Creating the Elevation view*

7. Select the building elevation line at **p1** as shown in Figure 12.69, right-click, and choose **Properties** to open the **Properties** palette. Select the **Design** tab and click the **Subdivisions** button (see Figure 12.69) to open the **Subdivisions** dialog box. Click the **Add Subdivision** button at the right, click twice on the default subdivision distance, and edit the distance to **15' [4000]** for the building elevation line as shown in Figure 12.69. Click **OK** to dismiss the Subdivisions dialog box.

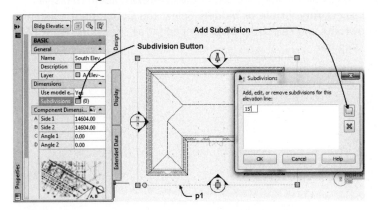

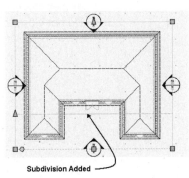

FIGURE 12.69 *Creating a subdivision for the elevation*

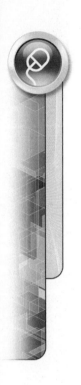

8. To view the elevations, select **Zoom Extents** from the Zoom flyout menu of the Navigation bar.

9. Select **Zoom Window** from the **Zoom** flyout menu of the Navigation bar. Respond to the workspace prompt as shown below to view the South Elevation:

> Specify first corner: *(Select a point near* **p1** *as shown in Figure 12.70.)*
>
> Specify opposite corner: *(Select a point near* **p2** *as shown in Figure 12.70.)*

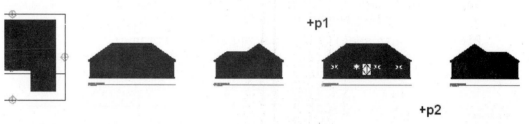

FIGURE 12.70 *Specifying points for the Zoom Window command*

10. To customize the display of the elevation using the subdivision and elevation styles, select the south elevation and choose **Copy Style** from the General panel of the 2D Section/Elevation tab. Edit the new style: Choose the **Display Properties** tab and click the **Edit Display Properties** button to open the **Display Properties (2D Section/Elevation Style Override-2D Section Style 96(2)) [100(2)]** dialog box. Select the **Layer/Color/Linetype** tab and then select Subdivision 2. Click in the **Linetype** column for Subdivision 2 to open the **Select Linetype** dialog box. Choose the **Hidden2** linetype. Click **OK** to dismiss all dialog boxes.

11. To view the changes of the elevation style modifications for subdivision 2, select the elevation and choose **Refresh** from the **Modify** panel. The components within subdivision 2 are displayed using the hidden linetype.

12. The elevation style can be modified to display a floor line with center linetype. The style will represent with centerlines the slab object based upon its color number, in this case color 192. Select the south elevation and choose **Edit Style** from the General panel.

13. Select the **Components** tab and click the **Add** button as shown in Figure 12.71. For Index 3, edit the Name to **Slab** and the Description to **slab line**.

14. Select the **Design Rules** tab. Click the **Add** button to add Rule 14. Select the **Color** column to open the **Select Color** dialog box, type **192** in the color field, and click **OK** to dismiss the Select Color dialog box. Select the **Context** column to display the drop-down list and select **Any visible or hidden**. Select the **Component** column to display the drop-down list and select the **Slab** component as shown in Figure 12.71. Do not close the dialog box.

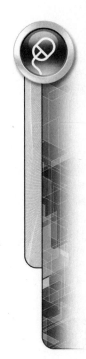

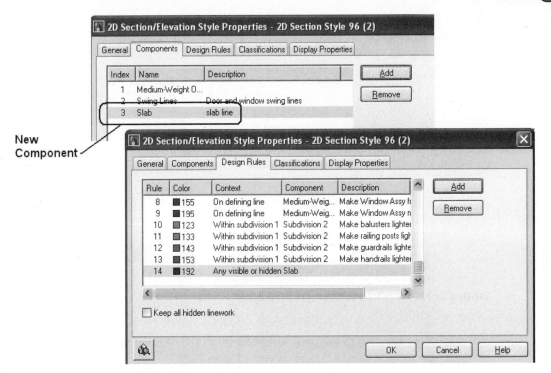

New Component

FIGURE 12.71 *Creating a new display component*

15. In this step the linetype and color of the slab display component will be defined. Select the **Display Properties** tab. Click the **Edit Display Properties** button. Select the **Layer/Color/Lineytpe** tab, scroll down the list of components, and select **Slab**. Click in the **Linetype** column for the Slab component and select **Center2** from the **Select Linetype** dialog box. Click **OK** to dismiss the **Select Linetype** dialog box. Select in the **Color** column and edit the color to color **7**. Click **OK** to dismiss the Select Color dialog box. Click **OK** to dismiss all dialog boxes.

16. To apply the elevation style, select the elevation and choose **Refresh** from the Modify panel of the 2D Section/Elevation tab to display the changes in the elevation.

17. In this step the display of surface hatch will be turned off to enhance editing. Select the South Elevation, right-click, and choose Properties. Choose the Display tab of the Properties palette. Click the **Select Component** button, shown in Figure 12.72, of the Display tab. Select the brick surface hatch of the elevation. In the Display component field toggle off the light bulb for the Surface Hatch Linework component as shown in Figure 12.72. Choose OK in the Modify Display Component at Style Override Level to apply the setting to the 2D Section/Elevation style.

18. Select **Zoom Window** from the Zoom flyout menu of the Navigation bar and respond to the workspace prompts as shown below:

    Specify first corner: *(Select a point near p1 as shown in Figure 12.72.)*

    Specify opposite corner: *(Select a point near p2 as shown in Figure 12.72.)*

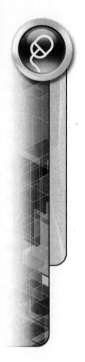

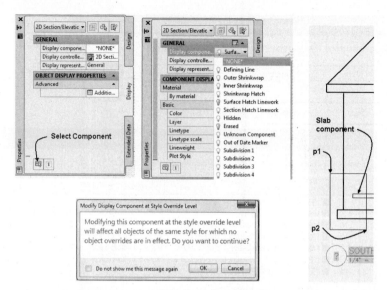

FIGURE 12.72 *Slab component displayed in elevation*

19. Select **Line** from the **Draw** panel in the Home tab and respond to the following work-space prompts to draw a line as shown in Figure 12.73:

    Specify first point: (Hold SHIFT down, right-click, and choose Endpoint. Select the end of the top of the slab at p2 as shown in Figure 12.73.)

    (Move the cursor left to set the direction.)

    Specify next point or [Undo]: **10' [3000]** ENTER (A line is drawn 10" [3000] long using direct distance entry.)

    Specify next point or [Undo]: ENTER (Press ENTER to end the command.)

20. Select the line at **A** as shown in Figure 12.73 to display the grips. Select the grip at the midpoint of the line, right-click, and choose **Copy**. Move the cursor up to display a polar angle of 90 and type **8' [2400]** in the workspace. Press ESC twice to end the selection and the grip edit.

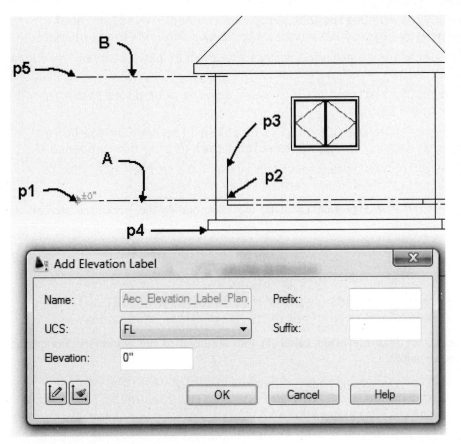

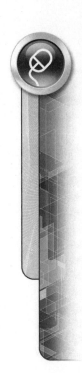

FIGURE 12.73 *Adding lines to an elevation*

21. Select the elevation and choose **Merge** from the **Linework** panel. Respond to the workspace prompts as shown below:

    ```
    Select objects to merge: (Select line at A as shown in
    Figure 12.73.)

    Select objects to merge: (Select line at B as shown in
    Figure 12.73.)

    Select objects to merge: ENTER (Press ENTER to end
    selection.)

    (Select the Slab component from the drop-down list of the
    Select Linework Component dialog box. Click OK to dismiss
    the Select Linework Component dialog box.)
    ```

22. Select **1/4″ = 1′-0″ [1:50]** from the **Annotation Scale** flyout menu and select High Detail display configuration from the **Drawing Window** status bar. Verify that **Automatically add scales** and **Annotation Visibility** are ON in the Drawing Window status bar.

23. Select the **Annotate** tab and double-click the Annotation Tool command of the Tools panel. Choose the Annotation palette and scroll down the tools to **Plan Elevation Label (1)**. Respond to the workspace prompts as shown below:

    ```
    Specify insertion point: (SHIFT + right-click, choose
    Endpoint object snap, and then select the line at p1 as shown
    in Figure 12.73.)
    ```

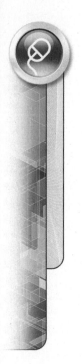

(Click the Define UCS *button of the* Add Elevation Label *dialog box and respond to the remaining workspace prompts.)*

Specify base point: *(Select the slab at* p2, *as shown in* Figure 12.73.*)*

Specify z-direction: *(Select a point near* p3 *as shown in* Figure 12.73.*)*

Enter name for UCS or [?]: FL ENTER *(The name of the UCS is defined and the* Add Elevation Label *dialog box reopened.)*

(Verify that the elevation value = 0 and then click OK *to dismiss the Add Elevation Label dialog box.)*

24. Select the **Plan Elevation Label (1)** tool. Respond to the workspace prompts as shown below:

    Specify insertion point: *(SHIFT + right-click, choose* Endpoint *object snap, and then select the end of Footing at* p4 *as shown in* Figure 12.73.*)*

    (Verify that UCS = FL and that the elevation value is -2'-0" [-406], and then click OK *to dismiss the dialog box.)*

25. Select the **Plan Elevation Label (1)** tool. Respond to the workspace prompts as shown below:

    Specify insertion point: *(SHIFT + right-click, choose* Endpoint *object snap, and then select the end of floor line at* p5 *as shown in* Figure 12.73.*)*

    (Verify that UCS = FL and that the elevation value is 8'-0" [2400], and then click OK *to dismiss the dialog box.)*

26. To turn on the display hatch, select the South Elevation, right-click, and choose Properties. Choose the Display tab. Select **Surface Hatch Linework** from the Display component flyout. Toggle the light bulb on for the Surface Hatch Linework. The **Modify Display Component at Style Override Level** dialog box opens. Choose OK to accept the style override. Press ESC to end selection. The surface hatch is displayed.

27. Select the elevation and choose **Edit** from the **Linework panel**. The surface hatch is turned off during the edit and the Edit In-Place: Linework tab is displayed. Select the footing lines shown at **p1**, **p2**, **p3**, and **p4** in Figure 12.74 and choose **Modify Component**. Select **Subdivision 2** from the list of display components. Click **OK** to dismiss the **Select Linework Component** dialog box. Select Finish from the **Edits** panel to save your changes.

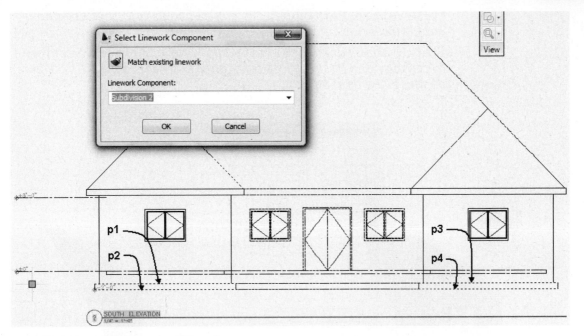

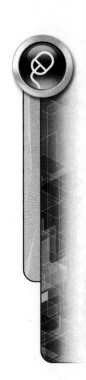

FIGURE 12.74 *Footing lines selected for edit*

28. In this step you will remove some of the slab lines to display a single line representing the top of slab. Select the elevation; choose **Edit** from the **Linework** panel. Select the lines representing the vertical edges and bottom of the slab using a crossing window from **p1** to **p2**, **p3** to **p4**, and **p5** to **p6** as shown in Figure 12.75, and then right-click and choose **Basic Modify Tools** > **Delete** from the shortcut menu. Select **Finish** from the **Edit In Place: Linework** tab.

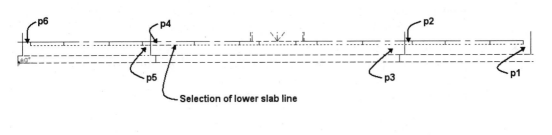

FIGURE 12.75 *Lower slab lines removed using Linework > Edit*

29. Select the Home tab. Select the **Polyline** command from Line flyout menu of the Draw panel. Draw a closed polyline as shown in Figure 12.76. Select the elevation; choose **Add** from the **Material Boundary** panel. Respond to the workspace prompts as shown below to modify the hatch pattern:

> Select a closed polyline for boundary: *(Select the polyline as shown at* **p1** *in Figure 12.76.)*

Erase selected linework? <No>: y *(Choose* Yes *to erase the polyline.)*

(Verify that Purpose = Limit, *Apply to =* Surface & Section Hatching, *and Material Selection =* All Materials. *Click* OK *to dismiss the dialog box.)*

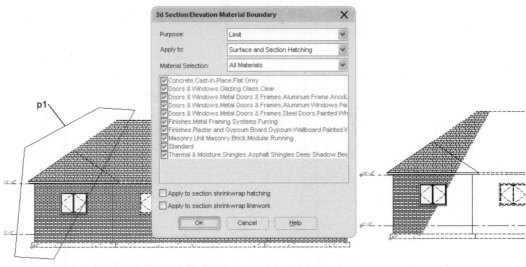

FIGURE 12.76 *Editing hatch of the elevation*

30. Select **View Top** from the **View** Cube. Save and close the Model.dwg drawing.

31. Select the **Constructs** tab of the **Project Navigator**. Double-click to open the **Floor 1** drawing. Select an existing double-casement window, right-click, and choose **Select Similar**. Edit Height to **5'-0"** [1500] in the Design tab of the **Properties** palette. Move the cursor from the Properties palette and choose ESC to clear selection.

32. Select an existing double-casement window, right-click, and choose **Add Selected**. Verify that Position along wall = **Offset/Center** and that Automatic offset = **3'-0"** [900]. Place additional windows as shown in Figure 12.77. Save and close the drawing.

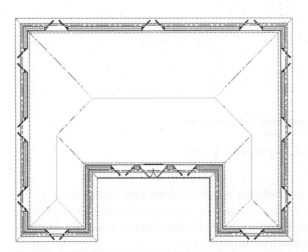

FIGURE 12.77 *Windows added to floor plan*

33. Click the **Views** tab and expand the **Exterior** category. Double-click the **Model** drawing. The changes on the Floor 1 construct are reflected in the model but not in the elevations.

34. Select **Zoom Extents** from the **Zoom** flyout menu of the Navigation bar. To apply the changes to all elevations, select the Home tab. Extend the Section & Elevation panel and choose Batch Refresh as shown in Figure 12.78 to open the **Batch Refresh 2D Section/Elevations** dialog box. Verify that the **Current Project** radio button is toggled ON and choose **Begin** to refresh all elevations. Click **Close** to dismiss the dialog box. The elevations are updated as shown in Figure 12.79.

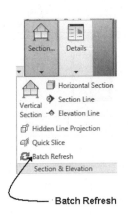

FIGURE 12.78 *Updating elevations using Batch Refresh*

FIGURE 12.79 *All elevations of the project are updated*

35. Click the Views tab of the Project Navigator. Expand the Model, double-click the **South Elevation** of the Model View drawing to view the revised elevation as shown in Figure 12.80.

36. Select the Project tab of the Project Navigator. Choose Close Current Project. Click Close all project files of the Project Browser—Close Project Files dialog box. Click Yes to save each drawing file.

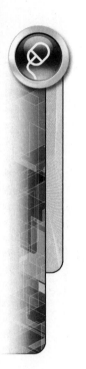

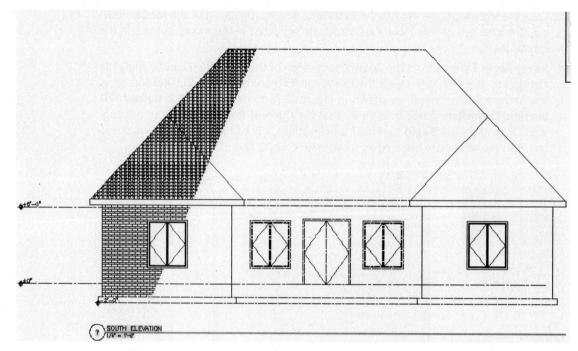

FIGURE 12.80 *The model space view of the South Elevation*

TUTORIAL 12-3: CREATING 2D SECTIONS AND A LIVE SECTION

1. Verify that the Accessing Tutor\Imperial or Accessing Tutor\Metric content for Chapter 12 of the CengageBrain http://www.cengagebrain.com has been downloaded to your Accessing Tutor folder on your computer as described in Organizing Tutorial Directories in the Preface.

2. Open Autodesk AutoCAD Architecture 2012. Choose Project Browser from the Quick Access toolbar. Use the Project Selector drop-down list to navigate to your Student\ Ch12 directory. Select New Project to open the Add Project dialog box. Type 23 in the Project Number field and type Ex 12-3 in the Project Name field. Check Create from template project, choose Browse, and edit the path to \Accessing Tutor\Imperial\ Ch12\Ex 12-3\Ex 12-3.apj or \Accessing Tutor\Metric\Ch12\Ex 12-3\Ex 12-3.apj in the Look in field. Click Open to dismiss the Select Project dialog box. Click OK to dismiss the Add Project dialog box. Click Close to dismiss the Project Browser.

3. Select the **Views** tab of the Project Navigator.

4. Select the **Views** category, right-click, choose **New Category**, and overtype **Sections**.

5. Expand **Exterior** and select the **Model** View drawing in the **Views** tab of the **Project Navigator**, then right-click, and click **Open**.

6. Toggle **ORTHO** ON on the status bar. Toggle SNAP OFF on the status bar. Choose **High Detail** from the **Display Configuration** flyout menu of the **Drawing Window** status bar.

7. Choose the Annotate tab. Select **Section (Sheet Number)** from the Section flyout of the **Callouts** panel and respond to the workspace prompts as shown below:

   Specify first point of section line: *(Select near point* A *as shown in* Figure 12.81.*)*

   Specify next point of line: *(Select near point* B *as shown in* Figure 12.81.*)*

Specify next point of line or: ENTER *(Press* ENTER *to end the line.)*

Specify section extents: *(Select a point near* C *as shown in* Figure 12.81 *and then edit the* Place Callout *dialog box as described in the next step.)*

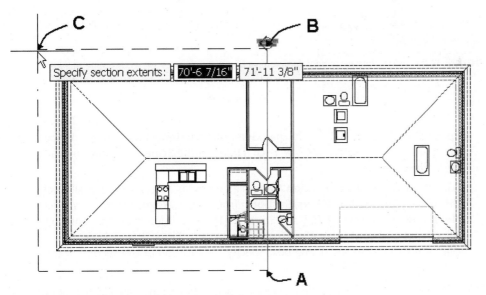

FIGURE 12.81 *Section line placement for Transverse view*

8. Edit the **Place Callout** dialog box as follows: type **Transverse** in the **New Model Space View Name** field, verify that **Generate Section/Elevation** and **Place Title-mark** are checked, select **1/4" = 1'-0" [1:50]** in the **Scale** drop-down list, and click **New View Drawing**.

9. Edit the **Add Section/Elevation View** dialog box as follows: Type **Transverse** in the name field, select **Views/Sections** from the **Category** drop-down list, click **Next**, and check Level **1 [G]** and Level **2 [1]** for division **1** on the **Context** page. Click **Next** and verify that **Constructs**, **Basement**, **Floor 2**, and **Stair** are checked. Click **Finish** to dismiss the dialog box. Specify the location of the section in the workspace prompts as follows:

    Specify insertion point for the 2D section result: *(Select a point near* p1 *as shown in* Figure 12.81.*)*

10. Select **Section (Sheet Number)** from the Section flyout in **Callouts** panel and respond to the workspace prompts as shown below:

    Specify first point of section line: *(Select near point* p1 *as shown in* Figure 12.82.*)*

    Specify next point of line: *(Select near point* p2 *as shown in* Figure 12.82.*)*

    Specify next point of line or: ENTER *(Press* ENTER *to end line.)*

    Specify section extents: *(Select a point near* p3 *as shown in* Figure 12.82 *and then edit the* Place Callout *dialog box as described in the next step.)*

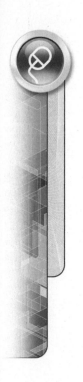

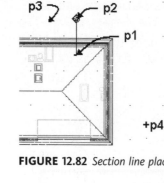

FIGURE 12.82 *Section line placement for Wall Section view*

11. Edit the **Place Callout** dialog box as follows: Type **Wall Section** in the **New Model Space View Name** field, verify that **Generate Section/Elevation** and **Place Title-mark** are checked, select **3/4" = 1'-0" [1:20]**, in the **Scale** drop-down list, and click **New View Drawing**.

12. Edit the **Add Section/Elevation View** dialog box as follows: Type **Wall Section** in the name field, select **Views/Sections** from the **Category** drop-down list, click **Next**, and check Level **1 [G]** and Level **2 [1]** for division **1** on the **Context** page. Click **Next** and verify that **Constructs**, **Basement**, **Floor 2**, and **Stair** are checked. Click **Finish** to close the dialog box. Specify the location of the section in the workspace prompt as follows:

    Specify insertion point for the 2D section result: *(Select a point near* p4 *as shown in* Figure 12.82.*)*

13. Select the Views tab of the Project Navigator. Select the **Sections** category and select **Refresh Project** from the toolbar at the bottom of the Views tab as shown in Figure 12.83. Expand the Sections category and verify that the Transverse and Wall Section views are now displayed in the Sections category.

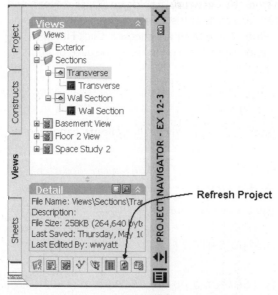

FIGURE 12.83 *Section View drawings of the Project Navigator*

14. Expand the **Transverse** View drawing and double-click the **Transverse** model space view. The view title has been inserted in the View. The View number is a question mark since the view has not been placed on a sheet. The view number is field data and will be linked to the sheet when placed. Edit the Display Configuration to **High Detail**.

15. Note that the hatch patterns assigned in the section include hatching of the wood walls as shown in Figure 12.84. To refine the section, you can turn off the section hatch component of the **Standard** material definition to remove this hatch. This step modifies the Standard material definition. Select the **Manage** tab and choose **Style Manager** from the Style & Display panel to open the **Style Manager**. Expand the Multi-Purpose Objects category for the **Transverse** drawing in the left pane. Expand the Material Definitions category. Select the **Standard** material definition to open the editing tabs of the right pane. Verify that the High Detail is current. Select the **Display Properties** tab and choose the **General High Detail** display representation. Click the **Edit Display Properties** button and select the **Layer/Color/ Linetype** tab. Toggle **Off** the **Section Hatch** component. Click **OK** to dismiss all dialog boxes.

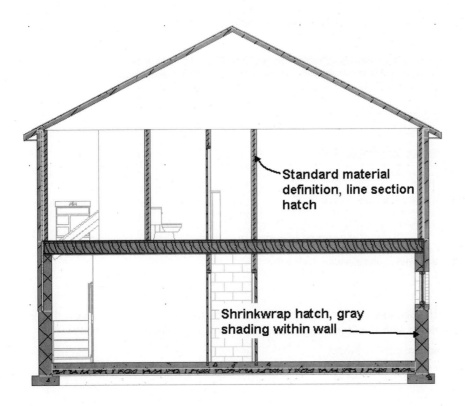

FIGURE 12.84 *Transverse section of the building*

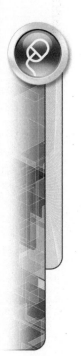

16. Select the section and choose **Edit Style** from the General panel. Select the **Display Properties** tab and click the **Edit Display Properties** button. Check **By Material** for the Section Hatch Linework. When By Material is selected, the Standard Material Definition edit of step 15 will be applied. Toggle OFF **Shrinkwrap hatch** as shown in Figure 12.85. Click **OK** to dismiss all dialog boxes.

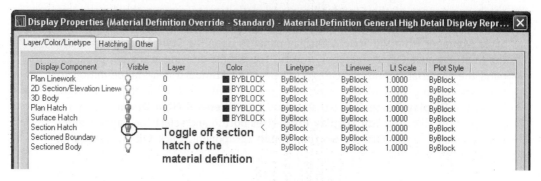

Edit Material Definition

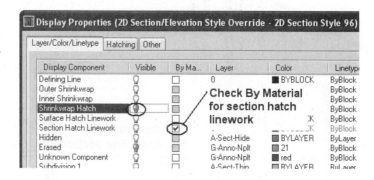

Edit 2D Section/Elevation Style

FIGURE 12.85 *Material definition and elevation style edit*

17. Select the section, right-click, and choose **Refresh**. The hatching is removed from components with the Standard material definition as shown in Figure 12.86.

18. Save and close the drawing.

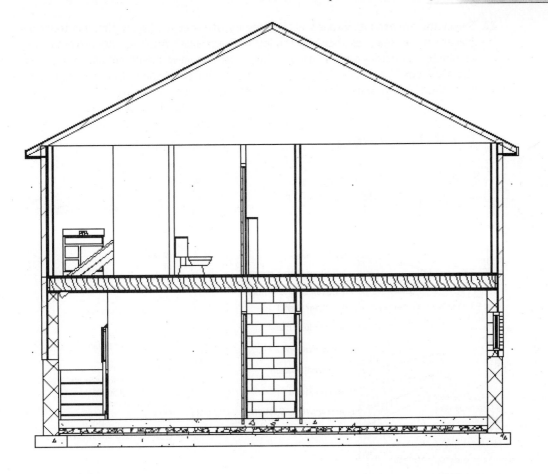

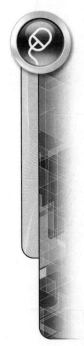

TRANSVERSE
1/4" = 1'-0"

FIGURE 12.86 *Hatch pattern removed from objects with the Standard material definition*

19. Expand the **Wall Section** View drawing and double-click the **Wall Section** view in the Views tab. In this step the section hatching and shrinkwrap hatching will be turned off to prepare the section for the development of additional detail components. Select the section object and choose **Edit Style** from the General panel of the 2D Section/Elevation tab. Choose the **Display Properties** tab. Click the **Edit Display Properties** button at the right. Select the Layer/Color/Linetype tab and toggle OFF **Shrinkwrap Hatch** and **Section Hatch Linework**. The detail components within the boundaries of the section can be added from the Detail Components Manager. Click **OK** to dismiss all dialog boxes.

20. Save and close all project drawings.

21. Open the Project Navigator and select the **Sheets** tab. Expand the Architectural category. Select the **Sections** category, right-click, and choose **New > Sheet** to open the **New Sheet** dialog box. Type **A-14** in the **Number** field and **Sections** in the **Sheet title** field. Choose the *Open in drawing editor* checkbox. Click **OK** to dismiss the dialog box and open the new sheet.

22. Select the **Views** tab. Select the **Transverse** View drawing and drag the drawing that includes the model space view onto the A-14 sheet. Select the **Wall Section** drawing and drag the drawing onto the A-14 sheet. Save and close project drawings.

23. Select the **Sheets** tab. Expand each category of sheet to display the sheets views placed on the sheets as shown at the left in Figure 12.87. Although the project is not complete, the names and numbers of the sheet and view should be adjusted. Select the A-2 Floor Plan 2 sheet, right-click, and choose Delete Sheet. Choose Yes to dismiss the Confirm Sheet Remove dialog box. Select the remaining sheets or views, right-click, and choose **Rename and Renumber**. Edit the sheets and views as shown in Figure 12.87. Check **Rename drawing file to match** the sheet title and check **Prefix with sheet number**.

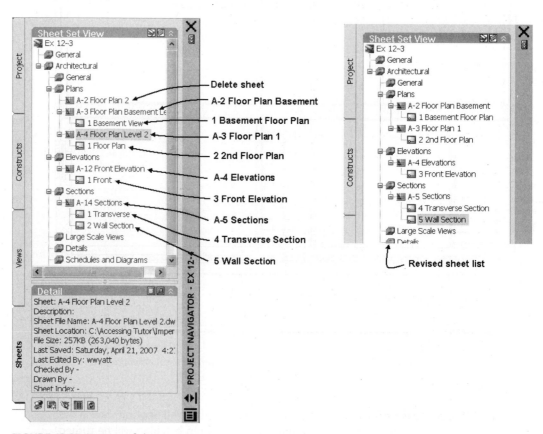

FIGURE 12.87 *Revision of sheets*

24. To add callouts to each plan sheet, double-click the **A-3 Floor Plan 1** to open the sheet drawing from the Sheets tab. Double-click inside the viewport. Click the Lock in the Drawing Window status bar to unlock the viewport. Click Annotation Visibility to toggle annotation visibility ON. Verify that **1/4″ = 1′-0″ [1:50]** is the Annotation Scale from the **Drawing Window** flyout menu.

25. Select the Annotate tab. Choose the **Section (Sheet Number)** from the **Callouts** palette. Respond to the workspace prompts as follows:

    Specify first point of section line: *(Select a point near* p1 *as shown in* Figure 12.88.*)*

    Specify next point of line: *(Select a point near* p2 *as shown in* Figure 12.88.*)*

    Specify next point of line: *(Press* ENTER *to end selection.)*

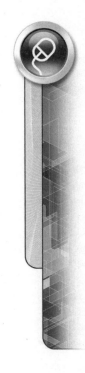

Specify section extents: *(Select a point near* p3 *as shown in* Figure 12.88. *Since you are only placing a callout the extents of the section is not defined by p3. The* `Place Callout` *dialog box opens; choose* `Callout only`. *)*

(Select the callout using a crossing window from point p5 *to* p4 *as shown in* Figure 12.88. *Click the selected geometry, continue to hold down the left mouse button, and drag the callout over to the* 4 Transverse Section *view as shown in* Figure 12.88. *The Callout is changed after a brief time to reflect the sheet and view. Click the Lock Viewport in the Drawing Window status bar.)*

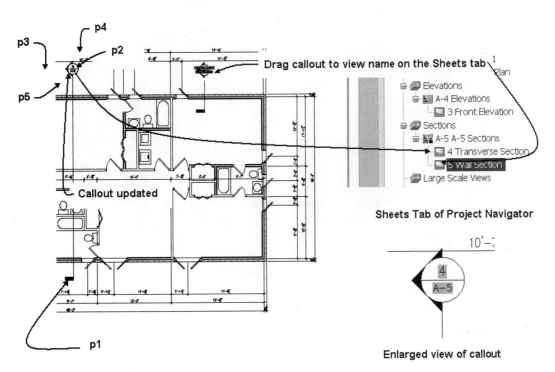

FIGURE 12.88 *Link of fields to views*

26. Repeat step 24 to place a callout for the **5 Wall Section** view. Save and close A-3 Floor Plan 1.

27. Open the **A-2 Floor Plan Basement** sheet. Double-click inside the viewport. Click the lock to unlock the viewport. Verify Display Configuration = **High Detail** and Annotation Scale = **1/4" = 1'-0" [1:50]**, Global Cut plane Cut Height = **5'-0" [1500]**.

28. Repeat the procedure in step 24 to place callouts for the **4 Transverse Section** and **5 Wall Section** views on the A-2 Floor Plan Basement sheet. Choose the lock in the Drawing Window status bar to lock the viewport. Save and close A-2 Floor Plan Basement.

29. Select the **Views** tab, select the **Transverse** View drawing of the **Project Navigator**, right-click, and choose **Copy**. Select the **Sections** category, right-click, and choose **Paste**. A copy of the transverse View drawing is listed in the Sections category. Select the **Transverse (2)** drawing, right-click, and click **Open**. Select the **SE Isometric** view in the **View Cube**.

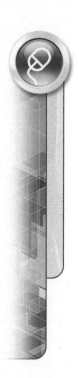

30. Select the section line and choose **Enable Live** Section from the Live Section panel of the Building Section Line tab. Choose **Realistic** from the **Visual Styles** flyout menu of the Visual Styles panel in the View tab. The live section is shown in Figure 12.89.

31. Save and close all project drawings.

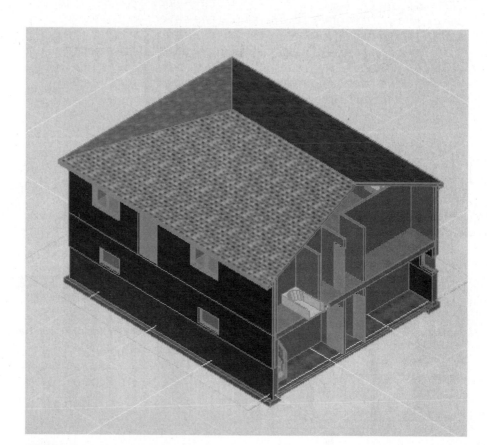

FIGURE 12.89 *Live section enabled*

32. Select the Views tab of the Project Navigator. Select the **Exterior** category, right-click, and choose **New View Dwg > General**. Edit the Add General View as follows: Name = **Presentation** and Category = **View/Exterior**. Click **Next** and check level **1 [G]** and **2 [1]** of the **Context** page. Click **Next** and verify that **Floor2**, **Basement**, and **Stair** constructs are checked. Choose Open in drawing editor. Click **Finish** to dismiss the dialog box and open the Presentation view drawing.

33. To create a presentation elevation, choose **Front** from the **View Cube**. Choose **Realistic** from the **Visual Styles** flyout menu of the Visual Styles panel in the View tab. Toggle OFF Grid Display in the **Application** status bar.

34. Choose the View tab. Choose the View Manager from the flyout View menu of the Appearance panel to open the View Manager. Click the New button to open the **New View/Shot Properties** dialog box. Edit the New View/Shot Properties as follows: View name = **Front Presentation**. Choose **Define view window** and choose a window from points **p1** to **p2** as shown in Figure 12.90. Press ENTER to end window specification. Check **Save layer snapshot with view** checkbox. Set UCS = **World** and Visual Style = **Current**. Click **OK** to dismiss all dialog boxes.

35. Select **Refresh Project** from the toolbar of the **Views** tab in the **Project Navigator**. Expand the Exterior category. Expand the Presentation drawing. Double-click the **Front Presentation** view name to display the new view. Save and close the drawing.

36. Select the **Sheets** tab. Expand Architectural and choose the **3D Representations** category. Right-click and choose **New > Sheet**. Type **A-6** in the **Number** field and **Presentation Elevations** in the **Sheet title**. Click the Open in drawing editor checkbox. Click **OK** to dismiss the New Sheet dialog box.

37. Select the **Views** tab. Select the **Front Presentation** view, right-click, and choose **Place on Sheet**. Right-click, choose **1/4" = 1'-0" [1:50]**, and select a location on the sheet.

38. Select the Project tab. Click **Close Current Project**. Click Close all project files. Click Yes to save changes to open drawings.

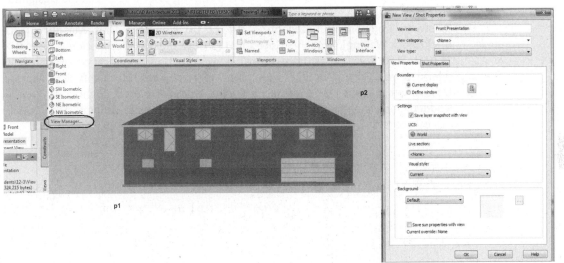
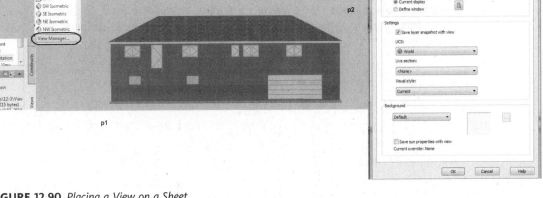
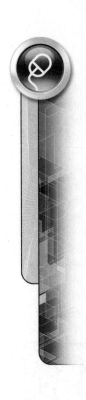

FIGURE 12.90 *Placing a View on a Sheet*

TUTORIAL 12-4: CREATING DETAILS AND KEYNOTING

1. Verify that the Accessing Tutor\Imperial or Accessing Tutor\Metric content for Chapter 12 of the CengageBrain http://www.cengagebrain.com has been downloaded to your Accessing Tutor folder on your computer as described in Organizing Tutorial Directories in the Preface.

2. Select Project Browser from the Quick Access toolbar. Verify that the Project Selector folder location is set to your Student\Ch12 folder. Click the New Project button located in the lower-left corner of the Project Browser to open the Add Project dialog box. Type **24** in the Number field and **Ex 12-4** in the Name field. Check Create from template project, choose Browse, and edit the path to Accessing Tutor\Imperial\Ch12\Ex 12-4\Ex 12-4.apj or Accessing Tutor\Metric\Ch12\Ex 12-4\Ex 12-4.apj in the Look in field. Click **Open** to dismiss the Select Project dialog box. Click OK to dismiss the Add Project dialog box. Click **Close** to close the Project Browser.

3. Select the **Views** tab of the **Project Navigator**. Expand the Sections category and double-click **Wall Section** to open the file. To update all views of the project, choose the Home tab. Expand the Section & Elevation panel and choose **Batch Refresh** to open the **Batch Refresh 2D Sections/Elevations** dialog box. Verify that Current

Project is toggled ON and choose **Begin**. Refreshing the Sections/Elevation process can take several minutes. Click the **Close** button upon completion of the update.

4. Right-click the Object Snap button of the status bar, select **Settings**, check **Endpoint** object snap, and toggle OFF all other object snap modes. Click the **Options** button in the lower-left corner and verify that **Display polar tracking vector**, **Display full-screen tracking vector**, and **Display Auto Track tooltip** are checked in the **AutoTrack Settings** section of the **Drafting** tab. Verify that the **Display AutoSnap tooltip** is checked in the **AutoSnap Settings** section. Click **OK** to dismiss the Options dialog box. Click **OK** to dismiss the Drafting Settings dialog box. Toggle **ON** Object Snap, Polar Tracking, Object Snap Tracking, and Dynamic Input in the Application status bar.

5. Select the Home tab and click the Tools button of the Build panel. Right-click the titlebar of the tool palette and choose the Detailing palette set. Select **04-Masonry [F-Masonry]** from the **Basic** palette of the **Detailing** palette set. Click the Design tab of the Properties palette. Click in the fields of the **Properties** palette and specify the following: Component: Category = **04 22 00 Concrete Unit Masonry [F10-Brick/Block Walling]**, Type = **3 Core CMU [3 Core Blocks]**, Description = **12 × 8 × 16 CMU [140 × 200 × 400mm-3Core]**, View = **Section**; Specifications: Hatch CMU [block] = **yes**, Grout fill cores = **No**, Show mortar = **yes**; Mortar: Type = **Mortar**, Description = **Type "S" Mortar**, Left joint type = **Rodded**, Right joint type = **Rodded**, and Hatch mortar = **yes**. Respond to the workspace prompts as follows to place 7 courses:

> Start point: *(Press the down arrow of the workspace prompt and choose the* Count *option.)*
>
> Number of courses <1>: 7 ENTER *(Enter the number of courses.)*
>
> Start point: *(Select a point near* **p1** *as shown in* Figure 12.91.*)*
>
> Pick point for direction or: *(Select point* **p2** *shown in* Figure 12.91.*)*
>
> Start point or: *(Press* ESC *to end the command.* Metric only: *Select the 7 courses of the block, right-click, and choose* Basic Modify Tools > Copy. *Copy the wythe 155 mm to the left as shown at the right in* Figure 12.91.*)*

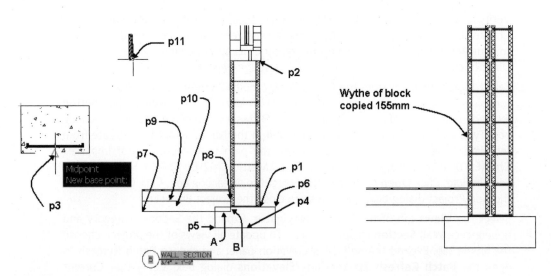

FIGURE 12.91 *Start point of a detail component*

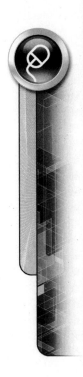

6. Right-click the tool palette titlebar and select the **Ex 12-4** palette set. Right-click the tool palette titlebar, choose **New Palette**, and overtype **My Detail** in the palette field. Select the Home tab and select **Detail Components** from the Details panel of the Home tab. Verify that the Current detail component database is *{Vista—C:\ProgramData\Autodesk\ACA 2012\enu\Details\ Details (US)\ AecDtlComponents (US).mdb}* or *{Vista—C:\ProgramData\Autodesk\ACA 2012\enu\Details\Details (UK)\ AecDtlComponents (UK).mdb}*. Expand Division 03-Concrete, 03 31 00 Structural Concrete > Strip Footings [E-In Situ Concrete/Large Precast Concrete > E10 Mixing/ Casting/ Curing In Situ Concrete > Strip Footings]. [Caution: metric readers should choose Strip Footings, not Strip Footing located above.] Click the **Selector** of the **18 × 12 Concrete Footing [600 mm × 300 mm Concrete Footing]** tool and drag it to the **My Detail** tool palette. Close the **Detail Component Manager**.

7. Right-click the **18" × 12" Concrete Footing** [600 mm × 300 mm Concrete Footing] tool on the tool palette and choose **Properties** to open **Tool Properties** as shown in Figure 12.92. Edit and verify the following in the **Tool Properties** dialog box: Component: Category = **03 31 00—Structural Concrete [E10—Mixing/Casting/Curing In Situ Concrete]**, Type = **Strip Footings**, Description = **18 × 12 Concrete Footing [600 mm × 300 mm Concrete Footing]**, View = **Section**; Specifications: Hatch footing = **Yes**, Show reinforcing = **Yes**; Rebar Longitudinal: Type = **ACI Reinforcing Bar [BS Reinforcing Bar]**, Description = **#4 Rebar [R8 Rebar]**, Draw type = **Solid**, Bars = **2**, Bars on outside = **No**; Rebar-Lateral: Type = **ACI Reinforcing Bar [BS Reinforcing Bar]**, Description = **#4 Rebar [R8 Rebar]**, Draw type = **Solid**, Edges = **Edge 2**, and Edge offset = **1 1/2" [38]**.

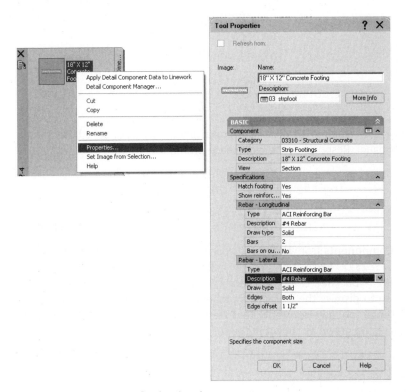

FIGURE 12.92 *Settings of a detail tool*

8. Click **OK** to dismiss the Tool Properties dialog box.

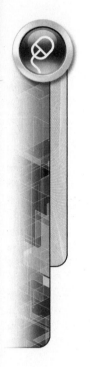

9. Select the **18" × 12" Concrete Footing [600 mm × 300 mm Concrete Footing]** tool from the **My Detail** palette and place the footing below the CMU wall as shown in the following workspace prompt sequence:

    Insert point or: *(Type B to choose the Base point option. SHIFT + Right-click, choose the Midpoint, and select the midpoint of the bottom edge of the footing detail component at p3 as shown in Figure 12.91.)*

    Insert point or: *(SHIFT + Right-click, choose the Midpoint mode, and select the midpoint of the footing in the section at p4 shown in Figure 12.91. Press ESC to end the command.)*

10. Select the footing detail component placed in the previous step, right-click, and choose **AEC Modify Tools > Trim**. Respond to the workspace prompts as follows:

    Select the first point of the trim line or ENTER to pick on screen: *(Select using the Endpoint object snap p6 as shown in Figure 12.91.)*

    Select the second point of the trim line: *(Select using the Endpoint object snap p1 as shown in Figure 12.91.)*

    Select side to trim: *(Select a point above p1 as shown in Figure 12.91.)*

11. Select the footing detail component, right-click, and choose **AEC Modify Tools > Merge**. Respond to the workspace prompts as follows:

    Select object(s) to merge or NONE to pick rectangle: *(Press ENTER to choose the NONE option.)*

    Specify first corner: *(Select point p5 as shown in Figure 12.91.)*

    Specify opposite corner: *(Select a point near p6 as shown in Figure 12.91.)*

12. A closed polyline is required to place the gravel under the slab. Select **Rectangle** from the Draw panel. Respond to the workspace prompts as follows:

    Specify first corner point or: *(Select point p7 as shown in Figure 12.91 using the Endpoint object snap mode.)*

    Specify other corner point or: *(Select point p8 as shown in Figure 12.91 using the Endpoint object snap mode.)*

13. Select **Detail Components** from the Details panel of the Home tab. Expand Division 31-Earthwork > 31 23 00 Excavation and Fill > Backfill [D-Groundwork > D20 Excavating and Filling > Backfill]. Drag the selector for **Gravel** to the **My Detail** tool palette. Click **Close** to dismiss the Detail Component Manager. Select the **Gravel** tool from the My Detail tool palette. Verify that Description = **Gravel** in the Properties palette and place the gravel as shown in the following workspace prompts:

    Select objects to form the backfill boundary: 1 found *(Select the rectangle at p9 as shown in Figure 12.91.)*

    Select objects to form the backfill boundary: ENTER

    Select a point within a boundary to backfill: *(Select a point near p10 inside the rectangle as shown in Figure 12.91. The border of the rectangle will display red prior to selecting the interior point.)*

    Select objects to form the backfill boundary: ESC *(Press ESC to end command.)*

14. Select the gravel placed under the slab, right-click, and choose **AEC Modify Tools >
Obscure**. Respond to the workspace prompts as follows:

> Select obscuring object(s) or NONE to pick rectangle:
> *(Select the concrete hatch of the footing.)*
>
> Select obscuring object(s) or NONE to pick rectangle: ENTER
> *(Press ENTER to end selection.)*

15. Select the rectangle created in Step 12 located at **p9** as shown in Figure 12.91, right-click, and choose **Basic Modify Tools > Delete**. Select the green line at **A** as shown in Figure 12.91 and choose **Edit** from the Linework panel. Select the lines at **A** and **B**, right-click, and choose **Basic Modify Tools > Delete**. Choose **Finish** from the **Edit In-Place:Linework** tab.

16. Select **Detail Components** from the Details panel. Expand Division 07-Thermal and Moisture Protection > 07 95 00 Expansion Control > 07 95 13 Expansion Joint Cover Assemblies > Fiber Expansion Joint [J-Waterproofing > J-21 Mastic Asphalt Roofing/Insulation/Finishes > Fiber Expansion Joint. Choose **1/2" [12 mm] Expansion Joint** and drag this tool to the **My Detail** tool palette. Click **Close** to close the Detail Component Manager. Click the **1/2" Expansion Joint** tool from the **My Detail** tool palette. Respond to the workspace prompts as follows:

> Insert point or: *(Type* B *to choose the* Base point *option.*
> *Select the lower-right corner of the expansion joint symbol
> at* p11 *as shown in* Figure 12.91.*)*
>
> Insert point or: *(Select a point at the end at* p8 *as shown in
> Figure 12.91. Press ESC to end the command.)*

17. Select the **18" × 12" Concrete Footing** [600 mm × 300 mm Concrete Footing] tool of the My Detail tool palette, right-click, and choose **Detail Component Manager**. Expand Division 03-Concrete > 03 31 00 Structural Concrete > Slabs with Optional Haunch [E In Situ Concrete/Large Precast Concrete > E10 Mixing/Casting/Curing In Situ Concrete > Slabs with Optional Haunch] of the **Detail Component Manager**. Choose the **Selector** of the **4" Slab with Haunch [100 mm Slab with Haunch]** and drag to the My Detail tool palette. Click **Close** to close the Detail Component Manager.

18. Right-click the **4" Slab with Haunch [100 mm Slab with Haunch]** tool and choose **Properties**. Edit the **Tool Properties** as follows: Component: Category = **03 31 00 Structural Concrete [E10 Mixing/Casting/Curing In Situ]**, Type = **Slabs with Optional Haunch**, Description = **4" Slab with Haunch [100 mm SLAB WITH HAUNCH]**, View = **Section**; Specifications: Use custom size = **No**, Hatch slab = **Yes**, Show reinforcing = **No**, and Draw haunch = **None**. Click **OK** to dismiss the **Tool Properties** dialog box.

19. Select the **4" Slab with Haunch [100 mm SLAB WITH HAUNCH]** tool from the My Detail tool palette and respond to the workspace prompts as shown below to place the slab.

> Start point or: *(Select a point at* p1 *as shown in* Figure 12.93.*)*
>
> Endpoint or: *(Select a point at* p2 *as shown in* Figure 12.93.*)*
>
> Start point or: ESC *(Press ESC to end the command.)*

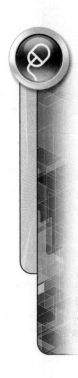

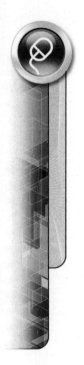

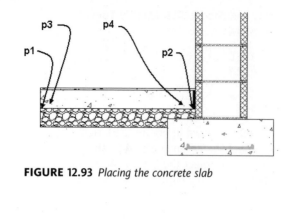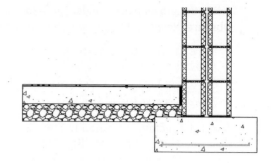

FIGURE 12.93 *Placing the concrete slab*

20. Right-click the **4″ Slab with Haunch [100 mm SLAB WITH HAUNCH]** tool of the My Detail tool palette and choose **Detail Component Manager**. Expand Division 03-Concrete > 03 22 00 Welded Wire Fabric Reinforcing > Wiremesh Reinforcing [E In Situ Concrete/Large Precast Concrete > E30 Reinforcement For In Situ Concrete > Wiremesh Reinforcing]. Choose **6x6 × W1.4 × W1.4 Welded Wire Mesh [150 × 150 WELDED WIRE MESH]** and drag this tool to My Detail tool palette. Click **Close** to close the Detail Component Manager.

21. Toggle OFF Object Snap. Select **6x6 × W1.4 × W1.4 Welded Wire Mesh [150 × 150 WELDED WIRE MESH]** from the My Detail tool palette. Respond to the workspace prompts as follows to place the welded wire fabric in the approximate position:

    Start point: *(Select a point at* p3 *as shown in* Figure 12.93.*)*

    Endpoint: *(Select a point at* p4 *as shown in* Figure 12.93. *Press* ESC *to end the command.)*

22. Verify that **3/4″ = 1′-0″ [1:20]** is set in the **Annotation Scale** flyout menu of the Drawing Window status bar. Toggle ON Object Snap in the Application status bar.

23. Select the Annotate tab and choose the **Annotation** tool palette from the Document palette set. Select **Reference Keynote (Straight) [Keynote (Straight)]** from the Leader flyout in the **Keynoting** panel and respond to the following workspace prompts to place the keynote. Select object to keynote or ENTER to select keynote manually: *(Select the* **12 CMU [3 Core Block]** *at* p1 *as shown in* Figure 12.94.*)*

    Select first point of leader: *(Select a point near* p1 *as shown in* Figure 12.94.*)*

    Specify next point of leader line: *(Select a point near* p2 *as shown in* Figure 12.94.*)*

    Specify next point of leader line: ENTER *(Press* ENTER *to end the leader.)*

    Select text width <0>: ENTER *(Press* ENTER *to accept the default text width.)*

    (Keynote information is placed as shown in Figure 12.94.*)*

FIGURE 12.94 *Keynotes placed*

24. Repeat the **Reference Keynote (Straight Leader) [Keynote (Straight Leader)]** command to place additional reference keynotes as shown in Figure 12.94. (**Metric only:** When you choose the fabric reinforcement, the **Select Keynote** dialog box opens. Choose **E In Situ Concrete Reinforcement for In Situ Concrete > E30/210 Fabric reinforcement**.)

25. Select **Reference Keynote Legend** from the Legend flyout menu in the Keynoting panel. Respond to the workspace prompts as shown below to create the legend:

> Select keynotes to include in the keynote legend or: *(Select a point at* **p3** *as shown in* Figure 12.94.*)*
>
> Specify opposite corner: *(Select a point at* **p4** *as shown in* Figure 12.94.*)*
>
> Select keynotes to include in the keynote legend or: ENTER
>
> Insertion point of table: *(Select a point near* **p5** *as shown in* Figure 12.94.*)*

26. Save and close the drawing.

27. In the previous steps you added content from the Detail Component Manager to create a detail from a section object. During the final steps of this tutorial you will develop a detail from the graphics of a Detail object. Double-click the Basement View drawing of the Views tab in the Project Navigator.

28. Select the Annotate tab and choose the **Detail (Sheet No. Tail)** from the Detail flyout menu in the Callouts panel of the Annotate tab. Respond to the command line prompts as follows:

> Specify first point of detail line: *(Select point* **p1** *as shown in Figure 12.95.)*
>
> Specify next point of line: *(Select point* **p2** *as shown in Figure 12.95.)*
>
> Specify next point of line or [Break]: *(Press ENTER to end selection.)*
>
> Specify side for tail: '_layer *(Select point* **p3** *as shown in Figure 12.95 to open the Place Callout dialog box.)*

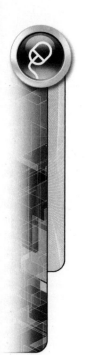

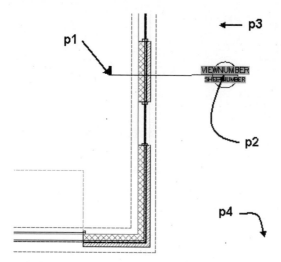

FIGURE 12.95 *Placing the Detail Mark A2T*

29. The Place Callout dialog box opens as shown in Figure 12.96. Type **Sill Detail** in the New Model Space View Name. Check Generate Section/Elevation and Place Title-mark checkboxes. Select **1″ =1′-0″ [1:10]** scale. Click the New View Drawing button to open the **Add Detail View** dialog box.

FIGURE 12.96 *Place Callout dialog box for the Detail Mark*

30. In the Add Detail View dialog box shown in Figure 12.97, type FOUNDATION DE-TAILS in the Name field and verify Category = Views. Click Next to move to the Con-text page and check Division 1 for Level 1 [G]. Click Next to move to the Content page. Verify that Constructs, Basement, and Stair are checked. Click Finish to dismiss the **Add Detail View** dialog box.

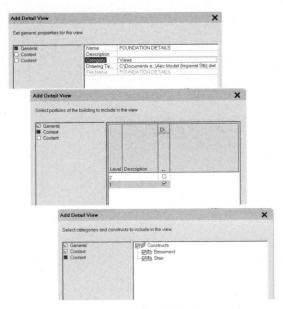

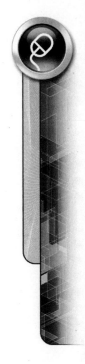

FIGURE 12.97 *Add Detail View dialog box*

31. To specify the location of the view drawing, respond to the following workspace prompt:

    Specify insertion point for the 2D section result: *(Select a point near* p4 *as shown in Figure 12.95.)*

32. Save and close the Basement View drawing.

33. Click the Views tab of the Project Navigator. Expand Foundation Details. Double-click the **Sill Detail**. Toggle OFF Snap Mode in the Application status bar.

34. Choose the Annotate tab and double-click Annotation Tools to display the Document palette group.

35. Choose the Annotation palette. Choose **Cut Line 1** from the Annotation tool palette. Respond to the workspace prompts to place the Cut Line 1.

    Specify first point of break line: *(Select point* p1 *as shown in Figure 12.98.)*

    Specify second point of break line: *(Select point* p2 *as shown in Figure 12.98.)*

    Specify break line extents: *(Select point* p3 *as shown in Figure 12.98. The lower portion of the wall is masked.)*

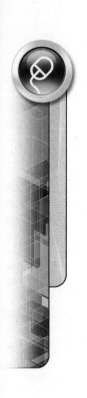

Sill Detail

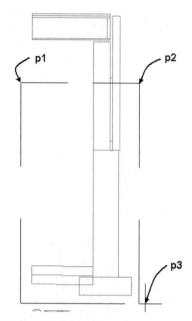

FIGURE 12.98 *Location of points to place a cutline*

36. Place the following detail components as shown in Figure 12.99.

Imperial:

Division 06 Wood, Plastics, and Composites > 06 11 00 Wood Framing > Nominal Cut Lumber place 2 × 8 sill plate and 2 × 10 rim joist

Division 04-Masonry > 04 22 00 Concrete Unit Masonry > 3 Core place two courses of 8 × 8 × 16 CMU

Division 04 Masonry > 04 21 00 Clay Unit Masonry > Bricks place Standard 3/8 JT courses

Division 05 Metals > 05 05 00 Common Work Results for Metals > 05 05 23 Metal Fastenings > Anchor Bolts Hook place 3/8 Hooked Anchor Bolt

Metric:

G-Structural/Carcassing Metal/Timber > G20 – Carpentry/Timber Framing/First Fixing>Timber Framing>Nominal Cut Timber place 50 × 200 sill plate and 50 × 250 joists

F-Masonry > F10 Brick/Block Walling > 3 Core Blocks place 2 courses of 190 × 225 × 450mm Concrete Block3 Core

F-Masonry > F10 Brick/Block Walling > Bricks place Standard 65mm – 10 mm Jt

G – Structural/Carcassing Metal/Timber > G20 – Carpentry/Timber Framing/First Fixing > First Fixing > Anchor Bolts Hook place M10 Hooked Anchor Bolt

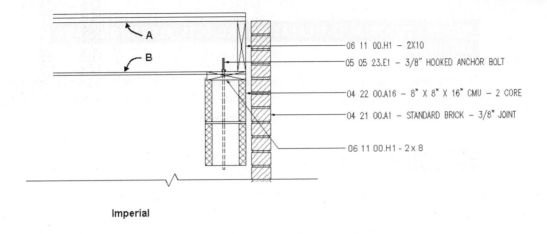

Imperial

06 11 00.H1 – 2X10

05 05 23.E1 – 3/8" HOOKED ANCHOR BOLT

04 22 00.A16 – 8" X 8" X 16" CMU – 2 CORE

04 21 00.A1 – STANDARD BRICK – 3/8" JOINT

06 11 00.H1 - 2 x 8

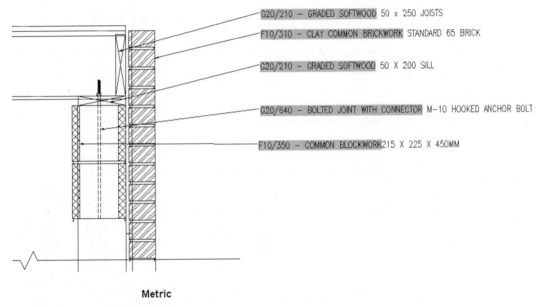

Metric

G20/210 – GRADED SOFTWOOD 50 x 250 JOISTS

F10/310 – CLAY COMMON BRICKWORK STANDARD 65 BRICK

G20/210 – GRADED SOFTWOOD 50 X 200 SILL

G20/640 – BOLTED JOINT WITH CONNECTOR M–10 HOOKED ANCHOR BOLT

F10/350 – COMMON BLOCKWORK 215 X 225 X 450MM

FIGURE 12.99 *Detail components*

37. The 2D Section/Elevation Style Properties—2D Section Style Background 2D Section/Elevation style places all content of the detail on the G-Anno-Nplt layer. Set the current layer to A-Detl-Wide [AEC-Detail-Wide]. Select Line from the Draw panel. Draw lines using the Endpoint Object Snap over lines A and B.

38. Save and close the drawing. Close the project in the Project Browser.

PROJECT TUTORIAL 12-5: CREATING PROJECTS, ELEVATIONS, AND SECTIONS

Create a new project using the *Accessing Tutor\Imperial\Ch12\Ex 12-5\Ex 12-5.apj* or *Accessing Tutor\Metric\Ch12\Ex 12-5\Ex 12-5.apj* as a template. Assign project number 25 to the project named Ex 12-5. Create left, rear, and right side elevations as a View drawing.

SUMMARY

- The Project Navigator allows you to attach Construct drawings as reference files to create a model as a View drawing.
- Floor elevation and floor to floor heights are defined in the Levels of the Project tab of the Project Navigator.
- Sections and elevations should be developed from a model that consists of the floor plans of the building attached as external reference files.
- Building elevation lines are created with the Building Elevation Mark A2 located on the Callouts tool palette. Additional elevation marks are located in the Documentation > Callouts folder of the Content Browser.
- Building section lines are created when you insert the Section Mark A2T from the Callouts tool palette. Additional section marks are located in the Content Browser.
- Changes made in the model can be reflected in the elevation and section objects with the Refresh (2DSectionResultRefresh) and Regenerate (2DSectionResultUpdate) commands.
- Editing the Object Display of the elevation and section object allows you to add and modify linetype, color, and lineweight of the lines of the object.
- Elevation styles can be created to modify the display content of an elevation.
- Lines of the elevation or section can be assigned to different display components when you select 2DSectionResultEdit and 2DSectionEditComponent commands.
- Subdivisions can be created in the building elevation line and building section line to add emphasis to objects within the subdivision.
- The surface hatch can be limited by a closed polyline when you select the Material Boundary > Add (2DSectionResultAddHatchBoundary) command.
- Live Sections can be created from orthographic and pictorial views of the model.
- The Hidden Line Projection (CreateHLR) command creates a 2D block from a view of the model.

 NOTE

Refer to e **Review Questions** folder for download of review questions for Chapter 12 of Student Companion Site of **CengageBrain** http://www.cengagebrain.com described in the Preface.

Drawing Commercial Structures

INTRODUCTION

The development of drawings for commercial building requires the insertion of structural grids and ceiling grids. Layout curves can be developed to insert columns or other AEC objects spaced uniformly along a layout curve. Structural grids, layout curves, and ceiling grids are presented in this chapter. In addition, the development of views and walk-throughs using a camera is presented.

OBJECTIVES

Upon completion of this chapter, you will be able to

- Create columns, beams, braces, and Custom Columns.

- Create structural member styles.

- Create and edit structural frames and column grids with structural members.

- Add, move, remove, mask, and clip column grid lines.

- Create Enhanced Custom Grids and Column Grids.

- Label and dimension a column grid.

- Use the LayoutCurveAdd command to place columns and other AEC objects symmetrically along a layout curve.

- Place a camera in the drawing using the Camera command.

- Create a video of animations.

INSERTING STRUCTURAL MEMBERS

Structural members including steel, concrete, and wood shapes can be inserted in a drawing as columns, beams, and braces. The components of the structural framework are inserted as 3D objects but can be viewed from the top view to create 2D plan views. Lines, arcs, or polylines can be converted to structural shapes by extruding member shapes along the geometry. Structural components are object styles created from the **Structural Member Catalog**. The style extracts the dimension and shape properties of the component from the catalog and you assign a name. Custom shapes can be created and applied to structural member styles. Two structural components can be combined to create a composite column or beam. Beams can be created that are tapered with different dimensional properties at each end of the beam and used as a rigid frame. Styles of column structural members can be defined for structural grids and anchored to the structural grid or inserted independently of a grid. Structural member styles can also be imported from other drawings in the Style Manager. The tools for inserting columns, beams, and braces are located on the Design tool palette. The member inserted as a column, beam, or brace is a style developed from the Structural Member Catalog.

Using the Structural Member Catalog

The **Structural Member Catalog** is a library of structural components, which includes Imperial and Metric shapes of concrete, steel, and wood materials. Access the **Structural Member Catalog** as shown in Table 13.1.

TABLE 13.1 *Structural Member Catalog command access*

Ribbon	Select Manage tab. Extend the Style & Display panel, choose Structural Member Catalog
Command prompt	AECSMEMBERCATALOG

When you select the **Structural Member Catalog** command (**AecSMemberCatalog**), the **Structural Member Catalog** window opens.

The left pane of the Structural Member Catalog lists a hierarchical tree view of the components available as shown in Figure 13.1. If you select a structural member in the left pane, a graphic preview of the category of members is shown in the upper-right pane and the text descriptions of the physical properties of the members are shown in the lower-right pane. The columns describing the physical properties of the beam usually exceed the width of the lower-right pane; therefore, you can use the scroll bar at the bottom of the pane to view the additional properties.

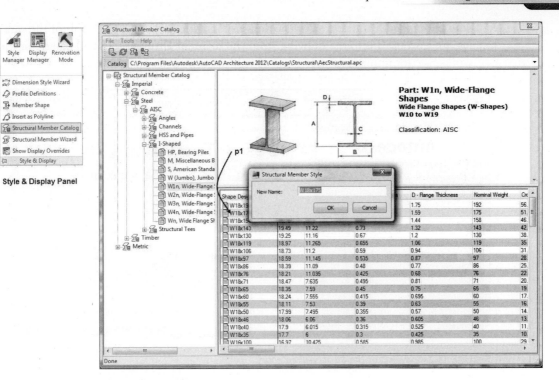

FIGURE 13.1 *Structural Member Catalog*

The contents of the Structural Member Catalog – Imperial folder are summarized in Table 13.2.

TABLE 13.2 *Structural member styles*

Material	Form	Components
Concrete	Cast-in-Place	Beams, Circular columns, Rectangular columns
Precast	Inverted Tee Beams, Joist Double T, Joist T, L Shaped Beams, Rectangular Beams, Rectangular Columns (Joist Double T and Joist T are not included in Structural Member Catalog – Metric)	
Steel	AISC (AISC, BS4, CISC & DIN included in Structural Member Catalog – Metric)	Angles, Channels, HSS and Pipes, I Shaped, Structural Tees
Timber	Glue Laminated Beams	Beams
Lumber	Heavy Timber, Nominal Cut Lumber, Rough Cut Lumber	
Plywood Web Wood Joist DIN (Metric only)	Plywood Joist Squared Timber Profiles	

In addition, open web steel bar joists can be added to the drawing by importing their styles from the AutoCAD Architecture Design Tool Catalog – Imperial or Design Tool Catalog – Metric\Structural catalog of the Content Browser. This catalog consists of the Steel Bar Joists shown in Figure 13.2 and other structural members.

The Design Tool Catalog – Metric does not include bar joists. The tools of the catalog are i-dropped into the drawing from the Content Browser to a tool palette or the workspace for immediate use.

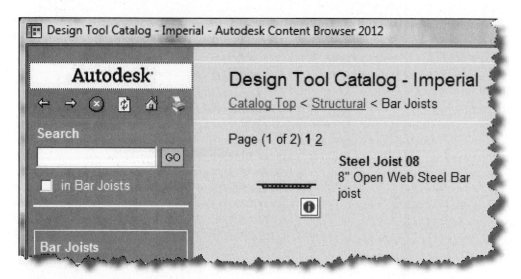

FIGURE 13.2 *Bar Joists of the Structural category*

Creating a Structural Member Style

Structural member styles are created from the components in the **Structural Member Catalog**. When you create a structural member style, you select a component from the Structural Member Catalog and assign a name to the style. The style has the physical properties of the component selected from the Structural Member Catalog. The styles created can be used as a column, beam, or brace.

Steps to Create a Structural Member Style

1. Select Manage tab. Choose **Structural Member Catalog** from the Style and Display panel to open the **Structural Member Catalog** window.
2. Expand a folder in the left pane of the desired structural member category.
3. Select the item in the left pane to display the contents in the lower-right pane.
4. Select the structural member in the lower-right pane.
5. Create a style name by one of the following five methods:

   Double-click the component name at p1, shown in Figure 13.1, in the lower-right pane.

   Select **Tools > Generate** Style from the menu bar of the **Structural Member Catalog**.

   Select a member and then type CTRL+G.

   Select the member, right-click, and choose **Generate Member Style**.

   Select a member and then select the **Generate Member Style** icon from the **Structural Member Catalog** toolbar.
6. Type a name in the **Structural Member Style** dialog box as shown in Figure 13.1.
7. Click **OK** to dismiss the **Structural Member Style** dialog box.

ADDING STRUCTURAL MEMBERS

Structural member styles can be applied to create beams, columns, and braces. The commands to add a beam, column, or brace are located on the Design tool palette and the Column Grid flyout menu in the Build panel as shown in Figure 13.3.

The styles created from the Structural Member Catalog are listed in the **Style** list of the Properties palette, shown in Figure 13.6 when you add a column, beam, or brace. When you place a member, the On object selection in the Properties palette can be Yes or No. To place a freestanding structural member, set On object to No. If On object is set to Yes, the member can be arrayed within slabs, roof slabs, walls, or column grids to represent the framing within the object. Columns can be anchored to the nodes or intersections of a column grid.

When you place a beam or brace, you specify the start point and endpoint that will create the length of an axis line. The associated cross-section of the structural member is extruded along the axis line. The length of the axis line is specified in the **Logical length** field of the Properties palette. When placing a freestanding structural member, you can position it with the aid of the elevation toggle located on the Application status bar. Click the **Elevation** toggle shown in Figure 13.3 to open the **Elevation Offset** dialog box and specify a Z value for your working plane.

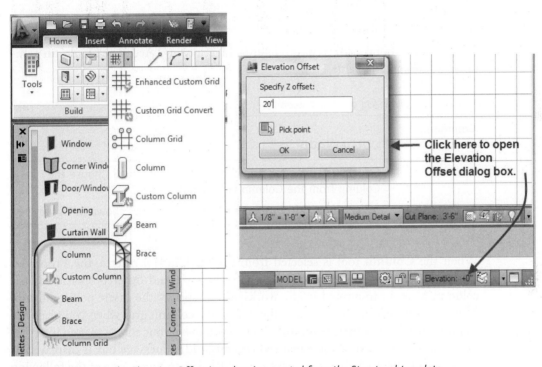

FIGURE 13.3 *Setting the Elevation Offset in a drawing created from the Structural template*

When placing a structural member, use the Node mode of object snaps to place members which begin and end at the justification lines of other structural members. When you place structural members using the Node object snap, the members are connected. The default edit mode for modifying existing members using grips is for the members to remain connected by their adjoining justification lines; therefore, movement of the column shown in Figure 13.4 will result in movement of the beam. You can press the CTRL key to toggle motion of the grip edit as connected or unconnected as shown in Figure 13.4. The axis line includes grips at the start, end, and midpoint. The **Node** object snap is active at the start, end, and midpoint of this axis line. The **Start offset**, **End offset**, **Roll**, and **Justify** values are all relative to the axis line.

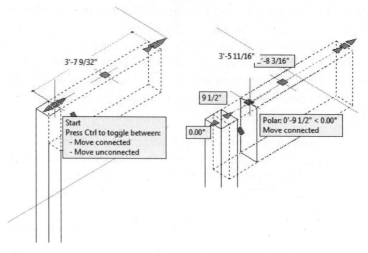

FIGURE 13.4 *Grip edit of structural members*

Inserting Columns

The **Column** tool is used to place columns in a drawing. Access the **Column** command as outlined in Table 13.3.

TABLE 13.3 *ColumnAdd command access*

Command prompt	COLUMNADD
Ribbon	Select the Home tab and choose the Build panel. Choose the Column from the Column Grid flyout.
Tool palette	Select Column from the Design Tool palette as shown in Figure 13.3.

When the **Column** tool is selected from the Design tool palette, the Properties palette opens, which allows you to specify the style and other properties of the column. You are prompted in the workspace to specify the insert point and roll. Multiple columns can be placed at the nodes of a column grid when you place the column grid. If **On object** is set to Yes when you move your cursor near an intersection of a column grid or the enhanced column grid, a message box is displayed as shown in Figure 13.5. You can press the CTRL key to toggle one of the following insertions: Add a column to a node, Pick or window to select nodes, or Add columns to all nodes. The structural shape of the column is inserted at the intersection of axis lines.

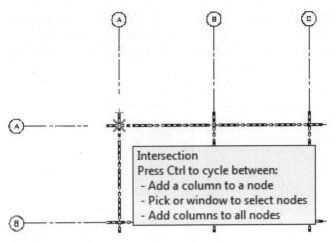

Intersection
Press Ctrl to cycle between:
- Add a column to a node
- Pick or window to select nodes
- Add columns to all nodes

FIGURE 13.5 *Placing structural members on the grid*

Inserting Beams

Beams are placed in the drawing by selecting the **Beam** tool from the Design tool palette (refer to Table 13.4 for command access).

TABLE 13.4 *BeamAdd command access*

Command prompt	**BEAMADD**
Ribbon	Select the Home tab. Select Beam from the Column Grid flyout of the Build panel.
Tool palette	Select Beam from the Design tool palette as shown in Figure 13.3.

When you select the **Beam** tool from the Design palette, the Properties palette opens, allowing you to specify the offsets, justification, and roll of the beam. The options of the Properties palette of a Beam are shown in Figure 13.6.

Inserting a Brace

Braces are placed in the drawing by selecting the **Brace** tool in the Design tool palette shown in Figure 13.3. The brace is added on the **S-Cols-Brce** layer, which has color 32 (a hue of brown) and a dashed linetype. Access the **Brace** tool as shown in Table 13.5.

TABLE 13.5 *BraceAdd command access*

Command prompt	**BRACEADD**
Ribbon	Select the Home tab and choose the Build panel. Select Brace from the Column Grid flyout.
Tool palette	Select Brace from the Design Tool palette as shown in Figure 13.3.

If you select the **Brace** tool from the Design tool palette, the Properties palette opens as shown in Figure 13.6. This palette allows you to specify the style and position of the structural member about the axis line. The content of the Properties palette for beams, columns, and braces is described below.

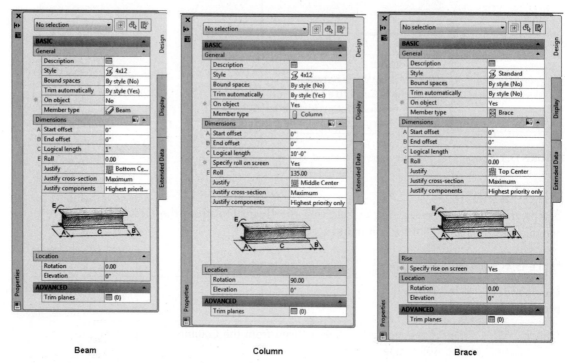

Beam   Column   Brace

FIGURE 13.6 *Properties palette of a structural member*

Design Tab—Basic
General

Description—The Description button opens a Description dialog box, which allows you to add text describing the column, beam, or brace.

Style—The Style drop-down list consists of the styles created in the drawing from the **Structural Member Catalog**.

Bound spaces—The column, beam, or space can be specified to restrict a space.

Trim automatically—Trim automatically can be toggled to **Yes** or **No**. If Trim automatically is toggled to **Yes**, the geometry of the member will be trimmed by the geometry of a connecting member.

On object—On object can be toggled to **Yes** or **No**. If On Object is **Yes**, the structural member can be arrayed within a slab, roof slab, wall, or column grid as shown in Tutorial 13.1. When On object is **Yes**, a Layout type option is added to the Dimensions section to specify placement as Fill or Edge. If On object is **No**, the structural member is added freestanding.

Member Type—The Member Type lists the structural member type, such as column, beam, or brace.

Dimensions

A-Start offset—The Start offset field allows you to specify the distance to adjust the start of the member relative to the axis line. A positive distance starts the member shape above the axis start point, as shown in Figure 13.7, while a negative distance extends the member shape below the axis start point.

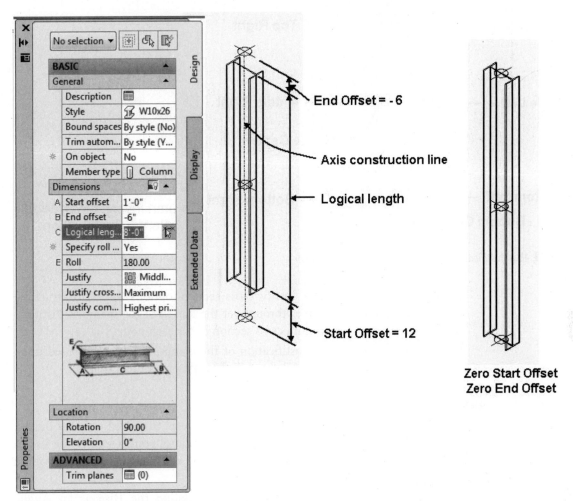

FIGURE 13.7 *Offsets and logical length of the column*

B-End offset—The End offset field allows you to specify the distance to the end of the member. A positive offset extends the shape above the axis line of a member and a negative offset shortens the member from the end of the axis line.

C-Logical length—The Logical length field allows you to specify the length of the column, beam, or brace axis line.

Specify roll on screen—The Specify roll on screen **Yes** option allows you to specify the **Roll** in the workspace or in the Properties palette.

E-Roll—The Roll field allows you to specify the rotation angle about the axis line. Positive angles rotate the shape about the axis line in a counterclockwise direction.

Justify—The **Justify** list allows you to specify the location of the axis line relative to the geometry of the shape as shown in Figure 13.8. In addition to the nine positions shown, the **Baseline** option is located on the centroid of the member, as shown at p1 in Figure 13.8.

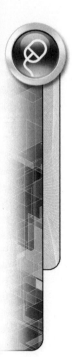

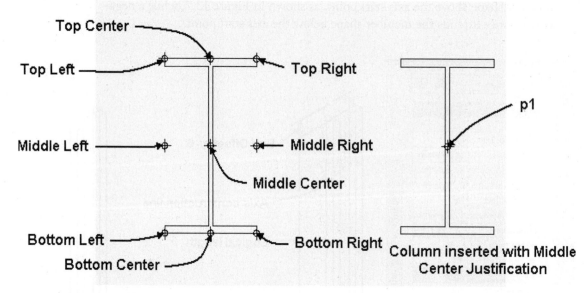

FIGURE 13.8 *Justification locations of a structural member*

Justify cross-section—The Justify cross-section options apply to structural members with multiple shapes throughout the length of the member. Justification can be set to the maximum cross section or for each node.

Justify components—The justification of the components can be based upon only the highest priority shape defined in the style or about all shapes.

Rise

Specify rise on screen—The Specify rise on screen, displayed only for Braces, includes **Yes** and **No** options. The **Yes** option allows you to specify the rise in the Z direction in the workspace. The No option activates the Rise fields of the Properties palette.

Method—The Method value allows you to define the Rise as an Angle, Distance, or Height.

Design Tab—Advanced

Trim planes—A trim plane can be specified for the ends of the structural member. The trim plane cuts the end of the member at an angle relative to the baseline of the member.

TUTORIAL 13-0: CREATING WOOD FRAMING

1. **Download Chapter 13 files from the Accessing Tutor\Imperial or Accessing Tutor\ Metric category of the Student Companion site of CengageBrain** http://www .cengagebrain.com described in the Preface to your Accessing Tutor folder. Open *Ex 13-0.dwg from your Accessing Tutor\Imperial\Ch13 or Accessing Tutor\Metric\Ch13* directory.

2. Select **Save As** > **AutoCAD Drawing** from the Application menu and save the drawing as **Lab 13-0** in your Ch 13 student directory.

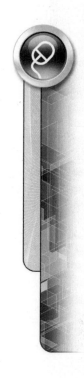

3. Right-click OSNAP in the Application status bar and choose Settings. Clear all object snap modes except the Node object snap mode. Click OK to dismiss the dialog box. Verify that the OSNAP toggle is pushed in.

4. Select the Manage tab. Choose **Structural Member Catalog** from the Style & Display panel extension.

5. Expand the tree in the left pane to **Imperial > Timber > Lumber > Nominal Cut Lumber** [**Metric > Timber > Lumber Metric**].

6. Double-click **2 × 10 [50 mm × 200 mm]** in the right pane to open the **Structural Member Style** dialog box. Click **OK** to accept the default style name and dismiss the dialog box. From the menu bar of the **Structural Member Catalog** select **File > Close** to dismiss the dialog box.

7. Select the Home tab. Select the **Beam** tool from the **Column Grid** flyout menu. Edit the **Design** tab of the **Properties** palette as follows: Style = **2 × 10 [50 mm × 200 mm]**, Trim automatically = **Yes**, On object = **Yes**, Start offset = **1 1/2" [50]**, End offset = **−1 1/2" [−50]**, Roll = **0**, Layout type = **Fill**, Justify = **Bottom Center**, Array = **Yes**, Layout Method = **Repeat**, Bay size = **16" [400]**. Move the cursor to the workspace and respond to the workspace prompts as follows:

Start point or: *(Move the cursor to* p1*, verify that the* Node *object snap tool tip is displayed, and left-click when the joists are dynamically displayed in a vertical position as shown in Figure 13.9.)*

Start point or: *(Press* ESC *to end the command.)*

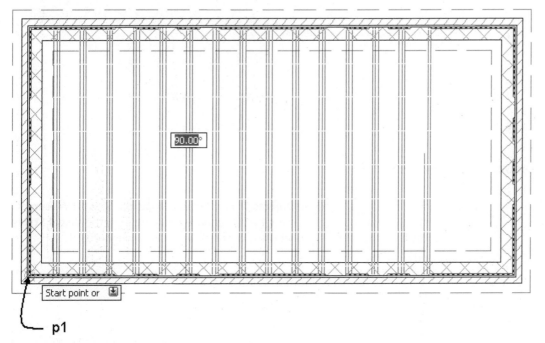

FIGURE 13.9 *Joists array in slab*

8. Select the Home tab. Select **Layer Manager** on the **Layers** panel. Thaw the layer A-Roof-Slab [A370G]. Close the Layer Properties Manager.

9. Select the ridge of the roof slab and choose **Select Similar** from the **General** panel of the **Roof Slab** contextual tab. Choose **Isolate Objects** from the **Isolate Objects** flyout in the General panel.

10. **Metric only step**: Select the 4 roof slabs and choose **Copy Style** from the General panel. The roof slab will be revised in this step for a 200 mm joist. Choose the

General tab and change the name to **200 – 25 × 200 Fascia + Soffit**. Choose the **Components** tab and edit Joist Thickness to **200**, Sheathing Thickness Offset to **200**, and Shingle Thickness Offset to **220**. Click **OK** to dismiss the Roof Slab Styles dialog box.

11. Right-click Snap Mode in the Application status bar and choose **Settings** to open the **Drafting Settings** dialog box. Toggle ON **Polar Snap** and set the **Polar distance** to **16″ [400]**. Click **OK** to dismiss the dialog box. Toggle ON Polar Tracking, Object Snap, Object Snap Tracking, and Dynamic Input in the Application status bar.

12. Verify that the Home tab is current. In this step rafters will be placed in one of the four roof slabs. Select the **Beam** tool from the Column Grid flyout of the Build panel. Edit the **Properties** palette as follows: Style = **2 × 10 [50 mm × 200 mm]**, Trim automatically = **Yes**, On object = **Yes**, Start offset = **0**, End offset = **0**, Roll = **0**, Layout type = **Fill**, Justify = **Bottom Center**, Array = **Yes**, Layout Method = **Repeat**, Bay size = **16″ [400]**. Move the cursor to the workspace and respond to the workspace prompts as follows:

> Start point or: *(Hover the cursor at* p1, *move the cursor from* p1 *to display the beams in the roof slab as shown in Figure 13.10. Click to specify the insertion of the beams. Press ESC to end the command.)*

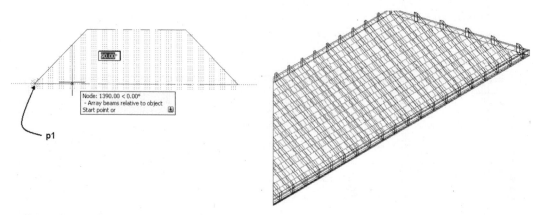

FIGURE 13.10 *Rafters placed in roof slab*

13. Right-click and choose **Isolate Objects > End Object Isolation**. Select **SW Isometric** from the **View** Cube. Select the View tab; choose **Realistic** from the **Visual Styles** flyout menu of the View tab to view the roof slab as shown in Figure 13.11.

14. Save and close the drawing.

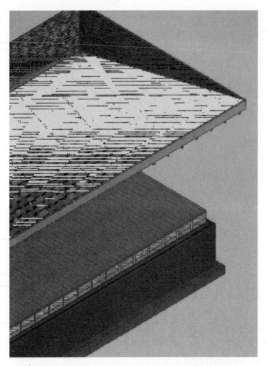

FIGURE 13.11 *Structural members placed objects*

Customizing Structural Members Using Styles

Styles created from the **Structural Member Catalog** can be customized in the style definition. The **Edit Member Style** command edits the properties of a structural member style. A style created from the Structural Member Catalog assigns a shape for the axis line. Materials, classifications, and display properties can be edited in the member style. You can assign a shape for each end of the structural component or combine two shapes for a structural component. Access the **Edit Member Style** command as shown in Table 13.6.

TABLE 13.6 *MemberStyleEdit command access*

Command prompt	MEMBERSTYLEEDIT
Ribbon	Select a structural member and choose Edit Style from the General panel.
Shortcut menu	Select a structural member, right-click, and choose Edit Member Style.

When you select a structural member and choose **Edit Style**, the **Structural Member Style Properties** dialog box opens as shown in Figure 13.12. The General, Materials, Classifications, and Display Properties tabs include content similar to other object styles, whereas the **Design Rules** tab includes options for defining the start and end position of shapes. The Design Rules tab consists of columns that define the shape name for the start and end segments of the structural member. The content of the tab is displayed when you click the **Show Details** button. The content of the Design Rules tab shown in Figure 13.12 indicates the W10 × 26 shape is applied along the axis line. The member shape can be positioned with an offset in the x, y, or z direction.

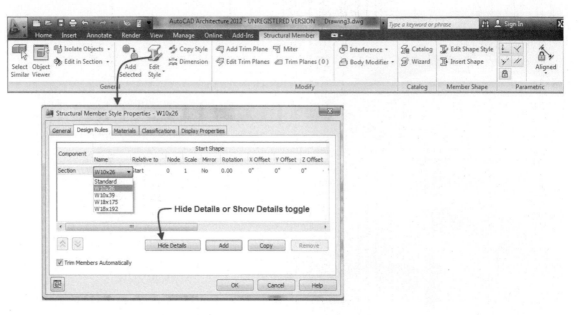

FIGURE 13.12 *Structural Member Style Properties dialog box*

Converting Linework to Structural Members

The **Column**, **Beam**, and **Brace** tools located on the Design tool palette include a shortcut menu option to convert linework to the structural member. Therefore, you can draw an arc, line, or polyline and convert this geometry to a structural member. The linework becomes the axis line for the structural member. Access the **ColumnToolToLinework** command as shown in Table 13.7.

TABLE 13.7 *ColumnToolToLinework command access*

Command prompt	**COLUMNTOOLTOLINEWORK**
Tool palette	Select Column from the Design Tool palette, right-click, and choose Apply Tool Properties to > Linework.

The ColumnToolToLinework command allows you to draw in elevation or pictorial views lines, arcs, or polylines and convert the geometry to a structural member. The following workspace sequence converted a line shown in the front view of Figure 13.13 to a column.

Select lines, arcs, or open polylines to convert into members: 1 found *(Select linework as shown in* Figure 13.13*.)*

Select lines, arcs, or open polylines to convert into members: ENTER *(Press* ENTER *to end the selection.)*

(The Convert to Column *dialog box opens; respond to the* Erase Layout Geometry *checkbox. Click* OK *to dismiss the Convert to* Column *dialog box.)*

(Edit the Properties palette to specify the column type as shown in Figure 13.13*.)*

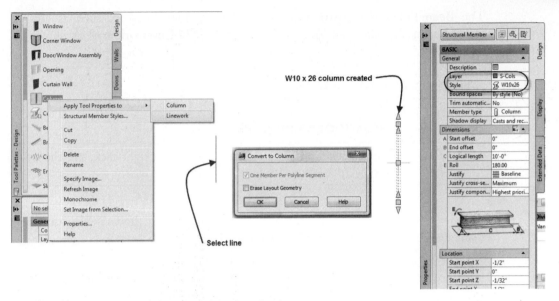

FIGURE 13.13 *Line converted to W10 × 12 structural style*

If you draw a polyline to represent the column, you can assign various structural shapes to each segment of the polyline. The polyline drawn in Figure 13.14 consists of a lower segment 12″ long and an upper segment 4′ long. The lower segment is converted at the right to a 12″ × 12″ concrete shape in the Design Rules tab of the Structural Member Style Properties dialog box shown in Figure 13.14. The upper segment is converted to a 4 × 4 wood shape. A node has been assigned to the vertex of the polyline. A shape can be defined to start and end at the node specified relative to the start or end of the polyline in the Design Rules tab as shown in Figure 13.14. The ColumnToolToLinework tool allows you to convert the polyline to a column that applies the shapes defined for the nodes. When creating the column, you must clear the **One Member Per Polyline Segment** checkbox of the **Convert to Column** dialog box shown in Figure 13.13.

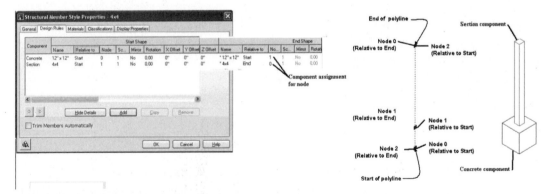

FIGURE 13.14 *Polylines converted to members and the Structural Member Style Properties dialog box*

The **BeamToolToLinework** tool allows you to draw linework such as arcs to create beams for canopies. Access the **BeamToolToLinework** tool as shown in Table 13.8.

TABLE 13.8 *BeamToolToLinework command access*

Command prompt	**BEAMTOOLTOLINEWORK**
Tool palette	Select Beam from the Design Tool palette, right-click, and choose Apply Tool Properties to > Linework.

The BeamToolToLinework command was applied to the arc drawn in elevation view to create the curved beam shown in Figure 13.15.

> *(Select the* Beam *tool of the Design tool palette, right-click, and choose* Apply Tool Properties to > Linework.*)*
>
> Select lines, arcs, or open polylines to convert into members: 1 found *(Select the arc at* p1 *as shown in Figure 13.15.)*
>
> Select lines, arcs, or open polylines to convert into members: ENTER *(Press* ENTER *to end the selection.)*
>
> *(The* Convert to Beam *dialog box opens; respond to the* Erase Layout Geometry *checkbox. Click* OK *to dismiss the Convert to Beam dialog box.)*
>
> *(Edit the Properties palette to assign the structural style.)*

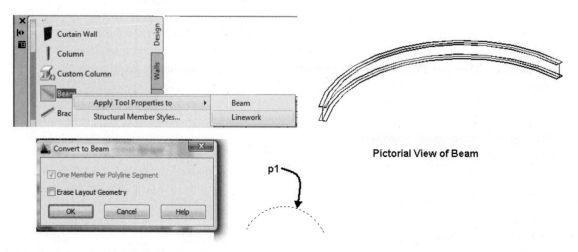

FIGURE 13.15 *Converting an arc to a Beam*

Creating Structural Members in the Style Wizard

Custom sizes for the shapes included in the Structural Member Catalog can be developed in the **Structural Member Style Wizard**. Access the **Structural Member Style Wizard** as shown in Table 13.9. The Structural Member Style Wizard shown in Figure 13.16 includes a tree listing of the shapes used in the Structural Member Catalog. The first page of the wizard allows you to select a shape from the tree list. Click the **Next** button at the bottom of the first page to open the next page and define the sizes for the shape as shown in Figure 13.16. The final page of the wizard allows you to specify the name for the style. A summary of properties of the style is listed as shown in Figure 13.16. The new shape can be specified as a shape for a component in the **Structural Member Style Properties** dialog box.

TABLE 13.9 *MemberStyleWizard command access*

Command prompt	MEMBERSTYLEWIZARD
RibbonMenu bar	Select the Style & Display panel of the Manage tab. Choose Structural Member Wizard from the Style & Display panel extension.

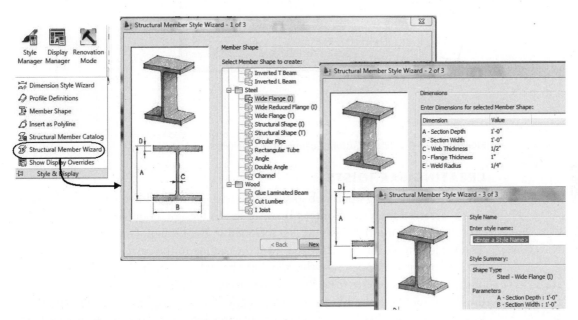

FIGURE 13.16 *Accessing the Structural Member Style Wizard*

Creating Custom Columns

The **Custom Column** tool located on the Design tool palette, as shown in Table 13.10, allows you to create columns with custom column shapes. The command allows you to use the geometry of polylines, splines, ellipses, or circles to create the shape.

TABLE 13.10 *AecCustomColumn command access*

Command prompt	AECCUSTOMCOLUMN
Tool palette	Select Custom Column from the Design tool palette.
Ribbon	Choose Custom Column from the Column flyout of the Build panel in the Home tab.

(Select the Custom Column tool of the Design tool palette; respond to the workspace prompts as follows.)

Select a closed polyline, spline, ellipse, or circle for column profile: *(Select the polygon at p1 shown in Figure 13.17.)*

> Insertion point or [Centroid] <Centroid>: *(Select the AutoCAD point shown at p2 in Figure 13.17.)*
>
> *(The Convert to Column dialog box opens; respond to the Erase Layout Geometry checkbox. Click **OK** to dismiss the Convert to Column dialog box.)*

FIGURE 13.17 *Creating a Custom Column from a polyline*

CREATING BAR JOISTS

The tools for creating open web joists are located in the Bar Joists category. This catalog is accessed from the Content Browser in the Design Tool Catalog - Imperial > Structural category. Bar joists are *not* included in the Design Tool Catalog - Metric. The bar joist tools can be i-dropped from the catalog to the current drawing. When a bar joist tool is i-dropped into the drawing, the style will be included in the **Style** list of the Properties palette for a Beam. Therefore, additional joists can be added using the **Beam** tool with the Bar Joists style assigned in the Properties palette. The **Baseline** justification should be used when placing bar joists, because the baseline is located at the bearing surface of the bar joist, as shown in Figure 13.18.

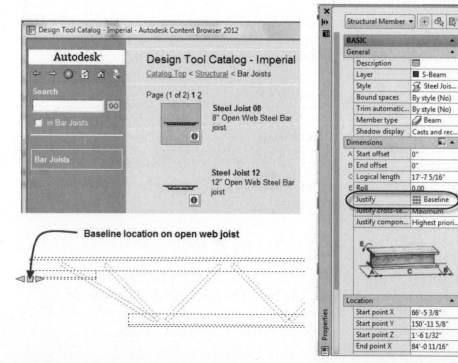

FIGURE 13.18 *Baseline location of a bar joist*

Steps to Insert a Bar Joist

1. Select **Content Browser** from the **Tools** panel extension of the Home tab.
2. Double-click the Design Tool Catalog - Imperial.
3. Select the **Structural > Bar Joists** catalog as shown in Figure 13.18.
4. Click and drag the steel joist from the catalog to the drawing.
5. The **BeamAdd** command is current, and the bar joist style is the current style. Verify that the justification is **Baseline**. Select the start and endpoint locations of the joist.

The Bar Joist tool can be i-dropped from the catalog to a tool palette for future use.	NOTE

CREATING STRUCTURAL GRIDS

The Column Grid and the Enhanced Custom Grid are two types of structural grids available. The column grid creates a layout for the placement of columns. The enhanced custom grid introduced in the 2012 release allows you to create a grid by entering size parameters and annotation in the Custom Grid dialog box.

Column grids are developed to uniformly place Columns in the plan view. The size and orientation of the structural grid are dependent upon the floor plan and maximum span of the structural components. The column grid is created on the S-Grid [A030G] layer, which has the color 191 (a hue of purple) with Center2 linetype. The column grid can be created with structural components from the Structural Member Catalog. The structural component is assigned in the Style field of the Properties palette as shown in Figure 13.19. The column shape is inserted at each grid intersection. The **Column Grid** command (**ColumnGridAdd**) is used to create a column grid. The grid can be rectangular or radial in shape. See Table 13.11 for **Column Grid** command access.

TABLE 13.11 *Column Grid command access*

Command prompt	COLUMNGRIDADD
Ribbon	Select the Home tab. Choose the Column Grid tool from the Build panel.
Tool palette	Select Column Grid from the Design tool palette.

When you select the **Column Grid** tool, the Properties palette opens, allowing you to specify the parameters of the column grid as shown in Figure 13.19.

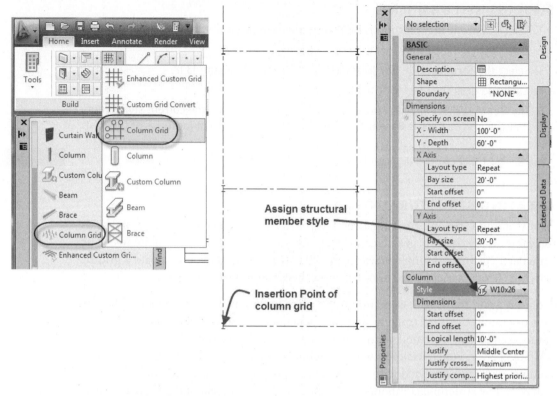

FIGURE 13.19 *Column Grid Properties palette*

The rectangular column grid allows you to specify the overall size of the grid according to the **X-Width** and **Y-Depth** dimensions. The X-Width and Y-Depth dimensions can be spaced evenly or repeated by a bay size dimension. The column grid is developed relative to its insertion point, which is in the lower-left corner as shown in Figure 13.19.

Specifying the Size of the Column Grid

The size and number of divisions of the rectangular column grid can be specified in the X-Width and Y-Depth fields of the Properties palette. The **Repeat** and **Space evenly** layout types allow you to type the number of bays or the bay size in the Properties palette. The bay size and number of bays can also be determined dynamically on screen. When the **Specify on screen** option is selected, you can specify the overall size of the column grid by selecting two diagonal points. Moving the cursor from the last diagonal point dynamically sets the number of divisions. As you move the cursor back **toward the direction of the insertion point**, the **number of bays increases**. Movement of the cursor in a horizontal direction toward the insertion point increases the number of divisions along the horizontal grid line. Vertical movement of the cursor in the direction toward the insertion point increases the number of divisions along the vertical grid line.

Steps to Size a Rectangular Column Grid on Screen

1. Select the **Column Grid** tool from the **Build** panel.
2. Select the **Rectangular** shape from the **Shape** list.
3. Select **Yes** for the **Specify on screen** option of the **Properties** palette.
4. Select **Space evenly** Layout type for the **X-Axis** and **Y-Axis** dimensions.

5. Respond to the workspace prompts as follows to place the column grid:

> Insertion point or: *(Select the insertion point at* **p1** *as shown in Figure 13.20.)*
>
> New size or: *(Select the diagonal corner of the column grid at* **p2** *in Figure 13.20.)*
>
> New size for cell or: *(Move the cursor horizontally toward point* **p3** *as shown in Figure 13.20 and click to specify the number of* **X-Width** *divisions and* **Y-Depth** *divisions.)*
>
> Rotation or <0.00>: 0 ENTER *(Rotation specified.)*
>
> Insertion point or: ENTER *(Press ENTER to end the command.)*

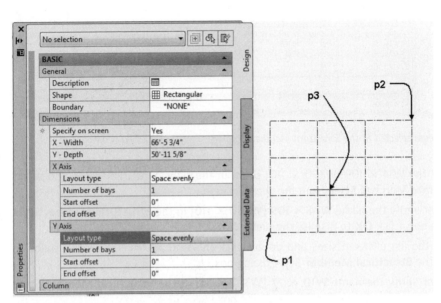

FIGURE 13.20 *Specifying size of grid and bay size on screen*

TUTORIAL 13-1: CREATING STRUCTURAL PLANS

1. Choose **New > Drawing** from the Application menu. Select the Structural Model (Imperial Stb).dwt [Structural Model (Metric Stb).dwt] template. Click **Open** to dismiss the **Select Template** dialog box.

2. Select **Save As** from the menu bar and save the drawing as **Lab 13-1** in your Ch 13 student directory. Toggle OFF **Allow/Disallow Dynamic UCS** in the Application status bar.

3. Select the Manage tab. Choose **Structural Member Catalog** from the Style & Display panel extension.

4. Expand the tree in the left pane to Imperial > Steel > AISC > I-Shaped > W1n, Wide-Flange Shapes [Metric > Steel > AISC > I-Shaped > Wide Flanges > W1nn, Wide-Flanges] as shown in Figure 13.21.

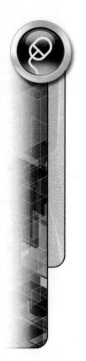

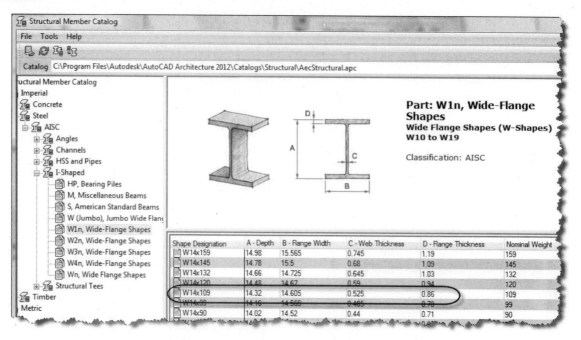

FIGURE 13.21 *Wide Flange beam selected in the Structural Member Catalog*

5. Select the W14 × 109 [W360 × 110] beam in the right pane, right-click, and select **Generate Member Style** from the shortcut menu.

6. Type or verify the name **W14 × 109 [W360 × 110]** in the **Structural Member Style** dialog box. Click **OK** to dismiss the dialog box.

7. Scroll the list of beam sizes and double-click the W10 × 77 [W310 × 52] shape to open the **Structural Member Style** dialog box.

8. Type or verify the name **W10 × 77 [W310 × 52]** in the **Structural Member Style** dialog box. Click **OK** to dismiss the dialog box.

9. From the menu bar of the **Structural Member Catalog**, select **File > Close** to close it.

10. Toggle OFF the Grid Display in the Application menu. Select the Home tab. Select the **Column Grid** from the structural flyout of the Build panel.

11. Edit the **Properties** palette as follows: Shape = **Rectangular**, Boundary = **None**, Specify on screen = **No**, X-Width = **72' [21600]**, Y-Depth = **36' [10800]**; X Axis: Layout type = **Repeat**, Bay size = **18' [5400]**, Start offset = **0**, End offset = **0**; Y Axis: Layout type = **Repeat**, Bay size = **18' [5400]**, Start offset = **0**, End offset = **0**; Column Style = **W14 × 109 [W360 × 110]**; Column Dimensions: Start offset = **0**, End offset = **0**, Logical length = **12' [3600]**, Justify = **Middle Center**, Justify cross-section = **Maximum**, and Justify components = **Highest Priority Only**. Move the cursor to the workspace and type the insertion point in the workspace prompts shown below:

    Insertion point or: 15', 15' [5000,5000] ENTER *(Specify the insertion point location.)*

    Rotation or <0.00>: ENTER *(Press ENTER to accept 0 rotation.)*

    Insertion point or: ENTER *(Press ENTER to end the command.)*

12. Right-click the Object Snap toggle of the status bar and choose **Settings**. Clear all object snap modes, then toggle ON the **Node** object snap. Check **Object Snap On (F3)** and click **OK** to dismiss the Drafting Settings dialog box.

13. Verify that the Model tab is current and choose **SW Isometric** from the View Cube. Choose **Elevation value** in the **Application** status bar to open the **Elevation Offset** dialog box. Choose **Pick point** and use the Node object snap to select the top of the column. Verify Elevation to **12' [+3600]**. Click **OK** to dismiss the Elevation Offset. Set the **Replace Z value with current elevation** toggle ON as shown at the right in Figure 13.22.

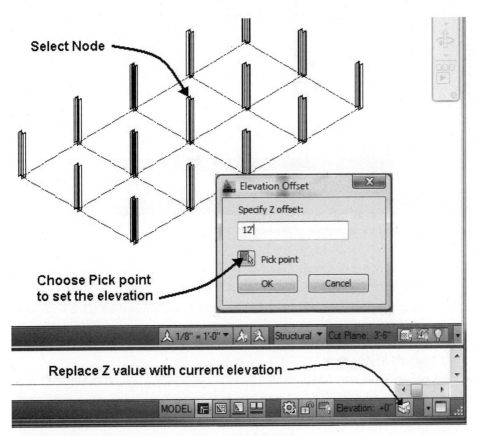
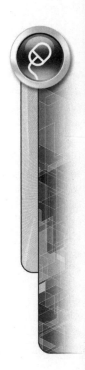

FIGURE 13.22 *Setting elevation in the Application status bar*

14. Select the Home tab. Choose Design Tools to display the Design tool palette as shown in Figure 13.23.

15. Select the **Beam** tool from the Design palette. Edit the **Properties** palette for the Beam as follows: Style = **W10 × 77 [W310 × 52]**, Trim automatically = **Yes,** Start Offset = **0,** End Offset = **0,** Roll = **0,** Layout type = **Edge,** Justify = **Top Center,** Justify cross section = **Maximum,** and Justify components = **Highest Priority Only** as shown in Figure 13.23.

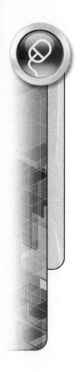

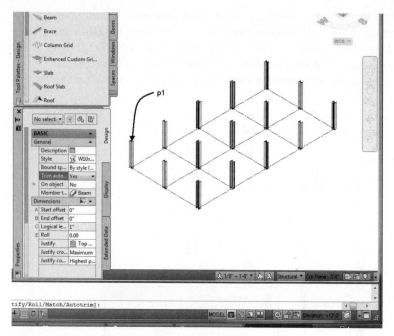

FIGURE 13.23 *Placing the start point of the beam*

16. Respond to the workspace prompts to place the beam as shown below:

> Start point or: *(Select the node of the column at* **p1** *as shown in Figure 13.23.)*
>
> Endpoint or: *(Select the node of the column at* **p2** *as shown in Figure 13.24.)*
>
> Start point or: *(Move the cursor from the end of the last beam, verify that the Node object snap tip is displayed, and click to start the beam at* **p2** *as shown in Figure 13.24.)*
>
> Endpoint or: *(Select the node of the column at* **p3** *as shown in Figure 13.24.)*
>
> Start point or: ENTER *(Press ENTER to end the command.)*

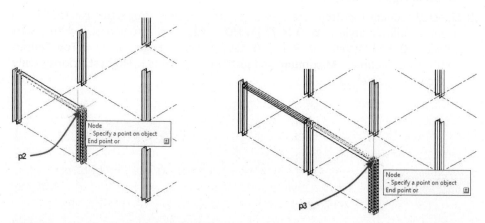

FIGURE 13.24 *Placing the endpoint of the beam*

17. Continue to place the beams as shown in Figure 13.25.

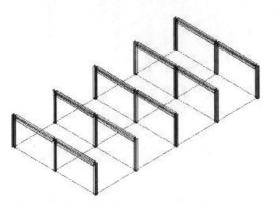

FIGURE 13.25 *Beams placed between columns*

18. Right-click the titlebar of the tool palettes and choose New Palette. Overtype **Structural** as the name of the new palette. Select Top View from the View Cube.

19. Select the **Content Browser** from the Tools flyout menu in the Build panel. Select the Design Tool Catalog – Imperial > Structural > Bar Joists [Design Tool Catalog – Metric > Structural Members > Castellated Beam W530 × 66] as shown in Figure 13.26. Select the i-drop for the **Steel Joist 20** [Castellated Beam W530 × 66] and drag it from the catalog to the Structural tool palette.

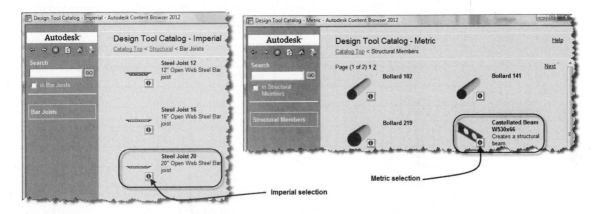

FIGURE 13.26 *Steel Joist 20 [Castellated Beam W530 ×× 66] i-dropped into the drawing*

20. Select the **Steel Joist 20** [Castellated Beam W530 × 66] tool and verify the settings of the Properties palette as follows: Style = **Steel Joist 20** [Castellated Beam W530 × 66], Trim automatically = **No**, On object = Yes, Start Offset = **1″ [25]**, End Offset = **−1″ [−25]**, Roll = **0**, Layout type = **Fill**, Justify = **Baseline [Bottom Center]**, Array = **Yes**, Layout method = **Repeat**, Bay size = **3′-0″ [900]**. Note that Baseline justification will place the bar joist on top of the beam or column.

21. Move the cursor over the grid and press CTRL to toggle the display of the joist so that all bays are filled as shown at left in Figure 13.27. Click to accept the display and place the joist throughout the column grid. Press ESC to end the command.

22. Choose SW Isometric from the View Cube to view the joist as shown at right in Figure 13.27.

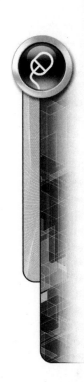

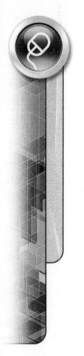

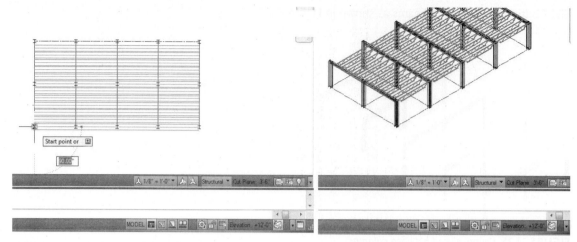

FIGURE 13.27 *Insert joists using Fill Layout*

23. Choose **Top View** from the **View** Cube. The Structural display representation depicts each beam as a line as shown in Figure 13.28.

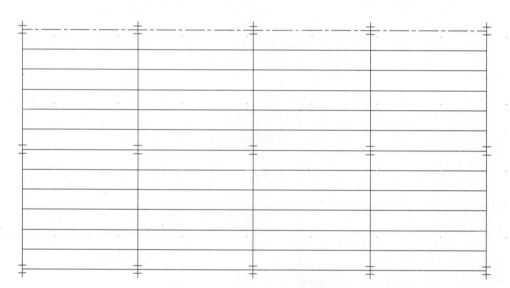

FIGURE 13.28 *Structural display representation of beams and columns*

24. Select **Medium Detail** from the **Display Configuration** flyout menu of the **Drawing Window** status bar to view the structure as shown in Figure 13.29.

25. Save and close the drawing.

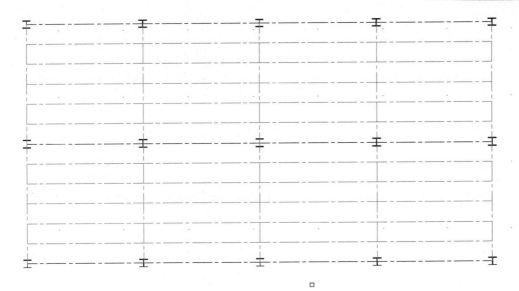

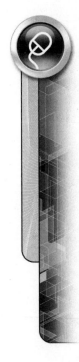

FIGURE 13.29 *Medium Detail display of structural members*

Clipping the Column Grid

The column grid can be clipped by a closed polyline. The LayoutGridClip command can be used to restrict a column grid to a closed polyline or to create and remove holes placed in the column grid. Refer to Table 13.12 for **LayoutGridClip** command access.

Clipping a column grid allows you to see only that part of the grid inside or outside the closed polyline. The clip can be applied when the grid is inserted or added to an existing grid. To clip a column grid upon insertion, select the **Boundary** field of the Properties palette, choose the **Select object** option, and select a closed polyline. The column grid will be restricted to inside a closed polyline as shown in Figure 13.30. The clip can be removed by selecting the column grid and editing the **Boundary** field to **None** in the Properties palette.

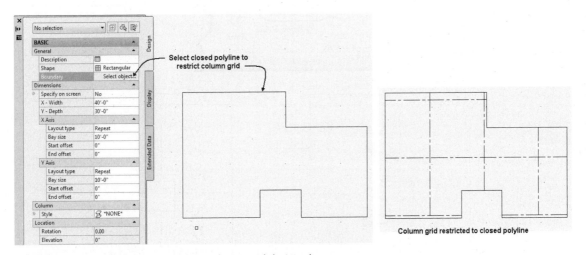

FIGURE 13.30 *Set Boundary applied to Column Grid during placement*

The LayoutGridClip command includes three options Set boundary/Add hole/Remove hole. The Set Boundary option of the LayoutGridClip command allows you to restrict the column grid to a closed polyline after the grid has been placed. The Set boundary option will mask the grid to only include the column grid inside the closed polyline. The Set boundary option clips the grid in the same manner as selecting a **Boundary** in the Properties palette. Access the LayoutGridClip command as shown in Table 13.12.

TABLE 13.12 *LayoutGridClip command access*

Command prompt	LAYOUTGRIDCLIP
Ribbon	Select a column grid and choose Set Boundary, Add Hole, or Remove Hole from the Clipping panel of the Column Grid tab.
Shortcut menu	Select a column grid, right-click, and choose Clip.

Holes can be placed in a column grid with the **LayoutGridClip** command. When you select a grid and choose **Add Hole** from the Clipping panel, you are prompted in the workspace to select layout grids and the closed boundary to create the hole. The Add Hole option removes the grid from within the closed polyline. The Remove hole option removes the clip from the column grid. A hole was placed in the column grid shown in Figure 13.31 as shown in the following workspace sequence:

(Select a column grid; choose Add Hole from the Clipping panel of the Column Grid contextual tab.)

Select layout grids: 1 found *(Select the column grid at* **p1** *as shown in Figure 13.31.)*

Select layout grids: ENTER *(Press ENTER to end the command.)*

Select a closed polyline for hole: *(Select the polyline at* **p2** *as shown in Figure 13.31.)*

Press ESC to end the command and view the column grid as shown at the right in Figure 13.31.

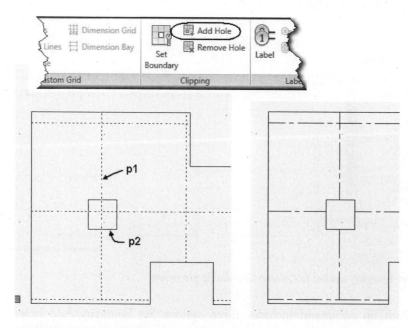

FIGURE 13.31 *Hole clipped in column grid*

Defining the Column for the Grid

Columns can be inserted at each intersection of the Column Grid or an Enhanced Custom Grid by the toggling options as shown in Figure 13.5. When you place a column with the ColumnAdd command with **On object = Yes**, you can press the CTRL key to toggle insertion to all nodes of the Column Grid or an Enhanced Custom Grid. A structural member style included in the drawing can be assigned in the Properties palette when a Column Grid is placed as shown in Figure 13.32.

The content of the drop-down list includes the structural member styles created from the Structural Member Catalog and used in the drawing. The column style will be inserted at each intersection of the column grid; however, you can erase selected columns from the grid. Columns inserted in the column grid can be edited independent of the remaining columns by selecting the column and changing the style or other properties in the Properties palette.

The Properties palette can be used to globally edit all columns of a column grid by selecting the grid and columns and editing the filter at the top of the Properties palette, as shown on the right in Figure 13.32. In addition you can select a column, right-click, and choose **Select Similar** to select all columns of the same style for edit in the Properties palette.

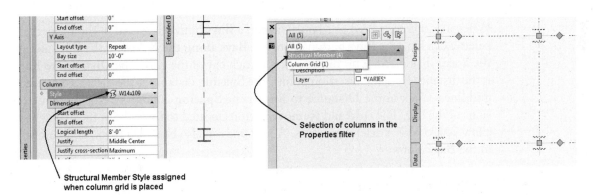

FIGURE 13.32 *Filtering objects in the Properties palette*

The Properties palette of an existing column grid does not include a column style field. Therefore, if this field is not specified when the column grid is placed, you must use the **Column** tool to insert columns in the grid. The **Node** and **Intersection** object snap modes can be used to insert the column at the intersection of the column grids.

REFINING THE COLUMN GRID

The Properties palette allows you to edit most features of an existing column grid. The **Layout type** of the column grid determines how the grid will be edited. Column grids placed with **Space evenly** and **Repeat** include grips at the four corners. The **X-Width and Y-Depth** dimensions of the column grid can be changed if you select one of the four corner grips and stretch the grip to a new location. Column grids created with **Space evenly** or **Repeat** layouts can be converted to **Manual** by selecting the **Manual** layout type in the Properties palette. Column grids converted to Manual have grips for each grid line, as shown on the right in Figure 13.33. A grid line can be relocated by selecting one of the grips and stretching it to a new location.

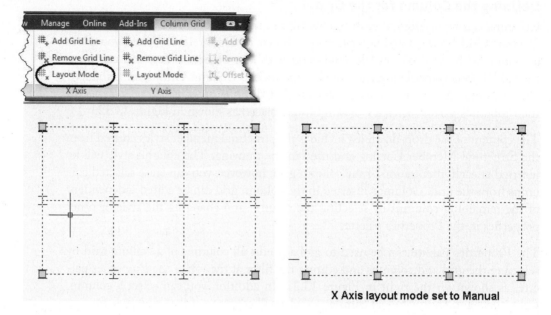

FIGURE 13.33 *Grips of the rectangular column grid*

When a grid is converted to Manual layout, a **Bays** button is added in the Properties palette. Clicking the **Bays** button opens the **Bays along the X Axis** or the **Bays along the Y axis** dialog box. This dialog box lists each bay of the grid. The **Distance to line** is the distance from the insertion point. The **Spacing** distance is the distance between grid lines. Click in the **Distance to line** or the **Spacing** column to edit the dimensions of the bay. The **Add** and **Remove** buttons located to the right of the dialog box allow you to add or delete grid lines. A grid line can also be added if you double-click in the **Bay** column.

Adding and Removing Column Grid Lines

The commands to add, remove, and change the layout type of a column grid are located on the shortcut menu of the column grid. The shortcut menu of the grid includes **X Axis** and **Y Axis** cascade menus. The cascade menu includes the following options: **Add Grid Line**, **Remove Grid Line**, and **Layout Mode**. These options allow you to quickly change the column grid mode and add or remove column grid lines for the X axis or the Y axis.

The **Layout Mode** command on the shortcut menu allows you to define the spacing system used in the column grid. See Table 13.13 for **Layout Mode** command access.

TABLE 13.13 *ColumnGridXMode or ColumnGridYMode command access*

Command prompt	COLUMNGRIDXMODE or COLUMNGRIDYMODE
Ribbon	Select the column grid and choose Layout Mode from the X Axis or Y Axis panels.
Shortcut menu	Select the column grid, right-click, and choose X Axis > Layout Mode or Y Axis > Layout Mode from the shortcut menu.

Choosing the **Layout Mode** command from the shortcut menu allows you to select one of the following modes for the column grid: **Manual**, **Repeat**, or **Space Evenly**. Selecting the **Manual** option allows you to add or remove individual column grid lines at specific locations. The **Repeat** option allows you to specify in the command line the bay size to be repeated in the column grid. The **Space Evenly** option allows you to specify the number of equal spaces in the column grid. These options allow you to edit the column grid by entering values in the command line. The layout mode can also be changed in the **Layout type** field of the Properties palette for existing column grids.

Adding Column Grid Lines

Additional column grid lines can be added to the grid at specific locations if the column grid is set to the **Manual** mode as shown in Figure 13.34. The **Space evenly** layout type allows you to add grid lines uniformly spaced within the grid. However, **Repeat** layout does not allow you to add or remove grid lines with the **Column-GridXAdd** or **ColumnGridYAdd** command.

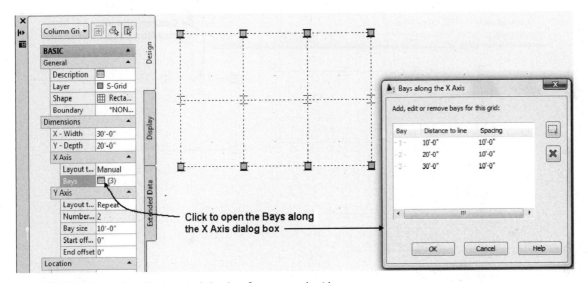

FIGURE 13.34 *Bays along the Y Axis dialog box for a manual grid*

You can add column grid lines for the X axis or the Y axis by choosing the column grid and choosing **Add Grid Line** from the X axis or Y axis panels of the Column Grid contextual tab as shown in Table 13.14.

TABLE 13.14 *ColumnGridXAdd and ColumnGridYAdd command access*

Command prompt	COLUMNGRIDXADD or COLUMNGRIDYADD
Ribbon	Select the column grid and choose Add Grid Line from the X Axis or Y Axis panel.
Shortcut menu	Select the column grid, right-click, and choose X Axis > Add Grid Line or select the column grid, right-click, and choose Y Axis > Add Grid Line.

When you select a grid and choose **Add Grid Line** from the **X Axis** or **Y Axis** panels, you are prompted to specify the location for the new line. The new line can be specified in the command line or selected with the cursor. The distance inserted in the command line is relative to the insertion point of the column grid. The new column grid line is combined with the remainder of the column grid. The procedure for adding a new column grid line is shown in the following workspace sequence:

(Select the column grid with Manual layout type, as shown in Figure 13.35 then Add Grid Line *from the* X Axis *panel of the Column Grid tab. Move the cursor in the right direction as shown in Figure 13.35.)*

Enter X length <10'-0 [3000]>: 5' [1500] ENTER *(The column grid line added 5' [1500] from the insertion point along the X axis as shown at the right in Figure 13.35.)*

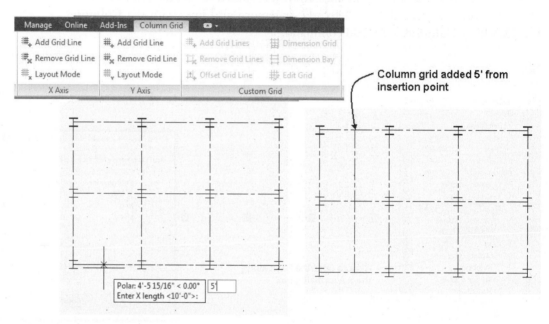

FIGURE 13.35 *Column grid prior to and after adding a grid line*

Removing Column Grid Lines

You can remove column grid lines by choosing the **ColumnGridXRemove** or **ColumnGridYRemove** command. See Table 13.15 for command access.

TABLE 13.15 *ColumnGridXRemove or ColumnGridYRemove command access*

Command prompt	COLUMNGRIDXREMOVE or COLUMNGRIDYREMOVE
Ribbon	Select the column grid and choose Remove Grid Line from the X Axis or Y Axis panels.
Shortcut menu	Select the column grid, right-click, and choose X Axis > Remove Grid Line or Y Axis > Remove Grid Line.

When you select a column grid, right-click, and choose **X Axis > Remove Grid Line** or **Y Axis > Remove Grid Line**, you are prompted in the workspace to specify the location of the grid line to remove. Selecting near the line with the cursor or typing the location as the distance from the insertion point specifies the location. The procedure for removing the column grid line is shown in the following workspace sequence:

> *(Select the column grid and choose* Remove Grid Line *from the X Axis panel of the Column Grid panel.)*
>
> Enter approximate X length to remove <10'-0>: *(Select the location of the column grid line at point* **p1** *as shown in Figure 13.36.)*
>
> *(The column grid is removed as shown in* Figure 13.36.*)*

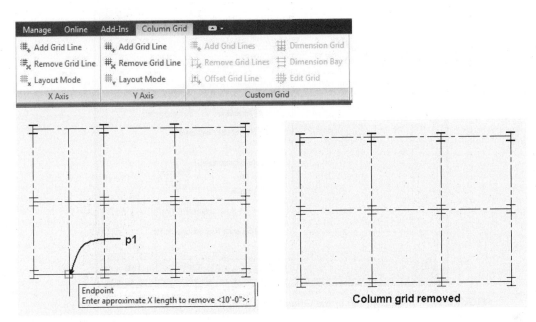

FIGURE 13.36 *Selected column grid line removed*

Labeling the Column Grid

The column grid is labeled with the **ColumnGridLabel** command. The column grid can be labeled automatically with numbers or letters, or each column grid line can be assigned a unique label. If the column grid labels are automatically generated, the numbers or letters can be assigned in an ascending or descending pattern generated from the column grid line passing through the insertion point. The column grid label components consists of annotative content; therefore, set the Annotation Scale and toggle ON Automatically add scales prior to placing label components. See Table 13.16 for **ColumnGridLabel** command access.

TABLE 13.16 *ColumnGridLabel command access*

Command prompt	COLUMNGRIDLABEL
Ribbon	Select a column grid and choose Label from the Label panel of the Column Grid tab.
Shortcut menu	Select a column grid, right-click, and choose Label.

When you select a column grid and choose **Label** from the **Label** panel, the **Column Grid Labeling** dialog box opens, as shown in Figure 13.37.

The Column Grid Labeling dialog box consists of the **X - Labeling** and **Y - Labeling** tabs. The column grid lines for each grid direction are labeled in the tab. The **Automatically Calculate Values for Labels** checkbox allows the numbering of the column grid lines to be assigned in ascending or descending order from the first column grid line. The number assigned to the first column grid line as shown at **p1** in Figure 13.37 defines the beginning character of the sequence. In the figure, if 1 is assigned at **p1** and Ascending order is toggled ON, the columns will be labeled 1, 2, 3. The options of the X - Labeling and Y - Labeling tabs are described below.

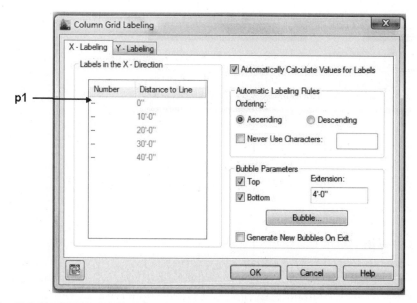

FIGURE 13.37 *X-Labeling tab of the Column Grid Labeling dialog box*

Labels in the X - Direction

The Labels in the X - Direction list box shows each grid label and the distance from the insertion point.

> **Number**—The Number column allows you to assign an alphanumeric character for each column grid line. If **Automatically Calculate Values for Labels** has been selected, the labels are generated from the label specified in the top field of the **Number** column. You can override the column grid labels generated with the **Automatically Calculate Values for Labels** checkbox by selecting the grid label and typing the character for the column grid in the **Number** column.
>
> **Distance to Line**—The Distance to Line column lists the distance for each column grid line from the insertion point of the column grid.
>
> **Automatically Calculate Values for Labels**—The Automatically Calculate Values for Labels checkbox automatically assigns the labels to the column grid from the beginning column grid in ascending or descending order.

Automatic Labeling Rules

> **Ascending**—The Ascending radio button assigns the column grid label in increasing numeric or alphabetic order.

Descending—The Descending radio button assigns the column grid label in decreasing numeric or alphabetic order.

Never Use Characters—The Never Use Characters checkbox allows you to identify characters to exclude when the labels are automatically generated.

Bubble Parameters

The Bubble Parameters section allows you to define the size and location of the bubbles used to identify the column grid.

Top—The Top checkbox creates bubbles at the top of the column grid.

Bottom—The Bottom checkbox creates bubbles at the bottom of the column grid.

Extension—The Extension field defines the distance from the grid to extend the grid centerline and place the bubble.

Bubble—The Bubble button opens the **MvBlockDefSelect** dialog box, which allows you to select the multi-view block for the bubble.

Generate New Bubbles On Exit—The Generate New Bubbles On Exit checkbox redisplays the position and size of the bubbles according to the changes. The size of the column bubbles is determined in the **Drawing Setup** dialog box (**AecDwgSetup** command).

> The *Y-Labeling* tab includes similar options for specifying the labeling of the column grid in the Y direction.

NOTE

Dimensioning the Column Grid

The column grid can be automatically dimensioned with the **AEC Dimension** command (**DimAdd**). The dimensions will be updated to reflect any changes in the column grid. Access the **AEC Dimension** command as shown in Table 13.17.

TABLE 13.17 *AEC Dimension for column grids command access*

Command prompt	DIMADD
Ribbon	Select the column grid and choose Dimension from the General panel of the Column Grid contextual tab.
Shortcut menu	Select the column grid, right-click, and choose AEC Dimension. Select AEC Dimension – Exterior from the Dimensions tool palette of the Document palette set.

When you select the column grid and choose **AEC Dimension**, the direction in which you move the cursor determines whether you create a horizontal or vertical dimension. In the example shown in Figure 13.38 the cursor was moved right from the grid. In the Properties palette, specify a dimension style for the AEC Dimension. Verify that an AEC Dimension style is current and then select a location with the cursor to specify the location of the dimension string. Shown below is the workspace prompt sequence for dimensioning a column grid:

```
(Select AEC Dimension - Exterior from the Dimensions tool
palette, right-click, and choose Import Exterior - Center of
Opening AEC Dimension Style.)
```

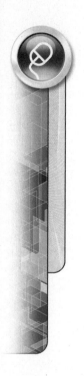

(Select the column grid at p1 as shown in Figure 13.38. Choose Dimension from the General panel. Select Exterior - Center of Opening from the Style drop-down list of the Design tab in the Properties palette.)

Specify insert point: *(Specify the location of the dimension string at p2 as shown in Figure 13.38.)*

1 added

(The column grid is dimensioned as shown in Figure 13.38.)

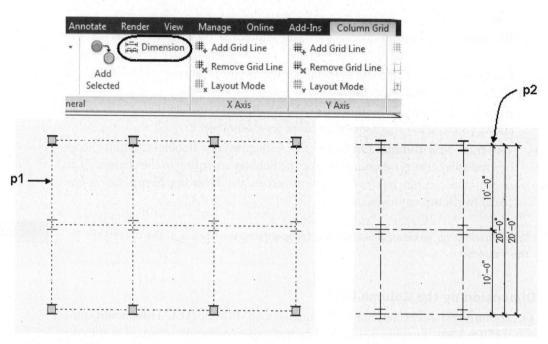

FIGURE 13.38 *Dimensioned column grid*

TUTORIAL 13-2: DIMENSIONING THE COLUMN GRID

1. Download Chapter 13 files from the Accessing Tutor\Imperial or Accessing Tutor\Metric category of the Student Companion site of CengageBrain http://www.cengagebrain.com described in the Preface to your Accessing Tutor folder.

2. Open Ex 13-2.dwg from your Accessing Tutor\Imperial\Ch13 or Accessing Tutor\Metric\Ch13 directory.

3. Select **Save As > AutoCAD Drawing** from the Application menu and save the drawing as **Lab 13-2** in your Ch 13 student directory. Verify Annotation Scale = 1/8" = 1'-0" [1:100] and Automatically add scales is toggled ON in the Drawing Window status bar.

4. Select the column grid line and choose **Label** from the Label panel of the Column Grid contextual tab. Edit the **X-Labeling** tab: Check **Automatically Calculate Values for Labels**; select the **Descending** radio button; for Bubble parameters, check **Top** and **Bottom**; set Extension = **12'-0" [3500]**; and select **Generate New Bubbles on Exit**. Click in the **Number** column and type **E** in the field as shown at **p1** in Figure 13.39.

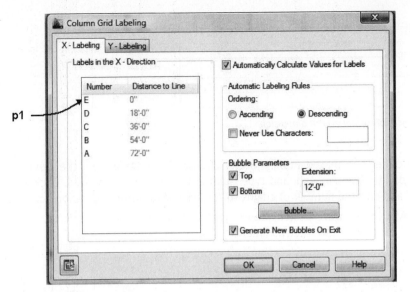

FIGURE 13.39 *X-Labeling of the column grid*

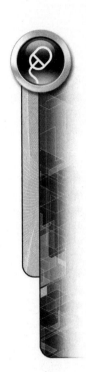

5. Select the **Y-Labeling** tab. Edit as follows: Check **Automatically Calculate Values for Labels**; select the **Descending** radio button; verify that the **Bubble Parameter Left** is checked and clear the **Right** bubble; set the extension to **12'-0" [3500]**; and select **Generate New Bubbles on Exit**. Type **3** in the top row of the **Number** column as shown at **p1** in Figure 13.40. Click **OK** to dismiss the Column Grid Labeling dialog box.

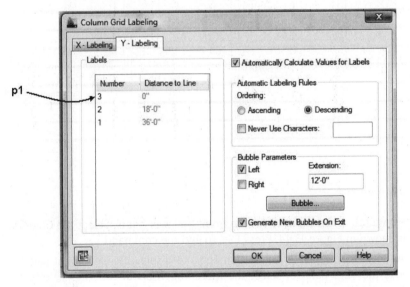

FIGURE 13.40 *Y-Labeling of the column grid*

6. Select the Annotate tab. Double-click the Annotation Tools button of the Tools panel to display the Document palette set. Choose the Dimensions tool palette. Select the AEC Dimension – Exterior tool from the Dimensions palette. Select the column grid and press ENTER. Select a point near **p1** as shown in Figure 13.41.

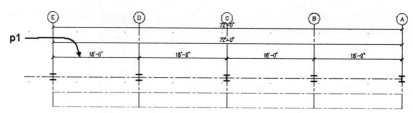

FIGURE 13.41 *Location of horizontal dimension*

7. Select the dimension placed in step 6 and choose **Copy Style** from the General panel. Choose the General tab, and type Column Grid in the Name field. Click the Chains tab and edit the number of chains to 2. Click OK to dismiss all dialog boxes.

8. Select the AEC Dimension placed in step 6, right-click, and choose Add Selected from the General panel. Add the vertical dimension as shown in Figure 13.42.

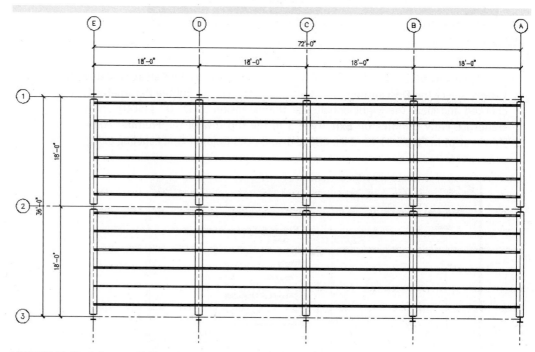

FIGURE 13.42 *Column grid dimensioned*

9. Select the **Home** tab. Choose **Layer Freeze** from the Layer panel. Select the Column grid at **p1** as shown in Figure 13.44. Press ESC to end the command.

10. Select the Annotate tab. Choose Annotation Tools to display the Document palette group. Choose the Tags palette. Select the More Tag Tools of the Tags palette to open the **Content Browser**. Double-click Structural Tags to open the Structural Tags category as shown in Figure 13.43. Drag the i-drop of the **Beam Tag by Style** tool to the workspace. Respond to the workspace prompts given in the following and in Figure 13.44 to place the tag:

> Select object to tag: *(Select the beam at* p2 *as shown in* Figure 13.44.*)*

> Specify location of tag <Centered>: *(Select a point near* p3 *as shown in* Figure 13.44.*)*

> *(Click* OK *to dismiss the Edit Property Set Data dialog box. Press* ESC *to exit the command.)*

11. Select Home tab. Extend the Layers panel and choose **Thaw All Layers** from the **Layers** panel extension. Save and close the drawing.

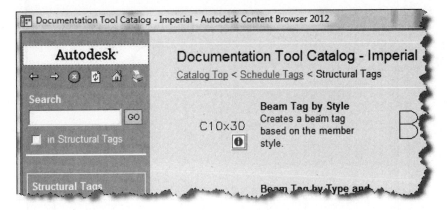

FIGURE 13.43 *Beam tag i-dropped from the Documentation Tool Catalog - Imperial*

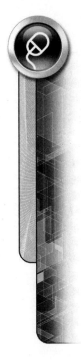

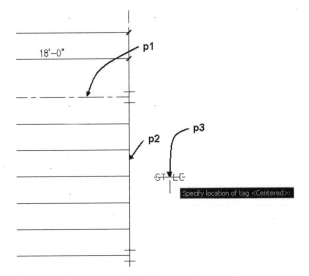

FIGURE 13.44 *Beam tag placed*

CREATING AN ENHANCED CUSTOM GRID

The Enhanced Custom Grid, accessed from the Build panel as described in Table 13.18, is developed from the Column Grid dialog box shown in Figure 13.45. Separate grid spacing and numbering or labeling of rectangular grids can be defined for the top, bottom, left, and right. Secondary grid lines can be added to the primary grid.

TABLE 13.18 *Custom Column Grid command access*

Command prompt	CUSTOMCOLUMNGRID
Ribbon	Select the Home tab. Choose Enhanced Custom Grid from the Build panel.
Tool palette	Select Enhanced Custom Grid from the Design Tool palette.

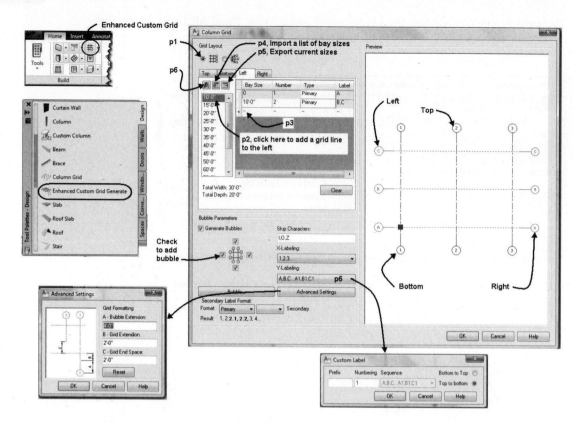

FIGURE 13.45 *Creating a Column Grid with the Enhanced Custom Grid command*

The Column Grid dialog box includes toggles for rectangular or radial grids as shown at p1 in Figure 13.45. The size of the grid can be specified for each of the following sides of the column grid: top, bottom, left, and right. In Figure 13.45 the tab for the Left side of the grid has been chosen, and the size for each grid line is defined by clicking at p2 in the left window or by clicking at p3 to specify the size for the Bay Size. The left pane includes a sample of bay sizes for the grid. You may click at p4 to import a list of preconfigured bay size values or choose p5 to export the current size list to an XML file. You may choose the Set From button to select an existing column grid for the development of a custom grid from the lines of a column grid.

The lower portion of the Column Grid dialog box includes Bubble Parameters that consist of toggles for setting the display and content of the column grid bubbles. Choose the Advanced Setting button at p6 to edit the bubble position from the column grid.

EDITING AN ENHANCED CUSTOM GRID

The Enhanced Custom Grid can be edited by the CustomColumnGridEdit command as described in Table 13.19. The CustomColumnGridEdit command opens the Column Grid dialog box as shown in Figure 13.46. The Column Grid dialog box allows you to change the labels and the positions of the grid lines.

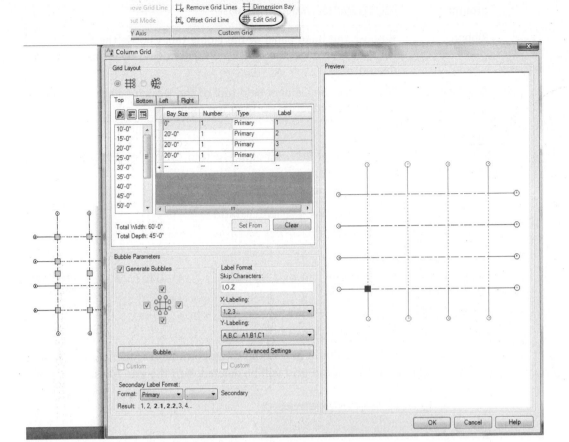

FIGURE 13.46 *Editing a custom column grid*

TABLE 13.19 *Edit Custom Grid command access*

Command prompt	CUSTOMCOLUMNGRIDEDIT
Ribbon	Select a custom grid; choose Edit Grid from the Custom Grid panel of the Column Grid tab.
Shortcut menu	Select a custom grid, right-click, and choose Edit Grid.

CONVERTING LINEWORK OR A COLUMN GRID TO A CUSTOM GRID

The Custom Grid Convert (CUSTOMCOLUMNGRIDADD) command, described in Table 13.20, will convert lines, arcs, or an existing column grid which was placed using the ColumnGridAdd tool into a Custom Grid. The conversion to a Custom Grid allows you to edit the column bubbles using the CustomColumnGridLabelEdit. The CustomColumnGridEdit command presented in Table 13.19 cannot be used to edit converted custom grids.

TABLE 13.20 *Custom Grid Convert command access*

Command prompt	CUSTOMCOLUMNGRIDADD
Ribbon	Choose the Build panel of the Home tab. Choose the Custom Grid Convert from the Enhanced Custom Grid flyout as shown in Figure 13.47.
Tool Palette	Right-click the Column Grid tool of the Design palette, and choose Apply tool properties to from the menu.

Select a column grid, lines, or arcs. Choose Custom Grid Convert from the Enhanced Custom Grid flyout of the Build panel in the Home tab of the ribbon. Respond to the workspace prompts as shown below.

```
Command: _AecCustomColumnGridAdd
Select linework: (Select a column grid or linework) 1 found
Select linework: (Press ENTER to end selection.)
Enter label extension or [No labels] <4'-0">: (Press ENTER to
accept label extension distance.)
Erase selected linework? [Yes/No] <No>: (Press ENTER to
retain selected linework.)
```

If you choose to erase the column grid in the above command sequence, any columns that were inserted in the column grid will be deleted with the grid lines.

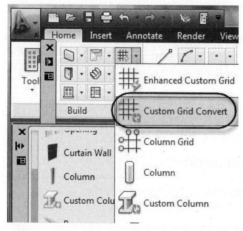

FIGURE 13.47 *Accessing the CustomColumnGridAdd command*

APPLYING CONSTRAINTS TO GRIDS

Column Grids and Enhanced Custom Grids can be positioned using constraints. The following constraints are available from the Parametric panel of the Column Grid tab: Coincident, Collinear, Fix, Vertical, Horizontal, Aligned, and Angular. The position of walls can be constrained to the column grid. Tutorial 13.3 includes an application of the constraints for the column grid.

TUTORIAL 13-3: CREATING AN ENHANCED CUSTOM GRID

1. Download Chapter 13 files from the Accessing Tutor\Imperial or Accessing Tutor\ Metric category of the Student Companion site of CengageBrain http://www .cengagebrain.com described in the Preface to your Accessing Tutor folder.

2. Open *Accessing Tutor\Imperial\Ch 13\Ex13-3.dwg* or *Accessing Tutor\Metric\Ch13\ Ex 13-3.dwg*.

3. Save the file to your Ch 13 student directory as **Lab 13-3**.

4. Choose the Enhanced Column Grid Generate from the Build panel to open the Column Grid dialog box. Verify that the Top tab is current, click at **p1** shown in Figure 13.48 to add a 20' [6000] column grid.

5. Click at **p2** as shown in Figure 13.48, to add a 20' [6000] primary grid. Continue to click in the Bay Size field until the Total Width is 80'-0" [24000].

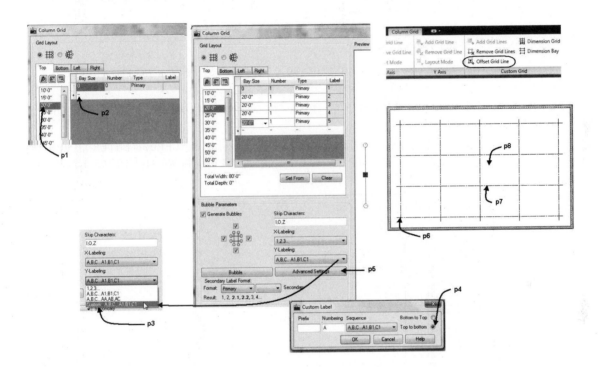

FIGURE 13.48 *Modifying the Column Grid dialog box*

6. Choose the X-Labeling flyout of the Column Grid dialog box. Choose **A,B,C, ...** if necessary to change the X-Labeling pattern to A,B,C,...

7. In this step you will modify the Y-Labeling to numbers and descending. Choose the **Custom** option from the Y-Labeling flyout as shown at p3 as shown in Figure 13.48 to open the **Custom Label** dialog. Overtype **1** in the Numbering field and choose the **Top to bottom** radio button shown at **p4**. Click OK to dismiss the dialog box.

8. Choose the Advanced Setting button at **p5** shown in Figure 13.48; verify that the A-Bubble Extension is 14'-4" [4300] and B-Grid Extension and C-Grid End Space are 2'-0" [600]. Click OK to dismiss the dialog box.

9. Choose the Left tab in the Column Grid dialog box, click 15' [4500] three times in the left bay list to create a Total Depth of 45' [13500].

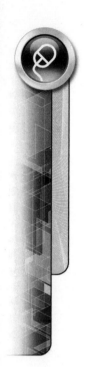

10. Click OK to dismiss the Column Grid dialog. Respond to the workspace prompts as follows:

Specify insertion point or [Rotate]: *(Choose a point inside the four walls near p6 as shown in Figure 13.48.)*

Specify rotation angle <0.00>: ENTER

Specify insertion point or [Rotate]: ENTER

11. In this step you will add a custom grid 6' [1800] to the right of the A column grid. Select the column grid at **p7** to display the Column Grid tab. Choose Offset Grid Line of the Custom Grid panel in the Column Grid tab. Respond to the workspace prompts as follows:

Command: _AecCustomColumnGridLineOffset

Select a reference grid line for offset...

Select a custom column grid line: (Select the grid shown at **p7** in Figure 13.48.)

Specify offset distance or [Through] <10'-0" [1000]>: 6'[1800] ENTER

Specify point on side to offset: *(Select a point at p8 as shown in Figure 13.48.)*

Is the added grid line a secondary grid line? [Yes/No] <No>: **y** ENTER

Select a custom column grid line: ENTER (Press ENTER to end the command. Custom grid line created as shown in Figure 13.49.)

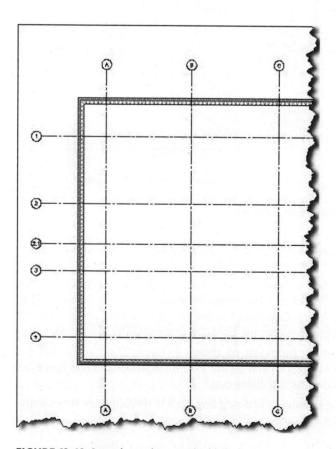

FIGURE 13.49 *Secondary column grid added*

12. Right-click over the Object Snap toggle. Choose only the Intersection object snap mode from the flyout menu.

13. Choose the Column tool. Edit the Properties palette as follows: Style = **Standard** and On object = **Yes**. Move the cursor near an intersection of the column grid to display the column placement tip; press CTRL twice to display the node marker at each of the grid intersections. Left-click to place a column at each of the intersections.

14. Select the Column Grid at p1 as shown in Figure 13.50. Choose **Collinear** from the Parametric panel of the Column Grid contextual tab. Respond to the workspace prompts as follows:

>   Select first object or [Multiple]: *(Select column grid at p1 as shown in Figure 13.50.)*
>
>   Select second object: *(Select the edge of the concrete masonry wall at p2 as shown in Figure 13.50.)*

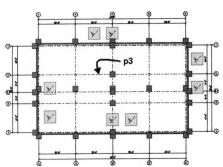
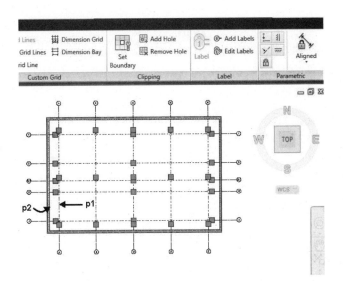
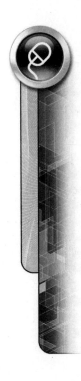

FIGURE 13.50 *Constraining walls to a column grid*

15. Repeat step 14 to position the remaining exterior walls to the column grid as shown at right in Figure 13.50.

16. Select the column grid at p3 as shown in Figure 13.50 to display the Column Grid tab. Choose **Dimension Grid** from the Custom Grid panel. Respond to the workspace prompts as follows:

>   Enter offset distance <11'-4">: ENTER *(Press ENTER to accept the default distance and place the dimension as shown in Figure 13.50.)*

17. Save and close the drawing.

TUTORIAL 13-4: CONVERTING LINEWORK TO A CUSTOM COLUMN GRID

1. Download Chapter 13 files from the Accessing Tutor\Imperial or Accessing Tutor\Metric category of the Student Companion site of CengageBrain **http://www.cengagebrain.com** described in the Preface to your Accessing Tutor folder.

2. Open Accessing Tutor\Imperial\Ch 13\Ex13-4.dwg or Accessing Tutor\Metric\Ch13\Ex 13-4.dwg.

3. Save the file to your Ch 13 student directory as Lab 13-4.

4. Choose **Custom Grid Convert** from the Enhanced Custom Grid flyout of the Build panel in the Home tab. Respond to the workspace prompts as follows:

> Select linework: *(Select a point near* p1 *as shown in Figure 13.51.)*
>
> Specify opposite corner: *(Select a point near* p2 *as shown in Figure 13.51.)*8 found
>
> Select linework: *(Press ENTER to end selection.)*
>
> Enter label extension or [No labels] <4'-0">: *(Press ENTER to accept the default value.)*
>
> Erase selected linework? [Yes/No] <No>: **y** *(Type Y to erase existing linework.)*

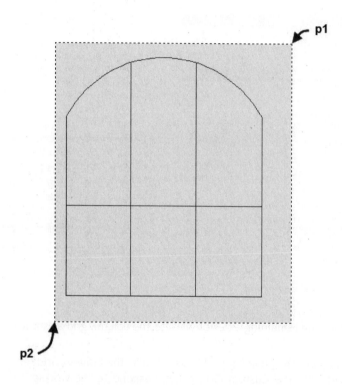

FIGURE 13.51 *Selection of linework for enhanced column grid*

5. Select the custom grid at p1 as shown in Figure 13.52, and choose **Edit Labels** of the Labels panel of the Column Grid tab.

> Select grid line(s) to edit label...
>
> Pick the first custom column grid line: *(Select the column grid at p2 as shown in Figure 13.52.)*
>
> Pick the next custom column grid line or: *(Select the column grid at p3 as shown in Figure 13.52.)*
>
> Pick the next custom column grid line or: *(Select the column grid at p4 as shown in Figure 13.52.)*
>
> Pick the next custom column grid line or: *(Select the column grid at p5 as shown in Figure 13.52.)*

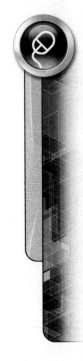

Pick the next custom column grid line or: *(Press* ENTER *to end selection of column grids.)*

Enter a label value or <->: **A** *(Type A to specify the column bubble value for the first column grid.)*

Pick the first custom column grid line: *(Select the column grid at* p6 *as shown in Figure 13.52.)*

Pick the next custom column grid line or [Edit] <Edit>: *(Select the column grid at* p7 *as shown in Figure 13.52.)*

Pick the next custom column grid line or [Edit] <Edit>: *(Select the column grid at* p8 *as shown in Figure 13.52.)*

Pick the next custom column grid line or [Edit] <Edit>: *(Press* ENTER *to end selection.)*

Enter a label value or <->: **1** (Type 1 to specify the column bubble value for the first column grid.)

Press ESC to end the command.

6. Save and close the drawing.

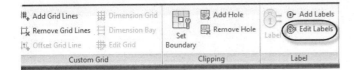

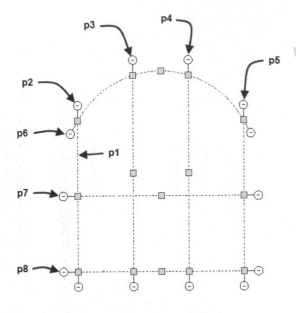

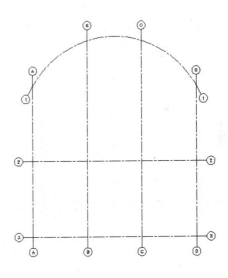

FIGURE 13.52 *Applying labels to column grid lines*

CREATING CEILING GRIDS

Ceiling grids represent the building components of the suspended ceiling. The ceiling grid is created as an object with properties similar to the column grid. The ceiling grid is developed in a rectangular shape; when it is inserted, its size should exceed the maximum dimensions of the room. If the ceiling grid is attached to a space or closed polyline as a boundary, only that part of the ceiling grid inside the room will be displayed. The ceiling grid should be anchored to a space or a polyline; however, it can be placed freestanding. The command to place a ceiling grid is **CeilingGridAdd**. See Table 13.21 for **CeilingGridAdd** command access.

TABLE 13.21 *CeilingGridAdd command access*

Command prompt	CEILINGGRIDADD
Ribbon	Select the Home tab. Choose Ceiling Grid from the Build panel.
Tool palette	Select Ceiling Grid from the Design Tool palette.

Ceiling grids and spaces are displayed in Reflected and Reflected Screened display configurations. Therefore, select the **Reflected** or **Reflected Screened** display configuration prior to selecting the **Ceiling Grid** tool from the Design tool palette. When you select the **Ceiling Grid** command, the Properties palette opens, allowing you to specify the size and properties of the ceiling grid as shown in Figure 13.53.

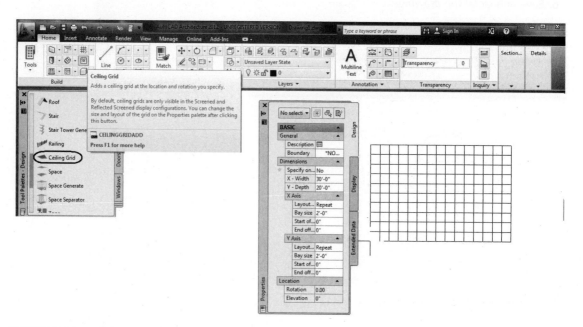

FIGURE 13.53 *Properties palette for inserting a ceiling grid*

The ceiling grid is placed as a rectangle regardless of the shape of the room. If the ceiling grid is attached to a boundary, the ceiling grid outside the boundary is not displayed. When the **Select Object** option is chosen from the **Boundary** field, you can select a closed polyline or space to attach to the ceiling grid. The closed polyline or space creates a boundary for the ceiling grid, which turns off the display of the ceiling grid outside the boundary. The ceiling grid shown in Figure 13.53 was attached to a Space object. Ceiling grids attached to any closed polyline will be bound by the polyline and placed at the elevation of the polyline. Using polylines and spaces as the boundaries for the ceiling grid creates ceiling grids that fit irregularly shaped spaces.

TIP

When you attach a ceiling grid to a space, the grid can be attached to the floor or ceiling boundaries. If a ceiling grid has been attached to the polyline of the floor, view the drawing in pictorial, choose **Select object** for the **Boundary** option in the Properties palette, and select the ceiling object.

When the **Ceiling Grid** command is selected, the default width and depth are retained from the last ceiling grid insertion. The grid size can be determined in the workspace if you select **Yes** in the **Specify on screen** option in the Properties palette. The **Specify on screen** option allows you to select a location for the insertion point (**p1**) and a diagonal point (**p2**) that is inclusive of the room geometry as shown in Figure 13.54. Selecting two points equal to or greater than the room size creates a ceiling grid that will cover the room. When **Specify on screen** is selected, you are prompted in the command line to select an insertion point, and the new size is determined by selecting the diagonal location as shown in the following command line sequence and Figure 13.54. The ceiling grid shown in Figure 13.54 and in the following workspace prompt sequence is **not attached to a boundary**; therefore the ceiling grid is displayed beyond the walls of the room.

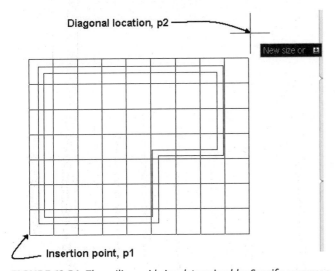

FIGURE 13.54 *The ceiling grid size determined by Specify on screen*

> Insertion point or: *(Select a point near* **p1** *as shown in Figure 13.54.)*
>
> New size or: *(Prior to selecting the 2nd point, verify that* **Specify on screen** *in the Properties palette is set to* **Yes** *and then select a point near* **p2** *as shown in Figure 13.54.)*
>
> Rotation or <0.00>: ENTER *(Press ENTER to accept 0° rotation.)*
>
> Insertion point or: ESC *(Press ESC to exit the command.)*

If the ceiling grid is attached to a boundary, the ceiling grid can be centered about the boundary by selecting the SNap to center option from the shortcut menu when the cursor is positioned to size the ceiling grid. The SNap to center option centers the ceiling grid about the boundary as shown in the following steps.

NOTE	You can use the *Space Generate* tool of the Design palette to create spaces for ceiling grids of floor plans.

Steps to Center and Attach a Ceiling Grid to a Space Object

1. Select **Reflected** or **Reflected Screened** display configuration from the **Drawing Window** status bar.
2. Select the **Ceiling Grid** tool from the **Design** palette.
3. Choose **Select object** from the **Boundary** field of the **Properties** palette. Select a space at **p1** as shown in Figure 13.55.
4. Set **Specify on screen** to **Yes**, move the cursor to the workspace, and respond to the following command line prompts:

> Insertion point or: *(Select a point near* p2 *as shown in Figure 13.55.)*
>
> New size or: *(Move the cursor near* p3, *right-click, and choose* SNap *to center.)*
>
> Rotation or <0.00>: 0 (Type 0 to specify rotation)
>
> Insertion point or: ENTER *(Press ENTER to end the command.)*

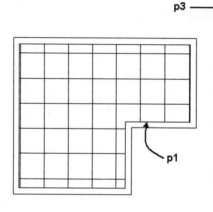

FIGURE 13.55 *Ceiling grid sized and attached to a boundary*

CHANGING THE CEILING GRID

After the ceiling grid is placed, the layout type, bay size, start offset, and end offset can be changed in the Ceiling Grid contextual tab as shown in Figure 13.56. The layout mode and clipping panels are identical to the column grid editing tools presented earlier in this chapter.

The grips of the Repeat and Space evenly layouts are located at the four corners of the ceiling grid as shown at the left in 13.56. The grips located at the corners of the ceiling grid allow you to resize the X-Width and Y-Depth of the grid. If you select a ceiling grid and edit the **Layout type** to **Manual** for the X axis and the Y axis in the Properties palette, the grips are located on each ceiling grid line. The grips of the Manual layout are shown at the right in Figure 13.56.

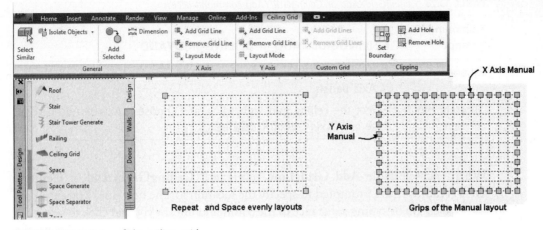

FIGURE 13.56 *Grips of the ceiling grid*

The Ceiling Grid contextual tab includes the following commands for modifying the ceiling grid: **Add Grid Line**, **Remove Grid Line** (for both the X and Y axes), **Layout Mode**, and **Clip**. These commands allow you to change the dimensions of the ceiling grid and add or remove sections of the grid as needed.

Adding and Removing Ceiling Grid Lines

The tools of the Ceiling Grid contextual tab allow you to modify the layout of the ceiling grid. **Layout Mode** command of the shortcut menu allows the ceiling grid to be set to **Manual** and ceiling grid lines can then be added or removed from the grid. The Layout Mode command (**CeilingGridXMode** or **CeilingGridYMode**) allows editing of each ceiling grid line. You can edit grid lines in the X axis direction by selecting the **Layout Mode** command for the X axis (**CeilingGridXMode**). Edit the ceiling grid lines in the Y direction by selecting the **Layout Mode** command for the Y axis (**CeilingGridYMode**). See Table 13.22 for **Layout Mode** command access.

TABLE 13.22 *CeilingGridXMode or CeilingGridYMode command access*

Command prompt	CEILINGGRIDXMODE or CEILINGGRIDYMODE
Ribbon	Select the ceiling grid and choose Layout Mode of the X Axis or Y Axis panels.
Shortcut menu	Select the ceiling grid, right-click, and choose X Axis > Layout Mode or Y Axis > Layout Mode.

When you select the **Layout Mode** command for either the X or Y axis, the ceiling grid is released from the automatic mode, and additional ceiling grid lines can be added or removed. Selecting the **Manual** option is equivalent to selecting **Manual** in the Properties palette to edit an existing ceiling grid.

Adding Ceiling Grid Lines

After the layout mode has been toggled to **Manual**, you can add ceiling grid lines with the **Add Grid Line** command (**CeilingGridXAdd** or **CeilingGridYAdd**). See Table 13.23 for **Add Grid Line** command access.

TABLE 13.23 *CeilingGridXAdd or CeilingGridYAdd command access*

Command prompt	**CEILINGGRIDXADD or CEILINGGRIDYADD**
Ribbon	Select the ceiling grid and choose Add Grid Line from the X Axis or Y Axis panels.
Shortcut menu	Select the ceiling grid, right-click, and choose X Axis > Add Grid Line or Y Axis > Add Grid Line.

When you select the **Add Grid Line** command (**CeilingGridXAdd** or **Ceiling-GridYAdd**), you are prompted to specify the location for the new line. You can specify the location by typing a distance in the command line, or you can click to select the location. If a distance is entered in the command line, it is measured relative to the insertion point of the grid. The procedure for adding a ceiling grid line is shown in the following workspace sequence:

NOTE	Layout mode must be set to *Manual* prior to grid lines being added.

(Verify that ORTHO is toggled ON in the status bar and grid layout mode is Manual.)

(Select the ceiling grid Add Grid Line *from the X Axis panel.)*

Enter X length <10'-0" [3000]>: *(Select a point near* p1 *in Figure 13.57.)*

(Ceiling grid line added as shown in Figure 13.57.)

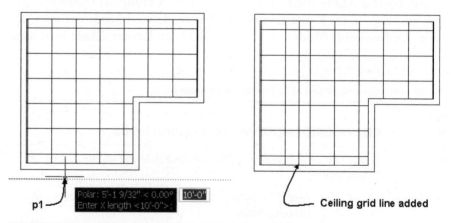

p1

Polar: 5'-1 9/32" < 0.00°   10'-0"
Enter X length <10'-0">:

Ceiling grid line added

FIGURE 13.57 *Specifying the location to add a ceiling grid line*

Removing Ceiling Grid Lines

Ceiling grid lines can be removed if the layout mode has been set to **Manual** for the axis of the ceiling grid and you select the **CeilingGridXRemove** or **CeilingGridY-Remove** command. See Table 13.24 for **Remove Grid Line** command access.

TABLE 13.24 *CeilingGridXRemove or CeilingGridYRemove command access*

Command prompt	**CEILINGGRIDXREMOVE or CEILINGGRIDYREMOVE**
Ribbon	Select the ceiling grid and choose Remove Grid Line from the X Axis or Y Axis panels.
Shortcut menu	Select the ceiling grid, right-click, and choose X Axis > Remove Grid Line or Y Axis > Remove Grid Line.

When the layout mode of the ceiling grid is set to **Manual**, you can remove selected ceiling grid lines. The following command line sequence removes a ceiling grid line.

(Verify that ORTHO is toggled ON on the status bar.)

(Select the ceiling grid at p1 *in Figure 13.58 and choose* Remove Grid Line *from the X Axis panel of the Ceiling Grid contextual tab.)*

Enter approximate X length to remove <10'-0" [3000]>: *(Click near the grid line at* p2 *in Figure 13.58 to remove the grid line. The grid line is removed as shown in Figure 13.58.)*

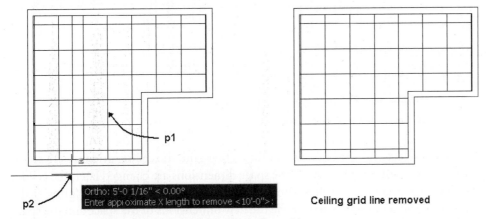

Ceiling grid line removed

FIGURE 13.58 *Ceiling grid line removed from the ceiling grid*

Adding Boundaries and Holes

Boundaries and holes can be added to existing ceiling grids with the **CeilingGrid-Clip** command. This command has three options: Set boundary, Add hole, and Remove hole. The command can be used to create a boundary for a ceiling that has not been bounded to a space or polyline. The Add hole and Remove hole options allow you to carve out an area from the ceiling grid for penetrations through the ceiling, such as columns, shafts, and plumbing chases. See Table 13.25 for CeilingGrid-Clip command access.

TABLE 13.25 *CeilingGridClip command access*

Command prompt	**CEILINGGRIDCLIP**
Ribbon	Select a ceiling grid and choose Set Boundary from the Clipping panel.
Shortcut menu	Select a ceiling grid, right-click, and choose Clip.

When you select a ceiling grid and choose Set Boundary, you are prompted in the workspace to select a closed polyline. If a grid has been placed without being attached to a boundary, the Set boundary option can be used to assign a boundary, which can be a space or closed polyline. The workspace prompt sequence for assigning a boundary to an existing ceiling grid is shown below:

> *(Select a ceiling grid at* **p1** *as shown at the left in Figure 13.59 and choose* Set Boundary *from the* Clipping *panel.)*
>
> Select a closed polyline or space object for boundary: *(Select the space at* **p2** *as shown at the left in Figure 13.59.)*
>
> Ceiling grid clip [Set boundary/Add hole/Remove hole]: ENTER *(Press* ENTER *to end the command and display the grid as shown at the right in Figure 13.59.)*

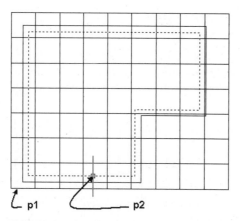 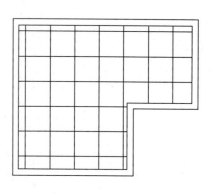

FIGURE 13.59 *Ceiling grid attached to boundary*

If a space is selected as the boundary of the ceiling grid, the ceiling grid is restricted to the boundary limits of the space. If the space dimensions are changed later, the ceiling grid does not stretch to the new dimensions of the space. However, the grips of the ceiling grid can be stretched to include the new dimensions of the space and the space will continue to serve as the boundary for the ceiling grid.

Creating Holes in the Ceiling Grid

The Add hole option of the **CeilingGridClip** command allows you to carve from the ceiling grid the shape of any closed polyline or AEC object. The polyline does not have to have the same elevation or intersect with the plane of the ceiling grid. The ceiling grid shown at the right in Figure 13.60 is trimmed by the closed polyline. The following workspace prompt sequence was used to edit the ceiling grid as shown in Figure 13.60.

> *(Select ceiling grid at* **p1** *as shown at the left in* Figure 13.60 *and choose* Add Hole *from the* Clipping *panel.)*
>
> Select a closed polyline or AEC entity for hole: *(Select the column at* **p2** *as shown at the left in Figure 13.60.)*
>
> Ceiling grid clip [Set boundary/Add hole/Remove hole]: ENTER *(Press* ENTER *to end the command. The ceiling grid is modified as shown at the right in Figure 13.60.)*

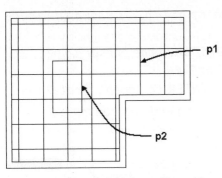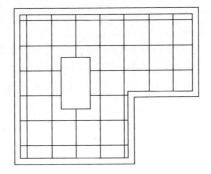

FIGURE 13.60 *Hole added to ceiling grid*

The Remove hole option allows you to edit the grid and remove any holes that have been created.

TUTORIAL 13-5: CREATING AND MODIFYING A CEILING GRID

1. Download Chapter 13 files from the Accessing Tutor\Imperial or Accessing Tutor\ Metric category of the Student Companion site of CengageBrain http://www .cengagebrain.com described in the Preface to your Accessing Tutor folder.

2. Open *Accessing Tutor\Imperial\Ch 13\Ex13-5.dwg or Accessing Tutor\Metric\Ch13\ Ex 13-5.dwg.*

3. Save the file to your Ch 13 student directory as **Lab 13-5**.

4. Select the **Reflected** display configuration from the **Drawing Window** status bar. Toggle OFF Snap Mode, Grid Display, Ortho Mode, Polar Tracking, Object Snap, and Object Snap Tracking in the status bar.

5. Select the Home tab. Select **Ceiling Grid** from the **Build** panel.

6. Edit the **Properties** palette as follows: Specify on screen = **No**; X Axis: Layout type = **Repeat**, Bay size = **2'-0 [600]**, Start offset = **0**, End offset = **0**; Y Axis: Layout type = **Repeat**, Bay size = **2'-0 [600]**, Start offset = **0**, End offset = **0**. To set the remainder of the ceiling grid dimensions, select Specify on screen = **Yes**. Choose **Select object** from the **Boundary** list and select the edge of the AEC Space for the room at **p1** as shown in Figure 13.61.

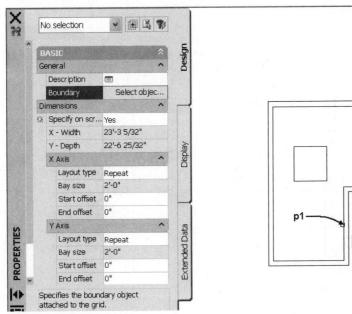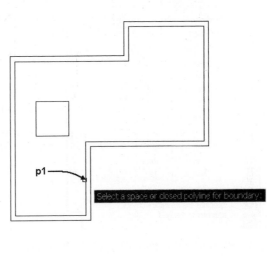

FIGURE 13.61 *Selecting the AEC Space boundary for the ceiling grid*

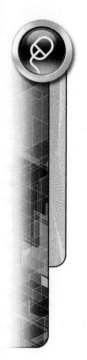

Insertion point or: *(Select a point near* **p1** *as shown in Figure 13.62.)*

New size or: *(Refer to Figure 13.62 move the cursor to* **p2**, *right-click, and choose* SNap *to center.)*

Rotation or <0.00>: ENTER *(Press* ENTER *to accept 0 rotation.)*

Insertion point or: ESC *(Press* ESC *to end the command.)*

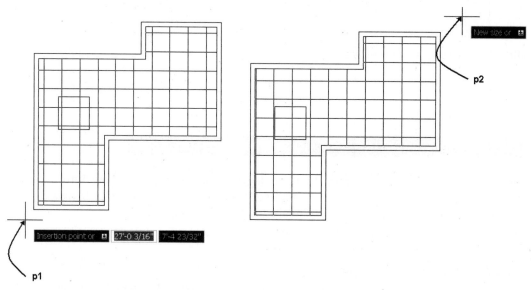

FIGURE 13.62 *Specifying the size of the ceiling grid*

7. Select **SW Isometric** view from the **View** Cube. Select the ceiling grid, choose **Select object** from the **Boundary** field of the Properties palette, and then select the ceiling boundary at **p1** as shown in Figure 13.63.

8. Select the **Top View** from the **View** Cube.

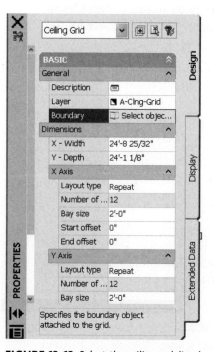

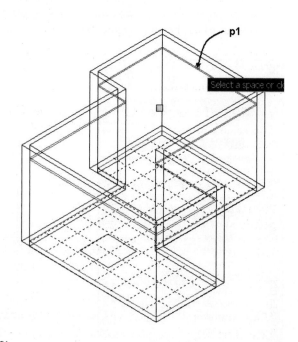

FIGURE 13.63 *Select the ceiling polyline boundary*

9. Select the ceiling grid and choose Add Hole from the Clipping panel. Respond to the workspace prompts as shown below to remove the ceiling grid from inside the rectangle shown in Figure 13.64.

> *(Select the ceiling grid and choose* Add Hole.*)*
>
> Select a closed polyline or AEC object for hole: *(Select the rectangle at* p1 *as shown in Figure 13.64.)*
>
> Ceiling grid clip [Set boundary/Add hole/Remove hole]: *Cancel* ESC *(Press* ESC *to end the command.)*

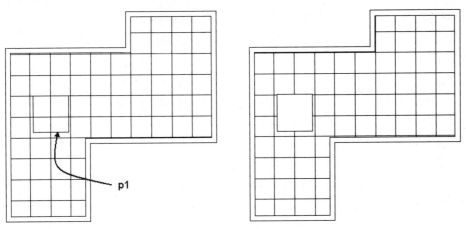

FIGURE 13.64 *Creating a hole in the ceiling grid*

10. Save changes and close the drawing.

USING LAYOUT CURVES TO PLACE COLUMNS

Placing columns spaced evenly along a centerline does not require the creation of a column grid line. Columns and other AEC objects can be placed along a layout curve. Isolated pier footings used in the foundations of residences can be placed on a layout curve. When a layout curve is created, nodes are defined on the curve, and AEC objects can be inserted using the Node object snap at the nodes on the layout curve. The nodes can be spaced evenly or at a specified distance apart. Layout curves are similar to the AutoCAD **Divide** and **Measure** commands; however, the layout curve provides additional enhancements.

Access the **Layout Curve** tool from the Content Browser; select **Stock Tool catalog > Parametric Layout and Anchoring**. See Table 13.26 for **LayoutCurveAdd** command access.

TABLE 13.26 *LayoutCurveAdd command access*

Command prompt	**LAYOUTCURVEADD**
Tool palette	Access the Content Browser, select Stock Tool catalog > Parametric Layout and Anchoring, and select Layout Curve.

When you select the **LayoutCurveAdd** command, you are prompted to select a curve. The curve can be a straight line, arc, polyline, or spline. A Wall can be used

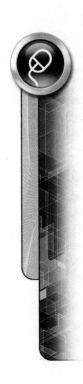

as the layout curve. The LayoutCurveAdd command allows you to use layout nodes equally spaced or at a specified distance apart. After selecting the entity to be used as a layout curve, you are prompted in the workspace to select the mode for the layout. The options for the layout mode are described below.

Manual—Requires that you specify the location of each node relative to the start point of the layout curve.

Repeat—Requires that you specify the starting offset distance, ending offset distance, and the spacing value along the curve to place the nodes.

Space evenly—Requires that you specify the starting offset distance, ending offset distance, and the number of nodes you want placed along the curve.

The LayoutCurveAdd command is used in the following workspace prompt sequence to place nodes along an arc. The nodes can then be used to locate the columns as shown in Figure 13.65.

```
Select a curve: (Select the arc at p1 as shown in Figure 13.65.)
Select node layout mode [Manual/Repeat/Space evenly]
<Manual>: S ENTER (Select the Space evenly option.)
Start offset <0>: 2' [600] ENTER
End offset <0>: 2' [600] ENTER
Number of nodes <3>: 5 ENTER
(The Column command can then be used to place columns along
the arc using the Node object snap.)
```

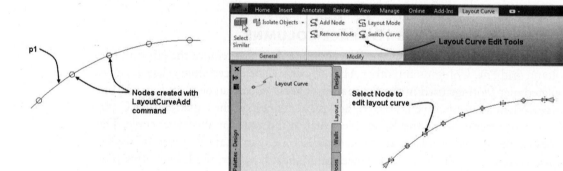

FIGURE 13.65 *Nodes placed on the arc with the LayoutCurveAdd command*

VIEWING THE MODEL

Most frequently, the designs of commercial buildings are presented to clients in the form of perspective drawings or a walkthrough. AutoCAD Architecture includes a camera to facilitate the view of the building model. The camera provides a quick visualization tool for presenting the model. The model can be viewed through one or more cameras placed in the plan view of the drawing. The still view or walkthrough presentation can be created with the camera. Access the camera from the Massing tool palette or the Ribbon as shown in Table 13.27.

TABLE 13.27 *Camera command access*

Command prompt	**CAMERA**
Ribbon	Select the Render tab. Choose Create Camera from the Camera panel.
Tool palette	Select Camera from the Massing tool palette of the Design palette set.

When the **Camera** command is selected, you are prompted to specify the insertion point and target location of the camera as shown in the following workspace sequence:

(Select Camera *from the* Massing *tool palette.)*

Specify camera location: *(Select a point near* p1 *as shown in Figure 13.66.)*

Specify target location: *(Specify the target select a point near* p2 *as shown in Figure 13.66.)*

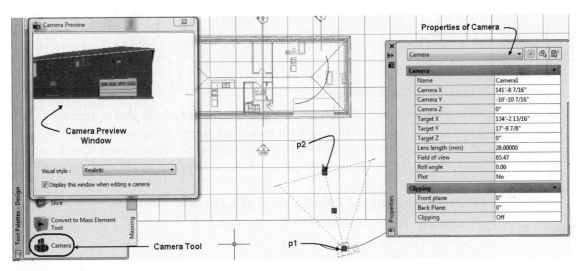

FIGURE 13.66 *Camera placement*

When you select an existing camera, the **Camera Preview** window and Properties palette opens. The Camera Preview window allows you to choose a visual style. You can view the camera image full screen in the workspace if you select the camera, right-click, and choose **Set Camera View**.

Creating a Walkthrough

You can create a walkthrough by inserting a camera and defining the path and target for the camera to follow. The camera path and target can be defined by a polyline or spline. The camera path, target path, and other animation settings are defined in the **Motion Path Animation** dialog box as shown in Figure 13.67. Access the Motion Path Animation dialog box as described in Table 13.28. The path for the camera and target is specified in the Motion Path Animation dialog box as shown in Figure 13.67. An animation is created in Tutorial 13.5.

TABLE 13.28 *Accessing the Motion Path Animation tool*

Command	ANIPATH
Ribbon	Choose Animation Motion Path from the Animations panel of the Render tab.

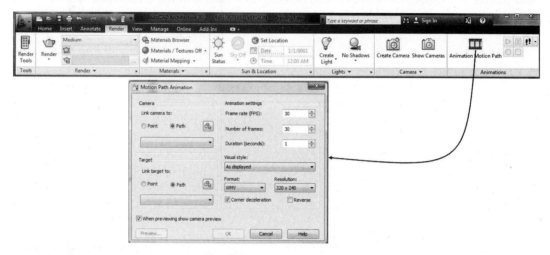

FIGURE 13.67 *Motion Path Animations dialog box.*

Creating a Cover Sheet

Near the completion of a project you can develop list of the sheets included in the project. The sheet list can be placed on a cover sheet or any sheet within the sheet set. The cover sheet may also include a pictorial view of the model. The following steps outline the procedure to create a pictorial view of the building model.

Steps to Creating a Pictorial View of a Model

1. Open the Project Navigator. Choose the Views tab.
2. Choose the Views category, right-click, and choose New View Dwg > General to open the Add General View dialog box.
3. Type **Pictorial** in the Name field of the General page.
4. Click Next to open the Context page.
5. Check all levels on the Context page.
6. Click Next to open the Content page.
7. Verify that Constructs and all construct drawing names are checked.
8. Click Finish to dismiss the Add General View dialog box.
9. Click the Sheets tab of the Project Navigator. Select a subset of the project, right-click, and choose New > Sheet.
10. Type a number in the Number field and Cover in the Name field. Click OK to dismiss the New Sheet dialog box.
11. Double-click the Cover name of the Sheets tab to open the Cover.dwg.
12. Click the Views tab. Click and drag the Pictorial.dwg view drawing onto the sheet. The Pictorial view drawing is displayed in Top view.

13. Toggle OFF Snap Mode. Select the viewport to display its grips. To prepare the viewport for the larger pictorial view, drag the grips of the viewport to increase the size of the viewport.

14. Click the lock in the Drawing Window status bar to Unlock. The viewport must be unlocked to change the view. Double-click inside the viewport. Select SW Isometric from the **View** Cube. Choose Realistic from the Visual Styles panel of the View tab.

TIP

To create a monochrome hidden view, create a monochrome plot style in your plot style folder. A monochrome plot style named Test1.stb is provided in Accessing Tutor\Ch13 folder in the Cengage Brain. Set the monochrome plot style current and assign all layers of the drawing to Style 1 (Black). Set the Shade Plot for viewport to Hidden.

After placing the pictorial view of building on the cover you can place the sheet list on the cover. The sheet list can be updated later in the project if additional sheets are added or removed. To update the sheet list select the table, right-click and choose Update Table Data Links to update the table to the current sheet contents of the project. The following steps outline the procedure to create a sheet list.

Steps to Creating a Sheet List

1. Open the project. Select the Sheets tab.

2. Select a category of the sheet set, right-click, and choose New > Sheet.

3. Type a number and sheet name in the New Sheet dialog box to create the cover sheet.

4. Double-click the cover sheet in the Sheets tab to open the drawing.

5. Select the Sheets tab. Select the Sheet Set name, right-click, and choose **Insert Sheet List Table** to open the Insert Sheet List Table dialog box as shown in Figure 13.68.

6. Type the project title in the Title Text field.

7. Click OK to dismiss the dialog box. Click a location on the cover sheet to specify the insertion point for the table.

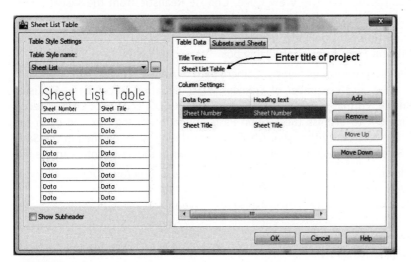

FIGURE 13.68 *Inserting Sheet List table*

TUTORIAL 13-6: CREATING AN ANIMATION

1. Download Chapter 13 files from the Accessing Tutor\Imperial or Accessing Tutor\ Metric category of the Student Companion site of CengageBrain http://www.cenga-gebrain.com described in the Preface to your Accessing Tutor folder. Open Autodesk AutoCAD Architecture and select **Project Browser** from the Quick Access toolbar. Use the **Project Selector** drop-down list to navigate to *Student\Ch13 directory*. Select **New Project** to open the **Add Project** dialog box. Type **26** in the Project **Number** field and type **Ex 13-6** in the Project **Name** field. Check **Create from template project**, choose **Browse**, and edit the path to *Accessing Tutor\Imperial\Ch13\Ex 13-6\Ex 13-6.apj* or *Accessing Tutor\Metricl\Ch13\Ex 13-6\Ex 13-6.apj* in the **Look in** field. Click **Open** to dismiss the Select Project dialog box. Click **OK** to dismiss the Add Project dialog box. Click **Close** to dismiss the Project Browser.

2. Select the **Views** tab and expand the **Exterior** category. Double-click the **Animation** view drawing to open the drawing.

3. Select the Render tab. Choose **Create Camera** from the Camera panel. Respond to the following workspace prompts to place the camera:

   Specify camera location: (Select a point near p1 as shown in Figure 13.69.)

   Specify target location: (Select a point near p2 as shown in Figure 13.69.)

4. Choose **Animation Motion Path** to open the **Motion Path Animation** dialog box as shown in Figure 13.69.

5. In this step you will link the camera to an existing polyline. The Z coordinate of the polyline drawn at **p4** is preset to 8'. Select the **Select Path** button at **p3**. Select the arc at **p4** to open the **Path Name** dialog box. Type **AAA-C** in the **name** field. Click **OK** to dismiss the dialog box.

6. In the **Target** section, select the **Select Path** button at **p5**. Select the arc at **p6** to open the **Path Name** dialog box. Type **AAA-T** in the **name** field. Click **OK** to dismiss all dialog boxes. Type **Accessing** in the **File name** field. Verify that your Ch 13 student directory is specified in the **Save as** window. Click **Save** to dismiss the dialog box.

7. Select **SE Isometric** from the **View** Cube. Choose Realistic from the Visual Styles panel of the View.

8. Choose **Animation Motion Path** from the Animations panel in the Render tab to open the **Motion Path Animation** dialog box.

9. Edit the animation settings as follows: Camera path = **AAA-C**, Target path = **AAA -T**, Frame rate (FPS) = **30**, Number of Frames = 150, Duration (seconds) = **5**, Visual style = **Realistic**, Format = **WMV**, Resolution = **800 × 600**, and check **Corner deceleration**. Click **OK** to dismiss the Motion Path Animation dialog box and open the **Save As** dialog box. Verify that your Ch 13 student directory is specified in the **Save in** window. Select Accessing .wmv. Select **Yes** to overwrite the new content to the file.

10. Open Windows Explorer and navigate to your student directory. Double-click the *Accessing .wmv* file.

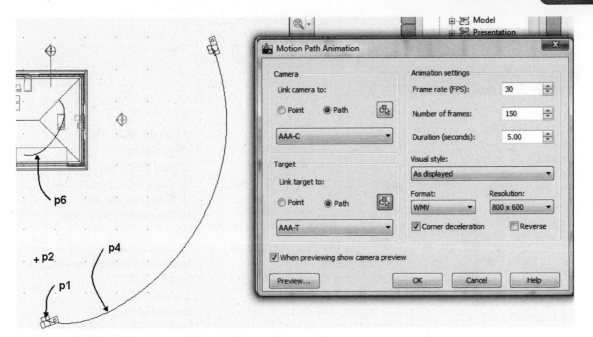

FIGURE 13.69 *Motion Path Animation Settings*

SUMMARY

- Styles of structural components are created in the Structural Member Catalog.
- The logical length of a structural member is the length of its axis line.
- Start offset and end offset values modify the extension of the column about the end of the axis line.
- ColumnToolToLinework, BeamToolToLinework, and BraceToolToLinework commands convert linework to the structural shape.
- Additional custom structural shapes are located in the Structural category in the Design Tool Catalog – Imperial and Design Tool Catalog – Metric of the Content Browser.
- Columns, beams, and braces are inserted in the drawing using the Column, Beam, and Brace commands of the Design palette.
- Column grids are placed in the drawing with the ColumnGridAdd and Custom-ColumnGrid commands of the Design tool palette.
- Set the Layout Mode to Manual for an axis to add column grid lines at specific locations along the axis.
- The start offset and end offset distances of the column grid are relative to the insertion point of the column grid.
- The column grid can be dimensioned with the AEC Dimensions.
- Ceiling grids are added to the drawing with the CeilingGridAdd command.
- The Clip option for ceiling grids and column grids allows you to trim the grid by a polyline or an AEC object.
- The LayoutCurveAdd command can be used to create nodes that are evenly spaced along a line, arc, or polyline.

Refer to the **Review Questions** folder at the Student Companion Site of **CengageBrain** http://www.cengagebrain.com described in the Preface to download review questions for Chapter 13.

NOTE

INDEX

D